Sourcebook in the

Mathematics of Medieval Europe and North Africa

Sourcebook in the

Mathematics of Medieval Europe and North Africa

Edited by

VICTOR J. KATZ

MENSO FOLKERTS
BARNABAS HUGHES
ROI WAGNER
J. LENNART BERGGREN

PRINCETON UNIVERSITY PRESS • PRINCETON AND OXFORD

Published by Princeton University Press, 41 William Street,
Princeton, New Jersey 08540
In the United Kingdom: Princeton University Press, 6 Oxford Street,
Woodstock, Oxfordshire OX20 1TR

press.princeton.edu

ISBN 978-0-691-15685-9

Library of Congress Control Number: 2016935714

British Library Cataloging-in-Publication Data is available

This book has been composed in Times Roman and Arial

Printed on acid-free paper. ∞

Printed in the United States of America

Typeset by Nova Techset Pvt Ltd, Bangalore, India

1 3 5 7 9 10 8 6 4 2

Contents

Appendices

Preface

In 2007, Princeton University Press published *The Mathematics of Egypt, Mesopotamia, China, India, and Islam: A Sourcebook*, of which I was the general editor. The book has been well received, and the reviews have been uniformly favorable. However, two of the reviewers wondered why we chose those five particular civilizations for the book, as there are other civilizations with mathematics that are not well represented in the general sourcebooks available. One of these reviewers specifically wondered why we did not include sources from Medieval European mathematics. The true answer to the reviewers' question was that the book was quite large already. But now that some time has passed, I decided that we could respond positively to the reviewers' question with a new sourcebook on Medieval European mathematics.

Four scholars whose work I know graciously agreed to edit the three main chapters of this new book, on mathematics in Latin, Hebrew, and Arabic. We worked together to choose the sources to include. We agreed that if there were already English translations of the chosen sources, we would use them, but as it turned out, the editors (and other scholars) translated the majority of our sources from the original language, especially those sources in Hebrew and Arabic. On occasion, we retranslated sources from an existing translation into French or German, but always with reference to the original language.

We have aimed this *Sourcebook* at readers with a reasonable background in college-level mathematics, although much of the mathematics can be understood by those having only secondary-level mathematics. Thus, we hope that this *Sourcebook* will prove valuable to mathematics teachers at both the secondary and college level who want to gain a stronger understanding of medieval mathematics for themselves as well as explore those ideas with their students. They will find here the beginnings of numerous mathematical ideas that have proved fruitful over the years and also, perhaps, some mathematical dead ends. But as much recent research has found, delving into original sources is a particularly valuable way of gaining a deeper understanding of mathematics. We also hope that historians of mathematics will find here valuable material for tracing the flow of ideas among medieval mathematicians writing in the three languages.

The editors thank the readers of the manuscript for their careful work and their numerous valuable suggestions. We also thank the scholars who contributed the material in some of the chapters or who helped with the translations. These people include Naomi Aradi,

Avinoam Baraness, José Bellver, Ahmed Djebbar, David Garber, Julió Samso, Stela Segev, Shai Simonson, Immo Warntjes, and Ilana Wartenberg. We also thank the two foremost current scholars of medieval Hebrew mathematics, Tony Lévy and Gad Freudenthal, for their generous assistance. Not only did they make valuable suggestions as to which sources to use in the book, but they also provided some translations on their own or helped check our translations. Furthermore, the editors thank Vickie Kearn, our editor at Princeton University Press for her help and support. In addition, we thank Betsy Blumenthal, our editorial assistant; Dimitri Karetnikov, our illustration specialist; Jenny Wolkowicki, our production editor; and Cyd Westmoreland, our copyeditor for their friendly and efficient handling of a difficult manuscript through the production process. Finally, the general editor thanks his wife, Phyllis, who, as always, has provided encouragement and love and lots more.

Victor J. Katz
Silver Spring, MD
November 2015

Permissions

The editors thank the publishers listed below for permission to reprint certain texts. The texts not listed were all translated directly from the original languages. For more details on the sources, see the sections on sources at the end of each chapter in the book.

CHAPTER 1

I-1-1: Reprinted from *Martianus Capella and the Seven Liberal Arts*, by William Stahl, Richard Johnson, and E. L. Burge (1977), by permission of Columbia University Press.

I-2-1 and I-3-1: Reprinted from *Boethian Number Theory*, by Michael Masi (1983), by permission of Brill Publishers.

I-2-2: Reprinted from *Bede, The Reckoning of Time*, by Faith Wallis (1999), by permission of Liverpool University Press.

I-3-4: Reprinted from "Pandulf of Capua's *De calculatione*: An Illustrated Abacus Treatise and Some Evidence for the Hindu-Arabic Numerals in Eleventh-Century South Italy," by Craig A. Gibson and Francis Newton, in *Mediaeval Studies* 57 (1995), by permission of the Pontifical Institute of Mediaeval Studies.

I-4-5: Reprinted from *Practical Geometry, attributed to Hugh of St. Victor*, translated by Frederick A. Homann, SJ (1991), by permission of Marquette University Press.

I-5-3: Reprinted from "The Puzzle of the Thirty Counters," translated by Gerard Murphy, in *Béaloideas* 12 (1942), by permission of University College Dublin.

II-1-1, II-2-3, II-4-6 (second part), II-6-4: Reprinted from *A Sourcebook in Medieval Science*, edited by Edward Grant (1974), by permission of Harvard University Press.

II-2-1: Reprinted from "Thus Spake al-Khwarizmi: A Translation of the Text of Cambridge University Library Ms. Ii.vi.5", by J. N. Crossley and A. S. Henry, in *Historia Mathematica* 17 (1990), by permission of Elsevier Publishers.

II-2-2 and II-3-2: Reprinted from *Fibonacci's Liber Abaci: A Translation into Modern English of Leonardo Pisano's Book of Calculation*, by Laurence Sigler (2002), by permission of Springer-Verlag.

II-2-5: Proposition IX-70 reprinted from "The Arithmetical Triangle of Jordanus de Nemore," by Barnabas Hughes, in *Historia Mathematica* 16 (1989), by permission of Elsevier Publishers.

II-2-6: Reprinted from "De Vetula, a Medieval Manuscript containing Probability Calculations," by D. R. Bellhouse, in *International Statistical Review* 68 (2000), by permission of John Wiley and Sons.

II-3-3: Reprinted from *Leonardo Pisano Fibonacci, The Book of Squares: An Annotated Translation into Modern English*, by L. E. Sigler (1987), by permission of Elsevier Publishers.

II-3-4: Reprinted from *Jordanus de Nemore, De numeris datis*, by Barnabas Hughes (1981), by permission of University of California Press.

II-3-5: Reprinted from "Part I of Nicole Oresme's *Algorismus proportionum*," by Edward Grant, in *Isis* 66 (1965), by permission of the University of Chicago.

II-3-6: Reprinted from *Nicole Oresme, De proportionibus proportionum and Ad pauca respicientes. Edited with Introductions, English Translations, and Critical Notes*, by Edward Grant (1966), by permission of University of Wisconsin Press.

II-4-1: Reprinted from *Archimedes in the Middle Ages, Vol. 1: The Latin Tradition*, by Marshall Clagett (1964), by permission of University of Wisconsin Press.

II-4-3 and II-5-2: Reprinted from *Fibonacci's De Practica geometrie*, by Barnabas Hughes (2008), by permission of Springer-Verlag.

II-4-4: Second section reprinted from *Archimedes in the Middle Ages. Vol. 3: The Fate of the Medieval Archimedes 1300–1565*, by Marshall Clagett (1978), by permission of the American Philosophical Society.

II-4-5: Reprinted from *Archimedes in the Middle Ages. Vol. 5: Quasi-Archimedean Geometry in the Thirteenth Century*, by Marshall Clagett (1984), by permission of the American Philosophical Society.

II-5-1: Reprinted from *Ptolemy's Almagest*, translated by G. J. Toomer (1998), by permission of Princeton University Press and Bloomsbury Publishing.

II-5-4: Second part reprinted from *Richard of Wallingford: An Edition of His Writings with Introductions, English Translation and Commentary*, by John D. North (1976), by permission of Clarendon Press.

II-5-6: Reprinted from *Regiomontanus On Triangles*, by Barnabas Hughes (1967), by permission of University of Wisconsin Press.

II-6-1: Second part reprinted from *Thomas Bradwardine, Geometria speculativa. Latin Text and English Translation with an Introduction and a Commentary*, by A. G. Molland (1989), by permission of Franz Steiner Verlag.

II-6-3: Reprinted from "Franciscans and Mathematics," by Barnabas Hughes, in *Archivum Franciscanum Historicum* 76 (1983), by permission of Archivum Franciscanum Historicum.

II-6-5: Reprinted from *Nicole Oresme, Questiones super geometriam Euclidis*, by H. L. Busard (2010), by permission of Franz Steiner Verlag.

II-7-1: Reprinted from *Robert Grosseteste and the Origins of Experimental Sciences 1100–1700*, by Alistair Crombie (1962), by permission of Clarendon Press.

II-7-2: Reprinted from *The Medieval Science of Weights (Scientia de Ponderibus)*, by Ernest A. Moody and Marshall Clagett (1960), by permission of University of Wisconsin Press.

II-7-3: Reprinted from *Thomas of Bradwardine. His Tractatus de Proportionibus. Its Significance for the Development of Mathematical Physics*, by H. Lamar Crosby (1961), by permission of University of Wisconsin Press.

II-7-4 and II-7-5: Reprinted from *The Science of Mechanics in the Middle Ages*, by Marshall Clagett (1959), by permission of University of Wisconsin Press.

II-7-6: Reprinted from *Nicole Oresme and the Medieval Geometry of Qualities and Motions. A Treatise on the Uniformity and Difformity of Intensities known as Tractatus de configurationibus qualitatum et motuum*, by Marshall Clagett (1968), by permission of University of Wisconsin Press.

III-4-2: Reprinted from "The Earliest Vernacular Treatment of Algebra: *The Libro di Ragioni of Paolo Gerardi* (1328)," by Warren Van Egmond (1978), in *Physis* 20, by permission of Casa Editrice Leo S. Olschki.

III-4-3: Reprinted from *Jacopo da Firenze's Tractatus Algorismi and Early Italian Abbacus Culture*, by Jens Høyrup (2007), by permission of Springer-Verlag.

CHAPTER 2

I-6 and V-1a: Problems reprinted from "The Missing Problems of Gersonides: A Critical Edition," by Shai Simonson, in *Historia Mathematica* 27 (2000), by permission of Elsevier Publishers.

II-2: Reprinted from *Abraham Ibn Ezra: The Book of the World*, by Shlomo Sela (2010), by permission of Brill Publishers.

III-1: Reprinted from "'Sefer ha-Middot': A Mid-Twelfth-Century Text on Arithmetic and Geometry attributed to Abraham ibn Ezra," by Tony Lévy and Charles Burnett in *Aleph* 6 (2006), by permission of Indiana University Press.

III-5: Material from chapter 4 reprinted from *The Astronomy of Levi ben Gerson (1288–1344)*, by Bernard Goldstein (1985), by permission of Springer-Verlag. Material from chapter 49 reprinted from "Heuristic Reasoning: Approximation Procedures in Levi ben Gerson's Astronomy," by J. L. Mancha, in *Archive for History of Exact Sciences* 52 (1998), by permission of Springer-Verlag.

IV-1: Translation made from the French in "Gersonide, Commentateur d'Euclide: Traduction annoteé de ses gloses sur les *Éléments*," by Tony Lévy (1992), in *Studies on Gersonides: A Fourteenth Century Jewish Philosopher-Scientist*, edited by Gad Freudenthal, by permission of Brill Publishers.

IV-3: Reprinted from "Hebrew Texts on the Regular Polyhedra," by Tzvi Langermann (2014), in *From Alexandria, Through Baghdad: Surveys and Studies in the Ancient Greek and Medieval Islamic Mathematical Sciences in Honor of J. L. Berggren*, edited by Nathan Sidoli and Glen Van Brummelen, by permission of Springer-Verlag.

IV-4: Reprinted from "Immanuel ben Jacob of Tarascon (Fourteenth Century) and Archimedean Geometry: An Alternative Proof for the Area of a Circle," by Tony Lévy, in *Aleph* 12 (2012), by permission of Indiana University Press.

IV-6: Proposition 23 reprinted from "Medieval Hebrew Texts on the Quadrature of the Lune," by Tzvi Langermann, in *Historia Mathematica* 23 (1996), by permission of Elsevier Publishers.

CHAPTER 3

I-3: Translation made from the French in "Quelques methods arabes de construction des carrés magiques impairs," by Jacques Sesiano, in *Bulletin Societé Vaudoise des Sciences Naturelles* 83 (1994), by permission of Societé Vaudoise des Sciences Naturelles.

IV-2: Translation made from the French translation of *The Plane Sections of a Cylinder and the Determination of Their Areas* by Tony Lévy (1996), in *Les mathématiques infinitésimals du IXe au XIe siècle*, edited by Roshdi Rashed, by permission of Al-Furqān Islamic Heritage Foundation.

IV-4(3): Reprinted from "The Lost Geometrical Parts of the *Istikmāl* of Yusuf al-Muʾtaman ibn Hūd (11th century) in the Redaction of Sartaq (14th century): An Analytical Table of Contents," by Jan Hogendijk in *Archives Internationales d'Histoire des Sciences* 53 (2004) by permission of the author.

IV-4(4): Reprinted from "Four Constructions of Two Mean Proportionals between Two Given Lines in the Book of Perfection (*Istikmāl*) of al-Muʾtaman Ibn Hūd," by Jan Hogendijk, in *Journal for History of Arabic Science* 10 (1992), by permission of the journal.

IV-4(5): Reprinted from "Al-Muʾtaman's Simplified Lemmas for Solving 'Alhazen's Problem,'" by Jan Hogendijk (1996), in *From Baghdad to Barcelona: Studies in the Islamic Exact Sciences in Honour of Prof. Juan Vernet*, edited by Josep Casulleras and Julio Samsò, by permission of the University of Barcelona.

IV-5(1): Reprinted from "Simplicius on Euclid's Parallel Postulate," by A. I. Sabra, in *Journal of the Warburg and Courtauld Institutes* 32 (1969), by permission of the journal.

IV-5(2): Reprinted from "An Arabic Text on the Comparison of the Five Regular Polyhedra: 'Book XV' of the *Revision of the Elements* by Muhyi al-Dīn al-Maghribī, by Jan Hogendijk, in *Zeitschrift für Geschichte der arabisch-islamischen Wissenschaften* 8 (1994), by permission of Institut für Geschichte der Arabisch-Islamischen Wissenschaften.

V-2: Reprinted from "The Latin Version of Ibn Muʿādh's Treatise 'On Twilight and the Rising of Clouds,'" by A. Mark Smith, in *Arabic Sciences and Philosophy* 2 (1992), by permission of Cambridge University Press.

APPENDIX 1

Introduction: Reprinted from "Numeracy and Science," by Anne Tihon (2008), in the *Oxford Handbook of Byzantine Studies*, by permission of Oxford University Press.

1: Reprinted from "The Great Calculation According to the Indians, of Maximus Planudes," by Peter G. Brown, in *Convergence* 3 (2006), by permission of the Mathematical Association of America.

2: Reprinted from "The Magic Squares of Manuel Moschopoulos," by Peter G. Brown, in *Convergence* 2 (2005), by permission of the Mathematical Association of America.

3: Translation made from the French in "Le petit traité d'Isaac Argyre sur la racine carrée," by André Allard, in *Centaurus* 22 (1978), by permission of John Wiley and Sons.

4: Translation made from the German in *Ein Byzantinisches Rechenbuch des 15. Jahrhunderts: Text, Übersetzung und Kommentar*, by Herbert Hunger and Kurt Vogel (1963), by permission of J.B. Metzler'sche Verlagsbuchhandlung.

Sourcebook in the

Mathematics of Medieval Europe and North Africa

General Introduction

Medieval Europe, from around 800 to 1450, was a meeting place of three civilizations: the Latin/Christian civilization that was forming on the foundation of the defunct Western Roman Empire; the Jewish/Hebrew civilization, which witnessed great scholarly activity in every location where Jews resided; and the Islamic/Arabic civilization, whose European center was in Spain, but which had a close relationship with the Islamic civilization of North Africa. Although these three civilizations clashed at numerous times in the medieval period—with the Christians gradually pushing the Muslims out of Spain and with both Muslims and Christians attacking Jews at various times—a fertile intellectual exchange took place at this time that is reflected in the developments in mathematics. Thus, the aim of this sourcebook is not just to present original sources in mathematics from the three civilizations, but also to present sources that reflect this interchange. Thus, the reader will frequently find similar problems or methods discussed in texts originally in Latin, Arabic, or Hebrew, and we have provided internal references to similar texts in the various sections.

The original design of this book was to include sections on mathematics originally written in Europe in Latin, in Hebrew, and in Arabic. However, as the book developed, it became clear that confining the geographical locus to European soil was too restrictive. After all, for much of this period, the Islamic rulers of Spain also ruled in North Africa (the Maghrib); and whether the rulers of both areas were from the same dynasty or not, there was much interchange of ideas and people between the two areas. Furthermore, although most of the medieval Hebrew work in mathematics was accomplished in Spain and France, when Jews had to leave those areas, many migrated east to Constantinople, so it seemed important to include a mathematical treatise from the Byzantine and later Ottoman capital. And since we did include such a work, the question came up as to whether we should also include original sources from the Byzantine Empire itself. Unfortunately, precious few such works have been edited to date, so we made a decision to include just a limited number of excerpts of Byzantine work, translated from the Greek, in Appendix 1.

The editorial team also realized that it is nearly impossible to make a sharp distinction between Latin, Hebrew, and Arabic works. For example, some mathematical texts were originally written in Arabic, but no Arabic manuscript is extant. Fortunately, at some point during the medieval period, these manuscripts were translated into Latin or Hebrew. Similarly, there are Hebrew manuscripts that today only exist in a Latin translation made during that period. Thus, it is not always clear into which of the three chapters of this book to place a given piece of mathematics. In general, our decision was based on whether the original author was known. If he was, then we placed the work under the language in which the document was originally written. But if not, then it is placed under the language in which the extant manuscript is written. In any case, it is clear that mathematicians in the three cultures—the Latin, the Arabic, and the Hebrew—were aware of what was being written in the other cultures and somehow made use of these materials.

Although medieval texts that involve groundbreaking mathematics are rare, nevertheless, we decided that the excerpts presented would not, in general, be snippets but would be long enough for readers to get a solid understanding of the mathematical ideas the author was expressing. Thus, many of our sources are quite long and develop a mathematical idea in some detail. We hope that readers will come to appreciate the mathematical struggles of our medieval ancestors and the answers they found to the problems they posed. As will be clear from the frequent references to Euclid and other Greek authors, mathematicians writing in Latin, Hebrew, and Arabic had usually studied classical Greek mathematics, mostly in translation, and were trying both to understand that mathematics and, often, to develop it further. However, they also often had practical goals and thus ignored the Greeks when they needed to solve problems that were not part of the Greek mathematical enterprise. In both of these situations, however, these medieval mathematicians prepared the ground for the great advances in mathematics that were made in the Renaissance and later.

The study of the Latin mathematics of medieval Europe has a long history, reflected today in the large number of texts that have been translated from that language into English. By 1974, Edward Grant had edited *A Source Book in Medieval Science*, which included more than 100 pages of mathematics in its total of over 800 pages. We have, in fact, reused a few of the translations from that book, but it is interesting to note that many of the works that were excerpted there now have complete English editions available, including especially works of Nicole Oresme, Jordanus of Nemore, and Leonardo of Pisa (Fibonacci). Thus, in this book we have been able to use the work of the numerous scholars who contributed to this translation effort, supplementing the original translations by the two editors of our Latin section. We decided to include pieces of mathematics in this section that were originally written in Greek or Arabic but whose influence in Europe only occurred after they had been translated into Latin. For example, we have included excerpts from both the arithmetic and algebra of al-Khwārizmī, both of which appeared in longer excerpts in the earlier *Sourcebook*. And at the end of the Latin section are some sources originally in Italian that presage the development of Renaissance mathematics and yet are still intimately related to medieval material. In addition, the trigonometry section includes a substantial excerpt from Regiomontanus's *On Triangles*, which could be considered either the last medieval trigonometry text or the first one of the early modern period, but whose author evidently learned many of his ideas from his medieval ancestors.

Scholars have only been studying medieval Hebrew mathematics in relatively recent times. Such a study needs to begin with editions collated from extant manuscripts and then translated into modern languages, and major efforts in this direction are only a product of the late twentieth century. In particular, very few medieval Hebrew works have previously been translated into English, although some were translated into German early in the twentieth century and others have been translated into French more recently. Thus, the editor and several other young scholars contributed numerous original translations to the Hebrew section. And even when a previous translation into German or French existed, we retranslated the material into English, always checking with the Hebrew original when that was available.

The situation for Arabic was somewhat different from that of the other two languages. Although numerous Arabic works from the Muslim East were translated into German or French in the nineteenth century, it is only in recent years that medieval manuscripts from the Maghrib or from Spain (al-Andalus) have been studied, edited, and translated. Early in the twentieth century, some Arabic manuscripts from Spain were translated into Spanish, while the last decades of the twentieth century witnessed a steady flow of translations of North African manuscripts into French. We were able to make use of some of this material in the present work, again checking translations from French or Spanish against the original Arabic. And with the help of several other scholars, the editor of Chapter 3 on Arabic mathematics also translated some material into English directly from Arabic, material that has only recently become available.

One criticism of the earlier *Sourcebook* was that no attempt had been made either to compare the mathematics of the five civilizations whose sources were included or to speculate about any methods of transmission from one of these civilizations to another. At the time, too little information was available to even attempt the latter task, so each of the five sections stood independently. But in this *Sourcebook*, it is clear that the Latin, Hebrew, and Arabic mathematical cultures influenced one another. The reader can easily compare, for example, the works on measurement in the three cultures or the problem of men buying a horse or even some of the trigonometrical treatises. In some of these instances, it appears that one author simply borrowed from another, given that the numerical values and even the lettering of the diagrams are identical. In most cases, however, we can only speculate as to whether one author of an included work was actually familiar with the work of another. It is always possible that both borrowed from works that are no longer extant or even that each developed the particular idea or example independently. To help trace the flow of ideas, we have also included a few problems from Diophantus (see Appendix 2) and from Mahāvīra (Appendix 3) that are very similar to those considered in the main text. And as a further help, we have included a time line that includes all authors whose works we have used in the *Sourcebook*.We therefore hope that historians may be able to take our *Sourcebook* as a starting point for further research in the study of the flow of mathematical ideas during the medieval period.

NOTE TO THE READER

We have used different fonts to make clear to the reader the source of any given paragraph. Namely, the original sources are generally in Arial, while in cases where we have intermingled a source and a later commentary on the source, we have used Arial in the first case and **boldfaced Arial** in the second. Comments by the editors are generally in Times Roman. The numbering for the figures begins anew in each part of each chapter of the book. Thus, the figures have a four-digit identifier in the Latin chapter (Chapter 1) and a three-digit identifier in the other two chapters.

1

The Latin Mathematics of Medieval Europe

MENSO FOLKERTS AND BARNABAS HUGHES
with contributions by Immo Warntjes

INTRODUCTION

The mathematics that developed in Latin Catholic Europe, 800–1480, is the topic of this chapter. The period begins with the implementation of the *General Directive* (*Admonitio Generalis*) of Charlemagne in 793 that was directed to all bishops and abbots in his realm. Among the many imperatives he sought for improving life in his kingdom is edict 70. It states, among other things,

In every monastery and cathedral there shall be for boys who can read schools with well corrected Catholic books on the Psalms, written signs, chant, calendar, and grammar. [Mordek, 2012, p. 222]

This edict initiated what has been called the *Carolingian Renaissance*. Further, the kingdom became the Roman Empire in 800 with the coronation of Charlemagne in Rome as emperor. The ebb and tide of the boundaries of the Holy Roman Empire over the centuries does not concern us here.

What at the beginning of the ninth century people thought constituted mathematics introduces the section on the first period: "Latin Schools: 800–1140." For an understanding of this mathematics, Boethius (ca. 480–524) had crafted a fourfold model in his *De institutione arithmeticae*.

Arithmetic considers that multitude which exists of itself as an integral whole; the measures of musical modulation understand that multitude which exists in relation to some other; geometry offers the notion of stable magnitude; the skill of astronomical discipline explains the science of moveable magnitude. If a searcher is lacking knowledge of these four sciences, he is not able to find the true; without this kind of thought, nothing of truth is rightly known. [Masi, 1983, pp. 72–73]

The four—arithmetic, geometry, music, and astronomy—became known as the *quadrivium*, a word first used by Boethius;[1] and it provided the template for the curricula of

[1] Apparently Boethius translated the Greek word *methodos* from Nicomachus's *Arithmetic* (I.4.1–5) as *quadruvium*—whence *quadrivium*, "a place where four roads or ways meet;" [see Stahl, Johnson, and Burge, 1971, p. 156

the first period. Other elements are included, including some mathematical games and instruments.

The second period, discussed in the section "A School Becomes a University: 1140–1480," witnessed the birth of universities and the wealth of studies gathered and translated in Spain and Italy that would become much of the curricula for these institutions. In France a Guild of Masters separated from the Cathedral School to form the University of Paris. The Masters were all clerks, tonsured and dressed appropriately as befit members of the clergy. Logic, philosophy, and the liberal arts,[2] encompassing the *trivium* (grammar, rhetoric, and dialectic) and the quadrivium, were the original studies, to which was soon added the study of Sacred Scripture that came to dominate the university. The situation was different south of the Alps. Perhaps a year or two after Paris a Guild of Students founded the University of Bologna as a law school. Students from many nations gathered, organizing themselves into national groups (colleges) of students. The older Italian universities are still called *Università degli studi.* The commonality of the two universities was the required basic courses. To begin a career in law or theology, a student was expected to have completed successfully the study of the liberal arts. Further, literacy required proficiency in a common language, Latin. The curricula of these universities would soon expand.

About the same time that a Guild of Masters was beginning the University of Paris, translators in Toledo and in other parts of Spain were bringing the intellectual wealth of the Arabic Near East to scholars proficient in Latin. A hundred years later in Italy, similar wealth would be harvested from Greek sources. Translations from both sources, Arabic and Greek, would become the bases for courses in the universities that were arising throughout Europe. Further, from the translations would come novel subjects, such as trigonometry (but not yet under that name) and the use of mathematics in, of all places, theology. Among the mathematical sciences that originated here and would endure is mechanics. Not to be overlooked because it would become so popular was the challenging mathematical game *Rithmimachia.* By the end of this period, some dozens of universities had been established in Europe, all enriched by the translations from Spain and Italy and the creativity of their Masters.

The third period, which overlaps about half of the second, is discussed in the section "Abbacist Schools: 1300–1480." Just before 1300 the vernacular so necessary for economic survival began to replace Latin as the language of instruction in the arts. The word *arts* used here means mostly crafts, the tools of the working class. Giovanni Villani (ca. 1278–1348) in his *Nuova cronica* offered an overview of educational life in early fourteenth century Florence:

> We find that the boys and girls learning to read number from 8,000 to 10,000. From 1000 to 1200 boys were learning the abacus and algorism in six schools. And from 550 to 600 were learning grammar and logic in four large schools. [Villani, 1991, XII, sec. 94]

n. 39]. Masi remarked that the tenth-century Benedictine author, Sister Hrosvitha, in discussing moral philosophy, proposed this dialogue between Teacher and Student: "[S:] Quid est musica? [T:] Disciplina una de philosophiae quadruvio. [S:] Quid est hoc quod dicis quadruvium? [T:] Arithmetica, geometrica, musica, astronomia" [Masi, 1983, p. 35].

[2]The term "liberal arts" finds its origin in Greek education, particularly in Aristotle's *Politics* (VIII.1), where he wrote of the liberal sciences, a core of knowledge necessary for freemen to function freely. He considered as liberal any science that was pursued for its own sake and not for any outside advantage. Over time, their study came to be considered preparatory for university work.

The six schools, wherein students learned abacus and algorism, were called "abbacist schools," and in them students would also learn about foreign exchange, geometry, and algebra. This type of education would flourish and spread throughout Italy and into the rest of Europe, thus setting the stage for the explosion of mathematics in the Renaissance.

I. LATIN SCHOOLS, 800–1140

The mathematics available to Latin Europe at the beginning of the ninth century came from Roman sources and was written in Latin, the language of literacy. Since Roman times, mathematics was taught within the framework of the seven liberal arts. Four of these were mathematical: arithmetic, geometry, astronomy, and music (i.e., theory of music based on mathematical proportions). Arithmetic and geometry were also used for practical purposes (e.g., for calculation and to measure plots of lands). Astronomy provided material for astrologers. Musical theory was more an academic exercise than a tool for choirmasters. And churchmen and rabbis needed to compute the Easter and Passover dates. The principal writings from late antiquity and the early Middle Ages on the seven liberal arts were encyclopedic in nature. These books were copied in the monasteries, and they were used for teaching in monastic and cathedral schools. One reason for studying mathematics was a sentence from the Bible that God has ordered everything by measure, number and weight (Wisdom 11:20). In fact, Cassiodorus (sixth century), in his *Institutiones* identified the seven liberal arts as the disciplines that should be studied by Christian people, noting that the word *liberal* is derived from *liber* or book.

The principal sources available at the beginning of Period I, and exemplified here in chronological order, are the *Quadrivia* of Martianus Capella, Cassiodorus, and Isidore of Seville, all encyclopedic in nature. At the beginning of the ninth century, at least one new idea arose in mathematics: the arithmetic of signed numbers. Unfortunately, this idea fell into oblivion. Thereafter training in the use of Roman numerals is exemplified together with finger reckoning. The first occurrence of Hindu-Arabic numerals is coupled with the impact of Gerbert on the abacus. Organized instruction in arithmetic is from Boethius and in geometry from Gerbert (until the translations of Euclid from the Arabic appear). Next the issue of the quadrature of the circle is addressed, and the use of the astrolabe to measure heights is exemplified. The first section concludes with a selection of number puzzles and thought problems, including *Rithmimachia*.

I-1. Brief Selections

1. THE QUADRIVIUM OF MARTIANUS CAPELLA

The most influential source for ninth-century mathematics is a book well known throughout this period, *The Marriage of Philology and Mercury* by Martianus Felix Capella (ca. 370–ca. 440). He lived in Carthage, on the Mediterranean coast of North Africa. As remarked in the work itself, he wrote this wonderful account as an old man, perhaps as something with which to occupy himself, having retired as an advocate. This masterpiece was certainly "one of the most popular books of Western Europe for nearly a thousand years."[1] It contributed immensely to the development of the seven liberal arts in medieval Europe, for it offered basic information about each of the arts in separate chapters. Capella wrote in a style entirely different from what we are accustomed to, one described as a dialogue or even a monologue in places with a narrator giving directions, all in a florid style. He begins with an introduction to geometry that describes an unexpected instrument.

Now the lady over there, prepared to perform equally important functions, as you observe, is called Paedia [Learning], a very well-to-do woman, who regards the wealth and treasures of Croesus or a Darius as contemptible alongside her own. . . . That object which the women (her attendants) brought in is called an abacus board, a device designed for delineating figures; upon it the straightness of lines, the curves of circles, and the angles of triangles are drawn. The board can represent the entire circumference and the circles of the universe, the shapes of the elements, and the very depths of the earth; you will see there represented anything you could not explain in words.

Immediately there came into view a distinguished-looking lady, holding a geometer's rod in her right hand and a solid globe in her left. . . . This tireless traveler was wearing walking shoes, to journey through the world, and she had worn the same shoes to shreds in traversing the entire globe. . . . Geometry came to a halt. . . . finally (breaking) the spell of her rapt admiration of the glittering heavens. . . . A request was made that she begin disclosing the secrets of her knowledge, from the very beginning.

. . .

First, I must explain my name, to counteract any impression of a grimy itinerant coming into this gilded senate chamber of the gods and soiling this gem-bedecked floor with dirt collected on earth. I am called Geometry because I have often traversed and measured out the earth, and I could offer calculations and proofs for its shape, size, position, regions, and dimensions. There is no portion of the earth's surface that I could not describe from memory.

And so the Lady goes on, describing all the places on the earth through forty-four pages of geography, until toward the end she offers a six-and-a-half page list of geometric facts. This medieval idea, that geometry was for measuring the earth, would eventually be replaced by Euclid's *Elements*, but not for several centuries. After all, the word *geometry* does literally mean *earth measure*.

Arithmetic is

a lady of striking appearance. . . . from her brow a single, scarcely perceptible whitish ray appeared, and from it emanated another ray, the projection of a line, as it were, from its original source. Then came a third and fourth ray and on to a ninth and tenth, the first decad.

First she identifies

the monad being called sacred; numbers coming after it and associated with it have taught that before everything the monad is the original quickener. . . . Number takes its beginning from the dyad.

[1]For a thoroughly documented analysis of the influence of this treatise, see [Stahl, Johnson, and Burge, 1971, pp. 55–71].

All the while she describes the origin, properties, and significance of the numbers up to ten, for instance:

When the monad has extended itself in any direction, although an indivisible line is produced without any suggestion of breadth, it forms the dyad. The dyad, because it is the first offspring, is called Genesis by some. Because between it and the monad the first union and partnership occurs, it is called Juno or wife.

Thereafter follow the properties and relations of numbers found in elementary theories of number.

Next in Capella's order is Astronomy.

The time is now at hand to speak of the path of the starry sphere, the course of the poles and of the region where the hallowed planets trace their diverse and winding courses. I see the canopy of heaven gleam, now struck by a bolt of lightning from the sky. . . . Before their eyes a vision appeared, a hollow sphere of heavenly light, filled with transparent fire gently rotating and enclosing a maiden within. Several planetary deities, especially those which determine men's destinies, were bathed in its glare, the mystery of their behavior and orbits revealed.

Thus astrology is allied with astronomy, a relationship that persists to this day. Of more interest to modern readers, perhaps, is Capella's acceptance of the heliocentric theory regarding the orbits of Mercury and Venus, first proposed by Heraclides of Pontus (ca. 340 BCE):

Now Venus and Mercury, although they have daily risings and settings, do not travel about the earth at all; rather they encircle the sun in wider revolutions.

As Stahl, Johnson, and Burge [1971, p. 175] have pointed out, Capella produced a largely free, occasionally literal, translation of part of *De Adrasti Peripatetici in Platonis Timaeum Commentario* compiled by Theon of Smyrna (fl. 100 CE). Noteworthy also is Capella's use of a clepsydra, or water clock, an instrument that measures time by dripping water at a fixed rate, so that the volume of water dripped is proportional to the time elapsed. He used it to measure the diameter of the moon and its orbit. Martianus prefaced his computation of the orbit of the moon with information that compares the relative sizes of the moon and earth, as follows.

It often happens that an eclipse of the sun occurring at the latitude of Meroë darkens the entire orb, but at a nearby clime, that is, one passing through Rhodes, the obscuration is partial, and at the latitude of the mouth of the Borysthenes [Dnieper River] there is no obstruction and the full orb shines forth.[2] Since the correct distance in stadia at which the latitude of Rhodes is located is known, I have found that the breadth of the shadow, which the moon casts, is one-eighteenth part of the earth. Now since the body, which casts the conical shadow, is larger than the shadow itself, it has been ascertained from the latitudes on either side at which the sun was partially obscured, that the moon itself is three times as large as its shadow. Thus it has been determined by the foregoing calculations that the moon is one sixth as large as the earth.[3]

[2]These three locations are chosen to represent parallels of given maximum length of day, as was common from the time of Ptolemy. Meroë, on the Nile in what is now Sudan, has a longest day of 13 hours; Rhodes, off the coast of Turkey, has a longest day of $14\frac{1}{2}$ hours; while at the mouth of the Borysthenes in the Black Sea near Odessa, the longest day is 16 hours.

[3]The correct ratio of the moon's diameter to that of the earth is about 1:4.

That the moon's diameter is one six-hundredth of its orbit is determined by the use of clepsydras. Place two copper vessels in position, an empty one below, one full of water above. Mark the rising of the moon and that of a fixed star rising simultaneously with it. At the moment when the upper edge of the moon begins to appear above the horizon, quickly release the stopper from the upper vessel, the one containing the water, and let the water flow until the moon's entire orb appears. At this point remove the first vessel, into which the water has flowed, and put another in its place, into which the water may flow until on the following night, the very same star rises which rose together with the moon on the previous night. Remove the upper vessel, from which the water flowed out in the space of twenty-four hours. Compare it with the amount, which flowed out into the first vessel while the orb of the moon was rising, and you will find that the entire amount increased six hundredfold. Hence it is clear that the orbit of the moon surpasses the diameter of the moon by six hundred times.[4] We concluded, then, that the lunar orbit is one hundred times as great as the earth.

With this information Capella measured the orbits of the other planets.

The final chapter on *Harmony* (or *Music* as others would name it) begins by discussing at length for nearly half the text the supernatural and mystic powers of music,

to soothe the cares of the gods, gladdening the heavens with her songs and rhythms.

When at last Martianus focuses on music in itself, he presents his reader with a plethora of definitions and distinctions, for which he offers no mathematical representations. For instance,

I shall first discuss melodic composition. I maintain that whatever sounds strike the ear in the right proportion form either a whole tone, a half tone, or a quarter tone, the last being called *diesis*. A whole tone is an interval of appropriate size that lies between two mutually different sounds.[5]

2. THE QUADRIVIUM OF CASSIODORUS

In far southern Italy on the shore of the Ionian Sea at Vivarium (near modern Squillace), M. Aurelius Cassiodorus (490–583) had founded an abbey. He reasoned that the monks needed expertise in the trivium and quadrivium to understand Sacred Scripture. To educate them, he wrote *De Artibus ac Disciplinis Liberalium Litterarum*. This book had circulated somewhat among the nascent Benedictine abbeys and sundry monasteries. As preface to Chapter Four is a generous section titled "De Mathematica."

As a subject to be taught mathematics is the science that considers quantity in the abstract. This means that the intellect separates quantity from material things and their accidental qualities, for example, equal, unequal and characteristics of this sort treated by reasoning alone. The parts of mathematics are four: arithmetic, music, geometry, and astronomy. Arithmetic considers numerable quantity in itself. Music considers numbers as found in sound. Geometry concerns itself with the immobile, solid magnitude. Astronomy traces the course of heavenly stars and investigates with one's probing reason all the figures, the characteristics of the stars, by themselves and as related to the earth.

Cassiodorus described the common thinking about arithmetic.

[4] The correct ratio of the diameter of the moon's orbit to that of the moon's diameter is about 220:1.

[5] In Stahl, Johnson, and Burge [1971, pp. 202–227] the mathematical representations are supplied.

Writers on secular matters rank arithmetic in the first place of all the mathematical arts because music, geometry and astronomy need arithmetic to explain their own powers. For instance, the ratio of one to two found in music depends on arithmetic. Geometry needs arithmetic to describe triangles, rectangles and other figures. Astronomy uses arithmetic to describe the position of points in the movements of the stars. On the other hand arithmetic needs not music, nor geometry, nor astronomy to be understood. Therefore arithmetic is the mother of all. [As the prophet said in Wisdom 11:20,] "God created all in number, measure, and weight."

Similarly, he describes geometry.

The Latin word "geometry" means "measurement of the earth," because from various appearances of this discipline, as some say, the Egyptians were the first to partition their property. Teachers of this subject used to be called measurers. . . . Then the studious were prompted to understand what is invisible. They began to seek how much space does the earth contain, how far is the sun from the moon, how far is it to the vertex of heaven. These are the kind of things that the expert geometricians seek. Then they subject the measurements they make of the whole earth to probing reason. Thus the discipline of geometry takes its name and has kept it these many centuries.

Cassiodorus has music set a tone for conduct.

Music is the discipline that is spread through everything we do, for these reasons. First, we obey the command of our Creator and observe his statutes with pure minds. For whatever we say or whatever inward effect is caused by the beating of our pulse is joined by musical rhythms to the power of harmony. Music therefore is the science of correct modulation. If we observe the good way of life, we are always associated with this excellent science. But when we do what is wrong, music does not accompany our actions. Heaven and earth, all that happens above us, none of these are without the discipline of music. Pythagoras established that this world was created with music and can be governed by music.

Finally, he considers astronomy.

The divisions of astronomy are as follows: spherical position; spherical motion; the East; the West; the North; the South; the hemisphere above the earth; the hemisphere beneath the earth; the orbital number of the stars; the forward movement or progression of the stars; the backward movement . . . , the pause of the stars; the correction of a computation by addition; . . . by subtraction; the magnitude of the sun, moon, and earth; the eclipse; and other phases which occur in these bodies.

3. THE QUADRIVIUM OF ISIDORE OF SEVILLE

Isidore of Seville (ca. 560–636) produced an encyclopedia of universal knowledge in 448 chapters spread over 20 Books titled *Etymologies*. Identifying the quadrivium with mathematics, he assigned it to Book III. This is his understanding of mathematics:

In Latin mathematics is called knowledge of the doctrine that considers abstract quantity. A quantity is called abstract if the intellect can separate it from matter or some accident, for example, equal, unequal or such that we treat solely with reasoning. There are four kinds: arithmetic, music, geometry and astronomy. Arithmetic is the doctrine of numerable quantity in itself. Music speaks of numbers that are found in sound.

Geometry thinks about size and shape. Astronomy contemplates the movement of the heavenly stars and figures and all the properties of the stars.

Each of the four sections begins with remarks on the meaning of its name and its discovery. While there is much repetition of what might have been expected from the preceding sources, this paragraph is singularly interesting:

How many infinite numbers are there? It is certain that numbers are infinite, because regardless of whatever number you make the last, not only can it be increased by the addition of one, but regardless of its size and whatever huge multitude it contains, the doctrine of numbers can not only double it but can multiply it. Secondly: there are certain numbers that because of their properties have boundaries, for instance no number can be made equal to another number. Therefore, no matter how unequal among themselves and diverse they are, each individual number is finite, but all of them are infinite.

Five hundred years later, Hugh of St. Victor (1096–1141) paid tribute to the same division of mathematics:

Mathematica igitur dividitur in arithmeticam, musicam, geometriam, astronomiam. [Buttimer, 1939, chapter VI, 755C]

Such was his conclusion to the chapter focusing on *De quadrivio*. Thereafter he states that abstract quantity is the scope of mathematics, with much room for its visible aspect, magnitude. This has two parts: continuous (as described by dimensions) and discrete (as analyzed in multitude). The latter makes use of counting and computing. Then he continued that magnitude might also be mobile, like the spheres of the world (astronomy), or immobile, like the earth (geometry). Multitude itself is discussed in arithmetic, and relations between multitudes are the focus of music. One may well wonder whether he knew about the translations just beginning to appear in Spain as he watched the rise of the university in Paris. Regardless, what we have reviewed above represents selections from the mathematics that was available to students at the beginning of the 800s.

4. A NEW CREATION: RULES FOR ADDITION OF SIGNED NUMBERS

There is apparently no evidence that Hugh (or anyone else of significant influence) was aware of the most unique creation in mathematics that occurred in the early ninth century: the rules for the addition of signed numbers. The rules appear in five unrelated manuscripts of this period, written immediately after or close to three number puzzle problems (discussed in section I-5-1). This text shows how a "positive number" and a "negative number" of different values can be combined. Before the fifteenth century the concept of negative number was found only in Chinese and Indian mathematics, although Leonardo of Pisa encountered negative quantities in certain problems in his *Liber abbaci*, where he dealt with them as "debts." We have become so accustomed to our terminology for operating with positive and negative numbers (note the adjectives!), that we would do well to pause in the reading of the very literal translation; the first line is "Verum cum verum facit verum." Expressing a new idea was certainly difficult. Surely the author was assuming readers would discover for themselves what the words signify.

True with true makes true. Less (*minus*) with true makes true. True with less makes less. Less with less makes less. True signifies being; less means nothing. Put any sum of numbers you wish called true, that is being, and put another sum you wish called less by the adverb that I said signified nothing, and combine the two sums. That which was larger conquers the less and consumes it according to the quantity of its size.

For example: If two sums of numbers are joined, the one is VII called true from the name of being and the other is III called less from the name of an adverb. These two sums of numbers are joined together, that is, being and non-being. Because the sum of the true is greater than the sum called less as VII is greater than III, the true VII conquers the less III. But it can conquer not by the larger part that is IIII, for the VII takes the III less. So that with three removed from the VII, IIII remains. Therefore when the III less is joined to the VII, IIII remains. Similarly, if III true in name and VII less are joined, because the value of the not being is greater than (the value) of the being, the non-existing seven conquers the existing three, and its non-being consumes it. From itself remains the non-existing number IIII. This is what is said: join III and VII less; they make IIII less. If however you join III non-existing and VII similarly non-existing numbers, you show ten non-being. For as two existing true numbers such as VII and III true make a true, that is an existing number, namely ten, so two non-existing values of numbers so named as III less and VII less make X less. True means to make and less means not to make. Join III and VII; they are X. Join III less and VII; they are IIII. Join III and VII less; they are IIII less. Join III less and VII less; they are X less.

I-2. Numbering

1. ROMAN NUMERALS

In Latin Christian Europe, Roman numerals were used for writing numbers. They continued to hold a commanding place even after the twelfth century, when the so-called Hindu-Arabic numerals became known, and did not lose their influence before early modern times. Roman numerals had special symbols for the powers of 10 (I = 1, X = 10, C = 100, M = 1,000) and for the multiples of 5 between them (V = 5, L = 50, D = 500). Numbers were written by adding up to four symbols for the powers of 10 and, if necessary, one symbol for a multiple of 5 (e.g., MMMDCCLXXXXII = 3,792). The subtractive principle—that is, the rule that a smaller number placed to the left of a larger shall be subtracted from it (e.g., IX = 9)—was sometimes used during the Middle Ages, but one might see both IIII and XL in the same document (representing 4 and 40, respectively).

It is easy to add and subtract using Roman numerals. But they are not very convenient for multiplication and division. Therefore in Roman antiquity and in the Middle Ages, tables were used for multiplication. Such a table is given in Boethius's *Arithmetic* in the context of number theory and not for practical use.

Let there be such a diagram in which shall be placed in an order up to the number ten the natural order of continuous numbers, and in the second line the duplex order shall be extended, in the third the triple order, in the fourth the quadruple order, and this is done up to ten [Fig. I-2-1-1].

For more complex calculations, more elaborate multiplication tables were devised so that multiplications and divisions of integers and fractional numbers in the Roman system could be managed. The main problem came from the Roman fractions. The fractions were derived from the division of weights. The unit, which was called *as*, was divided into twelve parts. Every part had its own name and was denoted with a special sign. The smallest part (*uncia*)

Longitudo

Latitudo

I	II	III	IIII	V	VI	VII	VIII	VIIII	X
II	IIII	VI	VIII	X	XII	XIIII	XVI	XVIII	XX
III	VI	VIIII	XII	XV	XVIII	XXI	XXIIII	XXVII	XXX
IIII	VIII	XII	XVI	XX	XXIIII	XXVIII	XXXII	XXXVI	XL
V	X	XV	XX	XXV	XXX	XXXV	XL	XLV	L
VI	XII	XVIII	XXIIII	XXX	XXXVI	XLII	XLVIII	LIIII	LX
VII	XIIII	XXI	XXVIII	XXXV	XLII	XLVIIII	LVI	LXIII	LXX
VIII	XVI	XXIIII	XXXII	XL	XLVIII	LVI	LXIIII	LXXII	LXXX
VIIII	XVIII	XXVII	XXXVI	XLV	LIIII	LXIII	LXXII	LXXXI	XC
X	XX	XXX	XL	L	LX	LXX	LXXX	XC	C

Latitudo

Longitudo

Fig. I-2-1-1. From [Guillaumin, 1995, p. 54].

was subdivided again in twelve parts. It is obvious that the base 12 is not practical even with addition, let alone multiplication. Therefore, tables listing the multiples of integers and of Roman fractions were developed. The most important one is the treatise of Victorius of Aquitaine (fifth century), which was heavily commented by Abbo of Fleury (ca. 940–1004) [Fig. I-2-1-2].

The main content of the so-called Calculus by Victorius are 98 columns in which are given the multiples of the integers from 2 to 50 with integers (from 2 to 1,000) and with Roman fractions (from 1/144 on). For example, if 24 (= XXIIII) must be multiplied by 19 (= XVIIII), one goes to the table for 24 and notes how much the tenfold value is (CCXL) and how much the ninefold value is (CCXVI). Then the partial results must be added [Fig. I-2-1-3].

2. FINGER RECKONING

Since olden times it was very common to represent numbers by positions of fingers. Greek and Roman authors mention the custom to represent numbers below 100 on the left hand and the hundreds on the right hand. The term "finger reckoning" is somewhat misleading, because the fingers did not apparently serve the purpose of calculation; rather, different positions of the fingers represented the various numbers, and the system was simply a way of recording numbers. The oldest detailed description how this was done was given by Bede the Venerable (ca. 672–735), who belonged to the order of St. Benedict. In the first chapter of a treatise on calculating the date of Easter (*De temporum ratione*) he explains the *loquela digitorum* (i.e., the language of the fingers) and explains explicitly how by bending the fingers of the right and the left hand, all numbers less than 10,000 can be represented. The left hand is used for the units and the digits, the right hand for the hundreds and thousands. To present the numbers correctly, each finger joint must be bent and stretched independently from the others—a difficult task. At the end of this chapter Bede notes that even bigger numbers up to 1,000,000 can be represented by placing the hands on different parts of the body. In some manuscripts of Bede's treatise, illustrations show the different positions of the fingers [see Fig. I-2-2-1].

Calculating or Speaking with the Fingers. Before discussing the basics of the calculation of time, we have decided to demonstrate a few things, with God's help, about that very useful and easy skill of flexing the fingers, so that when we have conveyed maximum

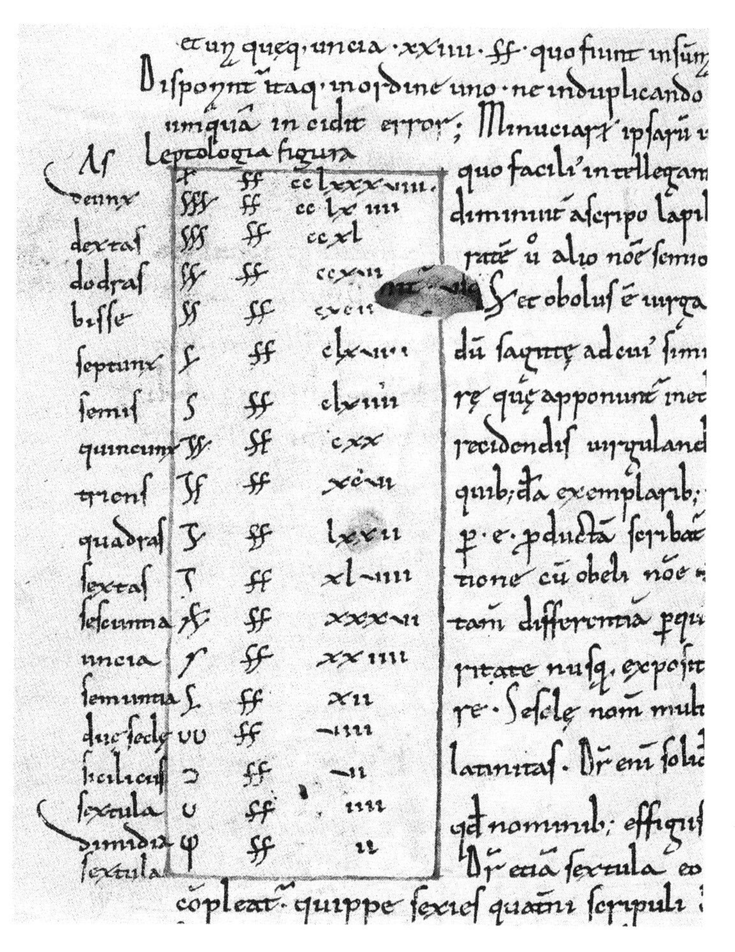

Fig. I-2-1-2. Roman fractions of 288, with their name, symbol, and value. For example, starting near the bottom, a *sextula* is 1/72, and so the value is 4; an *uncia* is 1/12, so its value is 24; a *triens* is 1/3, with value 96; a *semis* is 1/2, with value 144 (erroneously written as clxiiii = 164 instead of cxliiii); a *septunx* is 7/12, with value 168; and a *deunx* is 11/12, with value 264. Source: [Taegert, 2008, p. 25], reprinted by permission of Staatsbibliothek Bamberg.

facility in calculation, we may then, with our readers' understanding better prepared, attain equal facility in investigating and explaining the sequence of time through calculation. For one ought not to despise or treat lightly that rule with which almost all the exegetes of Holy Scripture have shown themselves well acquainted, no less than they are with verbal expressions. Many have said other things [on this topic], and even Jerome, that translator of the sacred narrative, says in his treatise on the evangelical precept (and [Jerome] did not hesitate to take up the aid of its discipline): "The thirty-fold, sixty-fold and hundred-fold fruit, though born of one earth and one seed, nevertheless differ vastly as to number. Thirty refers to marriage, for this conjunction of fingers depicts husband and wife, wrapped and linked (as it were) in a tender kiss. Sixty refers to widows, because their position is one of confinement and tribulation; hence they are pressed down against the upper finger,

XXIIII milia	mille
XXI milia DC	DCCCC
XVIIII milia CC	DCCC
XVI milia DCCC	DCC
XIIII milia CCCC	DC
XII milia[c]	D
VIIII milia DC	CCCC
VII milia CC	CCC
IIII milia DCCC	CC
II milia CCCC	C
II milia CLX	XC
mille DCCCCXX	LXXX
mille DCLXXX	LXX
mille CCCCXL	LX
mille CC	L
DCCCCLX	XL
DCCXX	XXX
CCCCLXXX	XX
CCXL	X
CCXVI	VIIII
CXCII	VIII
CLXVIII	VII
CXLIIII	VI
CXX	V
XCVI	IIII
LXXII	III
XLVIII	II
XXIIII	I
XXII	[illegible]
XX	[illegible]
XVIII	[illegible]
XVI	[illegible]
XIIII	[illegible]
XII	[illegible]
X	[illegible]
VIII	[illegible]
VI	[illegible]
IIII	[illegible]
III	[illegible]
II	[illegible]
I	[illegible]
[illegible]	[illegible]
[illegible]	[illegible]
[illegible]	[illegible]
[illegible]	[illegible]

Fig. I-2-1-3. This is the multiplication table for 24. Note the multiplication by fractions toward the end of the table. For example, 11/12 of 24 is 22, 1/12 of 24 is 2, and 1/72 of 24 is 1/3. Source: [Peden, 2003, pp. 19–20].

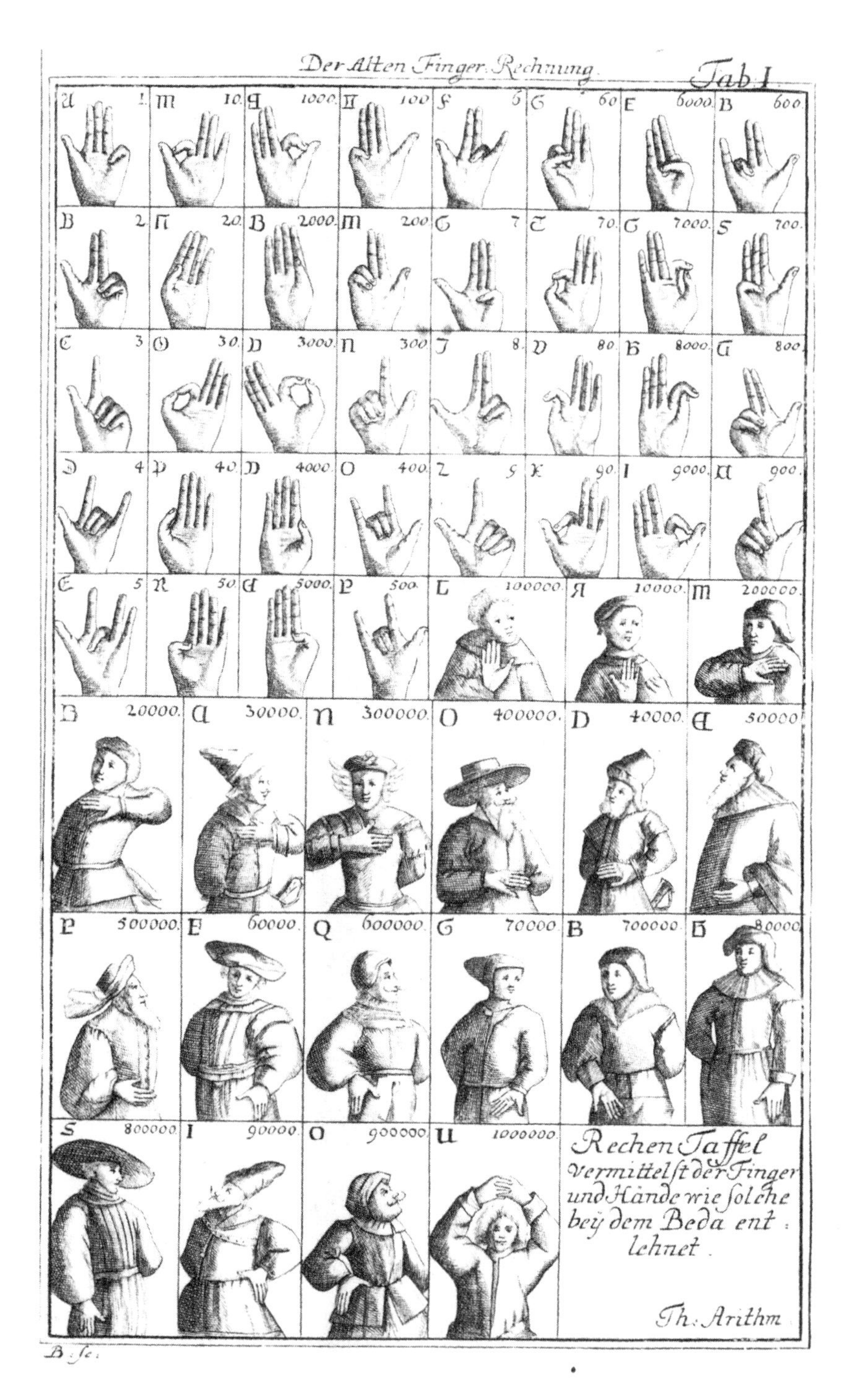

Fig. I-2-2-1. This is a page from Jacob Leupold's *Theatrum Arithmetico-Geometricum* (Leipzig, 1727). The author interchanges the hundreds and thousands compared with Bede's description.

for the more the will of a [sexually] experienced person suffers in abstaining from sin, the greater the reward. Finally the hundred-fold number (pay careful attention, reader, I pray!) is transferred from the left hand to the right, and symbolizes the crown of virginity by making a circle with the same fingers, but not on the same hand, by which marriage and widowhood are signified on the left hand."

So when you say "one," bend the little finger of the left hand and fix it on the middle of the palm. When you say "two," bend the second from the smallest finger and fix it on the same place. When you say "three," bend the third one in the same way. When you say "four," lift up the little finger again. When you say "five," lift up the second from the smallest in the same way. When you say "six," you lift up the third finger, while only the finger in between, which is called *medicus*, is fixed in the middle of the palm. When you say "seven," place the little finger only (the others being meanwhile raised), on the base of the palm. When you say "eight," put the *medicus* beside it. When you say "nine," add the middle finger.

When you say "ten," touch the nail of the index finger to the middle joint of the thumb. When you say "twenty," you insert the tip of the thumb between the middle joints of the index and middle fingers. When you say "thirty," you join the tips of the index and middle fingers in a gentle embrace. When you say "forty," you pass the under side of the thumb over the side or top of the index finger while holding both erect. When you say "fifty," you rest the thumb, bent at the last joint into the shape of the Greek letter *gamma*, against the palm. When you say "sixty," you carefully encircle the thumb, bent as before, by curving the index finger forward. When you say "seventy," you fill the index finger, bent as before, by inserting the thumb, with its nail upright, through the middle joint of the index finger. When you say "eighty," you fill the index finger, curved as before, with the thumb extended full length and its tip placed against the middle joint of the index finger. When you say "ninety," you place the tip of your bent index finger against the base of your upright thumb. So much for the left hand. You make one hundred on the right hand the way you make ten on the left, two hundred on the right the way you make twenty on the left, three hundred on the right the way you make thirty on the left, and the rest in the same manner up to nine hundred. You make one thousand on the right hand the way you make one on the left, two thousand on the right hand the way you make two on the left, three thousand on the right hand the way you make three on the left, and so forth up to nine thousand. Then when you say "ten thousand," you place your left hand flat on the middle of your chest, but with the fingers pointing upwards to the neck. When you say "twenty thousand," place the same hand, spread out sideways, on your chest. When you say "thirty thousand," place it flat but upright with the thumb on the breastbone. When you say "forty thousand," turn it on its back upright against the belly. When you say "fifty thousand," lay it flat but upright, with your thumb against your belly. When you say "sixty thousand," grasp your left thigh with your flattened hand. When you say "seventy thousand," turn [your hand] on its back on your thigh. When you say "eighty thousand," lay it flat on your thigh. When you say "ninety thousand," grasp your hip with your thumb turned towards the groin.

One hundred thousand, two hundred thousand and so forth up to nine hundred thousand, you perform in the same manner as we said, but on the right side of the body. When you say "one million," cross your two hands, linking your thumbs together.

3. ISIDORE OF SEVILLE, *LIBER NUMERORUM* (*BOOK OF NUMBERS*)

Many treatises on numerology were written during the medieval period in the Christian part of Europe. They connect arithmetic with theology by interpreting numbers allegorically. Terms like "mystery of numbers" (*mysteria numerorum*) or "sacred numbers" (*numeri sacrati*) indicate the topics that are dealt with. By explaining number as an allegorical symbol and uncovering the sacredness, the secrets can be made apparent. At the same time, numbers that have a specific mathematical meaning are also treated, especially in the theory of numbers according to the Pythagoreans and to Boethius. For instance, the number 6 is a (mathematically) "perfect number," because it is equal to the sum of its divisors. But other numbers can be "perfect" as well, because they have special relevance in the Bible. In this way, number and arithmetic serve as aids for understanding the Bible. For a modern reader, this unconnected coexistence of mathematical evidence and biblical number symbolism is difficult to understand, but in the Middle Ages it did not raise problems, as can be seen from the numerous texts that were popular at that time.

One of the earliest authors of such a treatise was Isidore of Seville. Later, in the twelfth century, many elaborate treatises on numerology were written by Hugh of St. Victor, Odo of Morimond, William of Auberive, Thibaut of Langres, and others. Here follows a short section from Isidore of Seville's *Liber numerorum qui in sanctis scripturis occurrunt* (*Book of Numbers That Occur in Sacred Scripture*). We only include the introduction and the sections dealing with the numbers 6 and 7. Noteworthy in this selection is the absence of numerical symbols, except in one place where Isidore used Roman numerals. Thus numbers are addressed conceptually by their names only, with no other symbols.

Paying attention to the meaning of numbers in Sacred Scripture is not a useless activity, because they have about themselves a certain teaching of knowledge and many mystical wonders. So, as you wished, I am pleased to make known briefly the rules for [understanding] certain numbers.

To begin with, number must be defined. Number is a collection of units or a multitude proceeding from one. Its size is indeed endless; it cannot be fixed.

An even number can be divided into two equal even numbers. An uneven number, however, cannot be divided into two equal parts, for one part in the middle will be either more or less than the other. An even times even number can be divided evenly into even numbers. And each part can be divided further into even numbers, until they reach unity that is indivisible. For example, consider sixty-four. Its half is thirty-two, then by halving, sixteen, whose half is eight, followed by four then two and finally one. One is the unique indivisible.

One is the least part of all numbers; it cannot be divided. However, one is the seed of numbers; but it is not a number. For from it the rest of the numbers flow or are procreated. It alone is the measure, the cause of any increase and decrease in any number. So, the increase of all numbers begins with it, and again they return to its unity.

. . .

Consider the number six. Six is the first perfect number because it is filled with its own parts. Six contains its sixth part, which is one, then its third part, which is two, and its half, which is three. The sum of these: one, two and three, is six. There is no number smaller than six, whose parts when divided sum to it. The perfection of this number is also clear in the work of the world, for in six days God completed His work. On the first day He created light, on the second day the firmament, on the third the sea and land, on the fourth the stars, on the fifth the fish and birds, and on the sixth day man and other animals. Then there are six days in which the world was perfected by ages. The first of these was from Adam to Noah, the second up to Abraham, third to David, the fourth up to the transmigration, the fifth up to the coming of Christ, and the sixth that is now, our era to the end of the world. The number six also consumes the year. The number sixty is the sixth part of days whose sum multiplied by the six of the first turn that is six sixties, is CCCLX. What is left are five days, to which if the four parts are joined, they make the number of sixes.

Indeed the fourth itself is made up of these. Likewise the number six, if it is associated with the square and solid number four, measures the hours of day and night, because four sixes make twenty-four. Moreover, the perfection of this number can be found in the ages of men and in the levels of things. The course of mortal life is sent over six ages, namely, infancy, childhood, adolescence, youth, adulthood, and old age. There are six levels of all things: nonliving like rocks, living like trees, sensitive like sheep, rational like men, immortal like the angels, and at the extraordinary and highest level that is above all things is God. Likewise there are six natural properties without which nothing can exist: magnitude, sight, figure, distance, state, and motion. There are even six different motions, because anything can be moved backwards or forward, to the left or right, up or down. Furthermore, many examples of the use of the number six are found in sacred preaching. For man was created in the image of God on the sixth day. Furthermore, during the sixth age of man the Savior became flesh. And on their sixth day in the desert the people were ordered to collect manna. Ezekiel told us that he saw a feather of six cubits on the right side of a man. He also wrote about the fountains of the temple being six cubits high. Likewise he commanded the prince to offer six lambs in sacrifice. In the Gospel there are six vases full of water that Christ changed into delicious wine. It was six days before his Passion that entering Jerusalem Christ was seated upon an ass. There are many other examples that are omitted here lest they tire the reader.

On the number seven. The number seven is born from no other number. It neither generates nor is generated. For all the numbers less than ten either produce others or are produced by others. Seven however neither generates nor is generated. Six and eight are only generated. Four and two both create and are created. Seven bears nothing nor is borne by another. Nevertheless, this number seven is legitimate. Wherever it is used as seventy or seven hundred, it contains so many sevens within itself. And among the wise persons of this world this number is considered perfect for this reason, because it is a combination of the first even number with the first odd. For the first odd number is three and the first even is four. From these two, seven was made. When these parts are multiplied then we have twelve. For whether by three fours or four threes, twelve is made.

But through three the mystery of the Trinity and through four the action of the virtues is illustrated. And by this in these parts[6] through the image of the Trinity the action of the

[6]Isidore is referring to what he wrote in an omitted passage that the numbers 3 and 4 illustrate the Trinity and virtues.

virtues is perfected, and through the representation of the virtues one comes to the notion of the Trinity. On the other hand, when they become twelve, this shows the twelve apostles made perfect by the seven-fold grace of the Spirit. For by their preaching with the strength of the four virtues, faith in the Trinity grew throughout the whole world. In Holy Scripture this number signifies peace during all time in this world. Sometimes it displays the unity of the Church. . . .

Seven very appropriately also signifies the Holy Spirit. According to the Law, sanctification belonged to the seventh day. For God did not sanctify His work on any day but only on the seventh, when he rested from his work. Therefore and by right that person carries the image of the Sevenfold Spirit, who through the fullness of divinity in Christ lives. Isaiah the prophet testified, "The spirit of the Lord shall alight upon him: a spirit of wisdom and understanding, a spirit of counsel and might, a spirit of knowledge and piety, and he shall be filled with the Spirit of the fear of the Lord" (11: 2).

4. HINDU-ARABIC NUMERALS

The decimal system with nine digits and zero comes from India. At first, distinct symbols were used for the units and the tens. A significant advance was made when practitioners adopted the same symbols for the tens, hundreds, and so forth as for the units. So began the "place-value" system for base 10. The system goes back at least to the first centuries CE. From the eighth century, inscriptions and manuscripts in India using these numbers become relatively numerous.[7] The decimal place-value system was also known outside India at least as early as the eighth century. In their numerous conquests beginning in the seventh century, the Arabs took over the forms of the numbers found in the conquered territories. Indian numerals with the zero were known to educated Muslims at least by 760. In the following centuries the Indian numerals became known throughout the Muslim empire from the eastern provinces to Spain. The form of the numerals was not uniform; in particular, two conventions were developed, one in the East (including Egypt) and the other in North Africa and Spain. They differ in the way some of the numerals were written.

Among the first who wrote mathematical works in Arabic was Muḥammad ibn Mūsā al-Khwārizmī (fl. 830), who worked in Baghdad. His treatises on arithmetic and algebra marked the beginning of the Arabic writings on these subjects. In the twelfth century, both his algebra and his arithmetic texts were translated into Latin, with the latter being the first in Latin describing the workings of the Hindu-Arabic system.

The Indian numerals themselves were known in the West much earlier. The oldest known examples are in two Latin manuscripts written in monasteries in Northern Spain in the tenth century: the *Codex Vigilanus* in 976 and the *Codex Emilianus* in 992. Both contain excerpts from Isidore of Seville's *Etymologiae* dealing with chronology and arithmetic. They are twin manuscripts (i.e., they go back to the same source). This source may have been a manuscript of the monastery of Ripoll in Catalonia, which was at that time a center of scholarship. One of the excerpts runs as follows.

[7] See [Plofker, 2009] for more details on the development of this system in India.

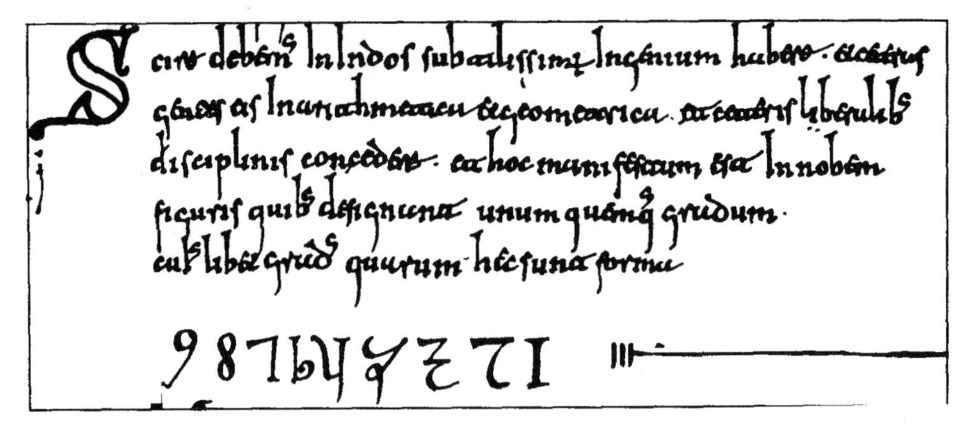

Fig. I-2-4-1. From the *Codex Vigilanus*. Source: [Burnam, 1912–1925, II, plate XXIII].

Also on the figures of arithmetic.

We must know that the Indians have a very subtle intellect and that the other nations give place to them in arithmetic, geometry, and the other liberal arts. This is obvious in the nine figures with which they denote any level of any number. The [numerals] have the following form: 9 8 7 6 5 4 3 2 1 [Fig. I-2-4-1].

The numerals from 1 to 9 appear here in the western Arabic form. The order of the numerals (from right to left) is in accordance with the Arabic way of writing. This form of the numerals became well known in monastic schools, because they were used on the calculation board (Latin: *abacus*). From the late tenth century until at least the mid-twelfth century the principal method of studying practical arithmetic in the schools of Western Europe was that of the abacus with marked counters. This originated with one person: Gerbert of Aurillac (ca. 945–1003). On a journey to the Spanish march now called Catalonia in 967, Gerbert studied mathematics and probably learned the Hindu-Arabic numerals as presented in the *Codex Vigilanus*. After his return to France he taught in the cathedral school at Reims. In 999 he became Pope Sylvester II. In his teaching Gerbert used an abacus. Richer, one of his students at Reims, informs us precisely about the form of this instrument:

In teaching geometry he gave himself no less trouble [than in the other disciplines]. As a preparation he introduced an abacus, i.e. a table of an appropriate size, with the help of a shield-maker. He divided its length in 27 parts and he placed marks, nine in number, to represent any number. In the likeness of which he had a thousand characters made out of horn, which he distributed through the 27 parts of the abacus, to designate the multiplication or division of each number: [the characters] divide or multiply the multitude of the numbers with such a shortening [of effort] that with the excessive multitude it is easier to understand than to describe in words. Whoever wishes to know fully the science of these [characters], may read his book which he wrote to Constantinus the grammarian. There he will find this enough and abundantly treated.

From this description, it follows that Gerbert's abacus was divided into 27 columns and that Gerbert used nine marks, each representing one of the nine digits. These marks were reproduced a thousand times, one each on a small piece of horn. There was no need for a zero. To represent a number such as 706, the mark for 6 was placed in the units' column and the mark for 7 in the hundreds' column. An empty tens' column needed no mark to signify that there are no tens in the number.

No extant manuscript of Gerbert shows what the marks looked like. But in manuscripts of the eleventh century we find pictorial representations of counting-boards, one of them with 27 columns, and there are texts of other scholars in which the abacus is described. From them we know that Gerbert indeed used the Hindu-Arabic numerals to mark the counters. Figure I-2-4-2 illustrates a counting board with the numerals represented in the second row. Moving from right to left are the numerals for 1, 2, 3, 4, 5, 6, 7, 8, and 9. The symbol to the left of the 9 resembles a zero but is not. Named *sipos* or *counter*, it is used as a placeholder in a nonmathematical sense: during a computation the person may wish to pause; so he places the *counter* at the place in the computation where he has paused. The first row in the accompanying figure lists the names of the numerals: *igin* (1), *andras* (2), *ormis* (3), *arbas* (4), *quinas* (5), *calctis* (6), *zenis* (7), *temenias* (8), and *celentis* (9). The names of the numerals 4, 5, and 8 were probably derived from the Arabic, but no widely accepted effort has explained the origin of the other words [Folkerts, 2001].

The numerals on Gerbert's abacus were placed at the top of counters and were not used for writing numbers. Therefore they do not really represent the decimal place value system. It was not comfortable to calculate with them, because during the operation a special counter had to be replaced many times by another one. Therefore, it is not surprising that Gerbert's abacus went out of use in the twelfth century, when the translations from the Arabic written calculation with the Hindu-Arabic numerals became known.

A consistency can be observed in the Latin forms of the numerals, from their earliest examples in the *Codex Vigilanus* (976) until the thirteenth century [Fig. I-2-4-3]. The * indicates forms not usually written by the scribe but done here with obvious difficulty. The figures 4, 5, and 7 were not changed to their present form before the end of the fifteenth century.

I-3. Arithmetic

1. BOETHIUS, *DE INSTITUTIONE ARITHMETICA* (*INTRODUCTION TO ARITHMETIC*)

The most important work on theoretical mathematics not only in the period treated in this section, but also all over Latin Christian Europe until the Renaissance, was the *Introduction to Arithmetic* (*De institutione arithmetica*), written by Anicius Manlius Severinus Boethius (ca. 480–524). It is a loose translation of the *Arithmetic* of Nicomachus (fl. first century CE) and is part of Boethius's translation program of around 500 CE, in which he tried to preserve Greek knowledge for the West—in a world where knowledge of Greek was rapidly vanishing. Boethius's *Arithmetic* is divided into two books and contains essential elements of Pythagorean theory of numbers, for example, the division of the (natural) numbers into even and odd; prime numbers; perfect numbers; and the classification of numerical ratios, which were used in the theory of music. He also treats the polygonal and polyhedral numbers (i.e., numbers represented by regular polygonal or polyhedral arrays of points), and finally the three means (arithmetic, geometric, and harmonic).

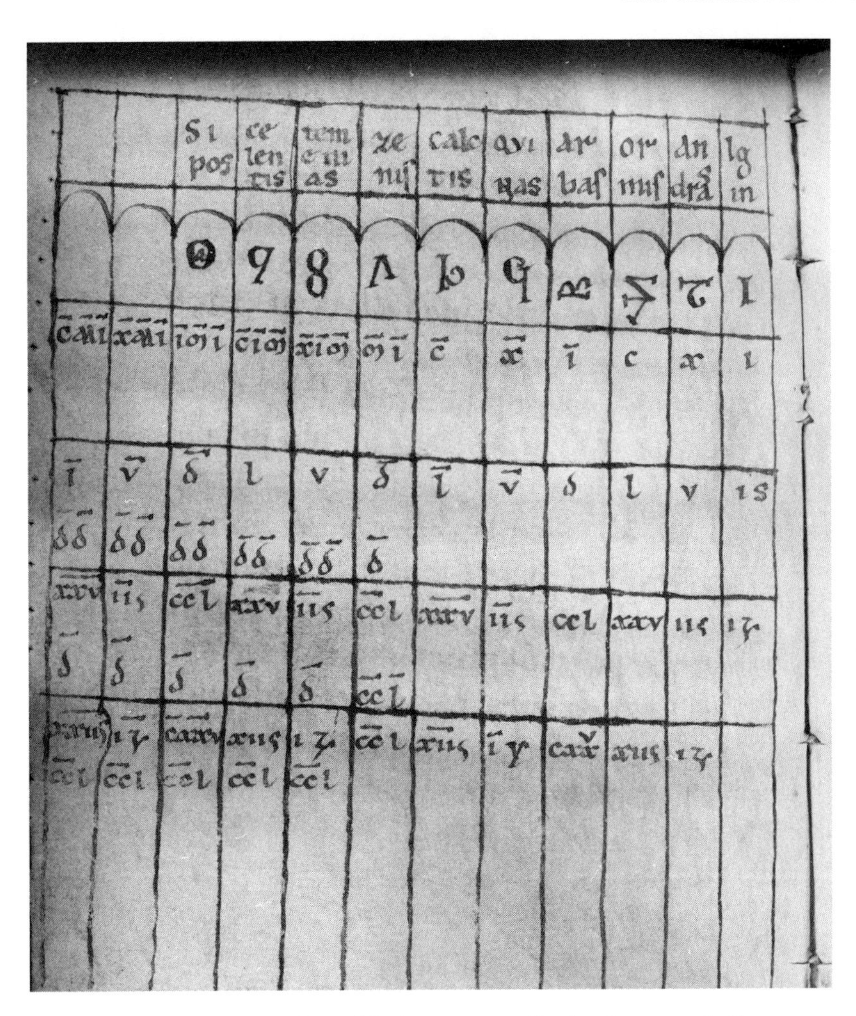

Fig. I-2-4-2. Hindu-Arabic numerals and their names on the top two rows in the representation of an abacus table (first half of the eleventh century); the remaining rows display Roman numerals representing column values. Source: [Folkerts, 2001]; reprinted by permission of Associazione culturale Amici di Archivum Bobiense.

Presented here are some selections from Book I. They refer to the definition of number, its division into even and odd and their properties (I 2–6); prime numbers (I 14); and perfect, imperfect, and superabundant numbers (I 19–20). It should be noted that, according to the Pythagoreans and to Boethius, 1 is not a number but is the unity from which numbers are developed. The idea of perfect numbers, being equal to the sum of their parts (including 1), was of special interest not only for Greek mathematics, but also for the Latin Middle Ages. In I 20 Boethius explains how perfect numbers can be found. In modern terms: if $1 + 2 + 4 + \ldots + 2^n = 2^{n+1} - 1 = p$ is a prime number, then $p \cdot 2^n$ is a perfect number. This theorem was proved in Euclid, *Elements* IX-36, and by its use, the first four perfect numbers

Arabic										
a) Rabat, Maktaba al-'āmma, k 222 (Ibn al-Yāsamīn)	9	8	7	6	5	4	3	2	1	
b) Univ. of Tunis, 2043, 1611, (ash-Sharīshī)	9	8	7	6	5	4	3	2	1	0
Latin										
a)* Escorial, d.I.2 (from Albelda, 976 A.D.)	9	8	7	6	5	4	3	2	1	
b) British Library, Harley 3595, s. XI^mid (Abacus forms)	9	8	7	6	5	4	3	2	1	
c) Munich Clm 18927 (unnamed row)	9	8	7	6	5	4	3	2	1	0
d) Cambridge, Trinity O.7.41	9	8	7	6	5	4	3	2	1	
e) Paris, BNF, lat. 16208 (*Liber ysagogarum*)	9	8	7	6	5	4	3	2	1	0
f) Vienna 275 (*Liber ysagogarum*)	9	8	7	6	5	4	3	2	1	0
g) Bodleian, Digby 51 (al-Battānī)	9	8	7	6	5	4	3	2	1	0
h) Hispanic Society of America, HC 397/726 (*Dixit Algorismi*)	9	8	7	6	5	4	3	2	1	0
i) Erfurt, Q 351 (John of Seville)		8		6	5	4	3	2	1	0
j) Erfurt, Q 365 (Raymond of Marseilles)			7	6		4	3		1	0
k)* Paris, BNF, lat. 16204 (*On the Great Conjunctions*)				6		4			1	
l)* Paris, BNF, lat. 14738 (Gerard of Cremona)	9	8	7	6	5	4	3	2	1	0
m)* Munich, Clm 18927 ('toletane f<igure>')	9	8	7	6	5	4	3	2	1	
n) Paris, BNF, lat. 15461 (*Liber Alchorismi*)	9	8	7	6	5	4	3	2	1	0
o) Raniero	9	8	7	6	5	4	3	2	1	0
p)* Florence, Bibl. naz. centrale, Magliabech. (Fibonacci I)				6	5	4	3	2	1	0
q) Florence, Bibl. naz. centrale, Magliabech. (Fibonacci II)	9	8	7	6	5	4	3	2	1	
Greek										
Vatican, gr. 184, s. XIII (*Psēphēphoria*)	9	8	7	6	5	4	3	2	1	0

Fig. I-2-4-3. Source: [Burnett, 2002, p. 265], reprinted by permission of Franz Steiner Verlag.

can be found easily:

$(2^2 - 1) \cdot 2 = 6$

$(2^3 - 1) \cdot 2^2 = 28$

$(2^5 - 1) \cdot 2^4 = 496$

$(2^7 - 1) \cdot 2^6 = 8128.$

From this, the (erroneous) conjecture came that perfect numbers ended alternately in 6 and 8 and that they were distributed in a regular way: always one perfect number between 10^n and 10^{n+1} ($n \geq 0$). The second part of this conjecture failed when the fifth perfect number, $(2^{13} - 1) \cdot 2^{12} = 33{,}550{,}336$, was discovered in the West by the thirteenth century and given by Johannes Regiomontanus in the middle of the fifteenth century [Folkerts, 2006, VIII, pp. 422–423]. The first part of the conjecture was disproved when it was discovered late in the sixteenth century that the sixth perfect number also ended in a 6.

I 2. The substance of number

From the beginning, all things whatever which have been created may be seen by the nature of things to be formed by reason of numbers. Number was the principal exemplar in the mind of the Creator. From it was derived the multiplicity of the four elements; from it were derived the changes of the seasons; from it the movement of the stars and the turning of the heavens. Since things are thus and since the status of all things is founded on the binding together of numbers, it is necessary that number in its own substance maintain itself evenly at all times, permanently, and that it not be composed of diverse elements. What substance would one join with number when the model of it itself holds all things together? It seems to have been composed of itself alone. Nothing can be considered as composed of similar parts or composed of things which are joined without reasonable proportion. Numbers are discrete of themselves and differ from every other substance and nature. But it becomes evident that number is composed of parts, not similar parts, nor of those things which adhere to each other without reasonable proportion. There are, therefore, first principles which join numbers together, which are in accord with its substance and which are always permanent. Nothing can be made from that which does not exist, and things from which something is made must be dissimilar but must possess the capacity of being combined. These then are the principles of which number consists: even and odd. These elements are disparate and contrary by a certain divine power, yet they come forth from one source and are joined into one composition and harmony.

I 3. The definition and division of number; the definition of even and odd

First we must define what number is. A number is a collection of unities, or a big mass of quantity issuing from unities. Its first division therefore is into even and odd. Even is that which is able to be divided into equal parts without one coming between the two parts. Odd is that which is unable to be divided into equal parts unless the aforesaid one should come between the parts. This kind of definition is common and well known.

I 4. The definition of even and odd number according to Pythagoras

Yet the definition of number is different according to the teaching of Pythagoras. An even number is that which can be divided by the same division into the biggest and the smallest [entities]: into the biggest in respect to the space [= magnitude], and into the smallest in respect to the quantity, according to the contrary properties of these two types.[8] An odd number is one to which this cannot happen but whose separation into two uneven parts is natural. Here is an example. If some given even number is divided, neither part is found

[8]This refers to the difference between magnitude and quantity. Concerning magnitude, halves are the greatest possible ones; concerning quantity, there are only two halves, the smallest possible quantity for division.

to be larger than the other; there is nothing but a separation into halves. No quantity is smaller when a division of each of these halves into equal parts is made. In this way the even number 8 is divided into 4 and 4. There cannot be any other division which would divide this term into a smaller number of parts since there are no fewer parts than two. Now when a whole is divided in a three-fold division, the total space within each part is diminished, but the number of the divisions is increased. What was said about the contrary natures of the two types of numbers is relevant to this kind of situation. We have previously explained that a quantity grows into infinite pluralities, and that a space, which may otherwise be called a magnitude, can be diminished into infinitely small sections and that this occurs in contrary ways. This is the case in the division of an even number: when the space is maximum, the quantity is minimum.

I 5. Another definition of even and odd, according to a more ancient method

According to a more ancient method, there is another definition of an even number. An even number is that which can be divided into two equal or two unequal parts, but in neither division is there an even number mixed with an odd number or an odd mixed with an even number, excepting alone the principal even binary number which cannot be divided into two unequal parts because it consists only of two unities. This number is the first equality, two. What I am saying is this: if one takes an even number, it can be divided into two even parts, as ten is divided into fives. It can also be divided into unequal parts, as when the same ten is divided into 3 and 7. In this manner, when one part of a division is even, the other part is even; and if one part is odd, the other part, which is also odd, when added to it does not make a total of more or less than ten. When ten is divided into fives, or into three and seven, and these portions are divided further, only odd numbers result. If, however, this number or any other even number is divided into even parts, as when eight is divided into 4 and 4, or in the same manner into odd numbers, as when the same eight is divided into 5 and 3, this is the result: in the first division both parts are even, and in the second both are odd. If one part of a division is even, the other cannot be odd, and if one part is odd, the other cannot be even. An odd number is that which is divided into other odd numbers by means of any division; numbers always show both types and neither type of numbers is ever able to exist without the other, but one must be understood as even, the other as odd. If you divide seven into three and four, one part is even, the other is odd. This same condition is found to exist in all odd numbers, and it can never be otherwise in the division of an odd number. These are the two types which naturally make up the power and substance of number.

I 6. The definition of even and odd in terms of each other

Now if these types must be defined in terms of each other, the odd number may be said to be that which differs from the even by a unity, either by increase or reduction. In the same way, an even number is one which differs from an odd number by a unity, either by increase or reduction. If you subtract one from an even number or add one to it, you make it odd; if you do the same thing to an odd number, an even number is created thereupon.

I 14. Prime and incomposite numbers

The prime and incomposite number is that which has no other factor but that one which is a denominator for the total quantity of that number so its fraction is nothing other than unity, and such are 3, 5, 7, 11, 13, 17, 19, 23, 29, 31. In these numbers, there is only one factor found, and each is a denominator of itself and that only once, as said above. In three is only one factor, that is, a third, which has a denominator of three, and unity is a third part of the number. In the same way, there is only a fifth part of five, and that is unity; this is found to be the case sequentially with each individual number. Such a number is called prime and incomposite because no number of this kind can be measured except through that factor which is the mother of all numbers, unity. A three cannot give name to 2, for if you compare just 2 alone to 3, it is smaller than 3, but if you double it, it becomes greater than 3, since it grows to 4. One number is a measure of another number as often as, either alone or doubled or multiplied by three, or however often that number is compared to another; its sum, neither smaller nor larger, comes exactly to the amount of the number it is compared with. So if you compare 2 to 6, the binary number is the measure of the larger number by three. Therefore, no other number can measure prime and incomposite numbers except unity alone, because it is put together from no other number, but is produced only from unities, increased and multiplied by them. Three times one is three, five times one is five, and seven times one is seven, and so all the others which we described above are generated in the same manner. These multiplied by themselves as prime numbers create other numbers, and you will find them to be the first substance and power of things; the origin of all numbers is generated from them as though they were elemental material because they are uncomposed and formed by simple generation. Into them all numbers are resolved since all numbers are drawn from them; these numbers are not produced from others nor can they be reduced to others.

I 19. Another division of even numbers according to perfect, imperfect, and superabundant

As much an introduction to odd numbers as brevity permits is thus finished. Of even numbers there is a second division. Some are abundant (*superflui*), others deficient (*diminuti*), according to their types of inequality. Now every type of inequality may be considered to be figured either in terms of larger numbers or in terms of smaller numbers. The larger numbers surpass by means of an immoderate plenitude, in terms of the numerosity of parts, the size of their own total bodies; the smaller numbers, as though needy and oppressed by poverty, suffer a certain slight lacking of their nature. The sum of their parts makes them such as they are. Those larger numbers, whose parts extend themselves more than is enough, are called abundant numbers, as 12 and 24. These, when compared to the sum of their parts factored out of the total body, are found to be smaller than that sum. Half of 12 is 6, a third part is 4, a fourth part is 3, and a sixth part is 2, a twelfth part is 1. The total sum [$6 + 4 + 3 + 2 + 1$] amounts to 16. This surpasses the total of the entire body. Again half of the number 24 is 12, a third is 8, a fourth is 6, a sixth is 4, an eighth is 3, a twelfth is 2, and a twenty-fourth is 1. All of these numbers add up to 36. In this matter, it is obvious that the sum of the parts is greater than and exceeds the size of the original number. And so this number, whose parts added together exceed the sum of the same number, is called abundant.

That number is called deficient whose parts, when put together in the same way, are exceeded by the multitude of the whole term, as 8 and 14. For 8 has a half, which is 4, and a fourth, which is 2, and an eighth, which is 1, all of which added together give 7, and this number is still confined within the total body of 8. Again 14 has a half which is seven, a seventh, which is two, and a fourteenth, which is one, the sum of which reaches the number 10. This is altogether smaller than the original term. So these numbers, those whose parts added together exceed the total, are seen to be similar to someone who is born with many hands more than nature usually gives, as is the case with a giant who has a hundred hands, or three bodies joined together, such as the triple-formed Geryon. Or this number is like some monstrosity of nature which suddenly appears with a multiplicity of limbs. The other number, whose parts when totaled are less than the size of the entire number, is like one born with some limb missing, or with an eye missing, like the ugliness of the Cyclops' face. Or the number is like one who is born naturally deficient in relation to some member, who emerges short of his total fullness.

Between these two kinds of number, as if between two elements unequal and intemperate, is put a number which holds the middle place between the extremes like one who seeks virtue. That number is called perfect and it does not extend in a superfluous progression nor is it reduced in a contracted reduction, but it maintains the place of the middle; the sum of its parts is not more than the total nor does it suffer from a lack in comparison with the total, as are 6 and 28. Six has a half which is three, and a third which is two and a sixth which is one, and these numbers, if brought together to form a total sum, are found to be equal to the original number. Indeed, the number 28 has a half of 14 and a seventh part of 4, and is not lacking in a fourth part, 7, and a fourteenth part, 2, and a twenty-eighth part, one, all of which numbers brought together into one total are equal in parts to the size of the original term. Joined together, all these parts equal 28.

I 20. Concerning the generation of the perfect number

There is in these a great similarity to the virtues and vices. You find the perfect numbers rarely, you may enumerate them more easily, and they are produced in a very regular order. But you find superfluous or diminished numbers to be many and infinite and not disposed in any order, but arranged randomly and illogically, not generated from a certain point. Within the first ten numbers there is only one perfect number, 6; within the first hundred, there is 28; within a thousand, 496; within ten thousand, 8,128. These perfect numbers always end in one of two numbers, 6 or 8, and these numbers always provide the final term in alternating fashion for the perfect numbers. First there is six, then 28, after this 496, which ends in 6 and which is the first number, then 8,128 which ends in 8 and that is the second number.

The generation and production of these numbers is fixed and firm and they cannot be achieved in any other way nor, if they are brought about in this way, is it at all valid to create them by another mode. Let the even times even numbers be disposed from one, and in order as far as you wish. Then you will add together the second with the first, and if a prime and incomposite number is made from that addition, you will then multiply the sum by the second number you have added on. If from that addition, a prime number does not emerge, but a secondary and composite number, skip this number and add to it the next one which follows. If then you do not yet emerge with a prime and incomposite number, add another number and see what emerges. But if you find a prime and incomposite number, then multiply it by the number added on from the last sum. Now the even times even numbers may be disposed in this manner: 1, 2, 4, 8, 16, 32, 64, 128. Then work in this fashion. Put down one and add two, and see what number is made from this addition; from it comes 3, which of course is prime and incomposite. After unity, you add the number two. If you then multiply three, which is the number achieved by this addition, by two, which is the last number added on to the sum, then without a doubt a perfect number is born. Twice 3 makes six, which has one part which is its factor, namely six, and 3 is its half, while 2, the second number added on, is its third and these two have been multiplied to give the product 6.

Twenty-eight is produced in the same way. If to one and two, which make three, you add the following even times even number, that is four, you arrive at seven. Then take the last number four, which you added in sequence, multiply the total by it, and a perfect number is produced. Seven times 4 is 28, which is a number equal to the sum of its parts; it has one as its factor, which is its twenty-eighth, and a half of 14, a fourth of 7, a seventh of 4, and a fourteenth of 2 and that corresponds to the medial term.

After these numbers are found, if you want to go on to discover the others, it is necessary that you pursue them by the same reasoning. It is necessary that you put down one, and after this 2 and 4, which add up to seven. The perfect number 28 showed itself a while ago by means of this process. The even times even which follows this number is 8; this is added on, increasing the previous number to 15. But this is not a prime and incomposite number, for it has another factor beyond the number by which it is named in itself, that is beyond the factor of one into fifteen. Because this number is secondary and composite, pass by it and add to the previous number the next even times even number, that is 16, which when added to 15, makes thirty-one. This is prime and incomposite. Multiply this number by the last number added on to the total, so that 16 times 31 yields 496. Now this sum is the perfect number within the order of 1000, and it is equal to the sum of its parts.

Unity is first in power and possibility but not in act and operation, and is itself perfect. If I take this first number from the proposed order of numbers I see that it is prime and incomposite because if I multiply it by itself, that same unity is produced for me. Also, the number one brings about only unity, which is equal only in potency to its parts, and is also like other perfect numbers in act and operation. Rightly therefore is unity perfect in its own strength, because it is prime and incomposite and when by means of itself it has been multiplied, it maintains itself through its own strength.

2. COMPUTUS

This section was prepared by Immo Warntjes

After the fall of the Western Roman Empire in the fifth century, the rapidly spreading monasteries replaced secular institutions as the centers of learning. As a consequence, the study of science turned decidedly Christian in character. It had two principal objectives, one general and one very specific: deciphering God's creation as manifested in the cosmos and regulating the liturgical calendar. A new subject was created called "computus," which ultimately embraced all scientific concepts only vaguely related to its core themes (from blood-letting to tidal theory). In the first three centuries of the emerging Latin West (ca. 500–800), science was synonymous with computus. In the Carolingian period, more nuanced scientific subjects like astronomy and geometry emerged, but computus retained its dominant position until the adoption of Arabic, Greek, and Hebrew science in the eleventh and twelfth centuries.

The centerpiece of computus was the calculation of Easter, the most important Christian feast. It was agreed that Easter should be celebrated on the first Sunday after the first full moon after the spring equinox (though the last criterion was rather fluid at the beginning). Therefore, the course of the sun and that of the moon had to be pressed into mathematical models, and these had to be combined. The sun's orbit was defined as $365\frac{1}{4}$ days; this constituted a solar year as represented in the Julian calendar, which determined a 28-year solar or weekday cycle. (For example, if a leap year started on a Sunday, it would take 28 years before the next leap year started on a Sunday; see the fourth column in Fig. I-3-2-1. The principal challenge was to align the moon phases of 29.5306 days (a synodic lunar month, the period from new moon to new moon) with the Julian calendar. The West preferred an 84-year lunar cycle, which had the advantage of coinciding with exactly three solar cycles ($84 = 3 \times 28$). The East, with its intellectual center in Alexandria, followed Babylonian and Greek tradition by applying a more accurate 19-year lunar cycle commonly ascribed to Meton (the Metonic cycle); since 19 and 28 have no common divisor, this led to an Easter cycle of $19 \times 28 = 532$ years. The inevitable controversy about the correct calculation of Easter provided much of the scientific stimulus in the early Middle Ages. By around 800, the Alexandrian system was the commonly accepted Easter reckoning throughout Christendom and remained so until the Gregorian calendar reform of 1582; in the Orthodox Church it prevails to the present day.

The Latin West became acquainted with Alexandrian practice through translation from Greek into Latin by Dionysius Exiguus (ca. 470–544) in 525, and therefore this system is commonly called the Dionysiac reckoning. His computistical writings include two letters, a translation of a letter by Proterius of Alexandria, an Easter table, and calendrical formulas (*argumenta*) for the calculation of its data. In his Easter table, he introduced AD as the preferred linear timeline, and AD forms the basis for the calendrical algorithms. As these formulas provide the data at the heart of computus, they proved extremely popular over the centuries. They were cited, refined, and added to in 562 (in the circle of Cassiodorus), 625, 675, and 689 (in the circle of Willibrord[9]). In the early eighth century, they were used by the Venerable Bede. Since the explosion of manuscripts in the Carolingian age, they can be found in almost every one of the thousands of medieval computistical codices.

Fig. I-3-2-1 reproduces a standard Dionysiac Easter table and the formulas (*argumenta*) to calculate its data as outlined in Cassiodorus's 562 recension of Dionysius's 525 original.[10] The last formula, for the calculation of the Julian calendar date of the Easter full moon and Easter Sunday, is not originally from Dionysius; the version translated here was possibly composed by Alcuin in 776.

[9]Willibrord (ca. 658–739) was a Northumbrian missionary saint, known as the "Apostle to the Frisians" in the modern Netherlands. He became the first Bishop of Utrecht and died at Echternach, Luxembourg.

[10]Dionysius's original is lost but can be reconstructed from the surviving evidence; see [Krusch, 1938, pp. 75–77], [Mosshammer, 2008, pp. 96–106], and [Warntjes, 2010].

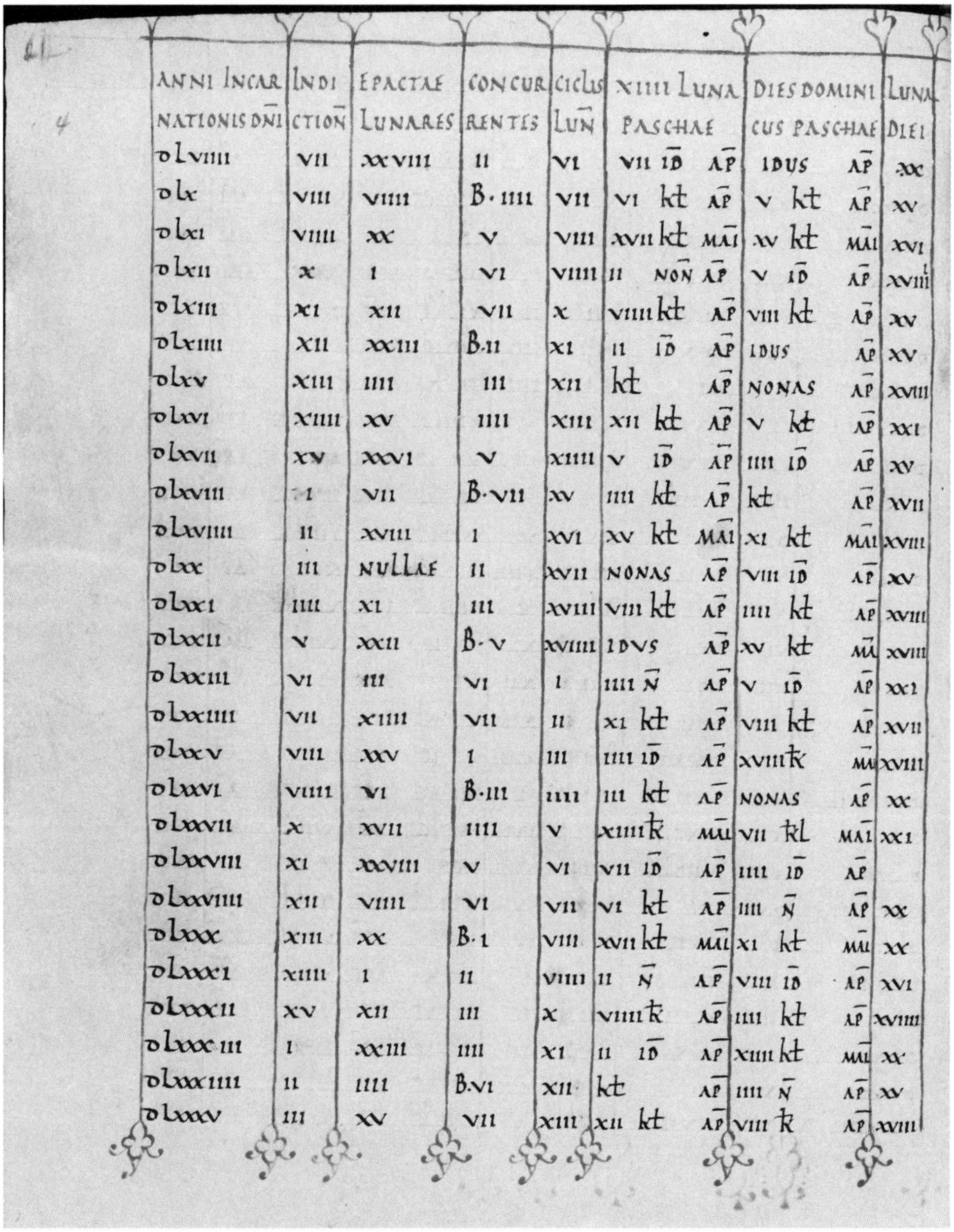

ANNI INCARNATIONIS DNI	INDICTION	EPACTAE LUNARES	CONCURRENTES	CICLUS LUN	XIIII LUNA PASCHAE		DIES DOMINICUS PASCHAE		LUNA DIEI
DLVIIII	VII	XXVIII	II	VI	VII ID	AP	IDUS	AP	XX
DLX	VIII	VIIII	B. IIII	VII	VI KL	AP	V KL	AP	XV
DLXI	VIIII	XX	V	VIII	XVII KL	MAI	XV KL	MAI	XVI
DLXII	X	I	VI	VIIII	II NON	AP	V ID	AP	XVIII
DLXIII	XI	XII	VII	X	VIIII KL	AP	VIII KL	AP	XV
DLXIIII	XII	XXIII	B. II	XI	II ID	AP	IDUS	AP	XV
DLXV	XIII	IIII	III	XII	KL	AP	NONAS	AP	XVIII
DLXVI	XIIII	XV	IIII	XIII	XII KL	AP	V KL	AP	XXI
DLXVII	XV	XXVI	V	XIIII	V ID	AP	IIII ID	AP	XV
DLXVIII	I	VII	B. VII	XV	IIII KL	AP	KL	AP	XVII
DLXVIIII	II	XVIII	I	XVI	XV KL	MAI	XI KL	MAI	XVIII
DLXX	III	NULLAE	II	XVII	NONAS	AP	VIII ID	AP	XV
DLXXI	IIII	XI	III	XVIII	VIII KL	AP	IIII KL	AP	XVIII
DLXXII	V	XXII	B. V	XVIIII	IDUS	AP	XV KL	MAI	XVIII
DLXXIII	VI	III	VI	I	IIII N	AP	V ID	AP	XXI
DLXXIIII	VII	XIIII	VII	II	XI KL	AP	VIII KL	AP	XVII
DLXXV	VIII	XXV	I	III	IIII ID	AP	XVIII KL	MAI	XVIII
DLXXVI	VIIII	VI	B. III	IIII	III KL	AP	NONAS	AP	XX
DLXXVII	X	XVII	IIII	V	XIIII KL	MAI	VII KL	MAI	XXI
DLXXVIII	XI	XXVIII	V	VI	VII ID	AP	IIII ID	AP	
DLXXVIIII	XII	VIIII	VI	VII	VI KL	AP	IIII N	AP	XX
DLXXX	XIII	XX	B. I	VIII	XVII KL	MAI	XI KL	MAI	XX
DLXXXI	XIIII	I	II	VIIII	II N	AP	VIII ID	AP	XVI
DLXXXII	XV	XII	III	X	VIIII KL	AP	IIII KL	AP	XVIIII
DLXXXIII	I	XXIII	IIII	XI	II ID	AP	XIIII KL	MAI	XX
DLXXXIIII	II	IIII	B. VI	XII	KL	AP	IIII N	AP	XV
DLXXXV	III	XV	VII	XIII	XII KL	AP	VIII KL	AP	XVIII

Fig. I-3-2-1. Page of an Easter table covering the years 559–585. Source: [St. Gallen, Stiftsbibliothek, 250, p. 4], reprinted by permission of Stiftsbibliothek St. Gallen.

First column—AD (*Anni Incarnationes Domini*): The linear timeline from the birth of Christ (*Anni Domini*, or AD) appears to have been invented in the third century. It was popularized through the Dionysiac Easter reckoning, replacing the count from the Passion of Christ (*Anni Passionis*, or AP) connected to the Victorian reckoning and the world years

(*Anni Mundi*, or AM) of chronicling tradition. Its accuracy was questioned in the tenth to twelfth centuries, but it proved persistent and still is the common era (CE) used today. The algorithms presented by Dionysius and Cassiodorus started with a formula (*Argumentum* 1) for calculating AD. In general, at this time, turning all the historical/chronological data available into AD was a major operation for chroniclers, since years were usually named in terms of reigns of kings or Roman consuls. It was only during the Carolingian period that AD became generally used. In any case, all of the following formulas are based on the knowledge of AD.

Second column—indiction (*Indictiones*): The indiction is a fiscal cycle of 15 years introduced in the Roman Empire in the late third century, which remained a popular dating tool throughout the Middle Ages. The indiction was calculated from AD as follows (*Argumentum* 2):

If you want to know the indiction ..., take the years from the incarnation of our Lord Jesus Christ, i.e. 562. To these always add 3, which makes 565. These divide by 15, and 10 remain: 10 is the indiction. Should, however, nothing remain, the indiction is 15.

In mathematical terms, indiction = (AD + 3) (mod 15); if 0, then use 15. For example, the indiction of 562 is (562 + 3) (mod 15) = 10. The constant 3 indicates that an indiction cycle started three years before the beginning of the AD count, since AD 1 = indiction 4.

Third column—epact (*Epactae Lunares*): The key problem for the calculation of Easter was aligning the moon phases with the Julian calendar. In the Dionysiac reckoning, a 19-year cycle was employed. This means that for any given Julian calendar date, the same lunar age recurs after exactly 19 years. A year within this cycle can therefore be defined by the lunar age of a set Julian calendar date. The date chosen was equivalent to 22 March (the earliest possible Easter Sunday), with the lunar age on this date being termed the epact of a given lunar year. The epact was calculated from AD as follows (*Argumentum* 3):

If you want to know the epact ..., take the years from the incarnation of the Lord, which are, say, 562. These divide by 19, 11 remain. These multiply by 11, which makes 121. These divide by 30, 1 remains. 1 is the epact.

In mathematical terms, epact = (AD (mod19) × 11) (mod 30). For example, the epact of 562 is (562 (mod19) × 11) (mod 30) = 1. The epact is ruled by the 19-year lunar cycle (i.e., the same value recurs after exactly 19 years); thus we calculate modulo 19. The epact increases by 11 every year, since the solar year of 365 days is 11 days longer than 12 lunar months of 29.5 days. Thus, we multiply the year number in the 19-year cycle by 11 to calculate the number of extra days beyond that number of 12-month lunar years. Because the epact can never exceed 30, the maximum length of a lunar month, we then take the remainder modulo 30 to calculate the lunar age.

Fourth column—concurrent (*concurrentes*): As Christ's resurrection occurred on a Sunday, Easter was set to fall on this weekday. Therefore, the 7-day weekday cycle had to be aligned with the Julian calendar of $365\frac{1}{4}$ days. Since it was preferred to assign an integer number of days to the calendar year, three common years of 365 days were followed by a bissextile (leap) year of 366 days. The weekday cycle of 7 days recurs 52 times in the 365 days of a calendar year, with a remainder of one day ($365 = 52 \times 7 + 1$). Therefore, the weekday of any given Julian calendar date increases by one in common years and by two in bissextile years. This leads to a solar or weekday cycle of 28 years. A year within this cycle can thus be defined by the weekday of a set Julian calendar date plus the additional information of whether this year is bissextile or not. The date chosen was equivalent to 24 March, calling the weekday on that date the "concurrent." The weekdays (*feriae*) were numbered from 1 (Sunday) to 7 (Saturday). The concurrent was calculated from AD as follows (*Argumentum* 4):

If you want to know ...the concurrent days of a week, take the years from the incarnation of the Lord, which are, say, 562. ...Always add the fourth part, here 140, which together make 702. To these add 4 [which makes 706]. These divide by 7, 6 remain. 6 is the concurrent.

In mathematical terms, concurrent = (AD + [AD/4] + 4) (mod 7); for example, the concurrent of 562 is (562 + 140 + 4) (mod 7) = 6. The concurrent increases by 1 every year and by an extra 1 in bissextile years every 4 years, thus we add AD to [AD/4]. But the year before the year 1 (i.e., 1 BCE) is considered to be bissextile and its concurrent is Wednesday (i.e. *feria* 4 in medieval terminology). Thus we also add the fixed parameter 4. Since the concurrent is ruled by the seven-day weekday cycle, it never exceeds 7, thus we calculate modulo 7.

The bissextile year was calculated from AD as follows (*Argumentum* 8):

If you want to know, when a bissextile day occurs, take the years of the Lord, say, 562. These divide by 4. If nothing remains, there is a bissextile day; if 1 or 2 or 3 remain, there is no bissextile day.

In mathematical terms: y = AD (mod 4); if $y = 0$, then AD is bissextile.

Fifth column—lunar cycle (*cyclus lunaris*): The Dionysiac reckoning is based on a 19-year lunar cycle. The start year of this cycle is a matter of convention. Dionysius continued the Alexandrian Easter table, which was appropriately arranged in sections of 19 years, beginning in 437 CE and 19-year equivalents (termed *cyclus decemnovenalis*). The Byzantines, however, started the 19-year lunar cycle three years later (termed *cyclus lunaris*). The year in the *cyclus lunaris* was calculated from AD as follows (*Argumentum* 6):

If you want to know the year in the *cyclus lunaris* ..., take the years of the Lord, say 562. Always subtract 2, which makes 560. These divide by 19, 9 remain. It is the 9th year in the *cyclus lunaris*. If nothing remains, it is the 19th.

Given that year 1 of the 19-year cycle is set at 440, which is congruent to 3 (mod 19), any other year congruent to 3 will also have year number 1. So to calculate the year number of year n, we must calculate its remainder modulo 19 and then subtract 2. The given algorithm simply reverses these two steps. That is, *cyclus lunaris* = (AD – 2) (mod 19), but if this is 0, we take the year number as 19. Thus, for the year 562, the *cyclus lunaris* is (562 – 2) (mod 19) = 9.

Sixth to eighth columns: With the information of columns 3 and 4 (the epact and the concurrent, respectively), it is straightforward to calculate the Julian calendar date and weekday of the Easter full moon (column 6). From the lunar age of 22 March (epact), one only has to count forward to the next lunar age 14; the date on which this falls is the Julian calendar date of the Easter full moon. With this date established, one only has to count the weekdays forward from 24 March (the concurrent) to determine the weekday on that date. The following Sunday is the Julian calendar date of Easter Sunday (column 7); its lunar age (column 8) is established by counting forward to this date from lunar age 14 of the Easter full moon. On the table, the dates are indicated in the Roman fashion: There are three marker days in each month, called the Calends (*KL-* in Fig. I-3-2-1), the Nones (*NON-* or *N-*), and the Ides (*ID-*), which indicate Roman feast days. The Calends are the first day of the month; in March, May, July, and October, the Nones are on the seventh of the month and the Ides on the fifteenth; in other months, the Nones are on the fifth and the Ides on the thirteenth. A date is indicated by the number of days before the next of these marker days, with the count including the days at both ends of the interval. So, for example, the sixth day before the Ides of April is 8 April.

We complete the example for the year 562, given that the epact is 1 and the concurrent is 6 (so 24 March is a Friday); also 562 is not a bissextile year. Since the lunar age of 22 March (epact) is 1, the Easter full moon will be 13 days later, that is, 4 April, which is the second day before the Nones of April. And since 24 March is a Friday, 4 April is a Tuesday and the following Sunday, 9 April, is Easter. This is designated as the fifth day before the Ides of April. Finally, since the lunar age of 4 April is 14, the lunar age of Easter Sunday is 19.

In the late sixth or early seventh century, an algorithm was invented to calculate the data of columns 6 to 8, known today as the pseudo-Dionysiac *Argumentum* 14; it was variously refined, most notably in the *Calculatio Albini* of 776, and remained popular throughout the Middle Ages:

From the epacts, you easily recognize whether the Easter full moon falls in March or April. If the epact is more than 15 or less than 5, the Easter full moon is computed in April. If, however, it is more than 5 and less than 15 the Easter full moon is considered in March. [Epact 5 does not exist; for epact 15, the Easter full moon falls in March.]

In April, take the 35 regulars and always subtract the epact of this year; and what remains is the day of the Easter full moon. E.g., in the third year of the *cyclus decemnovenalis* the epact is 22; deduct 22 from 35, and 13 remain. The Easter full moon occurs on the 13th day of April, i.e. the Ides.

If you are looking for the weekday of the Easter full moon, add the concurrent to the number that remained, say those 13 of the above, and also the 7 regulars in April. Take all of these together and then divide by 7. And what remains is the weekday of the Easter full moon. Like this you should easily get to Easter Sunday.

In March, on the other hand, take the 36 regulars and subtract the epact of this year, e.g., 11 in the second year of the *cyclus decemnovenalis*, which makes 25. The Easter full moon falls on the 25th day of this very month, i.e. the 8th Calends of April.

If, however, you are looking for the weekday of this day, add to the above given number the concurrent of this year and also the 4 regulars in March. Take all of these together and divide by 7 and what remains is the weekday of the Easter full moon. If nothing remains, it is *feria* 7 (Saturday).

Should 30 remain after deduction of the epact, this remainder still is the day on which you will find the Easter full moon. As in the year, in which the epact is 4, deduct 4 from 35, and 31 remain; deduct 30, and 1 remains. The Easter full moon occurs on the first day of the month, i.e. the Calends of April. If, however, only 30 remain after deduction of the epact, the Easter full moon falls on the 30th day of the month, which happens once in 19 years, when 6 is ascribed to the epact and the Easter full moon falls on 30 March (3rd Calends of April).

I should instruct you with examples for finding the weekday on which the Easter full moon falls: E.g., in the present year of AD 776, take epact 26 of this year; after these are deducted from the 35 regulars, 9 remain. And, indeed, the Easter full moon will be on the ninth day of the month, i.e. the 5th Ides of April. Combine the concurrent of the present year, i.e. 1, with these 9, which makes 10. To these add 7, which makes 17. These divide by 7: two times 7 is 14, and 3 remain. The Easter full moon (*luna* 14) will be on *feria* 3 (Tuesday), *luna* 15 on *feria* 4 (Wednesday), *luna* 16 on *feria* 5 (Thursday), *luna* 17 on *feria* 6 (Friday), *luna* 18 on *feria* 7 (Saturday), *luna* 19 on *feria* 1 (Sunday), which is Easter Sunday. The Easter full moon (*luna* 14) is on 9 April, *luna* 15 on 10 April, *luna* 16 on 11 April, *luna* 17 on 12 April, *luna* 18 on 13 April, *luna* 19 on 14 April.

In the year following this one, because epact 7 is more than 5 and less than 15, the Easter full moon falls in March, which you find thus: Take the 36 regulars of March, deduct 7 from these, and 29 remain. On the 29th day of March, i.e. on the 4th Calends of April, the Easter full moon will appear to you. In order to find the weekday, take these 29, and add to these the concurrent of that year, i.e. 2, which makes 31. To these also add the 4 regulars, and altogether there will be 35. Divide by 7: 5×7 is 35, and nothing remains. Thus, the Easter full moon will be on *feria* 7 (Saturday), 29 March; and *luna* 15 will be on *feria* 1 (Sunday), 30 March, which is Easter Sunday.

In mathematical terms, we have

(a) If $15 <$ epact or epact < 5, then the Easter full moon is in April. In fact, (35 – epact) (mod 30) $= y$ April for the Easter full moon. Also ($y +$ concurrent $+ 7$) (mod 7) = feria z for the Easter full moon; if 0, then use 7.

For example, in AD 776, the epact is 26 and the concurrent is 1. Therefore, $(35 - 26) \pmod{30} = 9$ April for the Easter full moon and this date falls on $(9 + 1 + 7) \pmod 7 =$ *feria* 3 (Tuesday).

or

(b) If $5 <$ epact ≤ 15, then the Easter full moon is in March. In fact, (36 – epact) $= y$ March for the Easter full moon. Also ($y +$ concurrent $+ 4$) (mod 7) = feria z for the Easter full moon; if 0, then use 7.

For example, in AD 777, the epact is 7 and the concurrent is 2. Therefore, $36 - 7 = 29$ March for the Easter full moon, and this date falls on $(29 + 2 + 4) \pmod 7 =$ *feria* 7 (Saturday).

These rules can be explained as follows:

(a) If the Easter full moon (*luna* 14) falls on y April and the epact on 22 March, then the Julian calendar day difference between y April and 22 March $(y + 9)$ is congruent modulo 30 to the lunar day difference between *luna* 14 and the epact (the congruence modulo 30 is necessary here, as the epact and the Easter full moon may belong not to the same, but successive lunar months, with the lunar month ending between them consisting of 30 days). So $(y + 9) \pmod{30} \equiv$ (14 – epact) (mod 30), or $y \pmod{30} \equiv$ (5 – epact) (mod 30), which is the same as $y =$ (35 – epact) (mod 30).

If *feria* z is the weekday of the Easter full moon on y April and the concurrent is the weekday on 24 March, then the Julian calendar day difference $y + 7$ between y April and 24 March is congruent modulo 7 to the weekday difference between *feria* z and the concurrent (the congruence modulo 7 is necessary here, because the weekday of the Easter full moon occurs in a different week than the concurrent). Thus, $(y + 7) \pmod 7 \equiv (z - \text{concurrent}) \pmod 7$, or $z = (y + 7 + \text{concurrent}) \pmod{30}$.

(b) If the Easter full moon (*luna* 14) falls on y March and the epact on 22 March, then $y - 22 = 14 - \text{epact}$, so $y = 36 - \text{epact}$.

If *feria* z is the weekday of the Easter full moon on y March and the concurrent the weekday of 24 March, then the Julian calendar day difference $y - 24$ between y March and 24 March is congruent modulo 7 to the weekday difference between *feria* z and the concurrent (the congruence modulo 7 is necessary here, because the weekday of the Easter full moon may occur in a different week than the concurrent). So $(y - 24) \pmod 7 \equiv (z - \text{concurrent}) \pmod 7$, and $z \equiv (y - 24 + \text{concurrent}) \pmod 7$, which is the same as $z = (y + 4 + \text{concurrent}) \pmod 7$.

Addendum: One of the key mathematical operations in computus was the modulo-operation, the division with remainder. The pseudo-Dionysiac *Argumentum* 16, which appears to have originated in the second half of the seventh century, provides a good insight into how this operation was executed with higher numbers, and it also is illuminating for multiplications. It consists of two main calculations: (a) establishing the number of hours in a year by multiplying 365 days with 2×12 hours ($= 8{,}760$ hours); (b) division of these 8,760 hours by 7. For a better understanding of the complexity, the Roman numerals are reproduced in the translation. (Note that a full day of 24 hours is here divided into 12 hours of day and 12 hours of night.)

First, calculate how many hours CCC days [of 12 hours] have; ten times three hundred makes $\overline{\text{III}}$;[11] two times three-hundred is six hundred; in three hundred days [of 12 hours] this makes $\overline{\text{III}}$ DC hours. Ten times sixty is DC, and two times sixty is CXX; thus, in sixty days there are DCCXX hours. Ten times five is L, and two times five is X; see, in five days [of 12 hours] you have LX hours. Together, in a full year, in CCCLXV days, there are $\overline{\text{IIII}}$ CCCLXXX hours; and there are as many in the night; together this makes $\overline{\text{VIII}}$ DCCLX hours of day and night in a full year.

Divide these in VII parts: First, seven times one thousand is $\overline{\text{VII}}$, $\overline{\text{I}}$ DCCLX remain. Seven times two hundred makes $\overline{\text{I}}$ CCCC, CCCLX remain. Seven times fifty makes CCCL, X remain. Seven times one is VII, III remain.

3. "GERBERT'S JUMP"

Boethius had argued in his *Arithmetic* that all inequality (i.e., plurality) can be reduced to equality (i.e., unity). The last chapter of the first book (I 32) has the title: "A demonstration of how every inequality proceeds from equality" and Book II starts with a chapter titled "How every inequality is reduced to equality." These chapters deal with the main proportions: the multiples, the "superparticulars" (numbers containing another number and one aliquot part),

[11] A bar over a Roman numeral indicates that the value is multiplied by 1,000.

and the "superpartients" (numbers containing another number and several aliquot parts). In I 32 Boethius had shown how these proportions can be produced from the unity and that the superparticulars and superpartients were derived from related multiples (e.g., 3:2 is derived from the double and 5:3 from the triple). In II 1 Boethius had claimed that there is a method of reduction that can be applied to all types of equality, especially to multiples, superparticulars, and superpartients, which will bring them to equality. He gave only one example, reducing the three quadruples 8, 32, and 128 into three eights. Generally speaking, his method was: Let the three multiples be $a_1 < b_1 < c_1$. Reduce them to a_2, b_2, c_2 with $a_2 = a_1$, $b_2 = b_1 - a_1$, $c_2 = c_1 - a_1 - 2b_2$. Do the same with a_2, b_2, c_2 and continue with $a_{n+1} = a_1$, $b_{n+1} = b_n - a_1$, $c_{n+1} = c_n - a_1 - 2b_{n+1}$, It can easily be shown that in every step the order of the multiple will be reduced by one and that this process will eventually lead to the identity a_1, a_1, a_1.

Boethius ended with: "If one directs his attention from here to the other types of inequality he will unwaveringly find the same accord." The medieval readers found this difficult. There were some attempts to give a general rule, but apparently only Gerbert succeeded. The text of his commentary to Boethius's *Arithmetic*, II 1, is given below. Moreover, there are some manuscripts in which not the complete text, but only the sets of number triples, are written in the margin.

Gerbert's idea is to combine the method given in Boethius II 1 for reducing multiples with the latter's explanation in I 32 of how inequality, especially the different kinds of proportions, proceeds from equality. His starting point is the triple (16, 20, 25). These numbers are *superparticulares sesquiquarti*, because $20 = (1 + \frac{1}{4}) \cdot 16 = 16 + \frac{1}{4} \cdot 16$ and $25 = (1 + \frac{1}{4}) \cdot 20 = 20 + \frac{1}{4} \cdot 20$. In part 1 of his commentary, he applies the formula of Boethius II 1 to the triple (16, 20, 25) and reduces it to (16, 4, 1). He puts the numbers into reverse order (1, 4, 16) and by using the formula of II 1 again, he reduces them to (1, 3, 9). Now the *quadrupli*, which are the basis of the triple (1, 4, 16), have become *tripli*. They are put into reverse order (9, 3, 1), and by using methods from I 32 we arrive at the *sesquitertii* (9, 12, 16). Gerbert emphasizes that one must proceed in an orderly manner, neither *confused* nor *inordinate*, "that is, they ought not suddenly to be resolved into *sesquitertii*, but in an orderly manner," by way of triples.

In part 2, Gerbert reduces the *tripli* to *sesquialteri*. The steps are: (9, 3, 1)= *tripli*; (1, 3, 9)= *conversi*; (1, 2, 4) = *dupli* (by applying I 32); (4, 2, 1) = *conversi*; (4, 6, 9) = *sesquialteri* (by applying II 1). Further in part 2, the *dupli* are reduced to units. The steps are: (4, 2, 1) = *dupli*; (1, 2, 4) = *conversi*; (1, 1, 1) = units (by applying II 1). This passage was later called *saltus Gerberti* (Gerbert's jump). The name came from Rudolf of Longchamps (de Longo Campo, ca. 1216), who in his commentary on Alan of Lille's *Anticlaudianus* wrote:

"Gerbert." He knew a great deal about arithmetic and by an admonition of a daemon he became Archbishop of Reims; afterwards he was Pope. He commented on *Arithmetic*. When he questioned a devil that he had locked in a golden bust about something he had doubts about, the demon explained a very difficult place in *Arithmetic* so badly that Gerbert skipped over the place. So it is that the unexplained place is called the *Leap of Gerbert*.

Evans gives another explanation of the name:

> The *saltus* was not a 'jump' over a difficult passage, but a leap of a quite different kind. ... The *saltus* would appear to consist in the consistent application of the principle of reversing the order of the terms in the series, so that the mathematician is repeatedly making his figures leap round to face the other way, as a dancer might jump round and back in a dance. This is a purposeful leaping, not a disorderly tumbling from one to the other. ... Far from leaping over the whole problem and ignoring it, as Rudolph of Longchamps describes him as doing, Gerbert has applied himself to it with remorseless determination to discourage the reader from unscholarly haste. [Evans, 1980, p. 266]

1. This passage, which, as some persons believe, has not been mastered, is solved in this way. Take [three] numbers which are *superparticulares sesquiquarti*, for example 16, 20, 25.[12] If you want to know how these *sesquiquarti* first can be resolved into *sesquitertii*, then into *sesquialteri*, and finally into three equal terms, arrange them in this way: 16, 20, 25. Take away from the middle [number] the smaller one and set it as the first [number]; that is, take away 16 from 20 and set these 16 as the first. And from 20 there remain 4: these put at the second place. From the third term, from 25, take away the first term, 16, once and the second term twice, i.e. two times 4, this is 8, and there remains 1. This 1 put at the third place, and there are: 16, 4, 1. You see how the *sesquiquarti* have been reduced to quadruples, from which they came. But this resolution cannot be done in a confused or inordinate way. That is, they ought not suddenly to be resolved into *sesquitertii*, but in an orderly manner, that is, into quadruples. Convert 16, 4, 1 and dispose them in this way: 1, 4, 16. Take away the smaller from the middle one, i.e., 1 from 4, and this 1 put at the first place and the 3, which remains from 4, at the second place. From the third term, 16, take away the first term, 1, once and the second term twice, i.e., 1 and two times three, and what remains from 16, i.e. 9, put at the third place and order them as: 1, 3, 9. You see that the quadruples have been reduced to triples, after the quadruples have been converted, from which they took their origin. These triples, 1, 3, 9, are converted and placed as 9, 3, 1. If you, according to the precepts of Boethius,[13] let the first, i.e. 9, be equal to the first; the second be equal to the first plus the second, i.e., 12; the third be equal to the first, twice of the second and the third, i.e., 16, you have the resolution of the *superparticularis sesquiquartus* to the *sesquitertius*, as Boethius teaches, not in a confused, but in an ordered way, such as at the beginning the numbers were created, by exactly using the precepts.

2. If you want to know how the *sesquiquarti* can be reverted in the second step to *sesquialteri*, take the converted triples 9, 3, 1 and bring them back to the former order in this way: 1, 3, 9. Then take away the smaller from the middle [number], i.e. 1 from 3, and put it [= 1] in the first place and, what remains from the 3 [= 2], in the second place. From the third, 9, take away the first, 1, and twice the second, i.e. two times 2, and what remains from 9, i.e., 4, put in the third place. Then the numbers are ordered as: 1, 2, 4. Therefore

[12]Of course, all numbers are given in the original in Roman numerals.

[13]This refers to the rules given in I.32 of Boethius's *Arithmetic*.

you see how the triples are reverted to doubles, from whence they came. Convert these doubles and order them in this way: 4, 2, 1. Do as before: the first, 4, is equal to the first; the second is equal to the first and the second; the third is equal to the first, twice the second and the third, i.e. 9. We have 4, 6, 9. The resolution of the *sesquiquartus* to the *sesquialterus* has been done, not in the first step, but in the second, as Boethius teaches, not in a confused, but in an ordered way.

3. If you still want to know how the *sesquiquarti* can be reverted in a final step to three equal terms, unroll the converted doubles, i.e. 4, 2, 1, and order them in this way: 1, 2, 4. Take away the smaller from the middle [number], i.e. 1 from 2, and put the 1 in the first place and what remains [from the 2], i.e. 1, put at the second place. From the third term, 4, take away the first, 1, and twice the second, i.e. two times 1, and one unit remains. This is 1, 1, 1. Therefore you see how the total quantity of the *sesquiquartus* has been reduced to three equal terms, i.e. units: 1, 1, 1, not in a confused, but in an ordered way, as it had been created in the beginning. Thus, this is the true nature of the numbers.

4. PANDULF OF CAPUA, *LIBER DE CALCULATIONE*

The following text on practical calculation on "Gerbert's" abacus is remarkable, because in contrast to most treatises on calculation written in the eleventh century, it does not give extensive and often exhausting lists of calculation rules. Its purpose is to explain the use of the abacus to beginners who do not already understand the basic terminology associated with the abacus, such as the use and the form of counters and the place-value system. The text starts with the basics, discusses the forms and values of the Hindu-Arabic numerals, the use of the counters, and thereafter gives two samples, one for multiplying (the conversion of a bigger coin, *solidus*, into a smaller one, *denarius*) and the other one for adding. Its author was Pandulf of Capua, a Benedictine monk of the abbey of Monte Cassino (southern Italy), who lived in the second half of the eleventh century.

> His originality lies in the skillful presentation of the sample problem and its solution, in the introduction of differently shaped *caracteres* to aid the user, and in the promotion (and perhaps invention) of a unique variety of Hindu-Arabic numerals whose forms he explains with a simple mnemonic device. Pandulf did not preach to the converted: in his detailed, step-by-step example of multiplication, Pandulf's goal was clearly to create a calculation device and method that could be used even by small children. . . . Pandulf is one of the few early abacists who demonstrate in writing the practical application of the abacus. . . . Pandulf himself does not assume that the student is familiar with the concept of the place-value system. He continually emphasizes the relationships among the different columns, clearly showing his educational aims. By working through a sample problem on the abacus and by reading (or listening to) Pandulf's explanations of the relative positions of different columns, the student should gain both a theoretical and a practical knowledge of the workings of the abacus. [Gibson and Newton, 1995, pp. 305, 307]

Pandulf is also original for his emphasis on memorizing the forms of the numerals. He uses shapes that are similar to, but not identical with, those in the tradition of Gerbert (see section I-2-4). But he gives an original explanation (chapters 4–7): "each of the first six numerals has

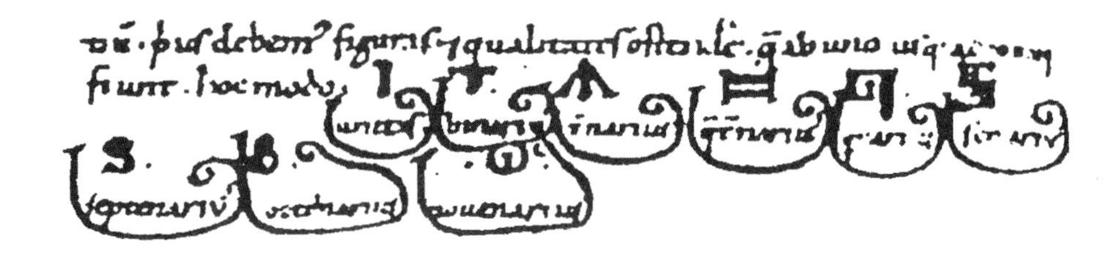

Fig. I-3-4-1. Source: [Gibson and Newton, 1995, p. 312], by permission of the Biblioteca Apostolica Vaticana.

a number of lines corresponding to its numerical value; the seven (*septenarius*) is indicated by a capital *S*; eight (*octonarius*) has the two *o*'s of *octo* stacked one on top of the other; and nine has a one inside a circle" [Gibson and Newton, 1995, p. 313]. Fig. I-3-4-1 shows the forms in the manuscript (a) and without the decorative frames (b).

Equally remarkable is an illustration following the treatise [Fig. I-3-4-2]. It is the earliest known illustration of an abacist at work. The figure is holding an object with some of the Hindu-Arabic numerals. This illustration is unique for its time.

(1) Whoever desires to have a full knowledge of computing (as much as human weakness permits), let him earnestly direct his attention to this page and, with the aid of him who grounded all things in measure and number,[14] we will explain it in as simple a style as we can, (to see) if perhaps through the grace of God even children may understand. (2) Indeed, so useful is this knowledge, that our studies of the *computus* and arithmetic and music, geometry also, and astronomy use it as if it were an alphabet.

(3) First, the rule and law of the whole abacus consists solely in multiplication and division, for which purpose we should first indicate the shapes and values of the *caracteres*, i.e., of the things that signify numbers, which go from 1 to 9 as follows: [1] *unitas*, [2] *binarius*, [3] *ternarius*, [4] *quaternarius*, [5] *quinarius*, [6] *senarius*, [7] *septenarius*, [8] *octonarius*, [9] *novenarius*.[15] (4) Six of these signify the number they represent by how many lines they have. (5) Seven (*septenarius*) is indicated by the first letter of its name. (6) Eight (*octonarius*) is indicated by the first and last letters, i.e., *o* and *o*, since *octo* begins with *o* and ends with *o*. (7) Nine has a one in a crown, signifying either how much it is superior to eight or how far it is from the perfection of ten. (8) Meanwhile, to be able to distinguish more clearly, we have set up three triangular and three quadrangular

[14] Citation from Wisdom 11: 20.

[15] See Fig. I-3-4-1.

Fig. I-3-4-2. Source: [Gibson and Newton, 1995, p. 334], by permission of the Biblioteca Apostolica Vaticana.

tabulae;[16] the remaining three we have fitted out on circles.[17] (9) It is a good idea for you to have in a little bag as many as ninety or more of these, as it were, "numbers," with which, when dice and chess have been put aside, you may play wisely on the abacus, and when these *caracteres* have been positioned on it, you will be able to signify every number.

(10) An actual *tabula* of the abacus is depicted in such a way that from one up to almost infinity, each individual *tabula* proceeding from the next—i.e., from the next from itself—is multiplied by 10; from the third, by 100; from the fourth, by 1000; and so on, like this: 1, 10, 100, 1000, 10,000, 100,000, 1,000,000, 10,000,000, 100,000,000. (11) For just as 10 times 1 is ten, and 100 times 1 is 100, and 1000 times 1 is 1000, thus also from whatever number of the *tabula* you like, the pattern remains the same. (12) Here is an example: 10 times 100 is 1000, and the thousands *tabula* is the second, i.e., the next from the hundreds. (13) So also 100 times 100 is 10,000; 1000 times 100 is 100,000, which are the third and fourth *tabulae* from 100.

[16]Normally, *tabula* is a synonym for *linea* (= column), but here it refers to counters.

[17]Pandulf's explanation is not fully developed. Perhaps he wants to say that three shapes of the numerals (1, 2, and 3) could easily be etched within the boundaries of triangular counters; 4, 5, and 6 are rectangular, and 7, 8, 9 have rather rounded shapes [Gibson and Newton, 1995, p. 310].

(14) Not only in this way but also in whatever way one is multiplied by something else, we must get the same result for the rest of the numbers, preserving the proportion between the "powers" of the *tabulae*. (15) For example: by whatever a unit—because it is first—is multiplied, you put it in the same by which it is multiplied, so also you put a number of tens value, because it is the second *tabula*, in the next from that which is multiplied or from which it multiplies; as 10 times 1 is 10, (one) multiplied by tens value ends up in the tens. (16) But if you were to say 10 times 100 is 1000, which is the fourth *tabula* from 1, and 100 times 1000 times 10 is 1,000,000 which is the seventh *tabula* from 1, just as 100,000 is the sixth tabula from 1, so also in the other cases, whatever happens in one case must also happen in the others.

(17) Moreover, by placing the aforementioned *caracteres* along the *tabulae*, you can represent any number you please. (18) But things placed in the "singular," i.e., the ones *linea*, signify "simply," as it has been said that they (the *caracteres*) signify. (19) In other cases, the *tabulae* by these very (*caracteres*) multiply a number; for instance, *ternarius* (three), when placed in the ones *linea*, means 1 times 3; in the tens, 10 times 3; in the hundreds, 300. (20) But because there is a method that is easier on the eyes, let us make a description of it in order, by means of which, when any *caracter* is placed in whatever *tabula*, you might be able to know and properly to relate what it signifies. (21) You must note clearly that, just as *caracteres* (placed) in the ones *linea*—as has been said—signify "simply," so the *unitas* (one) *caracter*, when placed in whatever *tabula*, "simply" signifies this (value)—which I have also said above—neither increasing nor decreasing in this way: 1, 10, 100, 1000, 10,000, 100,000. (22) For just as I have said that a unit puts its multiplication into that which it multiplies, we have said that the tens place puts its multiplication in the second, thus also the hundreds in the third and the thousands in the fourth, and so on in order with the remaining numbers of the *tabulae*.

(23) Wherefore, before you proceed further, take care to have and hold this little verse fixed in your memory: whatever a unit number is multiplying, put the "digit" in the same as what it is multiplying, the "article" in the next one.[18] (24) Likewise, whatever a tens place number is multiplying, put the "digit" in the next from that which it is multiplying, the "article" in the one after that; a hundreds in the third, a thousands in the fourth, a ten-thousands in the fifth, and so on with the rest. (25) The unit number is whatever *caracter* is placed in the ones *linea*; for example, the *binarius* (two) or the *ternarius* (three), because they signify merely twos or threes. (26) It, i.e., the *binarius* (two), will be of tens value, when placed in the next *linea*, because 2 times 10 signifies one unit of 20. In the same way also successively: in whatever *tabula* the *caracteres* are placed, they take on the appropriate "rule."

(28) There is also a difference among the terms "multiplier" and "multiplicand" and "multiplication"; as when we say "2 times 6 is 12," "2 times" is the multiplier, but 6 is the multiplicand or that which is multiplied by the first *binarius* (two). (29) But what proceeds from each of the two, i.e., the 12, is called "multiplication," or better "digit and article". (30)

[18]The terms *digitus* and *articulus* are common in all texts on calculation with the abacus, and the rule given here is given in almost all of them: if two numbers, both less than 10, are multiplied and the result is greater than 10, then the units of the product are placed in the column of the multiplier (*digiti*) and the tens are placed in the next column to the left (*articuli*).

Digits are the numbers from one to nine. (31) Articles are the numbers 10, 20, 30, etc., up to 100, which Bede, the most holy computator, demonstrates well in *Counting on the Hands or Talking with the Fingers*.[19] (32) For up to 9, he uses the bendings of fingers (*digiti*); in the rest he touches [sc. with his thumb] joints (*articuli*), i.e., the parts or tiny limbs and connections of fingers; as 12 is made with two fingers, but 10 is made with joints. (33) Multiplication is like this with the actual *caracteres*: just as many units as the one has, so many times let the other's number be duplicated; as when we say "2 times 6," we employ two *senarii* (sixes), but when we say "by one," we find once 6 to be only *senarius* (six).

(34) So, when you wish to multiply some number, you will place the multiplier on the upper part of the *tabula*, as when you say, "how many *denarii* are 50 *solidi*?"[20] the multiplier 50 is the number for which you will place a *quinarius* (five) in the tens *linea* on the upper part. (35) You will put the *solidus*, i.e., 12, which is the multiplicand, below, for which you will put a *binarius* (two) in the ones, a *unitas* (one) in the tens. (36) Afterwards you will multiply *binarius* (two) by *quinarius* (five) in the following fashion: twice *quinarius* (five) is 10, for which you will take a *unitas* (one) in your hands, and because *quinarius* (five) is the multiplier in the tens place—say the "tens place" rule, as has been said above[21]—and because the article which you hold in your hand is of tens value, not in the next but in the farther one—i.e., in that *linea* which is beyond the next one from that one where the *binarius* (two) is, (the two) that the *quinarius* (five) is multiplying (for from that *linea* where the *binarius* [two] is, the tens is the next, the hundreds is the one after that)—there, therefore, put what you have in your hand.

(37) But because there are two multiplicands, you will also multiply the other number saying "once" for the 1 in the following fashion: once 5 is 5. (38) So you will take a *quinarius* (five) in your hands and you will place it, keeping in mind the aforementioned rule that 5 is a digit, in that *linea* which is the third from that one in which the multiplicand stands, i.e., in the hundreds, where you have first put the *unitas* (one). (39) Add 5 and 1, which equals 6. (40) Then remove them and put down a *senarius* (six), which, when placed in the hundreds, makes 600. (41) Therefore 50 *solidi* have just so many *denarii*.

(42) But if there had been many multipliers and multiplicands, all the multiplicands likewise would be multiplied by all the multipliers. (43) Therefore, if you preserve the rules of placement, nothing keeps you from "taking a rule" from the multiplicand; as when I said above, "twice 5 is 10," if I might state a rule, whatever the unit number produces by multiplication, I would also place this product in the next from the five, holding the five for a multiplicand. (44) And *vice versa*, the one would likewise come out in the hundreds. (45) Generally it is better if you take a rule from that which is closer to the ones, because it holds the lesser number, and so, for example, place it in the same, or the next, or the third. (46) At the same time the term "multiplier" is more appropriately expressed through that which is smaller, as it is better to say "twice 7" than "7 times 2". (47) You must furthermore

[19] See the text in section I-2-2.

[20] One *solidus* is equal to 12 *denarii*. Therefore the multiplication $12 \cdot 50$ must be carried out.

[21] In §24 of this treatise.

note this, that you should not count *lineae* from where you take the rule,[22] as when I said above "twice 5," I took the rule from the *quinarius* (five), which the tens place is multiplying, and from the *binarius* (two), I counted both the next *linea* and the one after that. (48) Likewise, when many *caracteres* end up in the same *tabula*, add them together and, however many *denarii* you find, put that many units in the third *tabula*.

(49) And however many remain after the "tens business" [the calculations performed above with counters in the tens column, lit. "tens"], leave it so that there is a *senarius* (six) and an *octonarius* (eight). (50) Let them be added together in the first: they make fourteen. (51) Take away 10, for which place a one in the tens, which is the next from the ones *linea*, and leave a *quaternarius* (four); you will do likewise in all the *tabulae*. (52) Therefore there is no inconsistency in multiplication. (53) This is to be observed to such an extent that you should determine the multiplier or the multiplicand, whether they be one or many, as was mentioned in the above passages, and the articles or digits which result from the multiplication of these. (54) By however many *lineae* the individual multiplicand or multiplier among these is distant from the first, you should count and move it that many *lineae* from the second, and you should place it afterwards in the higher one. (55) So you should not count what you have just counted—namely, how far it is from the first *linea*—by using the same placement of digits or articles; rather you will always locate the exact *tabula* that this rule identifies by counting from the second, i.e., the article. (56) It remains for me to say that however many digits or articles are added, if there be many, let that be the result of the multiplication. (57) But now let these words suffice on multiplication.

I-4. Geometry

1. GERBERT, *GEOMETRY*

Until the twelfth century, Gerbert's *Geometry* was the most influential work on this topic. Chapter 1 refers to its name, the inventors of geometry and its use. Chapter 2 deals with its fundamental terms. The next chapters are on the units of measurement, the plane figures and especially the various forms of triangles. The end of the work is missing. Here, chapters 1 and 2 are given.

Chapter I

1. Among the four ordered mathematical disciplines, the tract on geometric speculation naturally holds the third place after arithmetic and music. The reason for this is that the highly distinguished Boethius, the most learned and knowledgeable commentator on the liberal arts, set it in the very principles of the theory of arithmetic. But the reason was concealed, as noted elsewhere, because of our more than simplicity. This discipline, if I may speak more simply, received its name from the Greeks as measure of land, since ΓΗ (*ge*) is the Greek word for land and METPON (*metron*) for measure.

2. The first inventors of this discipline are said to be the Egyptians. Because the rising of the Nile River very often flooded the boundaries of land, they discovered the usefulness

[22]This means that one should consider the position, and thus the value, of the multiplicand when trying to position the product correctly. For example, in our problem, 2 times 5 is really 2 times 50, so the product will end up not in the tens *tabula*, but rather in the hundreds.

of such a craft. This made it possible for anyone using it to be able to separate out very easily the area of a small piece of land. Although at first the craft and its vocabulary were developed for the utility of land measurement, later after diligently investigating its rationale, people adjusted speculation to other things, whether useful for thought or as a delightful exercise. Hence they adapted the word to this definition: Geometry is the study of magnitude and forms, which are investigated in their magnitude. Further, and unless I am mistaken, it can be defined differently: Geometry is the science of probable dimension reasonably investigated by the rationale of proposed magnitude.

3. All the lovers of wisdom consider the usefulness of this science the greatest. For it is the most subtle for sharpening one's intuition and exercising the powers of the mind and of the talents; it takes the greatest delight to investigate rationally many certain and true things, which to many seem miraculous and inexplicable; and it is fullest of subtle speculation in contemplating, admiring, and praising the wonderful vigor of nature and the power and ineffable wisdom of its Creator who made everything in number, measure and weight. Because of its rationale and rules we are about to collect some things from wherever according to the small means of our feeble intellect, so that we may lead the mind of a beginner along a more orderly path to subtle things. We want to start with the elements of this art which are called "boundaries."

Chapter II

1. The origin and the elements, as it were, of this craft are these: points, lines, planes, and solids. Boethius and others, lay persons as well as commentators on divine writings, disputed more than enough in many places in their own writings as did that most blessed and eloquent Doctor of the Church, Augustine. He discussed at great length in some of his books, especially in the one titled *On the Measure of the Soul,* that the eye of the mind, darkened by images of bodily things, can be cleansed and become more than a little sharp by exercising these arts to contemplate as much as possible spiritual and true things. If by chance the better educated have deigned to look on and are not too bored, I shall reverse the order of discussion and attempt to show briefly and simply the individual aspects of solid bodies of which human senses are more aware.

2. A solid body is whatever is set within three boundaries or dimensions. It is whatever is held by length, width, and height, such as anything that one can understand by sight or touch, like this tablet here on which I write. The Greek word is *stereon*. We call "surfaces" the boundaries or covering planes that the Greeks name *epiphaniae*. This must be so understood that nothing that is high or thick may be involved, but only what is contained by length and width spreads out. For if to these one adds height, it is no longer a surface but a part of the body, and therefore it is a solid body. The extremity or boundary of a surface is a "line" (*linea*), which is called in Greek *gramma*. You must mentally perceive that (the line), having no width, is promoted solely by the force of length; if width were added, it would not be a line, but a surface. The beginning and the end of a line is a "point" (*punctum*). It crowds together in such an intelligible way that only the end of the line exists and that (the point) contains in itself neither a quantity of a part nor of any magnitude.

3. So I would define each of the foregoing thus: a point is the smallest and indivisible sign that the Greeks call *symion* [= *semeion*]. Similar to the unit, which is the origin of all

Fig. I-4-1-1.

Fig. I-4-1-2.

numbers but is not a number, (the point) is the origin of measures but itself is neither a measure nor capable of being measured. A line is length without width, and it only admits division or cutting in its length. A surface has width without height. It has both length and width that can be divided, but it rejects a section in thickness.

4. But these three, point, line and surface cannot subsist in the nature of things without a body. Mentally however they are considered as incorporeal things having their existence [= *suum*] as though outside a body. The solidity defined above exists in solid bodies and can be perceived by the senses, and it is subject to being cut up in all kinds of ways by length, width, or height. Meanwhile let this brief account of the foregoing simpler items suffice as given. There is no need to detain the more knowledgeable with longer instruction in these matters.

5. Consequently whatever is proposed for reasonable measuring according to the three foregoing dimensions of a solid body is measured by a geometrical theorem under the guidance of reason. Or, length or width or certainly thickness, which geometricians usually call altitude, is investigated by measuring. Length is found in the lines that include the area of a figure or field, or in the length of a trip, or in the height of trees and buildings, or in measuring the linear distance of rivers or other objects to a certain end point, which is called "linear measurement." Width is found in the quantity of an area or surface, which is included by certain borderlines; it is called "floor" or "plane measure," and the Greeks call it *epipeda*. Altitude is found in the thickness or density of some structures with definite measure, or by the capacity of vases with defined quantity; it is called "solid measure." And hence it is that we are accustomed to call certain measures sometimes linear, other times floor-like or solid.

6. A linear foot is that by which we measure lines or some length, paying no attention to height or width, as shown here [Fig. I-4-1-1]. Floor or plane feet are those by which we measure surfaces or planes or areas bounded by lines, such as a square with equal length and width but lacking height, as shown here [Fig I-4-1-2]. A solid is used for something equally extended or squared in length, width, and height by which we measure solid bodies while keeping the form of a cube or tile. It cannot be clearly drawn on a surface, but it can be imaged mentally, easily formed from wax or wood or some other material like

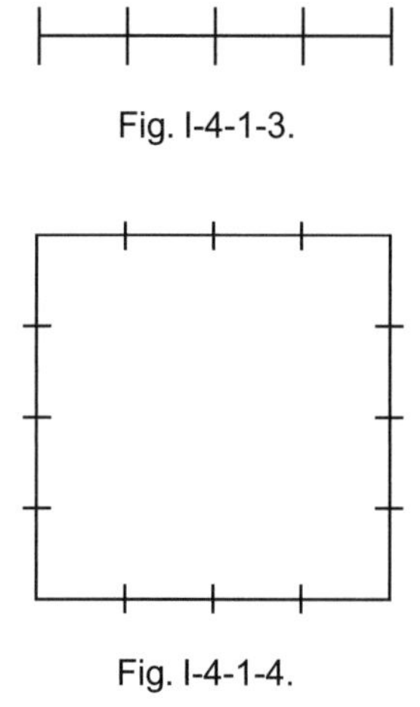

Fig. I-4-1-3.

Fig. I-4-1-4.

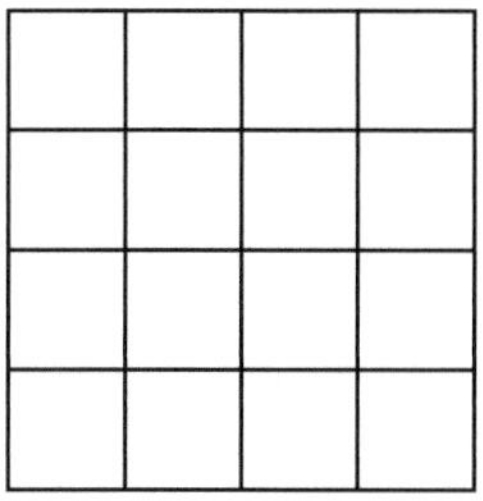

Fig. I-4-1-5.

these, although in his commentary on Plato's *Timaeus* Calcidius expounded in one way or another that a solid body is drawn on a plane (surface).

7. There are still other measures we shall later speak about, which are distinguished by three kinds, such as the aforementioned foot. We understand these things to be linear or floor-like or solid. It is particularly important to know that floor-like measures must be found by linear measures, and solid measures by floor-like and by linear measures. Floor measures are found by multiplying linear measures by themselves or among themselves. Then by multiplying a floor measure by a linear measure, the solid measure is certainly found.

8. An example easily clarifies this: let a line of linear foot be multiplied by a length, and that cut in four linear palms in this way [Fig. I-4-1-3]. Therefore having added a width of the same quantity and having sketched off this linear foot equally to a square, a floor of this kind is formed [Fig. I-4-1-4]. If therefore you multiply four linear palms of length by the same amount noted for width, that is four by four, you will find in this way that you have a floor of sixteen palms of plane feet [Fig. I-4-1-5]. But for making a solid foot, if you use the same foot and superimpose an equal height onto the length and width, then you multiply the sixteen palms of the floor by the four linear feet of the aforementioned height. Thus you shall find sixty-four solid palms in the solid. Anybody can understand this yet more easily than what I have described in [the] plane. By this way one linear foot equals four

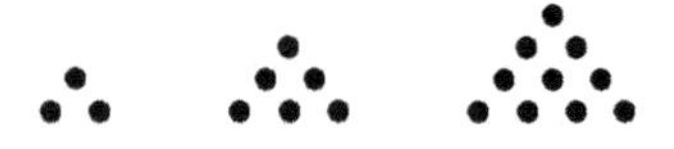

Fig. I-4-2-1.

linear palms, a floor (foot) sixteen floor units, and a solid (foot) sixty-four solid palms. The same reasoning for multiplying is found with the other measures according to the quantity of each (of them).

2. GERBERT: AREA OF AN EQUILATERAL TRIANGLE

In the time of Gerbert, two rules were used for calculating the area A of an equilateral triangle with side a. One of them, the so-called geometrical rule, was based on the general rule for the area of a triangle: $A = 1/2\, ah$ (a = base, h = height). The height of an equilateral triangle is: $h = (a/2)\sqrt{3}$, and, by using the approximation $\sqrt{3} \approx 12/7$, we have: $h \approx 6/7a$. The other, the so-called arithmetical rule, had to do with the "triangular numbers," which came from the Greek and were well known in the Middle Ages. The series of numbers 1, $1 + 2 = 3$, $1 + 2 + 3 = 6$, $1 + 2 + 3 + 4 = 10$, ... can be ordered in the form of triangles (see Fig. I-4-2-1). In antiquity and in the Middle Ages the area of an equilateral triangle with side a was sometimes identified with the corresponding triangular number $\mathrm{F}_3(a) = 1/2[a(a + 1)]$. It can easily be proved that the correct area of an equilateral triangle is less than the value obtained by the arithmetical rule. In an equilateral triangular with the side 7, by using the approximation of $\sqrt{3}$, the height h is 6 and, accordingly, the area A is $(7 \times 6)/2 = 21$. Using the arithmetical rule, the area A is $(7 \times 8)/2 = 28$. Gerbert was the first to explain these different results correctly.

Gerbert [offers to] Adelbold, whom he highly esteems now and forever, integrity and the constancy of integrity.

1. One of the geometrical figures which you have obtained from me was an equilateral triangle with side of 30 feet, height of 26 feet, and by combining side and height an area of 390 [square] feet. If you measure the same triangle without calculating the height according to the arithmetical rule, i.e., by multiplying one side by itself, adding to this product the measure of the [same] side, and taking half of this sum, the area will be 465. Do you see how these two rules differ? But the geometrical rule, which with the help of the height measured the area to be 390 feet, has been discussed by me quite carefully. I take the height to be only $25\frac{5}{7}$ [feet] and the area only $385\frac{5}{7}$ [square feet]. Let the universal rule for finding the height in every equilateral triangle be for you this: always take away from the side its seventh part and give the six parts which are left to the height.

2. In order that you understand this better, I will give an example with smaller numbers. I show you a triangle whose side has a length of 7 feet. I measure [its area] by the geometrical rule in this way: I take away the seventh part of the side and I give the six feet which are left to [be the measure of] the perpendicular. I multiply the side by saying this: six times seven is 42. The half of this, 21, is the area of the triangle. If you measure the same triangle with the arithmetical rule saying: seven times seven is 49. Now if you add the side, there are 56. And if you divide it (by 2) to reach the area, you will have 28 as result. So, in a triangle of the same magnitude there are different areas. This cannot happen.

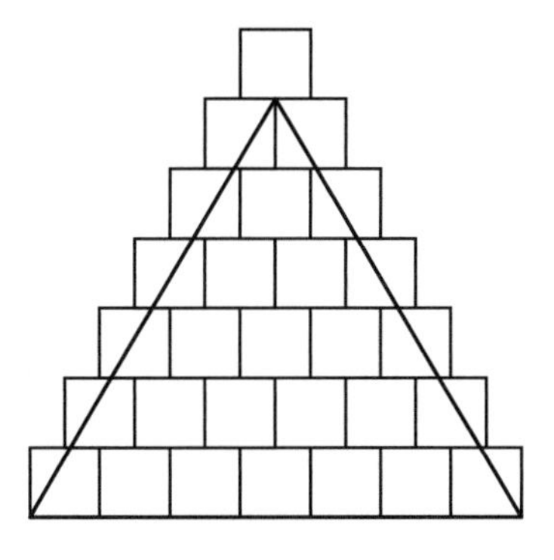

Fig. I-4-2-2.

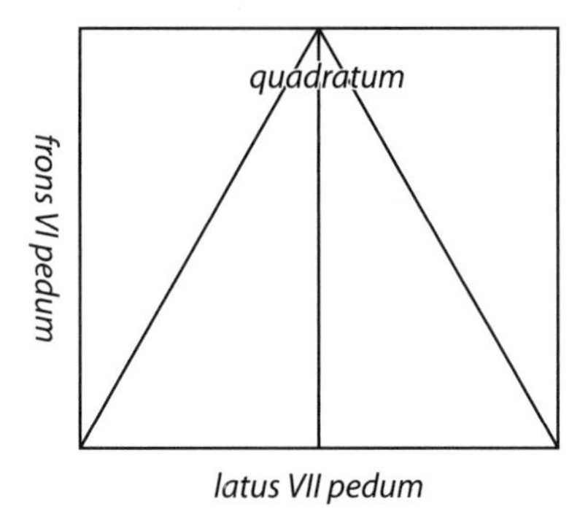

Fig. I-4-2-3.

3. But, in order that you no longer wonder, I shall show you the reason for this difference. I think that you know which feet are called long, which square, and which solid, and that for measuring areas we need only square feet. How little so ever part of them the triangle touches: the arithmetic rule computes them as whole numbers. I will draw it in order to make more clear what is said [Fig. I-4-2-2]. See in this small figure 28 feet but not integral. Therefore the arithmetic rule takes a part instead of the whole [measurement], and uses the diminished [length] with the integral one. The practice of geometry is to omit the small parts that exceed [integral lengths of] the sides. Having cut away the small parts, it combines what remains within the sides. And it computes only what is included by the lines. For in this small drawing is the part, the "seven" measures the sides and the "six" the perpendicular. If you multiply this by 7, you will complete almost a square, whose base is 6 feet and the side 7. And so you find as its area 42 feet. If you halve this, there remains a triangle of 21 feet [Fig. I-4-2-3]. To understand this more clearly, open your eyes and remember me forever.

3. UNITS OF MEASUREMENT

In the first half of the eleventh century, a handbook on geometry was compiled that included excerpts from Euclid's *Elements* (in the translation by Boethius), material from the Roman *agrimensores* (land surveyors), and a section on the abacus in the tradition of Gerbert. The book was attributed to Boethius, but we know now that this compilation was done by an

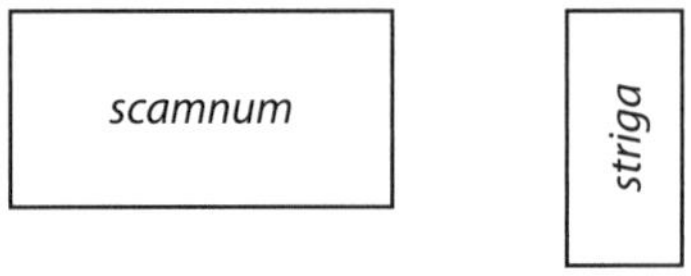

scamnum

striga

Fig. I-4-3-1.

unknown person living in Lorraine [Folkerts, 1969]. The author begins the second book by introducing his reader to the units of measurement and the concept of measure itself.

Even though the above volume has treated succinctly all theorems of the geometrical art, this book as an inquirer and without delay unravels and clarifies the knowledge of measuring by the foot and the fine details of all questions pertaining to it. The old surveyors were accustomed to measure each quadrilateral by dividing it into two parts, length and width. If the width were longer, they would name it a *scamnum*, if the length were longer, a *striga,* as the following figure shows [Fig. I-4-3-1].

The most cautious searchers of the old scholarship fixed twelve types of measures for measuring the areas of figures and fields. Their names are mile (*miliarium*), stade (*stadium*), act (*actus*), rod or perch (*decempeda* or *pertica*), pace (*passus*), grade (*gradus*), cubit (*cubitus*), foot (*pes*), half foot (*semipes*), palm (*palmus*), inch (*uncia*)[23] and digit (*digitus*). The mile has been set at 5000 feet in a line. A stade is 625 feet. There are three kinds of acts: the minimum, the square, and the double. The minimum act is 4 feet wide by 120 feet long. The square act is 120 feet on each side. The double act is 120 feet wide by 240 feet long. The rod is 10 feet, the pace 5 feet. The grade is 2.5 feet, the cubit 1.5 feet. The foot is 4 palms, the half foot 2. The palm is 4 digits wide. . . . The foot is called extended (*porrectus*) where the measurement in feet is for length only. The foot is called spread (*constratus*) when both length and width are considered. The foot is called square where all three dimensions are equal.

4. FRANCO OF LIÈGE, *DE QUADRATURA CIRCULI*

Franco of Liège (d. ca. 1083) was the head of the cathedral school in Liège. His main work is a treatise on squaring the circle. Franco started from a statement in the *Categoriae* of Aristotle that this is a problem that can be solved but of which the solution still is unknown. Franco did not know the writings of Archimedes and took it for granted that the proportion of circumference to diameter of a circle is equal to 3 1/7. Tacitly using this assumption, Franco first proved that a circle and a square cannot be equal in number. Because he was convinced that the areas of all figures can be compared, he tried to find a geometrical way to prove the equality of a square and a circle. He considered a circle with the diameter 14 and, accordingly, the circumference 44. Then he showed that the area of the circle was equal to that of an intermediate figure: a rectangle with sides 11 and 14. The last step was the conversion of the rectangle into a square. Here, Franco succeeded only partially, because he did not know the methods of the Greeks to find the geometrical mean between two terms. Franco's text is a

[23]Recall that *uncia* is the Roman fraction that denotes the twelfth part of the unit [Fig. I-2-1-2]. When used as a length measure, we translate it as "inch" (one-twelfth of a foot), but when used as a measure of weight, we translate it as "ounce" (still used for one-twelfth of a Troy pound).

document for the limits of mathematical knowledge before the translations from the Arabic were available. But it shows the awakening of original mathematical thinking.

Perhaps someone will object: Who has ever known how much the area of a circle comprehends, and who can therefore know what is equal to a circle or not equal? For, how can a person who does not know the correct quantity of something estimate the equality with it? There are good reasons to answer to these objections: I say that the experience in geometry to find the area of the circle gives the rule to multiply half of the diameter with half of the circumference. Because the diameter of this circle is 14 and the circumference is 44, and their halves are 7 and 22, it is evident that the circle comprises 154, because this is the result of the multiplication of 7 and 22. Since this is obvious, it is totally correct that the rectangle which consists of the whole diameter and its 11 parts is equal to the circle, because 7 times 22 is equal to 11 times 14. . . .

The learned geometers were moved by a long discussion about the meaning of the quantity of the area of a circle. They were not able to determine the circle in the same way by measures which they used for areas included by straight angles, such as inches, feet or any other measures, with which every measurement in length, width and height starts, with which a porch, a mile, the length of a stadium, the width of fields or rivers, and the height of walls and mountains can be compared. The curvature of the circumferential line did not allow by any means a computation of this kind. Nevertheless they wanted to know with absolute certainty which value the circular area had. They noted that its line, although curved and spreading around, had always the same distance from the center of the circle in all directions. Knowing this they divided the whole circumference into as many parts as they liked, namely into 44, and they marked them by the same number of points. They drew lines from these points to the center of the circle and, after the labor of cutting, they had 44 parts each of which having the length of 7 quantities, that is, half of the diameter having 14 quantities, and without delay they joined the parts with the parts and directed the apex to the width. In doing so, the result were 11 and 11 (parts) congruent to each other, and there were generated two figures out of the 44, each of which containing 22 of the 44 parts, i.e. one half of the circle, and having 7 in one side and 11 in the other. When they are joined with each other by this way, so that the length corresponds with the width as much as possible, a figure is formed with two sides having 11 units and two having 14 units.

In order that the situation becomes more evident by this example, I divide the given circle into 4 parts and I draw one diameter from A to B and another one from Γ to Δ. Thereafter I divide the four quarters of the circle into 11 parts, as the following figure shows [Fig. I-4-4-1].

Having done this dissolution, I combine the two sets of 11 parts by mixing them in such a way that the extremities of one part are directed to the heads of the other part, as given in this [same] diagram. When the circle has been decomposed and its parts have been rearranged in this way, has not the figure of a rectangle been produced? Having 11 in width and 14 in length, which, multiplied, produce 154, this figure evidently shows how much the circumference of the circular space comprises. Therefore no one can doubt that from this rearrangement insightful geometers have assembled the rule for finding the area of a circle, by which they advise that, either half of the circumference must be multiplied by half of the diameter, or the complete diameter by the fourth part of the circumference,

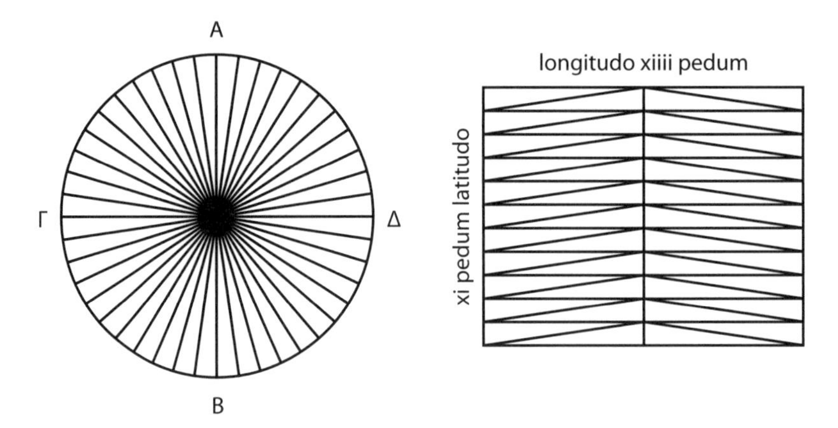

Fig. I-4-4-1.

or the complete circumference by the fourth part of the diameter, or by any other way which comes to the same result. We think that this rule has been invented to avoid always decomposing the circle whenever we want to find its area, especially because they knew that despite all due care there were some objects that were not easily decomposed. For this [decomposition] can not be done properly except with parchment and skin and other similar [devices].

Having satisfied the objections well enough we think that we proved at the same time how that middle figure[24] proves that the square is equal to the circle by being rearranged from the circle itself or its diameter. Now that a certain near foundation of the knowledge strengthens us, we confidently seek uncertainties from the certain, the unknown from the known. Then from the same quadrilateral into which a circle was rearranged, we move on to other kinds of squares. With care we will get close [to their areas]. We shall do this, if the previous rugged speech is lightened by some introductory words.

. . .

In the preceding book we have shown how a rectilinear figure can be produced from a circle. Now we undertake to produce a square from this figure. This is a difficult and laborious task and no less unattempted than squaring the circle: I think that we shall fail necessarily, if we are not supported by the help of geometry. Who has transmitted until now the method of squaring? Why was it passed over by all persons, if not by its difficulty, which has not been mastered? Therefore in this problem we are tortured by great difficulties and we are held back as by a closed door, wherever we try to go.

If we divide the excess of the length beyond the width, add its half to the width and leave the other half with the length, a truncated figure will occur, as though something angular were removed. This can be seen in the attached figure [Fig. I-4-4-2]. If we want to take the quantity of this area obliquely, would it by this modification become neither more

[24]The idea is to find a square equal to the circle. The rectangle found first is the "middle figure" between the circle and the square.

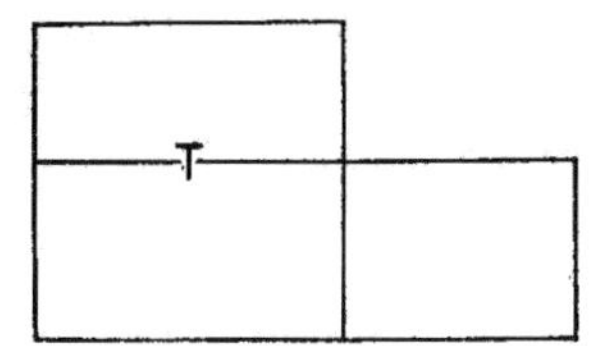

Fig. I-4-4-2. This sketch from a manuscript source does not accurately reflect the text; the lower left rectangle should be a square.

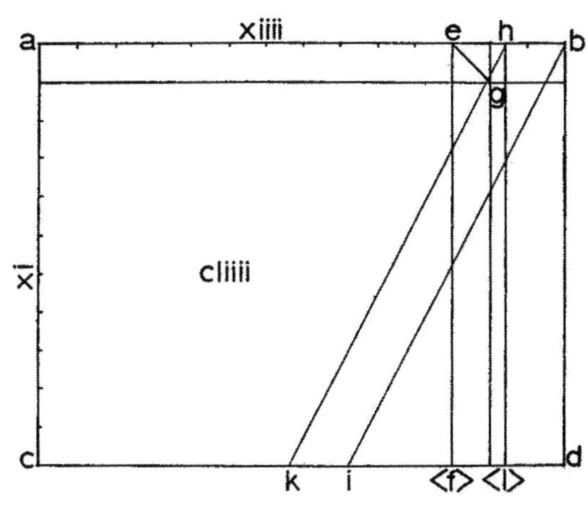

Fig. I-4-4-3.

nor less? I think that the difficulty has become clear. We are, however, now able to give a very short demonstration.

The side of the rectangle, *ab*, may be reduced by three parts at point *e* and also the side *cd* by three parts at point *f* [Fig. I-4-4-3]. The area between lines *ef* and *bd* will be divided in the following way: a line from *e* to *g* is drawn and between *e* and *b* the point *h* is assigned as having the same length [as *eg*].[25] The lines *hk* and *bi* are drawn having the same length as *ha*, [so $kibh = ldbh$]. If the included space is subtracted and added to the latitude, I say that a square has been produced from the rectangle.[26] By this way the [rectangle], which came from the circle, is changed into a square.

5. HUGH OF ST. VICTOR, *PRACTICA GEOMETRIAE*

Hugh of St. Victor (1096–1141) was a canon of the abbey of St. Victor in Paris and the head of the famous school attached to the monastery. In his writings, he made an explicit

[25] That is, $eg = \sqrt{2} = eh$, so $ah = 11 + \sqrt{2}$.

[26] Franco seems to be suggesting that the perpendicular distance between *hk* and *bi* is also $\sqrt{2}$. The actual value is approximately 1.405. In any case, it appears that Franco intends to cut away the strip *ldbh* and place its equal, the strip *kibh*, onto line *ah*, thus producing a square. His square then has side $11 + \sqrt{2}$. The square of this value is 154.11, a bit larger than the desired 154.

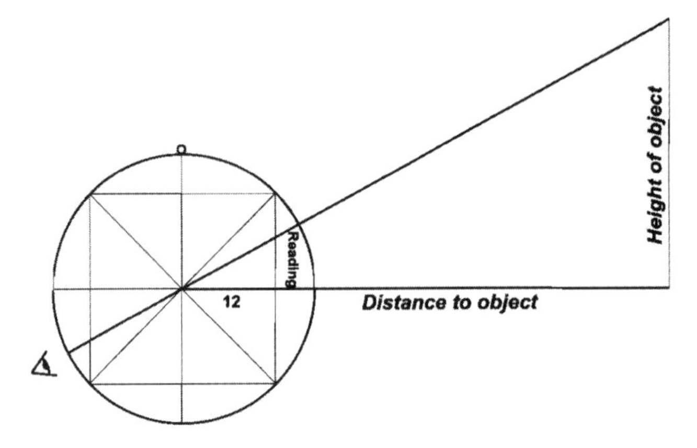

Fig. I-4-5-1.

distinction between theoretical geometry, which works only by rational argumentation, and practical geometry, in which instruments are used. His *Practica geometriae* is in the tradition of Gerbert. But for measuring angles, Hugh also made use of the astrolabe. This instrument had become known in the West from Arabic sources in the eleventh century but spread out in the Latin West only one century later with the translations of Arabic texts into Latin.

To understand the following text, one must know that on the back of the astrolabe is a ruler, called *medicline* or *alhidade*, which can rotate about the center of the astrolabe. It is moved to the line of sight from the eye to the top of the measured object. The angle of height is not measured by a scale divided into 360 degrees, but by a square on the back of the astrolabe, which is divided into twelve equal parts [Fig. I-4-5-1]. A reading of the height of the *medicline* is taken where it crosses the edge of the square. Then the height can be found by the proportion:

$$\text{reading}:12 = h : d \; (h = \text{height}, \, d = \text{distance}).$$

Of course, the height of the astrolabe must be added to the calculated height to find the height of the measured object.

If the distance to the object cannot be measured, because a river or other obstacle is between the person and the object, two measurements must be done. Let r_1 be the reading at one place having the distance d_1 from the object and r_2 be the reading at another place having the greater distance d_2. Then:

$$d_1 = 12h/r_1, \quad d_2 = 12h/r_2, \quad \text{and thus } d_1 - d_2 = (12/r_1 - 12/r_2)h.$$

Because r_1, r_2 and the difference $d_1 - d_2$ are known, the height h can be calculated.

To measure the height of an object in front of you without moving from your place, raise the astrolabe to the object. Adjust the medicline until you can see the top of the object through both apertures. Then compare the medicline degree reading with the whole side of the square. The ratio of the medicline reading to the whole side [i.e., to twelve] is the ratio of height to intervening space with surveyor's height added either proportionally or exactly.

If the intervening space is impassable because of an obstacle such as a river or a gorge, you can still get your result. Use the astrolabe where you are. Adjust the medicline to the top of the object until you can see it through both apertures. After this, note how many degrees of the side of the square below appear above the medicline. Compare them to the 12 degrees of the whole side. By rule, this is the ratio of height to intervening distance plus surveyor's height.[27]

Next, move back some distance to a second position. Take the astrolabe, and sight the top along the medicline. Record the medicline degrees on the square side and compare them to the whole side. The ratio is now that between height and intervening space plus surveyor's height. Then compare the first and the second base, to determine how much the second exceeds the first. Now compute the length of the first base by means of the difference between first and second, i.e., the distance between your first and second positions.

For example, suppose the medicline marker reading at the first station is four. Because twelve is the triple of four, the intervening space plus surveyor's height will be triple the object's height. Suppose the medicline marker reading at the second station is three. Because twelve is the quadruple of three, the space plus surveyor's height will be quadruple the object's height.

So suppose the first station separation plus surveyor's height (the first base) is triple the object height (the perpendicular), and the second station separation plus surveyor's height (the second base) quadruple the object height. Clearly, the second base is one and a third times the first. A third part of the first base will be the excess of the second over the first. Evaluate this distance, and take it as just one third of the first base.

A warning. This distance is not always that from first to second station. The surveyor's height adjustment is not the same at both. Rather the distance is measured from the end point of the first addition (where first base ends) to the end of the second addition (where the other ends). This gives the true difference between bases.

I-5. Recreational Mathematics

1. NUMBER PUZZLES

Number puzzles seem always to have been sources of delight and challenge. Present knowledge ascribes the three that follow to Bede, although they do not appear in his own list of his works. In the manuscripts they are titled *De arithmeticis propositionibus* (*Arithmetical Representations*). They are the first such problems to appear in Western Latin manuscripts, and there are many copies of them. In the first three problems a person has to find the number thought of by another. All of them are solved without algebraic manipulations, and they assume that only whole numbers are chosen. The treatment of fractions remaining at the end of a calculation varies.

[27]If h is the height of the object from the level of the astrolabe, m the height of the astrolabe, and H the actual height of the object, then $H = h + m$. So $\frac{r}{12} = \frac{h}{d} = \frac{H-m}{d} = \frac{H}{d+\frac{d}{h}m}$, where $\frac{d}{h}m$ is the surveyor's height "added proportionally."

How a number thought of by any mind can be recognized for what it is.

Problem 1: Take any number and triple it. Divide the triple into two parts. If both are equal, then take one of them without any difference and triple it again. If they were unequal, then take the larger and triple it. Then after the tripling consider how many VIIII can be found in it. Because however so many VIIII it has, the number taken is twice that number.

For example: if two had been thought of, being tripled it makes VI. Six divided is resolved into three and three. Three tripled becomes VIIII and nothing remains. Therefore it was two that had been mentally thought of.

Problem 2: Take any number and triple it. Divide the triple into two parts. Then ask the one who thought of the number if the division of the numbers is equal. If he answers that both parts are equal, then nothing is to be set aside. If however he said that they were unequal, then the diviner is to remember one for the first division. And again without differing triple one part from the division where both parts were equal. If however they were unequal, then triple the larger part. Then divide as before into two parts, and ask if the division made equal or unequal [parts]. If each was equal, then nothing is to be set aside. If unequal, two must be set aside for the second division. In the halving, if the [parts] are equal, without any difference between the parts, you must ask how many nines one of them contains. But if unequal, seek from the larger part how many nines can be found in it. The diviner then takes four times the quantity [and adds the numbers set aside for the number originally taken].

For example: if six had been thought of, when tripled it made XVIII. This is divided into VIIII and VIIII. Because this division was equal, nothing is set aside. Again the nine that was half of the division is tripled to make XXVII. This XXVII is divided into XIII and XIIII. Because the division is unequal, two is set aside. Then one must find out how many VIIII can be found in the greater part of the division, that is the XIIII. In XIIII there is one VIIII. The diviner takes from the VIIII, IIII, which is added to the two that had been set aside to make VI. Therefore the number six is set aside first in the mind. Now this must be noted as a reason that if both divisions are equal, nothing from them is set aside. If however the first [division] was unequal, one is set aside. If in the second, two is set aside. The nine holds the meaning for the four. What happens therefore in this situation seems to differ from what happened above, because here the tripling occurs twice and the dividing was twice, whereas above the tripling occurred twice but the dividing once. For this reason one nine means in this [problem] "four," but in the first [problem] it meant "two."

Problem 3: The same differently. How to divine at which day of the week a person has done something. First you must double whatever number you think of that corresponds to a day of the week. Then add five to the double. Next quintuple the sum that you collected. Then multiply the whole by ten. Afterwards from this subtract two hundred L. What remains is the number of the day.

For example: if the first day were chosen, then double one to become two. Add this to V to become VII. Multiply VII five times to become XXXV. Multiply XXXV ten times to become CCCL. Subtract CCL from this, and C remains. This stands for the monad or one, which represents the first day that had been set aside. In this computation it must be kept in mind that CCL is always taken from the entire sum and it must be diligently considered how many hundreds remain, because as was said above, a single hundred represents the first day, two hundreds the second, three hundreds the third, four hundreds the fourth, five hundreds the five, six hundreds the sixth, and seven hundreds the seventh [day of the week].

2. THOUGHT PROBLEMS

Alcuin of York (735–804) is credited with assembling a collection of problems, mostly mathematical but all requiring thought, whose purpose, as the title attests, was *Problems to Sharpen Young Minds*. There are at least fourteen manuscripts of the *Problems*, the earliest from the late 800s. While none of the manuscripts identify an author, Alcuin is usually recognized as such for two reasons. First, he loved problems of this sort, as suggested in a letter he wrote Charlemagne, sending him "puzzles of arithmetic subtlety for delight." Second, he is noted for his ability to organize. The *Problems* offer 56 challenges that can be categorized under six modern rubrics: linear problems with one unknown, linear problems with several unknowns, problems involving sequences and series, ordering problems, geometric problems (not recreational at all), and strictly thought, non-mathematical problems. While these kinds of problems have a long history going back to the *Hau* problems in ancient Egypt (e.g., see [Imhausen, 2007, pp. 26–29]), this particular collection seems to have had great influence. These problems are found in collections assembled in the later Middle Ages, in the writings of Fibonacci, and in the writings of the German *Rechenmeisters* from the early Renaissance. Some were studied intensely by Leonhard Euler in the eighteenth century. Thus the *Problems* form part of the tie that binds together Antiquity, the oriental and occidental Middle Ages, the Renaissance, and modern mathematics. A selection of problems follows.

(1) The Problem of the Snail

A snail was invited to breakfast somewhat less than a *leuva* away by a swallow. In a single day, however, it could travel no farther than one inch (*uncia*). Let someone say, if he will, how many days the snail took to travel.

Solution: A *leuva* consists of one thousand five hundred paces; that is, $\overline{\text{VII}}\text{D}$ feet [or] $\overline{\text{XC}}$ inches. The number of inches is equal to the number of days. They make CCXLVI years and CCX days.

(5) The Problem of the Buyer with C Dinars

A certain buyer said, "I want to buy C pigs with a hundred dinars, in this way: boars costs X dinars a piece, breeding sows cost V dinars a piece, and two piglets cost one dinar. Let someone say, whoever understands, how many boars, sows, and piglets must there be so that the cost is neither more nor less [than C dinars]."

Solution: Take VIIII sows and one boar for fifty-five dinars. Then 80 piglets for XL. Thus there are XC pigs. With the remaining V dinars buy X piglets, and you will have one hundred for both numbers.

(12) The Problem of the Father and his Three Sons

As he was dying a certain father willed to his sons XXX glass jars. Ten were full of oil, another ten were half full, and the third ten were empty. Let someone, whoever can, divide the oil and the jars, so that each of the sons receives the same [number of jars and the same] amount of oil.

Solution: There are three sons and XXX jars. Ten were full of oil, another ten were half full, and the third ten were empty. Now thrice ten is XXX. Each son is to have X jars. Divide by one third. That is, give the first son the X jars half full. To the second give V full and V empty, and likewise to the third son. And thus the oil and the jars have been divided equally among the three sons.

(17) The Problem of the Three Brothers Each with a Single Sister

There are three brothers, each having a sister, and they all have to cross a river.[28] Each one of the men desires the sister of one of [the other] brothers. When they reach the river, they discover only a small boat capable of carrying only two of them across at once. Let the person say, who can, how they cross the river without any one of them being soiled.

Solution: I cross in the boat with my sister. Having arrived I leave her there and return alone. Then the other two sisters, who had been left on the shore, enter the boat and go to the other side. There they leave the boat and my sister, who had gone over before, returns to me on the boat. Once she has disembarked the two brothers enter and go to the other side. Then one of them with his sister boards the boat and comes back to us. He who had just sailed and I go over leaving my sister behind. Once we have landed one of the two remaining ladies takes the boat back, obtains my sister and they return to the other side. Then he whose sister remained takes the boat, gets her, and returns. Thus the crossing has been made without any touch of evil.[29]

(42) Problem of the Stairway with One Hundred Steps

There is a stairway with C steps. On the first sits one pigeon, on the second two, on the third three, on the fourth IIII, and on the fifth V. And thus this continues up to the one hundredth step. Let the person say, who can, how many pigeons there are.

Solution: You will count this way. Take from the first step the one pigeon and it with the XCVIIII that are on the ninety-ninth step, and they are C. Likewise put [the two] on the second step with the [ninety-eight] on the ninety-eighth [step], and they are C, too. In this way at every step join one of the higher steps with another one of the lower steps in this order and you find always C for two steps. The fiftieth step however is alone with no equal. Likewise the one hundredth step remains alone. Join the two therefore and you will have $\overline{\text{V}}$L pigeons.

3. THE JOSEPHUS PROBLEM

In this well-known problem, men are placed around a circle and by counting, it is determined that every nth man has to die. In general, half of the persons will be killed. How must the men be ordered so that only certain specified individuals will survive? Such problems can be easily solved empirically. In the Renaissance, this problem was called the "Josephus problem" after a story related in Josephus's *De Bello Judaico*: when Josephus and 40 other Jews had been captured by the Romans, he was saved on the occasion of a choice of this kind.

[28] By "brother," the author simply means that the person has a sister. The problem clearly requires that the three brothers are unrelated.

[29] This puzzle and several similar ones are discussed with scholarly clarity by [Franci, 2002].

The oldest European trace of this problem is found in manuscripts of the ninth century, where it is given in metrical form. There are two groups of 15 persons, and one commander must decide which of them shall make the vigil. Every ninth has to do the job. The commander of group A orders everyone, so that only persons of group B are chosen. From the tenth century on, many other verses describe how the persons must be arranged. During that time the guise of the problem changed, too. In some Latin manuscripts of the twelfth century, which rephrase the problem to deal with soldiers, we find a note that this problem concerns 15 Christians and 15 Jews. Since the thirteenth century, the most common form in the Latin West was the version with Christians and Jews on a ship in distress. In the twelfth century the problem was also treated by Abraham ibn Ezra. In the fifteenth century this problem became widely known in the Latin West, not only with 30 persons out of whom every ninth has to be taken, but also with other numerical values. It is also present in later times and in non-European cultures.[30]

The version presented here probably originated in Ireland.[31]

Here is a pleasant pastime, if one knows it well:

On a certain night a black man named Dub, and another man [named] White, by chance entered beneath the same roofs. White brought along with him thrice five gleaming ones, and the black man had as many black men like [himself] in disposition and color. "White," the other had said, "who first of our men is to take charge of the watch? For I shall follow your suggestions." White answering replied as follows in a quiet voice: "I do not wish by a decision of mine to vex each man, lest by my fault a hitherto unknown quarrel provoke our company to arms. But I will not deprive you of my advice. I will arrange that the whole company recline in order, so that the ninth lot may pick them out for watching through the night. But let the white group sit mixed up with the black men, so that no one may think that I wish to play the men false."

This is how they sat: Four of exceeding whiteness, five blacklets, two whitelets, and a single black, three gleaming ones, a blacklet with darkened skin, here a single white, and two coal-colored ones; two glittering ones, three with darkened covering, after that a snowy one, and two dreadful ones; two whitelets gleaming with lovely skin; and following all these a single black. Thus, by this ingenious arrangement, the ninth lot fell on all the blacklets; the white group were untouched by the lot. Awake and unthankful the black leader kept watch alone with his dark soldiery till day. But all through the night the ingenious White enjoyed quiet sleep along with his men.

4. THE MOST IMPORTANT MEDIEVAL NUMBER GAME: THE *RITHMIMACHIA*

Rithmimachia (*Rithmomachia*) is a mathematical game invented in the eleventh century by monks of southern Germany. It was popular during the whole of the Middle Ages: only in the seventeenth century was the game abandoned. As a board game it competed with chess and in the Middle Ages it was even more popular. The name means "fight with ratios of numbers." *Rithmimachia* made it possible to learn, by playing, the arithmetic that normally had to be

[30] More information about this problem may be found in [Ahrens, 1918, pp. 118–169], [Ball and Coxeter, 1974, pp. 32–36], [Smith, 1958, pp. 541–544], and [Strecker, 1923, pp. 1117–1124].

[31] The leaders are called "Dub" and "Candidus." "Candidus" is the Latin word for "white". The margin of the oldest manuscript contains the note: "These two soldiers from Ireland, one called 'find,' the other 'dub,' were chasing. 'Find' means 'white,' 'dub' means 'black'."

	even				odd			
(1)	2	4	6	8	3	5	7	9
(2)	4	16	36	64	9	25	49	81
(3)	6	20	42	72	12	30	56	90
(4)	9	25	49	81	16	36	64	100
(5)	15	45	91	153	28	66	120	190
(6)	25	81	169	289	49	121	225	361

Fig. I-5-4-1.

learned the hard way in the cathedral and monastic schools as part of the quadrivium. The principal aim was the comprehension of arithmetic (i.e., the Boethian number theory; see section I-3-1), but it was also of service in learning the other subjects of the quadrivium (geometry, astronomy, and music). The game was played on a rectangular board with 16 rows and 8 columns, whose squares are alternately colored black and white as on a checkerboard. There are two armies, that of the "even" numbers and that of the "odd," each with 24 pieces. On the tops of the pieces are marked the following numbers [Fig. I-5-4-1]:

All numbers in the party of the "even" are generated by the smallest even numbers 2, 4, 6, 8, and the "odd" numbers are generated by the smallest odd numbers 3, 5, 7, 9. All elements of the game are derived from Pythagorean number theory as described in Boethius's *Arithmetic*. The numbers in row 2 represent the *proportio multiplex* $mn{:}n$, those in rows 3 and 4 the *proportio superparticularis* $(n+1){:}n$ and those in rows 5 and 6 the *proportio superpartiens* $(n+m){:}n$, with $n > m > 1$. In rows 2, 4, and 6 there are squares, and the numbers in rows 3 and 5 may be found by adding the two numbers directly above them. Some numbers are present in the pieces of both players, and the numbers 25 and 81 occur as many as three times.

Two numbers enjoy a special status: the 91 in the "even" army and the 190 in the "odd" army. Since $91 = 1^2 + 2^2 + 3^2 + 4^2 + 5^2 + 6^2$ and $190 = 4^2 + 5^2 + 6^2 + 7^2 + 8^2$, the 91 may be considered as a pyramid of six squares of sides $1, 2, \ldots, 6$ and the 190 as a truncated pyramid of five squares with sides $4, 5, \ldots, 8$. It is "truncated," because the first three squares (of sides 1, 2, and 3) are missing. These two numbers are described as pyramids built up from squares of corresponding size.

The pieces on which the numbers are written have various shapes according to the ratios that they represent: in the older descriptions the *multiplices* are symbolized by smaller squares, the *superparticulares* by larger squares, and the *superpartientes* by circles. Later the *multiplices* are symbolized by circles, the *superparticulares* by triangles, and the *superpartientes* by squares. The colors of the pieces in the early descriptions were not consistent, and only later was one color assigned for one team and another for the other—in general, black and white were chosen.

The pieces of the two armies were placed on the board in a way not specified in the earliest sources, but they appear variously in the later texts (see Fig. I-5-4-2).

They moved by rules that differed according to the classes assigned to the numbers. In most texts the round pieces could move to adjacent squares, the triangular pieces to a square

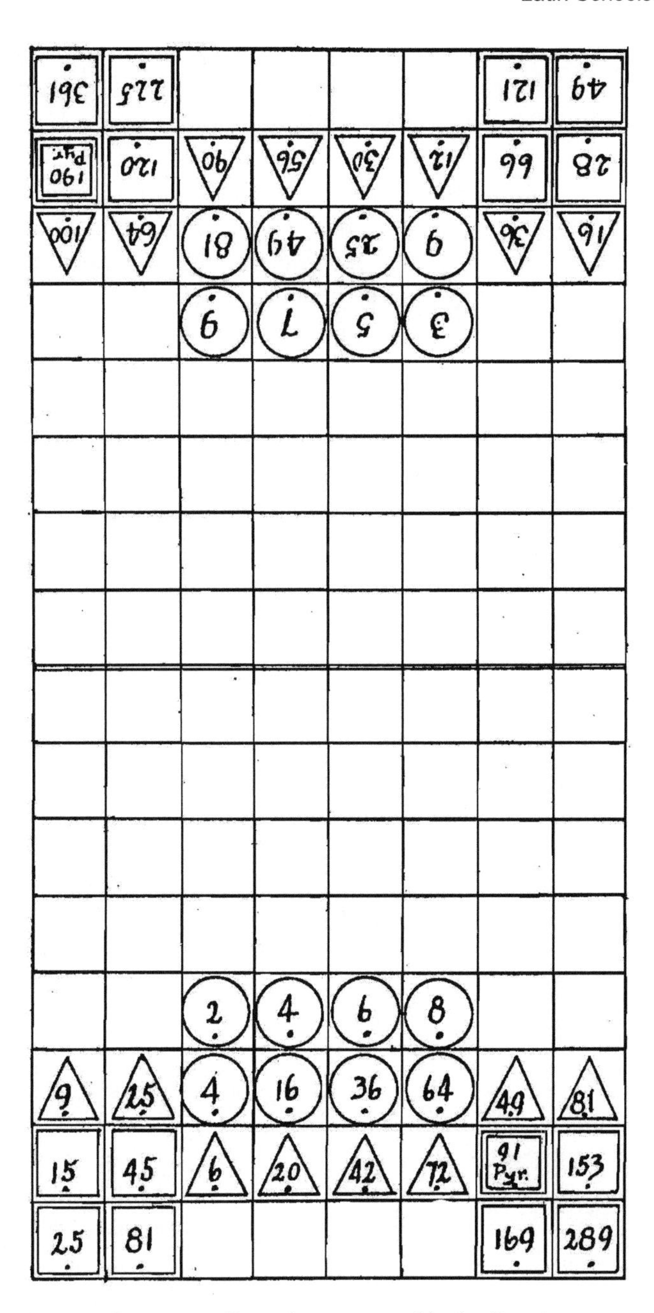

Fig. I-5-4-2. This is one of many configurations attested in the literature.

two squares away, and the quadrangular to a square three squares away.[32] A piece may not jump over other pieces. The pyramids move according to special and various rules.

[32]In some texts they move to the second, third, and fourth field, respectively, so that one, two, or three empty fields are between the offensive and the attacked counter.

For the capturing of enemy pieces there are typically four possibilities: "concourse" (*congressus*): a piece is placed on a square so that with its next move, it could reach the square occupied by the enemy piece holding the same number; "ambush" (*insidiae*): two pieces are placed on squares so that with their next moves, they could reach the square occupied by the enemy piece holding the number that is the sum of those of the two pieces; "eruption" (*eruptio*): a piece is placed on a square so that the product of the number on the piece and the number of squares between it and the enemy piece is equal to the number on the enemy piece; "siege" (*obsidio*): the enemy piece is surrounded on all sides, so that it cannot move; in this case the numbers are not involved. To capture a pyramid, it usually suffices to capture the square forming its lowest part.

To win a game, it is necessary first to capture the pyramid. After this, the player must contrive to put three or four pieces into a winning position in the half of the board originally occupied by the enemy: the pieces must be placed in a straight line and must present at least one of the three Pythagorean means (arithmetic, geometric, or harmonic). If two or more means are presented, the victory is correspondingly better. Since it was hardly possible to know all the winning positions by heart, tables were made, even in the earliest times.

There are more than 30 versions of this game, beginning around 1030 and ending in the seventeenth century.[33] We present here an early text, the so-called Bavarian compilation, which was written shortly after 1100 in southern Germany, perhaps in the monastery of Tegernsee.

(1) Let there be a table divided in the width into 8 and in the length into 16 fields. At both sides of this table there are disposed at the very ends the multiplex numbers, the *superparticulares* and the *superpartientes* up to the tenfold proportion,[34] in such a way that on one side the numbers are presented which have their denominations from the even and on the other side those which have them from the odd.

(2) The eight smaller numbers, [those] of the multiplex family, shall be placed: the doubles, fourfolds, sixfolds, eightfolds. Opposite to them the eight smaller numbers of the same [multiplex] family shall be placed: the triples, fivefolds, sevenfolds, ninefolds.[35]

(3) Behind the "even" *multiplices* there are eight greater [numbers] which belong to the family of the *superparticulares*, so that the *sesqualteri* adhere to the doubles, the *sesquiquarti* to the fourfolds, the *sesquisexti* to the sixfolds and the *sesquioctavi* to the eightfolds.[36] In the other party, i.e. the party of the "odds," behind the *multiplices*, i.e. the foot-soldiers, there are placed eight greater (*superparticulares*) in a similar way. They are ordered so that the *sesquitertii* are joined to the triples, the *sesquiquinti* to the fivefolds, the *sesquiseptimi* to the sevenfolds and the *sesquinoni* to the ninefolds.[37]

[33]Further discussions of the game may be found in [Evans, 1976], [Folkerts, 2003, XI], and [Smith and Eaton, 1911].

[34]This means: all numbers before 10, i.e., 2, 3, 4, ..., 7, 8, 9. ("One" is not considered a number but is the unit that creates the numbers.)

[35]These are the numbers of the first two lines in Fig. I-5-4-1. Every number in line (2) is created from the corresponding number in line (1) by multiplying it with itself. Evidently, they are square numbers.

[36]Line (3), left part: $6 = 3/2 \cdot 4$, $20 = 5/4 \cdot 16$, $42 = 7/6 \cdot 36$, $72 = 9/8 \cdot 64$. Line (4), left part: $9 = 3/2 \cdot 6$, $25 = 5/4 \cdot 20$, $49 = 7/6 \cdot 42$, $81 = 9/8 \cdot 72$. Evidently, the numbers in line (4) are square numbers.

[37]Line (3), right part: $12 = 4/3 \cdot 9$, $30 = 6/5 \cdot 25$, $56 = 8/7 \cdot 49$, $90 = 10/9 \cdot 81$. Line (4), right part: $16 = 4/3 \cdot 12$, $36 = 6/5 \cdot 30$, $64 = 8/7 \cdot 56$, $100 = 10/9 \cdot 90$. Evidently, the numbers in line (4) are square numbers.

(4) Behind the *superparticulares* there are in the party of the "even" the eight greatest [numbers] which belong to the family of the *superpartientes*, so that the *superbipartientes* are joined to the *sesqualteri*, the *superquatuorpartientes* to the *sesquiquarti*, the *supersexpartientes* to the *sesquisexti* and the *superoctopartientes* to the *sesquioctavi*.[38] And in the party of the "odds" behind the middle [numbers], i.e. the counts, there are eight [numbers] of the family of the *superpartientes*, so that the *supertripartientes* adhere to the *sesquitertii*, the *superquinquepartientes* to the *sesquiquinti*, the *superseptempartientes* to the *sesquiseptimi*, and the *supernonipartientes* to the *sesquinoni*.[39]

(5) Having disposed these [numbers] in this way, the permitted steps of both parties may be done, so that the *multiplices*, i.e. the foot-soldiers, may be moved to the next field, forward, backward, up, to the right, to the left, or at an angle. The *superparticulares* may leap over one field similarly, forward, backward, to the right, to the left, or at an angle. In the same way the *superpartientes* may leap two fields to the fourth from their places.

(6) If by a permitted step any number offends a number of the enemy party having the same quantity, it may take it away. If a number is surrounded by numbers of the enemy which, by multiplication or by addition, have the same value as the surrounded number, it may be taken away. And if one [number] faces a [number] of the contrary party and if the product of it with the number of the fields which lie between them is equal to the amount [of the adverse counter], it may remove it.

(7) Only the prime, the not composed numbers, can be moved safely, unless they are surrounded by enemies' [counters] so that they cannot escape by legitimate moves. That is to say, if they do not have a fellow of their party with them which is their guardian between aliens, they may be removed regardless.

(8) On the odd side is the pyramid, i.e. the queen, 190. It can be taken with 8 (units) over 8 fields, because 8 times 8 is 64, the base of this pyramid. Or it is robbed by 16 over 4 fields, because 4 times 16 are 64, or by 64, the proper base of it, if it attacks from the next field. This pyramid, 190, is called "three times shortened," because in the construction of this pyramid the first three numbers, 3, 2, and 1, are not counted. Indeed for its construction 8 times 8, which form its base, and 7 times 7, i.e. 49, and 6 times 6, i.e. 39, and 5 times 5, i.e. 25, and 4 times 4, i.e. 16, are sufficient. Therefore it is called "thrice cut" because of the cutting of the three numbers 3, 2, and 1.

(9) The pyramid on the reverse side, that is in the evens' party, i.e. 91, can be removed by 12 over 3 fields, because 3 times 12 is 36, the base of 91. Or it is robbed by 9 over 4 fields, because 4 times 9 is equally 36, or by the base itself, i.e. 36, in the third field, because it is a courtier.[40] One must also know: each pyramid falls by its base, when it serves as a soldier in the camp of the contrary party and when it is attacked by legitimate

[38]Line (5), left part: $15 = 5/3 \cdot 9$, $45 = 9/5 \cdot 25$, $91 = 13/7 \cdot 49$, $153 = 17/9 \cdot 81$. Line (6), left part: $25 = 5/3 \cdot 15$, $81 = 9/5 \cdot 45$, $169 = 13/7 \cdot 91$, $289 = 17/9 \cdot 153$. Evidently, the numbers in line (6) are square numbers.

[39]Line (5), right part: $28 = 7/4 \cdot 16$, $66 = 11/6 \cdot 36$, $120 = 15/8 \cdot 64$, $190 = 19/10 \cdot 100$. Line (6), right part: $49 = 7/4 \cdot 28$, $121 = 11/6 \cdot 66$, $225 = 15/8 \cdot 120$, $361 = 19/10 \cdot 190$. Evidently, the numbers in line (6) are square numbers.

[40]This means: if the 36 (in the party of the odds) is two fields away from the pyramid 91 (with the basis 36), the attacking 36 (being a *superparticularis* number) can reach the field of the pyramid and, according to the "concourse" rule, remove the basis of the pyramid and then the pyramid itself.

moves, and equally each base sinks down, if the pyramid offends the contrary pyramid at the fourth place.[41]

(10) If one wants to acquire the victory, so that the community of number cannot any longer rebel, he must try by all ways to achieve harmonical or arithmetical means beyond the line of separation, i.e. in the reign of the adversary. You must very cautiously think ahead to which place you do it. Occupy the place, where you want to place your own numbers to triumph. First, either blockade the place with your own men, or foresee that the enemies' will be harmless to you, so that none of the enemy [numbers] can creep between them. Only after thorough preparation you must indicate the [counter] which you want to place at first to gain the victory, whether it is a very small one, a medium, or a very big one. You have no license to move it from this place, but you must add the other two as fast as possible.

(11) I shall say what a numerical mean is. You must join three numbers together in such a proportion that the difference between the middle and the largest is the same as between the smallest and the middle, e.g., 4, 6, 8.[42] For between 8 and 6 the difference is 2, and the same 2 is between 6 and 4. The difference between 9, 12 and 15 is 3. But whenever you do not have a third number for two proper numbers to make any means, first take by robbery one which you can put to the place to fill the vacancy. For instance: if you want to make the means within the even group and you need the 12, you should steal it before by 6 or 4 or 2 and keep it, until you require it. And when you have placed 9 and 15, keep the middle position for it. Place the 12 there so potent is it as though it were yours.

II. A SCHOOL BECOMES A UNIVERSITY: 1140–1480

The transitional scholar, the shining intellectual in the early years of the Augustinian Abbey of St. Victor in Paris, was Master Hugh. Deeply familiar with the lofty minds of the past, from Aristotle, Plato, and Cicero, through Augustine of Hippo, Isidore of Seville, and the twelfth-century John the Scot, Hugh was well prepared to compose a *tour de force* that advocated and fostered what in twentieth century terms would be called a "Great Books" program.[1] He composed *Didascalicon*, as its name signifies,

to select and define all the areas of knowledge important to man and to demonstrate not only that these areas are essentially integrated among themselves, but that in their integrity they are necessary to man for the attainment of his human perfection and his divine destiny.

The text looks at the basic elements of learning on which a Christian should focus, both as to content and as to the manner of learning. Its purpose was

to relieve the physical weaknesses of earthly life and of restoring that union with the divine Wisdom for which man was made.

[41] This means: the pyramid cannot only be attacked (if its base is attacked), but the pyramid itself can attack the other pyramid or counters having the same value, because the pyramid is a *superpartiens* number and therefore can jump over two empty fields.

[42] Surprisingly, only the arithmetical mean, and not the geometrical or the harmonic mean, is explained here.

[1] For authors, titles, and cautions, see [Taylor, 1961, Book II Ch. 2 and Book IV Chs. 5, 13–15].

This single work provided resources and guidelines not only for contemporaries but also for writers in the following years and well through the fifteenth century.

As for mathematics, in Book II, Chapter 1, Hugh laid down the principles whereby a person in the twenty-first century could understand what an intellectual in the twelfth century understood by mathematics. In the second paragraph he stated,

Philosophy is the art of arts and the discipline of disciplines, namely, that toward which all arts and disciplines are oriented. . . . [It is therefore] divided into theoretical, practical, mechanical, and logical. . . . The theoretical can be divided into theology, mathematics, and physics.

Chapter 3 is titled *Concerning Mathematics*.

Since, as we have said, the proper concern of mathematics is abstract quantity, it is necessary to seek the species of mathematics in the parts into which such quantity falls. Now abstract quantity is nothing other than form, visible in its linear dimension, impressed upon the mind, and rooted in the mind's imaginative part. It is of two kinds: the first is continuous quantity, like that of a tree or a stone, and is called magnitude; the second is discrete quantity, like that of a flock or of a people, and is called multitude. Further, in the latter some quantities stand wholly in themselves, for example, "three," "four," or any other whole number; others stand in relation to another quantity, as "double," "half," "once and a half," "once and a third," and the like. One type of magnitude, moreover, is *mobile*, like the heavenly spheres; another, *immobile*, like the earth. Now, multitude that stands in itself is the concern of arithmetic, while that which stands in relation to another multitude is the concern of music. Geometry holds forth knowledge of immobile magnitude, while astronomy claims knowledge of the mobile. Mathematics, therefore, is divided into arithmetic, music, geometry, and astronomy.

Thereafter in four relatively short paragraphs Hugh offered origins of names.

The Greek word *ars* means *virtus*, or power in Latin; and *rithmus*, or number so that "arithmetic" means "the power of number." And the power of number is this—that all things have been formed in its likeness.

Music takes its name from the word water, or *aqua* because no euphony, that is, pleasant sound, is possible without moisture.[2]

Geometry or "earth-measure," for this discipline was first discovered by the Egyptians, who, since the Nile in its inundation covered their territories with mud and obscured all boundaries, took to measuring the land with rods and lines.

Astronomy and astrology differ in the former taking its name from the phrase "law of the stars," while the latter takes its from the phrase "discourse concerning the stars"—for *nomia* means law, and *logos*, discourse.

Thereafter Hugh expounded at varying lengths on the four pillars of the *quadrivium.* What he laid down throughout the *Didascalicon* was copied and recopied nearly a hundred times into the fifteenth century throughout Christian Europe. Hugh had set a norm for general school–level education, not to be replaced until the Italian Renaissance. His death marked the demise of the Carolingian Renaissance.

[2]Perhaps Hugh was thinking that a person could not sing well with a dry throat.

The twelfth century, however, saw a new renaissance; the universities were becoming significant. Well before the end of the century the Masters at the School of St. Victor, together with the Masters at the Schools of St. Geneviève and Notre-Dame de Paris, would construct the cradle of the University of Paris. Oxford arose from dissatisfied, mostly English, Masters and students who left Paris for their homeland; similarly, Cambridge was founded from Oxford. The origins of universities in other countries have their own histories, such as the earlier University of Bologna, formed by the students who hired the Masters. Some schools followed the English model with Masters in charge. Others followed the Italian model with students in charge. Regardless, if there are universities, then there must be students, Masters, and a curriculum. The new curriculum was the gift of the translators. As will be seen, the translations had their impact on all fields of knowledge. The impact in mathematics was felt in geometry, arithmetic, astronomy (the beginnings of trigonometry), the introduction of algebra, and a plethora of offshoots that are gathered under the rubric of mechanics, not to overlook its application in theology. Among the new topics were algebra and trigonometry, the latter not called by that name, but rather identified as the study of triangles.

II-1. Translations

1. TRANSLATORS AND TRANSLATIONS

For medieval mathematics in the West, the twelfth century was a turning point: this was the century in which a mass of new knowledge from Arabic sources suddenly became available. The greater part of this knowledge was from Greek works that had earlier been translated into Arabic, but also some Indian and original Arabic works were put into Latin. After the reconquest of Toledo (1085), a series of translators and compilers began their work principally in this city, but also in other parts of Spain. Scholars from all over Europe worked together; they were often supported in their endeavors by Jews, Mozarabs, or converted Muslims. By the second quarter of the twelfth century, particularly under the patronage of Raymond, Archbishop of Toledo from 1125 to 1152, these men gathered there to do their work. Names like Gerard of Cremona (1114–1187), Adelard of Bath (1080–1152), Robert of Chester (twelfth century), Hermann of Carinthia (1100–1160), and Plato of Tivoli (twelfth century) point up the international character of this movement. In the second half of the twelfth century, another group worked in southern Italy and in Sicily to translate mathematical and astronomical works directly from Greek into Latin, but these translations were not so widely disseminated.

Domingo Gundisalvo (fl. 1155), archdeacon at the cathedral in Segovia and one of the translators in Toledo, fashioned a magnificent tract, *De divisione philosophiae*. Its effect on the study of mathematics, and concomitantly on the university curriculum, was explosive.

The "end" of natural science is the cognition of natural bodies. ... Therefore, this science provides the principles of natural bodies and their accidents. The "instrument" of this science is the dialectical syllogism, which consists of truths and probables. Whence Boethius says, "It is necessary to be versed rationally in natural things." The "artificer" is the natural philosopher who, proceeding rationally from the causes to effect and from effect to causes, seeks out principles. This science, moreover, is called "physical," that is, "natural," because it intends to treat only of natural things which are subject to the motion of nature. Moreover, it is to be read and learned after logic....

On Mathematics

The same things are also to be sought concerning mathematics. It is defined thus: Mathematics is an abstractive science considering things existing in matter, but without the matter; for example, a line, a surface, a circle, a triangle, and similar things which do not exist except in matter. ... Whence it is defined by others thus: Mathematics is an abstractive science considering abstract quantity. We call abstract quantity that which we separate from matter in the intellect or from other accidents—for example: an even number, an odd number, and other things of this kind which we treat by the powers of reason alone. ... And so mathematics is called abstract, since it completely separates the things it treats from matter and its accidents. For since it concerns number or figure, it pays no attention to the matter, color, or position of it but considers it absolutely [or simply,] so that [for example] *all* numbers, *all* triangles, and so on, can be embraced equally in a definition.

The genus of mathematics is the second part of theoretical philosophy abstracted from matter and with motion. For however much things are abstracted from matter by the intellect, nevertheless such entities, which cannot exist without matter—and to the extent that they are not in the intellect—are surely not without motion in matter. The matter of mathematics is universally quantity but considered separately as *magnitude* and *multitude*. It treats most fully those things happening to magnitude and multitude.

Mathematics is also a universal science, since it contains seven arts under it: arithmetic, geometry, music, astrology, the science of aspects, the science of weights, and the science of devices (*ingenia*).

That last sentence enlarged the *quadrivium* into the *septrivium*, a natural science curriculum replacing astronomy with astrology and adding the sciences of optics, weights, and devices (engineering). Thus began the "Twelfth-Century Renaissance."[3]

By the third quarter of the twelfth century Gerard of Cremona had arrived in Toledo, to begin his masterful collection of translations [Haskins, 1927, pp. 9–16]. Among his scientific and mathematical titles are the following:

On Dialectic

1. Aristotle, *Posterior Analytics*
2. Themistius, *Commentary on the Posterior Analytics*
3. Alfarabi, *On the Syllogism*

On Geometry

4. Euclid, The Fifteen Books [of the *Elements*]
5. Theodosius, Three Books *On the Sphere*
6. Archimedes, [*On the Measurement of the Circle*]
7. [Aḥmad ibn Yūsuf], *On Similar Arcs*
8. Mileus [i.e., Menelaus], Three Books [*On Spherical Figures*]
9. Thābit [ibn Qurra], *On the Divided Figure*
10. Banū Mūsā [i.e., the Three Sons of Moses or the Three Brothers], [*On Geometry*]

[3]This renaissance is well described in [Benson and Constable, 1982].

11. Aḥmad ibn Yūsuf, [Letter] *on Ratio and Proportion*

12. [Abū ʿUthmān or Muḥammad ibn ʿAbd al-Bāqī], *The Book of Judeus*[4] *on the Tenth Book of Euclid*

13. Al-Khwārizmī, *On Algebra and Almucabala*

14. *Book of Practical Geometry*[5]

15. Anaritius [i.e., al-Nayrīzī], [*Commentary*] *on* [*the Elements of*] *Euclid*

16. Euclid, *Data*

17. Tideus [i.e., Diocles], *On* [*Burning*] *Mirrors*

18. Alkindi, *On Optics*

19. *Book of Divisions*

20. [Thābit ibn Qurra], *Book of the Roman Balance*

On Astronomy

21. Alfraganus [i.e., al-Farghānī], *The Book Containing XXX Chapters* (or *Elements of Astronomy*)[6]

22. [Ptolemy], The Thirteen Books of the *Almagest*

23. [Geminus of Rhodes], *Introduction to the Spherical Method of Ptolemy*

24. Geber [i.e., Jābir ibn Aflaḥ], *Nine Books* [*on the Flowers from the Almagest*]

25. Messehala, *On the Orb*

26. Theodosius, *On Habitable Places*

27. Hypsicles, [*On the Rising of the Signs*]

28. Thābit [ibn Qurra], *On the Exposition of Terms in the Almagest*

29. Thābit [ibn Qurra], *On the Forward and Backward Motion*

30. Autolycus, *On the Moving Sphere*

31. *Book of the Tables of Jaen with Its Rules*

32. [Abū Abdallāh Muḥammad ibn Muʿādh], *On the Dawn*[7]

Others were also having an impact on university curricula with their translations. The most prolific was the Dominican friar William of Moerbeke (ca. 1215–1286), who worked in the court of the Pope in Viterbo. In contrast to the people who worked in Spain and translated from Arabic into Latin, he translated directly from Greek. He provided the universities with

[4]The Latin text in the list of translations runs: "Liber iudei in decimum Euclidis". Indeed, the word "iudei" is curious. Normally in this list after "liber" follows the name of the author (in genitive). But no Arabic author with a name similar to "Iudeus" (at least, with the same consonants) is known. There are some possibilities: (1) Such a person has existed, but his name was not transmitted. (2) "Iudeus" means "Jew." This would be exceptional. (3) We do have two Latin commentaries of Euclid X which were translated from the Arabic by Gerard of Cremona: one written by Abū ʿUthmān, the other by ʿAbd al-Bāqī, and one of these translations is meant. Most historians of mathematics tend toward option 3.

[5]This seems to be the *Liber mensurationum* ascribed to Abū Bakr (see section II-4-2). In all Latin manuscripts this treatise is transmitted jointly with three texts on mensuration: the *Liber Saydi Abuothmi*, the *Liber Aderameti*, and the *Liber augmenti et diminutionis*.

[6]This is al-Farghānī's most influential work: a compendium of astronomy according to Ptolemy, divided into 30 books. It was translated twice in the twelfth century, and Gerard's translation was highly influential until the sixteenth century.

[7]See section V-2 in Chapter 3 for a translation of this work from the Arabic original.

the complete works of Aristotle, among his other contributions. Here follows a list of some of his mathematical and philosophical translations:

Alexander of Aphrodisias:

1. Commentary on the *Meteorologica* of Aristotle
2. Commentary on *On Sense and Sensible Objects* of Aristotle
3. *Short Work on Fate*
4. *To the Emperors on Fate*

Ammonius:

5. Commentary on *On Interpretation* of Aristotle

Archimedes:

6. *On Spiral Lines*
7. *On the Equilibrium of Plane Figures*
8. *Quadrature of the Parabola*
9. *Measurement of the Circle*
10. *On the Sphere and the Cylinder*
11. *On Conoids and Spheroids*
12. *On Floating Bodies*

Aristotle:

13. *Posterior Analytics*
14. *On the Soul*
15. *On the Heavens*
16. *Categories*
17. *Nicomachean Ethics*
20. *On Interpretation*
21. *Metaphysics*
22. *Meteorologica*
23. *On the Short Natural (or Physical) Treatises*
24. *Physics*
25. *Poetics*
26. *Politics*
27. *Sophistical Refutations*
28. *Rhetoric*
29. *On Colors*

Eutocius:

30. Commentary on *The Equilibrium of Plane Figures* of Archimedes
31. Commentary on *The Sphere and the Cylinder* of Archimedes

Heron of Alexandria:

32. *Catoptrics*
33. *Pneumatics*

John Philoponus:
34. Commentary on *On the Soul* of Aristotle

Proclus:
35. *Elements of Theology*
36. Commentary on the *Parmenides* of Plato
37. Commentary on the *Timaeus* of Plato

Ptolemy:
38. *Book on "Taking up"* [or the *Analemma*]

Simplicius:
39. Commentary on the *On the Heavens* of Aristotle
40. Commentary on the *Categories* of Aristotle

Themistius:
41. Paraphrase of *On the Soul* of Aristotle

A third translator, though not so extensive in his work as the others, was Robert Grosseteste (ca. 1175–1253), Master at Oxford and future Bishop of Lincoln. While the list of his translations is not so long as those of Gerard of Cremona—after all, he was principally a teacher—his work influenced the curricula at his home school and eventually those of schools on the Continent. Among his translations are these [Crombie, 1962, p. 136 n2; Rashdall, 1936, p. 240 n1]:

1. Aristotle, *On the Heavens*, Books 1–3
2. Pseudo-Aristotle, *On indivisible lines*
3. Simplicius, *Commentary on Aristotle's On the heavens*, Book 2

2. TRANSLATIONS: A PRACTICAL ILLUSTRATION

The important translations made from Arabic into Latin in twelfth-century Spain were the work of translators who did not understand Arabic even though translation was their purpose in coming to Spain. They did not know in advance which writings they would find, and sometimes they had to work with poor-quality manuscripts. Among the Arabic manuscripts that became available to the translators there were—besides texts by Arabic authors—many Arabic versions of Greek texts that had been translated into Arabic in the ninth and tenth centuries in the Arabic East. Many contained technical terms in Arabic unknown to the translators, and therefore some terms were retained in Latin transliteration of the Arabic (as, e.g., *algebra*). Sometimes an Arabic- or Hebrew-speaking person assisted the translator; or the Arabic text was first translated into Castilian and then into Latin. The overall results were Latin translations that differed considerably when compared among themselves or even with the Greek originals.

Here follow different versions of the same text in Latin translations and reworkings of the twelfth and thirteenth centuries: the initial theorem in Euclid's *Elements* (I Prop. 1). For comparison's sake, a translation from the Greek text is offered first (a). In what follows, efforts were made to keep the English translations as literal as possible; thereby the differences among the translations from the Arabic into Latin, indeed even twice removed from the Greek, are made evident.

In the twelfth century, different Arabic translations of the *Elements* were available in Spain. This makes uncertain which version(s) the Latin translators used. It is certain, however, that the *Elements* were translated independently three times. Adelard of Bath made the first translation in the second quarter of the twelfth century (b). Adelard did not always use technical terms consistently, and it seems that he was constantly seeking the right Latin expressions for the various Arabic terms. Possibly in the 1140s, Hermann of Carinthia made another translation. The extant text is not literal and may be a reworking (c). The most famous translator of all, Gerard of Cremona, made the third Latin translation, presumably in the third quarter of the twelfth century (d). Being closest to the Greek tradition it is superior to the other Latin translations. Gerard is also well known for his extreme literalness in translating Arabic texts; sometimes his translations look highly "Arabicized." In addition to the three twelfth-century translations of the *Elements* from Arabic into Latin, a translation was made directly from the Greek into Latin, shortly before 1175 in southern Italy or Sicily. This translation is very close to the translation from the Greek below (a), and therefore it is not presented here. At that time, Sicily was a meeting point of Greek and Latin cultures, when Greek, Latin, and Arabic civilizations lived side by side in peace and toleration, and the Sicilian kings encouraged the production of translations.

Even more important than the translations were the reworkings of the *Elements* done in the twelfth and thirteenth centuries, using the Arabic-Latin translations and including material from other sources. Two reworkings gained the greatest influence. The most important Latin Euclidean text in the West until the second half of the thirteenth century was made by Robert of Ketton (Chester) around 1140 (e). It does not contain detailed proofs but "directions for proof." These show how by the use of the axioms, postulates, and previous theorems the proof of another theorem can be arrived at. Shortly before 1259, Campanus of Novara (1220–1296) produced by far the most important medieval Latin text of the *Elements* (f). His work became the definitive text of the *Elements* until the sixteenth century. As a textbook it was used at all universities. The additions that Campanus made to the basic text are especially noteworthy.

a. Greek text

On a given finite straight line to construct an equilateral triangle.

Let AB be the given finite straight line. Thus it is required to construct an equilateral triangle on the straight line AB.

With center A and distance AB let the circle $B\Gamma\Delta$ be described; again, with center B and distance BA let the circle $A\Gamma E$ be described; and from the point Γ, in which the circles cut one another, to the points A, B let the straight lines ΓA, ΓB be joined [Fig. II-1-2-1].

Now, since the point A is the center of the circle $\Gamma\Delta B$, $A\Gamma$ is equal to AB. Again, since the point B is the center of the circle ΓAE, $B\Gamma$ is equal to BA. But ΓA was also proved equal to AB; therefore each of the straight lines ΓA, ΓB is equal to AB. And things which are equal to the same thing are also equal to one another; therefore ΓA is also equal to ΓB. Therefore the three straight lines ΓA, AB, $B\Gamma$ are equal to one another.

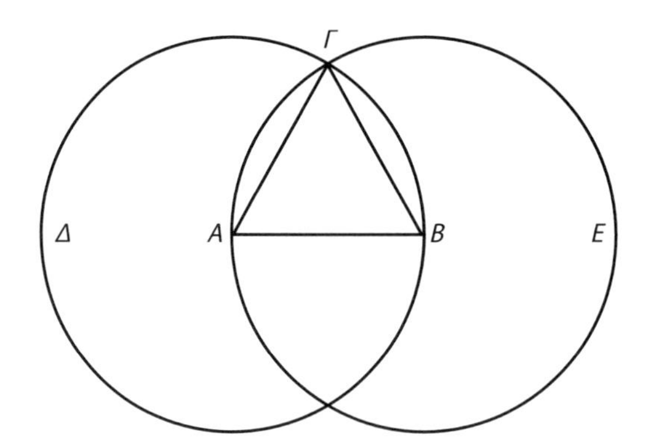

Fig. II-1-2-1.

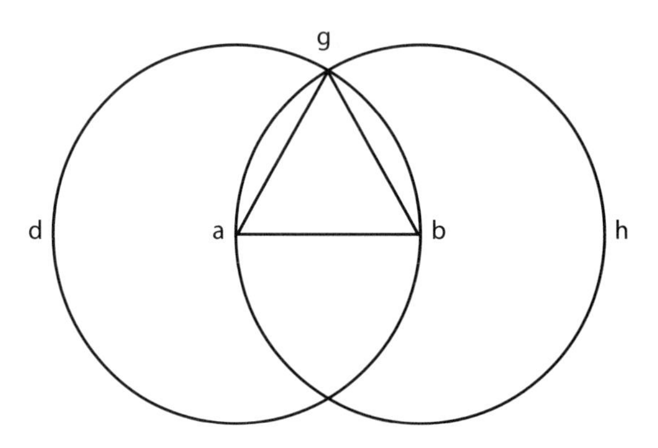

Fig. II-1-2-2.

Therefore the triangle *ABΓ* is equilateral; and it has been constructed on the given finite straight line *AB*. (Being) what it was required to do.

b. Translation by Adelard of Bath

Now it must be demonstrated how we can make a triangular surface with equal sides atop a straight line of assigned quantity.

Let *ab* be the designated line. Let a center be put on *a*, occupying the distance between *a* and *b* by a circle on which are *g d b*. Likewise let a center be put on *b*, occupying the distance between *a* and *b* by another circle on which are *g a h*. And from point *g*, where the circles cut themselves,[8] let go out two straight lines to point *a* and to point *b*; these will be *ga* and *gb*. I say that in this way we have made a triangle with equal sides atop the designated line *ab* [Fig. II-1-2-2].

The reason of this procedure: Because point *a* has become the center of the circle *gdb*, the line *ag* has become equal to the line *ab*. And because the point *b* is the center of the circle *gah*, the line *bg* has become equal to the line *ba*. And so each of lines *ga* and

[8]Literally: "on which there is the incision of the circles."

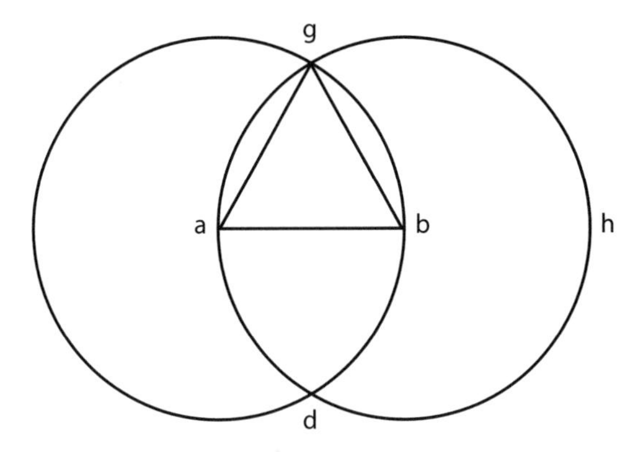

Fig. II-1-2-3.

gb is equal to the line *ab*. But things which are equal to the same thing are also equal to one another. Therefore the three lines *ag*, *ab* and *bg* are equal to each other. Therefore the triangle *abg* with equal sides has been made atop the designated line *ab*. And this is what we intended to show in this figure.

c. Translation by Hermann of Carinthia

First we place an equilateral triangle atop a straight line of defined quantity.

Given is the straight line between points *a* and *b*. Then the distance of the points, i.e. the space of the line, is taken, and fixing a compass on center *a* a circle *bgd* is made. The compass is firmly transferred, the same space is held, and another circle *agh* is made around center *b*. Then from point *g*, the point of the cut of the circles, straight lines shall fall on *a* and *b*. In this way there will be an equilateral triangle [Fig. II-1-2-3]. For *a*, the center of circle *bgd*, forces line *ag* to be equal to line *ab*. In a similar way, *b*, the center of circle *agh*, makes line *bg* coequal to line *ba*. For the center is the point from which all lines going out to the circle are equal to one another. It is true, however, if two things are equal to the same thing, each of the two is equal to the other. Therefore because the two *chateti* [= sides] are equal to the base, they are equal, too, to one another. And so an equilateral triangle has been placed atop a straight line.

d. Translation by Gerard of Cremona

To construct atop a straight line with definite quantity an equilateral triangle.

For instance: Let there be placed a straight line *ab* with definite quantity, and around the center *a* according to the quantity of the space between *a* and *b* a circle is drawn, on which are *g*, *d*, *b*. Another circle is drawn around the center *b* according to the quantity of the space between *a* and *b*, on which are *g*, *a*, *e*. Then from point *g*, in which one of the two circles cuts the other, two straight lines to the two points *a* and *b* are drawn. Let them be the lines *ga* and *gb*. I say that we now have made an equilateral triangle atop the given line *ab* [Fig. II-1-2-4].

This is its proof: Because the point *a* was made the center of the circle *gdb*, the line *ag* becomes equal to the line *ab*. Similarly, because point *b* was made the center of the circle *gae*, the line *bg* becomes equal to the line *ba*. Therefore each of these two lines *ga*

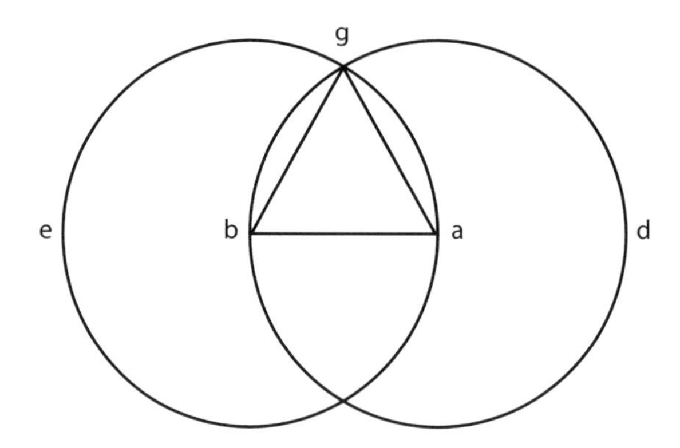

Fig. II-1-2-4.

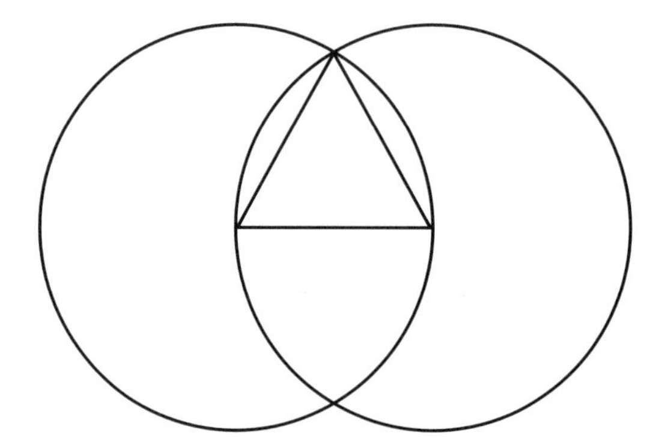

Fig. II-1-2-5.

and *gb* has been found equal to the line *ab*. But things that are equal to the same thing are also equal to one another. Therefore the three lines *ag*, *ab*, and *bg* are equal to each other. And so the triangle *abg* has equal sides and, as has been shown, is constructed atop the given line *ab*. And this is what we wanted to prove.

e. Reworking by Robert of Ketton

To place an equilateral triangle on a given straight line.

From the two endpoints of the given line describe with the compass by occupying [= using] this line two circles that cut each other. From the common point of intersection of the circles draw two straight lines to the two endpoints of the given line. Then draw an argument from the description [= definition] of the circle [Fig. II-1-2-5].

f. Reworking by Campanus of Novara

To construct an equilateral triangle atop a given line.

Let *ab* be the given straight line. I want to construct atop it an equilateral triangle. Upon one of its endpoints, namely point *a*, I place the immobile foot of a compass and I extend the other, mobile, foot to *b*, and I draw according to the quantity of the very given

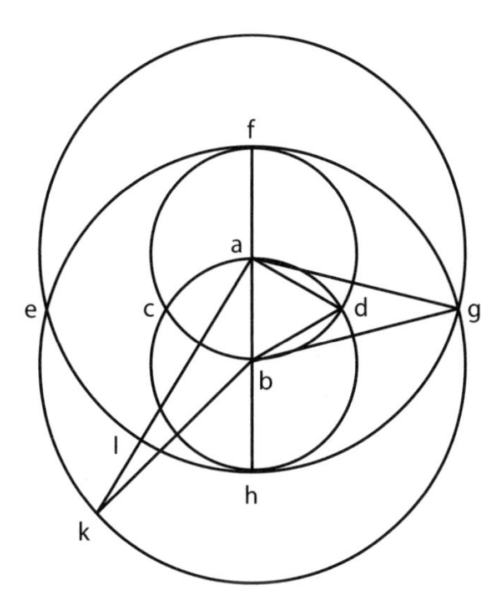

Fig. II-1-2-6.

line, by the second postulate,[9] circle *cbdf*. Again, I make its other endpoint, *b*, the center and, according to the same postulate and with the same quantity, I draw the line of the circle *cadh*. These circles intersect each other in points *c* and *d*. I join one of the two sections, say section *d*, with both extensities of the given line, having drawn lines *da* and *db* according to the first postulate [Fig. II-1-2-6].[10] Because from *a*, the center of the circle *cbd*, the lines *ad* and *ab* are drawn to its circumference, they are equal according to the definition of a circle. Similarly, because from *b*, the center of the circle *cad*, the lines *ba* and *bd* are drawn to its circumference, they too are equal according to the same definition. Because therefore each of the two lines, *ad* and *bd*, are equal to line *ab*, as has been proved, they are equal to each other according to the first common notion.[11] Therefore we have constructed an equilateral triangle atop the given line. This was proposed.

After that, Campanus has added two cases: to construct an isosceles triangle and to construct a triangle with three different sides atop a straight line. Accordingly, Fig. II-1-2-6 is more complex than those for the other translations.

II-2. Arithmetic

1. AL-KHWĀRIZMĪ, *ARITHMETIC*

Already by the end of the tenth century, the Hindu-Arabic numerals had become known in the Christian part of Western Europe, but they were used only on counters of the abacus in the tradition of Gerbert (see section I-2-4). They became more widely known in the twelfth century, beginning through the translation of the *Arithmetic* of al-Khwārizmī (780–850) from Arabic into Latin. This treatise describes the way to write numbers using the

[9]This postulate reads: "Let the following be postulated: . . . To describe a circle with any center and distance."
[10]This postulate reads: "Let the following be postulated: To draw a straight line from any point to any point."
[11]The first common notion reads: "Things which are equal to the same thing are also equal to one another."

"Arabic" numerals, to calculate with integers and fractions, and to extract square roots. The first chapter describes the forms of the numerals and states the principles of the place-value system. We present extracts from this chapter, showing how difficult it was at that time to explain the concepts of this system, and especially the concept of zero.[12]

Algorizmi said: Let us speak praises to God our guide and defender, worthy both to render Him His due and multiply His praise by increasing it, and let us entreat Him to guide us in the path of righteousness and lead us into the way of truth, and to help us in addition with goodwill in these things which we have decided to set out and reveal: concerning the numbering of the Indians by means of IX symbols, by which they set out their universal system of numbering, for the sake of its ease and brevity, so that this work, to be sure, might be made easier for the seeker after arithmetic, i.e., the greatest number as much as the smallest, and whatever there is in it as a result of multiplication and division, also addition and subtraction, etc.

Algorizmi said: since I had seen that the Indians had set up IX symbols in their universal system of numbering, on account of the arrangement which they established, I wished to reveal, concerning the work that is done by means of them, something which might be easier for learners if God so willed. If, moreover, the Indians had this desire and their intention with these IX symbols was the reason which was apparent to me, God directed me to this. If, on the other hand, for some reason other than that which I have expounded, they did this by means of this which I have expounded, the same reason will most certainly and without any doubt be able to be found. And this will easily be clear to those who examine and learn. So they made IX symbols, whose forms are these: 9 8 7 6 5 4 3 2 1. . . .

I have found, said Algorizmi, that everything that can be expressed in terms of number is also whatever is greater than one up to IX, i.e., what is between IX and one, i.e., one is doubled and two results, and likewise one is tripled and three results; and so on for the rest up to IX. Then X is put in the place of one and X is doubled and tripled, just as was done in the case of one; from its doubling results XX, from its tripling XXX, and likewise up to XC. After this C (a hundred) comes back in the place of one and is doubled there and tripled, just as was done in the case of one and X; and there will be produced from it CC and CCC etc. up to DCCCC (nine hundred). Again, a thousand is put in the place of one, and by doubling and tripling, as we have said, there result from it II thousand and III thousand etc. up to infinity according to this method. And I have found that the Indians worked according to these places. Of these, the first is the place of the units, in which is doubled and tripled whatever is between one and IX. The second is the place of the tens, in which is doubled or tripled whatever is from X to ninety. The third is the place of the hundreds, in which is doubled and tripled whatever is from C to DCCCC. Furthermore, the fourth is the place of the thousands, in which is doubled and tripled whatever is from a thousand to IX M. The fifth place is X thousand in the following way: every time the number rises, places are added. The arrangement of a number will be as follows: everything that will have been one in the higher place will be X in the lower, which is before it, and what will have been X in the lower will be one in the higher, which

[12]Here we have just included excerpts from the beginning of this text. More details can be found in [Berggren, 2007, pp. 525–530].

precedes it; and the beginning of the places will be on the right of the writer, and this will be the first of them and is itself placed there for the units. But when X was put in the place of one and was made in the second place, and its form was the form of one, they needed a form for the tens because of the fact that it was similar to the form of one, so that they might know by means of it that it was X. So they put one space in front of it and put in it a little circle like the letter o, so that by means of this they might know that the place of the units was empty and that no number was in it except the little circle, which we have said occupied it, and [thus] it is shown that the number that is in the following place was a ten and that this was the second space, which is the place of the tens. And they put after the circle in the aforesaid second place whatever they wished from the number of tens from what is between X and XC and these are the forms of the tens: the form of X is thus 10, the form of XX 20. And likewise the form of XXX is thus 30, and so on up to IX [tens] there will be, clearly, a circle in the first place and a character pertaining to the number itself in the second place. Moreover, one must know this, that the character that signifies one in the first place, in the second signifies X, in the third C and in the fourth M. And likewise the character that in the first place signifies two, in the second signifies XX and in the third CC and in the fourth two thousand and understand likewise about the rest. . . .

But when X or more is gathered in any of the places, it is to be raised to a higher place and from each X a one is to be produced in the higher place.[13] Again, if there is another number in the same place, at which a number arrives by increasing, it is to be added on and they are to be added together and if there is in it X or more, from each X a one is to be made and to be raised to a higher place, i.e., if ten is gathered in the first place, one is to be made from it and placed in the second place, and if in the same place there is likewise a number, it is to be added to it; and if there is X there, one is to be made from it and raised also to the third place. E.g., if in the first place, which is the place of units, you have X, make a one from it and place it in the second place. Moreover, in the first place put a circle just as we have said, so that it may be shown that there are two places.[14] But if there is XI, make a one from the X and put it in the second place as above and send down one into the first. But if you find some number in the second place, where you have placed the very number that you made from X, add it to that. And if there is X, or more, make from X a one and again place it in the third place; and what remains below X, let it remain in its own place. Moreover, what we say of more than ten holds for any large number. E.g., if there is in the second or third place a large number, such as if you find IX in the third place, which is the place of the hundreds, and if there is a X in the second place, make from the X a one and change it to the third place, and there add it to IX and there results X; make a one from the X itself and change it to the fourth place and there it will be a thousand. If, on the other hand, you found XX in the second place, you would also make two from it, and adding two to IX in the third place, XI would also result; you would again make a one from the X and change it to the fourth place where it would be a thousand: and there remains a one in the third place and therefore indicates X or more. And this must be known that, because you have changed your number and put it in the

[13]This is the carrying procedure. "Higher place" means to the left.

[14]If in adding you get 10, write down 0 and carry 1.

following place, you must put it by means of its own characters, i.e., if it is X, instead of it place the character that signifies one in the first place, and if it is XX, instead of it place the character that signifies two in the first place. And understand likewise for the rest. But if there remains in the same place, from which you have changed a number, something from the number, move it down likewise by means of its own characters, i.e., if there remains a one or two, move it down there by the character that signifies the same number, i.e., if there remains one, copy there the character of one, and if there remains two, copy there the character of two, and so on. But each form will have significance according to the place, i.e., in the first place it will signify units, in the second tens, in the third, hundreds etc., just as has been said above.

Moreover, if it is a large number and you wish to know which it is in numerical order or how many places are in it, so that you may write it in a book or talk about it, know that there is not in any place more than IX nor less than one unless there is a circle (i.e., o), which is nothing; when therefore you wish to know this, count the places beginning from the first, which will be on the right side, and this will be the place of the units. The rest of the places will be marked out by their succession toward the left side of the writer. Of these the second will be the place of the tens and the third of the hundreds and the fourth of the thousands and the fifth of the X thousands. Moreover, the sixth will be the place of the C thousands and the seventh of the thousand thousands. Again, the VIIIth will be the place of the X thousand thousands and the ninth of the C thousand thousands, and the tenth of the thousand thousand thousands in three stages and the eleventh of the X thousand thousand thousands in three stages, and the twelfth of the C thousand thousand thousands in three stages and the XIIIth of the thousand thousand thousand thousands in four stages, and likewise in every place you will add according to the places of the number in your utterance. But if beyond three places,[15] i.e., the places of the hundreds and tens and thousands there is a one left, there will be X thousand of the thousands themselves that have resulted for you in words. But if there remain two places there will be C thousand of the thousands themselves. And now I have put together an example for you, by which you will be able to know and prove by it whatever is added to a number or subtracted from it: and this is the form of the same: 1 180 703 051 492 863.

When you add two symbols according to what we have said about these signs, the number of thousands of those signs will be one thousand thousand thousand thousand of thousands in five stages according to the number of characters that are below[16] them, and one hundred thousand thousand thousand of thousands in four stages according to the number of characters that are below them, and eighty thousand thousand thousand of thousands in four stages according to what is from those characters. Next seven hundred thousand thousand of thousands in three stages according to the characters that are below them and three thousand thousand of thousands in three stages and fifty-one thousand of thousands in two stages, and four hundred thousand and ninety-two thousand and eight hundred and sixty-three.

[15]That is, to the right.

[16]That is, farther to the right.

2. LEONARDO OF PISA (FIBONACCI), *LIBER ABBACI* (*BOOK ON CALCULATION*)

Leonardo of Pisa (Fibonacci) (ca. 1170–after 1240) was a member of a family named Bonacci in Pisa. The young Leonardo lived with his father in Bejaia (Bujie in modern Algeria) and traveled from there around the Mediterranean coast to Constantinople and to other places in the East. The most authentic information about Leonardo's early life is to be found at the beginning of his *Liber abbaci*:

When my father, appointed by his homeland, held the post of public *scriba* (notary or representative) in the custom-house of Bejaia for the Pisan merchants frequenting it, he arranged for me to come to him when I was a boy. Because he thought it would be useful and appropriate for me, he wanted me to spend a few days there in the *abbaco* school, and to be taught there. Here I was introduced to that art (the *abbaco*) by a wonderful kind of teaching that used the nine figures of the Indians. Getting to know the *abbaco* pleased me far beyond all else and I set my mind to it, to such an extent that I learnt, through much study and the cut and thrust of disputation, whatever study was devoted to it in Egypt, Syria, Greece, Sicily and Provence, together with their different methods, in the course of my subsequent journeys to these places for the sake of trade. But I reckoned all this, as well as the algorism and the arcs of Pythagoras,[17] as a kind of error in comparison to the method of the Indians (*modus Indorum*). Therefore, concentrating more closely on this very method of the Indians, and studying it more attentively, adding a few things from my own mind, and also putting in some subtleties of Euclid's art of geometry, I made an effort to compose, in as intelligible a fashion as I could, this comprehensive book, divided into 15 chapters, demonstrating almost everything that I have included by a firm proof, so that those seeking knowledge of this can be instructed by such a perfect method (in comparison with the others), and so that in the future the Latin race may not be found lacking this [knowledge] as they have done up to now.

Leonardo returned to Pisa ca. 1200. There he composed for the next 25 years works on arithmetic, algebra, and geometry. His main work, the *Liber abbaci*, was written in 1202 and reworked in 1228. Leonardo was in contact with the scholars of the circle around the Hohenstaufen emperor Frederick II, and he was presented to the emperor when Frederick held court in Pisa around 1225. The last known reference to him is from 1240, when the republic of Pisa awarded him a yearly *salarium*.

Leonardo began his text with an introduction to the Hindu-Arabic place-value system, followed by several chapters discussing rules for computing with these new numerals. He then included numerous problems in such practical topics as calculation of profits, currency conversions, alloying of money, barter, determining values of merchandise, as well as many recreational problems. In chapter 8, he introduced the the rule of three,[18] a procedure he used frequently in the rest of the book. The longest chapter is the twelfth, which encompasses a wide variety of problems and solution methods. Among the problems in that chapter is his most famous problem, the Rabbit Problem, leading to the Fibonacci sequence $a_1 = 1, a_2 = 2$, $a_{n+1} = a_n + a_{n-1}$ $(n \geq 2)$, a problem tucked inconspicuously between a problem on perfect

[17] These were arcs used by Gerbert on his abacus that marked triples of columns.

[18] The rule of three is the method of solving the proportion $x{:}a = b{:}c$ for x by multiplying a by b and dividing the result by c.

numbers and one calculating the amount of money held by four men. We present here a small selection from chapter 12, dealing just with problems solved by arithmetic methods. In Section II-3-2, we will present a few additional problems from chapters 12, 13, and 15 that are solved by algebraic methods.

Early in chapter 12, Fibonacci introduced the "tree problem," exemplified in the problem below. He showed that many problems could be thought of as tree problems (including the following one on two serpents) and could be solved using either the rule of three or "another method" he introduced here, the method of false position.[19] In modern terms, such problems can be expressed as linear equations of the form $ax = b$. See [Hannah, 2007] for an extensive discussion of Fibonacci's use of this method.

From Chapter 12

[On Problems of Trees]

There is a tree 1/4 + 1/3 of which lies underground, and it is 21 palms; it is sought what is the length of the tree. Because the least common denominator of 1/4 and 1/3 is 12, you see that the tree is divisible into 12 equal parts. Three plus four parts are 7 parts, and 21 palms; therefore as the 7 is to the 21, so proportionally the 12 is to the length of the tree. And because the four numbers are proportional, the product of the first times the fourth is equal to the second by the third. Therefore, if you multiply the second 21 times the third 12, and you divide by the first number, namely by the 7, then the quotient will be 36 for the fourth unknown number, namely for the length of the tree; or because the 21 is triple the 7, you take triple the 12, and you will have similarly 36.

There is indeed another method that is used. Namely, you put for the unknown number some arbitrary number which is integrally divisible by the denominators in the fractions that are posed in the problem, and according to the posing of the problem, with the posed number you strive to find the proportion occurring in the solution of the problem. For example, the sought number of this problem is the length of the tree; therefore you put it to be 12, which is divided integrally by the 3 and the 4 which are under the fractions. And because it is said that 1/4 + 1/3 of the tree is 21, you take 1/4 + 1/3 of the put 12; there will be 7, and if it would be 21, we would fortuitously have the proposition, namely that the tree is 12 palms. But because 7 is not 21, it therefore happens proportionally that as the 7 is to the 21, so is the put tree to the sought value, namely as the 12 is to the 36. Therefore one says according to custom, I put 12 and there results 7; what shall I put so that 21 results? And as it is said, the extreme numbers are multiplied together, namely the 12 and the 21, and their product is divided by the remaining number.

On Two Serpents

There is a serpent at the base of a tower that is 100 palms high, and he ascends daily 1/3 of a palm, and he descends daily 1/4. At the top of the tower there truly is another serpent who descends daily 1/5 of a palm, and ascends 1/6; it is sought in how many days they will meet in the tower. You put it that they will meet in 60 days because 60 is the least common multiple of the 6 and 5 and 4 and 3; you see therefore how much the

[19]The method of false position can be found in Egyptian mathematics as far back as the early second millennium BCE. See [Imhausen, 2007, pp. 26–28] for some examples. It is also found in ancient Mesopotamian mathematics.

serpents approach each other in the 60 days. The lower serpent truly ascends 5 palms more than he descends in the 60 days. The upper serpent truly descends 2 palms more than he ascends in the 60 days. Therefore they are closer by 7 palms. Therefore it is said, for the 60 days that I put, they are closer by 7 palms; what shall I put so that they are 100 palms closer? You multiply the 60 by the 100; there will be 6000 that you divide by the 7. The quotient will be 857 1/7 days,[20] and in this amount of time they meet each other. If you will seek in what part of the tower they meet, you do thus: you multiply the 5, namely the ascent of the lower serpent, by the 100; there will be 500 that you divide by the 7; the quotient will be 71 3/7, and this is the amount the lower serpent ascends. And if you wish the descent of the upper serpent you will multiply the 2 by the same 100, and you will divide by the 7; then the quotient will be 28 4/7 palms from the summit for the place of the meeting.

Here is a problem on "finding a purse," a problem type that seems to have originated in India,[21] but it also appears in Islamic and numerous medieval and early modern European works. In this example, we have only two men with a purse, but in further problems there are more men and the situation is a bit more complicated.

On the Finding of a Purse

Two men who had *denari* found a purse with *denari* in it; thus found, the first man said to the second, If I take these *denari* of the purse, then with *denari* I have I shall have three times as many as you have. Alternately the other man responded, And if I shall have the *denari* of the purse with my *denari*, then I shall have four times as many as you have. It is sought how many *denari* each has, and how many *denari* they found in the purse. It is indeed noted that because the first, having the purse, has three times as many as the second, that if he has with the purse 3, then the second has 1; therefore among them both and the purse they have 4; therefore as the first with the purse has 3, he has 3/4 the entire sum of their *denari* and the purse. And for the same reason, as the second with the purse has four times as many as the first, it is necessary for him to have 4/5 of the same sum. Therefore you find the least common denominator of 4/5 and 3/4, and it will be 20. Therefore you put the sum of the *denari* to be 20, of which the first with the purse has 3/4, namely 15. And the second with the purse has 4/5, namely 16; therefore among them both with the purse counted twice they have 31; the difference between 31 and 20, namely 11, is truly the *denari* of the purse. Because the purse is counted twice, and as one should only count it once, the purse is therefore counted once more than it should be. Whence the *denari* difference between the 20 and the 31, namely 11, is one times that which is found in the purse. Therefore you subtract the 11 from the 15; there remains 4, and this many the first man has; next you subtract the 11 from the 16; there remains 5, and this many the second has; therefore the first has 4, and the second 5, which added to the 11 of the purse makes 20 which we can put for the sum.

[20] When writing mixed numbers, Leonardo writes the fraction before the integer. Thus, this value is written as 1/7 857. For ease of reading, we have converted Leonardo's notation to the modern one.

[21] See Mahavira in appendix 3. Also, an abstract version of this problem occurs in the work of Levi ben Gershon (see section I-6 in Chapter 2).

In the following purse problem, Leonardo finds that one of the men must have a negative number of *denari*, so he considers that amount to be a "debit."

It is proposed that there are 5 men, and the first having the purse proposes to have two and one half times as many as the others, and another, if he has the purse, proposes to have three and one third times the others. Also the third with the purse has four and one fourth times as many as the others; the fourth with the purse truly has five and one fifth times as many as the others; the fifth moreover with the same purse affirms that he has six and one sixth times as many as the others. According to the abovewritten material, the first with the purse has two and one half times as many as the others; therefore if he with the purse has 2 1/2, then all the others have 1; therefore if he has 5, then the others will have 2; therefore among them all they have 7; from this, as the first with the purse has 5, he is undoubtedly demonstrated to have 5/7 of the total sum of them all and the purse. Therefore undoubtedly there remains 2/7 of the same sum for the other IIII men. Therefore for the same reason you see the second man with the purse to have 10/13 of the entire sum, and you find 3/13 to remain for the others. For the same reason you show the third man to have with the purse 17/21 of the entire sum, and you know 4/21 of the same sum remains for the others. And you see the fourth man with the purse to have 26/31 and you do not doubt 5/31 to remain for the rest. By the same method you will take care to see the fifth man with the purse to have 37/43 of the entire sum, and you find there remains 6/43 of the same sum for the other IIII men.

Leonardo now determines that 363,909 is the least common multiple of all the denominators. He then calculates that 5/7 of that sum is 259,935; 10/13 is 279,930; 17/21 is 294,593; 26/31 is 305,214; and 37/43 is 313,131.

You add the 259,935 to the 279,930, and to the 294,593, and to the 305,214, and to the 313,131; there will be 1,452,803 from which you subtract the 363,909; there remains 1,088,894 that is the amount of *denari* in the purse. And because there are 5 men, the purse is counted four times more than it should be. Therefore, the 363,909 is multiplied by 4; there will be 1,455,636 that is the amount of the purse and the *denari* of the five men, and because the first has 5/7 of the entire sum, you take 5/7 of the 1,455,636, and is 1,039,740, and this many the first has with the purse. But because above more is found in the purse than is had between the purse and the first man, this posed problem will not be solvable unless the first man has a debit, namely that which is the difference between his amount plus the purse, and the entire amount of the purse, namely that which is the 1,088,894 minus the 1,039,740, that is 49,154. Also you take 10/13 of 1,455,636, that is 1,119,720, and the second man has this many between him and the purse; from this is subtracted the *denari* of the purse, namely 1,088,894; there remains 30,826, and the second man has this many.

(Similarly, Leonardo finds that the third man has 89,478, the fourth 131,962, and the fifth 163,630.)

Another problem appearing frequently in later works is that of men buying a horse. As a purely mathematical problem, this problem is solved by Diophantus in Book I, #24[22] of the *Arithmetica* and in the work of Levi ben Gershon (see section I-6 in Chapter 2). It is also

[22]See Appendix 2.

solved by al-Karajī in his *al-Fakhri*, part III, #26 [Woepcke, 1853, p. 95]. In addition, a version can be found in the work of Mizrahi (see section I-5 in Chapter 2). The problem here is one of many of the same type that Leonardo includes.

Three Men and One Horse When Each Takes from the Others in Order

There are three men having bezants who desire to buy a horse. And as none of them can buy it, the first proposes to take from other two men 1/3 of their bezants. And the second proposes to take 1/4 of the bezants of the other two men. And similarly the third proposes to take 1/5 of the others, and thus each proposes to buy the horse. The 1/5 1/4 1/3 are written in the first position, and the 1 that is over the 3 is subtracted from the 3; there remains 2 over which you put the 1 making the fraction 1/2. Also the 1 which is over the 4 is subtracted from the 4; there remains 3 over which you put the 1, making the fraction 1/3. Again you subtract the 1 which is over the 5 from the 5; there remains 4 over which you put the 1, making the fraction 1/4. After this you put in order 1/4 1/3 1/2, and this is called the second position. And you see what number is the least common denominator, namely 12, that you multiply by the 3 of the first position; there will be 36 that you divide by the 2 of the second position; the quotient will be 18 that you keep. Also you multiply the same 12 by the 4 of the first position, and you divide by the 3 of the second position; the quotient will be 16. Also you multiply the aforewritten 12 by the 5 of the first position, and you divide by the 4 of the second; the quotient will be 15 that you add to the 18, and the 16; there will be 49 that is the sum of the bezants of the three men. Next one man of the three is subtracted; there remains 2 which you multiply by the same 12; there will be 24 that you subtract from the 49; there remains 25 that is the price of the horse. After this you multiply the aforewritten 24 by the 1 which is over the 2 in the second position, and you divide by the 2; the quotient will be 12 that you subtract from the 25, namely the price of the horse; there remains 13, and the first has this many. Also you multiply the same 24 by the 1 which is over the 3, of the same second position, and you divide by the same 3; the quotient will be 8 which subtracted from the said 25 leaves 17, and the second has this many; also you multiply the 24 by the 1 which is over the 4 of the second position, and you divide by the 4; the quotient will be 6 which subtracted from the 25 leaves 19, and the third has this many.

Leonardo explained his reasoning in other, similar, problems. From a modern point of view, with x, y, and z representing the three men (or the amounts they possess); with h the price of the horse; and with $s = x + y + z$, the problem results in the equations $s - h = (2/3)(y + z) = (3/4)(x + z) = (4/5)(x + y)$. It follows that y and z give away 1/2 of what they keep; x and z give up 1/3 of what they keep; and x and y give 1/4 of what they keep. Since 12 is the least common multiple of the three denominators, Leonardo first assumes that what remains to any two men after they have given up part of their money is 12 (i.e., $s - h = 12$). So if y, z have 12, then they gave 6. Therefore $y + z = 18$ $(= (3/2)12)$. Similarly, if x, z have 12, then they gave 4, so $x + z = 16$ $(= (4/3)12)$. And if x, y have 12, then they gave 3, so $x + y = 15$ $(= (5/4)12)$. Therefore, $2s = 49$. But 2 does not divide 49, so we double all the numbers and find that $s = 49$. Since now we are assuming that $s - h = 24$, we find that $h = 25$. Then, since x takes from y and z one half of what they keep, and they keep 24, it follows that x takes 12, so $x = 13$. By similar arguments, $y = 17$ and $z = 19$.

In another problem from chapter 12 of the *Liber abbaci*, two men each sell fish for the same price. Further, they have to pay a tax to the customs agent. Leonardo gives three variants, two of which he notes are not solvable. In the third version, Leonardo makes a proposal to change the problem so that it will be solvable.

[On Two Men with Fish and the Customs Agent.]

One of two men had 12 fish, and the other had 13 fish, and all of the fish were of one price. The customs agent took away one fish and 12 *denari* for the office of the director. And from the other he took 2 fish and gave him back 7*denari*; the customs fee and the price of each fish are sought. Because for 12 fish there are given for the office of the director one fish and 12 *denari*, for each fish there are given 1/12 of one fish and 1 *denaro*. Therefore the second man must have given for the thirteen fish 13/12 of one fish and 13 *denari* for which he actually gave 2 fish minus 7 *denari*; therefore 13/12 of one fish and 13 *denari* are equal to two fish minus 7 *denari*. Therefore if 7 *denari* are commonly added, then there will be two fish equal to 13/12 fish plus 20 *denari*; there will remain 11/12 of one fish equal to 20 *denari*, that is 11/12 fish are worth 20 *denari*; therefore proportionally just as the 11 is to the 12, so is the 20 to the price of one fish. Therefore you divide the product of the 12 and the 20 by the 11; the quotient will be 21 9/11 *denari* for the price of one fish; 1/12 of this, namely 1 9/11, you add to one *denaro*; there will be 2 9/11 which is the customs fee for one fish. For example, the customs fee for 12 fish is 33 9/11 *denari* which result from the multiplication of the 2 9/11 and the 12 which is equal to the price of one fish and 12 *denari*. Similarly the customs fee for the 13 fish is 36 7/11 *denari* which is equal to the price of two fish minus 7 *denari*, as it must be.

And if it is proposed that for 12 fish the customs agent takes one fish, and gives back 12 *denari*, and for 13 fish he takes 2 fish minus 7 *denari*, then the problem is not solvable. You find indeed by a similar investigation that 13/12 of one fish minus 13 *denari* are equal to two fish minus 7 *denari*; therefore if the 13 *denari* are commonly added, then it results that 13/12 of one fish are equal to two fish and 6 denari which is inconsistent. And if for 12 fish one fish is taken minus 7 *denari*, and for 13 fish is taken 2 fish minus 12 *denari*, then 13/12 of one fish minus 7 7/12 *denari* are equal to 2 fish minus 12 *denari*. Therefore if commonly the 12 *denari* are added and the 13/12 fish are subtracted, then there will remain 11/12 of one fish equal to 4 5/12 *denari*. Therefore you multiply the 4 5/12 and the 12, and you divide this by the 11; the quotient will be 4 9/11 for the price of each fish. And because the 7 *denari* are minus, and they are given back to him by the customs agent who took the fish, it is seen that this problem is not solvable as the customs agent gives back more to him than he takes from him. But if it will be proposed that the excess that was given back to the first man is to the excess that is returned to the second, just as the fish of one to the fish of the other, that is just as the 12 is to the 13, then each fish is worth the found 4 9/11 *denari*.

How Many Pairs of Rabbits Are Created by One Pair in One Year?

A certain man had one pair of rabbits together in a certain enclosed place, and one wishes to know how many are created from the pair in one year when it is the nature of them in a single month to bear another pair, and in the second month those born to bear also. Because the above written pair in the first month bore, you will double it; there will be two pairs in one month. One of these namely the first, bears in the second month, and thus there are in the second month 3 pairs; of these in one month two are pregnant, and in the third month 2 pairs of rabbits are born, and thus there are 5 pairs in the month; in this month 3 pairs are pregnant, and in the fourth month there are 8 pairs, of which 5 pairs bear another 5 pairs; these are added to the 8 pairs making 13 pairs in the fifth month; these 5 pairs that are born in this month do not mate in this month, but another 8 pairs are pregnant, and thus there are in the sixth month 21 pairs: to these are added the 13 pairs that are born in the seventh month; there will be 34 pairs in this month; to this are added the 21 pairs that are born in the eighth month; there will be 55 pairs in this month; to these are added the 34 pairs that are born in the ninth month; there will be 89 pairs in this month; to these are added the 55 pairs that are born in the tenth month; there will be 144 pairs in this month; to these are added again the 89 pairs that are born in the eleventh month; there will be 233 pairs in this month. To these are still added the 144 pairs that are born in the last month; there will be 377 pairs, and this many pairs are produced from the above written pair in the mentioned place at the end of the one year. You can indeed see in the margin how we operated, namely that we added the first number to the second, namely the 1 to the 2, and the second to the third, and the third to the fourth, and the fourth to the fifth, and thus one after another until we added the tenth to the eleventh, namely the 144 to the 233, and we had the above written sum of rabbits, namely 377, and thus you can in order find it for an unending number of months.

3. JOHN OF SACROBOSCO, *ALGORISMUS VULGARIS*

Al-Khwārizmī's *Arithmetic* became the model for many subsequent texts on arithmetic in the West. The first words of this translation were: "Dixit algorizmi," the origin of the word "algorismus" that categorizes and titles these texts. A typical text of this kind contained the following: the forms of the numerals and the principles of the place-value system; the six basic operations (addition, subtraction, multiplication, division, and also doubling and halving); extraction of square and cube roots; and the summation of arithmetic and geometric progressions. These texts were not practical instructions for merchants; they were intended for students in the faculty of arts in universities. The most ubiquitous work, and even the most popular mathematical writing of the entire Latin Middle Ages, was written by John of Sacrobosco (1195–1236).[23] It became the most prominent textbook for learning arithmetic at the universities. Many copies of this work give evidence of this. Numerous manuscripts contain interlinear notes to explain special terms and longer comments in the margins. Very often the main text is surrounded by a detailed commentary that explains section by section the content of the work.

Everything that has proceeded from the earliest beginning of things has been formed in a pattern of numbers, and the manner in which these exist must be understood as follows: that in the whole comprehension of things, the art of numbering is concerned with combination [or union]. A certain philosopher named Algus[24] wrote this brief science

[23] Very little is known of Sacrobosco's life with certainty. See [Pedersen, 1985] for a detailed discussion of the possibilities for his birthplace, his education, and his dates. He is sometimes referred to as John of Holywood.

[24] "Algus" is a corruption of "al-Khwārizmī" (see section II-2-1).

of numbering, for which reason it is called Algorismus, which is understood to be the art of numbering or the introductory art into number.

A number is made known in two ways: materially and formally. Materially a number is a collection of units; formally it is the multitude of units extended [or spread out]. A unit is that by which anything is said to be one. Included among numbers are the digit, article, and composite number. A digit is every number smaller than 10; an article is every number divisible into ten equal parts with no remainder; a composite number is mixed, consisting of digit and article. And it must be understood that every number between [any] two proximate [or successive] articles is composite. Moreover, there are nine species of this art [of Algorism]: (1) numeration, (2) addition, (3) subtraction, (4) halving, (5) doubling, (6) multiplication, (7) division, (8) progression, and (9) extraction of roots, which is twofold, since [it applies] to square numbers and cube numbers. Among these, numeration will be considered first and then the others will follow in order.

On Numeration

The numeration of any number by any suitable figures is an artificial representation. However, figure, difference, place, and limit suppose the same thing but they are imposed for different reasons. "Figure" is so called because of the drawing of a line: "difference" is shown in this way: how the figure following differs from the figure preceding; "place" is so called by reason of the space in which a number is written; "limit" is the path [or system] that has been organized for the representation of any number. Therefore, it should be understood that alongside the nine "limits" are found nine significant "figures" representing nine digits, which are these:

9. 8. 7. 6. 5. 4. 3. 2. 1.

The tenth [figure], 0, is called teca, circle, cipher, or "the figure of nothing," since it signifies nothing. It holds a place and signifies for others, for without a cipher or ciphers, a pure article cannot be formulated. And so it happens that any number can be represented by these nine significant figures, with a cipher or ciphers added whenever needed, rendering it unnecessary to find more significant figures [than these]. Therefore, it must be noted that every digit must be written with only one figure that is appropriate to it; every article, however, must be represented by a cipher placed in the first position and a digit by which this article is denominated, since every article is denominated by some digit—as ten by unity, twenty by two, and so on. Indeed, every number that is a digit has to be placed in a first [category of] difference, every article in a second [category of] difference. For every number from ten to one hundred, excluding one hundred, must be written by two figures; and if it is an article, [it is written] by a cipher, placed in the first position, and a figure written to the left, which signifies the digit by which the article is denominated. If it is a composite number, the digit which is part of it is written first, and as before, the article is placed to the left.[25] Again, every number from one hundred to one thousand, excluding one thousand, is written by three ciphers or figures; likewise every number from one thousand to ten thousand is written by four numbers, and so on. It must also be noted that any figure placed in the first position signifies its digit; in the second

[25]This means that a number like 32 would be written by first writing 2, then to the left of it 3.

position ten times its digit; in the third [position] one hundred times its digit; in the fourth [position] one thousand times its digit; in the fifth [position] ten thousand times its digit; in the sixth [position] one hundred thousand times its digit; in the seventh [position] thousand times thousand times its digit, and so into infinity by multiplying these three [things]: ten, one hundred, and one thousand. All these things are contained in this maxim: Any figure posited in the place [or position] following signifies ten times as much as in the preceding position. And it must be understood that above any figure in the thousands place, one can conveniently put points for denoting that the last figure ought to represent as many thousands as there are points. Moreover, in this art we write toward the left in the Arabic or Jewish custom, for they are the inventors of this science; or [we do it] for this reason, an even better one, [namely] that in following the usual order in reading, we place a greater number before a smaller number.

On Addition

Addition is the adding of a number, or numbers, to a number. There are two rows of figures in addition—that is, at least two numbers are necessary, namely the number to which the addition is to be made and the number to be added. The number to which the addition ought to be made is the number that receives the addition and ought to be written above; the number to be added to the other ought to be written below. It is more fitting that the smaller number be written below and added to the greater than contrarily. But whether or not this is done, the same [result] is always produced.

Therefore, if you wish to add a number to a number, write the number to which the addition should be made in the upper row by its differences so that the first number of the lower row is under the first number of the upper row; the second under the second, and so on. After this has been done, the first figure of the lower row is added to the first figure of the upper row. From this addition, then, either a digit, an article, or a composite number results. If a digit, let the resulting digit be written in the place of the upper figure, which is crossed out; if an article, let a cipher be written in place of the upper deleted number and let the article be carried to the left and added to the next figure following—if there is a following figure. If, however, there is no figure following, then let it [the article that is carried] be placed in an empty place. However, if it happens that the figure following, to which the addition of the article is to be made, is a cipher [zero], the digit of the article should be written in place of it after it [the cipher] has been deleted. But if there is a figure of nine and a unit is to be added to it, a cipher should be written in place of this nine and the article written to the left of it, as before. After this has been done, the second [number] is added to the second number placed above and the same procedure is followed as before. It must also be noted that in addition and in all the following species [of algorism], when one figure is placed directly above [or over] another, it must be used in this figure and [this is different than] if it were placed by itself.

On Subtraction

When two numbers have been proposed, subtraction is the finding of the excess of the greater to the lesser term; or subtraction is the taking away of a number from a number so that the sum remaining may be seen. Furthermore, a smaller from a greater or an even from an even [number] can be subtracted; but a greater from a smaller never. That number is called greater which has more figures so long as the last is significant. If, however, there are as many [figures] in one number as in the other, [the greater] must be judged by the last, or next to last, [figures], and so on.

In subtraction two significant numbers are necessary, namely the number from which the subtraction is to be made and the number to be subtracted. The number from which the subtraction is to be made is written in an upper row by its differences;[26] the number to be subtracted is placed in the lower row by its differences, so that the first [figure] is under the first [figure], the second under the second, and so on. . . .

It must be understood, however, that in both addition and subtraction we can very well begin from the left and go toward the right; but, as was said before, it is more convenient to do it in the manner that has been described. Moreover, if you wish to prove whether you have done it properly or not, add the figures which you have found to the [figures in the] lower row, and if you have done this correctly, the same figures should result which you had before. Similarly in addition, [for] when you will have added all the figures, subtract what you added before, and if you have done this correctly, the same figures which you had before will be produced. For subtraction is the proof of addition, and conversely.

On Halving

Halving is the finding of half of any number proposed in order to see what and how much is this half. In halving there is only a single row of figures and only a single number is necessary, namely the number to be halved. Therefore, if you wish to halve any number, let this number be written by its differences and begin from the right, namely from the first figure toward the right, and take it toward the left side. . . .

On Doubling

Doubling is the adding of a proposed number to itself so that the resultant sum can be seen. In doubling, only one row of figures is necessary and it must be begun from the left side, namely from the greatest figure—that is, from the figure representing the greatest number. In the three preceding species, we began from the right side, namely from the smallest figure—that is, [the figure] representing the smallest number. [But] in this species and in all those that follow, we begin from the left side and with the greatest figure. Whence comes the verse:

> You subtract or add or halve from the right side,
> On the left side you divide and multiply, [and] extract a double root.

The reason for this is that if you begin to double from the first figure it would sometimes happen that the same number would be doubled twice; and although we could begin from the right side, the theory and operation will be more difficult.

If you wish to double any number, let it first be written by its differences and the last figure be doubled.[27] From this doubling there results either a digit, an article, or composite number. If a digit results, it is written in place of the first [figure], which is deleted; if an article results, a cipher is written in place of the first [figure], which is deleted and the

[26] That is, in order of the place value of the numbers.

[27] Since Sacrobosco orders his terms from right to left, this is the first number on the left.

article is carried toward the left; if a composite number results, the digit is written in place of the upper or first deleted figure (which is part of this composite number) and the article is carried toward the left. After this has been done, the next to last number must be doubled and the same thing must be done as before. If, however, a cipher occurs, it must be left untouched; but if a number is to be added to a cipher, the number to be added must be written in place of what was deleted. The same procedure must be followed for all other numbers until the whole number is doubled. . . . The proof of doubling is halving, and conversely.

On Multiplication

Multiplication is the finding of a third number from two proposed numbers which contains one of them as many times as there are units in the remaining number. In multiplication two numbers are principally necessary—the multiplier [or multiplying number] and the number to be multiplied. The multiplier is designated adverbially; the number to be multiplied receives a nominal appellation. A third number to be assigned, which is called the product, arises from leading one number into the other. It should be noted that a multiplying number can always be converted into a number to be multiplied, and conversely, with the product always remaining the same. And this is what is commonly said or alleged by the arithmeticians, namely that every number is converted when it is multiplied by itself.

There are six rules of multiplication:

The first is [for] when a digit multiplies a digit. [After] the smaller digit [has been multiplied] by the difference between the greater [given] digit and ten, it [the product of this multiplication] must be subtracted from the article of its denomination, which has been computed at the same time with ten. For example, if you wish to know how much is eight [multiplied by] four, see how many units there are between eight and ten, while at the same time computing with ten. It is obvious that there are two [units]. Therefore let eight be subtracted from forty, and thirty-two will remain, the product of the whole multiplication.

The second rule is this: When a digit multiplies an article, the digit must be multiplied by the digit by which the article is denominated—and this is done according to the first rule. Then any unit [or digit] will represent ten, and any [multiple of] ten [will represent] one hundred. And this is true whether the article or composite number increases.

The third rule is this: When a digit multiplies a composite number, the digit, namely the multiplier, must be multiplied into every part of the composite number—the digit by the digit in accordance with the first rule; the digit by the article in accordance with the second rule. The products are added and the sum of the whole multiplication will be obvious.

The fourth rule is this: When an article multiplies an article, the digit which denominates one of them must be multiplied by the digit which denominates the other; and any unit represents one hundred, any [multiple of] ten, a thousand.

The fifth rule is this: When an article multiplies a composite number, the digit of the article must be multiplied with each part of the composite number and [then] the products are added and the sum [of the products] will be obvious.

The sixth rule is this: When a composite number multiplies a composite number, each part of the multiplier must be multiplied by each part of the multiplying number, and the products are added; the sum [of the products] will then be obvious. . . .

On Division

When two numbers have been proposed, division is the distribution of the greater by the smaller into as many parts as there are units in the smaller [number]. In division, therefore, it must be noted that three numbers are necessary—the number to be divided [the dividend], the dividing number, or divisor, and the number denoting the quotient, or number that comes forth. If the division is to be made with whole numbers, the dividend ought always to be greater or at least equal to the divisor number.

Therefore if you wish to divide any number by itself or by another number, write the dividend number by [the order of] its differences[28] in the upper row; and write the divisor by [the order of] its differences in the lower row so that the last [figure] of the divisor would be under the last [figure] of the dividend; the next-to-last [figure] under the next-to-last, and so on, if it can be properly done.[29] For there are two reasons why a last [number] cannot be placed under a last [number]: (1) either because the last number of the lower row cannot be subtracted from the last number of the upper row because [the latter] is smaller than the [number in the] lower row;[30] or because (2) although the last [number] of the lower row could be subtracted from its [corresponding number in the] upper row, the remainder cannot be [subtracted] a certain number of times from the number placed above it; as if the last [number] was equal to the figure above it and the penultimate or antepenultimate [number] is greater.[31] In such a case, the last figure of the divisor must be placed under the penultimate [or next-to-last number] of the dividend.

After arranging the figures, the operation should begin from the last figure of the divisor number. One must see how many times this figure can be subtracted from the figure placed above it; and if something should be left, [see how many times it can be subtracted] from its remainder. But it must be noted that one cannot subtract more than nine times nor less than once. When it has been seen how many times the figures of the lower row could be subtracted from the numbers in the upper row, the number denoting this—that is, the quotient—must be written directly over that figure under which lies the first figure of the divisor, and that figure [indicates how often] the lower figures must be subtracted from their upper figures.

Now when this has been done, the figures of the divisor must be moved forward toward the right by a single place or difference and the same procedure followed as before. However, should it happen that after the moving forward of the divisor, its last figure cannot be subtracted from the figure above it, then directly above the figure under which the first figure of the divisor lies, a cipher, [or zero,] must be written in the row of numbers denoting the quotient and the figures [of the divisor] must be moved forward as before. The same thing must be done wherever it happens that the divisor cannot be subtracted from the number to be divided, [that is,] a cipher must be put in the row of numbers denoting the quotient and the figures [of the divisor] must be moved forward as before. . . .

[28] That is, in the order of their place value.

[29] Thus, to divide 2852 by 12, begin by writing $\begin{array}{l}2852\\12\end{array}$

[30] For example, we cannot use the arrangement from note 29 in dividing 1352 by 25; instead we have $\begin{array}{r}1352\\25\end{array}$

[31] For example, in dividing 1352 by 15, we would write $\begin{array}{r}1352\\15\end{array}$

On Progression

A progression is an aggregation of numbers taken according to equal excesses from unity or from two so that the sum of the whole or of the different numbers is had compendiously. Furthermore, some progressions are natural or continuous, others interrupted or discontinuous. A progression is natural or continuous when it is taken from unity and no number is omitted in its ascent, as 1, 2, 3, 4, 5, 6, . . . , and the number following always exceeds the preceding number by unity only. An interrupted progression occurs when some number is uniformly omitted, as 1, 3, 5, 7, 9, ... Similarly, it can begin from 2, as 2, 4, 6, 8, ... And the number following always exceeds the preceding by two units. It should be noted, moreover, that two rules are given concerning a natural progression.

The first is this: When a natural progression is terminated in an even number, multiply the next higher number by half [of the even number terminating the series] and you will obtain the total [of the series]. For example, 1, 2, 3, 4. Multiply five by two and you get ten, the sum of the whole progression. Whence the verse comes:

> Even, you will multiply a greater by half of an even number.

The second rule is this: When a natural progression is terminated in an odd number, multiply it by the greater part of it and obtain the total sum. For example, 1, 2, 3, 4, 5. Let five be multiplied by three, and so, three times five results in fifteen, the sum of the whole progression. Whence the verse:

> Let odd be multiplied by its greater part.

As for an interrupted progression, two rules are likewise given:

The first rule is this: When an interrupted progression is terminated in an even number, you multiply its half by the next number higher than its half, as [for example,] 2, 4, 6. Let four be multiplied by three, and three times four will result in twelve, the sum of the whole progression. Whence the verse:

> If even, multiply the [number] following by its half.

The second rule is this: When an interrupted progression is terminated in an odd number, multiply its greater part by itself. For example, 1, 3, 5. Let three be multiplied by itself; thus three taken three times produces nine, the sum of the whole progression. Whence the verse:

> In odd numbers, the greater mean part is multiplied by itself.

On the Extraction of Roots; First in Square Numbers

What follows concerns the extraction of roots, first in square numbers. Therefore, we must see what a square number is; what is the root of a square number; and what it is to extract a root. However, this division must be noted beforehand: some numbers are linear, some superficial [or plane], and some solid. A linear number is considered only according to process and not with regard to the multiplication of a number by a number; and it is called "linear" because it has only a single number, just as a line has only a single dimension, namely length. A superficial [or plane] number arises from the multiplication of one number by another number; and it is called "superficial" because it has two numbers measuring it, just as a surface has two dimensions, namely length and width. But it must be understood that a number can be multiplied by another number in two ways: either once or twice. If a number is multiplied by another number once, it is either [multiplied] by itself or by another [number]; if by itself, it is a square number and is called "square" because written separately by units it will have four equal sides in the manner of a square. If it is multiplied by another [number], it would be a superficial and non-square number, as two multiplied by three constitutes six, a superficial but non-square number. Thus it is obvious that every square number is superficial, but not conversely.

Moreover, a root of a square number is that number which when multiplied by itself produces the square number, as two taken twice is four. Therefore, four is the first square number and its root is two. However, if a number is multiplied twice by a number, it constitutes a solid number; and it is called solid, since it has three dimensions just as a solid body, namely length, width, and depth. Thus this number has three numbers producing it. But a number can be multiplied twice by a number in two ways: either by itself or by another [number]. If the number were multiplied by itself twice, or [multiplied] once by its square, which is the same thing, it becomes a cube number and is called cube from the name *cubus*, which is a solid. A cube is a certain body having six surfaces, as a die, eight [solid] angles, and twelve edges. If, however, any number is multiplied twice by another, a solid and non-cube number results, so that twice three twice constitutes twelve. And so it is obvious that every cube number is a solid, but not conversely.

From the aforesaid, it is also obvious that the same number is a root of a square and a cube [number], but the square and cube are not, however, the same. It is also obvious that every number can be the root of a square and cube, but not every number can be a square or cube [number]. Therefore, since the multiplication of a unit by itself once or twice produces nothing but a unit, Boethius says, in his *Arithmetic*, that the unit is every number potentially, but not in act.

It must be noted also that between any two successive square numbers there is a single mean proportional which is produced by the multiplication of one square root by the root of another; and between any two successive cube numbers there are two mean proportionals—a lesser and a greater. The lesser mean arises from the multiplication of the root of the greater cube [number] by the square of the [root] of the lesser [number]; the greater mean is produced if the root of the lesser cube [number] is multiplied by the square of the [root of the] greater [number]. Since, in the present art, no process occurs beyond the sum of solid numbers, only the nine limits of numbers are distinguished. For there is a limit to the continuous sequence of numbers of the same nature contained by extreme terms; hence the first limit is the continuous progression of the nine digits; the second [limit] is that of the nine principal articles;[32] the third is [the limit] of hundreds; the fourth of thousands. Three limits also result in compound [numbers] by the application of digits to any of the three [limits mentioned before]—[that is,] if one is placed before another. But by the multiplication of a final term once beyond itself in the manner of making squares, or twice in the manner of making solids, the penultimate and final limits are produced.

[32]That is, 10, 20, 30, . . . , 90.

The Extraction of the Square Root of a Number

To find the root of a square number is to find, with respect to some proposed number, the square root of it, if a square number was proposed. If it is not a square, to find the root of the greatest square contained under the proposed number. If, then, you wish to extract the root of some square number, write that number by its differences and count whether the number of figures [in that square number] is even or odd. If even, the operation is to begin under the penultimate [figure]; if there are an odd number [of figures], begin from the last figure. To put it briefly, with an odd number [of figures] always begin from the last [of the figures]. Under the last figure placed in the odd position, there must be found a certain digit which when multiplied by itself would offset [that is, equal] the whole number located above it or come as close to it as possible. When such a number has been found and multiplied by itself, it must be subtracted from the number above it. This digit must then be doubled and placed under the next figure before it to the right and its half must be placed under it. After this has been done, a certain digit must be found under the next figure preceding the doubled number, which multiplied by the doubled number may offset [that is, equal] the whole number located above the doubled number; then this number multiplied by itself offsets, [or equals]—or comes as close as possible to—the whole number located above it.

4. JOHANNES DE LINERIIS, *ALGORISMUS DE MINUCIIS*

From the thirteenth century on, special algorismus treatises for calculating with fractions were produced. They taught the basics of writing common fractions and sexagesimal fractions (which were used in astronomy) and the calculation with both kinds of them. The best-known textbook on fractions is the *Algorismus de minuciis* (*minutiae* = fractions) written by Johannes de Lineriis (also known as John of Lignères because of his French origin) around 1320. Johannes, an astronomer who studied and taught at the University of Paris early in the fourteenth century, wrote several other works, mostly on astronomical topics (see section II-5-3). The topics of the *Algorismus* are: to convert a whole number into a fraction, and its opposite; to bring two fractions to the same denominator; multiplication, division, addition, subtraction, doubling, halving; extracting square and cube roots; to find for two (or three) given numbers the third (or fourth) whole number proportional to them. Johannes de Lineriis's book is characterized by having no proofs, only rules and examples. Its general style is similar to the *Algorismus* of John of Sacrobosco and was a text that could easily be studied.

The text begins with the two kinds of fractions (common fractions and physical, i.e. sexagesimal, fractions). There follows how to bring two common fractions to the same denominator and to reduce fractions to integers (and vice versa).

1. The way to represent common fractions (*minucie vulgares*) and physical fractions (*minucie phisice*).

There are two numbers in fractions: the counting number (*numerus numerans*) and the denominating number (*numerus denominans*). Therefore we must know at first what the counting number and the denominating number are, then, what the common and the physical fractions are, and thereafter, how they are represented and how they are written. First we must know that the counting number contains the unit as many times as we want to represent the parts of the whole [number]. The denominating number is the number in which the unit is contained as many times as the denominated part is [contained] in the whole number. Secondly, we must know that "common fractions" are those which are counted by their numbering parts in a uniform way according to their order, i.e., "half" by 2, "third" by 3, "fourth" by 4, etc. The "physical fractions" are related to the integers when divided by 60; i.e., the degrees (*gradus*), which are the integers, are divided into 60 minutes (*minuta*), a minute into 60 seconds (*secunda*), the second into 60 *tertia*, etc. Thirdly, we must know that the common fractions are represented so that the denominator is written below and the numerator above. E.g.: if we want to write four sevenths, we do it in this way: 4/7. The physical fractions are represented so that they are indicated by the place where they stand. Only the numerator is written, and the place indicates the denominator. E.g.: 2 *signa*[33] 5 *gradus* 20 *minuta* 15 *secunda* 35 *tertia*. The first place is that of the *signa*, the second of the *gradus*, the third of the *minuta*, the fourth of the *secunda*, the fifth the *tertia*, etc. By proceeding from left to right, e.g. 4 *signa* 22 *gradus* 20 *minuta* 30 *secunda* etc., [we can represent] as much as we want.

2. The way to reduce fractions with different denominators to the same common denominator and the way to reduce whole numbers to fractions and *vice versa*.

We will reduce common fractions with different denominators to the same common denominator in this way: we multiply the denominator of one fraction with the denominator of the other one and the product is the common denominator. For instance: if we want to reduce 3/4 and 2/5 to the same denominator, we multiply 5, the denominator of the second fraction, with 4, the denominator of the first fraction. The result is 20, the common denominator of both fractions. . . . If you want to reduce more than two fractions with different denominators to the same denominator, then first arrange the first two fractions according to what I have said, and then multiply the common denominator, which comes from the first two denominators, with the [denominator of the] third fraction. After that, multiply the common denominator, which comes from the last [process], with the [denominator of the] fourth fraction, and so on, until you have reduced all fractions with different denominators to the same denominator. E.g., if we want to reduce 2/3, 2/4, and 2/5 to the same denominator, we multiply the denominator of the first fraction with the denominator of the second, i.e., 3 and 4. The result is 12, the common denominator of the first two fractions. This 12, the common denominator of the first two fractions, multiply with 5. The result is 60, which is the common denominator of all three fractions. Do the same, if you have more [than two] fractions with different denominators. If you have several fractions, some with different denominators and some with the same denominator, then reduce in the way as instructed only those with different denominators. E.g., 2/3, 2/3, 2/4, 2/5, 2/5 have the common denominator 60. Nevertheless, if you want, you can also reduce them as if they all had different denominators. If you want to know how many sixtieths are in each fraction, divide 60 by the denominator of the fraction, multiply the quotient with the numerator, and you have what you want. E.g., if you divide 60, the common denominator of all of them, by 3, the denominator of the first fraction, you receive 20. These 20 multiply with 2, the numerator of the first fraction, and you receive 40. So many sixtieths are in the first fraction, 2/3. Divide 60 by 4, the denominator of the second fraction, and you receive

[33]The word *signa* means "sign" (i.e., one of the signs of the zodiac); each sign corresponds to 30 degrees.

15. Multiply with 2, the numerator of the second fraction, and the result is 30. So many sixtieths are in the second fraction, 2/4. Divide further 60 by 5, the denominator of the third fraction, and the result is 12. Multiply them with 2, the numerator of the third fraction, and the result is 24. So many sixtieths are in the third fraction, 2/5.

[3. . . .]

You can reduce any integer to a common fraction with given denominator by multiplying the given denominator with the number of integers. Let the product be the numerator and the denominator of the fraction be the denominator. For instance, if you want to reduce 4 integers to fifths, multiply 4 with 5; the result is 20. Accordingly 4 integers are 20/5. Vice versa, you can reduce common fractions to integers by dividing the numerator by the denominator. The quotient is the number of integers, which are contained in the given fractions. If there is some remainder that cannot be divided, these are fractions of one integer that cannot perform as an integer. They have the same relation to the integer as the numerator has to the denominator. E.g., if you want reduce 37/7 to integers, divide the numerator 37 by the denominator 7. The quotient is 5 and 2/7 remains which cannot perform as an integer. It has the same relation to the integer as the numerator 2 to the denominator 7. Note that the numerator is always taken as an adjective and the denominator as a noun.

You can reduce any integer to a physical fraction by multiplying the number of integers with 60. The result is *minuta*. If you want to reduce these *minuta* to *secunda*, multiply them with 60, and you have *secunda*. If you multiply these *secunda* with 60, there are *tercia*, etc. For instance, if you want to reduce 2 integers to *minuta*, multiply 2 with 60, and you have 120 *minuta*. If you want to reduce these 120 *minuta* to *secunda*, multiply them by 60, and you have 7200 *secunda*. If you want to reduce these 7200 *secunda* to *tercia*, multiply them by 60, and you have 432,000, etc. *ad infinitum*. You must know that you can make physical fractions from any integer in astronomy, where the integers are called “degrees” (*gradus*) and after the degrees come the *minuta*. But the “signs” (*signa*) precede the degrees. Therefore, if you want to reduce any sign to degrees, multiply the number of signs with 30, because any sign has the value of 30 degrees, and the product is the number of degrees which are contained in the given signs. So, two signs have the value of 60 degrees and three signs of 90 degrees. If we have signs, degrees, *minuta* and *secunda* and want to reduce them to *tercia*, or just degrees, *minuta* and *secunda*, or just *minuta* and *secunda*, we must reduce first the signs to degrees by multiplying the number of signs with 30. To the result we add the degrees, which were after the signs, and reduce the whole to *minuta* by multiplying with 60. To these *minuta* we add the *minuta* which were after the signs and degrees, and we multiply the whole with 60 and we have *secunda*. To these we add the *secunda* which we have after the signs, degrees, and *minuta*, and we have *secunda*, and we multiply the whole with 60 and we have *tercia*. And this must be done with any given physical fractions.

Vice versa, you can reduce physical fractions to integers by dividing the numerator by 60, the quotient again by 60, and so on, until you come to integers. But if something remains after any division by 60, we must not throw it away, but mark it at its place, so that we can say how many integers, *minuta*, *secunda* etc. the given fractions have. For instance, if we want to reduce 432,579 *tercia* to integers, we first reduce them to *secunda* by dividing by 60. There are 7209 *secunda*, and 39 *tercia* left. They are kept at their place. We divide the 7209 *secunda* by 60. The quotient is 120 *minuta* and 9 *secunda* left. These are written before the *tercia* which were kept prior to that. Again, the 120 *minuta* are divided by 60. The quotient is 2 integers, i.e., 2 degrees in astronomy. Therefore we can say that 432,579 *tercia* have the value of 2 integers, i.e. degrees, 0 *minuta*, 9 *secunda*, et 39 *tercia*. If you want to reduce to signs a number of degrees greater than 30, divide the number of degrees by 30, and the quotient are the signs, and if anything is left, they are degrees which are kept at their place. It must be noted that, if in astronomy more than 360 degrees come forth, which is the number of degrees of the whole circle, one must always take away 360 and keep the remainder. Even if more than 12 signs come forth, we must always take away 12 and keep the remainder.

5. JORDANUS DE NEMORE (NEMORARIUS), *DE ELEMENTIS ARITHMETICE ARTIS*

Jordanus de Nemore was one of the most important mathematicians of the Western Middle Ages. He lived in the first half of the thirteenth century, but no details of his life are known. His writings comprise theoretical and practical arithmetic, algebra, geometry, the planisphere, and the science of weights. One of his major works is *De elementis arithmetice artis*, which became the standard source for theoretical arithmetic in the High Middle Ages. It was published in 1496 and 1514; in the sixteenth century it was still being used as the arithmetic textbook at universities.

As the title *De elementis arithmetice artis* indicates, the work was modeled after Euclid's *Elements*. It is founded on definitions, postulates, and axioms and is organized in ten books with more than 400 propositions. It comprises the whole domain of theoretical arithmetic, Euclidean as well as Boethian. Jordanus's *Arithmetica* contrasts with the nonformal and sometimes philosophical *Arithmetica* of Boethius, since the former generally contains proofs or, at least, indications necessary for developing a proof. In his proofs, Jordanus used letters of the alphabet to represent numbers without giving numerical examples.

The early chapters of the book generally follow Euclid's number theory. The two propositions that we include are from the later chapters. First is proposition VI 12: To find three square numbers q^2, v^2, and r^2, so that $q^2 - v^2 = v^2 - r^2$. (Note that this problem is closely related to the central problem of Leonardo's *Book of Squares* that appears in section II-3-3.) Jordanus's solution is equivalent to: $q = a^2/2 + ab - b^2/2$, $v = (a^2 + b^2)/2$, $r = b^2/2 + ab - a^2/2$. Because modern readers may have some problems in understanding his proof, an explanation is given after the literal translation.

VI 12: To find three square numbers so that their differences continuously taken are equal.

Let *c* be the double of any number. Add to it a number *b* to get *a*. Let *d* be the sum of *a* and *b*, *h* the product of *c* and *a*, *k* the product of *c* and *b*, and *e* and *f* the products of *a* and *b* with *d*. Let the sum of *g*, *l*, and *m* equal *e* and *g* equal *f*. Then the product of *c* and *d* is equal to the sum of *h* and *k*. Let *l* equal *h* and *m* equal *k*. I say that the sum of *g* and *l* can be divided into two equal parts [Fig. II-2-5-1].

Because the product of *b* and *d* is *g*, and because *d* is *c* plus twice *b*, likewise the sum of twice the square of *b* and the product of *b* and *c* is equal to *g*, and the product of *a* and *c* is *l*. Therefore the sum of *l* and *g* is equal to twice the square of *b* added to the product

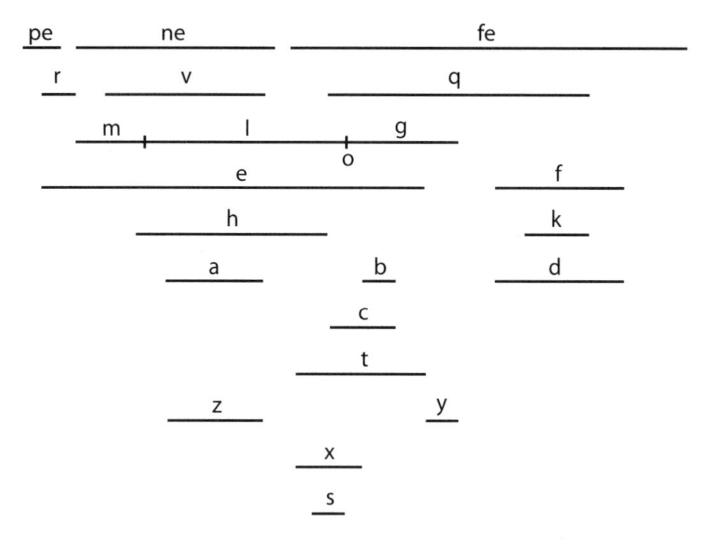

Fig. II-2-5-1.

of the sum of *a* and *b*, namely *d*, and *c*. Its half is *b* squared and the product of *d* by half *c*. Therefore the sum of *g* and *l* will be divided into two equal parts.

But *g* is not equal to *l*. If it were, the ratio of *d* to *a* would be equal to the ratio of *c* to *b*. Let *z*, *y* be the lowest terms in the ratio of *a* to *b* and let *x* be their difference and *t* their sum. Hence, the ratio of *t* to *z* equals the ratio of *x* to *y*. Let *s* be the difference of *x* and *y*. Then the ratio of *z* to *y* equals the ratio of *y* to *s*. But therefore, *z* and *y* are not the lowest terms, and thus there are none. And so *g* is not equal to *l*.

Divide the sum of *l* and *g* into equal parts to get *o*. Let *r* be the difference of this half and *g*. Let *v* be equal to the half, and let *q* be the difference of the half and the whole *g* and *l* and *m*. The squares of the proposed *r*, *v*, *q* are *pe*, *ne*, and *fe*. The differences of squares are equal to the product of the differences of the sides and the composed sides. The composed side of *r* and *v* is *g*, and their difference is *l*, because if one side is the difference, the other is the composed. Likewise the difference of *v* and *q* is *m* and the composed is the sum of *g*, *l*, *m*. The ratio of the sum *g*, *l*, *m* to *g* is equal to the ratio of *l* to *m*, because it is equal to the ratio of *a* to *b*. Therefore the products, which are differences of squares, are equal.

Explanation:[34]

Let $\boldsymbol{c}$ be an even number; $c+\boldsymbol{b}=\boldsymbol{a}$, $a+b=\boldsymbol{d}$, $c\cdot a=\boldsymbol{h}$, $c\cdot b=\boldsymbol{k}$, $a\cdot d=\boldsymbol{e}$, $b\cdot d=\boldsymbol{f}$. Let $e=\boldsymbol{g}+\boldsymbol{l}+\boldsymbol{m}$ with $g=f$. Hence $[e-f=a\cdot d-b\cdot d=(a-b)\cdot d=]c\cdot d[=c\cdot(a+b)]=h+k$. Let $h=l$ and $m=k$. I say that $g+l$ is even.

Proof: Since $g=b\cdot d$ and $d\ [=a+b]=2b+c$, hence $2b^2+b\cdot c=[b\cdot(2b+c)=b\cdot d=f=\ ]\ g$. And $a\cdot c\ [=h]=l$. Thus $l+g=[h+g=c\cdot a+2b^2+b\cdot c=]2b^2+(a+b)\cdot c=2b^2+d\cdot c$. Thus $l+g$ is even, since c is even.

[34] Terms not explicitly given in the text are added in square brackets. Letters occurring for the first time are marked in bold.

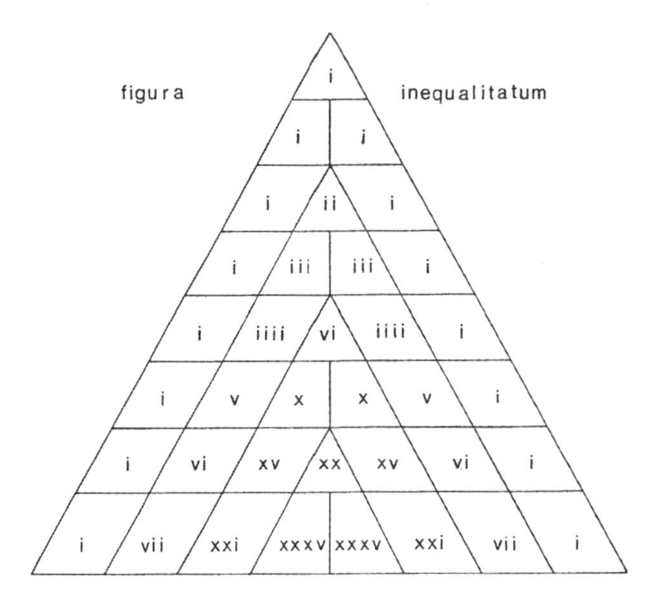

Fig. II-2-5-2.

But $g \neq l$. For assume $g = l$. Then [$b \cdot d = a \cdot c$ or] $d : a = c : b$. Let $\boldsymbol{z}, \boldsymbol{y}$ be the lowest terms in the ratio $a : b$ [with $a : b = z : y = d : c$.] Let $\boldsymbol{x} = z - y$, $\boldsymbol{t} = z + y$. Then [$(a + b) : a = (z + y) : z$ and $(a - b) : b = (z - y) : y$, or $d : a = t : z$ and $c : b = x : y$. Since $d : a = c : b$, we have] $t : z = x : y$. [Then $(t - z) : z = (x - y) : y$, or] $z : y = y : s$, with $\boldsymbol{s} = x - y$. [But from $a > b$ and $a : b = z : y$ follows: $z > y$, and from $z : y = y : s$ follows $y > s$.] Therefore, z and y are not the lowest terms in the ratio $a : b$, and thus $g \neq l$.

Let $\boldsymbol{o} = 1/2\ (l + g)$, $\boldsymbol{r} = g - o$, $\boldsymbol{v} = o$, $\boldsymbol{q} = (g + l + m) - o$, $r^2 = \boldsymbol{pe}$, $v^2 = \boldsymbol{ne}$, $q^2 = \boldsymbol{fe}$. Thus $v^2 - r^2 = (v + r) \cdot (v - r)$ [$= g \cdot (2o - g)$] $= g \cdot l$ and $q^2 - v^2$[$= (q + v) \cdot (q - v)$] $=$ $(g + l + m) \cdot m$ [$= e \cdot m$]. Thus $(g + l + m) : g$ [$= e : g = (a \cdot d) : (b \cdot d)$] $= a : b$ [$= (c \cdot a) : (c \cdot b) = h : k$] $= l : m$. [From $e : g = l : m$ follows:] $e \cdot m = g \cdot l$ and thus $q^2 - v^2 = v^2 - r^2$.

In Proposition IX 70, Jordanus displays, for the first time in a European work, the Pascal triangle. He uses it to construct series of terms in given ratios.

IX 70: Given several equal terms: Create double and triple ratios of any number of terms; then, from them produce as many near multiples of one another as you please in continued proportion; finally make from them equivalent superparticular numbers.

To do this, let us make a triangular arrangement of numbers as follows. Put 1 at the top and below two 1's. Then the row of two 1's is doubled so that the first 1 is in the first place and another 1 in the last place as in the second row; and 1.2.1 will be in the third row. The numbers are added two at a time, the first 1 to the 2 in the second place, and so on through the row until a final 1 is put at the end. Thus the fourth row has 1.3.3.1. In this way subsequent numbers are made from pairs of preceding numbers [Fig. II-2-5-2].

With this done, however, so many equal terms such as a, b, c, d are set down. They provide us with a way to find terms in double ratio. In general one has to know how many terms are in continued proportion and their ratio of proportionality. Then, considering all the terms: whatever is the ratio of the second to the first remains so for all [adjacent pairs

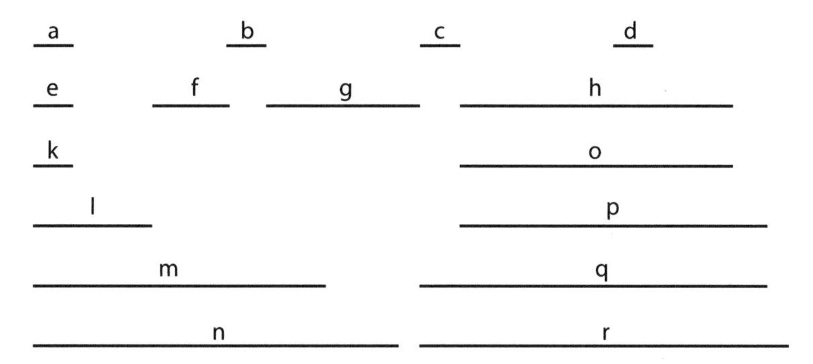

Fig. II-2-5-3.

of terms] from the second term on. For the same reason the terms ordered according to II 20,[35] with the ratio being the second to the first, have the same ratio for the second plus twice the third to the first plus twice the second. It is obvious that if a number is selected in any row of the triangle and when the number of consequent and antecedent terms is the same, the sum makes the next term of the same form. Since the first term of the antecedents is added to the second of the consequent terms, then the rest of the antecedents are added to the remaining consequent terms in the row until their last terms occupy the two end positions. This can be repeated in the arrangement of the figure.

It follows therefore by II 6[36] that, since the given terms are continually proportional and are multiplied according to the individual rows of the given figure, the individual numbers arising therefrom are continually proportional. Accordingly, take *a* once to become *e* the first term. Then add one *a* and *b* to get *f*. Now add one *a*, *b* twice and *c* once to make *g* the third term. And the sum of one *a*, thrice *b*, thrice *c*, and one *d* is *h* the fourth term. Therefore because *h*, *g*, *f* and *e* are continually proportional, and because *e* and *a* are equal, and because *f* exceeds *b* by *a* its equal, then *f* is twice *e*. For this reason all four terms are continually doubled [Fig. II-2-5-3].

In the same way multiply *e*, *f*, *g*, *h* to obtain *k*, *l*, *m*, *n* in continued proportion. And because *l* exceeds *f* by *e*, since *f* is a multiple of *e*, *l* is the next multiple with respect to *e*. The same can be said of the ratio of *l* to *k* because *k* and *e* are equal. Therefore *k*, *l*, *m*, *n* are continually tripled.

For the same reason you can obtain similar superparticular numbers from the multiples if the same multiplication begins with the larger term. For example: take *h* once to become *o*. Then add one *h* and one *g* to get *p*, and so on according to the rows of the given figure. Consequently *o*, *p*, *q*, *r* are continually proportional. Hence by IX 17[37] *p* is at once the sesquialter of *h* and *o*. For this reason all the terms are similarly sesquialters.

In this way therefore you can produce multiples from given multiples by beginning the multiplication with the least term. But if you begin with the largest, you will produce superparticular numbers. And so on [Fig. II-2-5-4].

[35]Proposition II 20 states that if $a : d = b : e = c : f$, then $(a + b + c) : (d + e + f) = a : d$.

[36]Proposition II 6 states that if $a : b = d : e$, then $(a + b) : b = (d + e) : e$.

[37]Proposition IX 17 states that if $a : b$ is a multiple ratio, then $(a + b) : a$ will be a similar superparticular ratio. That is, if $a : b = n$, then $(a + b) : a = 1 + 1/n$.

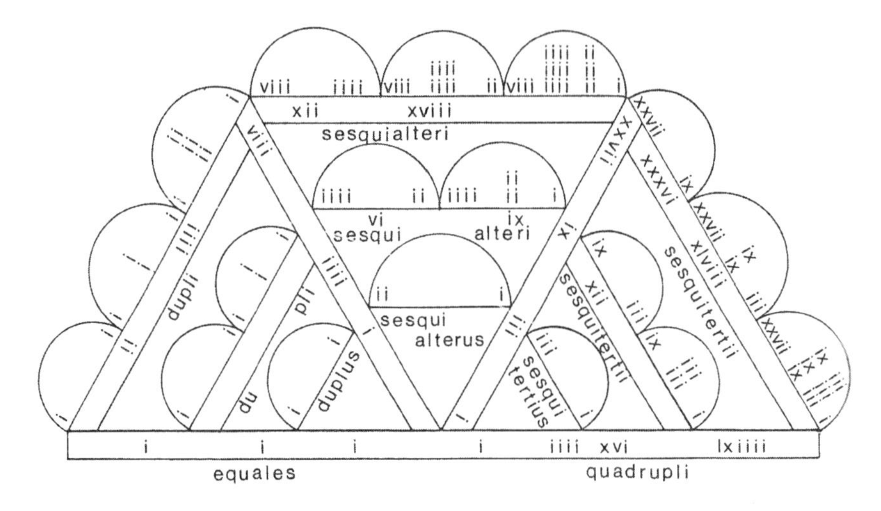

Fig. II-2-5-4. A diagram illustrating how to construct numbers in various ratios.

In the third paragraph, Jordanus constructs doubles as follows: If $a = b = c = d = 1$, then the numbers $e = 1a = 1, f = 1a + 1b = 2$, $g = 1a + 2b + 1c = 4$, and $h = 1a + 3b + 3c + 1d = 8$ form a continued proportion with ratio 2. In the next paragraph, $k = 1e = 1$, $l = 1e + 1f = 3$, $m = 1e + 2f + 1g = 9$, and $n = 1e + 3f + 3g + 1h = 27$, and we get a continued proportion with ratio 3. Finally, if we have $o = 1h = 8$, $p = 1h + 1g = 12$, $q = 1h + 2g + 1f = 18$, and $r = 1h + 3g + 3f + 1e = 27$, we get a continued proportion of ratio $1\frac{1}{2}$, the sesquialter.

6. COMBINATORICS AND PROBABILITY, *DE VETULA*

Virtually the only example of a combinatorial or probability calculation in medieval Catholic Europe is in the poem *De Vetula*, written in France in the mid-thirteenth century. The author is unknown, but some scholars have suggested that it is Richard de Fournival (1201–1260), chancellor of the Chapter of Notre Dame Cathedral in Amiens. The poem purports to be an autobiography of the Roman poet Ovid (43 BCE–17 CE). The first book describes his youth, which was devoted to love affairs and various other amusements, including dice playing and chess. The second book deals with a tragicomic love affair, which finally ended with Ovid being disillusioned with the lady, who becomes the *Vetula* (old woman) of the title. Finally, in the third book Ovid becomes convinced of the truths of Christianity. The section we reproduce, from the first book, gives the details of the results of throwing three dice.

> Perhaps, however, you will say that certain numbers are better
> Than others which players use, for the reason that,
> Since a die has six sides and six single numbers,
> On three dice there are eighteen,
> Of which only three can be on top of the dice.

These vary in different ways and from them,
Sixteen compound numbers are produced. They are not, however,
Of equal value, since the larger and the smaller of them
Come rarely and the middle ones frequently,
And the rest, the closer they are to the middle ones,
The better they are and more frequently they come.
These, when they occur, have only one configuration of pips on the dice,
Those have six, and the remaining ones have configurations midway between the two,
Such that there are two larger numbers and just as many smaller ones,
And these have one configuration. The two which follow,
The one larger, the other smaller, have two configurations of pips on the dice apiece.
Again, after them they have three apiece, then four apiece.
And five apiece, as they follow them in succession approaching
The four middle numbers which have six configurations of pips on the dice apiece.
The small table set out below will make these things easier for you:

18 666
17 665
16 664 655
15 663 654 555
14 662 653 644 554
13 661 652 643 553 445
12 651 642 633 552 543 444
11 641 632 551 542 533 443
10 631 622 541 532 442 433
9 621 531 522 441 432 333
8 611 521 431 422 332
7 511 421 331 223
6 411 321 222
5 311 221
4 211
3 111

These are the fifty-six ways for the numbers to fall,
And the number of them can neither be smaller nor larger.
For when the three numbers which make up the throw are alike,
Since six numbers can be matched up with one another,
There are also six configurations of pips on the dice, one for any number.
But, when one of them is not like the others,
And two are the same, the configurations of pips on the dice can vary in thirty ways,

Because, if you duplicate any of the six numbers,
After you have added any of the numbers which remain, then
You will come up with thirty, as if you multiply six five-fold.
But, if all three numbers are different,
Then you will count twenty configurations of pips on the dice
For this reason: Three numbers can be successive
In four ways and non-successive in just as many, but
If two are successive and a third non-successive,
You will discover from the one side twice three ways and from the other thrice two ways.
The figure set out below for your perusal makes this clear:
666 555 444 333 222 111 665
664 663 662 661 556 554 553
552 551 446 445 443 442 441
336 335 334 332 331 226 225
224 223 221 116 115 114 113
112 654 543 432 321 642 641
631 531 653 652 651 621 521
421 542 541 643 431 632 532
Again, if one looks more closely into the configurations of pips on the dice,
There are some which have only one way of falling,
And there are others which have three or six, since the ways of falling
Cannot be different when the three numbers in question
Are the same. But, if one of them should be unlike,
And two the same, three ways of falling emerge
After a different number turns up on top of any of the dice.
But if they are all unlike, you will discover
That they can vary in six ways, since,
When you give any position to one of the three, the remaining two change places,
Just as an alternation of the configuration of pips shows. And so
They vary in fifty-six ways in the configurations of pips on the dice,
And the configurations in two hundred and sixteen ways of falling.[38]
When these have been divided among the compound numbers which players use,
Just as they must be distributed among them,
You will learn full well how great a gain or a loss
Any one of them is able to be.

The table written out below can make this clear to you.
How many configurations of pips on the dice and ways of falling any compound number would have.

[38] $216 = 6\times1+30\times3+20\times6$.

3 18	Configurations of pips 1	Ways of falling 1
4 17	Configurations of pips 1	Ways of falling 3
5 16	Configurations of pips 2	Ways of falling 6
6 15	Configurations of pips 3	Ways of falling 10
7 14	Configurations of pips 4	Ways of falling 15
8 13	Configurations of pips 5	Ways of falling 21
9 12	Configurations of pips 6	Ways of falling 25
10 11	Configurations of pips 6	Ways of falling 27[39]

II-3. Algebra

1. AL-KHWĀRIZMĪ, *ALGEBRA*

Non-Arabic readers could whet their appetites for introductory Arabic algebra only with the translations of al-Khwārizmī's *Algebra* (ca. 830) by Robert of Chester, Gerard of Cremona, and William de Lunis (thirteenth century) and the compilation of Fibonacci, mid-twelfth to mid-thirteenth centuries (see section II-3-2). What is particularly interesting is how each translator/compiler understood al-Khwārizmī's introduction to his work.

Robert of Chester (ca.1145): In the name of God, tender and compassionate, begins the book of Restoration and Opposition of number put forth by Mohammed, the son of Moses Al-Khowarizmi [Karpinski, 1915, p. 81].

Gerard of Cremona (ca. 1170): The Book of Maumet Son of Moses Alchoarismi on Algebra and Almuchabala Begins [Hughes, 1986, p. 233].

Fibonacci (1228): Here Begins Part Three on the Solution of Certain Problems According to the Method of Algebra and Almuchabala, namely Proportion and Restoration [Sigler, 2002, p. 554].

William of Lunis (ca. 1250): Here begins the book called by the Arabs Algebra and Almucabala and named by us Book of Restoration and Opposition. And it was translated by Master Gerard of Cremona in Toledo from the Arabic into Latin [Kaunzner, 1986, p. 50].

Note that William's title combines the titles of the first two translations. Robert used "Restoration and Opposition," which in his eyes was a translation of what Gerard left untranslated from the Arabic text, "Algebra and Almuchabala." In short, if there were a title to William's efforts, it was lost, and some knowledgeable scribe or Master supplied a substitute title that covers the exigencies. Regardless, the translation is by William [Hughes, 1989, pp. 23–25]. Not yet settled is the issue of how Leonardo compiled the elementary algebra section of his text; the appearance of the word Restoration in his text and Robert's suggests the latter as a source. The version of Robert of Chester below comes from the reworking by Johann Scheubel (1494–1570), a German mathematician.

The tools for learning elementary Arabic algebra were available. Preliminary concepts, methods, and proofs are shown in this selection from the translation of Robert of Chester made by Louis Karpinski. He preferred to use modern symbols, such as x^2, where the Latin would have a noun, such as *substancia*. Here we present just a few excerpts from Robert's

[39]The configurations that give 10 are 442, 334, 226, 631, 541, and 532. The first three of these can each occur in three different ways, while the last three each in six different ways, giving a total of 27 "ways of falling."

translation. More of al-Khwārizmī's work, translated directly from the Arabic, can be found in [Berggren, 2007].

The Book of Algebra and Almucabola, Concerning Arithmetical and Geometrical Problems.

In the name of God, tender and compassionate, begins the book of Restoration and Opposition of number put forth by Mohammed, the son of Moses Al-Khowarizmi. Mohammed said, Praise God the creator who has bestowed upon man the power to discover the significance of numbers. Indeed, reflecting that all things which men need require computation, I discovered that all things involve number and I discovered that number is nothing other than that which is composed of units. Unity therefore is implied in every number. Moreover I discovered all numbers to be so arranged that they proceed from unity up to ten. The number ten is treated in the same manner as the unit, and for this reason doubled and tripled just as in the case of unity. Out of its duplication arises 20, and from its triplication 30. And so multiplying the number ten you arrive at one-hundred. Again the number one-hundred is doubled and tripled like the number ten. So by doubling and tripling etc. the number one-hundred grows to one-thousand. In this way multiplying the number one-thousand according to the various denominations of numbers you come even to the investigation of number to infinity.

Furthermore I discovered that the numbers of restoration and opposition are composed of these three kinds: namely, roots, squares, and numbers. However number alone is connected neither with roots nor with squares by any ratio. Of these then the root is anything composed of units which can be multiplied by itself, or any number greater than unity multiplied by itself: or that which is found to be diminished below unity when multiplied by itself. The square is that which results from the multiplication of a root by itself. Of these three forms, then, two may be equal to each other, as for example: Squares equal to roots, Squares equal to numbers, and Roots equal to numbers.

(CHAPTER I Concerning squares equal to roots)

The following is an example of squares equal to roots: a square is equal to 5 roots. The root of the square then is 5, and 25 forms its square which, of course, equals five of its roots. Another example: the third part of a square equals four roots. Then the root of the square is 12 and 144 designates its square. And similarly, five squares equal 10 roots. Therefore one square equals two roots and the root of the square is 2. Four represents the square.

In the same manner then that which involves more than one square, or is less than one, is reduced to one square. Likewise you perform the same operation upon the roots that accompany the squares.

Chapter II considers squares equal to numbers, while chapter III deals with roots equal to numbers. Al-Khwārizmī then continues:

Therefore, roots and squares and pure numbers are, as we have shown, distinguished from one another. Whence also from these three kinds which we have just explained, three distinct types of equations are formed involving three elements, namely a square

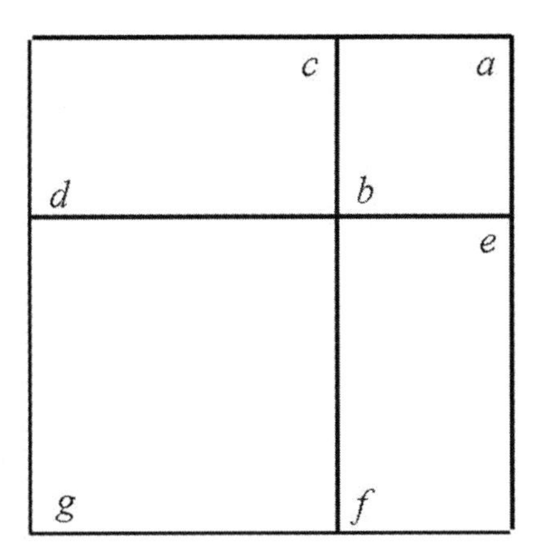

Fig. II-3-1-1.

and roots equal to numbers, a square and numbers equal to roots, and roots and numbers equal to a square.

(CHAPTER IV Concerning squares and roots equal to numbers)
The following is an example of squares and roots equal to numbers: a square and 10 roots are equal to 39 units. The question therefore in this type of equation is about as follows: what is the square which, combined with ten of its roots, will give a sum total of 39? The manner of solving this type of equation is to take one-half of the roots just mentioned. Now the roots in the problem before us are 10. Therefore take 5, which multiplied by itself gives 25, an amount which you add to 39, giving 64. Having taken then the square root of this which is 8, subtract from it 5, the half of the roots, leaving 3. The number three therefore represents one root of this square, which itself of course is 9. Nine therefore gives that square. Similarly, however many squares are proposed, all are to be reduced to one square. Similarly also, you may reduce whatever numbers or roots accompany them in the same way in which you have reduced the squares.

Al-Khwārizmī gives more examples; he then considers squares and numbers equal to roots in chapter V and roots and numbers equal to squares in chapter VI. He follows this description of algorithms by geometrical demonstrations. We present his second demonstration of the rule of chapter IV.

(Geometrical Demonstration)
Another method also of demonstrating the same is given in this manner: to the square *ab* representing the square of the unknown we add ten roots and then take half of these roots, giving 5. From this we construct two areas added to two sides of the square figure *ab*. These again are called (*cd*) and (*ef*). The breadth of each is equal to the breadth of one side of the square *ab* and each length is equal to 5 [Fig. II-3-1-1].

We have now to complete the square by the product of 5 and 5, which, representing the half of the roots, we add to the two sides of the first square figure, which represents the second power of the unknown. Whence it now appears that the two areas which we joined to the two sides representing ten roots together with the first square representing x^2 equals 39. Furthermore it is evident that the area of the larger or whole square is formed by the addition of the product of 5 by 5. This square is completed and for its completion 25 is added to 39. The sum total is 64. Now we take the square root of this, representing one side of the larger square and then we subtract from it the equal of that which we added, namely 5. Three remains, which proves to be one side of the square *ab*, that is, one root of the proposed x^2. Therefore three is the root of this x^2, and x^2 is 9.

2. LEONARDO OF PISA, *LIBER ABBACI* (*BOOK ON CALCULATION*)

In chapter twelve of the *Liber abbaci*, from which we have taken some problems in section II-2-2, Leonardo occasionally solves problems using what he calls the "direct method," a method we would characterize as using algebra. For example, he uses this to solve the following problem after first solving it by a more "arithmetic" method.

[From Chapter 12]

It is proposed that one man takes 7 *denari* from the other, and he will have five times the second man. And the second man takes 5 *denari* from the first, and he will have seven times the *denari* of the first.[40]

The goal is to find the number of denari with which each man begins.

In solving problems there is a certain method called direct that is used by the Arabs, and the method is a laudable and valuable method, for by it many problems are solved: if you wish to use the method in this problem, then you put it that the second man has the thing and the 7 *denari* which the first man takes, and you understand that the thing is unknown, and you wish to find it, and because the first man, having the 7 *denari*, has five times as much as the second man, it follows necessarily that the first man has five things minus 7 *denari* because he will have 7 of the *denari* of the second; thus he will have five whole things, and to the second will remain one thing, and thus the first will have five times it; therefore if from the first man's portion is added 5 to the second's that he takes, then the second will certainly have 12 *denari* and the thing, and to the first will remain five things minus 12 *denari*, and thus the second has sevenfold the first; that is because one thing and 12 *denari* are sevenfold five things minus 12 *denari*; therefore five things minus 12 *denari* are multiplied by the 7, yielding 35 things less 7 *soldi*[41] that is equal to one thing and one *soldo*; therefore if to both parts are added 7 *soldi*, then there will be thirty-five things equal to one thing and 8 *soldi* because if equals are added to equals, then the results will be equal. Again if equals are subtracted from equals, then those which remain will be equal; if from the abovewritten two parts are subtracted one thing, then there will remain 34 things equal to 8 *soldi*; therefore if you will divide the 8 *soldi* by the 34, then you will have $2\frac{14}{17}$ *denari* for each thing; therefore the second has $9\frac{14}{17}$ *denari*, as he has one thing and 7 *denari*. Similarly if from five things, namely from the product of the $2\frac{14}{17}$ by 5, are subtracted 7 *denari*, then there will remain $7\frac{2}{17}$ *denari* for the *denari* of the first man.

In the next example, Leonardo uses this method to solve the same problem on purses as exhibited in section II-2-2.

[40]There is a similar problem from Mizrahi in section I-5 of Chapter 2.

[41]One *soldo* equals 12 *denari*. The *soldo* and *denaro* were medieval Italian coins.

You put the first to have the thing; therefore with the purse he has the thing and the purse, which are triple the *denari* of the second; therefore the second has one third of the thing and one third of the purse. Therefore if he has the purse, he will have the purse and a third of a purse, and a third of the thing, which equal IIII things, namely quadruple the *denari* of the first, as the second with the purse has four times as many as the first. You therefore subtract from both parts one third thing; there will remain the purse, and a third of the purse, that are equal to IIII things minus one third thing. Therefore triple one and one third of a purse, namely 4 purses, are equal to triple IIII things minus triple one third of a thing, namely 11 things, and because four times 11 is equal to eleven times IIII, the proportion of *denari* of the purse to *denari* of the first man will be as 11 to 4. Whence if there are 11 *denari* in the purse, the first man has 4 *denari*, of which a number of *denari*, namely, 5, the second necessarily has, as the first with the purse has triple it.

In chapter 13, Leonardo introduces the method of double false position, a method that goes back at least to China at the beginning of our era. See [Dauben, 2007, pp. 269–274] for examples, as well as section II-1 in Chapter 3 and section V-3 in Chapter 2, although all of these relate to Fibonacci's second version of the procedure. In modern terms, double false position is a method for solving equations of the form $mx + r = c$ by making two guesses and then using an algorithm to produce the correct answer. After the examples, Fibonacci gives a proof of the method.

[From Chapter 13]
Indeed the Arabic *elchataym*[42] by which the solutions to nearly all problems are found is translated as the method of double false position. . . . Indeed the two false positions are put arbitrarily, when sometimes they both occur smaller than the true argument, sometimes greater, or sometimes one is greater and the other is smaller, and the true argument is found according to the proportion of the difference of one position to the other which is what occurs in the method of four proportionals in which three numbers are known and the fourth unknown, namely the true argument, is found. The first number is the difference between one false position and the other. The second is the difference between the approximations to the true value. The third is the difference between the second approximation and the true value. And we will first demonstrate how it is done with the method on hundredweights, so that the three differences are demonstrated subtly with hundredweights, and you will know how to understand the subtle solutions to other problems by *elchataym*.

Indeed the value of one hundredweight, namely 100 rolls, is 13 pounds,[43] and it is sought how much 1 roll is worth. We put arbitrarily that one roll is worth 1 *soldo*; therefore 100 rolls, namely one hundredweight, will be worth by the same rule 100 *soldi*, namely 5 pounds. But because the price of one hundredweight is 13 pounds, the first false position yields 5, and it differs from the true value by 8 pounds, namely the difference between the 5 pounds and the 13 pounds. Whence for the price of one roll we put 2, namely 1 *soldo* more than the first position; by the hundredweight rule it will be worth 200 *soldi*, namely

[42]The Arabic word *al-khaṭ*ʾ*-ayn* means "two errors."

[43]In this problem, Fibonacci assumes the reader knows that 1 pound is equal to 20 *soldi*, and 1 *soldo* is equal to 12 *denari*.

10 pounds; and this value similarly is false, and falls short of the true value by 3 pounds. Therefore with the difference between the first position and the second, namely 1 *soldo*, we approximated the true value more closely by 5 pounds, namely the difference from 8 to 3 pounds, and the approximation falls short by 3 pounds. Therefore you say: for the 12 *denari* that I added to the price of one roll, I approached the price of hundredweight more closely by 5 pounds. Therefore, what shall I add to the price of the same roll in order to decrease the difference of 3 pounds that resulted from the second position to the true price of the same hundredweight? You therefore multiply the extreme numbers, and divide by the middle one, as we demonstrated in the tree method and similar methods, namely the 12 by the 3, and you divide by the 5 which is the middle number. The quotient is 7 1/5 *denari* which is added to the 2 *soldi* which were put in the second position. You will have 2 *soldi* and 7 1/5 *denari* for the price of one roll.

. . .

There is indeed another *elchataym* method which is called the augmented and diminished method in which the errors are put below their positions; the first error is multiplied by the second position, and the second error is multiplied by the first position. And if both the errors are minus, or both are plus, the lesser product is subtracted from the aforesaid greater product, and the difference is divided by the difference of the errors, and thus the solution of the problem is found; and if one of the errors is plus and the other is minus, then their products are added together and the sum is divided by the sum of their errors. For example, we put above the proportion of one roll to be 1 *soldo* with which we erred by minus 8 pounds; therefore you put the 8 below the 1, and you will note minus above the 8, as it is minus. Next because we put 2 *soldi* in the second position for the price of the same roll, and we erred then by minus 3 pounds, you put the 2 *soldi* before the first position, and below this you put the error, namely the 3 pounds, above which you will note again minus as it is again deficient. And you will multiply the two *soldi* by the number of the first error; there will be 16 *soldi*. And you multiply the 1 *soldo* by the second error; there will be 3 *soldi*. And because both errors were minus, you subtract the lesser product from the greater, namely the 3 from the 16 leaving 13 *soldi*, which is divided by the error difference, namely by the 5, yielding 2 3/5 *soldi* [= 2 *soldi*, 7 1/5 *denari*], as we found above.

. . .

It is now demonstrated how this results. You take *.ab.* as the unknown number, namely the true solution to some problem that can be solved by *elchataym*. From this is taken the known number *.ag.* of the first position, for which the error is the number minus *.ez.*, and for the second position the number *.ad.*, similarly known, is taken again from the number *.ab.*; the error is the *.iz.*, similarly minus, and thus each of the numbers *.ez.* and *.iz.* are known [Fig. II-3-2-1]. Therefore, the difference between both errors, namely the number *.ei.*, is known; similarly the number *.gd.*, which is the difference between both positions is known, as the position numbers, namely *.ag.* and *.ad.* are known. But the number *.bd.* remains unknown, as the total *.ab.* is unknown; it must therefore be that if the problem is to be solved by *elchataym*, than as the known *.ei.* is to the known *.iz.*, so is the known *.gd.* to the unknown *.bd.* Therefore according to the first method, we multiplied the *.iz.* by the

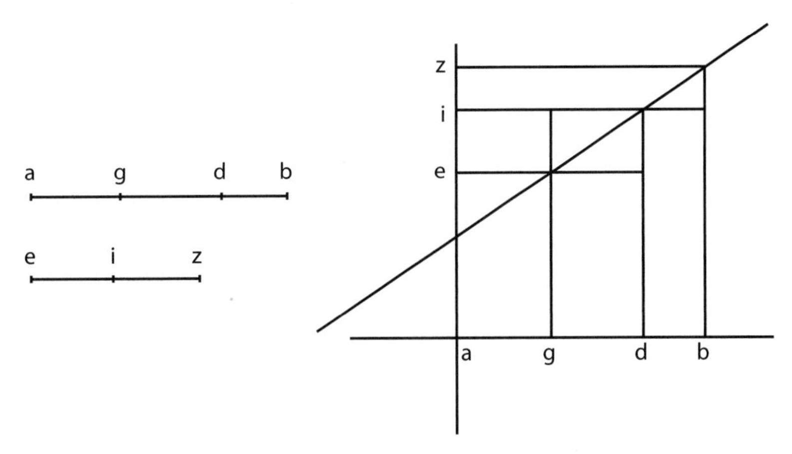

Fig. II-3-2-1. On the left is Fibonacci's diagram; on the right is a modern diagram justifying the procedure.

.gd., and we divided by the *.ei.*, namely, we multiplied the second error by the difference of the positions [and divided by the difference of the errors], and we had the known number *.db.* that we added to the second position, namely to the *.ad.*, and thus we had the known number *.ab.*, namely the solution to the posed problem.

But according to the other method, we multiply the first error by the second position, namely the *.ez.* by the *.ad.*, and we subtract the product of the second error by the first position, namely the number *.iz.* by the number *.ag.*, and we divide the residue by the number *.ei.*, and we have the entire number *.ab.* And this results because as the number *.ez.* is multiplied by the number *.ad.*, the numbers *.ei.* and *.iz.* are multiplied by the number *.ad.*; but as the number *.iz.* is multiplied by the number *.ad.*, the number *.iz.* is multiplied by the numbers *.ag.* and *.gd.*; therefore the number *.ez.* is multiplied by the number *.ad.*; the number *.ei.* is multiplied by the number *.ad.*, and the number *.iz.* by the numbers *.ag.* and *.gd.* But the product *.iz.* times *.gd.* is equal to the product *.ei.* times *.db.* Therefore, as *.ei.* is to *.iz.*, so is *.gd.* to *.db.* Therefore as *.ez.* is multiplied by *.ad.*, then *.ei.* is multiplied by the numbers *.ad.* and *.db.*, that is by the entire number *.ab.*, and the number *.iz.* by the number *.ag.* Whence if from the product of the numbers *.ez.* and *.ad.*, namely the first error by the second position, is subtracted the product of the numbers *.iz.* and *.ag.*, namely the second error by the first position, there remains the product of the numbers *.ei.* and *.ab.*, and if the product is divided by the same *.ei.*, namely by the difference in the errors, undoubtedly it is necessary that the number *.ab.* result, which had to be shown.

In chapter 15, Leonardo presents the Islamic methods for solving quadratic equations, along with geometric justifications. We just give here Leonardo's algorithm for solving squares (for which he uses the Latin word *census*) plus numbers equal roots.

[From Chapter 15]

And when it will occur that the *census* plus a number will equal a number of roots, then you know that you can operate whenever the number is equal to or less than the square

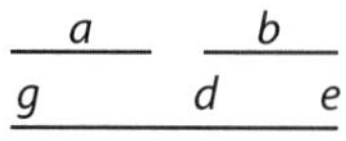

a b
g d e

Fig. II-3-2-2.

of half the number of roots. If it is equal, then half of the number of roots is had for the root of the *census*. And if the number which with the *census* is equal to the number of roots is less than the square of half of the number of roots, then you subtract the number from the square, and the root of that which will remain you subtract from half the number of roots. And if that which will remain will not be the root of the sought *census*, then you add that which you subtracted to the number from which you subtracted, and you will have the root of the sought *census*. For example, let the *census* plus 40 be equal to 14 roots; indeed half of the number of roots is 7. From the square of this, namely 49, you subtract the 40 leaving 9; the root of it, which is 3, you subtract from half the number of roots, namely 7; there will remain 4 for the root of the sought *census*. The *census* is 16, which added to the 40 makes 56, that is, 14 roots of the same *census*, as the root of 16 multiplied by the 14 yields 56. Or you add the root of 9 to the 7. There will be 10 for the root of the sought *census*, and thus the *census* will be 100, which added to the 40 makes 140, that is, 14 roots of 100, as the multiplication of the root of 100 by the 14 yields 140. And if this problem is not solved with subtraction, it thus is without doubt solved with addition.

Following the discussion of the rules for solution, Leonardo gave his readers about 100 problems with which to practice. Many of the problems were taken from the works of such Islamic authors as al-Khwārizmī, Abū Kāmil, and al-Karajī. The following problem is taken from Abū Kāmil's *Algebra* [Levey, 1966, p. 150], but it illustrates well Leonardo's expertise in algebraic manipulation. See also [Rashed, 2012].

I separated 10 into two parts, and I divided the greater by the lesser, and the lesser by the greater, and I added quotients, and the sum was the root of 5 *denari*. Let one of the parts be .*a*. and the other be .*b*.; and .*b*. is divided by .*a*. yielding .*gd*.; and .*a*. by .*b*. yielding .*de*. [Fig. II-3-2-2]. I say first that the multiplication of .*a*. by .*b*. times .*ge*. is equal to the sum of the squares of the two numbers .*a*. and .*b*.; for example, when .*b*. is divided by .*a*., then there results .*gd*.; if .*gd*. is multiplied by .*a*., then there results .*b*.; therefore if the product of .*gd*. and .*a*. is multiplied by .*b*., then it will be the same as .*b*. times itself. Again because when .*a*. is divided by .*b*. there results .*de*., if .*de*. is multiplied by .*b*., then .*a*. results; therefore if the product of .*de*. and .*b*. is multiplied by .*a*., then .*a*. times itself results. Because of this, if .*a*. is multiplied by .*b*. and the product is multiplied by .*ge*., then the result is the sum of the squares of the numbers .*a*. and .*b*.

And because this is so, you put the thing for .*a*., and 10 minus the thing for .*b*., and you multiply the .*a*. by itself yielding the *census*, and you multiply 10 minus the thing by itself yielding 100 plus the *census* minus 20 things; you add it to the *census*; there will be 2 *census* plus 100 *denari* minus 20 things. Next you multiply .*a*. by .*b*., namely the thing by 10 minus the thing; the product will be 10 things minus the *census*; and the total you multiply by .*ge*., the root of 5 *denari*. This yields the root of 500 *census* minus the root of 5 *census census* that is equal to 2 *census* plus 100 *denari* minus 20 things. You therefore restore the 20 things and the root of 5 *census census* to both parts. There will be the root of 5 *census census* plus 2 *census* plus 100 *denari* equal to 20 things plus the

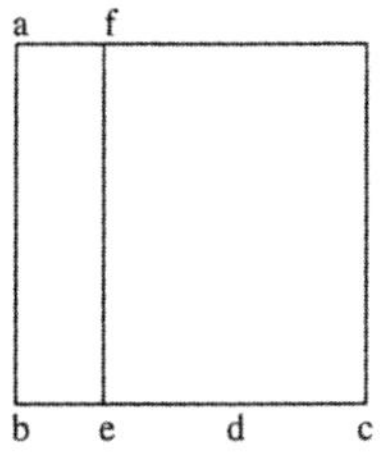

Fig. II-3-2-3.

root of 500 *census*. You reduce this to one *census*; you multiply them all by the root of 5 minus 2 *denari*, and the multiplication of the root of 5 *census census* plus two *census* by the root of 5 minus 2 yields one *census*, because when the root of 5 plus 2 *denari* is multiplied by the root of 5 minus 2, then 1 results. And the multiplication of the 100 by the root of 5 minus 2 yields the root of 50000 minus 200 *denari*; the multiplication of 20 things plus the root of 500 census by the root of 5 minus 2 *denari* yields as many as 10 things. Because the multiplication of the root of 5 by the root of 500 *census* yields the root of 2500 *census*, namely, 50 things, and the multiplication of the minus 2 by the 20 things yields minus 40 things, the difference between them and the 50 things just found is 10 things. Indeed the multiplication of the 20 things by the root of 5, that is plus, we cancel, as it is equal in quantity to the multiplication of the root of 500 *census* by the minus 2; and thus the multiplication of the 20 things plus the root of 500 *census* by the root of 5 minus 2 yields 10 things that are equal to one *census* plus the root of 50000 minus 200, and thus a number of roots are equal to the *census* plus a number, and we put this into a figure in order to see clearly that which we wish to say.

Let the side *.ab.* of a rectangle be the thing, and let *.bc.* be the 10, and thus the area *.ac.* holds 10 things, and because the 10 things are equal to one *census* plus the root of 50000 minus 200, we subtract from the area *.ac.* the square *.ae.*, which is the *census*; there will remain of the area *.ac.* the *.fc.* that is the root of 50000 minus 200, that is *.fe.* times *.ce.*, that is *.be.* times *.ec.* [Fig. II-3-2-3]. Thus the line segment *.bc.* is separated into two equal parts by the point *.d.* and into two unequal parts by the point *.e.*. Therefore, if from the square of the number *.bd.*, that is 25, we subtract the multiplication of *.be.* by *.ec.*, that is the root of 50000 minus 200, then there will remain 225 minus the root of 50000 for the square of the number *.de.*. Therefore if the root of it, that is the number *.de.*, is subtracted from *.bd.*, that is from 5, then there will remain 5 minus the root of the difference between 225 and the root of 50000 for the number *.be.*; and this is one thing, namely one of the two parts of the 10. The other truly is the number *.ec.* which is 5 plus the root of the difference between 225 and the root of 50000.

And if you wish to find the root of the difference of the 225 and the root of 50000, then you multiply the 225 by itself; there will be 50625 from which you subtract the 50000. There remains 625; the root of it, that is 25, you halve, yielding 12 1/2, which you subtract from half of the 225, that is 112 1/2. There will remain 100, and you add the 12 1/2 to the 112 1/2; there will be 125. You take the roots of the two numbers, and you subtract the lesser from the greater; there will remain the root of 125 minus the 10, which is the root of the difference between the 225 and the root of 50000, and this is the number *.ed.*.[44] If we add *.dc.*, namely the 5, then for the total *.ec.* will be had the root of 125 minus 5, that is the greater part, and if we shall subtract *.ed.* from *.bd.*, namely the root of 125 minus the 10 from the 5, then there will remain for the lesser part, namely for the number *.bc.*, 15 minus the root of 125.

3. LEONARDO OF PISA, *BOOK OF SQUARES*

If nowhere else, Fibonacci's skill as a theoretical mathematician appears in his *Book of Squares* (1225). Prompted by an invitation from Master John of Palermo to "find a square number from which when five is added or subtracted, always arises a square number," he created twenty-four propositions. Eight of these are theorems and fourteen are problems, for which he found the need to prove several less obvious results.

What is initially interesting about the treatise is that Fibonacci informs the reader of how he began the trek toward the solution of the challenge, now in Proposition 17.

I thought about the origin of all square numbers and discovered that they arise out of the increasing sequence of odd numbers; for the unity is a square and from it is made the first square namely 1; to this unit is added 3, making the second square, namely 4, with root 2, if to the sum is added the third odd number namely 5, the third square is created, namely 9, with root 3; and thus sums of consecutive odd numbers and a sequence of squares always arise together in order.

The musing led him to his first proposition, to find two square numbers that sum to a square number. So his solution begins.

I shall take any odd square and I shall have it for one of the two said squares; the other I shall find in a sum of all numbers from unity up to the odd square itself. For example, I shall take 9 for one of the mentioned two squares, the other will be had in the sum of odd numbers which are smaller than 9.

And so from $(1+3+5+7)+9=(1+3+5+7+9)$ to $4^2+3^2=5^2$, he would argue that $(ab+bc+cd+de)+ef=(ab+bc+cd+de+ef)$, and thus $ae+ef=af$, where ae, ef, and af are all square numbers.

Thereafter through Proposition 11, Fibonacci delights in posing problems about squares under different conditions. Then comes Proposition 12.

If two numbers are relatively prime and have an even sum, and if the triple product of the two numbers and their sum is multiplied by the number by which the greater number exceeds the smaller number, there results a number which will be a multiple of twenty-four.

Rather than the proof, the final sentence concerns us here:

This obtained number, namely the multiple of twenty-four, is called *congruous*.

In symbols, the statement of the theorem says that if m and n are relatively prime and both are odd numbers, then $mn(n+m)(n-m)$ is a multiple of 24. In the proof, Fibonacci also notes that if one of the two numbers is even, then $4mn(n+m)(n-m)$ is also a multiple of 24. In either case, these products are called congruous numbers. They will be crucial for the solution of Proposition 17, the occasion for the writing of *The Book of Squares*.

[44] Fibonacci is using the identity $\sqrt{a-\sqrt{b}}=\sqrt{\frac{a}{2}+\frac{\sqrt{a^2-b}}{2}}-\sqrt{\frac{a}{2}-\frac{\sqrt{a^2-b}}{2}}$.

Three propositions precede this proposition and provide information for its solution. They involve squares and congruous numbers. Of the first two,

Proposition 14: Find a number which added to a square number and subtracted from [the same] square number yields always a square number

presents a more cumbersome development and proof. Three squares, x^2, y^2, and z^2 must be found, along with what turns out to be a congruous number c, such that $x^2 + c = y^2$ and $y^2 + c = z^2$. Thus adding c to y^2 gives a square, and subtracting it from y^2 also gives a square. The full solution together with Sigler's explanation covers 20 pages of the translation [Sigler, 1987, pp. 53–74], and is not summarized here. The second of the two propositions,

Proposition 15: If some congruous number and its congruent squares are multiplied by another square, the number made by the product of the congruous number and the square will be a congruous number; the remaining squares will be congruent with this congruous number

is more reader friendly. Briefly, if c is a congruous number in $x^2 + c = y^2$ and $y^2 + c = z^2$, then by multiplying through by u^2, we have $u^2x^2 + u^2c = u^2y^2$ and $u^2y^2 + u^2c = u^2z^2$. Consequently for ux, uy, and uz, u^2c is a congruous number.

The third preparatory proposition and solution to be used in the challenge problem is

Proposition 16: I wish to find a congruous number which is a square multiple of five.

Let one of two given numbers be 5; let the other be a square number so that the sum of them is a square number and so that the lesser subtracted from the greater leaves a square number. For that reason there will be then a square 4, which added to five makes 9, a square; and 4 subtracted from 5 leaves 1. I say these two given numbers yield a congruous number which will be a square multiple of five.

There results certainly a congruous number from the products of one times twice five times twice four times nine; this is multiplying the double product one times ten by the double product eight times nine, which is 10 times 72. But the product of four times nine is a square number, for both are squares. Therefore the product of 8 times 9 is twice a square, and the multiplication of this double square by twice five results in four times a square times five. But four times a square is a square number; therefore four times a square times five is five times a square, and five times this square times 1, a square, is again five times a square. Therefore the congruous number made from these will be a square multiple of five.

Proposition 17, the problem that prompted Fibonacci's musing on square numbers, follows.

Proposition 17: Here is the question mentioned in the prologue of this book. I wish to find a square number, which increased or diminished by five, yields a square number.

Take a congruous number, a square multiple of five. 720 will be one, of which the fifth part is 144, by which divide the same 720 and the congruent squares, of which the first is 961, the second is 1681, the third indeed is 2401 [$31^2 + 720 = 41^2$ and $41^2 + 720 = 49^2$]. The root of the first square is 31, the second 41, third 49. There is for the first square $6\frac{97}{144}$ with root $2\frac{7}{12}$, which results from dividing 31 by the root of 144, which is 12, and there is for the second, which is the sought square, $11\frac{97}{144}$, with root $3\frac{5}{12}$, which results

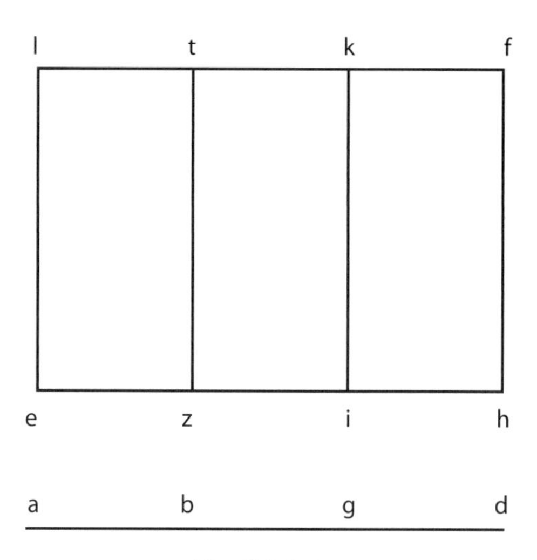

Fig. II-3-3-1.

from dividing 41 by 12, and there is for the last square $16\frac{97}{144}$ with root $4\frac{1}{12}$. [That is, $(41/12)^2 + 5 = (49/12)^2$ and $(41/12)^2 - 5 = (31/12)^2$.]

Fibonacci also enjoyed the phenomenon of one problem suggesting another. The following problem is not too different from the wording of Proposition 14 above, and it clearly leads to Proposition 20.

Proposition 19: I wish to find a square number of which the sum of it and its root is a square number and for which the difference of it and its root is similarly a square number.

Let a congruous number similarly be given with its three squares, numbers *.ab.*, *.ag.* and *.ad.*. Then the congruous number will be the numbers *.bg.* and *.gd.*. Each of the squares *.ab.*, *.ag.*, *.ad.* is divided by the congruous number *.bg.*, yielding numbers .ez., *.ei.*, *.eh.*. And construct on *.ei.* a square *.ek.* and complete the area *.lh.*; and place *.z.* so that the segment *.zt.* is parallel to the segments *.ik.* and *.el.*; and then the number *.ez.* results from the division of the number *.ab.* by *.bg.*; and the number *.ei.* results from the division of the number *.ag.* by the congruous number *.bg.*; in fact, the number *.zi.* results from the division of *.bg.* by itself. Therefore, *.zi.* is 1. Similarly, because *.ad.* is divided by the congruous number *.gd.*, that is by *.bg.*, there results the number *.eh.*, and from the division of *.ag.* by *.gd.* results *.ei.*. Therefore, *.ih.* results from the division of *.gd.* by itself. Then *.ih.* is similarly 1; therefore, *.hi.* is equal to *.iz.*. And because on the segment *.ei.* the square *.ek.* is constructed, and *.hi.* is 1, the area of *.kh.* or *.kz.* is therefore the root of the area. *ek.* [Fig. II-3-3-1]. Therefore, to the area of *.ek.* is added its root, namely the area of *.kh.*, and the result is the area of *.lh.*; and if from the square *.ek.* is subtracted the root, that is *.kz.*, there will remain the area *.zl.*. And because from the division of the numbers *.ab.* and *.ag.* and *.ad.* by some number, namely *.bg.*, there result numbers *.ez.*, *.ei.*, *.eh.*, therefore, as *.ab.* is to *.ag.*, so is *.ez.* to *.ei.*. The numbers *.ab.* and *.ag.* are certainly square. Therefore, the ratio of the number *.ez.* to the number *.ei.* is as the ratio of a square number to a square number. Therefore, the multiplication of *.ez.* by *.ei.* results in a square number. But the segment *.el.* is equal to the segment *.zt.*, which is equal to the segment *.ik.*. The area *.ek.* is that of a square; therefore, the area *.et.* is a square number.

Similarly, because the ratio of *.ei.* to *.eh.*, that is *.le.* to *.eh.*, is as a square number to a square number, the product of *.eh.* and *.le.* will be a square, that is the area *.lh.*. A square number, *.ek.* in fact, is found with added root, that is *.kh.*, which makes square number *.lh.*; and if from the square number *.ek.* is subtracted its root, there will remain the square number *.et.*; this is what had to be done.

Briefly, as Sigler says, the problem is to find numbers B, A, and C so that $B^2 - B = A^2$ and $B^2 + B = C^2$. With x, y, z, and the given congruous number c as in Proposition 14, he has $x^2 + c = y^2$ and $y^2 + c = z^2$. His geometric development is equivalent to deriving $(y^2/c)^2 - y^2/c = (xy/c)^2$ and $(y^2/c)^2 + y^2/c = (yz/c)^2$, so the solution is $A = xy/c$, $B = y^2/c$, and $C = yz/c$. With slight modifications the same equations may be used to provide a solution to the following.

Proposition 20: Similarly, a square number must be found which when twice its root is added or subtracted always makes a square number.

Given squares, *.ab.*, *.ag.*, *.ad.* are next divided by half the congruous number *.bg.* and there results the numbers *.ez.*, *.ei.*, *eh.*; and the number *.zi.* will be 2, which will be equal to the number *.ih.* Therefore, each of the areas, *.kh.* and *.kz.* will be equal to twice the root of the square number *.ek.*; and similarly, the ratio *.ez.* to *.zt.* will be as the square *.ab.* to the square *ag.*. Therefore the product of the numbers *.ez.* and *.zt.*, that is the area *.et.*, is a square; and because of this, the area, namely the number *.lh.*, is a square. Therefore, a square *.ek.* is found so that with the addition of two roots, namely *.kh.*, a square number *.lh.* results. And when from the same square *.ek.* two roots are subtracted, namely *.kz.*, there remains the square number *.zl.*. [See Fig. II-3-3-1.]

Again, the same result holds for three or more roots added or subtracted.

And to have the result in numbers, let the square *.ab.* be 1, and the square *.ag.* be 25. Also, let the square *.ad.* be 49. Therefore, the congruous number *.bg.* or *.gd.* will be 24; which is divided into 1 and 25 and 49 to yield *.ez.* 1/24; and *.ei.* will be 1 1/24. Also, the number *.eh.* will be 2 1/24. From the product of *.ei.* with itself, moreover, results the square *.ek.*, which is 625/576. If to it is added *.ki.* times *.ih.*, namely 1 1/24, there will result 1225/576, of which the root is 35/24, that is 1 11/24. Similarly, if one subtracts 1 1/24, that is 600/576, from 625/576, that is the number *.kz.* from the number *.ke.*, there will remain for the area *.et.*, 25/576, of which the root is 5/24. And if to some square it is proposed to add and subtract two roots, you will indeed double the numbers *.ez.*, *.ei.*, *.eh.*; there will result 1/12 and 2 1/12 and 4 1/12; which also will result if one divides 1, 25, 49, by half the congruous number, namely by 12; and thus the root of the sought square will be 2 1/12. And also the result holds for three or more roots added or subtracted.

The last sentence in Fibonacci's proof above reinforces the thought that both Propositions 19 and 20 might well be considered as corollaries to Proposition 14, which asks to find a number, which added to a square number and subtracted from a square number, always yields a square number. Each of the two propositions does ask for a specific number to be added or subtracted from a square number with the same result. However, Fibonacci's remark, in his letter to Emperor Frederick II,

Beyond this question (Prop. 14), the solution of which I have already found, I saw, upon reflection, that this solution itself and many others have origin in the squares and the numbers which fall between the squares.

suggests instead a series of interesting problems for the intellectual amusement of his liege lord, rather than a well-organized set of theorems and corollaries in the Euclidean format for fellow mathematicians. Thus it remains.

4. JORDANUS DE NEMORE, *DE NUMERIS DATIS* (*ON GIVEN NUMBERS*)

Jordanus de Nemore, the author of *De numeris datis* (*On Given Numbers*), was a scholastic unknown [see section II-2-5]. Recall that he had compiled *The Elements of the Arithmetic Art*, a title that echoes the greatest work in elementary geometry, *The Elements* of Euclid, so in parallel to Euclid's *Data*, he perhaps titled this work *De numeris datis*. But any analysis of the treatise uncovers its Arabic foundation.

Knowing full well the definitions, axioms, and postulates in these two works of Euclid, Jordanus very possibly saw no reason to duplicate what would be readily available to any scholar. Hence, the treatise begins with a thin set of definitions:

1. A number is given whose quantity is known.
2. A number is given in relation to another where the ratio of the one to the other is given.
3. A ratio is given whose denomination is known.

There is a significant play on the words "given" and "known." Jordanus developed a proof and an example for each proposition, beginning with something given and ending with a statement of fact. For instance, the example for Proposition I-6 gives the numbers 68 (the sum of the squares of two parts of an unknown number) and 36 (the square of the difference of the two parts), manipulates these, and concludes that the number is 10 and its parts are 8 and 2. In short, the numbers were always "known" because they were "given," not "found." This is exemplified fully in the very first proposition of the treatise that also uses Jordanus's first definition. Here and occasionally elsewhere we use the Roman numerals that Jordanus used throughout the text. For easier reading, however, we sometimes use modern numerals in the translation.

I-1. If a given number is separated into two parts whose difference is known, then each of the parts is given.

Since the smaller part and the difference equal the larger, the smaller with another equal to itself together with the difference make the given number. Subtracting therefore the difference from the total, what remains is twice the lesser. Halving this yields the smaller and, consequently, the greater part.

For example, separate x into two parts whose difference is ii. If that is subtracted from x, viii remains, whose half is iiii. This is the smaller number and the other is vi.

Here the given quantities are 10 and 2. The separation of 10 into two parts to form their difference is the condition of the problem. Hence, the parts exist. The second definition, "A number is given in relation to another where the ratio of one to the other is given," undergirds the following.

II-1. If four numbers are in proportion of which three are given, then the fourth is given.

By cross multiplication the products are equal. Taking them alternately then, multiply the two given numbers and divide by the third to produce the number not given at first.

For example, let xx be to some number as v is to iiii. Now by multiplying the given antecedent by the known consequent of the other, that is xx times iiii, one obtains lxxx. Divide this by v to get xvi, which is the not given consequent of the aforementioned xx.

The third definition, "A ratio is given whose denomination is known," is understood in the following.

II-21. If one part of a given number (or even both parts) has a known ratio to a constant so that a new sum with the other part is given, then both parts are given.

Divide *ab* into *a* and *b*. Let the ratio of *c* to *b* be given, and also *ac* be given. Let the difference of *ac* and *ab* be known as *e*, which is equivalent to the difference of *c* and *b*. Since the ratio of *c* to *b* is known, likewise is known the ratio of *e* to *b*. But since *e* is known, *b* and *a* are known.

For instance, separate 12 into two parts so that one part plus a third of the second is 6. Let the difference of this and 12 be 6, which is also the difference of the third and its whole. Since 6 is two thirds of the whole or 9, what remains is 3 or a third of 9. This with 3 makes 6.

In contrast with Euclid, who used single letters to represent lines that represented numbers, proposition I-3 employs single letters to represent numbers directly. Further, an expression such as *abc* does not signal the multiplication of three quantities, but rather their addition; *abc* means $a + b + c$.

I-3. If a given number is separated into two parts such that the product of the parts is known, then each of the parts is necessarily given.

Let the given number *abc* be separated into *ab* and *c* so that *ab* times *c* is given as *d*. Moreover, let the square of *abc* be *e*, and the quadruple of *d* be *f*. Subtract this from *e* to get *g*, which will then be the square of the difference of *ab* and *c*. Take the square root of *g* and call it *h*. *h* is also the difference of *ab* and *c*. Since *h* is given, then *c* and *ab* are given.

For example, separate 10 into two numbers whose product is 21. The quadruple of this is 84, which subtracted from the square of 10, namely from 100, yields 16. 4 is the root of this and also the difference of the two parts. Subtracting this from 10 yields 6, which halved produces 3, the lesser part; the greater is 7.

III-5. If the middle term of three numbers in continued proportion and the sum of the other two are known, then each of these is given.

Let the ratio of *a* to *b* equal the ratio of *b* to *c*, with *b* and *ac* given. Since the square of *b* equals the product of *a* and *c*, with this product given each of its parts is given.

For example, let the middle term be xii and the sum of the extremes be xxvi. This squared is dclxxvi. Square xii and subtract it four times from dclxxvi. What remains is c, whose root is x, the difference of the extremes. Therefore the extremes are viii and xviii.

The next two problems illustrate how Jordanus developed a subsequent proposition from its predecessor. Proposition 1-12 leads to Proposition 1–13.

I-12. If the sum of the squares of the two parts of a given number together with the square of their difference are given, then both parts are given.

If the sum is subtracted from the square of the given number, what remains is twice the product of the two parts less the square of their difference. This becomes the sum of the squares of the parts less twice the square of their difference, and finally it is the given sum less thrice the square of the difference. When, therefore, the remainder is subtracted from the given sum, take one third of what is left. The root of this is the difference that was sought. Hence, all are given.

For example, square the two parts of 10, and adding them to the square of their difference yields 56. Subtract this from 100 to get 44, which in turn is subtracted from 56. The remainder is 12, whose third is 4. The root of this is 2, the difference of the parts. Therefore the larger number is 6 and the smaller is 4.

I-13. If the sum of the product of two parts of a given number and the square of their difference are given, then each part is given.

By doubling the given sum you obtain the sum of the two parts squared together with the square of their difference. Since these are given, what is sought is given.

For example, let the product of the parts and the square of their difference be 28. Doubling this produces 56, which are the three squares as above. The rest follows as before.

The two propositions that follow testify to the sources that Jordanus had at hand. Both problems are clearly borrowed from some translation of Problems 2 and 3 in the set of various problems in al-Khwārizmī's *al-jabr*. For instance, in Gerard of Cremona's translation of the *al-jabr*, problem two states:

I have divided ten in two parts and I multiplied each of them into itself. And I subtracted the smaller from the larger and 40 remained. [Hughes, 1986, p. 218]

By rewording this and generalizing its answer, Jordanus produced the following proposition.

I-14. If the difference of the squares of two parts of a given number is given, the two parts are given.

By subtracting the given difference from the square of the given number what remains is twice the square of the lesser part together with twice the product of the parts. Halving this yields just the square of the lesser added to the product of the two parts; this is equal to the product of the given number and its lesser part. By dividing the product by the given number, the lesser part emerges and the greater part is given.

For example, take the difference of the squares of the two parts of 10, and let this equal 80. Subtracting this from 100 leaves 20, whose half is 10. Dividing 10 by 10 produces 1 for the smaller number, and 9 is the greater.

Similarly, Problem 3 in Gerard's translation states:

I divided ten in two parts and I multiplied each of them into itself. And I joined them. And moreover I added to them the excess that was between each of the sections before they were multiplied into themselves. And that produced the entire 54. [Hughes, 1986, pp. 250–251]

Jordanus modified problem 3 to become

I-15. If the sum of the squares of the two parts of a given number added to their difference is given, then the two parts are given.

If the sum is subtracted from the square of the given number and then the remainder is subtracted from the given sum, it is obvious that what remains is twice the sum less the square of the given number, that is to say, the product of the difference and itself increased by 2. Since this is known, the difference is given.

For example, let the sum of the squares of the parts increased by their difference equal 62. Subtract this from 100 to obtain 38, which in turn is subtracted from 62. The 24 that results equals the square of the difference of the parts increased by twice the difference. Doubling and redoubling this produces 96. To this add the square of 2 to reach 100. Its root is 10. Take 2 from this and halve the remainder to obtain 4, the difference of the parts, namely 7 and 3.

The following problem is a generalization of a problem of Abū Kāmil, but it also occurs in the work of Fibonacci. Jordanus's method, however, is quite different from those of his predecessors.

I-20. If the sum of the quotients of the two parts of a given number each divided by the other is given, then the parts are given.

Let the quotient of a by b be c, and of b by a be d. Let one be added to c and to d to produce e and f, and let the product of a and b be g. Since the product of a and f is ab [i.e., $a + b$] and the product of b and e is ab, it follows that e is to f as a is to b. Thus ef is to f as ab is to b, and alternately, ab is to ef as b is to f. Since a times the sum of b and f equals the sum of g and ab, we have that g is to ab as b is to f. Therefore g is to ab as ab is to ef. Consequently, the square of ab divided by the given ef produces g as given. Therefore a and b are given [by I-3].

For example, divide 10 into two parts and let the sum of the respective quotients be 2 1/6. Add two to this to get 4 1/6, by which 100 [10^2] is divided. The result is 24, which is the product of the two parts. Multiply this by four and subtract the product from 100. The remainder is four, of which the root is two. This is the difference of the two parts, so they are 6 and 4.

Manuscripts were not copied indiscriminately but rather from a belief in their worth. Jordanus's treatise was copied at least 15 times from the time of the autograph into the sixteenth century. Further, three digests were made in the later centuries, and Johann Scheubel (1494–1570), sometime professor of mathematics at the University of Tübingen, completely reedited the text and saw to its publication in 1545.

5. NICOLE ORESME, *ALGORISMUS PROPORTIONUM* (*ALGORITHM OF RATIOS*)

Nicole Oresme (1320–1382) seems to have become acquainted with ideas of what we would call exponents through the work of Thomas Bradwardine (ca. 1290–1349) (see sections II-6-2 and II-7-3 later in this chapter). To expand on this notion, Oresme dealt with compound proportions, such as the double proportion of a/b, which in modern terms is $(a/b)^2$, and the half proportion, which is $(a/b)^{1/2}$. In this terminology, the "addition" of two proportions means their multiplication and the "subtraction" means their division.

Oresme's primary focus was on the ratios of ratios, based ultimately on the notion of the commensurability of exponential parts. In his earliest known work, the *Algorismus proportionum* written shortly after 1350, he developed a mathematical theory of compound ratios. In writing irrational ratios, Oresme used a consistent pattern, although he wrote the rational bases with some variation. The work was dedicated to Philippe de Vitry, Bishop of Meaux from 1351 to 1361, the same musical scholar who commissioned Levi ben Gershon's *On Harmonic Numbers* (see section II-4 in Chapter 2). What follows is the beginning of the *Algorismus proportionum*.

One half is written as 1/2, one third as 1/3, and two thirds as 2/3, and so on. The number above the crossbar is called the numerator, that below the crossbar the denominator.

A double ratio [2/1] is written as 2^p, a triple ratio [3/1] as 3^p, and so forth. A sesquialterate ratio [3/2] is written as 1^p 1/2, and a sesquitertian ratio [4/3] as 1^p 1/3. A superpartient two-thirds ratio [5/3] is written as 1^p 2/3, a double superpartient three-fourths [11/4] as 2^p 3/4, and so on. Half of a double ratio [$(2/1)^{1/2}$] is written as 1/2 2^p and a fourth part of a double sesquialterate [$(5/2)^{1/4}$] as 1/4 2^p 1/2, and so on.[45] But sometimes a rational ratio is written in its least terms or numbers just as a ratio of 13 to 9, which is called a superpartient four-ninths [1 4/9]. Similarly, an irrational ratio such as half of a superpartient two-thirds [$(5/3)^{1/2}$] is written as half of a ratio of 5 to 3.

Every irrational ratio—and these shall now be considered—is denominated by a rational ratio in such a manner that it is said to be a part or parts of the rational ratio, as [for example] half of a double [$(2/1)^{1/2}$], a third part of a quadruple [$(4/1)^{1/3}$], or two-thirds of a quadruple [$(4/1)^{2/3}$].[46] It is clear that there are three things [or elements] in the denomination of such an irrational ratio: [1] a numerator, [2] a denominator, and [3] a rational ratio by which the irrational ratio is denominated, that is, a rational ratio of which that irrational ratio is said to be a part or parts, as, [for example,] in half of a double ratio [$(2/1)^{1/2}$] the unit is the numerator [of the exponent], or represents the numerator, two is the denominator [of the exponent], and the double ratio is that by which the irrational ratio is denominated.[47] And this can easily be shown for other ratios.

There follow nine rules for calculation with rational and irrational proportions. We consider rules one and five.

Rule One: To add a rational ratio to a rational ratio.

Assuming that each ratio is in its lowest terms, multiply the smaller term, or number, of one ratio by the smaller number of the other ratio; and then multiply the greater [number of one ratio] by the greater [number of the other], thereby producing the numbers or terms of the ratio composed of the two given ratios. In this way, three or any number [of ratios] can be added by adding two of them at a time and then adding a third to the whole composed of those two; then, if you wish, add a fourth ratio and so on.

For example, I wish to add sesquitertian [4/3] and quintuple [5/1] ratios. The prime numbers of a sesquitertian are 4 and 3, of the other 5 and 1. And so, as already stated, I shall multiply 3 by 1 and 4 by 5 obtaining 20 and 3, which is a sextuple superpartient two-thirds ratio [$6\frac{2}{3}$]. In this way a ratio may be doubled, tripled, and quadrupled, as many times as you please.

. . .

Rule Five: To add an irrational ratio to a rational ratio.

Let B, an irrational ratio, be added to A, a rational ratio, and assume that B is a part of rational ratio D, and in accordance with previous rules, this constitutes the most proper

[45] This means that the exponent is placed first, followed by the ratio expressing the rational base.

[46] The irrational ratio $(4/1)^{1/3}$ is a "part" of the rational ratio 4/1 because $(4/1)^{1/3} < 4/1$ and the exponent, 1/3, is a unit fraction. The irrational ratio $(4/1)^{2/3}$ is "parts" of 4/1 because $(4/1)^{2/3} < 4/1$ and the exponent, 2/3, is a proper fraction, where both integers are greater than 1.

[47] When an irrational ratio has a rational base, the latter is said to denominate the former. Thus if $(A/B)^{p/q}$ is irrational, A/B rational, and p, q are integers with $p < q$, then A/B will denominate $(A/B)^{p/q}$.

denomination of B. Then, by the first rule, add A to ratio D a number of times equal to the value of the denominator of B, and let C be the total result. I say, therefore, that the ratio composed of B and A will be part of C and will be denominated by the same denominator which denominated the part which B was of its denominating ratio, namely, D.

For example, let one-third of a double ratio [$(2/1)^{1/3}$] be added to a sesquialterate ratio [3/2]. Now join three sesquialterate ratios with a double ratio to yield a sextuple superpartient three-fourths ratio [$6\frac{3}{4}$], which is a ratio of 27 to 4. The ratio produced from [the addition of] one-third of a double ratio and a sesquialterate ratio is one-third of the ratio of 27 to 4 [$(27/4)^{1/3}$], which is written as 1/3 6^p3/4.

6. NICOLE ORESME, *DE PROPORTIONIBUS PROPORTIONUM* (*ON THE RATIO OF RATIOS*)

In another, apparently later, treatise, *De proportionibus proportionum*, Oresme dealt not only with rational but also with irrational exponents. However, he did not use the term "exponent," but the terminology of Euclid's *Elements*. Euclid had proved that commensurable magnitudes have to one another the ratio that a number has to a number (*Elements* X.5), that is: given two commensurable quantities, A and B, then $A{:}B = m{:}n$, where m and n are integers (in fractional terms, $A/B = m{:}n$). Oresme generalized this notion. His idea can be symbolized as follows: two ratios of quantities, C/D and E/F, are commensurable, if $C/D = (E/F)^{m/n}$, where m and n are integers. By applying these terms, he distinguished three types of ratios in this excerpt from Chapter 1.

Every rational ratio is immediately denominated by some number, either with a fraction, or fractions, or without a fraction. The determination of these denominations will be shown below.

An irrational ratio is said to be mediately denominated by some number when it is an aliquot part or parts of some rational ratio, or when it is commensurable to some rational ratio, which is the same thing, as [for example], the ratio of the diagonal [of a square] to its side is half of a double ratio.

I say, therefore, that it does not seem true that every irrational ratio is commensurable to some rational ratio. And the reason is that every ratio is just like a continuous quantity with respect to division, which is obvious by the last supposition.[48] Therefore, by the ninth proposition of the tenth [book of Euclid, every ratio] can be divided into two [ratios] any of which is incommensurable to the whole.[49] Thus there will be some ratio which will be part of a double ratio and yet will not be half of a double, nor a third part, or fourth part, or two-thirds part, etc., but it will be incommensurable to a double and, consequently [incommensurable] to any [ratio] commensurable to this double ratio (by the comment on the eighth proposition of the tenth book of Euclid).[50] And further, by the same reasoning there could be some ratio incommensurable to a double and also to a triple ratio and to

[48]Any ratio is as a continuous quantity in the sense that it is divisible into infinity just like a continuous quantity.

[49]This refers to the addition given by Campanus to *Elements* X.9 (= X.15 of the Greek text): "If two incommensurable magnitudes be added together, the whole will also be incommensurable with each of them; and if the whole be incommensurable with each of them or with one of them, it will also be incommensurable with the other and the original magnitudes will be incommensurable."

[50]This comment, which was added by Campanus to *Elements* X.8 (= X.12 in the Greek text), runs: "If two magnitudes be commensurable, and the one of them be commensurable with any magnitude, the remaining one will also be commensurable with the same; and if the one of them be incommensurable with any magnitude, the remaining one will also be incommensurable with the same."

any ratio commensurable to these, as [for example,] half of a sesquitertian ratio. And the same may be said of other ratios.

And there might be some irrational ratio which is incommensurable to any rational ratio. Now the reason for this seems to be that if some ratio is incommensurable to two [rational ratios], and some [ratio is incommensurable] to three [rational ratios], and so on, then there might be some [ratio] incommensurable to any [rational ratio] whatever, though this does not follow from the form of the argument as [it does follow when we say that] some continuous quantity is incommensurable to all quantities of one [geometric] series. However, I do not know how to demonstrate this; but if the opposite should be true, it is indemonstrable and unknown.

Oresme seems to have understood each one of these categories in the following way:[51]

(1) Rational ratios are always immediately denominated by some number or numbers. If, for example, we have a ratio of commensurable quantities, A/B, then $A/B = n$, where n is some integer or ratio of integers.
(2) Irrational ratios that, though incapable of immediate denomination by any number or numbers, can be mediately denominated by some number. This is possible in all cases where the irrational ratio is a part or parts of some rational ratio. Given an irrational ratio $(A/B)^{p/q}$, where p and q are integers with $p < q$ and A/B is a rational ratio, then the rational ratio A/B *immediately* denominates the entire irrational ratio $(A/B)^{p/q}$. The most likely interpretation of Oresme's text is that the exponent p/q *mediately* denominates the irrational ratio $(A/B)^{p/q}$.
(3) Irrational ratios that have no numerical denomination. Such a ratio is not an aliquot part or parts of the rational ratio that is supposed to denominate it immediately. Thus if A/B is a double ratio, namely 2/1, then an irrational ratio of this type cannot be expressed in the form $(2/1)^{p/q}$, where p and q are integers. In other words, p/q is itself irrational and, consequently, is not a number or ratio of numbers. Hence p/q, the exponent, cannot *mediately* denominate the irrational ratio. Oresme has, therefore, distinguished two types of irrational ratios, those where the exponent is rational and those where it is irrational. The former kind can be mediately denominated by a number; the latter is incapable of such denomination.

This important distinction reveals that Oresme had a clear grasp of the concept of an irrational exponent. This is, perhaps, the most significant contribution emerging from the *De proportionibus proportionum*. It seems to have been original with Oresme; at least, nothing similar is known from his predecessors and contemporaries. This may be seen to presage an extension of the concept of number. In chapter 3, Oresme states an interesting corollary of these definitions:

Proposition X: It is probable that two proposed unknown ratios are incommensurable because if many unknown ratios are proposed it is most probable that any [one] would be incommensurable to any [other].

As found in the first chapter, there are three types of ratios. Some, indeed, are rational ratios, others irrational with denominations, i.e., commensurable to rationals, and perhaps

[51] What follows is the interpretation given in [Grant, 1966, pp. 31–34].

there is a third type, namely irrational ratios which have no denomination because they are not commensurable to any rationals.

Let there be two unknown ratios. Now each might be rational, namely of the first type, and the argument would proceed as follows: with any number whatever of rational ratios forming one or more series of denominations, those which are mutually commensurable are much fewer than those which are incommensurable; and therefore it is likely that any two proposed unknown ratios are incommensurable.

The antecedent is [now] demonstrated. Let 100 ratios of the multiple genus be taken according to the sequence of their denominations, as 2/1, 3/1, 4/1, 5/1, and so on, up to 101/1, and let them be as 100 terms mutually compared. Then by comparing any one of these terms to any other of them, there would be 4950 ratios which are ratios of ratios and, of these 25—and no more—are rational and all the others are irrational, as I shall show afterwards.[52] Now if more rational ratios were taken as terms, [for example] 200 or 300, and if ratios of these were taken, the ratio of irrational to rational ratios would be much greater....

Therefore, if any unknown ratio of ratios were sought, it is probable that it would be irrational and its ratios incommensurable. And all this applies if the unknown ratios which are sought should be rational.

Now if, perchance, the proposed ratios were of the second type, namely irrational with denominations and commensurable to rationals, the same thing will be proved. ... And if it should happen that the proposed ratios belonged to the third type of ratio, if there are any such ratios with no denominations—and it is probable there are, but if not they can be imagined—the same applies to them as the others with respect to this, namely that among the ratios of these ratios, rationals are fewer than irrationals; and thus it would be probable that the proposed ratios would be incommensurable....

And so it is clear that with two proposed unknown ratios—whether they are rational or not—it is probable that they are incommensurable, which was proposed in the first place. Therefore, if many [unknown ratios] are proposed, it is [even] more probable that any one of them would be incommensurable to any other, which was proposed in the second instance.

II-4. Geometry

1. BANŪ MŪSĀ IBN SHĀKIR, *THE BOOK OF THE MEASUREMENT OF PLANE AND SPHERICAL FIGURES*

Three brothers of the ninth century, Muḥammad, Aḥmad, and al-Ḥasan, are known as the Banū Mūsā ibn Shākir. Educated at the House of Wisdom, they lived in Baghdad, where they devoted themselves to scientific inquiry and translation. They composed a text in mathematics in Arabic that in translation is known as *The Words of the Sons of Mūsā the*

[52] The number of ways of comparing two of the ratios from 2:1 up to 101:1, always comparing a greater to a smaller, is simply $\binom{100}{2} = 4950$. The rational ratios come from the following relationships: $(2{:}1)^2 = 4{:}1$; $(2{:}1)^3 = 8{:}1$; $\ldots (2{:}1)^6 = 64{:}1$; $(3{:}1)^2 = 9{:}1$; $\ldots (3{:}1)^4 = 81{:}1$; $\ldots (10{:}1)^2 = 100{:}1$; $(4{:}1)^{3/2} = 8{:}1$; $\ldots (8{:}1)^{4/3} = 16{:}1$; $(32{:}1)^{6/5} = 64{:}1$. A careful count gives 25 of these rational ratios.

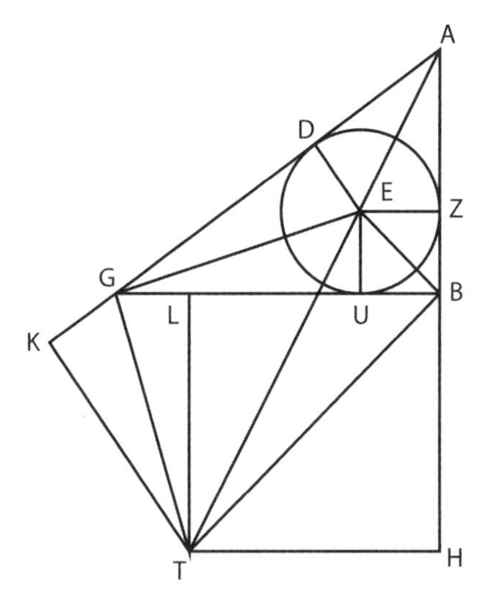

Fig. II-4-1-1.

Son of Sekir (*Verba filiorum Moysi filii Sekir*).[53] It was translated by Gerard of Cremona in the second half of the twelfth century in Toledo. An intellectual gold mine for the work of the schoolmen and particularly for Fibonacci, it proved of great assistance to Western scholars. Among other works, it contains the most accurate and popular translation of Archimedes's *On the Measurement of the Circle* together with early knowledge of some conclusions from Archimedes's *On the Sphere and the Cylinder*. Also included in the text is an original work containing a proof of Heron's formula for finding the area of a triangle, as offered here.

I wish to demonstrate that, if the excess of one half of the perimeter of every triangle over each of its sides is taken, and if one of the three excesses is multiplied by the second, and then that product is multiplied by the third excess, and further if the product of the three excesses is multiplied by one half of the perimeter of the triangle, then the final product is equal to the square of the area of the triangle.

For example let there be triangle ABG. I say therefore that assuming the excess of one half of the sum of AB, BG, and GA over each of the sides AB, BG, and GA, then the product of one half the sum of AB, BG, and GA less AB and one half the sum of AB, BG, and GA less BG and one half the sum of AB, BG, and GA less GA and one half the sum of AB, BG and GA equals the square of the area of triangle ABG. The proof follows.

I shall inscribe in triangle ABG the greatest circle that can fit in it, namely, the circle DZU. Let its center be E. I shall project from the center lines ED, EU, and EZ to the points of tangency of the sides of the triangle with the circle. I shall draw line AE. I shall show, therefore, that DA equals AZ, ZB equals BU, and UG equals GD [Fig. II-4-1-1]. When lines tangent to a circle meet in a single point, they are equal. For angle EDA equals angle EZA, both being right angles, and DE, EA equal ZE, EA. Therefore, DA equals AZ. In the same

[53] See [Berggren, 2007, pp. 520–523] for another excerpt from the work of the Banū Mūsā.

way it is known that ZB equals BU and that UG equals GD. And it is known from what we have recounted that DA equals AZ equals one half of the sum of AB, BG, and GA less BG and that ZB equals BU equals one half the sum of AB, BG, and GA less GA, and that DG equals GU equals one half the sum of AB, BG, and GA less BA. Then we shall extend line AE to T and line AB to H, and posit AH equal to one half the perimeter of triangle ABG. Therefore it is shown from what we have recounted that line HB equals line DG equals line GU. And we shall extend AG to K and posit line AK equal to line AH. Therefore it is shown that line GK equals line ZB equals line BU. And I shall draw from point H line HT at a right angle to line AH. And I shall draw from point K line KT at a right angle to line AK. Therefore, it is evident that KT equals HT. And I shall take from line BG a line equal to BH, and this will be line BL. I shall draw TL. Hence it is evident that TL is perpendicular to line BG. For when we have drawn the two lines BT and TG, it is therefore evident that BT squared minus TG squared equals BH squared minus KG squared. But KG equals LG, and BH equals BL. Hence, BT squared minus TG squared equals BL squared minus LG squared [or, BT squared minus BL squared equals TG squared minus LG squared]. Accordingly, TL is perpendicular to BG. And LT equals TH because BL equals BH, line BT is common, and the two angles BLT and BHT are right angles. And because of this, angle LBT equals angle TBH and angle LTB equals angle BTH. And line BH is a rectilinear continuation of line AB. Hence angle UBZ plus angle HBU equals two right angles. And angle LBH plus angle LTH equals two right angles. Hence, angle LTH equals angle ZBU.

But angle EBU equals one half of angle ZBU, and angle BTH equals one half of angle LTH. Hence, angle EBU equals angle BTH. And there remains of triangle BTH, angle TBH which is equal to angle BEU of triangle BEU. Therefore, triangle EBU is similar to triangle TBH. Therefore, the ratio of EU to UB is as the ratio of BH to HT. But ZB equals BU and EU equals ZE. Therefore, the ratio of EZ to ZB is as the ratio of BH to HT. Therefore the product of EZ and HT equals the product of ZB and HB. But the ratio of EZ squared to the product of EZ and HT equals the ratio of EZ to HT. And the ratio of EZ to HT is as the ratio of AZ to AH [by similar triangles]. Therefore, the ratio of AZ to AH is as the ratio of EZ squared to the product of EZ and HT. And the product of EZ and HT equals the product of ZB and HB. Therefore the ratio of EZ squared to the product of ZB and BH is as the ratio of AZ to AH. Therefore, the product of EZ squared and AH equals the product of HB, BZ, and ZA. But lines AZ, ZB, and BH are the excesses of one half the sum of AB, BG, and GA over each of the sides AB, BG, and GA. And line EZ is the radius of the circle DUZ. And line AH equals one half of the sum of AB, BG, and GA.

Therefore, it has now become evident that the multiplication of one of the excesses of one half the perimeter of the triangle over each of its sides by the second excess, followed by the multiplication of this product by the third excess, is equal to the multiplication of the square of the radius of the greatest circle falling in the triangle by one half the perimeter of the triangle. But the multiplication of the radius of the largest circle falling in the triangle by one half the perimeter of the triangle is equal to the area of the triangle. And the multiplication of this product again by the radius of the circle equals the multiplication of the square of the radius by one half the perimeter of the triangle. Then the further multiplication of this product by one half the perimeter of the triangle equals the square of the area of the triangle, since the multiplication of the radius of the largest circle falling in the triangle by one half the perimeter of the triangle is equal to the area of the triangle, as we demonstrated in those things that have been presented before. Therefore, the product of EZ squared and AH equals the product of the product of EZ and AH by EZ. Therefore, the product of EZ and AH by EZ equals the product of BH, BZ, and AZ. Hence EZ has now been multiplied by two numbers: by AH, and this is the area of the triangle [that is, the product of EZ and AH equals the area of the triangle], and by the area of the triangle itself, and this was equal to the product of BH, BZ and AZ [that is, the product of EZ and the area of the triangle equals the product of BH, BZ, and AZ]. Therefore the ratio of AH to the area of the triangle is as the ratio of the area of the triangle to the product of AZ, ZB, and BH. Therefore the multiplication of one of the excesses of one half the perimeter of the triangle over each of the sides of the triangle by the second excess, followed by the multiplication of this product by the third excess, and then followed once more by the multiplication of this product by one half the perimeter of the triangle, is equal to the triangle squared. And this is what we wished to show.

We can demonstrate the truth of what we have recounted by another method. When this has been done, then what we have stated before in regard to this proposition is evident. And this is as follows: since the ratio of EZ to ZB is as the ratio of BH to HT, hence the quantities EZ, ZB, BH, and HT are proportional terms. Therefore the ratio of the first term to the fourth is equal to the ratio of the first to the second multiplied by the ratio of the first to the third, since when the second term is placed as a mean between the first and the fourth, there results the ratio of the first to the second multiplied by the ratio of the second to fourth [which compound ratio] equals the ratio of the first to the fourth. But the ratio of the second to fourth is as the ratio of the first to the third. Therefore, the ratio of EZ to HT is as the product of the ratio of EZ to ZB and the ratio of EZ to BH. And since the ratio of EZ to HT is as the ratio of AZ to AH by similar triangles, therefore the ratio of AZ to AH is as the product of the ratio of EZ to ZB and the ratio of EZ to BH. Hence, the product of AZ, ZB, and BH equals the product of EZ squared and AH. And then we complete the demonstration as before.

2. ABŪ BAKR, *LIBER MENSURATIONUM* (*ON MEASUREMENT*)

Among the treatises that Gerard of Cremona translated in the twelfth century from Arabic into Latin is *On Measurement* by Abū Bakr. The Arabic text has not survived, and we do not have certain information about the author. We have several other Arabic treatises written by persons with the name "Abū Bakr/Bekr," but none of them can have been our author. The first half of the text deals with squares, rectangles, and rhombuses and represents a tradition that goes back to Old Babylonian mathematics. (See [Robson, 2007] for examples, particularly pp. 104–107.) The six problems given below belong to this section, with the first three dealing with squares, the next two with rectangles, and the last with a rhombus. In each problem, two solutions are given. The first, geometrical, is like a recipe, without giving an explanation. The second solution is algebraic; it uses the special terms for the unknown (*res*) (which we translate as "thing") and its square (*census*) (which we leave untranslated).

[3] If somebody has said to you: I have aggregated the side and the area, and what resulted was 110. How much is then each side?

The working in this will be that you take half of the side as the half and multiply it by itself, and one fourth results; add this then to 110 and it will be 110 1/4, whose root you then take, which is 10 1/2, from which you subtract the half, and there remains 10 which is the side. Understand!

There is also another way according to algebra (*aliabra*), which is, that you set the side as thing, and multiply it by itself, and what results will be the *census*, which is the area. This you add to the side according to what you have proposed, what results will be a *census* and a thing equal to 110. Do what you were told above in algebra, which is, that you halve the thing and multiply it by itself, and what results you add to 110, and you take the root of the sum, and subtract from that half of the root. What remains will be the side.

[6] If somebody has said to you: I subtracted its sides from its area and 60 have remained, how much thus is each side?

In this you must halve the sides which will be two. Multiply this by itself and add to it 60. Take the root of the sum, which is 8. To this add the half of the number of the sides. The result is 10, which is the side.

But its working according to algebra is that you set the side as thing. Multiply it by itself, and a *census* results which is the area. From it subtract its four sides, which are four things. Thus a *census* minus 4 things remains which equals 60. Restore and oppose, that is, restore the *census* by the four things that were subtracted and join to it 60, and thus you will have a *census* which equals 4 things and 60 *dragmae*. Do thus according to what was preceded in the sixth question,[54] i.e., halve the *radices* (roots), multiply them by themselves, join them to the *numerus* (number), and take its root, and the result is 8. To this add the half and 10 results, which is the side.

[9] If somebody has said to you: I subtracted its area from its sides and 3 have remained, how much thus is each side?

In this you must halve the number of the sides which will be two. Multiply this by itself and what results is 4. From it subtract 3 and the result is 1. Take the root of it, which is 1. If you add it to 2, you will have the side 3. If you subtract it from 2, there remains 1. This will be each side of it, according to the augmentation and diminution.

But its working according to algebra is that you set the side always as thing. Multiply it by itself, and the result is a *census*. Subtract it from the sides of the square, which are four things. Four things minus a *census* remain which equal 3. Restore and oppose, and after the opposition you will have a *census* and 3 *dragmae* which equal 4 things. Do thus according to what was preceded in the fifth question of the algebra.[55]

[30] If someone says to you: add the long side and the diameter [of a rectangle] and this produces 18, and the short side is 6; what is therefore the diameter and how much is the long side?

A way to find it: multiply the six by itself and this will be 36; then divide this by 18 and you will get 2. Then add this to the 18 above and you will get 20, of which half will be the

[54]This is the sixth type of equation given in al-Khwārizmī's *Algebra*: $bx + c = ax^2$ (in Latin: *radices et numerus equantur census*).

[55]This is the fifth type of equation given in al-Khwārizmī's *Algebra*: $ax^2 + c = bx$ (in Latin: *census et numerus equantur radicibus*).

diameter. If you want the side, subtract 2 from 18 and take half the remainder; this will be the long side.

Its working by algebra is that one sets the [long] side as thing and subtracts it from 18 and the remainder is the diameter: 18 less thing. Then multiply the thing by itself and add the product of the [short] side by itself, and there will be *census* plus 36 *dragmae* equal to the product of 18 less thing by itself, which is 324 less 36 things plus *census*. Restore and oppose, and after restoring 324 by 36 things and subtracting 36 from 324 and subtracting *census* from *census*, there remains after this that 36 things are equal to 288 *dragmae*. So thing is equal to 8, which is the [long] side. The diameter is the difference between 18 and this, so is 10.

[43] If someone says to you: I have aggregated the four sides and the area [of a rectangle] and the result is 76. And one side exceeds the other by two. How much is then each side?

A way to find it: multiply the increase of one side over the other always by 2, and what results will be 4. Then subtract this from 76 and what remains in 72. Next add the number of sides of the rectangle, which is 4, and join it to the increase of one side over the other, and what results is six. Then take its half, which is three, and multiply it by itself, giving nine, which you join to the 72 and the result is 81. Then take its root, which is nine, and subtract from this half of six, which is three, and what remains is the short side, which is 6. Then add two and this will be the long side, 8. Understand!

Its working by another method, algebra, is that one sets the short side as thing. Then the long side is thing plus two. Multiply thing by thing and by two and this will be the area: *census* and two things. Then add the four sides, which are 4 and 4 things. Add this to *census* and two things and the result is *census* plus 6 things plus 4, which equals 76. Then subtract 4 from 76, and there remains 72, which is equal to *census* plus six things. Do thus according to what came before in the fourth question of algebra.[56]

[62] If someone says to you: you have divided the longer diameter [of a rhombus] by the shorter one and the result is one and a third, and the area is 96. How large is each diameter?

A way to find it: double the area, the result is 192. Multiply it by one and a third: it is 256. Its (square) root is the longer diameter. To find the shorter diameter by this way, divide the double of the area by one and a third and extract the (square) root of what is collected.

There is another way by using algebra. Put the shorter diameter as the thing and the longer as the thing and its third, because he said: you have divided the longer diameter by the shorter one and the result is one and a third. Multiply the thing by the thing and its third. The result is one *census* and its third. This equals the double area, i.e. 192. Therefore one *census* is 144 and the thing is its root, i.e.12. This is the shorter diameter, and the longer diameter is 16.

The second half of *On Measurements* contains problems with trapezoids, triangles, circles, circular sections, and solids; the methods of solving them reflect Hellenistic mathematics. The following problem (#125) is the calculation of the area of an acute-angled triangle if the lengths of the sides are known. For solving it, the section p of the base that is cut off by

[56]This is the fourth type of equation given in al-Khwārizmī's *Algebra*: $ax^2 + bx = c$ (in Latin: *census et radices equantur numeris*).

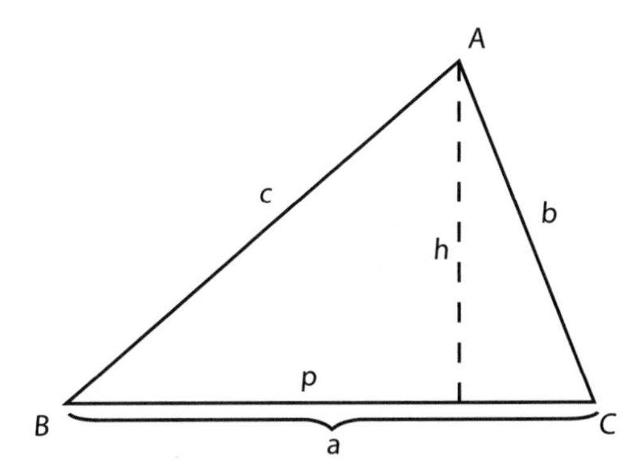

Fig. II-4-2-1.

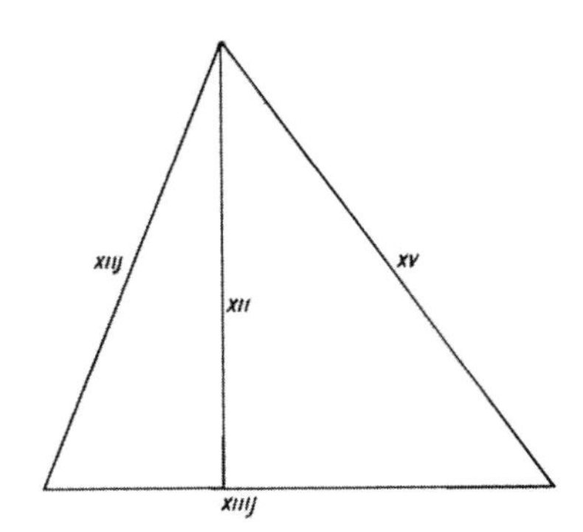

Fig. II-4-2-2.

the height must be found. The author uses the relation $p = (c^2 - b^2)/2a + a/2$, which follows from Euclid's *Elements* II.13, equivalent to the law of cosines and the formula $b^2 = a^2 + c^2 - 2ap$ (see Fig. II-4-2-1). This problem with the same numerical values was already treated by Heron (fl. ca. 60 AD) and it occurs, too, in Abraham Bar Ḥiyya's work (see section III-2 in Chapter 2).

[125.] A triangle with different sides and acute angles has one side with (length) 13, the second (with length) 15 and the base 14 [Fig. II-4-2-2]. How much is its area?

One cannot come to the area without producing its perpendicular. I shall mention this, when I have finished the doctrine of the triangles, God willing.[57] The rule of knowing how its perpendicular can be produced consists of the drawing of its sections. This is the rule of knowing how long the drawing of the sections is: multiply the first side by itself and the second side by itself and subtract the smaller from the larger. The remainder is 56. Its half is 28. Divide it by the base. The result is 2. Add it to the half of the base, i.e. to 7, and the longer section is 9. If you want the shorter section, subtract 2 from 7, and the remainder 5 is the shorter section.

[57]The announced explanation does not exist in the transmitted text. The author does not mention here the so-called Heron's formula.

If you want to know the perpendicular, multiply any of the sections by itself and the adjacent side by itself. Subtract the smaller from the larger. Find the root of the remainder. It is the perpendicular, and it is 12. If you want to know the area, multiply half of the base by the perpendicular. The result is the area, namely, 84.

Another way to know the drawing of the sections is: multiply 14 by itself—it is 196—and 15 by itself—it is 225. Their sum is 421. Subtract the product of 13 by itself. The remainder is 252. Its half is 126. Divide it by the base. The result is 9, the longer section, and the remainder of the base is the smaller section, namely 5.

3. LEONARDO OF PISA, *DE PRACTICA GEOMETRIE* (*PRACTICAL GEOMETRY*)

During his tour of the Mediterranean rim, Fibonacci acquired a profound knowledge of current mathematics. The major result of his practical education was *Liber abbaci* (*Book on Calculation*). This was followed by a small volume on geometry (1202) that in 1220 was enlarged into *Practical Geometry* (*De practica geometrie*). Intended for craftsmen of all kinds, the book enables its readers to measure the area of fields of every type, assist partners in dividing such fields, find the dimensions of solid bodies with the tools for measuring the height, depth, and lengths of planes, all aided by rules for calculating square and cube roots with a nod to algebra. The selections here display his adroitness. The first set of problems is taken from chapter 3, which deals with measuring fields of different shapes. Although the problems are geometrical, Leonardo generally uses algebra as part of the solution procedure. Many of the problems are closely related to problems found in the work of Abū Bakr (above), Abraham bar Ḥiyya (section III-2 in Chapter 2), Abraham ibn Ezra (section III-1 in Chapter 2) and Ibn Abdūn (section IV-1 in Chapter 3).

[18] Given acute scalene triangle *abc* with side *ab* 13 rods, side *ac* 15 rods, and base *bc* 14 rods. In this triangle the perpendicular cannot be found until first the place on the base is found where the perpendicular would fall. There are three ways to find that place. The first way is to add the square on one side to the square on the base, and to subtract the square on the other side from their sum. Divide half the remainder by the length of the base. The quotient will be the segment for that part from which was added the square on the side with the square on the base. For instance, in the figure [Fig. II-4-3-1], the square on the base of length 14 is 196. Add this to the square of 13 or 169. Subtract from their sum the square on the remaining side *ca* (225) to get 140. Divide half of this or 70 by the base or 14 to get 5. This is the length of segment *bd*. What remains is 9 rods, the length of *dc*, the difference between 14 and 5.

Fibonacci then discusses two other methods and concludes:

Once the place has been found, if you want to find the perpendicular, subtract the square (25) of the shorter segment from the square on side *ab* (169) to get 144. Its root is 12, the length of the perpendicular *ad*. ... The multiplication of the perpendicular by half the base or conversely yields an area of 84 rods for the entire triangle *abc*.

In the following two problems dealing with the area of a square and the sum of its four sides, note that the sides are transformed into rectangles of width 1.[58]

[58] This problem, in which a side is converted to a rectangle of width 1, is found in many of the algebraic works discussed in this sourcebook. For a detailed discussion, see [Høyrup, 1996].

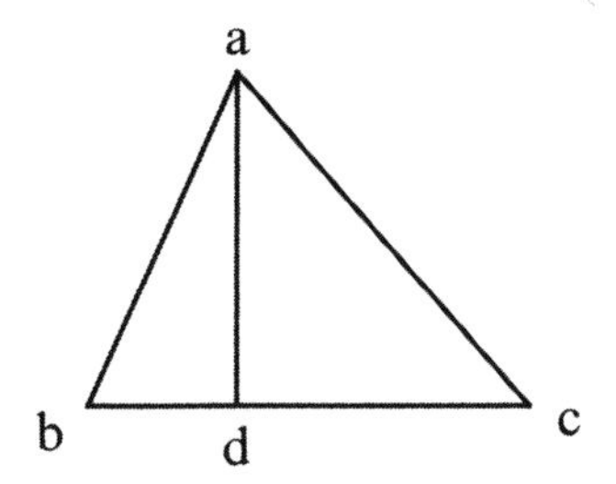

Fig. II-4-3-1.

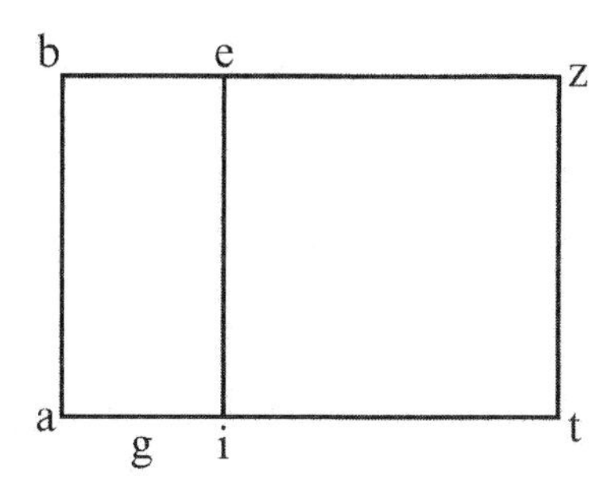

Fig. II-4-3-2.

[89] Given an area and four sides that make 140; and you wish to evaluate the sides from the area. Add the rectangular surface *ae* to a square *ezit* [Fig. II-4-3-2]. Let *ai* be extended to line *it* and *be* extended to *ez*. Let each of the lines *be* and *ai* be 4 to represent the number of sides of the square. Whence the surface *ae* equals four sides of the square *et* since the side *ei* is one of the sides of the surface *ae*. And the surface *et* contains the area of the square *zi* and its four sides. Therefore the surface *za* is 140. And that is what we had said, namely a *census* with four roots equals 140. Let the square *et* be the *census* and the surface *ae* be the four roots. Divide the line *ai* in two equal parts at point *g*. Because line *ti* was added to line *ai*, the surface of rectangle *it* by *at* with the square on line *gi* equals the square on line *gt* [*Elements* II.6]. But the surface *it* by *at* equals the surface of *zt* by *at*, since *it* equals *tz*. Therefore the surface *zt* by *at* with the square on line *gi* equals the square on line *gt*. But *zt* by *at* is the surface *za* that is 140. To this add 4, the square on line *gi*, to get 144 for the square on line *gt*. Whence *gt* is 12, the root of 144, as required. Whence if *gi* (2) is taken from *gt*, what remains is *it* (10), the side of square *et*. Its area is 100. If the four sides are added, then there is the required 140.

[93] If you subtract the area of a square from the sum of its four sides or four roots, 3 rods remain. Consider square *ge* with each side measuring less than 4 rods [Figs. II-4-3-3 and II-4-3-4]. Let there be added line *de* to line *ad* so that the whole length *ae* is 4. Divide line *ae* in two equal parts at point *b*. Draw line *az* equal to and equidistant from [i.e., parallel to] line *dg*. Extend line *fg* to point *z*. Because line *ae* is 4 and *ef* is the side of square *ge*, then the product of *fe* by *ea* (the surface *ze*) is equal to the four roots (the sides of square *ge*). If you take square *ge* from these, what remains is surface *ga* of 3 rods. But square *ge* and surface *ga* equal surface *ze*. Therefore the four roots equal the *census* and the three rods. And so it is necessary for us to find the *census* and its root. Because line *ae* equal

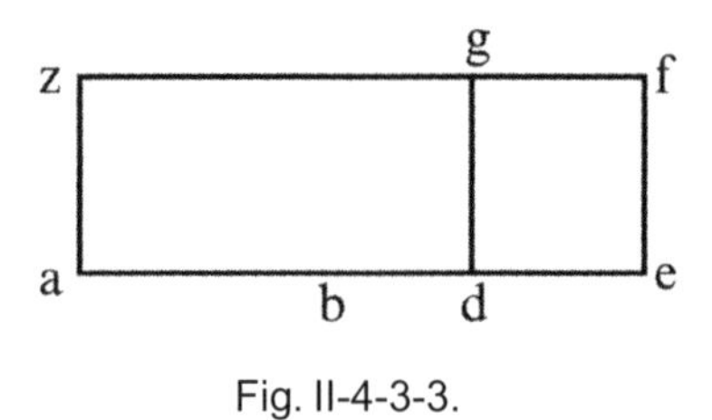

Fig. II-4-3-3.

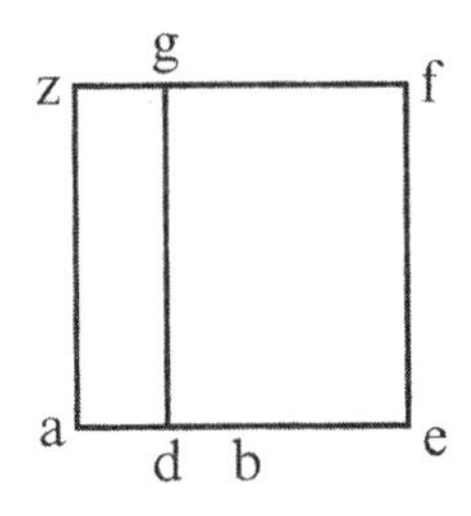

Fig. II-4-3-4.

to 4 was divided into two equal parts at point *b* and in two unequal parts at point *d*, the product of *ed* by *da* with the square on line *bd* equals the square on line *be* described as half the line *ae* [*Elements* II.5]. But the product of *ed* by *da* produces the surface *ga* which is 3, since *dg* equals *de*. Therefore surface *gd* by *da* with the square on line *bd* equals the square on line *be*, that is, 4. Therefore the square on *bd* is 1 with root 1. If it is taken from line *be* what remains from 1 is the side of square *ge*, the area of which is the desired *census* or 1 [Fig. II-4-3-3]. And if half the line *ae* were between *d* and *e* at point *b*, as seen in [Fig. II-4-3-4], add line *bd* to line *eb* (2) to make *de* equal 3, as we had foretold.

Such is the procedure for all such problems in which roots equal a square and a number: subtract the number from the square of half the roots; then add and subtract the remainder from the aforementioned half. Thus you have the root of the desired square.

[113] Again, I have joined the diameter [i.e., diagonal] [of a rectangle] with one side and found the sum to be 16, and the other side is 8. What is the diameter and what is the side joined to it? Multiply 16 by itself to get 256. From this subtract 64, the square on the given side, to leave 192. Divide this by twice 16 to reach 6, which is the side joined to the diameter. Subtract this side from 16 and 10 remains for the diameter. For example, consider the adjacent quadrilateral *bgde* [Fig. II-4-3-5]. Let the sum of the diameter *bd* and one side *bg* be 16 rods. Let side *dg* be 8. Made side *bg* a root so that the diameter *db* is 16 less the root. Squaring this produces 256 and one square less 32 roots for the square on the diameter. But the square on diameter *bd* equals the two squares on sides *bg* and *gd*. Therefore the squares on sides *bg* and *gd* equal 256 and the square on side *bg* less 32 roots. Restore the roots. And the squares on lines *bg* and *gd* with 32 roots equal 256 and the square on side *bg*. After subtracting the square on *bg* from both parts, what remains is the square on side *gd* and 32 roots equal to 256. Since the square on side *gd* equals 64, subtract it from both sides to leave 32 roots equal to 192. So we simplify the solution of this problem to one of the six paradigms mentioned above, namely that of

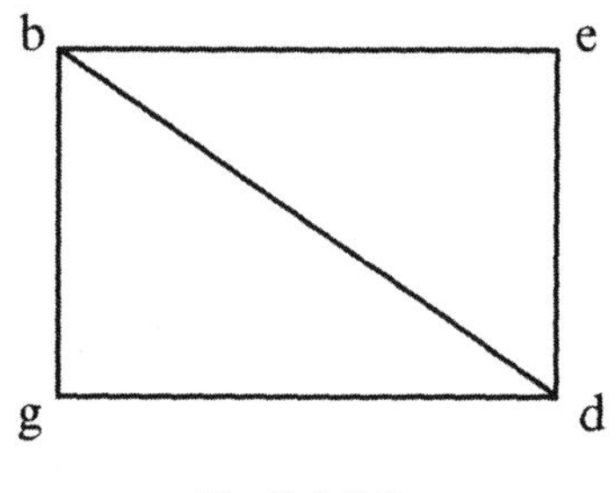

Fig. II-4-3-5.

roots equal to a number. Whence divide 192 by 32 to get 6 from the root equal to side *bg*. Subtract this from 16 to get 10 for the diameter *bd*, as I said.

[122] If the sum of the four sides and the area of a rectangle is 76, and the long side is two more than the short side, then put thing for the short side and thing and 2 for the long side. Multiply thing by thing and two to obtain a *census* and 2 roots equal to the area. Add to this 2 roots for the 2 short sides and two more and 4 (by which the long sides exceed the short sides) for the two long sides. Then you have a *census* and 6 roots and 4 dragmas equal to 76. Take 4 from both sides to leave a *census* and 6 roots equal to 72. Add the square of half the roots to 72 to yield 81. From its root or 9 take half the roots to leave 6 for the short side.

[138] Again: let the long side [of a rectangle] be 7 more than the short side and the diameter be 13. How much therefore is each side? Subtract the square of the excess from the square on the diameter, that is 49 from 169, to leave 120. Half of this is 60 for the area. For example: let the line *ab* equal the two sides, with *bg* the small side and *ga* the long side [Fig. II-4-3-6]. Further let *ac* be the 7 by which the long side *ag* exceeds the small side *bg*. Consequently *gc* equals *gb*. On line *ab* construct square *ad* as shown in the figure. The side of square *gk* is line *gb* or *gi*. The side of square *hf* is line *hi* or *if* that is equal to line *ag*. The supplement *ai*, therefore, is equal to the given rectangle whose dimensions are *ag* by *gi*. Now these are equal to the two sides of the given rectangle, and the supplement *id* equals the supplement *ai*. Therefore the supplements *ai* and *id* are twice the area of the given rectangle [*ai*]. Therefore twice the area with the squares *hf* and *gk* equal the square *ad*. But the two *hf* and *gk* are equal to the square on the diameter. Therefore the supplements *ai* and *id* with the square on the diameter of the given rectangle equal the square *ad*. But square *ad* with the square on line *ac* is described to be twice the squares which are described by lines *ag* and *gb*. But those which are described by the squares on lines *ag* and *gb* are equal to the square on the diameter. Therefore twice the square on the diameter is equal to square *ad* and the square on line *ac*. But one square on the diameter is equal to the squares *hf* and *gk*. Therefore the square on the diameter equals the supplements *ai* and *id* together with the square on the line *ac*. Whence if we subtract 49 the square on line *ac* from the square on the diameter, what remains for us is 120 for the supplements *ai* and *id*. Their half is 60 for the area *ai* of the quadrilateral equal to the given quadrilateral. Therefore the area of the given quadrilateral is 60, as I said. In order to find the sides, let the area be 60 and the long side 7 more than the short side. Then do as we taught you before. Or, let the short side be thing and the

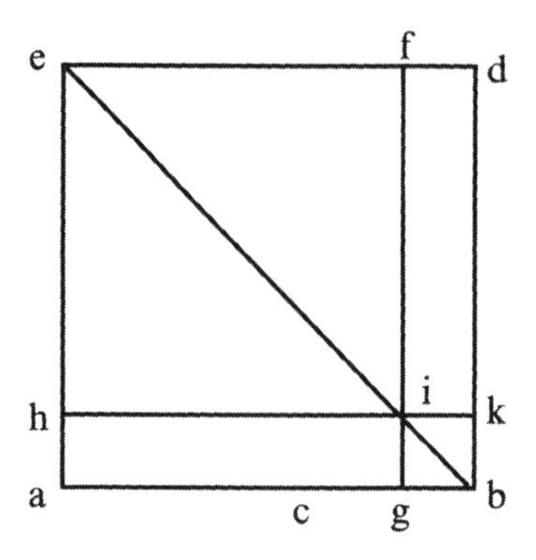

Fig. II-4-3-6.

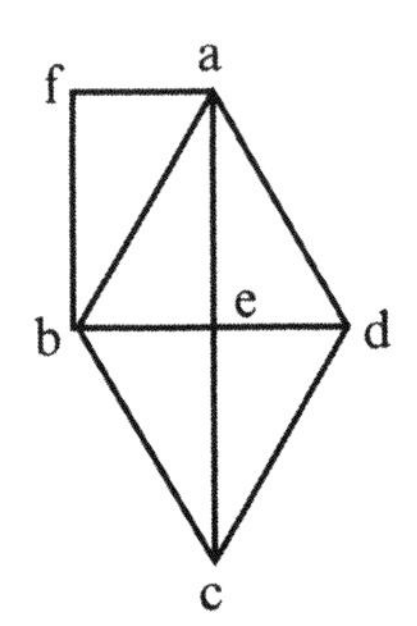

Fig. II-4-3-7.

long side be thing and 7. Add together the squares of thing and of thing and 7. The sum is two *census*, 14 roots, and 49 dragmas. Collect terms with the square on the diameter. And you have what you were looking for.

[160] Again: I divided the major diameter [of a rhombus] by the minor to get 2 2/5, and the area of the rhombus is 120. What is the measure of each diameter? Because the ratio of whole to whole is the same as part to the same part, so the ratio of the major diameter to the minor is as half the major to half the minor. And half the major is *ea*, the long side of the quadrilateral *ef* [Fig. II-4-3-7], and half the minor line *eb*, the short side. Now all numbers which have one and the same ratio, if the larger are divided by the smaller, always come out the same in the division process. Therefore, if we divide the long side of the quadrilateral *ef* by the short side, we still get 2 2/5. Therefore this is the problem: the area of the quadrilateral is 60, half the area of the rhombus. I divided the long side by the short side, and got 2 2/5. So multiply 1 by 2 2/5, then by 60 to reach 144. Its root is 12. Divide it by the numbers in the ratio, that is, by 1 and next by 2 2/5, to get first 12 and then 5 for the sides of the quadrilateral or halves of the diameters. Therefore the major diameter is 24 and the minor is 10.

Fibonacci continued in chapter 3 by showing how to measure the circle and its parts.

[188] Assuming a field you wish to measure is circular, you must know its diameter. Multiply this by 3 1/7; or multiply it by 22 and divide by 7. Either way you have the circumference that contains the circle. To find the area, multiply half the circumference

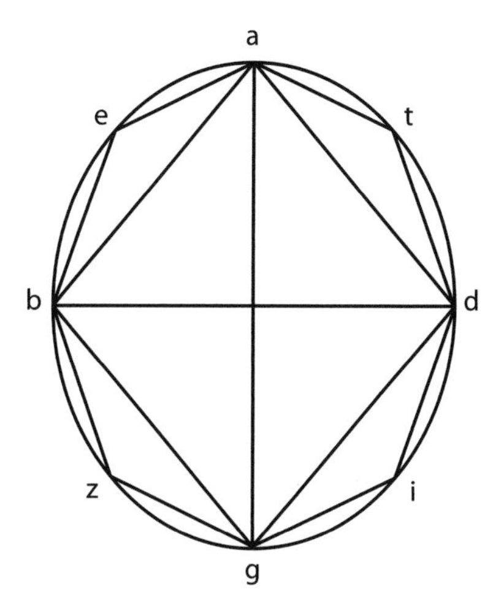

Fig. II-4-3-8.

by half the diameter; or take eleven fourteenths of the square on the diameter. Either way you have the area of the circle.

Fibonacci next gives various classical arguments as to why this method is correct and then applies his results to measure various parts of circles. To measure the area of a sector, however, he needs what we would think of as trigonometric results (section II-5-2). Then he continues.

[220] Since you know how to find arcs by their chords and chords by their arcs by what we have said and you want to find the area of a sector of a circle, then find its arc, multiply its half by the semi-diameter of the circle and the result is the area of the sector.

But if a figure was not bounded by lines and circular arcs, Leonardo taught that one should calculate its area by filling the region with smaller and smaller known areas until the difference between it and the known areas is as small as you choose, or, as Leonardo puts it, "so that nothing perceptible remains." He was clearly influenced by the so-called method of exhaustion used by both Euclid and Archimedes.

[225] Again: suppose there is a field described as *elana* or oblique which is circumscribed by only one line. It resembles a circle but with unequal diameters *ag* and *bd* that intersect each other at right angles [Fig. II-4-3-8]. In order to measure this field, try to use straight lines. First, the larger quadrilateral is bounded by straight lines *ba*, *bg*, *da*, and *dg*. What remain are four segments of the circle: the first contained by straight line *ab* and curved line *aeb*, the second by straight line *bg* and curved line *bzg*, the third by straight line *gd* and curved line *gid*, and the fourth by straight line *da* and curved line *dta*. Now if we removed the rectilinear triangles *eab*, *zbg*, *igd*, and *tad* from the four segments, not much remains in the figure apart from the little bit contained by the 8 segments. And if in any of them a triangle is drawn, try to proceed in the same way for the remaining segments. Thus the entire oblique figure can be resolved into rectilinear figures, so that nothing perceptible remains. If we can possibly find the area of just one of all the rectilinear figures, then we undoubtedly have the area of the whole figure.

In chapter 4, Fibonacci shows how to divide fields. The following problem displays an ingenious solution to a question with what may appear as an impossible condition, to divide

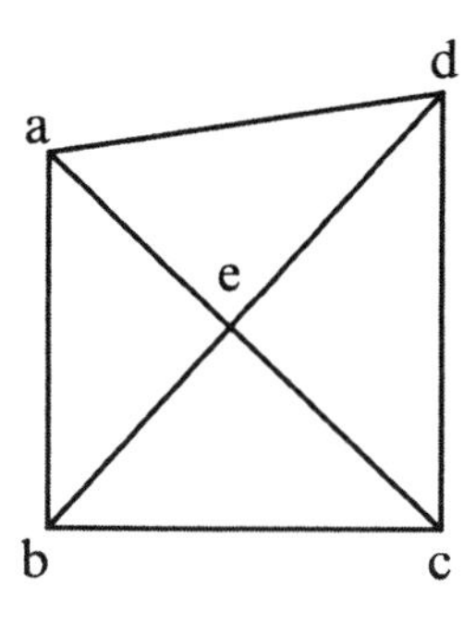

Fig. II-4-3-9.

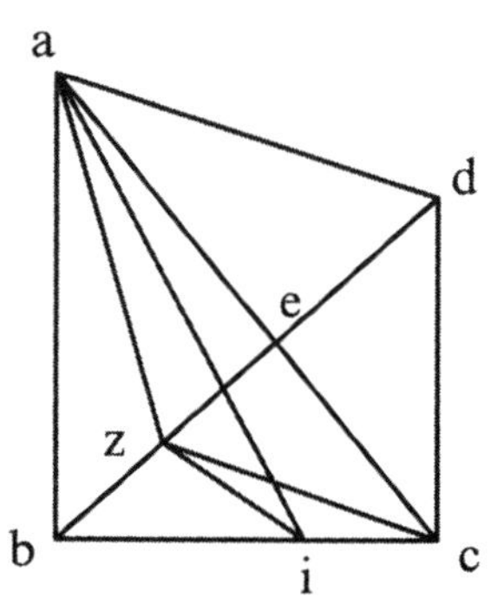

Fig. II-4-3-10. (not to scale).

equally among two partners an irregular four-sided piece of property from a required corner of the property. Compare this to the work of Abraham bar Ḥiyya in section III-2 of Chapter 2.

[60] To divide a quadrilateral with all sides unequal in length in two equal parts from a given angle.

Given quadrilateral *abcd* that I wish to divide in two equal parts from a given angle *a*. First I draw diameter *bd* opposite angle *bad*. Then I intercept line *bd* by diameter *ac* at point *e* [in Figure II-4-3-9]. Lines *be* and *ed* are either equal or they are not. Assume first that they are equal. Then because *be* and *ed* are equal, triangle *abe* equals triangle *ade*. Also triangle *ebc* equals triangle *ecd*. Whence the entire triangle *abc* equals the entire triangle *acd*. Quadrilateral *abcd* has therefore been divided in two equal parts by diameter *ac* drawn from angle *a*, as required.

On the other hand, assume that lines *be* and *ed* are not equal. Let line *bz* equal line *zd*. So I draw line *zi* equidistant from diameter *ac*, as seen [in Figure II-4-3-10]. And I join *ai*. I say again that quadrilateral *abcd* has been divided into two equal parts by line *ai* drawn from the given angle *a*. The equal parts are triangle *abi* and quadrilateral *aicd*. The proof follows. Draw lines *az* and *cz*. Thus triangles *azd* and *dzc* equal triangles *abz* and *bzc*. Whence quadrilateral *azcd* is half of quadrilateral *abcd*. And because triangles *aci* and *acz* share the same altitude and lie within the equidistant lines *ac* and *zi*, they equal each other. If triangle *acd* is added to both, then quadrilateral *aicd* equals quadrilateral *azcd*. But quadrilateral *azcd* is half of quadrilateral *abcd*. Therefore quadrilateral *aicd* is half of quadrilateral *abcd*, as required.

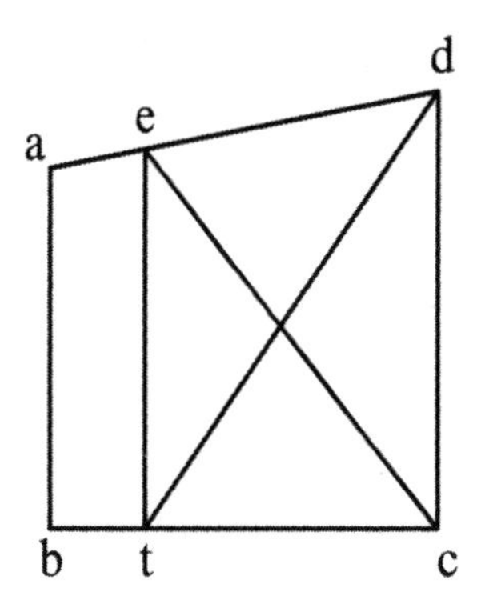

Fig. II-4-3-11.

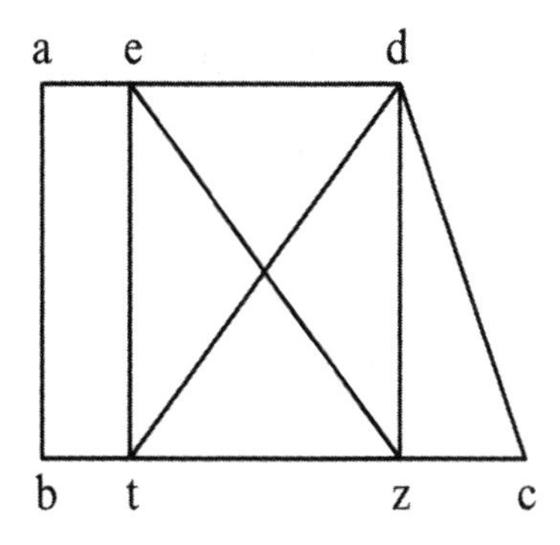

Fig. II-4-3-12.

To complete the picture, Fibonacci divides equally the same piece of property but from an arbitrary point on a side.

[61.1] To divide a quadrilateral in two equal parts by a line drawn from a point on one of the sides.

Given quadrilateral *ac* that you wish to divide from a given point *e* on side *ad*. First, divide quadrilateral *ac* in two equal parts by line *dt* drawn from angle *d*. Draw line *et*. Now line *et* is either equidistant from line *dc* or it is not. First, suppose that lines *et* and *dc* are equidistant, as appears in Fig. II-4-3-11. Draw line *ec*. I say that quadrilateral *ac* has been divided in two equal parts by line *ec* drawn from the given point *c*. The proof follows. Because lines *et* and *cd* are equidistant, triangles *ecd* and *tcd* are equal. But triangle *tcd* is half of quadrilateral *ac*. Whence triangle *ecd* is half of quadrilateral *ac*. Therefore quadrilateral *ac* has been divided into two parts by a line drawn from the given point *e*, as required.

But, suppose that lines *et* and *dc* are not equidistant. Then I draw line *dz* equidistant from *et*, as appears in Fig. II-4-3-12, and draw line *ez*. Quadrilateral *abcd* has been divided in two equal parts by line *ez*. The proof follows. Triangles *dze* and *dzt* are equal,

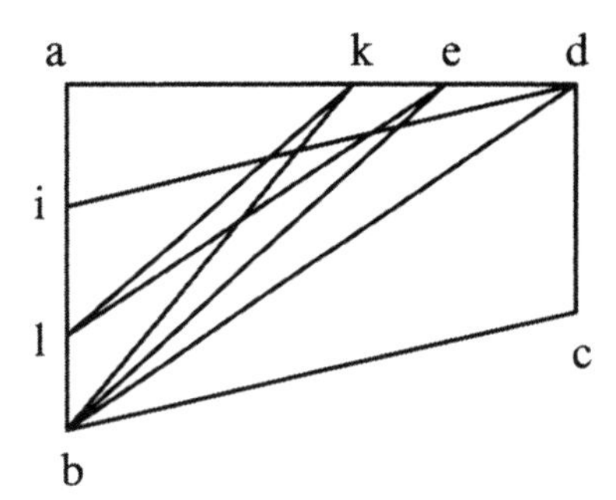

Fig. II-4-3-13.

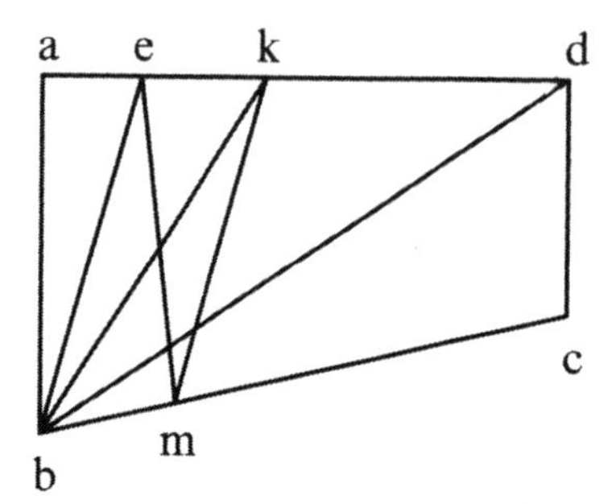

Fig. II-4-3-14.

because they share the same base and lie within the same equidistant lines. Now if triangle *dcz* is added to both, quadrilateral *ezcd* equals triangle *dct* that is half of the whole quadrilateral *ac*.

[61.2] It must be noted that if the diameter *bd* divides quadrilateral *ac* in two equal parts, what we said above about two figures pertains here. But if the division falls on side *ab* by line *di* that divided quadrilateral *abcd* in two parts, then another procedure is required [Fig. II-4-3-13]. I will divide quadrilateral *ac* by line *bk* drawn from angle *b*. And then if *k* is a given point on side *ad* from where dividing line *kb* must originate, then it divides quadrilateral *ac* in two equal parts. But if *k* is not the given point, then let the given point lie between points *k* and *d* or between points *k* and *a*. First then, let the given point *e* lie between points *k* and *d*. Draw line *be*. Through point *k* extend line *kl* equidistant from line *eb*. Draw line *el*, and it divides quadrilateral *ac* in two equal parts. The proof follows from what has been said before. Again, let the given point *e* lie between points *k* and *a*. Likewise, draw line *eb*. Through point *k* draw the equidistant line *km*. Draw line *em*, and it divides quadrilateral *ac* in two equal parts, quadrilaterals *am* and *ec*, shown in Fig. II-4-3-14. Again, the proof follows from what been said before.

Somewhat later, Fibonacci divides a region bounded by two straight lines and an arc of a circle into two equal parts.[59]

[59]If the two straight lines are both radii of the circular arc, then this region is what today is called a sector. But Fibonacci is interested in a more general region, as the proof shows.

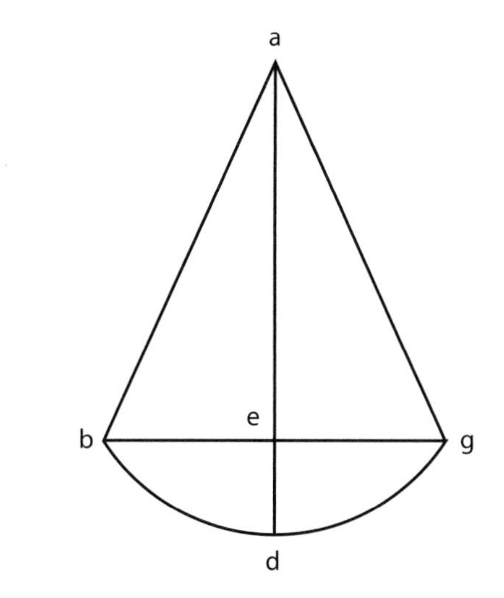

Fig. II-4-3-15.

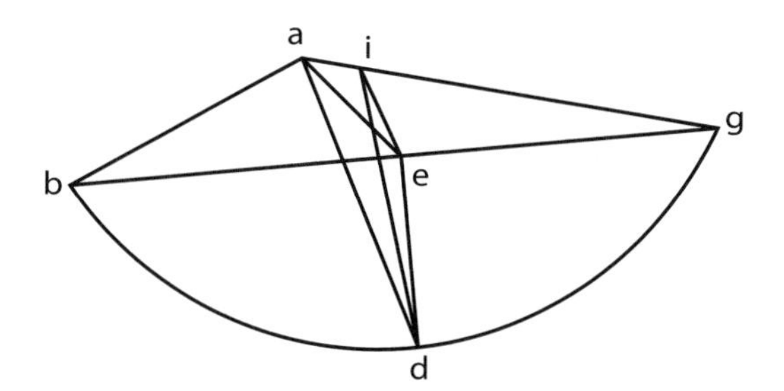

Fig. II-4-3-16.

[86] To divide a figure bounded by two straight lines and the arc of a circle into two equal parts.

Given triangle[60] *abdg* whose two sides *ab* and *ag* are straight lines, the remaining side *bdg* being an arc of a circle. First draw *bg* and divide it in two equal parts at point *e*. From here draw line *ae*. Then on line *bg* and from point *e* given on it, draw line *ed* at right angles. Lines *ae* and *ed* are either in a straight line or they are not. Assume that they are [Fig. II-4-3-15]. Then figure *abdg* has been divided into two equal parts by line *ad* because line *ae* has divided the rectilineal triangle *abg* in two equal parts. Thus line *ed* divides part *bdg* of the circle in two equal parts. On the other hand, lines *ae* and *ed* may not be in a straight line, as can be seen in this other figure [Fig. II-4-3-16]. Then draw line *ad*. Next, through point *e* draw line *ei* equidistant from line *ad*. Draw *di*. I say that *di* divides the whole figure *abdg* in two equal parts. The proof follows. Because lines *ad* and *ei* are equidistant and between them are [rectilineal] triangles *dai* and *dea* upon base *da*, these triangles are equal. Add to them the figure they have in common described by lines *ab* and *ad* and arc *db*. Thus the figure bounded by lines *di, ai*, and *ab* and arc *db* equals the figure bounded by lines *de*, *ea*, and *ab* and arc *bd*. This is half of triangle *abdg*, as required.

[60]Fibonacci uses the Latin word *trigonum* here to mean a region bounded by three curves, some of which could be straight lines. He uses *rectilineum trigonum* when all the curves are straight lines.

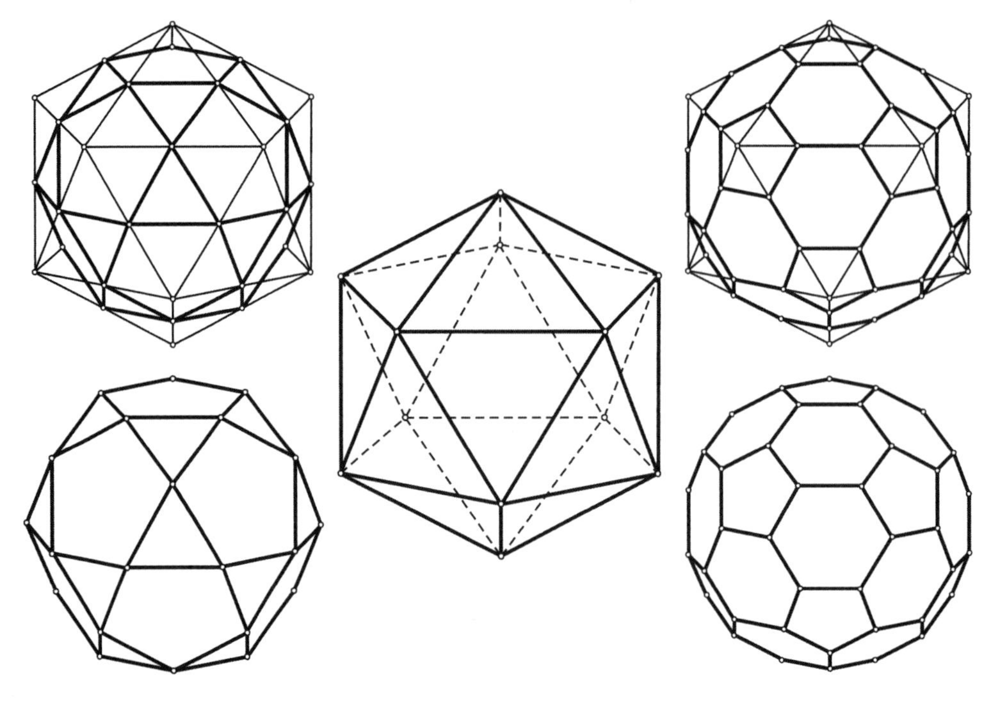

Fig. II-4-4-1.

4. JOHN OF MURS, *DE ARTE MENSURANDI*

John of Murs (Johannes de Muris) was active in science from 1317 until at least 1345. His works, mostly written in Paris, are on the theory of music, astronomy, and mathematics. His *Quadripartitum numerorum*, which takes its name from its division into four books, deals with arithmetic, algebra, and practical applications of arithmetic (including music and mechanics). Around 1344 he completed the *De arte mensurandi* in twelve chapters. The first four chapters and the beginning of chapter 5 were written by another, unknown, author; the rest is by John. The book starts with the mathematical knowledge necessary for astronomy (operations on sexagesimal fractions and trigonometry). What follows is not very original, being based mostly on contemporary writings, including the translation of Archimedes's works by William of Moerbeke.

Perhaps the most interesting part is chapter 11, which deals with semi-regular polyhedra. It seems unique for its time, because the source from which John could have taken it is unknown. Archimedes had discovered thirteen semiregular polyhedra. By not requiring their polygonal faces to be of the same kind, the semiregular polyhedra differ from the regular. The faces are still regular polygons, but they do not all have the same number of sides, nor are all the polyhedral angles equal. Since Archimedes's text has not survived,[61] many details are not known. But we do know, at least partly, the method by which the polyhedra were produced: Archimedes started from the five regular polyhedra and applied to them different procedures of truncation. While there were some Arabic writings on this topic (e.g., instructions for the

[61] For details, see [Dijksterhuis, 1956, pp. 405–408].

construction of the five regular and five semiregular polyhedra by Abū'l-Wafā), it is unlikely that John borrowed the idea from this or a similar work. He distinguished *corpora regulariter irregularia* (semiregular polyhedra) from *corpora irregulariter irregularia* (totally irregular polyhedra). He knew two methods for building them: by dividing the edges of the regular polyhedra into two or into three equal parts. Applying both methods to the icosahedron, he generates two of the so-called Archimedean polyhedra, the icosidodecahedron and the truncated icosahedron [Fig. II-4-4-1]. John tells that a stonecutter has made for him in his presence models of regular and semiregular polyhedra. This too is remarkable for that time. Before 1500 in the Latin West only one other person is known who dealt with the mathematics of the semiregular polyhedra: the famous painter, Piero della Francesca (ca. 1420–1492). Since Piero's procedure in his *Trattato d'abbaco* and in his *De quinque corporibus regularibus* differs from that of John; it is unlikely that Piero knew about John's treatise.[62]

The 11th chapter promised the measurement of irregular solids. I shall set irregular solids of two kinds: some are regularly irregular and others irregularly irregular solids. Without doubt there are numerically four regularly irregular solids which are generated from four regular solids, because from any regular solid except the tetrahedron one regularly irregular solid can be drawn. I thought it to be proper to start in reverse order with the icosahedron. For, if any side of each base of an icosahedron is halved, if straight lines are drawn to the points of section and if the pentagonal pyramids are cut by a plane—it is necessary that there are 12, just as the number of solid angles—from any equilateral triangle the fourth part, which is an equilateral triangle, will remain. And therefore in the newly formed solid there are 32 faces or bases, 12 of which being equilateral and equiangular pentagons and 20 equilateral triangles, and always between three pentagons one triangle is included. By another division of the side of any base you would create infinite irregular solids. For, if each side (of the base) is divided into three equal parts, from any triangle situated in front (of you) an equilateral and equiangular hexagon (and three equilateral triangles) will arise and as many in the other faces, and when the pentagonal pyramids are cut away, there would be as well 32 bases, and all pentagonal bases would be similar, but not equal, to each other; 20 would be larger and 12 smaller.[63] In this way it could be done in any regular solid. But we will first speak about those irregular solids that are generated by the first division, i.e. by halving; the other division can be transformed to the same way. . . .

Note that irregularity has come from regularity, just as inequality has resulted from equality. And from one irregular solid many other, infinite indeed, can be created: there is no limit to this. But perhaps the imagination cannot keep easily the form of solids like this. Therefore I consider to have constructed the five regular solids with sensible material in such a way as I have them drawn by my own hand, and I have arranged that later by a stone-cutter these solids will be made in stone in my presence to avoid that the stone will become faulty because of the great number of lines. At that day, sense and intellect will be in harmony and become friends. But now the measurement of irregular solids must be treated.

[62]For Piero's treatment, see [Clagett, 1978, pp. 398–406].

[63]This is not correct; there are 12 pentagons and 20 hexagons. The author seems to think that there are 20 bigger pentagons.

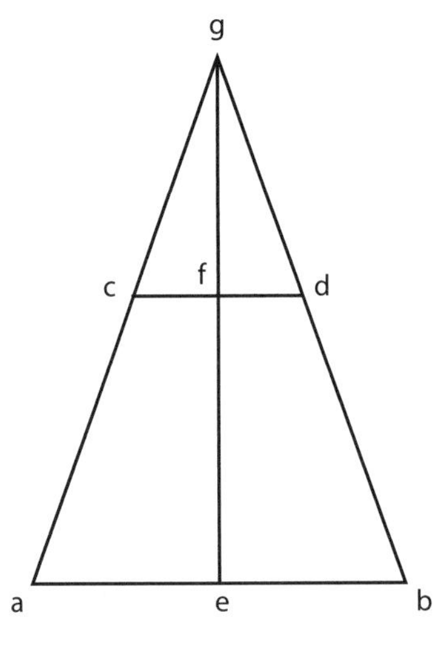

Fig. II-4-4-2.

Chapter 10 of *De arte mensurandi* concerns pyramids, cones, prisms, cylinders, spheres, rhomboidal bodies, and segments of these various figures. Inter alia, John cites here theorems from Archimedes's *On the Sphere and the Cylinder* and from *On Conoids and Spheroids*. What follows below is John's treatment of a truncated isosceles cone.

[Prop. 21] To comprehend the volume of a truncated cone.

If an (isosceles) cone is cut by a plane parallel to the base, that which is included between the section and the apex does not lose its form. But that which is left from the section down to the base is called a truncated cone. Its volume will be known by taking away the small cone, which is included from the section to the apex (and is known by the sixth [proposition] of this [chapter]),[64] from the quantity of the whole cone (known by the same [proposition]).

Example: Let the aforesaid [truncated] cone be *ad* [Fig. II-4-4-2], with its base circle about diameter *ab* and its upper surface a circle parallel to the base and about diameter *cd*, while its axis is *ef*. With sides *ac* and *bd* continued directly until they necessarily meet in *g* (by the fourth postulate [of Book I of *Elements*]),[65] hence there will be sides *cg* and *dg*. And hence whole side *ag* [and whole side] *bg* are known by the forty-first [proposition] of the second part of the fifth [chapter] of this [work].[66] And because the radius *ae* of the base is also known, therefore the axis *eg* is known by the penultimate [proposition] of [Book] I.[67] Therefore, the whole volume of cone *abg* will be known by the sixth [proposition] of this [chapter]. If from this whole cone is taken the volume of the small cone *cdg* (known by the same [proposition]), the volume of the truncated cone remains known to you; which

[64]In Chapter 10, proposition 6, the formula of the volume of a cone is given by reference to Euclid, *Elements* XII.9.

[65]This is the so-called parallel postulate, numbered as the fourth one in the Campanus version of the *Elements*.

[66]Chapter 5.2, proposition 41 reads: "If (in a trapezoid) the top, the bottom and both sides are known, to find the sides (of the produced triangle) to the vertex" (*Dato capite, nota basi, unoquoque laterum numerato, quantum distat usquoque concurrant latera reperire*).

[67]This is the Pythagorean theorem (Euclid, *Elements* I.47).

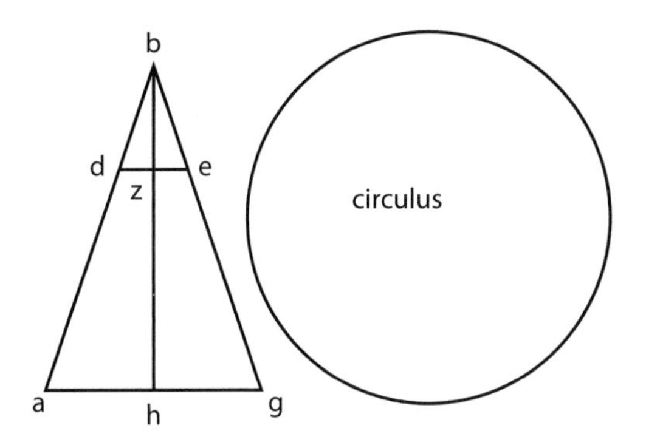

Fig. II-4-4-3.

was that proposed. And if you want to measure the surface of this [truncated] cone: With the cone completed, from its surface (known by the fifth [proposition] of this [chapter])[68] take the surface of the small cone (known by the same [proposition]) and the surface of the truncated cone remains.

[Argue] for the same [proposition] by [I.]16 of Archimedes in the place [cited] above:[69] Let there be a cone whose triangle through the axis is *abg* and which is cut by a plane parallel to the base [passing] through line *de* [Fig. II-4-4-3]. Moreover, the axis of the cone is *bh*. And let a circle be posited whose radius is the mean proportional between (1) *ad*, the side of the truncated cone, and (2) the sum of the radii of the bases, *dz* and *ah*; this circle will be equal to the surface of the truncated cone. Therefore, with the circle measured by VI.1 of this [work], the proposition will be had.[70]

5. JORDANUS DE NEMORE, *LIBER PHILOTEGNI*

One of the most significant mathematicians in medieval Western Europe was Jordanus de Nemore. His writings on arithmetic and algebra have already been excerpted above [sections II-2-5 and II-3-4]. His main work on geometry is *Liber philotegni*. Its objective is to determine and compare regular and irregular polygons, with some propositions for the division of triangles. One of the division problems follows. To assist understanding, numbers in square brackets have been added to mark the relations that are used later in the proof.

[Prop. 22] With a triangle given and a point marked outside [of the triangle], to draw a line proceeding through the point which bisects the triangle.

Let the triangle be *ABC* [Fig. II-4-5-1], and let the point *D* be taken outside of the triangle between lines *AEF* and *HBL*, which lines bisect [two] sides of the triangle and the triangle itself, and [let that point be] as far outside as you like. For if the point were to fall on one of those [lines], the conclusion would be given by inspection. Therefore, with the point placed between those [lines], let us draw from this [point] a line parallel to line *AC* until it

[68] Chapter 10, proposition 5 reads: "To find the surface of an isosceles cone."

[69] Proposition I.16 of Archimedes's *On the Sphere and the Cylinder* reads: "If an isosceles cone be cut by a plane parallel to the base, the portion of the surface of the cone between the parallel planes is equal to a circle whose radius is a mean proportional between (1) the portion of the side of the cone intercepted by the parallel planes and (2) the line which is equal to the sum of the radii of the circles in the parallel planes."

[70] Chapter 6, proposition 1 reads: "To find the area of a given circle."

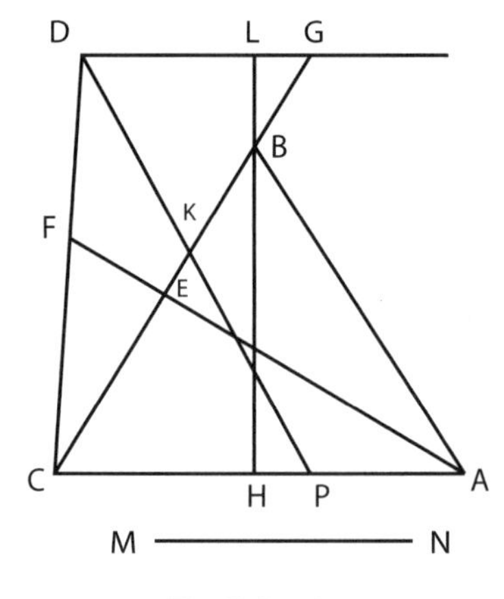

Fig. II-4-5-1.

meets with *CB*, however far that line need be extended, and let the point of juncture be *G*. And let line *DC* produce a triangle which is related to triangle *AEC* (which is one half of the given triangle) as line *CG* is related to some [line], and let that [line] be *MN* [1].[71] Also let *GC* be divided, by the argument of the 14th [proposition of this work],[72] into *GK* and *KC* so that $GK : KC = KC : MN$ [2], and let line *DK* be extended to meet line *AC* in *P*. And triangles *DGK* and *KPC* will be similar. And tri. *DGK* : tri. $KPC = (GK : KC)^2$ [3],[73] i.e., tri. *DGK* : tri. $KPC = GK : MN$ [4].[74] Also $GK : KC = DK : KP$ [5] because the triangles are similar, and $GK : KC = KC : MN$. Therefore, $DK : KP = KC : MN$ [6].[75] But $DK : KP =$ tri. *DCK* : tri. *CKP* [7]. Therefore, $KC : MN =$ tri. *CKD* : tri. *CKP* [8].[76] But it has been proved earlier that $GK : MN =$ tri. *DGK* : tri. *CKP*. Therefore, whole line $CG : MN =$ tri. *DCG* : tri. *CKP* [9].[77] But, by hypothesis, tri. *DGC* : tri. $CEA = CG : MN$, and *CEA* is one half the given triangle. Therefore *CKP* is one half the given triangle.[78] And this is what is proposed.

Shortly after Jordanus had composed *Liber philotegni*, an unknown author produced a new version with a different title, *Liber de triangulis Iordani*. He added geometrical material from other sources. One proposition deals with the construction of the regular heptagon. In a treatise that is only known from Arabic sources, Archimedes had proved the following properties of the intersection points of two long diagonals of the regular heptagon on the third: $BC \cdot BD = AD^2$ and $CD \cdot AC = BC^2$, or $a\,(a+b) = c^2$ and $b\,(b+c) = a^2$, and he had shown that it is possible to construct a regular heptagon if the segment *AB* is divided by points

[71] Take a line *MN* such that tri. *DCG* : tri. $AEC = CG{:}MN$.

[72] This proposition runs: "With two lines given, to divide either of them so that one of the segments is to the remaining segment as that same remaining segment is to the other of the proposed lines."

[73] This is evident from Euclid VI.19: "Similar triangles are to one another in the duplicate ratio of the corresponding sides."

[74] This follows from [2] and [3].

[75] This follows from [2] and [5].

[76] This follows from [6] and [7].

[77] This follows by adding the proportions in [4] and [8].

[78] By [1] and [9], tri. *DGC* : tri. $AEC =$ tri. *DGC* : tri. *CKP*, that is, tri. $AEC =$ tri. *CKP*. And since triangle *AEC* is one half the given triangle, so also is triangle *CKP*.

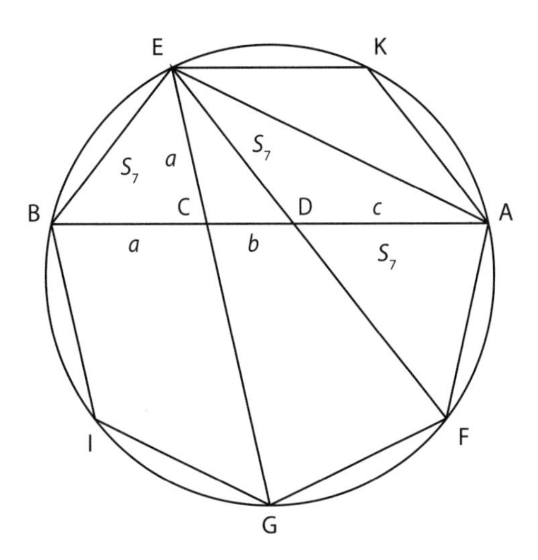

Fig. II-4-5-2.

D and *C*, so that these conditions hold [Fig. II-4-5-2]. The essentials are also given by the unknown author in *Liber de triangulis Iordani.*

[Prop. IV.23] When a line is divided into three segments and (1) the ratio of the last of these [segments] to the first is as the ratio of the first to the rest of the line (i.e., [to the sum] of the middle and last [segments]) and (2) the ratio of the middle to the last is as the ratio of the last to the rest [of the line] (i.e., [to the sum of] the middle and first [segments]), there results from these three segments [as sides] a triangle whose greatest angle is double its middle [angle], and its middle [angle] is double its smallest [angle] (4α, 2α, α). When therefore a circle contains this triangle, the arc which is subtended by the smallest angle is the arc of a chord [which is the side] of the [regular] heptagon falling in that circle.[79]

The demonstration of this is elicited as follows. Let *AED* be the triangle [composed] out of [sides equal to lines] *AD*, *DC*, and *CB* [with $AE = DC$ and $ED = BC$], constituted as the proportion demands,[80] and let *EF* be protracted equal to *ED* (or $EF = CB$), and line $GA = AE$ (or $GA = DC$) [Fig. II-4-5-3]. Therefore, by VI.16 [= Greek VI.17] and II.1 [both] of Euclid, $AD^2 = ED^2 + AE \cdot EF$.[81] Therefore, with line *DH* drawn perpendicular to *AF*, $AD^2 = DH^2 + EH^2 + AE \cdot EF$. Therefore, by the penultimate [proposition] of [Book] I [of Euclid] and with the common area subtracted, $AH^2 = HE^2 + AE \cdot EF$ because I shall [have] bisect[ed] *AF* in point *H* by the converse of II.5 of Euclid, and thus from I.4 [of Euclid] lines *AD* and *DF* are equal. Conclude therefore by I.32 and I.6 [both of Euclid] that ang. $AED = 2$ ang. *DAE*. In completely the same way it may be demonstrated that ang. $DAE = 2$ ang. *ADE* after *EK* has been protracted perpendicularly to *DG*, *EK* necessarily falling inside of triangle *DEA* since each angle on the base is acute.

[79]This text of the proposition, but not of the proof, is also found in a collection of short mathematical texts that were translated by Gerard of Cremona from the Arabic; see [Clagett, 1984, p. 599].

[80]That is, $BC{:}AD = AD{:}DB$, or $AD^2 = BC \cdot DB$ and $DC{:}CB = CB{:}AC$, or $CB^2 = AC \cdot DC$. Here and in what follows, we have replaced the Latin words by modern symbols to help readers understand the mathematics.

[81]$AD^2 = BC \cdot DB = BC(BC + DC) = BC^2 + BC \cdot DC = ED^2 + EF \cdot AE$.

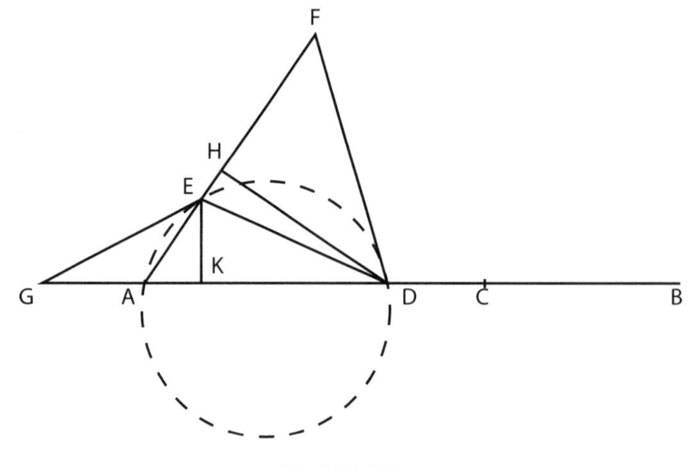

Fig. II-4-5-3.

6. DOMINICUS DE CLAVASIO, *PRACTICA GEOMETRIAE*

Dominicus of Clavasio was born near Turin, Italy, and was active at the University of Paris from around 1340 until around 1360, when he died. He started as a Master of Arts in the school of John Buridan and Nicole Oresme. Later he earned a similar degree in medicine. In 1346 he wrote a *Practica geometriae*. It is divided into an introduction and three books. The introduction contains arithmetical rules and the description of an instrument used in astronomy and mensuration. Book 1 deals with problems of measurement, Book 2 with the area of two- and three-dimensional figures, and Book 3 with the volumes of solids.

In the translation below the phrase "right shadow" (*umbra recta*) identifies the side of the quadrant attached to point *d*, whereby the length *de* is measured. The entire line called "right shadow" is divided into a number of points used for finding the necessary ratio that will be used in the computation described here. The line *oa* is the height of the person who measures the distance. Strangely, Dominic employed the word *planum* where the context clearly requires the object to be a line.

Book 1, 9th Exercise: To find with a quadrant the length of a line whose endpoint can be seen.

Let there be line *ab* whose endpoint *b* can be seen. Take the quadrant and look at point *b* through both openings of the small tables, having positioned to the eye the corner of the quadrant where is fixed the string of the perpendicular. Now see how many points of the right shadow the string or perpendicular cuts off. Compare these to 12. As is the ratio of the points cut off by the perpendicular to the number 12, such is the ratio of your height [= *oa*] to the length of the line [= *ab*] [Fig. II-4-6-1].

This is evident because the two triangles, namely triangles *oab* and *cde*, are equiangular. For *a* is a right angle and *edc* is a right angle and so they are equal. Angle *dec* equals angle *o* because lines *ob* and *ed* are equidistant and line *ce* falls on them. Therefore angle *dec* equals angle *ecb*, and angle *ecb* equals angle *o*. Because lines *ec* and *oa* are

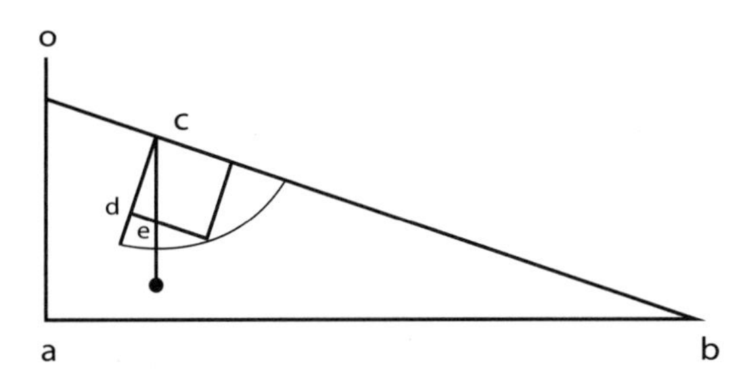

Fig. II-4-6-1.

equidistant and line *ob* falls on both, therefore by the second part of Euclid I.29,[82] angle *ecb* equals angle *o*, because *ecb* is an exterior angle and *o* is an interior angle. Then by Common Notion: things equal to the same thing are equal to one another, angles *aob* and *dec* are equal since both are equal to angle *ecb*. Therefore by the second part of Euclid I.32,[83] angle *dce* equals angle *b*. But if triangles are equiangular, then the [corresponding] sides are proportional. And so whatever the ratio of *ed* [the number of points on the right shadow] to *dc* [the given 12], such is the ratio of *oa* [your height] to *ab* [the length of the line]. But the first proportion is known, because both lines are known, and in the second [proportion] line *oa* is known, and so line *ab* will be known.

In Book 3 Dominicus gives 17 "constructions" with the rules for the volume of the following solids: a cube, *n*-sided prism, pyramid, prismoid, sphere, hemisphere, cylinder, and cone. At the end of Construction 12, he explains what will follow:

From what has been said, the manner of determining the volume of cisterns and vessels can be had quite satisfactorily. In order to show this in particular some special conclusions are posited.

The first of these conclusions is as follows.

Book 3, Thirteenth Construction. To obtain the volume of a cistern (*puteus*) according to a given bucket (*situla*).

Since a bucket ordinarily has one end (*fundum*) greater than another, make the greater end equal to the smaller—thus if the diameter of one end were 6 and the other 4 [Fig. II-4-6-2], add 6 and 4 to get 10, of which you take half and then you have a common [= average] end. After determining this, see how many times [a bucket] is contained in the area of the mouth of the cistern by the seventeenth [construction] of the second book.[84] Then you can know the depth of the cistern by the third construction of the first book[85] and you can know how many times the height of the bucket is contained in the

[82]This proposition runs: "A straight line falling on parallel straight lines makes the alternate angles equal to one another, the exterior angle equal to the interior and opposite angle, and the interior angles on the same side equal to two right angles."

[83]Proposition I.32: "In any triangle, if one of the sides be produced, the exterior angle is equal to the two interior and opposite angles, and the three interior angles of the triangle are equal to two right angles."

[84]This construction reads: "To find the circular area of a proposed circle according to a given measure." Erroneously, the Latin text refers to Book 2, Construction 18.

[85]This construction reads: "To measure the length of a valley when the eye is on the mountain and able to see the foot of the mountain."

Fig. II-4-6-2.

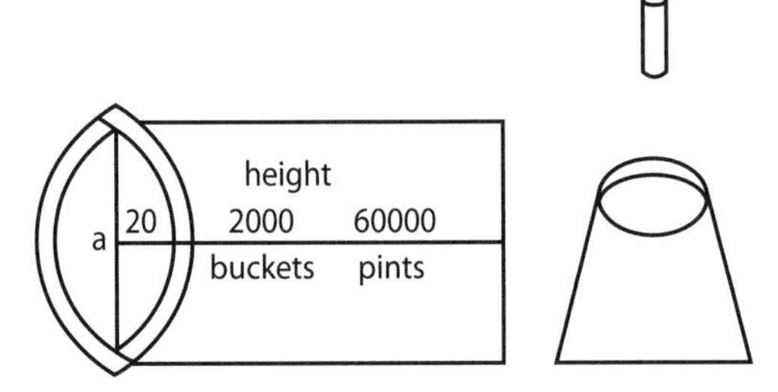

Fig. II-4-6-3.

[line representing the] depth of the cistern. Now you multiply the number representing the number of times the average area of the end of the bucket is contained in the area of the mouth of the cistern by the number representing the number of times the height of the bucket is contained in the depth of the cistern, and the product is the number denoting how many times the cistern contains the bucket. Thus, if the average area of the end of the bucket is contained in the area of the mouth of the cistern 20 times and the height of the bucket is contained in the depth of the cistern 100 times, you multiply 20 by 100 and get 2000 [Fig. II-4-6-3]. I say that the cistern holds 2000 buckets of water. And if you wish to know how many pints [there are in the cistern], you operate in the same way. See how many pints the bucket holds and multiply that number by the number denoting how many buckets the cistern holds. The product is the number which denotes how many pints a cistern holds. Thus, if the bucket which the cistern contains 2000 times contains 30 pints, multiply 30 by 2000 to get 60,000, the number of pints in the cistern. You proceed in the same way if you wish to know how many glasses there are in the cistern; or how many round measures there are, however small they may be.

II-5. Trigonometry

The origin of indirect measurement is lost in antiquity. While shadow reckoning today is easily taught to children, learning the complexities and use of trigonometric functions requires more sophisticated motivation. Trigonometry began in the medieval world when

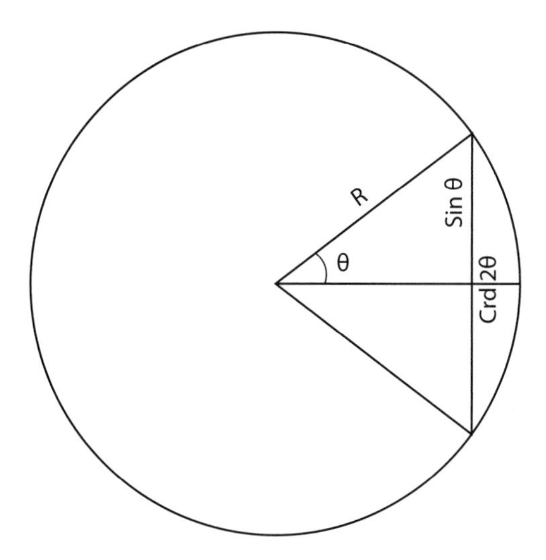

Fig. II-5-1-1.

translators, including Gerard of Cremona, brought Ptolemy's *Almagest* as well as Arabic texts on the mathematical basics of astronomical calculation to the Western world. Greek trigonometry was based on the chord. Accordingly, Ptolemy explained how the length of a chord corresponding to its arc can be calculated. In India, and later in the Arabic-speaking parts of the world, the chord was replaced by the Sine. There is a simple relation between the chord and the Sine, the Sine of an angle being half the chord of the double angle in a circle of radius R (see Fig. II-5-1-1). In the Latin Middle Ages, both concepts were used side by side. (In what follows, we use "Sine" to represent this medieval concept of a line segment in a circle of given radius, reserving "sine" to represent the modern concept.)

Our term *sinus* is derived from the word *jīvā*, which means "chord" in Sanskrit. The Arabs added it to their technical vocabulary. However, there was a genuine Arabic word with the same consonants: *jayb* (= "the opening in a garment for the head to be put through; breast"). So it is understandable that the translators from Arabic into Latin understood the Arabic term for "sine" in this way and gave its equivalent not as *chorda*, but as *sinus*.

Scholars in Spain, France, and England developed tables for astronomers. Here we can appreciate some of what Ptolemy accomplished; see how Leonardo of Pisa used these ideas to measure fields; consider the Sine tables created by a Parisian professor; learn how an English Abbot taught the functions known to him (they were not called functions, of course); and study the use of markings on the back of an astrolabe, as described by an English poet.

1. PTOLEMY, *ON THE SIZE OF CHORDS IN A CIRCLE*

Claudius Ptolemaeus, or Ptolemy as he is usually called, was born around 100 CE, probably in Egypt, and he certainly spent most of his life there, dying around 170. Nothing more is known of his personal life. Fortunately, his professional writings are well documented. His major works include *Almagest* and *Planisphaerium* in astronomy; *Apotelesmatika* and *Karpos* in astrology; and the voluminous *Geography*, which explains map projection. The treatises *Optics* and *Music* are among his minor works.

Ptolemy's most influential work is today called the *Almagest*. Ptolemy's own Greek title for it was (in English) *The Mathematical Collection*, later called *The Great Collection* in the

Muslim world, that is, *Kitāb al-majistī* in Arabic, from which the familiar title comes. This is a manual covering the whole of mathematical astronomy as the ancient world conceived it. He assumes in the reader nothing beyond a knowledge of Euclidean geometry and an understanding of common astronomical terms. Starting from first principles, he guides the reader through the prerequisite cosmological and mathematical apparatus to an exposition of the theory of the motion of those heavenly bodies known to the ancients: the sun, the moon, Mercury, Venus, Mars, Jupiter, Saturn, and the fixed stars. For each body, Ptolemy describes the types of phenomena that have to be accounted for, proposes an appropriate geometric model, derives the numerical parameters from selected observations, and finally constructs tables enabling one to determine the motion or phenomenon in question for a given date. He was careful to acknowledge his debts to Hipparchus and others. Yet all their work would have had little effect had Ptolemy not composed his *Almagest*. It must be considered an original work. Indeed, the Ptolemaic system was named after the right man. Among his contributions to trigonometry are the equivalents of identities, as illustrated below. He derived them from the seven theorems and one corollary that he used in compiling his table of chords. The table of chords enabled him to plot the distance between the horizon and a star and between stars.

To compile his table of chords, Ptolemy assumed that the circumference of a circle was divided in 360 parts (this from the Babylonians) and its diameter in 120 parts (this from Hipparchus). Furthermore, the length of the chords for 60°, 90°, and 120° arose from obvious Euclidean constructions. His first three theorems are presented here.

For the user's convenience, then, we shall subsequently set out a table of their [the chords'] amounts, dividing the circumference into 360 parts, and tabulating the chords subtended by the arcs at intervals of half a degree, expressing each as a number of parts in a system where the diameter is divided into 120 parts. [We adopt this norm] because of its arithmetical convenience, which will become apparent from the actual calculations. But first we shall show how one can undertake the calculation of their amounts by a simple and rapid method, using as few theorems as possible, the same set for all. We do this so that we may not merely have the amounts of the chords tabulated unchecked, but may also readily undertake to verify them by computing them by a strict geometrical method. In general we shall use the sexagesimal system for our arithmetical computations, because of the awkwardness of the [conventional] fractional system. Since we always aim at a good approximation, we will carry out multiplications and divisions only as far as to achieve a result which differs from the precision achievable by the senses by a negligible amount.

First, then, let there be a semicircle ABC on the diameter ADC and around center D [Fig. II-5-1-2]. Draw DB perpendicular to AC at D. Let DC be bisected at E, and EB be joined; and let EF be made equal to EB, and let FB be joined. I say that FD is the side of a [regular inscribed] decagon, and BF the side of a [regular] pentagon.

For since the straight line DC is bisected at E and a straight line DF is added to it, rect. CF, FD + sq. ED = sq. EF, [*Elements* II.6]. But sq. EF = sq. BE, since BE = EF, and sq. ED + sq. DB = sq. BE [*Elements* I.47]. Therefore rect. CF, FD + sq. ED = sq. ED + sq. DB. And, subtracting the common square on ED, rect. CF, FD = sq. DB = sq. DC. Therefore CF is cut at D in extreme and mean ratio [*Elements* VI, def. 3]. Now since the side of the hexagon and the side of the decagon, when both are inscribed in the same circle, make up the extreme and mean ratios of the same straight line [*Elements* XIII.9], and since DC,

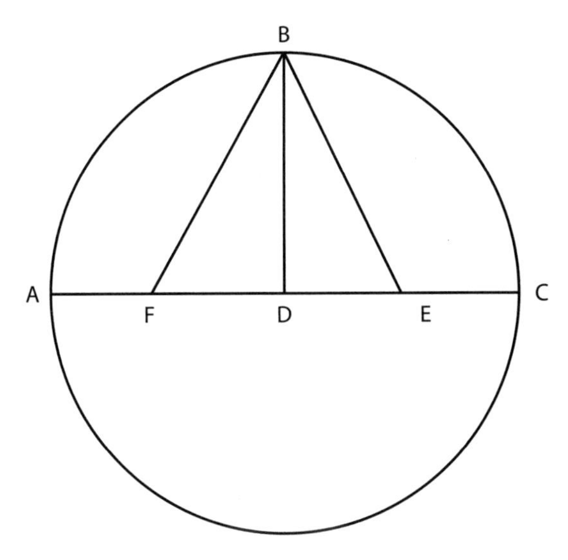

Fig. II-5-1-2.

being a radius, represents the side of the hexagon [*Elements* IV.15 corollary], therefore FD is equal to the side of the decagon.

Similarly, since the square on the side of the pentagon is equal to the square on the side of the hexagon together with the square on the side of the decagon when all are inscribed in the same circle [*Elements* XIII.10], and since in the right triangle BDF, sq. BF = sq. DB + sq. FD where DB is the side of the hexagon and FD the side of the decagon, it follows that BF equals the side of the pentagon.

Since, then, as I said, we set the diameter of the circle as 120 parts, it follows from the above, since ED is half the radius, that ED = 30^p, and sq. ED = 900; also rad. DB = 60^p, and sq. DB = 3600. Then sq. BE = sq. EF = 4500. So EF = $67;4,55^p$, and by subtraction FD = $37;4,55^p$. Therefore the side of the decagon, subtending an arc of 36°, has $37;4,55^p$, where the diameter has 120^p. Again, since FD = $37;4,55^p$, sq. FD = $1375;4,14^p$ and sq. DB = 3600^p, so sq. BF = sq. FD + sq. DB = 4975;4,14 and therefore BF = $70;32,3^p$. Therefore the side of the pentagon, which subtends 72°, contains $70;32,3^p$, where the diameter has 120^p.

Thus, Ptolemy has derived the lengths of the distance in parts between two heavenly objects that are either 36° or 72° apart. Assuming as usual the distance for 60°, he then derived the lengths of chords subtending arcs of 90°, 120°, and 144°.

It is immediately obvious that the side of the hexagon, which subtends 60° and is equal to the radius, contains 60^p. Similarly, since the side of the [inscribed] square, which subtends 90°, is equal, when squared, to twice the square on the radius, and since the side of the [inscribed equilateral] triangle, which subtends 120°, is equal, when squared, to three times the square on the radius, and since the square on the radius is 3600^p, we compute that the square on the side of the square is 7200^p and the square on the side of the [equilateral] triangle is 10800^p. And so the chord of an arc of 90° is $84;51,10^p$, and the chord of an arc of 120° is $103;55,23^p$, where the diameter is 120^p.

We can, then, consider the above chords as established individually by the above straightforward procedures. It will immediately be obvious that if any chord be given, the

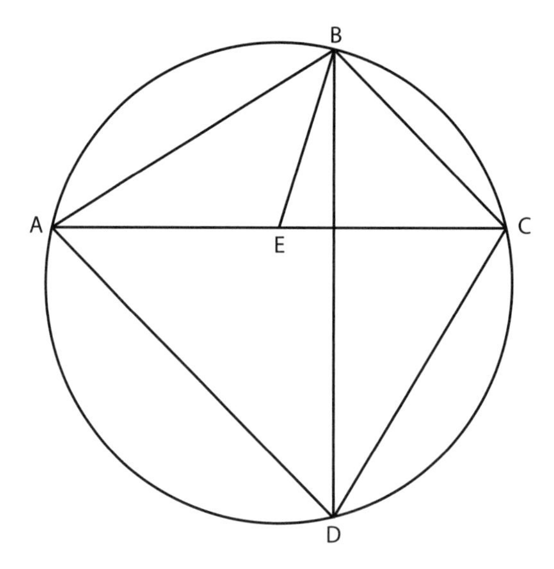

Fig. II-5-1-3.

chord of the supplementary arc is given in a simple fashion, since the sum of their squares equals the square on the diameter. For instance, since the chord of 36° was shown to be 37;4,55ᵖ, and the square of this is 1375;4,15ᵖ, and the square on the diameter is 14400ᵖ, the square on the chord of the supplementary arc (which is 144°) will be the difference, namely = 13024;55,45ᵖ, and so the chord of 144° equals 114;7,37. Similarly for the other chords [of supplements].

We shall next show how the remaining individual chords can be derived from the above [chords], first of all setting out a theorem that is extremely useful for the matter at hand.

Let there be a circle with an arbitrary quadrilateral ABCD inscribed in it, and let AC and BD be joined. We must prove that rect. AC,BD = rect. AB,DC + rect. AD,BC [Fig. II-5-1-3]. For let angle ABE be made equal to angle DBC. If then we add the common angle EBD, angle ABD = angle EBC. But also angle BDA = angle BCE, since they subtend the same arc [*Elements* III.21]. Then triangle ABD is equiangular with triangle BCE. Hence BC:CE = BD:AD [*Elements* VI.4]. Therefore rect. BC,AD = rect. BD,CE [*Elements* VI.16]. Again, since angle ABE = angle CBD and also angle BAE = angle BDC, therefore triangle ABE is equiangular with triangle BCD. Hence AB:AE = BD:CD. Therefore rect. AB,CD = rect. BD,AE. But it was shown that rect. BC,AD = rect. BD,CE. Therefore, by addition, rect. AC,BD = rect. AB,DC + rect. AD,BC.

In high school geometry books, this last theorem is often called "Ptolemy's Theorem," unfortunately without reference to its use. First change Figure II-5-1-3 into Figure II-5-1-4 by removing line BE, raising line AD to become the diameter of the circle with center at O, and drawing lines BO and CO. Then let angle AOB be $\angle\alpha$ and angle AOC be $\angle\beta$. To apply Ptolemy's Theorem, we have $AB = \text{chord } \alpha$, $AC = \text{chord } \beta$, $BD = \text{chord } (180^\circ - \alpha)$, $AD = 120$, $CD = \text{chord } (180^\circ - \beta)$, and $BC = \text{chord } (\beta - \alpha)$. Then, by substitution into

$$AD \cdot BC = AC \cdot BD - AB \cdot CD,$$

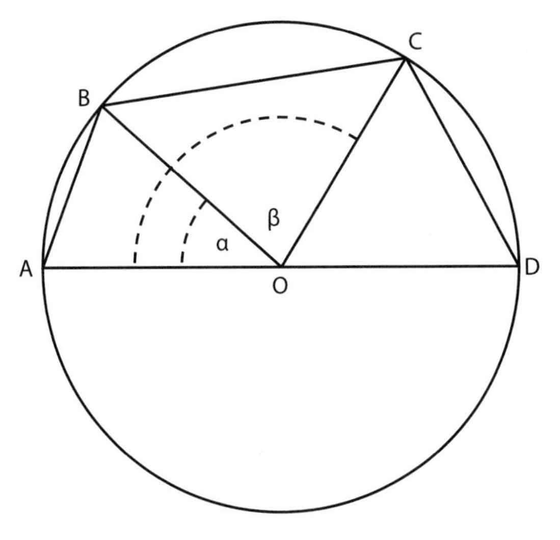

Fig. II-5-1-4.

we obtain the identity

$$120\,\text{chord}(\beta-\alpha) = (\text{chord}\,\beta)(\text{chord}(180^\circ-\alpha)) - (\text{chord}\,\alpha)(\text{chord}(180^\circ-\beta)),$$

which is equivalent to

$$\sin(\beta-\alpha) = (\sin\beta)(\cos\alpha) - (\sin\alpha)(\cos\beta).$$

Given this theorem, another theorem equivalent to our half-angle formula for the sine, and some clever approximation procedures, Ptolemy was able to produce the table he wanted, a table of chords to given arcs from 0° to 180° in a circle of radius 60, at intervals of one-half degree.[86]

2. LEONARDO OF PISA, *DE PRACTICA GEOMETRIE* (*PRACTICAL GEOMETRY*)

In chapter 3 of the *Practical Geometry*, Leonardo wanted to calculate lengths of arcs of circles as well as areas of circular segments. He was aware of Ptolemy's work in developing the chord table and, in fact, reproduced Ptolemy's proof of "Ptolemy's Theorem." But rather than use Ptolemy's table for the relationship between chords and arcs, he created his own, based on a radius of 21 and, therefore, a semicircumference of 66 ($= (22/7) \times 21$).

[211] For those who wish to find the measure of arcs by known chords, I have composed the following table. I constructed the ordinates of 66 known arcs in it. Further I have paired each with its chord in terms of rods, feet, inches, and points. Recall that one rod is 6 feet, a foot is 18 inches, and an inch is 20 points; or, a rod is 108 inches or 2160 points.[87] Be it understood that those 66 arcs are within a semicircle whose diameter is 42 rods, and each chord drawn in the semicircle is the chord of two equal arcs if the chord is the diameter, or of unequal arcs if the chord is not the diameter. Therefore I have set aside two arcs for each chord, as is shown in the following table [Fig. II-5-2-1].

[86] See [Toomer, 1998, pp. 57–60].

[87] These are Pisan measures.

Table of Arcs and Chords

Arcus pertice	Arcus pertice	Corde pertice	Ar pedes	Cun vncie	In puncta	Arcus pertice	Arcus pertice	Corde pertice	Ar pedes	CV vncie	VM puncta
1	131	0	5	17	17	34	98	30	2	6	17
2	130	1	5	17	13	35	97	31	0	8	5
3	129	2	5	17	4	36	96	31	4	8	7
4	128	3	5	17	2	37	95	32	2	5	15
5	127	4	4	12	10	38	94	33	0	1	9
6	126	5	5	16	7	39	93	34	3	13	0
7	125	6	5	14	5	40	92	35	1	4	15
8	124	7	5	12	9	41	91	35	4	12	10
9	123	8	5	8	16	42	90	36	2	0	0
10	122	9	5	7	8	43	89	36	5	3	5
11	121	10	5	4	2	44	88	37	2	4	6
12	120	11	4	17	18	45	87	37	5	3	2
13	119	12	4	13	6	46	86	38	1	17	15
14	118	13	4	7	16	47	85	38	4	12	13
15	117	14	4	1	0	48	84	38	1	4	0
16	116	15	3	11	18	49	83	39	3	11	15
17	115	16	3	3	12	50	82	39	5	17	2
18	114	17	2	12	8	51	81	40	2	2	1
19	113	18	2	0	15	52	80	40	4	2	10
20	112	19	1	8	12	53	79	40	0	0	11
21	111	20	0	13	18	54	78	40	1	14	5
22	110	21	0	0	0	55	77	41	3	7	8
23	109	21	5	2	16	56	76	41	4	16	2
24	108	22	4	4	5	57	75	41	0	4	12
25	107	23	3	4	8	58	74	41	1	8	1
26	106	24	2	3	2	59	73	41	2	9	0
27	105	25	1	0	6	60	72	41	3	7	14
28	104	25	5	16	2	61	71	41	4	9	2
29	103	26	4	8	0	62	70	41	4	15	10
30	102	27	3	0	3	63	69	41	5	6	9
31	101	28	1	9	7	64	68	41	5	12	17
32	100	28	5	16	4	65	67	41	5	6	14
33	99	29	4	3	9	66	66	42	0	0	0

Fig. II-5-2-1.

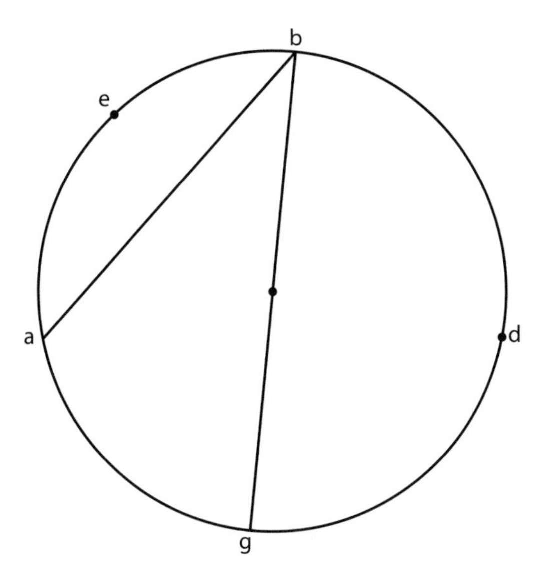

Fig. II-5-2-2.

[214] If you wish to find the arc by its chord when you know both the chord and the diameter of a circle, multiply the chord by 42, the diameter in the table, and divide the product by the diameter of the known circle. What results will be a chord in the table similar to the given chord. Take its arc from the table and multiply by the diameter of the given circle. Divide the product by the diameter of the table, and you will have the arc of the chord you want. . . .

[215] Let the chord *ab* be 8 rods, 3 feet, and 16 2/7 inches, the diameter *bg* being 10. So as we said: multiply 8 rods, 3 feet, and 16 2/7 inches by 42, and divide the product by 10 to get 36 rods and 2 feet for a chord in the table corresponding to the given chord *ba* [Fig. II-5-2-2]. [This is adjacent to] 42. Multiply 42 by 10, the diameter of the circle, and divide the product by 42 to obtain 10 for the measure of minor arc *aeb*. If you wish to know the measure of the major arc *bdga*, multiply the 90 [adjacent to 36 rods and 2 feet] by 10 and divide by 42, the diameter for the table, to obtain 21 3/7 rods for the measure of the major arc.

3. JOHANNES DE LINERIIS, *CANONES*

Among the astronomical writings of Johannes de Lineriis (section II-2-4) are *De Saphea* and *Aequatorium*, treatises on the use of certain astronomical instruments. In addition, there are the *Canones* to his astronomical tables (i.e., instructions for using the tables). These *Canones* were so well received that by 1405, students studying astronomy at the University of Bologna were required to read them during their second year [Malagola, 1888, p. 276]. In the selection that follows, one word may seem out of place: "equation." It certainly does not signify an equation in the algebraic sense. Rather, in astronomy, the word means the reduction of a number to a useful value.

The Canons Begin of the Tables of the Prime Mobile of Master John de Lineriis.

I Given any arc to find its right Sine.

The *right Sine* is half of the chord of the double arc.

The *versed Sine* is the part of the diameter contained between the arc and the given chord, passing through middle of the chord and cutting it perpendicularly.

The arc therefore, whose Sine you seek, is either more or less than 180 degrees. If it is more, then subtract 180 degrees from it and work with the remainder. If however it is less, then work with it. Look for the arc in the *Table of Half Chords* [Fig. II-5-3-1] that increases by half a degree. If you find it exactly, then directly opposite is the right Sine of the given arc.

If however you cannot find the arc exactly—this happens when degrees and minutes of the given arc have more or less than 30 minutes, then in the aforementioned *Table* take the Sine of the arc that is closest and less than it directly across and set it aside. Then take the Sine of the arc that is closest and more than it and write it below the other. Afterwards find the difference between the first and the second by subtracting the smaller from the larger. From this difference take the proportional part according to the ratio of minutes contained in the given arc, either from less than 30 to 30, or from more than 30 to 30, according to Proposition One of Book I of this treatise or to Proposition Two of Book II. Add the proportional part to the equation of the Sine that you found before, if it is less than the second reading, or subtract it if it is greater. Thus you will have the equation of the Sine of the given arc.

Be aware that 30 is the first number. The [number of] minutes contained in the given arc less or more than 30 is the second number. And the difference of the two entry numbers is the third number.

II Given any right Sine to find its arc.

Look for the given Sine in the *Tables*. If it is there exactly, then take that which is in the first line of the two lines of the number, because it is the arc of the given Sine.

But if you cannot find the exact given Sine, then use the Sine that is closest and less than it and take the arc on its line. Subtract this Sine from the given Sine, and set aside the difference because it is the difference between the smaller Sine found in the *Table* and the given Sine. This is the second number. Then take the difference, which is between the smaller Sine closer to the given one in the *Table* and the larger Sine that exceeds the given one in the *Table*, and subtract the less from the greater. This is the first number. And the thirty minutes, by which the *Table* is increased, is the third number. And so take that part of 30 minutes according to the ratio of the second number to the first, and multiply the second number by the third and divide the product by the first number. Add the result to the [measure of the] arc that you had set aside, and you will have what you sought.

III Given any arc to find its versed Sine.

If the number of degrees of the given arc is less than 90, then subtract it from 90 and find the right Sine of the remainder by the Proposition I above. Then subtract it from 60, which is the whole right Sine. What remains is the versed Sine of the given arc or the versed chord.

If the given arc is more than 90, take the difference between it and 90, and you know the right Sine by the Proposition I above. Add it to 60 (which is half the diameter), and the sum is the versed Sine of the given arc.

Incipiunt tabule sinuum et cordarum, ascensionum signorum, nec non eclipsium et aliorum quamplurium, quas composuit magister Johannes de Lineriis, Picardus, dyocesis Ambianensis anno domini M° ccc° 22° [1])

Sequitur Tabula Sinus.[2])

Arcus augmentati per dimidium gradum				Corde mediate			Arcus augmentati per dimidium gradum				Corde mediate		
Gr.	M^a	Gr.	M^a	Gr.	M^a	2^a	Gr.	M^a	Gr.	M^a	Gr.	M^a	2^a
0	30	179	30	0	31	25	75	30	104	30	58	5	20
1	0	179	0	1	2	50	76	0	104	0	58	13	4
1	30	178	30	1	34	14	76	30	103	30	58	20	32
2	0	178	0	2	5	37	77	0	103	0	58	27	44
2	30	177	30	2	37	2	77	30	102	30	58	34	40
3	0	177	0	3	8	25	78	0	102	0	58	41	21
3	30	176	30	3	39	45	78	30	101	30	58	47	44
4	0	176	0	4	11	7	79	0	101	0	58	53	51
4	30	175	30	4	42	28	79	30	100	30	58	59	53
5	0	175	0	5	13	46	80	0	100	0	59	9	18
5	30	174	30	5	45	2	80	30	99	30	59	10	38
6	0	174	0	6	16	20	81	0	99	0	59	15	40
6	30	173	30	6	47	32	81	30	98	30	59	20	27
7	0	173	0	7	18	44	82	0	98	0	59	24	50
7	30	172	30	7	49	53	82	30	97	30	59	29	12
8	0	172	0	8	21	1	83	0	97	0	59	33	13
8	30	171	30	8	52	7	83	30	96	30	59	36	51
9	0	171	0	9	23	10	84	0	96	0	59	40	16
9	30	170	30	9	54	10	84	30	95	30	59	43	25
10	0	170	0	10	25	8	85	0	95	0	59	46	18
10	30	169	30	10	56	3	85	30	94	30	59	48	54
11	0	169	0	11	26	54	86	0	94	0	59	51	14
11	30	168	30	11	57	43	86	30	93	30	59	53	17
12	0	168	0	12	28	29	87	0	93	0	59	55	4
12	30	167	30	12	59	11	87	30	92	30	59	56	35
13	0	167	0	13	29	49	88	0	92	0	59	57	49
13	30	166	30	14	0	24	88	30	91	30	59	58	49
14	0	166	0	14	30	55	89	0	91	0	59	59	37
14	30	165	30	15	1	22	89	30	90	30	59	59	52
15	0	165	0	15	31	45	90	0	90	0	60	0	0

Fig. II-5-3-1.

IV Given the versed Sine to find its arc.

If the versed Sine is less than 60, subtract it from 60 and you have the Sine of the complement of the arc. Find the arc by Proposition II above. Subtract this from 90 and what is left is what you want.

But if the versed Sine is more than 60, subtract 60 from it, and you have the Sine of the difference of the arc and 90. Find the arc by Proposition II above. Add this to 90 degrees and the sum is the arc of the given Sine.

4. RICHARD OF WALLINGFORD, *QUADRIPARTITUM*

Richard was born in Wallingford, a few miles southeast of Oxford in England, probably in 1291. Ten years later he was orphaned by the death of his father but was soon adopted by the prior of the nearby Benedictine monastery. This relationship may explain his attendance at the Benedictine college at Oxford for six years to study grammar and philosophy. He left there at the age of 22 to enter the premier Benedictine Abbey at St. Albans. Ordained a priest in 1317, he returned to Oxford for more philosophy and theology. Knowing his interest in natural philosophy, it is difficult to imagine that he did not visit with the scholars at Merton College, where the Calculators were in sway. Nine years later, he was licensed to lecture on Peter Lombard's *Sentences*. He became Abbot of St. Albans in 1327 and died in 1336, hardly 45 years of age, of a skin disease. Early on he was known as the "Clockmaker" because of his tract *A Treatise on the Astronomical Clock*, showing the movements of the planets.

Sometime before 1326 he composed the *Quadripartitum: Four Treatises on Right and Versed Chords and on the Proving of Sines*. The tract was a systematic development of propositions rather than haphazard insertions of theorems where needed. His basic template was Ptolemy's *Almagest*. "Richard had produced what was to remain the most comprehensive trigonometric corpus written in the Latin West before the final decades of the fifteenth century" [North, 1976, vol. I, pp. 3–19]. The Prologue and three results from Book I follow.

Here begins the treatise on the demonstration of Sines.

[Prologue] Because the Canons[88] do not completely pass on information about the Sine or its use, I intend to provide this in this four-part work. In the first part I shall speak plainly about the relationship of a circle to its diameter and about any arc of a circle to its chord. In the second part I will show the ratio of any Sine to another Sine. I shall manifest it with given numbers, because a ratio of geometric magnitudes cannot be known except by [the theorems of] the fifth book of Euclid borrowed from arithmetic. In the third tract I shall show that what is proved of commensurable numbers is true also of lines. Thus each chapter of Part II shall have a corresponding chapter in Part III, because there is one principle of proof common to both. In the fourth part I shall turn to proportion and show that the ratio of arcs is related to the ratio of their chords so that knowing the ratio of the chords no one shall be ignorant of the ratio of the arcs. And these are the complete contents.

And so we have the right Sine, the versed Sine, and the double Sine. The double Sine is a straight line, the ends of which lie on the two ends of an arc of a circle, and they are related as a chord to its arc. The right Sine is half the double chord with respect to half of its arc. The versed Sine is always that part of the diameter that cuts both the double Sine and its arc. It is like an arrow in relation to the chord and its arc.

For example, let there be circle *ABD* with center *O*, in which is drawn diameter *AC*. Upon it at point *E* falls the perpendicular *BD* [Fig. II-5-4-1]. And so I say that line *BED* is the double Sine of the very arc *BAD*, and line *BE* is the right Sine of arc *BA*, and line *EA* is the versed Sine at once of arc *BA* and arc *AD*, as is obvious in the first circle.

With the definitions of terms clear, Richard transits to his propositions. Numerous sentences employ the word "and" to indicate addition of the quantities so joined.

Proposition I: Knowing the right Sine of any arc leads to knowing the versed Sine of the arc.

Let circle *CGPE* and right Sine *FE* of arc *CE* be known to me. I say that from this information I can know *CF* that is the versed Sine of the same arc [Fig. II-5-4-2].

Proof: Because the product of *EF* by itself (by Euclid III.34)[89] equals the product of *CF* by *FP*, therefore (by II.1) it equals the products of *CF* by line *FD* and of *CF* by *DP*. And

[88] Richard was possibly referring to his own *Canons to the Tables* of John Maudith [North, 1976, vol. I, pp. 1–20].
[89] In Heath's edition, this is *Elements* III.35.

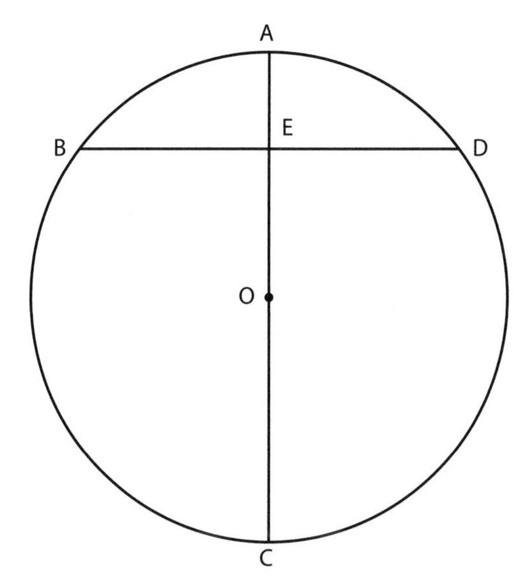

Fig. II-5-4-1.

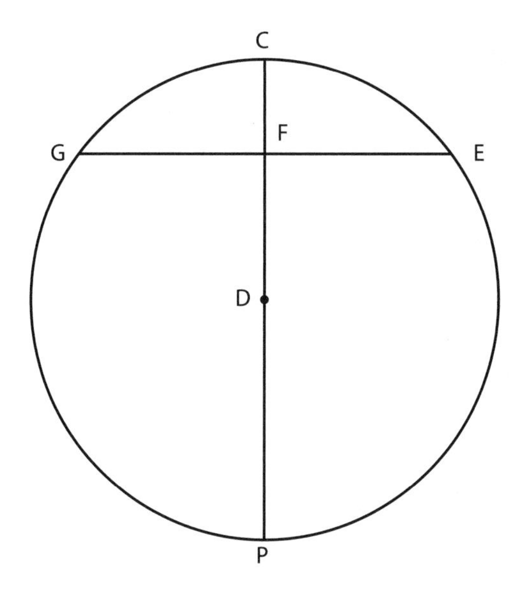

Fig. II-5-4-2.

because of this it equals the products of *CF* by *FD* and of *CF* by *CD*, since *CD* and *DP* are equal. But the product of *CF* by *CD* (by the same II.1 or by II.3) equals the products of *CF* by itself and of *CF* by *FD*. Therefore the product of *EF* by itself equals the product of *CF* by itself and twice the product of *CF* by *FD*. But the product of *CD* by itself (by II.4) equals the products of *CF* by itself and of *FD* by itself and twice the product of *CF* by *FD*. Therefore the product of *FE* by itself is less than the product of *CD* by itself by the product of *FD* by itself.

Therefore, since *CD* is known to me (for it is the whole Sine of 150 minutes, since it is half the diameter), I shall take therefore the square of *CD* and subtract from it the square of *FE* that is known to me; and the square of *FD* shall remain. And so I shall find the root of this square to be the quantity of line *FD*. Therefore I shall subtract from line *CD* the line *FD* now known to me; and line *FC* shall remain known to me. This is what was proposed: knowing *FE* the right Sine leads to knowing *CF* the versed Sine of the same arc.

Since Richard begins Proposition II with the versed Sine given, on studying the beginning of the proof, the reader will understand that *FC* is the versed Sine and *FE* the right Sine without Richard having to identify each as such.

Proposition II: Knowing the versed Sine of any arc you must not be ignorant of its right Sine.

This proposition is the converse of the preceding. Keep the arrangement of the figure as before. Therefore I say that if the versed Sine *FC* were known to me, then the Sine *FE* will necessarily be known. This is evident from what follows.

I will subtract *CF* from all of *CD*, and *FD* will remain. Therefore, since it has been demonstrated [in Proposition 1] that the square on *FE* is less than the square on *CD* by the square on *FD*, I will subtract the square on *FD* from the square on *CD*, and look for the root of what remains. This is necessarily the quantity of line *FE*, the right Sine of arc *CE*. And this is what I took care to show.

In the next several propositions, Richard follows Ptolemy as he proposes to construct a table of Sines. Thus, he gives equivalents to the half-angle formula and the sum and difference formulas and even uses Ptolemy's own approximation technique to calculate the Sine of 1°. But then he proposes two other methods, with the third probably being unique to him.

Thirdly, and most accurately, and without error such as would be detected by the mind's eye, you will proceed as follows for all [the calculations instanced in] this [treatise]. Having earlier found Sin $3/16^\circ$ by I-3 [the half-angle formula], you will find Sin $3/32^\circ$, the Sine of half the arc, [and Sin $3/64^\circ$]. Now find $\text{Sin}(3/64 + 15/16)^\circ$ by I-7 [the sum formula], which—as I am quite sure—is to say find $\text{Sin}(63/64)^\circ$. Lastly, halve the arc of $3/64^\circ$, giving $3/128^\circ$, whose Sine you will find by I-3. Next you will halve the arc of this. By I-3, seek Sin $3/256^\circ$. Find therefore $\text{Sin}(63/64 + 3/256)^\circ$, by I-7, that is to say, you will find $\text{Sin}(1 - 1/256)^\circ$. Again resume the work, take Sin $3/256^\circ$, and by I-3 seek Sin $3/512^\circ$. Likewise by I-3 seek the Sine of half this arc, that is of $3/1024^\circ$. By I-7 you will find $\text{Sin}(255/256 + 3/1024)^\circ$, or $\text{Sin}(1 - 1/1024)^\circ$. And proceed in this way even to the 9000th part of a degree, or even to the infinitely small, if by working minutely you wish to do so. This method of working is the most precise, but the first is the more to be recommended.

5. GEOFFREY CHAUCER, *A TREATISE ON THE ASTROLABE*

In medieval Europe, the most common instrument for indirect measurement, whether in the heavens or on earth, was the astrolabe. An exceptional medieval treatise describing this is from an unexpected source. Better known for his *Canterbury Tales*, Geoffrey Chaucer (ca. 1343–1400) also composed a scientific book, *A Treatise on the Astrolabe*. It describes the parts of the astrolabe and how to use it. In itself, this treatise is a unique piece of scientific writing, outstanding for its clarity and completeness. The part that interests us describes a numbered U-shaped figure on the back of the astrolabe called the "shadow square." It hangs as it were

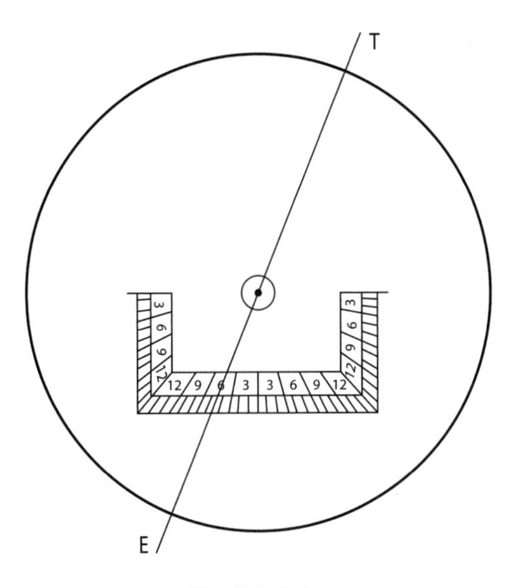

Fig. II-5-5-1.

from a horizontal diagonal through the center of the face of the instrument [Fig. II-5-5-1]. The two vertical arms of the U represent the *umbra versa*. The base, twice as long as either upright, consists of two *umbra recta* reflected away from one another. The four *umbrae* are each numbered 3, 6, 9, 12 for a total of twelve "points" along each *umbra*. The diagonal line through the figure represents a ruler, along which a person at E sights through holes on the ruler to T the top of, say, a tower. The mathematics is explained after the excerpt.[90]

41. If it so be that you would work by *umbra recta* and you may come to the base of the tower, in this manner you shall work. Take the altitude of the tower by both holes so that your rule lies even in a point. For example: I see it through at the point of 4. Then I measure the space between me and the tower, and I find it 20 feet. Then notice how 4 is to 12 and you shall find that it is the third part of 12. Right so is the space between you and the tower the third part of the height of the tower. Then thrice 20 feet is the height of the tower with adding of your person to your eye. And this rule is general in *umbra recta* for the point of 1 to 12. And if the rule falls upon 5, then 5 is 12 parts the height of the space between you and the tower, with adding of your person to your eye.

The problem posed above leads to a simple proportion. Someone is standing, say, $d = 20$ feet from the base of a tower. The sighting to the top of the tower has the ruler at point $r = 4$ on the horizontal or *umbra recta* scale. Hence $4{:}12 = 20{:}h$. Or in general, $h = 12d/r$. Of course, the height of the observer's eye above the ground is added to h for the true height of the tower.

Not so simple is the situation where the distance to the base of the tower cannot be measured. While the problem can be solved using the *umbra recta,* the *umbra versa* is employed here. To make sure that his reader is following the sequence of steps, Chaucer summarizes all of them briefly at the end of his instructions.

[90]For similar calculations with an astrolabe, see section III-1 in Chapter 2 as well as section I-4-5 above.

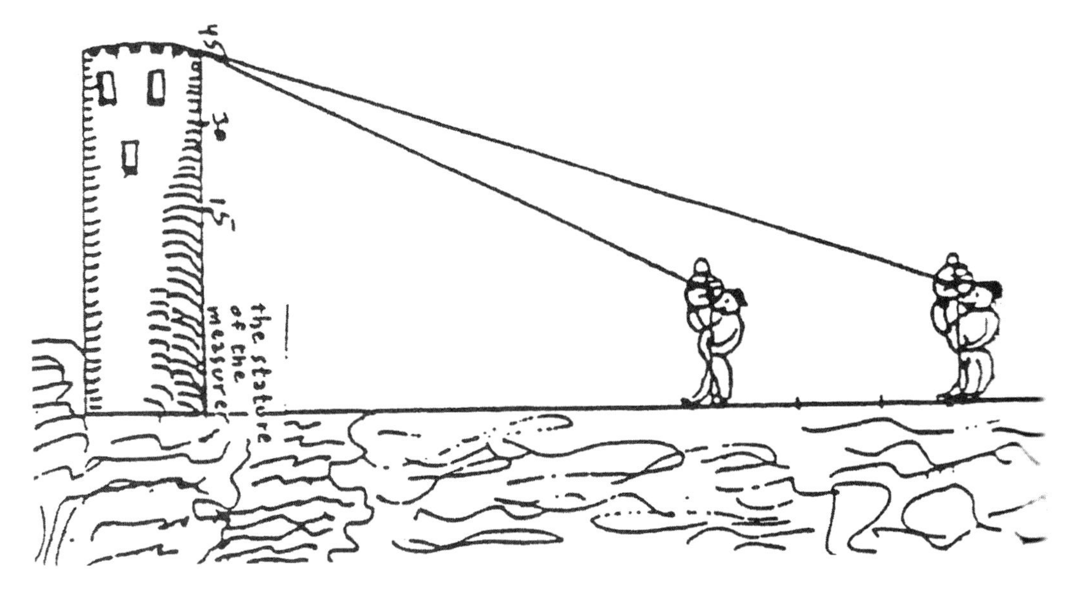

Fig. II-5-5-2.

42. A way of working by *umbra versa*. If it so be that you may not come to the base of the tower, I see it through [the holes] at the number of 1. I set there a prick at my foot. Then go I near to the tower, and I see it through the point of 2, and there set I another prick. And then behold how 1 has it to 12, and you shall find that it has it 12 times. Then you shall know that 12 passes 6 [by] the number 6. Right so the distance between your two pricks is the distance of 6 times your height [of the tower]. And note that at the first height of 1, you set a prick. And afterward when you see it at 2, there you set another prick. Then you find between these two pricks 60 feet. Then you shall find that 10 is the 6th part of 60. And then 10 feet is the height of the tower.

Since the distance to the base of the tower is inaccessible, two sightings r_1 and r_2 with the ruler on the vertical scale together with d the distance between the sightings are required to find h the height of the tower. In general $h = d/[(12/r_1) - (12/r_2)]$, where it is necessary to add the height of the observer's eye from the ground for the true height of the tower. (Note that the numbers in Fig. II-5-5-2 do not match the numbers in the problem, the difference arising from different manuscripts, one for the text and the other for the illustration.)

For more information on the use of the astrolabe as described by Fibonacci in his *Practica geometrie*, see [Hughes, 2008, pp. 349 ff].

6. REGIOMONTANUS, *ON TRIANGLES*

The term *trigonometry* would not enter the vocabulary until 1595, when Bartholomew Pitiscus (1561–1613) titled his book, *Trigonometria sive de solutione triangulorum tractatus brevis et perspicuus*. However, as we have seen earlier, the ideas of trigonometry were used by Ptolemy in his *Almagest* and were well thought through by Islamic scholars. These ideas were then translated from the Arabic to be reworked in the West until studied by Johannes Müller (1436–1476). Müller, better known as Regiomontanus, was born in Königsberg in Franconia. He wrote *Five Books on Triangles of Every Kind* (*De triangulis omnimodis libri quinque*)

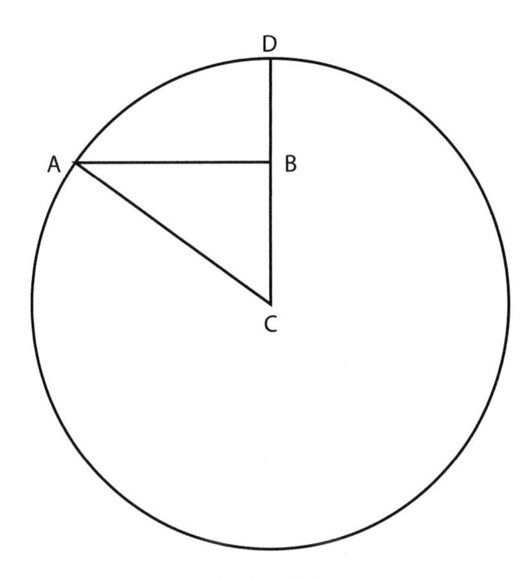

Fig. II-5-6-1.

for the most part during 1462–1464,[91] a work he conceived of as a prerequisite to the study of the *Almagest* or any other book on astronomy. We can consider the author as being on the cusp between medieval and Renaissance mathematics, given that his work was written at the end of the Middle Ages but was not published until 1533, well into the Renaissance. He was greatly influenced by his Islamic predecessors but certainly used his own creativity when writing the first book in Catholic Europe devoted to the study of both plane and spherical trigonometry as a purely mathematical subject. Among his noteworthy accomplishments was using algebra to solve trigonometric problems and explicitly formulating the Law of Sines. The basis of Regiomontanus's trigonometry is a circle with center C and two radii CA and CD with semichord AB [Fig. II-5-6-1]. The Sine or Right Sine of angle ACB is the line AB; its Sine complement is the line CB; and its Versine is the line BD. For calculations, Regiomontanus provided a table giving the Sines in a circle of radius 60,000.

We present here a sampling of Regiomontanus's work, both in plane and in spherical trigonometry. The Law of Sines begins Book II and is applied in Theorem II.4, as follows.

Theorem II.1. In every rectilinear triangle the ratio of [one] side to [another] side is as that of the Right Sine of the angle opposite one of [the sides] to the Right Sine of the angle opposite the other side.

As we said elsewhere, the Sine of an angle is the Sine of the arc subtending that angle. Moreover these Sines must be related through one and the same radius of the circle or through several equal [radii]. Thus if triangle ABG is a rectilinear triangle, then the ratio of side AB to AG is as that of the Sine of angle AGB to the Sine of angle ABG; similarly, that of side AB to BG is as that of the Sine of angle AGB to the Sine of angle BAG [Fig. II-5-6-2].

[91] The passages offered here are from the first printed edition. It differs in a number of places from Regiomontanus's surviving autograph, which is in the archive of the Academy of Science in St. Petersburg.

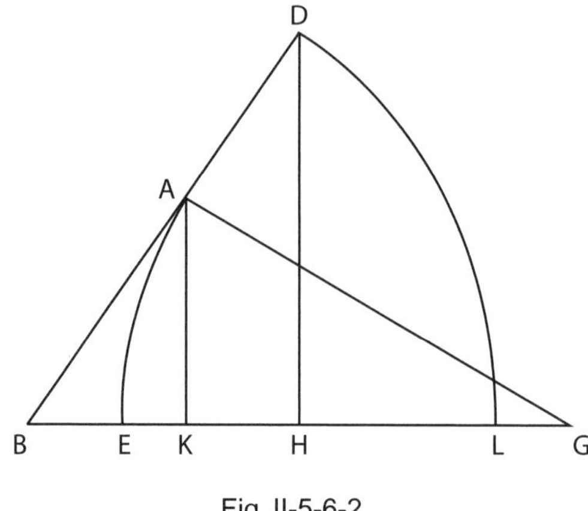

Fig. II-5-6-2.

If triangle ABG is a right triangle, we will provide the proof [directly] from Th. I.28 above.[92] However if it is not a right triangle yet the two sides AB and AG are equal, the two angles opposite [the sides] will be equal, and hence their Sines will be equal. Thus from the two sides themselves, it is established that our proposition is verified. But if one [of the two sides] is longer than the other—for example, if AG is longer—then BA is drawn all the way to D, until the whole [line] BD is equal to side AG. Then around the two points B and G as centers, two equal circles are understood to be drawn with the lengths of lines BD and GA as radii [respectively]. The circumferences of these circles intersect the bases of the triangle at points L and E, so that arc DL subtends angle DBL or ABG, and arc AE subtends angle AGE or AGB. Finally two perpendiculars AK and DH from the two points A and D fall upon the base. Now it is evident that DH is the Right Sine of angle ABG and AK is the Right Sine of angle AGB. Moreover, by VI.4 of Euclid,[93] the ratio of AB to BD, and therefore also to AG, is that of AK to DH. Hence what the proposition asserts is certain.

Theorem II.4. If in any scalene triangle two angles are given individually with any one of its sides, the other sides are easily measured.

If any two angles of triangle ABG, having three unequal sides, are given together with one of its sides—for example, AB—then the other two sides can be found [Fig. II-5-6-3].

By the application of I.32 of the *Elements,*[94] the third angle also becomes known. And since by an earlier theorem [Th. II.1 above] the ratio of the Sine of known angle AGB to the Sine of known angle ABG is as that of side AB to side AG, and since three of these quantities are known, then, except through utter ignorance of this, the fourth quantity—namely, side AG—may be found. In the same way the same will be found concerning side BG. By this method one may judge that it was unnecessary to draw any perpendicular.

Another innovation perhaps creditable to Regiomontanus is the use of algebra to solve problems in geometry. Consider this example, also from Book II, in which the words

[92] When the ratio of two sides of a right triangle is given, its angles can be ascertained.

[93] In equiangular triangles the sides about the equal angles are proportional.

[94] In any triangle . . . the three interior angles . . . are equal to two right angles.

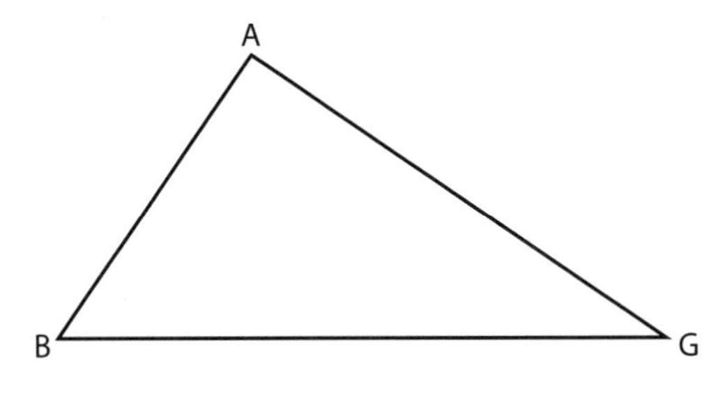

Fig. II-5-6-3.

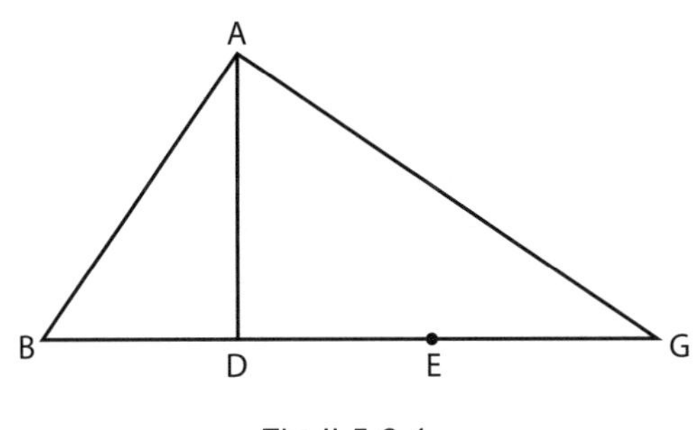

Fig. II-5-6-4.

res (thing) and *census* (square) are equivalent to our x and x^2. Note, too, the phrase he used that we call *algebra*.

Theorem II.12. If the perpendicular is given and the base and the ratio of the sides are known, each side can be found.

This problem cannot be proven by geometric means at this point, but we will endeavor to accomplish it by the craft of *res* and *census*. Thus, if triangle ABG has perpendicular AD and base BG known, and if the ratio of sides AB and AG is given, we seek each of these [Fig. II-5-6-4].

For example, let the ratio of AB to AG be as 3 to 5, so that side AB is shorter than side AG. From this, it follows, next, that segment BD is shorter than segment DG, as no one can deny. Therefore let DE be equal to BD, and let the perpendicular AD be given as 5 [feet] and the base BG as 20 feet. Take line EG as 2 *res*; hence, line BE will be 20 *minus* 2 *res*, and its half BD will be 10 *minus* 1 *res*, while the rest [of the line] will be 10 & 1 *res*. Square BD and 1 *census* & 100 *less* 20 *res* results, to which add the square of the perpendicular, namely 25, and 1 *census* & 125 *less* 20 *res* is obtained. Similarly, DG[95] squared makes 1 *census*, 20 *res* & 100, to which add the square of the perpendicular, 25, and 1 *census*, 20 *res* & 125 is obtained. Thus you will have the two squares of the lines AB and AG, whose ratio is as 9 to 25—namely, the squared ratio of 3 to 5, which was the ratio of the sides. Therefore, since the ratio of the first square to the second square is as 9 to 25, then, if the first square is multiplied by 25 and the second square by 9, the products

[95]The printed book has erroneously *BG*.

are equal. Restoring the deficits, as is customary, and subtracting equals on both sides [of the equation], we obtain 16 *census* & 2000 equal to 680 *res*. Whereupon what remains [to be done] the rules of the craft show. Therefore line GE, which was taken to be 2 *res*, will be known; hence the rest of the base, BE, and its half BD will be known. [BD] together with perpendicular AD reveals side AB. Whence, at last, side AG is pronounced known, as was promised.

The ratio the craft produced can be expressed in modern notation as

$$\frac{x^2 - 20x + 125}{x^2 + 20x + 125} = \frac{9}{25};$$

consequently, $x^2 + 125 = 42\frac{1}{2}x$, and the problem can be solved. Interestingly, Regiomontanus does not give the actual solution $x = 21\frac{1}{4} - \sqrt{326\frac{9}{16}}$, which leads to $AB \approx 8.45734$ and $AG \approx 14.09557$.

Book IV is on spherical trigonometry. The four theorems discussed here are central to the development of that subject. However, they have lengthy proofs; hence, only representative parts of the proofs are given here. To begin, Theorem IV.15 considers the relationship between the arcs of two great circles formed by the intersection of the circles. Specifically and in view of the accompanying figure, what this theorem considers are the arcs, *BR* and *GS*, which join the circumferences of two intersecting great circles. The theorem proves that the ratio of the Sine of arc *AB* to the Sine of arc *BR* is the same as the ratio of the Sine of arc *AG* to the Sine of arc *GS*. This theorem is generally called the "rule of four quantities." If for no other reason, this theorem is interesting because Regiomontanus addressed his reader.

Theorem IV.15. If, in a sphere, two great circles are inclined toward each other, and if two points are marked on the circumference of one of them or one point [is marked] on the circumference of each, and if a perpendicular arc is drawn from either one of the points to the circumference of the other circle, then the ratio of the Sine of the arc that is between one of those points and the point of intersection of the circles to the Sine of the perpendicular arc extended from that [same marked point] to the other circle is as the ratio of the Sine of the arc found between the other [marked] point and the point of intersection to the Sine of the arc drawn from that [second marked] point.

I pray that the present verbose and at first sight intricate theorem should not frighten you away, for in things mathematical you will scarcely make the language be clear enough, not to say graceful. Truly you will pluck the sweetest fruit from this tree, however unyielding; and when you have savored this [fruit], you will understand almost the entire present book.

Thus, if two great circles ABGD and AED in a sphere are inclined toward each other [and] their circumferences intersect each other at points A and D, and if on the circumference of circle ABGD two points B and G are marked, from which two perpendicular arcs BR and GS descend to the circumference of circle AED, then the ratio of the Sine of arc AB to the Sine of arc BR is as that of the Sine of arc AG to the Sine of arc GS [Fig. II-5-6-5].

From points B and G let two rectilinear perpendiculars apiece be drawn, one [pair], BM and GN, to the line of intersection of the circles—namely, line AD—and the other [pair], BK and GL, to the surface of circle AED. When lines KM and LN are drawn, and since the

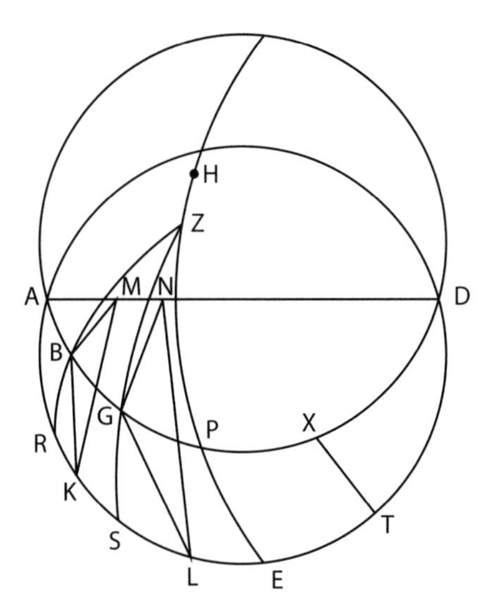

Fig. II-5-6-5.

two lines MB and BK, [which are] joined at an angle, are parallel to the two [lines] NG and GL, [which are] joined at an angle, because BM is parallel to line GN by [*Elements*] I.[not given] and BK to GL by [*Elements*] XI.[not given], then angle MBK will be equal to angle NGL. Moreover, each of the angles BKM and GLN is a right [angle] from the definition of a line perpendicular to a surface. Therefore by [*Elements*] I.32 the two triangles MBK and NGL are similar, and thus, by [*Elements*] VI.4 with [its] reasoning rearranged, the ratio of MB to BK is as the ratio of NG to GL. Moreover, MB is the Right Sine of arc AB by [*Elements*] III.[not given] and [by] definition, while NG is the Sine of arc AG. Similarly, BK is the Sine of arc BR and GL is the Sine of arc GS. Thus the ratio of the Sine of arc AB to the Sine of arc BR is as that of the Sine of arc AG to the Sine of arc GS. Therefore we have part of the proposition true when the two points are marked on one circumference.

Finally, if we mark these [points] on the two circumferences of the mentioned circles, and if B is the one on the circumference of circle ABGD and T is the other on the circumference of circle AED, [and] if a perpendicular arc TX descends from point T to the circumference of circle ABGD, then the ratio of the Sine of arc AB to the Sine of arc BR is as the ratio of the Sine of arc DT to the Sine of arc TX.

Let the poles of the two circles be H and Z, through which a great circle HZPE passes, intersecting the circumferences of the mentioned circles at points P and E. Then on the circumference of circle ABGD two points B and P are marked, from which two perpendiculars BR and PE descend. Therefore the ratio of the Sine of arc AB to the Sine of arc BR is as that of the Sine of arc AP to the Sine of arc PE. Again, since, on the circumference of circle AED, the two points E and T are marked, from which the two perpendiculars EP and TX originate, then the ratio of the Sine of arc DT to the Sine of arc TX will be as that of the Sine of arc DE to the Sine of arc EP. Moreover, this ratio of the Sine of arc DE to the Sine of arc EP is as that of the Sine of arc AP to the Sine of this same arc PE, for each of the arcs AP and DE is a quadrant. Therefore, the ratio of the Sine of arc AB to the Sine of arc BR is as that of the Sine of arc DT to the Sine of arc TX.

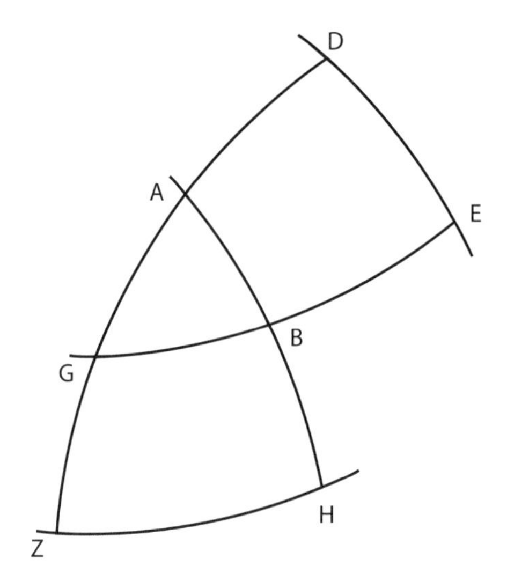

Fig. II-5-6-6.

In Theorem IV.16, Regiomontanus proves the spherical Law of Sines theorem for right triangles. He actually offers more than the theorem states: not only does he discuss triangles with two right angles or just one (both cases shown below), but he also addresses triangles with two acute, two obtuse, or just one of each type of angle.

Theorem IV.16. In every right triangle the ratio of the Sines of all the sides to the Sines of the angles which [the sides] subtend is the same.

If ABG is a triangle with right angle B, then the ratio of the Sine of side AB to the Sine of angle AGB is the same as the ratio of the Sine of side BG to the Sine of angle BAG and as the ratio which the Sine of side AG has to the Sine of angle ABG, which we will prove thus [Fig. II-5-6-6].

It is necessary that each of the angles A and G be a right angle or that only one of them [be a right angle] or that none [of them be a right angle]. If each of them is a right angle, then, by the hypothesis and Th. [2] above,[96] point A will be the pole of circle BG, B will be the pole of circle AG, and G will be the pole of circle AB. Therefore by definition each of the three mentioned arcs will determine the size of the angle opposite it. Therefore the Sine of any one of the three sides and [the Sine] of the angle opposite that [side] will be the same, and thus the Sines of all the sides to the sines of the angles opposite them will have the same ratio—namely, [that of] equality.

But if only one of the angles A and G is a right [angle], let that [angle] be angle G, for example. Moreover, the hypothesis gave angle B [to be] a right [angle]. Therefore, by Th. [2] above, A is the pole of circle BG, and, by Th. [1] above,[97] each of the arcs BA and

[96]If, from some point on the arc of a great circle, a great quadrant emerges orthogonally, its end [point] will be the pole of the circle from which the quadrant emerged; with this [fact], it may be stated that the meeting point of two arcs emerging orthogonally from a third arc is the pole of the circle containing that [third] arc.

[97]If you were to drop a great arc from the pole of a great circle in a sphere to its circumference or to an arc of [the circumference], that lowered arc will be a quadrant perpendicular to the circumference, forming two right angles upon the arc to which it falls.

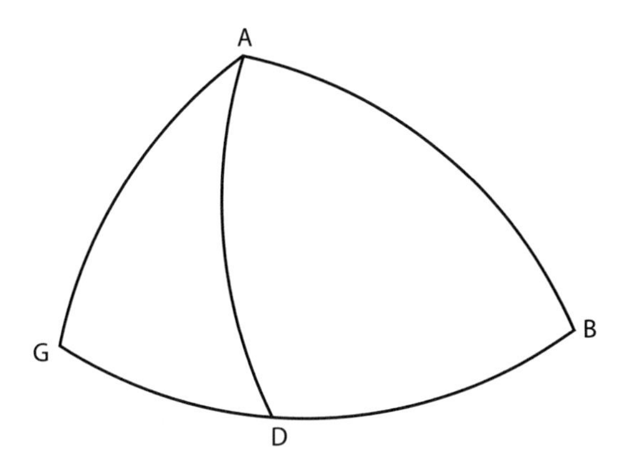

Fig. II-5-6-7.

AG is a quadrant of the great circumference. Then by definition each of the arcs AB, BG, and GA will determine the size of the angle opposite itself, and, by the converse of the definition of the Sine of an angle, the Sine of any side and the angle opposite it will be the same. Hence, finally, it will be established that the Sines of all the sides to the Sines of the angles opposite them have the same ratio—namely, [that] of equality. . . .

The next theorem continues the Law of Sines for nonright triangles with the assistance of the preceding theorem.

Theorem IV.17. In every nonright triangle the Sines of the sides to the Sines of the angles opposite those [sides] have the same ratio.

What proof the preceding [theorem] indicated concerning right triangles, the present [theorem] states also concerning nonright triangles. Thus if ABG is a triangle having no right angle, then the Sine of side AB to the Sine of angle G, and the Sine of side AG[98] to the Sine of angle B, and the Sine of side BG to the Sine of angle [A, all] have one and the same ratio [Fig. II-5-6-7].

From point A let perpendicular AD be dropped, falling upon arc BG if [AD] is found [to be] within the triangle, or meeting arc BG, necessarily extended, if [AD] falls outside the triangle. [AD] cannot coincide with either of the arcs AB and AG adjacent to it, for in that case one of the angles B and G would be a right [angle], which our hypothesis gives [to be] a nonright [angle].

Thus, first, let [it] fall within the triangle, determining two right triangles ABD and AGD. Then, by the preceding [theorem], with the terms changed around, the ratio of the Sine of AB to the Sine of AD is as that of the Sine of right angle ADB to the Sine of angle ABD. Furthermore, by the same previous [theorem], the ratio of the Sine of AD to the Sine of AG is as that of the Sine of angle AGD to the Sine of right angle ADG. And since the Sine of angle ADG is the same [as the Sine] of angle ADB because each of those [angles] is a right [angle], then, by indirect equal proportionality, the Sine of AB to the Sine of AG is as the Sine of angle AGB to the Sine of angle ABG; and, by a rearrangement, the Sine of side AB to the Sine of angle AGB is as the Sine of side AG to the Sine of angle ABG. Finally, you may infer the same ratio for the Sine of side BG to the Sine of angle BAG

[98]The printed book has erroneously *AB*.

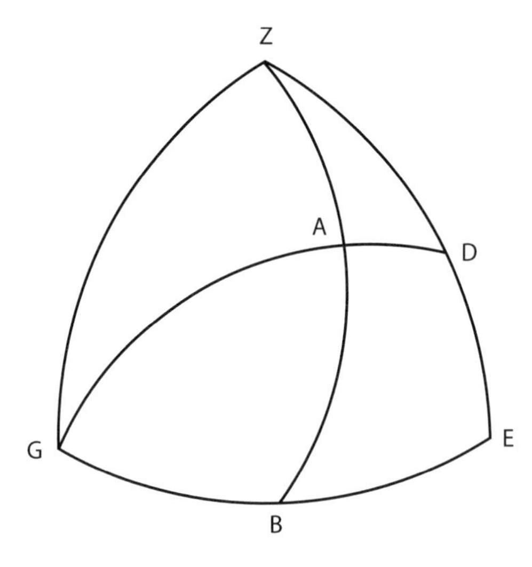

Fig. II-5-6-8.

if first you would drop a perpendicular arc from one of the vertices B and G to the side opposite it. . . .

In the following theorem, the cosine is used under its original name, sine of the complement of an angle.

Theorem IV.18. In every triangle having only one right angle, the ratio of the Sine of a nonright angle to the Sine of the right angle is as the Sine of the complement of the remaining angle to the Sine of the complement of the side subtending that [angle].

If ABG is a triangle whose angle B is a right [angle] but neither [of whose other] two, A and G, is a right [angle], then the ratio of the Sine of angle BAG to the Sine of right angle B is as that of the Sine of the complement of angle AGB to the Sine of the complement of side AB. Similarly, the Sine of angle AGB to the Sine of right angle B is as the Sine of the complement of angle BAG to the Sine of the complement of arc BG, which we shall prove in this way [Fig. II-5-6-8].

Because neither of the angles A and G is a right [angle], each of them will be either acute or obtuse, or one of them will be acute and the other obtuse. First, let each be acute; whereupon by Th. [3] above[99] each of the sides AB and BG will be less than a quadrant, and thus by Th. [6] above[100] arc AG will also be less than a quadrant. Then let GA be extended beyond point A until it becomes quadrant GD. Let a circle be drawn around G as the pole, intersecting arc GB, sufficiently extended, at point E. Next let arc BA, extended beyond point A, meet the circumference of the circle that was drawn around

[99] In every right triangle [the length of] the sides including the right angle compared to [the length of] a quadrant of the circumference and [the size of] the angles opposite these [sides] compared to [the size of] a right angle will have similar relationships. That is, if an angle which is opposite one of the sides including the right angle were equal to a right angle, that side will be equal to a quadrant; if greater than a right angle, the side will be greater than a quadrant; if less than a right angle, that side will be less than a quadrant.

[100] In every right triangle, if one of the two angles which the side opposite the right angle bears is a right [angle], that side will be a quadrant of the circle. If each of those [two angles] is either obtuse or acute, the mentioned side will be less than a quadrant. But if one of them is obtuse while the other is acute, that side is concluded to be greater [than a quadrant].

pole G at point Z. From Th. [2] above, this [point Z] is established to be the pole of circle GBE. Moreover, by Th. [1] above, each of the arcs ZB and ZE will be a quadrant and angle ZEG will be a right [angle]. Therefore, by definition, arc AZ will be the complement of side AB and arc DZ will be the complement of arc DE. And since arc DE determines the size of angle AGB because point G is the pole of circle ZDE, arc ZD will determine the size of the complement of angle AGB, which you may easily show when arc GZ is drawn. Moreover, the ratio of the Sine of angle BAG to the Sine of right angle ABG is as the ratio of the Sine of BG to the Sine of GA by the converse reasoning of Th. [16] above. Next, the Sine of BG to the Sine of GA is as the Sine of DZ to the Sine of ZA by Th. [15] above, because, on the circumferences of the two circles GAD and ZAB, two points G and Z are marked, from which, in turn, two perpendicular arcs descend to the circumferences of these circles—GB [being perpendicular] because angle ABG is a right [angle] by hypothesis and ZD [being perpendicular] because angle ZDG is a right [angle] since G is the pole of circle ZD. Furthermore, the Sine of ZD is the Sine of the complement of angle AGB and arc AZ is the complement of side AB. Thus the ratio of the Sine of angle BAG to the Sine of right angle ABG is as that of the Sine of the complement of angle AGB to the Sine of the complement of the side subtending it. . . .

The previous four theorems may be compared with the theorems of the *Almagest* by Jābir ibn Aflaḥ[101] in section V-5 of Chapter 3. Gerard of Cremona had translated this work into Latin, and it was available to Regiomontanus.

For practical applications of his theorems, Regiomontanus did not offer a plethora of examples. There were just enough to encourage readers to construct their own examples. What we find, however, are two types of examples: one numerical (as immediately below) and the other logical (as in Theorem V.3).

Theorem IV.26. When three angles of a right triangle are known, all its sides may be disclosed.

If ABG is a triangle whose three angles are known, then all its sides may be found. Either two of these [angles] are right or one is that size. If two are, let them be B and G, for example [Fig. II-5-6-9]. Then, by Th. [2] above, point A will be the pole of circle BG, and by Th. [1] above, each of the arcs AB and AG will be known [to be] a quadrant. Furthermore, arc BG determines the size of known angle BAG, and hence BG itself will be known. Thus we have found the three sides of the triangle.

We omit the proof of the case where only one angle is right, but the method follows directly from Theorem IV.18. To measure an arc opposite a given acute angle in a triangle with only one right angle, find the Sine of the complement of the desired arc by multiplying the Sine of the complement of the acute angle opposite the desired arc by the whole Sine and then dividing the product by the Sine of the other acute angle. Then the measure of the desired arc can be found by subtracting from 90° the measure just found of the complement of the arc. Regiomontanus presents two examples, one where two of the angles are right and one where just one is:

For example, let each of the angles B and G be 90° and let angle A be 40°. Each of the arcs AB and AG will be 90° and arc BG will be 40°. But [now] let angle B be a right [angle] while angle A is 50° and angle G is 70°. It is desired to find arc AB. The Sine of 50° is

[101] Known as "Geber" in the Latin West.

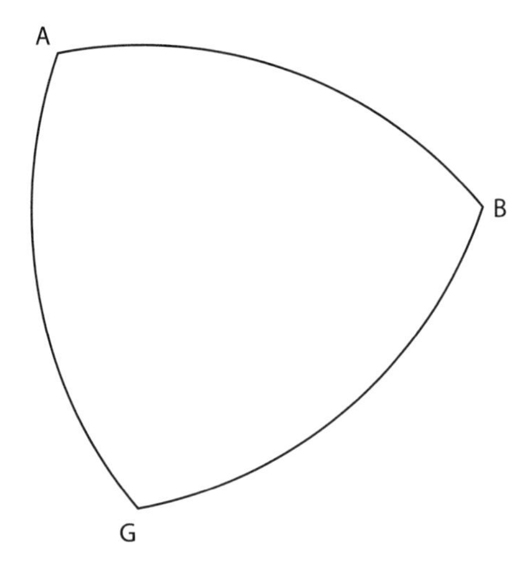

Fig. II-5-6-9.

45963. The Sine of the complement of 70° is 20521, which is multiplied by the whole sine [60000], giving 1231260000. This is divided by 45963, leaving 26788—namely, the Sine of the complement of arc AB. Its arc—namely, the complement itself—is 26°31′. This is subtracted from a quadrant, because arc AB must be less than a quadrant since angle G is acute, and 63°29′ are left. Arc AB will be considered [to be] this value. On the other hand, for side BG to be measured, the Sine of angle G, which is 70°, is 56382. The Sine of side AB, which we just now found [to be] 63°29′, is 53688. This is multiplied by the Sine of angle A, which was 45963, giving 2467661544. That [product], divided by 56382, leaves 43767—namely, the Sine of side BG. The smaller arc of this [Sine] is 46°50′, and this value is arc BG because angle BAG was given acute. Similarly, we may find arc AG, with no variation at all except that we take right angle ABG in place of angle BAG.

As far as available information indicates, Regiomontanus was the first mathematician to state the spherical Law of Cosines, albeit in phrasing quite unfamiliar to modern readers.

Theorem V.2. In every spherical triangle that is constructed from the arcs of great circles, the ratio of the versed Sine of any angle to the difference of two versed Sines, of which one is [the versed Sine] of the side subtending this angle while the other is [the versed Sine] of the difference of the two arcs including this angle, is as the ratio of the square of the whole right Sine to the rectangular product of the Sines of the arcs placed around the mentioned angle.

We omit the detailed proof, which involves drawing many lines into and across the sphere on which the given spherical triangle is situated. To translate the theorem into modern terms, since this theorem deals with ratios, we can substitute ordinary sines for Regiomontanus's Sines and can assume that the whole right Sine is equal to 1. Thus, if we assume the triangle is ABC and if we use a, b, and c to represent the sides subtending angles A, B, and C, respectively, we can write the conclusion as

$$\frac{\operatorname{vers} A}{\operatorname{vers} a - \operatorname{vers}(b-c)} = \frac{1}{(\sin b)(\sin c)}$$

or, in modern symbols,

$$\frac{1 - \cos A}{(1 - \cos a) - (1 - \cos(b - c))} = \frac{1}{(\sin b)(\sin c)},$$

which transforms to $\cos A \sin b \sin c = \cos a - \cos b \cos c$ or

$$\cos A = \frac{\cos a - (\cos b)(\cos c)}{(\sin b)(\sin c)}.$$

Immediately, Regiomontanus chose to offer his readers practice with using this law under the rubric of a theorem, although the example is in logical form. Note his understanding of the phrase "to solve a spherical triangle."

Theorem V.3. When three given sides of a spherical triangle are constructed from the arcs of great circles, all the angles of this [triangle] may be measured.

Although one may execute this proposition by Theorem [IV.34] above,[102] nevertheless since contemplation of the truth is more pleasant when one arrives at the same goal by several and different ways, the preceding theorem may be applied to our theorem. Therefore let ABC be such a triangle that is constructed from the arcs of great circles. The purpose is to find its angle BAC, or any other [angle].

We have taken its three sides [to be] unequal, for if any two sides of [the triangle] were equal, the procedure would be like the instruction of Th. IV.34 above [the second method]. Therefore, since from the preceding the ratio of the square of the whole right sine to the product of the right sines of the two arcs AB and AC is as the ratio of the versed sine of the desired angle BAC to the difference of the two versed sines, one of which is that of arc BC while the other is that of the difference of the two arcs AB and AC, and since three of these quantities are known from the hypothesis, then the fourth may be found—namely, the versed sine of angle BAC. Hence its arc, which determines the size of angle BAC, and therefore the angle itself, will be measured. Furthermore, we will teach nothing new for the identification of the other two angles, because, [with] angle BAC already known together with side BC opposite it and with the other sides known, we may find what is left by the reasoning of Th. [IV.30,[103] combined with Th. IV.8[104]] above.

In short, Regiomotanus developed further the foundations and articulated the basic theorems for the science that would come to be called trigonometry.

II-6. Mathematics of the Infinite

Scholars at the medieval universities studied intensively the writings of Aristotle and his Arabic commentators. These had become available by translations from Arabic into Latin.

[102]In any triangle that has three known sides, the three angles may be found. In chapter IV, this theorem was proved by drawing perpendiculars and using just Sines.

[103]When two sides of a non-right triangle are known together with the angle opposite one of them, [and] if it were established by what law the perpendicular opposite the given angle falls, then the remaining side and the other angles may be found.

[104]If any spherical triangle has two acute angles or two obtuse [angles], the arc passing from the vertex of the third angle to meet the side opposite that [third] angle perpendicularly will be found within the triangle. But if one of these [angles] is acute and the other obtuse the [arc] necessarily falls outside [the triangle].

The natural philosophy contained in Aristotle's *Physics* and in his other writings was not only important for laying the foundations of metaphysics and theology, but also perhaps even more important for studying the entire world of nature. One of the general questions in philosophy as well as in mathematics coming out of Aristotle's work was the concept of infinity. Can a magnitude (e.g., a line or a time interval) be divided without end, or do smallest units exist? And if the latter, are there an infinite number of entities ("atoms," "indivisibles") or not? Problems like these were discussed in the context of studying Euclid's *Elements*, as well as in dealing with notions of velocity and distance.

1. ANGLE OF CONTINGENCE

Since classical antiquity, mathematicians and philosophers have interested themselves in two angles: the "angle of a semicircle" and the "remaining angle." The first is the angle between the diameter of a circle and its circumference. The second is the angle between the circumference and the tangent at the point of tangency with the diameter. The two angles constitute the obvious right angle formed by the diameter and tangent. Because these two angles were formed by a straight line and a curved line, thereby differing from normal rectilinear angles, the ancient Greeks discussed them. They usually named the remaining angle, a "horn-like angle." Different from rectilinear angles, two horn-like angles do not have a ratio to one another, and one is not capable of being multiplied until it exceeds the other. Although the "angle of a semicircle" and the "horn-like angle" were of no use in elementary mathematics, they are mentioned in Proposition III.16 of Euclid's *Elements*:

The straight line drawn at right angles to the diameter of a circle from its extremity will fall outside the circle, and into the space between the straight line and the circumference another straight line cannot be interposed; further, the angle of the semicircle is greater, and the remaining angle less, than any acute rectilinear angle.

In the Latin West, from the thirteenth century on, the last part of this proposition gave rise to discussions about the horn-like angle, which normally was called the "angle of contingence" (*angulus contingentiae*). Does "less than any acute rectilinear angle" mean that the horn-like angle has no value? Johannes Campanus (d. 1296) (often known as Campanus of Novara), who reworked Euclid's *Elements* shortly before 1259, added to the text and to the proof of *Elements* III.16 (III.15 in his numbering) a lengthy comment:

Note that from this [proposition] the following argument does not hold: What transits from the less to the greater, [transits] through all intermediate [quantities], and therefore through the equal. And this, too, is not [true]: It is possible to find [something] greater than it and smaller than the same; thus it is possible to find [something] which is equal. This is [made] evident by this way [Fig. II-6-1-1]: Let *ab* be a circle with the center *c* and the diameter *acb*. From its end point *a* an orthogonal line *ad* is drawn which will touch the circle according to the corollary of this [proposition].[105] Another circle *bed* is described around point *a* with semi-diameter *ab*. Imagine that line *ab* [of the circle] is moved around point *a* on the circumference of the arc *bed* so that point *b* passes all points of the arc *bed*, until it comes to the line *ad* and covers it. Because the angle *bad* is a right angle, it is not

[105]Corollary to III.16: From this it is manifest that the straight line drawn at right angles to the diameter of a circle from its extremity touches the circle.

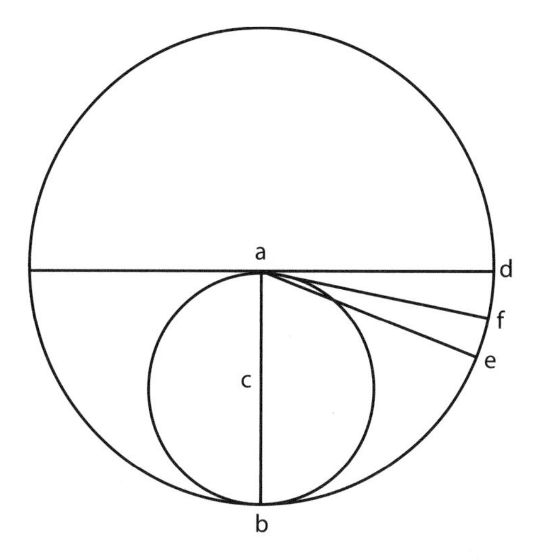

Fig. II-6-1-1.

possible to find any acute angle equal to it [enclosed] by the line *ad* and the diameter *acb* of the smaller circle. Therefore [the moved angle] transits to the right angle by passing through all acute angles. It is evident that some of them are smaller than the angle of the semicircle, which is enclosed by the semi-circumference *ab* and the diameter *acb*, and that the right angle is greater than the angle [of the semicircle].[106]

I argue that no angle in the transit from the smaller acute angles to the greater right angle is equal to it [= the angle of the semicircle]. If there were such an [angle], we suppose that this is formed by the line *ab*, when point *b* is in point *e* of the arc *bed*. Because angle *eab* equals the before-mentioned angle of the semicircle, and because this angle of the semicircle is the greatest of all acute angles according to the last part of this proposition, the angle *eab* is the greatest of all acute angles. This angle *ead* may be halved, as proposition I.9 [of Euclid's *Elements*] proposes,[107] by drawing the line *af*, and according to Conception 8[108] the angle *fab* is greater than the angle *eab*. Therefore something is greater than the greatest. This is impossible.

Another [argument]: The angle *eab* is equal to the angle of the semicircle, as it is proved, and the angle of the semicircle together with the angle of contingence is equal to a right angle. Similarly, the angle *eab* together with the angle *ead* is equal to a right angle. Therefore the angle *ead* is equal to the angle of contingence. Because, according to part 3 of this [proposition], the angle of contingence is the smallest of all acute angles, the angle *ead* being equal to it is equally the smallest of all acute angles. But the angle *eaf* is smaller than it, according to Conception [8]. Therefore, something is smaller than

[106]This implies tacitly that one angle is greater than another, if on superposition it includes the other in some neighborhood of the vertex. If the diameter is moved about its extremity until it takes the position of the tangent, then it makes an acute angle less than the angle of the semicircle as long as it cuts the circle; but if it ceases to cut, it makes a right angle greater than the angle of the semicircle.

[107]Proposition I.9: To bisect a given rectilineal angle.

[108]The Latin text has erroneously "per 9 conceptionem." Conception 8 (= Book 1, Common notion 5, in the Greek text) runs: The whole is greater than the part.

the smallest. This is impossible. Therefore no rectilinear angle is equal to the angle of the semicircle. And because there is a transit from the smaller to the greater, but not through the equal, and because it is possible to find [something] greater than it and smaller [than the same, but not something that is equal], there is an evident instance against both of the mentioned arguments.

Campanus was right to argue that angles of semicircles and angles of contingence do not obey the principle of continuity as formulated in the postulate of continuity. They are non-Archimedean quantities, and the theory of proportions cannot be applied to them. The remarks of Campanus soon became of basic authority for almost all later scholastic discussions of these unusual angles. They are found in many mathematical and philosophical works of the fourteenth and fifteenth centuries. Thomas Bradwardine (1295–1349), one of the Oxford Calculators at Merton College, addressed this problem in his *Geometria speculativa*, although his treatment only diverges from Campanus's in minor points. He defined the angle of contingence and then proved a major result about it:

The angle of contingence lies between the tangent to a circle and part of the circumference at the point of tangency.

[Proposition:] An angle of contingence is [1] less than any rectilinear angle, and yet [2] it is infinitely divisible. From which it is manifest that an angle of contingence is so much the greater as the circle is less and so much the less as the circle is greater.[109]

The first part is shown thus. Let line AD be touching circle ABC in point A, which is the end of diameter AC. I say that the angle which the tangent line makes with the circumference, which is called the angle of contingence, is less than any rectilinear angle, that is any angle contained by two straight lines. And the means to proving this is that between lines containing an acute rectilinear angle, however small, there can be taken a straight line dividing the angle, and that between a tangent line and the circumference it is impossible to take a straight line.

The first supposition is from the first and the last postulate. For let there be two equal lines AB and AC containing the given angle, and let the base BC be drawn, which let be divided in point D, because it is manifest that a line can be divided, and let line AD be drawn according to the first postulate [Fig. II-6-1-2]. I say then that line AD divides angle A. For either it is a third line distinct from line AB or AC, or it is the same as one of them. If it is a third line distinct from them, then, since it is applied not straightly to each of them in a surface, it will constitute two angles with them, by the definition of plane angle, which is what was proposed. If it be said to be the same as one of them, let it then be posited to be the same as AC, and it will then be that two straight lines, namely line DC and line AC, enclose a surface, which is contrary to the last postulate.

The second is evident, because, if there can be taken a straight line between the tangent and the circumference, let it be AG, to which let EF be drawn perpendicularly, for EA cannot be perpendicular to it, because it is perpendicular to AD, making two right angles with AD, and consequently angle GAE is acute [Fig. II-6-1-3]. Let therefore EF be perpendicular to AG, and angle EFA will be right, by the definition of right angle, wherefore, by the third conclusion of the chapter on triangles, AE will be the greatest side in triangle AEF. Therefore EF will be less than EA, and consequently it will be less than EB, which is

[109]For a discussion of the contents of Bradwardine's text, see [Molland, 1978, pp. 163–164].

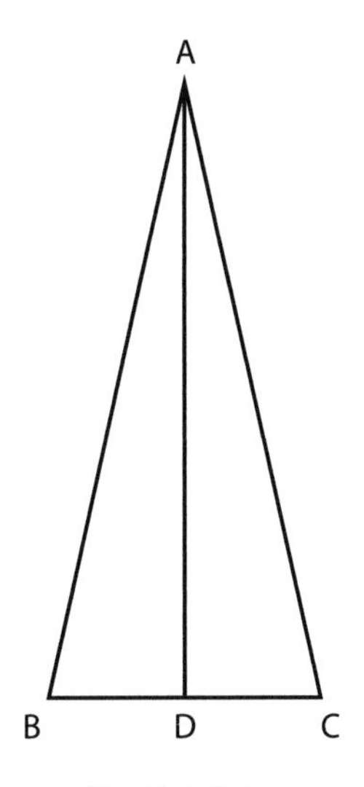

Fig. II-6-1-2.

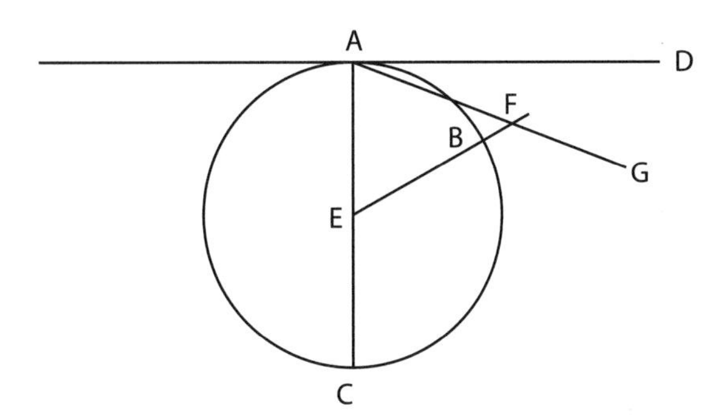

Fig. II-6-1-3.

equal to EA, as was argued in the preceding conclusion, which is impossible. It is therefore established that line AG cuts the circle, and the perpendicular [line, which is EF,] falls on part of the [aforementioned] line that is contained inside the circle.

The second part is evident, namely, that an angle of contingence is infinitely divisible. Although it cannot be divided by a straight line, it can still be divided by a curved line, such as an arc of a circumference, and this is evident by producing diameter AE continuously and in a straight line and on different centers situated on it describing different circles, all touching each other in point A [Fig. II-6-1-4]. For circumference AH described around center C divides the angle of contingence FAG, and the circumference AI described around center D divides angle FAH, and so on infinitely, by descending in the diameter AE and describing circles touching each other.[110] On account of this, Campanus says in the first comment on the tenth book of *Geometry* that any rectilinear angle is infinitely greater than any angle of contingence.

[110]The division is not into two angles of contingence, but into an angle of contingence and an angle between two arcs. Bradwardine does not remark on this [Molland, 1978, p. 164].

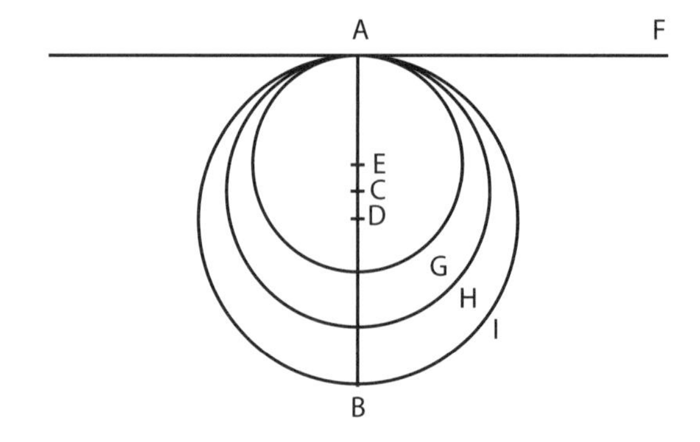

Fig. II-6-1-4.

The corollary is evident, because the tangent line AF constitutes the greatest angle FAG with the least circumference, and the least FAI with the greatest.

Bradwardine's presentation became a second standard source for the knowledge about the angle of contingence and the angle of the semicircle. His *Geometry* was printed several times between 1495 and 1530, and it influenced the discussion on this topic in the sixteenth century. A great advance was made by Jacques Peletier in his edition of Euclid's *Elements* (1557). He held that the angle of contingence was no angle at all, that the contact of a straight line with a circle is no quantity, and that angles contained by a diameter and a circumference of a circle are right angles and are equal to rectilinear angles.

2. THOMAS BRADWARDINE, *TRACTATUS DE CONTINUO* (*ON THE CONTINUUM*)

Bradwardine's *Tractatus de continuo* deals with continuity and discreteness, a border area between philosophy, mathematics, and physics. Bradwardine mentions five different opinions presented by scholars living in his time and in the past:

[Conclusion 31] If one continuum is composed by indivisibles in any way, then every continuum is composed in this way. And if one is not composed by atoms, then none is.

To understand this conclusion, one must know that the old and modern philosophers have five famous opinions about the composition of the continuum. Some of them, such as Aristotle, Averroes,[111] Algazel,[112] and most of the moderns, argue that the continuum is not composed of atoms, but of parts that can be divided without end. Others say that it is composed of two kinds of indivisibles, because Democritus had assumed that the continuum consists of indivisible bodies. Others say that it consists of points, and this [assumption is divided] into two parts: Pythagoras (the father of this position), Plato, and our contemporary Walter[113] assume that it is composed of a finite number of indivisibles, but others say that [it is composed] of an infinite number. This group, too, is divided into two parts. Some such as our contemporary Henry[114] say that it is composed of an infinite

[111]Known in Arabic as Ibn Rushd (1126–1198).
[112]Known in Arabic as al-Ghazalī (1058–1111).
[113]Walter Chatton (1290–1343) studied and wrote at Oxford from 1315 to 1332.
[114]Henry of Harclay (1270–1317) was chancellor of Oxford from 1312 to his death.

number of indivisibles that are directly joined. But others such as Lincoln,[115] say [that it is composed] of an infinite number [of indivisibles] that are indirectly joined to one another. Therefore the conclusion is this: "If one continuum is composed of indivisibles in some way" (the "way" includes any of the precedent ways), it then follows that "any continuum is composed of indivisibles according to a similar way."

Thus Bradwardine produced the following classification:

I. The nonindivisibilist position: continua composed of parts divisible without end
II. The indivisibilist position:
    A. The corporeal indivisibilist: there are extended indivisibles
    B. The incorporeal indivisibilist: there are nonextended indivisibles (i.e., points)
        1. Composition out of a *finite* number of points
        2. Composition out of an *infinite* number of points:
            a. Points are *immediately conjoined* to one another
            b. Points are *mediately conjoined* to one another

Bradwardine discusses all these hypotheses by giving examples and by listing the pros and contras. He rejects the position that the continuum is composed of a finite number of points.

[Conclusion 56] The assumption that the continuum is composed out of a finite number of indivisibles is adverse to all sciences. It attacks all of them and therefore is opposed uniformly by all.

One of the arguments against this assumption is the following.

[Conclusion 73] If this [is true], then the circumference of a circle is double of its diameter.

This is: half of the circumference is equal to its diameter. From the different points of the diameter, [assuming that] they are 10, ten perpendiculars are drawn directly to different points on half the circumference. It follows that there are 10 points on half the circumference, because only one point on half the circumference corresponds [to] a perpendicular. Therefore equally, there are the same number of points on half the circumference as are on the diameter. Therefore according to the second conclusion: half the circumference equals the diameter.

Then Bradwardine rejects two objections given by Walter Chatton and Henry of Harclay.

From this conclusion follows without difficulty that half of the circumference is equal to the diameter, and from this follows that the circumference is equal to twice the diameter, as proposed. But the third proposition of Archimedes' *Measurement of the Circle* does not admit this.[116] Bricklayers, carpenters and all craftsmen are convinced by experience that this is false. But someone could object weakly that it does not follow from the preceding that the circumference is double of the diameter, because the end points of the diameter are also points on the circumference. But then follows the even more impossible situation that the circumference is less than double the diameter.

[115]Robert of Lincoln, also known as Robert Grosseteste (1175–1253), taught at Oxford from 1229 to 1235 and then became bishop of Lincoln.

[116]The ratio of the circumference of any circle to its diameter is less than 3 1/7 but greater than 3 10/71.

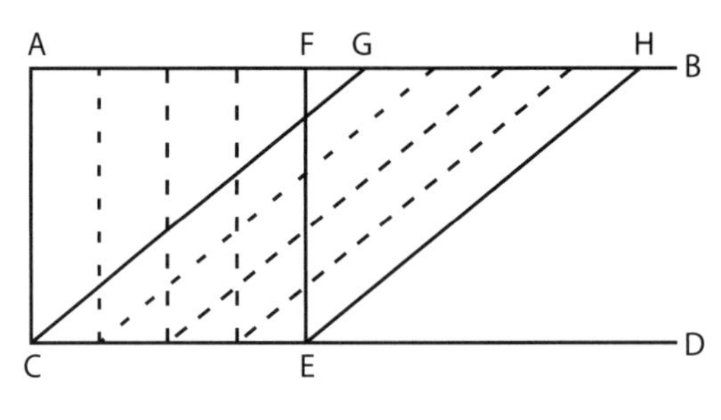

Fig. II-6-2-1.

The largest portion of *De continuo* is concerned with the deduction *ad absurdum* conclusions from the hypotheses of one or another form of indivisibilism. Some of the paradoxes that Bradwardine derives are original. Here is one of his counterexamples.

[Conclusion 135] If this is true [i.e., if each continuum is composed of an infinite number of indivisibles], a terminated surface can exceed another surface equal to it by any finite proportion.

Let AB and CD be parallel lines [Fig. II-6-2-1]. Atop base CE a right-angled parallelogram AFCE is constituted, and atop the same base another parallelogram CGHE is constituted with sides that are as much longer as you want than the sides of the parallelogram AFCE. Then all lines of CGHE which are drawn from all points of CE to the opposite points of GH are equal in number to those points, and consequently to all perpendiculars of AFCE which are drawn from the same points to the opposite points. But they are longer than those [lines]. Therefore, by [Conclusion] 133,[117] CGHE is larger than AFCE. But according to I.36 of Euclid's *Elements*,[118] the parallelograms are equal. The same can be argued with triangles made in the same way, according to Euclid I.37.[119]

From this and from other examples, Bradwardine announces his true view of the composition of continua:

[Conclusion 141 and Corollary]: No continuum is made up (*integrari*) of atoms. From here follows and elicits: Every continuum is composed of an infinite number of continua of the same species as it, . . . that is, every line is composed of an infinite number of lines, every surface composed of an infinite number of surfaces, and so on concerning other continua.

3. JOHN DUNS SCOTUS, INDIVISIBLES AND THEOLOGY

The use of strictly mathematical arguments to establish theological conclusions may strike modern readers as a mixture of water and oil: while mathematical analogies may be tolerated, arguing from mathematics seems almost pagan (shades of Plato!). Yet, one finds such chains of reasoning among the best of the Scholastics, here in a lecture of John Duns Scotus (ca. 1266–1307). The relevance of the mathematics becomes clear from the background to the

[117] Surfaces which are composed by lines that are equal in number and in length, are equal. But if they are composed by lines equal in number but not in length, the surface composed by the longer lines is greater.

[118] Parallelograms which are on equal bases and in the same parallels are equal to one another.

[119] Triangles which are on the same base and in the same parallels are equal to one another.

problem: can an angel be moved by a continuous motion? The Greco-Arabic cosmology developed from the Aristotelian synthesis assigned to "intelligences" the duty of moving the spheres on which the planetary system rotates, to provide harmony in the celestial hierarchy. When this philosophy was brought into Christian thinking, the "intelligences" became "angels," since the latter were already well known from the Bible. The Scholastics in turn discussed all aspects of angels to fit them into the ordered panoramas they sought for theology. Having located the angels in prescribed places, they addressed the problem. Scotus attacked the position that a line is composed of indivisible points; hence, an angel cannot be moved by a continuous motion. The argument selected considers a square with diagonal drawn.

I shall prove this by Euclid X, 5 or 9. Prop. 5 says, "the ratio of any commensurable quantity to one like itself is as the ratio of some number to another number." And consequently as Prop. 9 says, "If some lines are commensurable, then their squares have a ratio to one another as the square of some number is to the square of another number." However the square of a diameter does not have a ratio to the square of a side like the square of a number has a ratio to the square of another number. Consequently the line, which is the square of a diameter, is not commensurable with the side of that square. The minor of this follows from the Pythagorean Theorem: Because the square of the diameter is twice the square of the side, that is to say, it is equal to the sum of the squares of two sides, there is no square number that is twice some other square number. This is obvious from a survey of a table of squares found by multiplying roots by themselves. The conclusion therefore follows: the diameter is asymmetric to its side; that is, [it is] incommensurable.

On the other hand if these lines are composed of points, then they are not incommensurable (for points on one line would have a ratio to the points on another line in some numerical ratio). It follows therefore that not only are these lines commensurable but that they would be equal. And that is against common sense. The proof of this follows. Select two adjacent points on one side of a square and the corresponding points on the opposite side. Draw lines connecting each pair of opposite points, making them parallel to the base of the square. These lines cut through a diameter of the square. I want to know whether the points of intersection are adjacent to one another or not. If they are adjacent, then there are no more points in the diameter than in the side. If they are not adjacent, then I can find a point midway between the two non-adjacent points on the diameter (it falls between the two lines, as given). Moreover I can draw another line between the two parallel lines that is parallel to each by Euclid I.31. This line is drawn continuously equidistant and straight by Euclid I, Postulate 2. It cuts the sides without touching either of the two adjacent given points and lies equidistantly between them. (Otherwise the line would be concurrent with one of the other lines, with which it was drawn equidistant. And this contradicts the definition of equidistant that is the final definition in Euclid I.) Therefore between those two points, which were placed adjacent to one another on the side, there is a point between them. This follows from what was said about the point placed between the points located on the diameter. Therefore by contradicting the consequent, I contradict the antecedent. Therefore and so on.

Furthermore and in general: the whole of Book X of Euclid destroys the idea that a line is composed of points. If it were, then there could be no surd or irrational line.

Fig. II-6-4-1.

Therefore, since the line joining two points is a continuum, the angels can move continuously from one point on it to its endpoint, without being able to jump from one point to the next, only to discover that there is another point between the two, a conundrum that prevents all motion whatsoever.

4. DOES LIGHT TRAVEL INSTANTANEOUSLY OR OVER TIME?

The notions of infinity and infinitesimal also occurred in the discussion in the Middle Ages of whether light travels instantaneously or with a finite speed. By the thirteenth century, the major treatises on optics from Greek antiquity, such as Ptolemy's *Optica*, and medieval Islam, such as the *Perspectiva* of Alhazen (Ibn al-Haytham), had been translated into Latin, and the study of the subject surged dramatically. The only issue we consider here, however, is the question of the speed of light.

Witelo (ca. 1230–1300) held that light moved instantaneously. He presented his thinking on instantaneous movement succinctly in Book II of his major optical treatise *Perspectiva*.

Theorem 2: Unimpeded light is necessarily moved in an instant through the whole of a medium proportioned to it.

Let there be a line proportioned to the diffusion of a strong light, as the diameter of the world in the case of solar light. Let this line be ABCD, and let there be a strongly luminous body at point A [Fig. II-6-4-1].

If it should be said that the light is moved through line ABCD in time and not in an instant, then in part of that time it is moved through line AB, and in the least sensible time through the least sensible line AB; for if it were moved through an insensible space in a sensible time, it would follow that a sensible space is composed of insensible [parts], just as the time measured after that space [AB] is composed of partial times that are sensible.[120] Therefore, light is moved in the least sensible time through the least sensible space. But in the same time, the form of a luminous body weaker than that more strongly luminous body is moved through the same space, since there is no sensible space smaller than the least sensible space and no sensible time less than the least sensible time.

[120]That is, as the sum of the sensible partial times is equal to the whole time required to traverse space *AD*, so the sum of the insensible spaces corresponding to those partial times must equal the sensible space *AD*; but, as Witelo implies, it is absurd to hold that a sensible space is composed of insensible parts.

Therefore, stronger and weaker lights are of equal strength,[121] which is impossible, since they involve contradictories. Therefore, it is impossible for light to be diffused in time through a medium proportioned to it, and thus it is necessary for this diffusion to take place in an instant, which is the proposition.

Of all the major thinkers at that time, only Alhazen held that light travels with finite speed. While Roger Bacon (ca. 1219–1292) agreed, he found it necessary to discuss and refute the opinions of al-Kindi (ca. 801–873) and Aristotle (and others, he wrote), who held that light is diffused instantaneously. His thinking appears in the preface to his argument in *Opus maius* I, 9, iii.

Al-Kindi endeavors to demonstrate in his book, *De aspectibus*, that a ray traverses [some space] in an altogether indivisible instant, and he adduces quite a curious and probable argument when he says that if a certain species, such as the light of the sun when it rises, is produced in time in the first part of the air, then if that time is doubled in the second equal part and tripled in the third part, when it reaches the west it would be a time composed of many parts, great in proportion to the first time; and although the first time would be insensible, nevertheless the whole time will be sensible because of its magnitude, which is nearly incomparable with respect to the first time. And Aristotle says in the second book of *De anima* that although the multiplication of light over a small distance can be hidden from our sense, it cannot be hidden in the case of a distance as great as that between east and west. Therefore, if a species were produced in a certain time, this would be perceptible to sense. But we do not perceive it; therefore it is not produced in time but in an instant. . . .

And all authors express the same view except Alhazen, who tries to refute this view in his second book.

Bacon describes Alhazen's arguments, but then refutes them. He then continues as follows.

Nevertheless, an irrefutable argument for the opinion of Alhazen can be drawn from his statements in the seventh book. For there he teaches that from the same origin a perpendicular ray arrives sooner at the end of the space that a non-perpendicular ray. But sooner and later do not exist without time, as Aristotle maintains in the fourth and sixth books of *Physics*. And this is demonstrated without possibility of contradiction, for no finite power acts in an instant, as Aristotle says in the sixth book of *Physics*; and he proves this, since if it did, a greater power would act in less than an instant, which is impossible. But the power of the eye and of its species and of all created things is finite; therefore it cannot act in an instant. And at the end of the eighth book of *Physics*, he holds that a finite and an infinite power cannot act in the same or equal durations, since then they could have equal effects, and thus they would be equal. But it is the property of an infinite power to act in an instant. Therefore, a finite power cannot do anything in an instant but only in time.

Moreover, as an instant is to time, so is a point to a line. Therefore, by permutation, as an instant is to a point, so is time to a line. But the traversal of a point occurs in an instant; therefore the traversal of every line occurs in time. Therefore, species traversing a

[121] That is, if any light (or form) is moved through the least sensible space in the least sensible time (the view Witelo is attempting to refute), then strong and weak lights move with equal rapidity; but this, Witelo apparently thinks, violates the assumption of unequal strengths.

linear space, however small, traverse it in time. Furthermore, prior and posterior in space are the cause of prior and posterior in traversal of the space and in duration, as Aristotle says in the fourth book of *Physics*. Therefore, since the space through which a species is moved has a prior and posterior [part], its traversal must have a prior and posterior part in itself and in its duration; but prior and posterior in duration do not exist without time, since they cannot exist in an instant. And if it should be declared that this is true of those things that are measured by parts of space but that this excludes species (as is supposed), still it is nothing, since this second statement is applicable only to spiritual existence. Since the species of a corporeal thing has a real corporeal existence in the medium and is a true corporeal thing, as we have shown above, it must necessarily be dimensional and therefore adapted to the dimensions of the medium.

5. NICOLE ORESME, *QUESTIONES SUPER GEOMETRIAM EUCLIDIS* (*QUESTIONS ON THE GEOMETRY OF EUCLID*)

Nicole Oresme addressed the idea of infinity in several places in his works. In particular, he dealt with infinite series in his *Questions on the Geometry of Euclid*.[122] In Question 2, he found the sum of the geometric series with ratio 1/2:

$$1 + 1/2 + 1/4 + 1/8 + \ldots + 1/2^n + \ldots = 2,$$

and also found the sum of the geometric series with ratio 1/3:

$$1 + 1/3 + 1/9 + 1/27 + \ldots + 1/3^n + \ldots = 3/2.$$

In fact, he remarked that if a magnitude is increased an infinite number of times with proportional parts in a *proportio minoris inaequalitatis*, that is, in a ratio of less to greater (e.g., 1/4), the whole will never become infinite. But by adding nonproportional parts in such a ratio, the whole may become infinite. As example, he examined the harmonic series

$$1 + 1/2 + 1/3 + 1/4 + \ldots + 1/n + \ldots.$$

He proved its divergence by pointing out that the sums of the third and fourth, of the fifth through the eighth, of the ninth through the sixteenth "*et sic in infinitum*," are all greater than 1/2. It seems that Oresme was the first to prove the divergence of the harmonic series, 300 years earlier than Pietro Mengoli (1626–1686). Oresmes's proof is given here.

The third conclusion is: it is possible that an addition of non-proportional parts to a quantity according to a proportion of minor inequality can be done, but the whole may become infinite; if it were done of proportional parts, it would be finite, as it has been said. For instance: Given a quantity of one foot. To it half of one foot may be added in the first proportional part of an hour, then one third in the second [part], then one fourth, then one fifth etc. *in infinitum*, according to the order of the numbers. I argue that the whole will become infinite. This is proved by this way: there exist an infinite number of parts each of which being greater than half of a foot, and therefore the whole will become infinite. The first is evident, because the fourth part and the third part are more than one half, the same [is true] with the fifth up to the eighth part and with the ninth to the sixteenth part, etc. *in infinitum*.

[122] See section II-7-6 for his consideration of infinite series in the context of motion.

II-7. Statics, Dynamics, and Kinematics

Medieval philosophers and theologians used geometry to understand old ideas and develop new ones. Perhaps the most outstanding geometric exercise of this period was to lay the foundation for mathematical physics. On the one hand the philosophers were encouraged to describe in geometric terms the world in which they lived and moved; on the other hand they did not think of nature as physics in the modern sense. Instead, the medieval mind focused on a tripartite mechanics of statics, dynamics, and kinematics. The impetus for thinking along these lines is due to a major schoolman of the first half of the thirteenth century, Robert Grosseteste.

1. ROBERT GROSSETESTE, *DE LINEIS, ANGULIS ET FIGURIS* (*ON LINES, ANGLES AND FIGURES*)

Beginning in 1214 and before he became bishop of Lincoln in 1235, Robert Grosseteste was one of the major teachers at Oxford. His influence was particularly strong in the science and mathematics that provided the foundation for the study of philosophy and theology. His writings found their way to the Continent. Grosseteste considered mathematics as the vehicle that brings to students an understanding of the world they experience. He cautioned them to keep mathematics and the world separate and distinct, because the latter needed the former to be understood. The caution is described in the following selection from his *De lineis, angulis et figuris* (*On Lines, Angles and Figures*).

The usefulness of considering lines, angles, and figures is the greatest, because it is impossible to understand natural philosophy without these. They are efficacious throughout the universe as a whole and its parts, and in related properties, as in rectilinear and circular motion. They are efficacious also in cause and effect (*in actione et passione*), and this whether in matter or in the senses, and in the latter whether in the sense of sight, where their action properly takes place, or in other senses, in the operation of which something else must be added on top of those which produce vision. . . . For all causes of natural effects have to be expressed by means of lines, angles, and figures, for otherwise it would be impossible to have knowledge of the reason (*propter quid*) concerning them. This is clear in this way: a natural agent propagates its power (*virtutem*) from itself to the recipient, whether it acts on the senses or on matter. This power is sometimes called species, sometimes a similitude, and is the same whatever it may be called; and it will send the same power into the senses and into matter, or into its contrary, as heat sends the same thing into the sense of touch and into a cold body. For it does not act by deliberation and choice, and therefore it acts in one way, whatever it may meet, whether something with sense perception or something without it, whether something animate or inanimate. But the effects are diversified according to the diversity of the recipient. For when received by the senses this power produces an operation in some way more spiritual and more noble; on the other hand when received by matter, it produces a material operation, as the sun by the same power produces diverse effects in different subjects, for it cakes mud and melts ice. Hence these rules and principles and fundamentals having been given by the power of geometry, the careful observer of natural things can give the causes of all natural effects by this method. And it will be impossible otherwise, as is already clear in respect of the universal, since every natural action is varied in strength and weakness through variation of lines, angles and figures. But in respect of the particular this is even clearer, first in natural action upon matter and later upon the senses, so that the truth of geometry is quite plain.

The reason or *propter quid* for all this was based on Grosseteste's understanding of physical nature itself. All physical beings (animal, vegetable, and mineral) have "bodiness" or "corporality" in common. In other words, everything having a body exists in three dimensions and is capable of being moved either by itself or by something else, thus leading to the creation of the science of mechanics. The tool for describing this was geometry. In short, for Grosseteste and the scholars who would succeed him at Merton College, Oxford, geometry was the language that led to the understanding of nature. Their focus (as well as that of scholastics on the Continent) had turned early to mechanics in the field of statics.

2. JORDANUS DE NEMORE, *DE RATIONE PONDERIS* (*ON THE THEORY OF WEIGHTS*)

Contemporaneous with the yet-to-become bishop of Lincoln was Jordanus de Nemore. Easily hailed as a polymath because of his writings in so many fields, yet, as noted in sections II-2-5 and II-3-4, totally unknown as to home or kin, Jordanus composed *On the Theory of Weights* (*De ratione ponderis*) early in the thirteenth century. This is a crucial tract for statics that contains a proof of the law of the lever different from that of Archimedes as well as the first correct proof for the equilibrium of a bent lever. Some question the attribution of this treatise to Jordanus, but the evidence favors him [Moody and Clagett, 1960, pp. 171–172].

If the arms of a balance are proportional to the weights suspended, in such manner that the heavier weight is suspended from the shorter arm, the weights will have equal positional gravity.

Let the balance beam be ACB and the suspended weights *a* and *b* [Fig. II-7-2-1], and let the ratio of *b* to *a* be as the ratio of AC to BC. I say that the balance will not move in either direction. For let it be supposed that it descends on the side of B; and let the line DCE be drawn obliquely to the position of ACB. If the weight *d*, equal to *a*, and the weight *e*, equal to *b*, are suspended, and if the line DG is drawn vertically downward and the line EH vertically upward, it is evident that the triangles DCG and ECH are similar, so that the proportion of DC to CE is the same as that of DG to EH. But DC is to CE as *b* is to *a*; therefore DG is to EH as *b* is to *a*. Then suppose CL to be equal to CB and to CE, and let *l* be equal in weight to *b* ; and draw the perpendicular LM. Since then LM and EH are shown to be equal, DG will be to LM as *b* is to *a*, and as *l* is to *a*. But, as has been shown, *a* and *l* are inversely proportional to their contrary [upward] motions. Therefore, what suffices to lift *a* to D will suffice to lift *l* through the distance LM. Since therefore *l* and *b* are equal, and LC is equal to CB, *l* is not lifted by *b* ; and consequently *a* will not be lifted by *b*, which is what is to be proved.[123]

[123]As Moody remarks [Moody and Clagett, 1960, pp. 390–391], this proof and the following one are based on the principle of work, given in an earlier result: If a weight w, descending through a vertical distance d, can lift a counterweight x through a certain vertical height h, then another weight kw, by descending through the vertical distance d/k, can raise the same counterweight x the same height h.

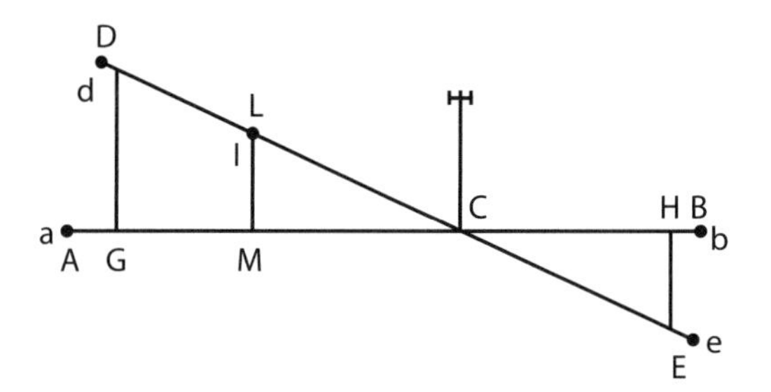

Fig. II-7-2-1.

If the arms of a balance are unequal, and form an angle at the axis of support, then, if their ends are equidistant from the vertical line passing though the axis of support, equal weights suspended from them will, as so placed, be of equal heaviness.

Let the axis be C, the longer arm AC, and the shorter arm BC. And draw the vertical line CEG; and let the lines AG and BE perpendicular to this vertical be equal. When therefore equal weights are suspended at A and B, they will not change from this position. For let AG and BE be extended by a distance equal to their own length to K and to Z; and on them let the arcs of circles MBHZ and KXAL be drawn [with centers at C]; and let the arcs AX and AL be equal to each other and similar to the arcs MB and BH. And let the arcs AY and AF also be equal and similar [on a circle through A and K with radius equal to CB] [Fig. II-7-2-2]. If then *a* is heavier in this position than *b*, let it be supposed that *a* descends to X and that *b* is raised to M. Then draw the lines ZM, KXY, KFL; and let MP be erected perpendicularly on ZBP, and XT and FD on KAD. And because MP is equal to FD, while FD is greater than XT—on account of similar triangles—MP will also be greater than XT. Hence *b* will be lifted vertically more than *a* will descend vertically, which is impossible, since they are of equal weight. Again, let it be supposed that *b* descends to H and lifts *a* to L; and let HR fall perpendicularly on BZ, and LN and YO on KAN. Then LN will be greater than YO, and consequently greater than HR; so that in the same way the impossible will result.

To make this more evident, let us draw a different figure, as follows: Let there be a vertical line YKCNZ; and around the center C let there be drawn two semicircles, YAEZ and KBDN; and let the lines AFE and BD be drawn at equal distances from the diameter; and from these let there be drawn the equal perpendiculars BL and CF [Fig. II-7-2-3]. Then draw the lines CB, CA, CE, and CD. If we then suppose that equal weights are suspended at A, B, D, E, and F, they will be of equal positional gravity. For if the lines BA, BXF, BE, DA, DF, and DE are drawn, all of them will be bisected by the diameter, as for instance BXF. For since BL and CF are equal, and the triangles BLX and CFX are similar, BX will be equal to XF. And in the same manner the others will be divided at their midpoints. Therefore, since the midpoints of all the lines are placed at a common center, so likewise the weights are placed; therefore they will be of equal heaviness.

A more subtle variant may, however, be added, if we suppose that *a* is heavier than *b*, *b* heavier than *f*, *f* heavier than *d*, and *d* heavier than *e*. Yet *d* is not able to lift *e*; for the segment of the line DE on the side of E would immediately become greater. But if *a* is given an impulse downward, it is able to raise *b*; and similarly *b* can raise *a*; and

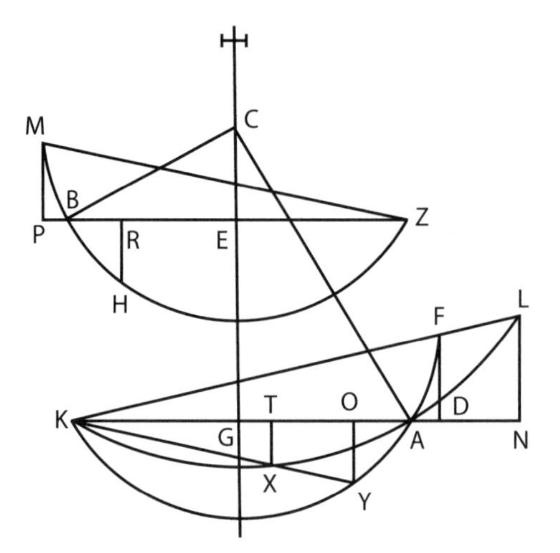

Fig. II-7-2-2.

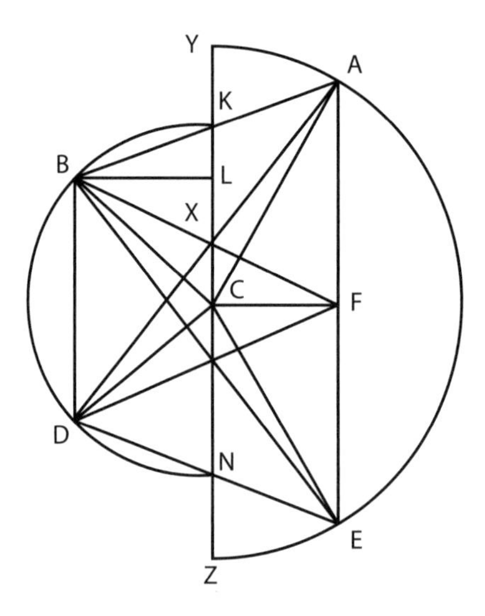

Fig. II-7-2-3.

d can raise *a* and *a* can raise *d*; and *b* can raise *f* and *f* can raise *b*; until they make a complete revolution, and hang in such manner that the angle with the axis is beneath them. For when *b* is moved downward, the segment of the line BA on the side B will become steadily longer, and *b* will become heavier.

While kinematics was an item of interest in ancient Greek philosophy, the philosophers at Merton College, Oxford, moved the science forward to include both kinematics and dynamics. These are like the two sides of a rolling wheel. The first looks only at the spatial or temporal changes in movement, as the second considers only the forces that produced the changes; in short, the former is effect and the latter its cause. An understanding of certain terminology created at Merton College is essential when reading tracts emanating from its scholars:

Motion is simply movement without implying vectors.
Velocity is speed.
Quality of motion or intensity of speed is usually instantaneous speed.
Quantity of speed or motion is average speed over time, measured by the distance traversed in that time.
Latitude of velocity or motion considers positive or negative changes of velocity.
Uniform velocity is constant speed.
Uniformly difform velocity or motion is constant acceleration [Clagett, 1959, pp. 210–211].

First we consider a work by Bradwardine.

3. THOMAS BRADWARDINE, *TRACTATUS DE PROPORTIONIBUS*

Of the half dozen or so medieval mathematical works that changed the way mathematicians thought, one is Thomas Bradwardine's *Tractatus de proportionibus* (1328). Here, in the context of the study of dynamics, he introduced the concept and use of what moderns would call exponential notation. He begins his discussion with a well-known principle, "each successive motion is proportional to another with respect to speed." After explaining the meanings of definitions and axioms applied to technical terms, he investigated four accepted theories regarding relationships among speed, force, and resistance. First he explained and demolished the thinking of Aristotle in *On the Heavens and Earth*: the proportion of the speeds of motions varies as the amount that the power of the mover exceeds the resistance offered by the thing moved. Then he explained and rejected Averroes's Comment 36 on Aristotle's *Physics Book VI*: the proportion of the speeds of motions varies with the proportion of the excesses whereby the moving powers exceed the resisting powers. Next he destroyed a generalization built on remarks in Aristotle's *Physics* and *On the Heavens* and from *On Weights*: with the moving power remaining constant, the proportion of the speeds of motions varies with the proportion of resistances, and with the resistance remaining constant, it varies with the proportion of moving powers. Finally he took apart Comment 79 on Aristotle's *Physics VIII* by Averroes: there is neither any proportion nor any relation of excess between motive and resistive powers. Then he began his own contribution.

Now that these fogs of ignorance, these winds of demonstration, have been put to flight, it remains for the light of knowledge and of truth to shine forth. For true knowledge proposes a fifth theory, which states that the proportion of the speeds of motions varies in accordance with the proportion of the power of the mover to the power of the thing moved. . . . Furthermore, there does not seem to be any theory whereby the proportion of the speeds of motions may be rationally defended, unless it is one of those already mentioned. Since, however, the first four have been discredited, therefore the fifth must be the true one. We therefore arrive at the following theorem:

Theorem I. The proportion of the speeds of motions follows the proportion of the force of the mover to that of the moved, and conversely. Or, to put it another way, which means the same thing: The proportion of motive to resistive power is equal to the proportion of their respective speeds of motion, and conversely. This is to be understood in the sense of geometric proportionality.

Thus the proof of this theorem is that it alone survives after all other explanations of the relationship of speed to the ratio of motive to resistive forces have been shown to be incorrect; therefore it must be true. As Crosby points out, it is more of a conclusion than a theorem. The remaining eleven theorems, however, are proved mathematically as will appear in the proofs for Theorems II to VI. The references are to Theorem I and other theorems developed and proved in the first part of the *Tractatus*, which discussed succinctly the theory of proportions. Symbolically, the first theorem can be expressed as $V = \log_n(\frac{F}{R})$ or $n^V = \frac{F}{R}$. That is to say, doubling the velocity squares the ratio of motive power to resistance, tripling the one cubes the other, and so on. For the formula to be correct universally, n is necessarily a constant equal to $\frac{F}{R}$ when $V = 1$ [Crosby, 1961, p.189 n. 90]. The remaining eleven theorems follow with some directions for proofs.

Theorem II. If the proportion of the power of the mover to that of its mobile is that of two to one, double the motive power will move the same mobile exactly twice as fast. This may be demonstrated by means of an example. Let A be a motive power that is twice B (its resistance), and let C be a motive power that is twice A. Then (by Theorem I, Chapter I[124]) the proportion of C to B is exactly the square of that of A to B. Therefore (by the immediately preceding theorem) C will move B exactly twice as fast as A does. This is what was to be proved.

Theorem III. If the proportion of the power of the mover to that of its mobile is two to one, the same power will move half the mobile with exactly twice the speed. This may be demonstrated by an argument like that used for Theorem II.

Theorem IV. If the proportion of the power of the mover to that of its mobile is greater than two to one, when the motive power is doubled the motion will never attain twice the speed. This may be demonstrated by means of Theorem IV, Chapter I[125] and Theorem I [above].

Theorem V. If the proportion of the power of the mover to that of its mobile is less than two to one, when the resistance of the mobile is halved, the motion will never attain twice the speed. This may be demonstrated by means of Theorem III, Chapter I[126] and Theorem I [above].

Theorem VI. If the proportion of the power of the mover to that of its mobile is less than two to one, when the power moving this mobile is doubled, it will increase the speed to more than twice what it was. This is likewise easily demonstrable, from Theorem VI, Chapter I[127] and Theorem I [above].

[124]If a proportion of greater inequality between a first and a second term is the same as that between the second and a third, the proportion of the first to the third will be exactly the square of the proportions between the first and the second, and the second and the third.

[125]If the first term is twice the second and the second is more than twice the third, the proportion of the first to the third will be less than the square of the proportion of the second to the third.

[126]If the first term is more than twice the second and the second is exactly twice the third, the proportion of the first to the third will be less than the square of the proportion of the first to the second.

[127]If the first term is twice the second and the second is less than twice the third, the proportion of the first to the third will be greater than the square of the proportion of the second to the third.

Theorem VII. If the proportion of the power of the mover to that of its mobile is less than two to one, when the same mover moves half that mobile, the speed of the motion will be more than doubled. . . .

Theorem VIII. No motion follows from either a proportion of equality of one or lesser inequality between mover and moved. ([The proof requires] the addition of the following axiom independently known: Axiom I. All motions of the same species may be compared to each other with regard to slowness and fastness.)

Theorem IX. Every motion is produced by a proportion of greater inequality, and from every proportion of greater inequality a motion may arise. . . .

Theorem X. Given any motion, one twice as fast and one twice as slow can be determined. . . . ([The proof requires] the help of the following axiom, independently known: Axiom 2. A proportion of greater inequality of mover to moved may be halved or doubled indefinitely.)

Theorem XI. An object may fall in the same medium both faster, slower, and equally with some other object that is lighter than itself. . . .

Corollary 1. It is manifest from the foregoing that the fastness and the slowness of any pure body and the slowness of any mixed body may be doubled indefinitely, but that the fastness of a mixed body may not be so doubled by rarefaction of the medium. . . .

Theorem XII. All mixed bodies of similar composition will move at equal speeds in a vacuum. . . .

Corollary 2: If two heavy mixed bodies of unequal weight but similar composition were balanced on a scale within a vacuum, the heavier would descend.

4. WILLIAM HEYTESBURY, *REGULE SOLVENDI SOPHISMATA* (*RULES FOR SOLVING SOPHISMS*)

Easily one of the most powerful theorems created for natural philosophy by fourteenth-century scholastics at Merton College is the kinematical result named the Mean Speed Theorem. That is: a body that moves with uniformly accelerating speed traverses in a given time the same distance as a body that in the same time moves with a constant speed equal to the accelerating body's speed at the middle instant. While it is difficult to assign this result to a single member of the College's group of profound philosophers, William Heytesbury (ca. 1313–1373), one of the renowned Calculators at Oxford, is surely a candidate. In part VI of his *Rules for Solving Sophisms* (*Regule solvendi sophismata*), he offered a clear exposition of the theorem, preceded by some basic definitions dealing with velocity and acceleration. (See paragraph III.vii of section II-7-6 for Oresme's proof of this result.)

Part VI: Local Motion

There are three categories or generic ways in which motion, in the strict sense, can occur. For whatever is moved, is changed either in its place, or in its quantity, or in its quality. And since, in general, any successive motion whatever is fast or slow, and since no single method of determining velocity is applicable in the same sense to all three kinds of motion, it will be suitable to show how any change of this sort may be distinguished from another change of its own kind, with respect to speed or slowness. And because local motion is prior in nature to the other kinds, as the primary kind, we will carry out our intention in this section, with respect to local motion, before treating of the other kinds.

Although change of place is of diverse kinds, and is varied according to several essential as well as accidental differences, yet it will suffice for our purposes to distinguish uniform motion from nonuniform motion. Of local motions, then, that motion is called *uniform* in which an equal distance is continuously traversed with equal velocity in an equal part of time. *Nonuniform motion* can, on the other hand, be varied in an infinite number of ways, both with respect to magnitude and with respect to time.

In uniform motion, then, the velocity of a magnitude as a whole is in all cases measured by the linear path traversed by the point which is in most rapid motion, if there is such a point. . . . In nonuniform motion, however, the velocity at any given instant will be measured by the path which would be described by the most rapidly moving point if, in a period of time, it were moved uniformly at the same degree of velocity with which it is moved in that given instant, whatever instant be assigned. . . .

With regard to the acceleration and deceleration of local motion, however, it is to be noted that there are two ways in which a motion may be accelerated or decelerated: namely, uniformly, or nonuniformly. For any motion whatever is *uniformly accelerated* if, in each of any equal parts of the time whatsoever, it acquires an equal increment of velocity. And such a motion is *uniformly decelerated* if, in each of any equal parts of the time, it loses an equal increment of velocity. But a motion is *nonuniformly accelerated or decelerated*, when it acquires or loses a greater increment of velocity in one part of the time than in another equal part. . . .

In this connection, it should be noted that just as there is no degree of velocity by which, with continuously uniform motion, a greater distance is traversed in one part of the time than in another equal part of the time, so there is no latitude (i.e., increment, *latitudo*) of velocity between zero degree [of velocity] and some finite degree, through which a greater distance is traversed by uniformly accelerated motion in some given time, than would be traversed in an equal time by a uniformly decelerated motion of that latitude. For whether it commences from zero degree or from some [finite] degree, every latitude, as long as it is terminated at some finite degree, and as long as it is acquired or lost uniformly, will correspond to its mean degree [of velocity]. Thus the moving body, acquiring or losing this latitude uniformly during some assigned period of time, will traverse a distance exactly equal to what it would traverse in an equal period of time if it were moved uniformly at its mean degree [of velocity].

For of every such latitude commencing from rest and terminating at some [finite] degree [of velocity], the mean degree is one-half the terminal degree [of velocity] of that same latitude.

From this it follows that the mean degree of any latitude bounded by two degrees (taken either inclusively or exclusively) is more than half the more intense degree bounding that latitude.

From the foregoing it follows that when any mobile body is uniformly accelerated from rest to some given degree [of velocity], it will in that time traverse one-half the distance that it would traverse if, in that same time, it were moved uniformly at the degree [of velocity] terminating that latitude. For that motion, as a whole, will correspond to the mean degree of that latitude, which is precisely one-half that degree which is its terminal velocity.

It also follows in the same way that when any moving body is uniformly accelerated from some degree [of velocity] (taken exclusively) to another degree inclusively or exclusively, it will traverse more than one-half the distance which it would traverse with a uniform motion, in an equal time, at the degree [of velocity] at which it arrives in the accelerated motion. For that whole motion will correspond to its mean degree [of velocity], which is greater than one-half of the degree [of velocity] terminating the latitude to be acquired; for although a non-uniform motion will likewise correspond to its mean degree [of velocity], nevertheless the motion as a whole will be as fast, categorematically, as some uniform motion according to some degree [of velocity] contained in this latitude being acquired, and, likewise, it will be as slow.

To prove, however, that in the case of acceleration from rest to a finite degree [of velocity], the mean degree [of velocity] is exactly one-half the terminal degree [of velocity], it should be known that if any three terms are in continuous proportion, the ratio of the first to the second, or of the second to the third, will be the same as the ratio of the difference between the first and the middle, to the difference between the middle and the third; as when the terms are 4, 2, 1; 9, 3, 1; 9, 6, 4. For as 4 is to 2, or as 2 is to 1, so is the proportion of the difference between 4 and 2 to the difference between 2 and I, because the difference between 4 and 2 is 2, while that between 2 and 1 is 1; and so with the other cases.

Let there be assigned, then, some term under which there is an infinite series of other terms, which are in continuous proportion according to the ratio 2 to 1. Let each term be considered in relation to the one immediately following it. Then, whatever is the difference between the first term assigned and the second, such precisely will be the sum of all the differences between the succeeding terms. For whatever is the amount of the first proportional part of any continuum or of any finite quantity, such precisely is the amount of the sum of all the remaining proportional parts of it.

Since, therefore, every latitude is a certain quantity, and since, in general, in every quantity the mean is equidistant from the extremes, so the mean degree of any finite latitude whatsoever is equidistant from the two extremes, whether these two extremes be both of them positive degrees, or one of them be a certain degree and the other a privation of it or zero degree.

But, as has already been shown, given some degree under which there is an infinite series of other degrees in continuous proportion, and letting each term be considered in relation to the one next to it, then the difference or latitude between the first and the second degree—the one, namely, that is half the first—will be equal to the latitude composed of all the differences or latitudes between all the remaining degrees, namely those which come after the first two. Hence, exactly equally and by an equal latitude that second degree, which is related to the first as a half to its double, will differ from that double as that same degree differs from zero degree or from the opposite extreme of the given magnitude.

And so it is proved universally for every latitude commencing from zero degree and terminating at some finite degree, and containing some degree and half that degree and one-quarter of that degree, and so on to infinity, that its mean degree is exactly one-half its terminal degree. Hence this is not only true of the latitude of velocity of motion commencing from zero degree [of velocity], but it could be proved and argued in just the same way in the case of latitudes of heat, cold, light, and other such qualities.

With respect, however, to the distance traversed in a uniformly accelerated motion commencing from zero degree [of velocity] and terminating at some finite degree [of velocity], it has already been said that the motion as a whole, or its whole acquisition, will correspond to its mean degree [of velocity]. The same thing holds true if the latitude of motion Is uniformly acquired from some degree [of velocity] in an exclusive sense, and is terminated at some finite degree [of velocity].

From the foregoing it can be sufficiently determined for this kind of uniform acceleration or deceleration how great a distance will be traversed, other things being equal, in the first half of the time and how much in the second half. For when the acceleration of a motion takes place uniformly from zero degree [of velocity] to some degree [of velocity], the distance it will traverse in the first half of the time will be exactly one-third of that which it will traverse in the second half of the time.

And if, contrariwise, from that same degree [of velocity] or from any other degree whatsoever, there is uniform deceleration to zero degree [of velocity], exactly three times the distance will be traversed in the first half of the time, as will be traversed in the second half. For every motion as a whole, completed in a whole period of time, corresponds to its mean degree [of velocity]—namely, to the degree it will have at the middle instant of the time. And the second half of the motion in question will correspond to the mean degree of the second half of that same motion, which is one-fourth of the degree [of velocity] terminating that latitude. Consequently, since this second half will last only through half the time, exactly one-fourth of the distance will be traversed in that second half as will be traversed in the whole motion. Therefore, of the whole distance being traversed by the whole motion, three-quarters will be traversed in the first half of the whole motion, and the last quarter will be traversed in its second half. It follows, consequently, that in this type of uniform intension and remission of a motion from some degree [of velocity] to zero degree, or from zero degree to some degree, exactly three times as much distance is traversed in the more intense half of the latitude as in the less intense half.

5. GIOVANNI DI CASALI, *DE VELOCITATE MOTUS ALTERATIONIS* (*ON THE VELOCITY OF MOTION OF ALTERATION*)

The apex in the development of the Merton theory of speed was reached perhaps by the two-dimensional geometric representation of Giovanni di Casali. A Franciscan friar, he had been teaching at the Franciscan *studium* or college for friars in Assisi (1335–1340) when he was called to be the forty-ninth lector at the Cambridge *studium* (ca. 1340–1341). During his stay there, he had ample time to visit with the friars in Oxford. At one of these two places he learned about the Merton Mean Speed Theorem. On returning to Italy, he was lector at the Bologna *studium* (1346–1352), where in 1346 he wrote *On the Velocity of Motion of Alteration* (*De velocitate motus alterationis*). The passage cited here shows the first use of a geometric diagram to represent what we would think of as a function. Compare this to Oresme's use of diagrams in section II-7-6.

1. *Latitude of calidity*[128] is [1] calidity uniformly difform, or [2] the qualitative distance between degrees of calidity, by which distance the motion of alteration of calidity is measured, or [3] the distance from a summit degree to zero degree. A summit degree (*gradus summus*) of calidity is calidity which throughout is equally and maximally distant from zero degree. An "intended" degree of calidity is calidity which throughout, or at its more intense extreme, is distant from zero degree by a latitude. A "remitted" degree of calidity is [1] a calidity which in its most intense part is truly distant from its summit degree by [some] indivisible latitude, or [2] a remitted degree is spoken of where in its latitude there is, or can be, another degree more intense than it, or [3] [it] is that which contains less intension or qualitative distance, or [4] [it is] that which is less distant from zero degree. A degree in latitude is nothing else than a part of the latitude, which is more intense or more remiss than another part of it. A degree of calidity is calidity which is uniform, [difform,] or uniformly difform. A degree of uniform calidity is a calidity no part of which is more intense or more remiss than another part of it. A difform calidity is [for example,] one whose first part is hot, whose second part is hotter than the first, whose third part is hotter than the second, and so on. A degree of calidity uniformly difform is a difform calidity, wherein, when any two immediate parts are taken, the most intense uniform degree, which is not in the one part, is the most remiss degree, which is not in the other part. According to some, calidity uniformly difform is one whose second part exceeds its first by [some] degree and whose third exceeds its second [by the same degree], and so on. The most intense uniform degree which is not in A is the uniform degree not in A such that some degree more remiss than any such uniform more intense degree is in A. And in such a degree, [there is] no degree more remiss or as remitted [as any one] in A. The most remiss degree which is not in A is the uniform degree not in A such that any degree more intense than any such more remiss degree is in A. And in such a degree [there is] no degree as intense [as any one] in A. . . .

2. The third conclusion is that any uniformly difform latitude is precisely as much as a latitude uniformly intense at the mean degree. Or, anything at all uniformly nonuniform in hotness is precisely as hot [quantitatively] as something uniformly hot at the mean degree of the uniformly nonuniform hotness.

Then Giovanni defined uniformly difform hotness as well as uniform hotness.

One can exemplify these qualitative things [Fig. II-7-5-1]. For something uniformly hot is throughout like a rectangular parallelogram constructed between two parallel lines [AB and CD]. Then any part of such a rectangle is equally wide (*lata*) as any other part, because the latitude of any such part is measured by the base.

Similarly a uniformly difform hotness is in every way like a right triangle [ABC]. This would be a uniformly difform hotness terminated in one extreme [point A] at zero. [This correspondence of uniformly difform hotness and a right triangle] holds because one quarter [ADE] of such a triangle has a line [FG] which is just as distant from one extreme of the quarter as from the other (i.e., the latitude at A is 0, and the latitude at D is DE, and there is a mean latitude FG), and it is just as distant from either extreme as the middle line [LJ] of another small quarter [EIC] is distant from its extremes. But when we speak

128 "Calidity" is an archaic name for "heat."

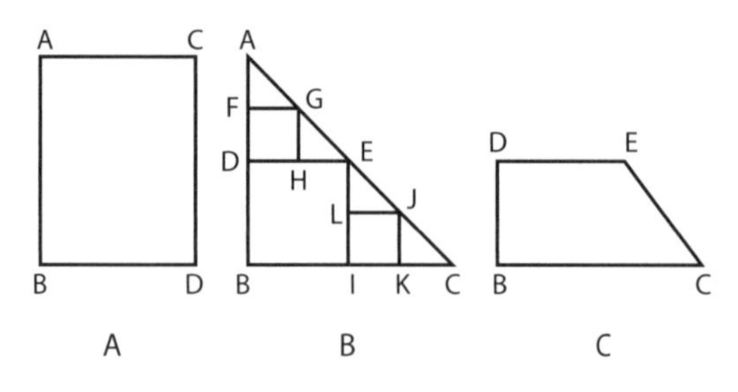

Fig. II-7-5-1.

of distance [we speak in terms] of arithmetic proportion. So the smaller any latitude of some part toward the base is, the more distant is the part it is in, [i.e.,] in any more distant part the latitude is smaller, and no part is uniformly wide. Similarly, any latitude at all is imaginable between the latitude corresponding to the base and a point corresponding to some (the highest?) point of that triangle, because it is possible to find in such a triangle any line that is less than the base of the triangle.

Any latitude uniformly difform terminated at both extremes at some degree is similar to a quadrangle [BDEC] produced by a line [DE] cutting off the apex of the triangle mentioned above, so that it would be terminated at a degree in the more remiss extreme which would be continually less as the line approaches the apex. [Thus DE forms the less intense extreme of the uniformly difform quality represented by BDEC, and obviously as we let DE be closer to A it becomes smaller.]

This passage contains two additional points of interest. First is that Casali applied the Merton Mean Speed Theorem to heat. The second is that he subsequently offered a proof for the Mean Speed Theorem that is more rhetorical than geometric.

Having seen these things, the proposed conclusion is proved this way. Let B be the midpoint of a uniformly hot object A, and C that of another. Apply two agents at the midpoint of A or at its endpoints, I do not care, in which one of them remits C uniformly difformly, so that at the end of an hour C ends without any degree or at a degree below which it was uniform before. Let the agent intend B uniformly difform in the same proportion in which C was remitted. By always calling this an arithmetic proportion, so that the degree in which A was uniformly hot is marked as 4, then whatever midpoint of C is remitted as 2, a point corresponding to itself in the midpoint of B is intended as 2 beyond that which it had before, so that regardless that that point is assigned 6 proportionally. When some point of C is remitted as one (or some degree to which one was assigned), then that point corresponding to it in B is intended, so that that point is assigned as 7, and so on with the others. But C is remitted though every degree among the degrees below which it had and no degree among the other points.

The argument goes this way. If the exact amount taken from some point of C were added to a similar point in B, and thus that point so added to intends to be in the same proportion, to which [point] is added as taken, remits the point from which it is taken, then we are speaking of an arithmetic proportion. Then precisely at the end of such an alteration there is as much latitude as there was at the beginning, because it is the same and nothing has been taken from it. But now, after such an increasing and decreasing made in the parts of A, it is everywhere as it was. If precisely that which is taken from some point of C and decreases that point is added to a similar point B and increases so much proportionally as it lessens the point from which it was taken, because already the median of A is lessened, as happened by such a removal, and another median is so increased by such an addition. Therefore now at the end of the alteration the latitude is not greater or less than it was at the beginning. It is as much as it was at the beginning. In the beginning the latitude was uniform according to the degree of the middle of the latitude, already uniformly difform. Therefore now the latitude is as much as is one uniform latitude at its middle degree. That was what had to be proved.

Similarly through all else: we argue from a uniformly difform latitude terminated at a certain degree of extreme lessening, as is obvious to anyone paying attention.

6. NICOLE ORESME, *DE CONFIGURATIONIBUS QUALITATUM ET MOTUUM* (*ON THE CONFIGURATIONS OF QUALITIES AND MOTIONS*)

Part I of Nicole Oresme's treatise, *De configurationibus qualitatum et motuum*, describes figures that characterize physical or psychological occurrences. As a whole they make up a toolbox for possibly explaining numerous changes. Each description represents a graph. What may be called the horizontal axis (the base of the accompanying picture) represents that time during which the subject acts. The shape of the figure above the base line is a composite of all the changes in intensity of a particular quality over time. We can think of this quality as speed, but Oresme thought about qualities more generally. We excerpt sections from many of the chapters of part I, so that the reader can understand the basics of Oresme's use of figures to represent "qualities."

I.i On the continuity of intensity

Every measurable thing except numbers is imagined in the manner of continuous quantity. Therefore, for the mensuration of such a thing, it is necessary that points, lines, and surfaces, or their properties, be imagined. For in them, as [Aristotle] has it, measure or ratio is initially found, while in other things it is recognized by similarity as they are being referred by the intellect to them [the geometrical entities]. Although indivisible points, or lines, are nonexistent, still it is necessary to feign them mathematically for the measures of things and for the understanding of their ratios. Therefore, every intensity which can be acquired successively ought to be imagined by a straight line perpendicularly erected on some point of the space or subject of the intensible thing, e.g., a quality. For whatever ratio is found to exist between intensity and intensity, in relating intensities of the same kind, a similar ratio is found to exist between line and line, and vice versa. . . . Therefore, the measure of intensities can be fittingly imagined as the measure of lines, since an intensity could be imagined as being infinitely decreased or infinitely increased in the same way as a line.

Again, intensity is that according to which something is said to be "more such and such," as "more white" or "more swift." Since intensity, or rather the intensity of a point, is infinitely divisible in the manner of a continuum in only one way, therefore there is no more fitting way for it to be imagined than by that species of a continuum which is initially divisible and only in one way, namely by a line. And since the quantity or ratio of lines is better known and is more readily conceived by us—nay the line is in the first species of

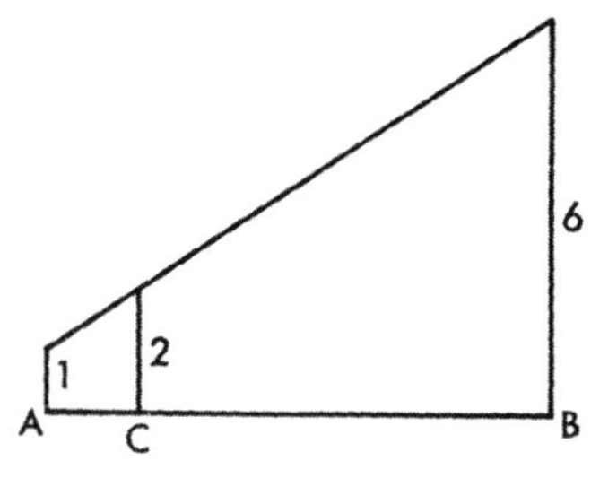

Fig. II-7-6-1.

continua, therefore such intensity ought to be imagined by lines and most fittingly by those lines which are erected perpendicularly to the subject. The consideration of these lines naturally helps and leads to the knowledge of any intensity. . . . Therefore, equal intensities are designated by equal lines, a double intensity by a double line, and always in the same way if one proceeds proportionally. . . .

I.iv On the quantity of qualities

The quantity of any linear quality is to be imagined by a surface whose length or base is a line protracted in a subject of this kind . . . and whose breadth or altitude is designated by a line erected perpendicularly on the aforesaid base. . . . And I understand by "linear quality" the quality of some line in the subject informed with a quality. . . .

I.vi On the clarification of the figures

. . . No linear quality is imagined or designated by any figure except the ones in which the ratio of the intensities at any points of that quality is as the ratio of the lines erected perpendicularly in those same points and terminating in the summit of the imagined figure.

For example, let line AB be divided in point C in any way such that the intensity in point C is double that in point A; and in point B let it be triple that in point C [Fig. II-7-6-1]. Therefore, by the first chapter the line imagined as rising perpendicularly above point C and denoting the intensity at that point is double the line imagined as rising above point A, and the line imagined as rising above point B is three times the line imagined as rising above C. Therefore, this quality can be imagined only by the figure which at point C is twice as high as at point A or whose summit at C is double that at point A, and whose summit at point B is triple that at point C—with the further stipulation however that the figure of this sort could be varied in altitude according to the ratio of intensities in the other points of line AB. . . .

I.vii On the suitability of the figures

Any linear quality can be designated by every plane figure which is imagined as standing perpendicularly on the linear [extension of the] quality and which is proportional in altitude to the quality in intensity. Moreover, a figure erected on a line informed with a quality is said to be "proportional in altitude to the quality in intensity" when any two lines perpendicularly erected on the quality line as a base and rising to the summit of the surface or figure have the same ratio in altitude to each other as do the intensities at the points on which they stand. . . .

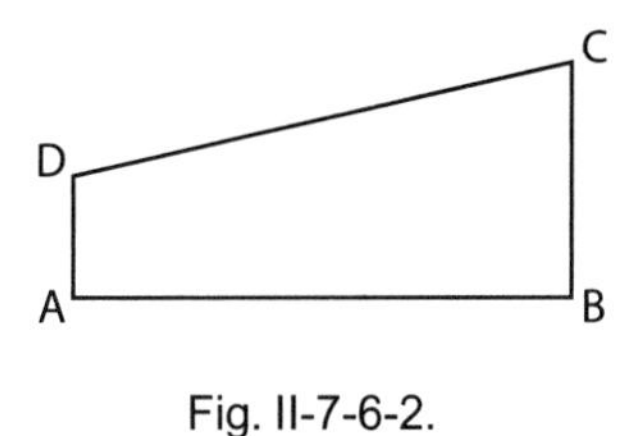

Fig. II-7-6-2.

I.viii On a right-triangular quality

Every quality which is imaginable by a triangle having a right angle on the base can be imagined by every triangle having a right angle on the same base; and by no other figure can it be imagined. That some quality is imaginable by such a triangle is evident . . . because some quality can be proportional in intensity to such a triangle in altitude. This quality is that which is commonly called a "uniformly difform quality terminated at zero degree." However, more properly it can be called a "quality uniformly unequal in intensity" just as the triangle to which it is proportional is uniformly unequal in altitude. . . .

I.x On quadrangular quality

A certain quality is imaginable by a rectangle—in fact by any such rectangle constructed on the same base—and by no other type of figure can it be designated. . . . Moreover, any such quality is said to be "uniform" or "of equal intensity" in all of its parts.

Again it ought to be known that some quality is imaginable by a quadrangle having two right angles on the base and the other two angles unequal, e.g., by quadrangle ABCD [Fig. II-7-6-2]. . . . Moreover, any such quality is spoken of as "uniformly difform terminated in both extremes at some degree," so that the more intense extreme is designated in the acute angle C and the more remiss in the obtuse angle D. The superior line, e.g., line CD, is called "the line of summit," or in relation to quality it can be called "the line of intensity" because the intensity varies according to its variation.

I.xi On uniform and difform quality

And so every uniform quality is imagined by a rectangle and every quality uniformly difform terminated at zero degree is imaginable by a right triangle. Further, every quality uniformly difform terminated in both extremes at some degree is to be imagined by a quadrangle having right angles on its base and the other two angles unequal. Now every other linear quality is said to be "difformly difform" and imaginable by means of figures otherwise disposed according to manifold variation. Some modes of the "difformly difform" will be examined later. The aforesaid differences of intensities cannot be known any better, more clearly, or more easily than by such mental images and relations to figures, although certain other descriptions or points of knowledge could be given which also became known by imagining figures of this sort: as if it were said that a uniform quality is one which is equally intense in all parts of the subject, while a quality uniformly difform is one in which if any three points [of the subject line] are taken, the ratio of the distance between the first and the second to the distance between the second and the third is as the ratio of the excess in intensity of the first point over that of the second point to the excess of

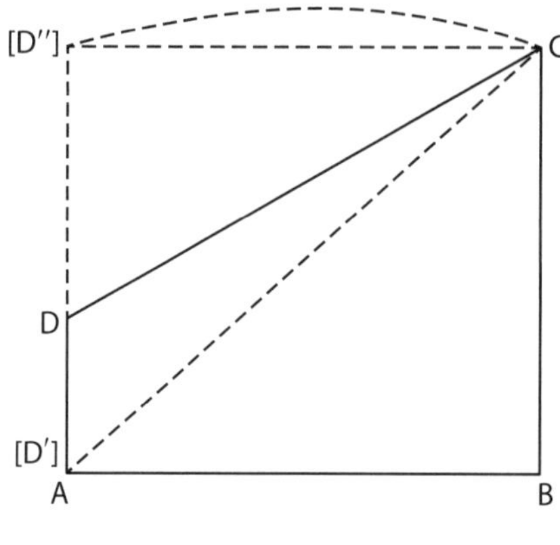

Fig. II-7-6-3.

that of the second point over that of the third point, calling the first of those three points the one of greatest intensity. . . .

I.xiii On these same [qualities considered] in still another way

The previous things can be discriminated in still another way by letting the superior line of the figure by which the quality is imagined be called the line of intensity or line of summit, as was said in chapter ten. An example is line DC in quadrangle ABCD [Fig. II-7-6-3]. If, therefore, the line of summit of the figure by which a quality is imagined is parallel to the base, e.g., to base AB, the quality imaginable by such a figure is simply uniform. If it is not parallel to the base but still is a straight line, then the quality is uniformly difform, so that if the aforesaid line is joined to the base in one extreme that uniform difformity is terminated at zero degree [e.g., D′] and if it is joined to the base in neither extreme, the quality or difformity is terminated in both extremes at [some] degree. And since such a line cannot be joined to the base in both extremes—for it is a straight line and thus would form a single line with the base which also is a straight line—it is clear that there cannot be a quality uniformly difform terminated in both extremes at zero degree. Further, if the line of intensity or summit line is a curve or is composed of several lines rather than one, then the quality imaginable by that figure will be difformly difform, and it can be that it is terminated in both extremes at some degree [e.g., D″], or in both extremes at zero degree, or at some degree in one extreme and at zero degree in the other.

I.xiv On simple difform difformity

We now treat of difform difformity; there are two modes of such difformity: simple and composite. We must first talk of the simple mode. Simple difform difformity is that which can be designated by a figure whose line of summit or line of intensity is a single line, i.e., not composed of several lines. It is necessary, therefore, that the line be a curve; because if it were straight, then it would be simply a uniformity or uniform difformity, as is clear from the preceding chapter. Furthermore, it is necessary that the curvature of the summit line does not attain that of a circular segment greater than a semicircle so that the angle on the base is greater than a right angle. . . . However, it can happen that the angle on the base is less than a right angle by any amount you please.

. . .

I.xv On four kinds of simple difform difformity

Therefore, every simple difform difformity either (1) is imaginable by a figure which is not a segment of a circle nor proportional in altitude to some segment of a circle but whose summit is determined by an irrational curvature, or (2) is imaginable by a figure whose summit is determined by a rational curvature, namely, by a circular figure or one proportional to it in altitude. And each of these two kinds of figures can be either convex or concave. According to these four differences, then, there are four kinds of simple difform difformity, namely (1) rational convex, (2) rational concave, (3) irrational convex, and (4) irrational concave.

In addition to these four [essential] differences there are other accidental differences, such as whether the qualities are terminated at [some] degree or at zero degree. Therefore, both rational and irrational convex difformity can be determined in both extremes at [some] degree, as in this figure ; or in one extreme at [some] degree and in the other extreme at zero degree, as in this figure ; or in both extremes at zero degree, as here . On the other hand, concave [difformity], whether rational or irrational, cannot be terminated in both extremes at zero degree, but it can be terminated in both extremes at [some] degree, as here ; or in one of the extremes, as here .

The overarching significance of the following section is Oresme's use of a method for determining the number of combinations of 6 things taken k at a time; in modern notation

$$\binom{6}{k} = \frac{6!}{(6-k)!k!}$$

All the results of the computations are common to the manuscripts, including the few errors corrected in the interpolated parentheses below. For the appearance of this method in Arabic and Hebrew mathematics of the time, see sections III-2 and III-3 in Chapter 3 and sections II-2 and II-3 in Chapter 2, respectively.

I.xvi On composite difformity and how it has sixty-two (63) species

Setting aside the variation of difformity arising from accidental differences dependent on whether the termination is at [some] degree or at zero degree—since we have said enough about this before—there are beyond the four simple kinds of quality figuration posited in the preceding chapter two others which were treated earlier: simple uniformity and uniform difformity. Thus there are six kinds of simple figuration of qualitative intensity. Further, since composite difform difformity can be effected from several simple figurations, either of one kind, or two, or three, or four, or five, or six, it follows by arithmetical rules that from each simple kind some combination or composition can be formed, and so we have six species of composite difformity. Then taking two simple kinds at a time, up to fifteen combinations and composite species are formed. Also, taking three at a time, twenty combinations are formed; and four at a time, fifteen combinations are formed; five at a time, five (six) combinations; and finally taking all of them simultaneously, one combination is formed. And so in summary there are sixty-two (63) species of composite difform difformity. And in each species one can make a combination of two simple figurations, or three, or four, and so on to infinity. And so accordingly in each species there can be infinite variation by reason of the number and by reason of the order or disposition of the simple figures of which these species are composed.

So that this [whole matter] might be seen more clearly, let me put forth examples of certain of the species. There can be a difform difformity composed out of several items of one simple kind, for example out of two or more uniform qualities, as here 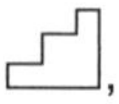 , and this can be called a graduated quality or difformity. A further example of a composition from another simple kind is one composed of two or more uniformly dlfform [qualities] thus:  , and also this one:  . A further example is a composition of two convex rationals, such as this one:  . And so in the same way one could run through the six kinds [of simple figuration] described above. Then again a composition can be formed from two [different] kinds, as for example from a uniform and a uniformly difform [quality] thus:  ; or from more of these two kinds thus:  . An example of a combination made from two other kinds is one formed from a uniform and a concave rational [difformity], thus:  ; or from more of these two kinds. In the same fashion there could be a mixture of two other kinds and again of still other pairs up to fifteen mixtures or species. Further, there can be mixtures of three kinds up to twenty species, and so on for the other [combinations] as was stated above. But examples of these can be sufficiently understood if one makes use of what we have already said. And so we have 62 (63) species of composite difform difformity plus four simple species or kinds, as is evident in the preceding chapter. Thus there are 66 (67) species or kinds of difform difformity, one of uniform difformity, and one of simple uniformity. It is also clear enough from what has been said that we cannot [very] well arrive at a knowledge of the diversity of these species of the difformity of qualities or other things except by assimilating them to, and imagining them by, figures.

In part III of *De configurationibus qualitatum et motuum*, Oresme continued his analysis of his graphical representations of qualities, in particular, speed, by determining in various cases the quantity, or total quality, over a given time period, which we interpret as distance traveled in that time, under various configurations of the speed. In some of these configurations, Oresme actually summed infinite series. But note that Oresme was not just concerned with speed. He intended his configurations to represent any type of quantity that varied in intensity. As before, we give excerpts from several of the chapters.

III.i How the acquisition of quality is to be imagined

Succession in the acquisition of quality can take place in two ways: (1) according to extension and (2) according to intensity. . . . And so extensive acquisition of a linear quality ought to be imagined by the motion of a point flowing over the subject line in such a way that the part traversed has received the quality and the part not yet traversed has not received the quality. An example of this occurs if point *c* were moved over line AB so that any part traversed by it would be white and any part not yet traversed would not yet be white. . . .

III.ii Application of the prior statements to uniformity and difformity

As has been said in chapter ten of the first part, the superior line of the figure by which a linear quality is imagined is called in our proposal "the line of summit." . . . So long as the summit . . . is parallel to the base, then the quality is uniform. Further, so long as it is right but is not parallel to the base, . . . then the quality is uniformly difform. But so long as the summit is related [to the base] in another way, then the quality or intensity is difformly difform. . . .

III.v On the measure of uniform qualities and velocities

The universal rule is this, that the measure or ratio of any two linear . . . qualities or velocities is as that of the figures by which they are comparatively and mutually imagined. . . . Therefore, in order to have measures and ratios of qualities and velocities, one must have recourse to geometry. Nothing more of our subject remains for treatment except to show the application of the mensuration of figures to the measure of qualities and velocities by means of certain examples and means of a certain exercise.

And so I say first that the ratio of all uniform qualities of equal degrees [of intensity] is as [that of] their subjects, just as the ratio of all rectangles of equal altitude is as [the ratio of] their base lengths. Further, the ratio of all uniform qualities whose subjects are equal is as the ratio of their intensities, so that if the subjects are equal and one is twice as intense as another, the one that is more intense will be double the other. However, the ratio of intensities is not so properly or so easily attainable by the senses as is the ratio of extensions. . . .

III.vi Further consideration of the same subject

Again returning to the principal subject, I say that in the case of uniform qualities which are unequal in both extension and intensity, if the one that is more extended is more intense, then the ratio of the greater to the lesser is composed of the ratio of extension to extension and the ratio of intensity to intensity. If, however, the one that is more extended is less intense, then the ratio of extension to extension is to be taken from [i.e., divided into] the ratio of intensity to intensity or vice versa, and the result is the ratio of the one quality to the other. But that quality will be greater which was either in extension or intensity the term of the larger ratio. For example, let A be a quality three times as extensive as quality B, and similarly twice as intense. There the double ratio is to be [multiplied] by the triple ratio, and the compound ratio will be a sextuple ratio, i.e., six to one. Therefore, quality A exceeds quality B in such a ratio. But let it be posited that A is twice as extensive as B, while B is three times as intense as A. Then the double ratio is to be taken away from [i.e. divided into] the triple ratio and a sesquialterate ratio results, i.e., a ratio of 3 to 2. And since B was the term of the larger ratio, i.e., the triple ratio, therefore quality B is greater than quality A, in the amount of the aforesaid ratio, i.e., in a sesquialterate ratio.

III.vii On the measure of difform qualities and velocities

Every quality, if it is uniformly difform, is of the same quantity as would be the quality of the same or equal subject that is uniform according to the degree of the middle point of the same subject. . . . Let there be a quality imaginable by triangle ABC, the quality being uniformly difform and terminated at zero degree in point B [Fig. II-7-6-4]. And let D be the middle point of the subject line. The degree of this point, or its intensity is imagined by line DE. Therefore, the quality which would be uniform throughout the whole subject at degree DE is imaginable by rectangle AFGB, as is evident by the tenth chapter of the first part.

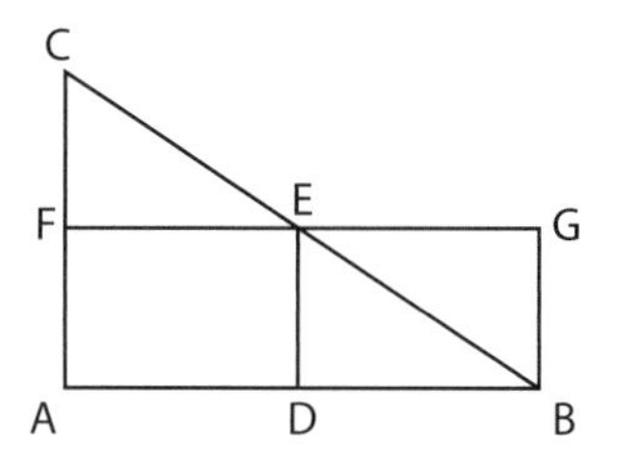

Fig. II-7-6-4.

Therefore, it is evident by I.26 of the *Elements* of Euclid that the two small triangles EFC and EGB are equal. Therefore, the larger triangle BAC, which designates the uniformly difform quality, and the rectangle AFGB, which designates the quality uniform in the degree of the middle point, are equal. Therefore the qualities imaginable by a triangle and a rectangle of this kind are equal. And this is what has been proposed. . . .

Further, if a quality or velocity is difformly difform, and if it is composed of uniform or uniformly difform parts, it can be measured by its parts, whose measure has been discussed before. Now, if the quality is difform in some other way, e.g. with the difformity designated by a curve, then it is necessary to have recourse to the mutual mensuration of the curved figures, or to [the mensuration of] these [curved figures] with rectilinear figures; and this is another kind of speculation. Therefore what has been stated is sufficient.

In section III.viii, Oresme summed the infinite series $\frac{1}{2}\cdot 1+\frac{1}{4}\cdot 2+\frac{1}{8}\cdot 3+\cdots+\frac{1}{2^n}\cdot n+\cdots$ and found that the result was 2.

III.viii On the measure and intension to infinity of certain difformities

A finite surface can be made as long as we wish, or as high, by varying the extension without increasing the size. For such a surface has both length and breadth and it is possible for it to be increased in one dimension as much as we like without the whole surface being absolutely increased so long as the other dimension is diminished proportionally. . . . For example, . . . let there be a surface of one square foot in area whose base line is AB; and let there be another surface, similar and equal to it, whose base line is CD [Fig. II-7-6-5]. Let the latter surface be imagined to be divided on line CD to infinity into parts continually proportional according to the ratio 2 to 1, with its base divided in the same way. Let E be the first part, F the second, G the third, and so on for the other parts. Therefore, let the first of these parts, namely E, which is one half the whole surface, be taken and placed on top of the first surface towards the extremity B. Then upon this whole let the second part, namely F, be placed and again upon the whole let the third part, namely G, be placed, and so on for the others to infinity. When this has been done, let base line AB be imagined as being divided into parts continually proportional according to the ratio of 2 to 1, proceeding towards B. And it will be immediately evident that on the first proportional part of line AB there stands a surface one foot high, on the second a surface two feet high, on the third one three feet high, on the fourth four feet high, and so on to infinity, and yet the whole surface is only the two [square] feet previously given, without augmentation. And consequently the whole surface standing on line AB is precisely four times its part standing on the first proportional part of the same line AB. Therefore, that

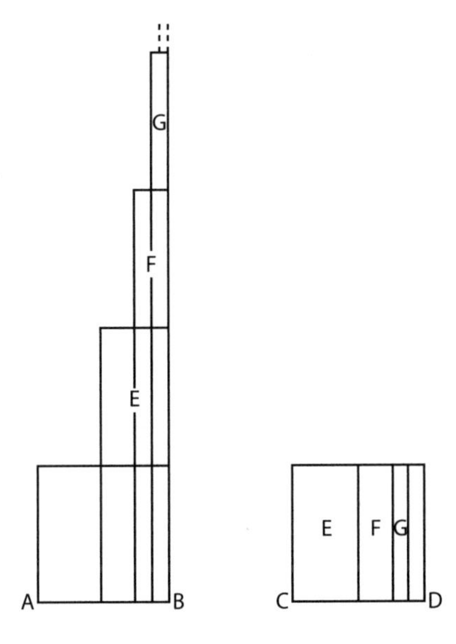

Fig. II-7-6-5.

quality or velocity which would be proportional in intensity to this figure in altitude would be precisely four times the part of it which would be in the first part of the time or the subject so divided. . . .

For example, if some mobile were moved with a certain velocity in the first proportional part of some period of time, divided in such a way, and in the second part it were moved twice as rapidly, and in the third three times as fast, in the fourth four times, and increasing in this way successively to infinity, the total velocity would be precisely four times the velocity of the first part, so that the mobile in the whole hour would traverse precisely four times what it traversed in the first half of the hour; e.g., if in the first half or first proportional part it would traverse one foot, in the whole remaining period it would traverse three feet and in the total time it would traverse four feet.

In chapter III.x, Oresme presented a more complicated example, in which a body moves partly uniformly and partly difformly in different sections of time which decrease infinitely. As Fig. II-7-6-6 shows, the line AB is divided successively into parts in the ratio of 2:1. Further, the heights (intensities) of the odd-numbered sections (1°, 3°, 5°) are uniform but successively twice the preceding odd section. In the even sections (2°, 4°, 6°) the heights vary from the preceding odd section to the height of the following odd section. Hence, the two sets of sections can be represented by these infinite series:

$$\text{Odd} : 1 + \frac{1}{2} + \frac{1}{4} + \ldots + \frac{1}{2^{n-1}} + \ldots = 2 = \frac{4}{2}.$$

$$\text{Even} : \frac{3}{2} \cdot \frac{1}{2} + \frac{6}{2} \cdot \frac{1}{8} + \ldots + 3\left(\frac{2^{n-1}}{4^n}\right) + \ldots = \frac{3}{2}.$$

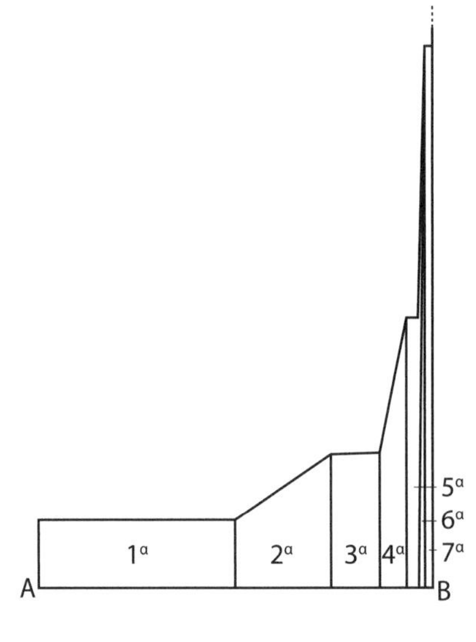

Fig. II-7-6-6.

The sum of the two series is obviously $\frac{7}{2}$. And so the text continues as follows.

Let line AB be divided in infinity by the designation of parts proportional according to a ratio of two to one, as is the common custom, so that the first proportional part is one-half of the whole, the second is half of the remainder, and so on to infinity. And let (1) the quality of the first part be uniform in a certain degree, (2) the quality of the second part be uniformly difform from that same degree to a degree twice it, (3) the quality of the third be uniform in that double degree, (4) the quality of the fourth be uniformly difform from that double degree to a degree twice the double degree, proceeding continually and alternately with the other parts in a manner similar to that of the figure placed here [Fig. II-7-6-6]. Therefore, I say that the quality of the whole will be triple sesquialterate to the quality of the first part of it, so that the ratio of the whole to that first part will be as 7 to 2. . . .

Proof of the conclusion: Let us take the parts denominated by odd numbers, namely, the first, third, fifth, seventh, and so on. Then it is evident from the case [posited] that the first part is quadruple the third in extension and similarly the third is quadruple the fifth in extension, and the fifth the seventh, and the seventh the ninth, and so on for the others. Again it is evident from the case that contrariwise the quality of the third is double that of the first in intensity, and the quality of the fifth similarly is double that of the third in intensity, and so on to the end. Therefore, by the prior statements made in the sixth chapter of this part, the quality of the first part is double the quality of the third, the quality of the third is double the quality of the fifth, and so on for the other [odd-numbered parts]. Therefore the total aggregate of all [the parts] denominated by an odd number is precisely double the first part which is denominated by unity, unity occupying the place of an odd number in a series of numbers. By the same method it can be shown that the quality of the second part is double the quality of the fourth, and the quality of the fourth is double that of the sixth, and so on for the other parts denominated by the even numbers.

But let us proceed in still another way, for the quality of the second part is uniformly difform. Therefore, by the seventh chapter of this part its quantity is just as great as if it were uniform at the degree of the middle point. And from the case it immediately follows that that degree of the middle point is sesquialterate the degree of the first part, i.e., as 3 to 2. Therefore, the quality of the second part is just as great as if it were uniformly more intense than the first part according to a ratio of 3 to 2, and it is one-half as extended as the first. Therefore, by the sixth chapter of this [part], the quality of the first part is sesquitertiate the quality of the second part, so that it is related to it as 4 to 3. But as the first part is to the second, so is the third to the fourth, the fifth to the sixth, and so on for the others. Therefore, the whole aggregate of the parts demonstrated by odd numbers is sesquitertiate the whole aggregate of the parts denominated by even numbers. If therefore, for example, the first part of this quality is as 2, then the whole aggregate of it and the other odd numbers is as 4, as was demonstrated before. Therefore, the whole aggregate of all the even numbers is as 3, as was just seen. Therefore, the whole quality is as 7. And so the ratio of the whole quality to its first part is as 7 to 2, i.e., it is a triple sesquialterate ratio, which was proposed. A similar [argument] can be easily applied to velocity, as was said [previously].

III. ABBACIST SCHOOLS: 1300–1480

For centuries, particularly after the mandate of Charlemagne, education outside the home had been in the hands of monks and clergy. The monasteries and cathedrals provided the venue where students—usually from the upper class—received their primary education (mostly in Latin) for a life in politics or religion. While there had been individuals not of the clergy who taught privately, such as Fibonacci, the first inroad on the established paradigm was the *scuola d'abbaco*, which were first established in Italy during the late thirteenth century and which held lectures and recitations in the vernacular. The most important locale for these abbacist schools was Florence, but similar schools existed in other merchant centers in northern Italy. The growth of business required an educated merchant class. The abbacist schools provided the necessary training. These schools were also businesses, each operated by a craftsman.

Giovanni Villani's description of the student population of schools in Florence was noted above in the introduction to this chapter. Children entered the first school at age 7 for three to four years of work on learning to read. Then, the mathematics school (the abbacist school itself) had a two-year curriculum, consisting mostly of arithmetic, algebra, and geometry for practical purposes (business and surveying). The grammar school kept the best students for two to three more years. After that they either entered business at a level befitting their education or went on to a university to study law, medicine, theology, or some combination of these topics. The outstanding mathematical achievement in this period was a solution of some cubic equations.

The purpose of the abbacist schools was to train students in everything necessary to maintain and foster the merchant economy that had come to make Florence, the banking capital of Europe, famous throughout the European world. (At that time, the word "abbacus" meant "calculation," and "abacus" referred to the tablet or instrument.) Florence had displaced Flanders in the production of cloth. Its gold florin had become the standard for currency in most of Northern and Western Europe. Individuals engaged in business needed to know the basic arithmetic operations related to commercial practice, such as problems of pricing, calculation of interest, computation and division of profits, exchange of currency, and the writing of the Hindu-Arabic numerals. The mathematics curriculum provided the tools. The typical content of an *abbacus* text, no longer written in Latin but in some form of Italian, has nine subjects:

1. Symbolic and finger representation of numbers;
2. Multiplication and division of whole numbers;
3. Addition, subtraction, multiplication, and division of fractions and mixed numbers;
4. Finding square and cube roots;
5. Basic geometric concepts;
6. The rule of three;
7. The Florentine monetary system;
8. Algebraic concepts, paradigms, processes; and
9. Problems both theoretical and practical to exemplify all of the foregoing.

Easily seen in this list is the influence of a similar outline in Fibonacci's *Liber abbaci* of nearly a hundred years before. While there is evidence of the use of the Hindu-Arabic numerals before Fibonacci's treatise, writing numbers in the place-value system became paramount after the dissemination of this work. This included understanding the necessity of the concept of zero and its double meaning: on the one hand it signified "nothing," on the other hand it had the power to increase the value of a numeral to which it was appended by tenfold.[1] Some Masters prepared a single text for algebra and another for practical geometry. We consider here the teaching of foreign exchange, geometry, and algebra.

III-1. Foreign Exchange

Several of the city-states in the Italian peninsula had their own coinage, so that the ability to exchange money from one currency system to another was crucial for doing business in Italy. Consider this example, which uses the rule of three [Van Egmond, 1976, p. 237]:

If 11 bolognini are worth 17 pisani, how much are 29 pisani worth?

This problem in foreign exchange had to be matched with the paradigmatic rule:

If we are given a problem in which there are given three things, we must multiply the thing that we want to find by that which is not the same, and divide by the other.

The problem is then solved:

We take the thing which we want to find, which is the value of 29 pisani, and multiply it by that number in the problem which is not of the same quantity, the 11 bolognini, and divide by the remaining quantity, 17 pisani. Thus 29 times 11 divided by 17 is 18 13/17.

Theory was closely tied to practice. Application of the theory was always directed to problems drawn from everyday life or at least related to it.

[1] The time would come when the Hindu-Arabic numerals would be written with periods fore and aft, lest anyone change their values. But that is another story, and here we are dealing with the late 1200s and early 1300s.

III-2. Geometry

The geometry taught to students was geared toward land measurement. This by definition is practical geometry. Codices from this period and later reflect classes in land measurement as early as the end of the thirteenth century. In the following example, an early maestro teaches how to calculate the area of a circle by multiplying half the diameter by half the circumference. He then states that the area of a circle inscribed in a square is 11/14 of the area of the square, and that the area of a square inscribed in a circle is 7/11 of the area of the circle. The results are exact if we consider π to be equal to its Archimedean value of 22/7. A linear *braccia* is a little more than half a meter. Note that volume is expressed in "square braccia." The author applies these methods to an example.

There is a round tank of circumference 220 braccia and 24 braccia deep, and it is half filled with water. A marble column has fallen in this tank. The circumference of this column is 44 braccia and its length is 4 braccia. We ask, how much has the level of the water increased?

This is its rule: we must know how many square braccia are [contained in] the whole tank. Thus we must divide 220 braccia by $3\frac{1}{7}$. The result is 70 braccia for the diameter. To know how many braccia this yields, we must take the half of the circumference, that is, 110 braccia, and half of its diameter, that is, 35 braccia, and multiply 35 by 110 to make 3850 square braccia. . . .

And to know how many braccia the water increased from the submerged column of circumference 44 braccia and of diameter 14 braccia, we must know the volume of this column in this way. We must take 22 braccia for half the circumference and 7 braccia for half the diameter and multiply 7 by 22 to make 154 braccia. Because it is 4 braccia long, we must take 4 times 154 to make 616 braccia, the square braccia of the whole column. So to know by how much the water has increased, we must divide 616 braccia by 3850. The result is 616/3850 braccia, the amount the water has increased. And to make finer the problem we must simplify the numbers and we obtain 4/25 braccia. And in this way do similar problems.

Towers with a fountain lying between them and a bird atop each is the setting for several problems in the practical geometry books used in the schools. One such problem follows.

There are two towers, one being 60 braccia high and the other 40 braccia high. They are 50 braccia away from one another. On the top of each tower there is one pigeon. They go to drink at a fountain that lies between the towers. Both flying at the same speed, they start at the same time and arrive at the same time. I want to know how far the fountain is from each tower.

You must do this: let the shorter tower be 40 braccia high. Calculate 40 by 40 to get 1600. The towers are 50 braccia apart. So calculate 50 by 50 to get 2500. Add this to 1600 to have 4100. The taller tower is 60 braccia high. Calculate 60 by 60 to reach 3600. Subtract it from 4100; 500 braccia remains. Now divide 500 braccia by twice the distance between the two towers or 100. Thus divide 500 braccia by 100 to get 5 braccia. Thus the taller tower is 5 braccia from the fountain. The shorter will be what remains from the 50 braccia; that is 45 braccia. We are done. Do likewise with similar problems.

The distance of the fountain from the higher tower is calculated by applying the following algorithm: square the height of the lower tower and add the square of the distance between the towers, subtract it from the square of the height of the higher tower and divide the result by twice the distance between the towers. This algorithm can be easily explained using algebra. Let x be the distance of the fountain from the higher tower, h_1 and h_2 the heights of the towers, d_1 and d_2 the distances of the fountain from the top of the towers, and d the distance between the towers. Then we have $d_1^2 = h_1^2 + x^2$ and $d_2^2 = h_2^2 + (d - x)^2$. Because $d_1^2 = d_2^2$ we have $h_1^2 + x^2 = h_2^2 + d^2 + x^2 - 2\,dx$ and finally $x = \frac{(h_2^2+d^2)-h_1^2}{2d}$.[2]

III-3. Algebra

Algebra began to develop in Italy, based on its Islamic roots, during the late Middle Ages. A chapter on algebra appears in perhaps one out of four surviving abbacist books; we can call them lecture notes testifying to the ingenuity or simply the interest of the teacher to explain what they understood. Examples follow of how the algebra, introduced in translations from the Arabic and by Fibonacci in his *Liber abbaci*, was applied and developed.

1. GILIO DA SIENA, *A LECTURE IN INTRODUCTORY ALGEBRA*

Gilio da Siena (fl. 1374–1407) taught in an abbacist school. In preparation for this duty, he wrote a series of lecture notes, finishing one set on March 11, 1384. In the translation that follows, three words are left in Italian for the sake of "flavor": *cosa* for "thing," *censo* for "square," and *cubo* for "cube." What follows is one day's lesson.

The Beginning of the Rules of the *Cosa*

Cosa times number makes *cosa.*

Cosa times *cosa* makes *censo.*

Cosa times *censo* makes *cubo.*

Cosa times *cubo* makes *censo* of *censo.*

Censo times *censo* makes *censo* of *censo.*

Now we have spoken of multiplication of the *cosa.* It would be good to multiply more today, but that would become too tiresome. However, for what we have multiplied above, we exemplify here below, dividing every multiplication by itself.

The first rule is: when the *cosa* is equal to a number, we must partition that number by the *cosa*, which is there; that is, reduce to one *cosa*. And what results is the value of the *cosa* which you want. And this problem is solved by a number. I give you an example and say thus:

Find one number which multiplied by 6 and this total partitioned by 3 yields 10.

Proceed thus. Let the number be 1 *cosa,* which multiplied by 6 makes 6 *cose.*

Divide by 3 to get 2 *cose.* We have 2 *cose* equal to 10. Reduce to one *cosa.* Partition 10 by 2 to get 5. And 5 is that number.

When *censo* is equal to *cosa* and number, it is said: reduce to 1 *censo* and halve the *cosa* and multiply the half by itself and join it to the number, and the root of that sum and

[2]For a thorough history of this problem, including its possibly Hindu roots, see [Dold-Samplonius, 1996].

the other half which you took is the value of the *cosa*, which you sought. That problem is solved by the root of the number and the number. For example:

Find 1 number which multiplied by itself and from that total take the number itself and the remainder is 20.

Make that number 1 *cosa*, which multiplied by itself makes 1 *censo*. Remove 1 *cosa*. There remains 1 *censo* less 1 *cosa* and that is equal to 20. Halving 1 *cosa* produces 1/2 *cosa,* which multiplied by itself makes 1/4 *censo.* Join with 20 to make 20 1/4. And the root of 20 1/4 and 1/2 is the value of the *cosa* that you sought. That much is the number. This is proved by the number because the root of 20 1/4 is 4 1/2, which joined to 1/2 makes 5, which multiplied by itself makes 25 from which 5 is taken and 20 remains. And that is correct.

Over time, the ability to express ideas improved, as seen in the work of a Florentine visitor to Montpellier.

2. PAOLO GIRARDI, *LIBRO DI RAGIONI*

Besides being a university city, Montpellier in southern France was a center for merchants, particularly from southern Europe. One of the many visitors was the Florentine Paolo Girardi (fl. 1325–1335). On January 30, 1327 in Montpellier he completed his copy of *Libro di Ragioni*, substantially an abbacist manual to which he appended theoretical explanations of rules for solving equations, at the same time offering practical examples. Consider these two forms of the quadratic equation with illustrative story problems.

[4] When the *censi* and the *cose* are equal to the numbers, we must divide by the *censi* and then divide the *cose* in half and multiply this half by itself and add it onto the numbers, and the root of this sum minus half of the *cose* is equal to the *cosa.*

A man loaned 20 *lire* to another for two years at compound interest. When the end of 2 years came he gave me 30 *lire*. I ask you at what rate was the *lire* loaned per month?

Let us suppose that the *lira* was loaned at one *cosa* a month.[3] The *lira* earns 12 *cose* per year. Take 1/20 of the 12 *cose* and say this: 1/20 of the 20 *lire* is also one *cosa* and I have for the first year 20 lire and one *cosa*. Now do the second year and say, 1/20 of 20 *lire* is also one *cosa* and 1/20 of one *cose* of the first year is 1/20 of a *censo*, and you now have 20 *lire* and 2 *cose* and 1/20 of a *censo.* And we have that 20 *lire* and 2 *cose* and 1/20 of a *censo* are equal to 30. Take 20 *lire* from 30, leaving 10. Then we have the 2 *cose* and 1/20 of a *censo* are equal to 10 *lire*. We must divide by the *censi.* We must change it to one *censo* by multiplying by 20 and say, 20 times 1/20 of a *censo* makes 1 *censo*, and 20 times 2 *cose* makes 40 *cose*, and 20 times 10 makes 200. Now we have that 40 *cose* and one *censo* is equal to 200. We must halve the *cosa*. We say that half of 40 is 20. We say 20 times 20 makes 400. Add it onto 200 to make 600. Then we say that the root of 600 minus 20, which was half of the *cose,* equals the *cosa.* And you supposed that the *lira* was loaned at one *cosa* a month. Thus the *lira* was loaned at this much per month, that is at the rate of the root of 600 minus 20 *denari.*

[3]If we let x represent the *cosa*, then x is the number of *denari* earned by 1 *lira* in one month, where there are 240 *denari* in a *lira*. Thus 1 *lira* earns $12x$ *denari* per year or $(1/20)x$ *lire*. So 20 *lire* become $20(1 + (1/20)x) = 20 + x$ *lire* after one year and $(20 + x)\ (1 + (1/20)x) = (1/20)x^2 + 2x + 20$ *lire* after two years.

[5] When the *cose* are equal to the *censo* and to the number, you must divide by the *censi* and then halve the *cose* and multiply it by itself and take away the number, and the root of what remains plus half of the *cosa* equals the *cosa*. Or half of the *cosa* minus the root of what remains after the number has been taken away from the multiplication of half the *cosa*, and this much will equal the *cosa*.

There is a man who went on 2 voyages. On the first voyage he earned 12 *denari*. On the second voyage he earned at the same rate that he made on the first voyage, and at the end he found [he had] 100 *denari*. I ask you with how many *denari* did he leave?

Let us suppose that he left with one *cosa* and earned 12 *denari*. Therefore he had at the [end of the] first voyage 1 *cosa* and 12 *denari*. We come to the other voyages. If on the first he made 1 *cosa* and 12 *denari* for one *cosa*, what would he make from one *cosa* and 12? Therefore multiply 1 *cosa* and 12 *denari* times 1 *cosa* and 12 *denari*, and then divide by one *cosa*. Now you say this: one *cosa* and 12 *denari* times one *cosa* and 12 *denari* makes 1 *censo* and 24 *cose* and 144 *denari*. You must divide it by one *cosa* and there results 100 *denari*. But 100 times one *cosa* makes 100 *cose*. You have that 100 *cose* are equal to one *censo* and 24 *cose* and 144 *denari*. Take the *cose* from the *cose*. Take 24 *cose* from 100 *cose*, leaving 76 *cose*, which half is 38. Multiply it by itself making 1444. Subtract the number 144, and it leaves 1300, and the root of 1300 plus half of the *cosa*, which was 38, equals the *cosa*. And you supposed that he carried one *cosa*. Therefore he must have at first the root of 1300 plus 38 in number. And it is done, and in this way do similar problems.

3. JACOBO DA FIRENZE, *TRACTATUS ALGORISMI*

A set of equations, same equations in the same order, appearing in two late fourteenth-century manuscripts became part of five fifteenth-century manuscripts, thus forming a special textual tradition [Van Egmond, 2008, pp. 312 ff.]. According to Van Egmond, a copy of the equations was added to a fifteenth-century copy of the 1307 *Tractatus algorismi* by Jacobo da Firenze (fl. 1305–1325), originally composed in Montpellier.[4] What is particularly interesting is the gradual exposition of the solution of a quadratic equation. First is the succinct solution of the quadratic in the (modern) form $bx = ax^2 + n$. Then follows a numerical example. Finally, a "real life" problem concludes the exposition.

When the things are equal to the *censi* and to the number, one shall divide in the *censi*, and then halve the things and multiply by itself and remove the number, and the root of that which remains and then the halving of the things is worth the thing. Or indeed the halving of the things less the root of that which remains.

Example to the said rule. And I want to say thus, make two parts of 10 for me, so that when the larger is multiplied against the smaller, it shall make 20. I ask how much each part will be. Do thus, posit that the smaller part was a thing. Hence the larger will be the remainder until 10, which will be 10 less a thing. Next one shall multiply the smaller, which is a thing, by the larger, which is 10 less a thing. And we say that it will make 20. And therefore multiply a thing times 10 less a thing. It makes 10 things less one *censo*, which

[4]Høyrup argues that these algebra problems come from the original by Jacobo and were not added later [Høyrup, 2007, pp. 5–25].

multiplication is equal to 20. Restore each part, that is, you shall join one *censo* to each part, and you will get that 10 things are equal to one *censo* and 20 numbers. Bring it to one *censo,* and then halve the things, from which 5 results. Multiply by itself, it makes 25. Remove from it the number, which is 20, 5 remains, of which seize the root, which it is manifest that it does not have precisely. Hence the thing is 5, that is, the halving, less root of 5. And we posited that the part, that is the smaller, was a thing. Hence it is 5 less root of 5. And the second is the remainder until 10, which is 5 and added root of 5. And it goes well.

Somebody makes two voyages, and in the first voyage he gains 12. And in the second voyage he gains at that same rate as he did in the first. And when he had completed his voyages he found himself with 54, gains and capital together. I want to know with how much he set out. Posit that he set out with one thing, and in the first voyage he gained 12. Hence when the first voyage was completed, he found himself with one thing and 12. Hence you see manifestly that from each one thing in the first voyage he makes a thing and 12. How much will it be at that same rate in the second voyage? It suits you to multiply a thing and 12 times a thing and 12, which makes one *censo* and 24 things and 144 numbers, which, according to what the rule says, one shall divide by a thing, and 54 shall result from it. And therefore multiply 54 times a thing. It makes 54 things, which equal one *censo* and 24 things and 144 numbers. Restore each part, that is, you shall remove 24 things from each part. And you will get that 30 things are equal to one *censo* and 144 numbers. Divide in one *censo,* the same results. Halve the things, 15 remain. Multiply by itself, they make 225. Detract from it the numbers, which are 144, so 81 is left. Find the root, which is 9. Detract it from the halving of the things, that is from 15, 6 is left, and as much is worth the thing. And we said that he set out with one thing. Hence you see manifestly that he set out with 6. And if you want to verify it do thus. You say that in the first voyage he gained 12, and with 6 he set out, one has 18. So that in the first voyage he found himself with 18. And therefore I say thus, if from 6 I make 18, what will I make from 18 at the same rate. Multiply 18 by 18, it makes 324. Divide by 6, and 54 results from it, and it goes well. And thus the similar computations are done.

4. MASTER DARDI, NEW EQUATIONS SOLVED

Even before Gilio's time, a unique Italian algebraist was solving problems thought impossible to solve. Known only as Master Dardi, he completed his text in 1344; others copied it several times during the fifteenth century.[5] His text is the earliest vernacular treatise entirely devoted to algebra. It describes 198 different types of equations, from first to fourth degree. Some equations are set in radicals, which Dardi resolves, so that effectively he uses powers up to the twelfth degree. Where some radicals are expanded, an equation becomes equivalent to an earlier one of lower degree. All the rules except for two are correct. These two err perhaps because the author had neither the terminology to express nor the techniques to find fifth and seventh roots of numbers. Most importantly, he developed methods for solving cubic and quartic equations that are uniquely original. Dardi used these abbreviations: R for root (square

[5] A translation was made into Hebrew by the renowned scholar Mordecai Finzi in 1473, who without any proffered justification attributed it to one Dardi da Pisa. For more information, see [Van Egmond, 1983] and [Hughes, 1987].

root if unmodified), C for *cosa* (or x), and Ç for *censo* (or x^2). The first selection is a general solution and example for $ax^2 + \sqrt[3]{n} = \sqrt{bx^4}$. In modern notation, Dardi's algorithm becomes $x = \sqrt{\sqrt[3]{\frac{n}{(\sqrt{b}-a)^3}}}$. The particular equation that Dardi solves is $4x^2 + \sqrt[3]{10} = \sqrt{20x^4}$. However, it is not clear how he gets his ultimate solution for x, expressed as the sum of a mixed number and five square roots. In fact, his answer differs from the correct one by 1 in the third decimal place.

158. When the Ç and the cube R of number are equal to the R of Ç of Ç, you take the Ç from each of the parts. What remains from one part is the cube R of number and from the other the R of Ç of Ç less Ç. Now reduce the R of the quantity of the Ç of Ç less the quantity of the Ç to a cube, and divide the numbers of the cube R by that multiplication, that is by the above named cubic product. The cube R of what comes will be the Ç, and the R of the Ç will become the C.

Here is the example. Find me 2 numbers which, when multiplying the first into itself makes the second, and, when multiplying the first by 2 and multiplying that multiplication into itself and adding the cube R of 10 makes as much as the second number multiplied by the R of 20. I ask how much each number comes to be.

This is its rule. Suppose that the first number is 1C, which number 1C multiply into itself. It gives Ç, and so much is the second number. Now multiply the first number, that is 1C, by 2; it makes 2C. And this product, that is 2C, multiply into itself; it makes 4Ç. Add this to the cube R of 10, and you will have 4Ç and the cube R of 10. Save this for one part of the equation. Now multiply the second number, that is 1Ç, by the R of 20; it makes the R of 20 Ç of Ç, which are equal to 4Ç and to the cube R of 10 that you saved. Now proceed according to the rule given above.

Take the Ç, that is 4Ç from each of the parts and there will remain from one part the cube R of 10 and from the other the R of 20 Ç of Ç less 4Ç. Now reduce the R of the quantity of the Ç of Ç less the quantity of the Ç to cube, that is, you must reduce the R of 20 less 4 to a cube in this way. Multiply the R of 20 less 4 times the R of 20 less 4. It gives 36 less the R of 1280, and this multiplication, that is, 36 less the R of 1280, multiply times the R of 20 less 4. It gives the R of 25920, and the R of 20480 less 144 less the R of 25600, and save. Now divide the number having the cube R, that is 10, by the multiplication that you saved, that is by the R of 25920, and the R of 20480 less 144 and less the R of 25600, and the cube R of what comes from it is the Ç, and the R of the Ç will be the C. Therefore the square R of the cube R of that is the C, that is the square R of the cube R of the sum 23 $\frac{2111}{2178}$ with the R of 564 and $\frac{1753356}{2070721}$ and with the R of 158 $\frac{876082}{2070721}$ and with the R of 157 $\frac{32578973}{33131536}$ and with the R of 125 $\frac{359875}{2070721}$ and with the R of 124 $\frac{1711096}{2070721}$, and so much becomes the first number. And you supposed the second to be 1Ç. Therefore the second number becomes the square of the first, that is, the cube R of the said division of which the first number is the square R of its cube R, as was said above.

An example of how Dardi solved one of the many types of cubic equations that he posed is illustrated by problem 41 of the text. In this case, the general equation is $n = ax^3 + \sqrt{bx^3}$, and the solution algorithm becomes $x = \sqrt[3]{\left(\sqrt{\frac{n}{a} + \frac{b}{4a^2}} - \sqrt{\frac{b}{4a^2}}\right)^2}$. The specific equation that Dardi solves here is $12x^3 + \sqrt{12x^3} = 342$.

41. When the numbers are equal to cubes and to R of cubes, you want to divide the numbers by the quantity of the cubes not having a R, and save what comes. Then multiply the quantity not having a R into itself and divide the quantity of the cubes having a R by that multiplication. A $\frac{1}{4}$ of what comes add to the division that was saved. The R of that sum less the R of the $\frac{1}{4}$ that you added, that is the R of the $\frac{1}{4}$ which came from dividing the cubes having a R by the multiplication of the cubes not having a R, becomes the R of the cubes. [When you] multiply this R into itself, it becomes the cube, and the cube R of this multiplication becomes the C.

Do this problem. Find me 2 numbers that are in the ratio of 2 to 3, such that the first multiplied by itself and this multiplication multiplied by the second and this multiplication added with its R makes 342. I ask how much each number is.

This is the rule. Suppose that the first number is 2C and the second 3C. Now multiply the first number, that is 2C, into itself. It makes 4Ç, which 4Ç you multiply by the second number, that is by 3C. It makes 12 cubes. [12 cubes added with] the R of 12 cubes becomes equal to 342.

Now proceed according to the rule given above. Divide the numbers by the quantity of the cube, that is 342 by 12. It becomes $28\frac{1}{2}$, and save it. Then multiply the quantity of the cube not having a R into itself, which is 12. It makes 144. Divide the quantity of the cube having a R, that is the other 12, by 144. It becomes $\frac{1}{2}$. Take $\frac{1}{4}$ of this $\frac{1}{12}$, which is $\frac{1}{48}$, and add it to $28\frac{1}{2}$, which you saved, and you will have $28\frac{25}{48}$. The R of $28\frac{25}{48}$ less the R of that $\frac{1}{4}$, that is less the R of $\frac{1}{48}$, becomes the R of the cube. Multiply this R into itself, that is R of $28\frac{25}{48}$ less R of $\frac{1}{48}$ into itself. It makes 27, and this much becomes the cube. And its cube R, which is 3, becomes the C.

And you supposed that the first number was 2C. Therefore multiply 2 times the value of the C, that is 2 times 3. It makes 6, and this becomes the first number. You supposed the second to be 3C. Therefore multiply 3 times 3. It makes 9, and this becomes the second number.

Several remarks are appropriate. By substituting y^2 for x^3, Dardi was able to treat the problem like a quadratic, which he solved by completing the square. When he found the value of y, he had to square it to produce a cube! This was his trick. For many years Dardi had a method that no one else had. The method was simple: at a certain stage treat the cubic equation like a quadratic, solve the quadratic, then turn the solution back into a cubic that is easily solved.

As we study his methods, we need to remember that mathematicians of this age always set their equations equal to some number greater than zero. They could not imagine anything being equal to zero. As with all general statements, there is an exception. Dardi does consider a simple linear equation set equal to zero. When reading the translation, be aware that C is Dardi's abbreviation for the unknown or thing (*cosa* in Italian), and N represents a known whole number. Similarly for him an *adequation* is the same as an *equation*. Further, he was experimenting with abbreviations. Hence, his $\frac{5}{C}$ is our $5x$. Similarly, $\frac{15}{N}$ represents 15, a known whole number. His complete explanation and example follow.

Note that there may be C and numbers on one side equal to nothing, as can be seen in some adequations mostly characterized, as we have been saying, by differences or other things. You must take the numbers from each side, and you will have C equal to nothing minus the number. Then you divide the number that is minus by the quantity of the C. However much results that much you debit to the C. Let there be $\frac{5}{C}$ and $\frac{15}{N}$ equal to nothing. Divide 15 by 5, and there comes 3. And so much is debited to the thing.

Dardi.s explanation follows the pattern used throughout his text. First he gives the general equation, $ax + b = 0$. Then follows the rule for solving the general equation: subtract the number from both sides and divide by the coefficient of the unknown. Finally, he gives an example completely solved, $5x + 15 = 0$. The "minus the number" in the above excerpt should be interpreted as a subtraction rather than as a negative number. In Dardi's time, what today we refer to as negative numbers were treated as *debts* to be subtracted from whatever a person had, even if it were nothing. This idea was at least discussed by Fibonacci. Nonetheless, Dardi certainly took a step toward setting equations equal to zero and considering negative roots.

In the foregoing selections from the world of scholars in the late Middle Ages, we have witnessed some of the major accomplishments that awaited further development in the next several generations. This task would be met by mathematicians like Christoph Rudolff and Michael Stifel in Germany, Pedro Nuñez in Portugal, Gerolamo Cardano and Rafael Bombelli in Italy, and François Viète in France. They might well have said what Newton did: "If I have seen so far, it is because I have stood on the shoulders of giants."

SOURCES

I. Latin Schools: 800–1140

I-1. BRIEF SELECTIONS

1. Martianus Capella, *Quadrivium*, from William Stahl, Richard Johnson, and E. L. Burge. 1977. *Martianus Capella and the Seven Liberal Arts. II. The Marriage of Philology and Mercury*. New York: Columbia University Press, pp. 217–220, 274, 276–278, 316–317, 333–335, 349, 359.
2. Cassiodorus, *Quadrivium*, translated by Barnabas Hughes from the Latin from R.A.B. Mynors. 1961. *Cassiodori Senatoris Institutiones. Edited from the manuscripts*, second edition. Oxford: Clarendon Press, 1203C, 1204C, 1208D, 1213B.
3. Isidore of Seville, *Quadrivium*, translated by Barnabas Hughes from the Latin from W. M. Lindsay (ed.) 1911. *Isidori Hispalensis Episcopi Etymologiarum sive Originum libri XX*. Oxford: Oxford University Press, 153D, 161B.
4. *A New Creation: Rules for Addition of Signed Numbers*, translated by Menso Folkerts from the Latin from Menso Folkerts. 2003. "*De arithmeticis propositionibus*. A Mathematical Treatise Ascribed to the Venerable Bede," in *Essays on Early Medieval Mathematics. The Latin Tradition*, chapter II. Burlington, VT: Ashgate/Variorum, pp. 29–30.

I-2. NUMBERING

1. Roman numerals; source translation from Michael Masi. 1983. *Boethian Number Theory: A translation of the De Institutione Arithmetica*. Amsterdam: Rodopi, I 26, 2.
2. Bede on finger reckoning, from Faith Wallis. 1999. *Bede, The Reckoning of Time. Translated, with Introduction, Notes and Commentary by Faith Wallis*. Liverpool: Liverpool University Press, pp. 9–11.
3. Isidore of Seville on numerology, translated by Barnabas Hughes and Menso Folkerts from the Latin from Jacques Paul Migne. 1862. *Patrologiae cursus completus*. Series Latina, vol. 83. Paris. cols. 179–180, 184–186.

4. Hindu-Arabic numerals. The first source is translated by Menso Folkerts from the Latin from John Burnam. 1920. "A Group of Spanish Manuscripts," *Bulletin Hispanique* 22, 229–233. The second source is translated by Menso Folkerts from the Latin from Nicolaus Bubnov (ed.) 1899. *Gerberti Opera Mathematica* (*972–1003*). Berlin. Reprint Hildesheim: Georg Olms, 1963.

I-3. ARITHMETIC

1. Boethius, *De institutione arithmetica*, translation from Michael Masi. 1983. *Boethian Number Theory: A Translation of the De institutione arithmetica*. Amsterdam: Rodopi, pp. 75–78, 89–90, 96–100.
2. Computus, translated by Immo Warntjes from the Latin. Cassiodorus's 562 recension is from Paul Lehmann. 1959. *Erforschung des Mittelalters: Ausgewählte Abhandlungen und Aufsätze*, vol. 2. Stuttgart: Hiersemann, pp. 52–55. The material from 776 is from Arno Borst. 2006. *Schriften zur Komputistik im Frankenreich von 721 bis 818*, 3 vols. Hannover: Hahnsche Buchhandlung, vol. 2, pp. 704–711. There is also a German translation with commentary in Kerstin Springsfeld. 2002. *Alkuins Einfluß auf die Komputistik zur Zeit Karls des Großen*. Stuttgart: Franz Steiner. The final section on calculation in Roman numerals is from Bruno Krusch. 1938. "Studien zur christlich-mittelalterlichen Chronologie: Die Entstehung unserer heutigen Zeitrechnung," *Abhandlungen der Preußischen Akademie der Wissenschaften, Jahrgang 1937, philosophisch-historische Klasse* 6, 80–81.
3. Gerbert's jump, translated by Menso Folkerts from the Latin of Gerbert's commentary on Boethius from Nicolaus Bubnov (ed.) 1899. *Gerberti Opera Mathematica* (*972–1003*). Berlin. Reprint Hildesheim: Georg Olms, 1963, pp. 391, 32–35.
4. Pandulf of Capua, *Liber de calculation*, translation from C. A. Gibson and F. Newton. 1995. "Pandulf of Capua's *De calculatione*: An Illustrated Abacus Treatise and Some Evidence for the Hindu-Arabic Numerals in Eleventh-Century South Italy," *Mediaeval Studies* 57, 293–335, pp. 331–335.

I-4. GEOMETRY

1. Gerbert, *Geometry*, translated by Menso Folkerts from the Latin from Nicolaus Bubnov (ed.) 1899. *Gerberti Opera Mathematica* (*972–1003*). Berlin. Reprint Hildesheim: Georg Olms, 1963, pp. 48–57.
2. Gerbert: area of an equilateral triangle in Gerbert's *Geometry*, translated by Menso Folkerts from the Latin from Nicolaus Bubnov (ed.) 1899. *Gerberti Opera Mathematica* (*972–1003*). Berlin. Reprint Hildesheim: Georg Olms, 1963, pp. 43–45.
3. Units of measurement, translated by Menso Folkerts from the Latin from Menso Folkerts. 1970. *"Boethius" Geometrie II. Ein mathematisches Lehrbuch des Mittelalters.* Wiesbaden: Franz Steiner, pp. 144–146.
4. Franco of Liège, *De quadratura circuli*, translated by Menso Folkerts from the Latin from Menso Folkerts and A.J.E.M. Smeur. 1976. "A Treatise on the Squaring of the Circle by Franco of Liège, of about 1050," *Archives Internationales d'Histoire des Sciences* 26, 59–105, 225–253, pp. 70–72.
5. Hugh of St. Victor, *Practica geometriae*, translation from Frederick A. Homann. 1991. *Practical Geometry Attributed to Hugh of St. Victor.* Milwaukee, WI: Marquette University Press, pp. 46–47.

I-5. RECREATIONAL MATHEMATICS

1. Number puzzles, translated from the Latin by Menso Folkerts from Menso Folkerts. 2003. *Essays on Early Medieval Mathematics. The Latin Tradition*, chapter III. Burlington, VT: Ashgate/Variorum, pp. 22–28.
2. Thought problems, translated from the Latin by Menso Folkerts from Menso Folkerts. 2003. *Essays on Early Medieval Mathematics. The Latin Tradition*, chapter V. Burlington, VT: Ashgate/Variorum.
3. The Josephus Problem, translation from Gerard Murphy. 1942. "The Puzzle of the Thirty Counters," *Béaloideas* 12, 3–28.
4. The most important medieval number game: the *Rithmimachia*, description translated by Menso Folkerts from the Latin from Arno Borst. 1986. *Das mittelalterliche Zahlenkampfspiel*. Heidelberg: Carl Winter, pp. 415–418.

II. A School Becomes a University: 1140–1480

INTRODUCTION

Sources translated by Barnabas Hughes from the Latin from Charles Henry Buttimer. 1939. *Hugonis de Sancto Victore Didascalicon de Studio Legendi. A Critical Text*, Book II. Washington, DC: Catholic University Press.

II-1. TRANSLATIONS

1. Translators and Translations: translation from Edward Grant. 1974. *A Source Book in Medieval Science*. Cambridge, MA: Harvard University Press, pp. 65–66, 36–37, 39–41.
2. Translations: A Practical Illustration: a. Translation from Thomas L. Heath. 1926. *The Thirteen Books of Euclid's Elements*, 3 vols., second edition. Cambridge: Cambridge University Press (reprint: New York: Dover, 1956). b. Translated by Menso Folkerts from H.L.L. Busard (ed.) 1983. *The First Latin Translation of Euclid's Elements Commonly Ascribed to Adelard of Bath. Books I–VIII and Books X.36–XV.2*. Toronto: Pontifical Institute of Mediaeval Studies. c. Translated by Menso Folkerts from H.L.L. Busard (ed.) 1968. *The Translation of the Elements of Euclid from the Arabic into Latin by Hermann of Carinthia (?)*. Leiden: E. J. Brill. d. Translated by Menso Folkerts from H.L.L. Busard (ed.) 1983. *The Latin Translation of the Arabic Version of Euclid's Elements Commonly Ascribed to Gerard of Cremona. Introduction, Edition and Critical Apparatus*. Leiden: New Rhine Publishers. e. Translated by Menso Folkerts from H.L.L. Busard and Menso Folkerts (eds.) 1992. *Robert of Chester's (?) Redaction of Euclid's Elements, the So-Called Adelard II Version*, vols. I, II. Basel: Birkhäuser Verlag. f. Translated by Menso Folkerts from H.L.L. Busard (ed.) 2005. *Campanus of Novara and Euclid's Elements*. Stuttgart: Steiner.

II-2. ARITHMETIC

1. Al-Khwārizmī, *Arithmetic*, translation from J. N. Crossley and A. S. Henry. 1990. "Thus Spake al-Khwārizmī: A Translation of the Text of Cambridge University Library Ms. Ii.vi.5," *Historia Mathematica* 17, 103–131, with some corrections by Menso Folkerts from his 1997 Latin edition of another, complete, manuscript of this work, *Die älteste lateinische Schrift über das indische Rechnen nach al-Ḫwārizmī*. München: Verlag der Bayerischen Akademie der Wissenschaften.
2. Leonardo of Pisa, *Liber abbaci*, translation of the introduction from Charles Burnett. 2003. "Fibonacci's 'Method of the Indians,'" *Bollettino di Storia delle Scienze Matematiche* 23, 87–97; translation of problems in L. E. Sigler. 2002. *Fibonacci's Liber abbaci*, New York: Springer, pp. 268–269, 274, 317–318, 320–322, 357–358, 396–397, 404–405.
3. John of Sacrobosco, *Algorismus vulgaris*, translation from Edward Grant. 1974. *A Source Book in Medieval Science*. Cambridge, MA: Harvard University Press, pp. 94–101.
4. John de Lineriis, *Algorismus de minuciis*, translated by Menso Folkerts from the Latin from H.L.L. Busard. 1968. "Het rekenen met breuken in de middeleeuwen, in het bijzonder bij Johannes de Lineriis," *Mededelingen van de Koninklijke Vlaamse Academie voor Wetenschappen, Letteren en Schone Kunsten van België. Klasse der Wetenschappen* (Brussels) Jaargang 30(7), 21–25.
5. Jordanus de Nemore, *De elementis arithmetice artis*: Proposition VI-12 translated by Menso Folkerts from the Latin from H.L.L. Busard. 1991. *Jordanus de Nemore, De elementis arithmetice artis. A Medieval Treatise on Number Theory*, 2 vols. Stuttgart: Franz Steiner. Translation of proposition IX-70 from Barnabas Hughes. 1989. "The Arithmetical Triangle of Jordanus de Nemore," *Historia Mathematica* 16, 213–223.
6. De Vetula, translation from D. R. Bellhouse. 2000. "De Vetula: A Medieval Manuscript Containing Probability Calculations," *International Statistical Review* 68, 123–136.

II-3. ALGEBRA

1. Al-Khwārizmī, *Algebra*, translation from Louis Charles Karpinski. 1915. *Robert of Chester's Latin Translation of the Algebra of al-Khowarizmi*. New York: Macmillan, pp. 67–71, 81.

2. Leonardo of Pisa, *Liber abbaci*, translation from L. E. Sigler. 2002. *Fibonacci's Liber abbaci,* New York: Springer, pp. 291–292, 318, 447–450, 557, 587–588.
3. Leonardo of Pisa, *Book of Squares*, translation from L. E. Sigler. 1987. *Leonardo Pisano Fibonacci, The Book of Squares. An Annotated Translation into Modern English.* Boston: Academic Press, pp. 4–5, 48, 52, 74, 76–78, 84–88.
4. Jordanus de Nemore, *De numeris datis,* translation slightly modified from Barnabas Hughes. 1981. *Jordanus de Nemore, De numeris datis. A Critical Edition and Translation.* Berkeley: University of California Press, pp. 127–128, 131–135, 140, 148, 157.
5. Nicole Oresme, *Algorismus proportionum*, translation from Edward Grant. 1965. "Part I of Nicole Oresme's *Algorismus proportionum*," *Isis* 66, 327–341.
6. Nicole Oresme, *De proportionibus proportionum*, translation from Edward Grant (ed.) 1966. *Nicole Oresme, De proportionibus proportionum and Ad pauca respicientes. Edited with Introductions, English Translations, and Critical Notes*. Madison: University of Wisconsin Press, pp. 161–163, 247–253.

II-4. GEOMETRY

1. Banū Mūsā ibn Shākir, *The Book of the Measurement of Plane and Spherical Figures*, translation from Marshall Clagett. 1964. *Archimedes in the Middle Ages*. Vol. 1: *The Arabo-Latin Tradition*. Madison: University of Wisconsin Press, pp. 279–289.
2. Abū Bakr, *On Measurement*, translated by Menso Folkerts from the Latin in H.L.L. Busard. 1968. "L'Algèbre au moyen âge: Le 'Liber mensurationum' d'Abû Bekr," *Journal des savants* (Avril–Juin), 65–124, pp. 87–88, 101, 114–115.
3. Leonardo of Pisa, *Practical Geometry*, translation slightly modified from Barnabas Hughes. 2008. *Fibonacci's De practica geometrie*. New York: Springer, pp. 73, 109–112, 119, 122, 127–128, 135, 151, 163, 165–166, 231–235, 252–253, 354–357.
4. John of Murs, *De arte mensurandi*. The first selection was translated by Menso Folkerts from the Latin from H.L.L. Busard. 1998. *Johannes de Muris, De arte mensurandi. A Geometrical Handbook of the Fourteenth Century*. Stuttgart: Steiner, pp. 337–339. The second selection is from Marshall Clagett. 1978. *Archimedes in the Middle Ages*. Vol. 3: *The Fate of the Medieval Archimedes 1300–1565*. Philadelphia: American Philosophical Society, pp. 117–118.
5. Jordanus de Nemore, *Liber philotegni*, translation from Marshall Clagett. 1984. *Archimedes in the Middle Ages*. Vol. 5: *Quasi-Archimedean Geometry in the Thirteenth Century*. Philadelphia: American Philosophical Society, pp. 269, 473.
6. Dominicus de Clavasio, *Practica geometriae*, first excerpt translated by Menso Folkerts from the Latin from H.L.L. Busard. 1965. "The Practica Geometriae of Dominicus de Clavasio," *Archive for History of Exact Sciences*, 2, 520–575; translation of second excerpt from Edward Grant. 1974. *A Source Book in Medieval Science*. Cambridge, MA: Harvard University Press, p. 187.

II-5. TRIGONOMETRY

1. Ptolemy, *On the Size of Chords in a Circle*, translation from G. J. Toomer. 1998. *Ptolemy's Almagest*. Princeton, NJ: Princeton University Press, pp. 48–51.
2. Leonardo of Pisa, *Practical Geometry*, translation slightly modified from Barnabas Hughes. 2008. *Fibonacci's De practica geometrie*. New York: Springer, pp. 354, 356–357.
3. John de Lineriis, *Canones*, translated by Barnabas Hughes from the Latin from Maximilian Curtze. 1900. "Urkunden zur Geschichte der Trigonometrie im christlichen Mittelalter," *Bibliotheca Mathematica*, 3.1, 321–416, pp. 391–393.
4. Richard of Wallingford, *Quadripartitum*, translation from John D. North. 1976. *Richard of Wallingford: An Edition of His Writings with Introductions, English Translation and Commentary*, 3 vols. Oxford: Clarendon Press, vol. I, pp. 24–28.
5. Geoffrey Chaucer, *Treatise on the Astrolabe*, translated by Barnabas Hughes from the Latin from Sigmund Eisner (ed.) 2002. *A Treatise on the Astrolabe*, Variorum edition of the *Works of Geoffrey Chaucer*, Vol. VI: *The Prose Treatises, Part One*. Norman: University of Oklahoma Press, pp. 312–315.

6. Regiomontanus, *On Triangles*. Translation from Barnabas Hughes. 1967. *Regiomontanus on Triangles*. Madison: University of Wisconsin Press, pp. 109–111, 113, 219–229, 247–249, 271–275.

II-6. MATHEMATICS OF THE INFINITE

1. Angle of contingence (Campanus and Bradwardine). Campanus source translated by Menso Folkerts from the Latin from H.L.L. Busard (ed.) 2005. *Campanus of Novara and Euclid's Elements*. Stuttgart: Steiner, pp. 121–122. Translation of Bradwardine source from A. G. Molland. 1989. *Thomas Bradwardine, Geometria speculativa. Latin Text and English Translation with an Introduction and a Commentary*. Stuttgart: Franz Steiner Verlag, pp. 66–71.
2. Thomas Bradwardine, *On the Continuum*, translated by Menso Folkerts from J. E. Murdoch. 1957. "Geometry and the Continuum in the Fourteenth Century: A Philosophical Analysis of Thomas Bradwardine's Tractatus de Continuo." Dissertation, University of Wisconsin, pp. 379–380, 400–401, 413–415, 456–457, 459.
3. John Duns Scotus, Indivisibles and Theology, translation from Barnabas Hughes. 1983. "Franciscans and Mathematics," *Archivum Franciscanum Historicum* 76, 98–128.
4. Speed of light (Witelo and Bacon). Translations by David Lindberg from Edward Grant. 1974. *A Source Book in Medieval Science*. Cambridge, MA: Harvard University Press, pp. 395–397.
5. Nicole Oresme, *Questiones super geometriam Euclidis*, translation from H.L.L. Busard (ed.) 2010. *Nicole Oresme, Questiones super geometriam Euclidis*. Stuttgart: Steiner, pp. 105–106.

II-7. STATICS, DYNAMICS, AND KINEMATICS

1. Robert Grosseteste, *On Lines, Angles and Figures*, translation from Alistair Crombie. 1962. *Robert Grosseteste and the Origins of Experimental Sciences 1100–1700*. Oxford: Clarendon Press, p. 110.
2. Jordanus de Nemore, *On the Theory of Weights*, translation from Ernest A. Moody and Marshall Clagett (eds.) 1960. *The Medieval Science of Weights (Scientia de ponderibus)*. Madison: University of Wisconsin Press, pp. 169–172.
3. Thomas Bradwardine, *Tractatus de proportionibus*, translation from H. Lamar Crosby. 1961. *Thomas of Bradwardine. His Tractatus de Proportionibus. Its Significance for the Development of Mathematical Physics*. Madison: University of Wisconsin Press, pp. 111–117.
4. William Heytesbury, *Rules for Solving Sophisms*, translation from Marshall Clagett. 1959. *The Science of Mechanics in the Middle Ages*. Madison: University of Wisconsin Press, pp. 270–272.
5. Giovanni di Casali, *On the Velocity of Motion of Alteration*, translation from Marshall Clagett. 1959. *The Science of Mechanics in the Middle Ages*. Madison: University of Wisconsin Press, pp. 382–384.
6. Nicole Oresme, *De configurationibus qualitatum et motuum*, translation from Marshall Clagett. 1968. *Nicole Oresme and the Medieval Geometry of Qualities and Motions. A Treatise on the Uniformity and Difformity of Intensities Known as Tractatus de configurationibus qualitatum et motuum*. Madison: University of Wisconsin Press, pp. 165–207, 393–425.

III. Abbacist Schools: 1300–1480

III-2. GEOMETRY

The first source was translated by Barnabas Hughes from the Italian in Florence, Biblioteca Riccardiana, Ms. 2404. The second source was translated by Barnabas Hughes from the Italian in Lucca, Biblioteca Comunale, Ms. 1754.

III-3. ALGEBRA

1. Gilio da Siena, *A Lecture in Introductory Algebra*, translated by Barnabas Hughes from the Italian from Raffaella Franci (ed.) 1983. *Maestro Gilio. Questioni d'algebra dal Codice L.IX.28 della Biblioteca Comunale di Siena*. Siena: Università di Siena, pp. 22–25.
2. Paolo Girardi, *Libro di Ragioni*, translation from Warren Van Egmond. 1978. "The Earliest Vernacular Treatment of Algebra: *The Libro di Ragioni of Paolo Gerardi* (1328)," *Physis* 20, 155–189, pp. 176–177.

3. Jacobo da Firenze, *Tractatus algorismi*, translation from Jens Høyrup. 2007. *Jacopo da Firenze's Tractatus Algorismi and Early Italian Abbacus Culture*. Basel: Birkhäuser Verlag, pp. 313–315.
4. Master Dardi, *Algebra*, translated by Warren Van Egmond from the Italian in Tempe, Arizona State University, Ms. UCM-35. The final paragraph is translated by Barnabas Hughes from the Italian from Barnabas Hughes. 1987. "An Early 15th-Century Algebra Codex: A Description," *Historia Mathematica* 14, 167–172.

REFERENCES

Ahrens, Wilhelm. 1918. *Mathematische Unterhaltungen und Spiele*, Vol. 2, second edition. Leipzig: B. G. Teubner.

Ball, W. W. Rouse, and H. S. M. Coxeter. 1974. *Mathematical Recreations and Essays*, twelfth edition. Toronto: University of Toronto Press.

Benson, Robert, and Giles Constable (with Carol D. Lanham). 1982. *Renaissance and Renewal in the Twelfth Century*. Cambridge, MA: Harvard University Press.

Berggren, J. Lennart. 2007. "Mathematics in Medieval Islam," in Victor J. Katz (ed.), *The Mathematics of Egypt, Mesopotamia, China, India, and Islam: A Sourcebook*. Princeton, NJ: Princeton University Press, pp. 515–675.

Burnam, John M. 1912–1925. *Palaeographia Iberica*, 3 vols. Paris: Champion.

Burnett, Charles, 2002. "Indian Numerals in the Mediterranean Basin in the Twelfth Century with Special Reference to the 'Eastern Forms,'" in Yvonne Dold-Samplonius, Joseph W. Dauben, Menso Folkerts, and Benno van Dalen (eds.), *From China to Paris: 2000 Years Transmission of Mathematical Ideas*. Stuttgart: Franz Steiner, pp. 237–288.

Buttimer, Charles Henry. 1939. *Hugonis de Sancto Victore Didascalicon de Studio Legendi. A Critical Text*. Washington, DC: Catholic University Press.

Clagett, Marshall, 1959. *The Science of Mechanics in the Middle Ages*. Madison: University of Wisconsin Press.

———. 1978. *Archimedes in the Middle Ages*, Vol. 3: *The Fate of the Medieval Archimedes 1300–1565*. Philadelphia: American Philosophical Society.

———. 1984. *Archimedes in the Middle Ages*, Vol. 5: *Quasi-Archimedean Geometry in the Thirteenth Century*. Philadelphia: American Philosophical Society.

Crombie, Alistair. 1962. *Robert Grosseteste and the Origins of Experimental Sciences 1100–1700*. Oxford: Clarendon Press.

Crosby, H. Lamar, Jr. 1961. *Thomas of Bradwardine. His* Tractatus de proportionibus. *Its Significance for the Development of Mathematical Physics*. Madison: University of Wisconsin Press.

Dauben, Joseph. 2007. "Chinese Mathematics," in Victor J. Katz (ed.), *The Mathematics of Egypt, Mesopotamia, China, India, and Islam: A Sourcebook*. Princeton, NJ: Princeton University Press, pp. 187–384.

Dijksterhuis, E. J. 1956. *Archimedes*. Copenhagen: Ejnar Munksgaard.

Dold-Samplonius, Yvonne. 1996. "Problem of the Two Towers," in Raffaella Franci, Paolo Pagli, and Laura Toti Rigatelli (eds.), *Itinera Mathematica. Studi in onore di Gino Arrighi per il suo 90° compleanno*. Siena: Centro Studi sulla Matematica Medioevale Università di Siena, pp. 45–70.

Dold-Samplonius, Yvonne, Joseph W. Dauben, Menso Folkerts, and Benno van Dalen (eds.) 2002. *From China to Paris: 2000 Years Transmission of Mathematical Ideas*. Stuttgart: Franz Steiner.

Evans, Gillian R. 1976. "The *Rithmomachia*: A Mediaeval Mathematical Teaching Aid?" *Janus* 63, 257–273.

——— 1980. "The Saltus Gerberti: The Problem of the 'Leap,'" *Janus* 67, 261–268.

Folkerts, Menso, 1969. *Boethius' Geometrie II, ein mathematisches Lehrbuch des Mittelalters*. Wiesbaden: Steiner.

———. 2001. "The Names and Forms of the Numerals on the Abacus in the Gerbert Tradition," in Flavio Nuvolone (ed.), *Gerberto d'Aurillac da Abate di Bobbio a Papa dell'Anno 1000*. Bobbio/Pesaro: Associazione culturale Amici di Archivum Bobiense, pp. 245–265.

———. 2003. *Essays on Early Medieval Mathematics. The Latin Tradition*. Burlington, VT: Ashgate/Variorum.
———. 2006. *The Development of Mathematics in Medieval Europe. The Arabs, Euclid, Regiomontanus*. Burlington, VT: Ashgate/Variorum.
Franci, Raffaella, 2002. "Jealous Husbands Crossing the River: A Problem from Alcuin to Tartaglia," in Yvonne Dold-Samplonius, Joseph W. Dauben, Menso Folkerts, and Benno van Dalen (eds.) 2002. *From China to Paris: 2000 Years Transmission of Mathematical Ideas*. Stuttgart: Franz Steiner, pp. 289–306.
Gibson, C. A., and F. Newton. 1995. "Pandulf of Capua's *De calculatione*: An Illustrated Abacus Treatise and Some Evidence for the Hindu-Arabic Numerals in Eleventh-Century South Italy," *Mediaeval Studies* 57, 293–335.
Grant, Edward (ed.) 1966. *Nicole Oresme, De proportionibus proportionum and Ad pauca respicientes. Edited with Introductions, English Translations, and Critical Notes*. Madison: University of Wisconsin Press.
Guillaumin, Jean-Yves (ed.) 1995. *Boèce, Institution arithmétique*. Paris: Les Belles Lettres.
Hannah, John. 2007. "False Position in Leonardo of Pisa's *Liber abbaci*," *Historia Mathematica* 34, 306–332.
Haskins, Charles Homer. 1927. *Studies in the History of Mediaeval Science*. Cambridge, MA: Harvard University Press.
Høyrup, Jens. 1996. "The Four Sides and the Area: Oblique Light on the Prehistory of Algebra," in Ronald Calinger (ed.), *Vita Mathematica: Historical Research and Integration with Teaching*. Washington, DC: Mathematical Association of America.
———. 2007. *Jacopo da Firenze's Tractatus algorismi and Early Italian Abbacus Culture*. Basel: Birkhäuser Verlag.
Hughes, Barnabas. 1981. *Jordanus de Nemore,* De numeris datis. *A Critical Edition and Translation*. Berkeley: University of California Press.
———. 1986. "Gerard of Cremona's Translation of al-Khwārizmī's *al-Jabr*: A Critical Edition," *Mediaeval Studies* 48, 211–263.
———. 1987. "An Early 15th-Century Algebra Codex: A Description," *Historia Mathematica* 14, 167–172.
———. 1989. *Robert of Chester's Latin Translation of al-Khwārizmī's al-Jabr: A New Critical Edition*. Stuttgart: Franz Steiner Verlag.
———. 2008. *Fibonacci's De practica geometrie*. New York: Springer.
Imhausen, Annette. 2007. "Egyptian Mathematics," in Victor J. Katz (ed.), *The Mathematics of Egypt, Mesopotamia, China, India, and Islam: A Sourcebook*. Princeton, NJ: Princeton University Press, pp. 7–56.
Karpinski, Louis Charles. 1915. *Robert of Chester's Latin Translation of the Algebra of al-Khowarizmi*. New York: Macmillan.
Kaunzner, Wolfgang. 1986. "Die lateinische Algebra in Ms Lyell 52 der Bodleian Library, Oxford, früher Ms Admont 612," *Österreichische Akademie der Wissenschaften, Philosophisch-historische Klasse. Sitzungsberichte* 475, 47–89.
Krusch, Bruno. 1938. "Studien zur christlich-mittelalterlichen Chronologie: Die Entstehung unserer heutigen Zeitrechnung," *Abhandlungen der Preußischen Akademie der Wissenschaften, Jahrgang 1937, philosophisch-historische Klasse* 6, 80–81.
Levey, Martin. 1966. *The Algebra of Abū Kāmil, Kitāb fī al-jābr wa'l-muqābala, in a Commentary by Mordecai Finzi*. Madison: University of Wisconsin Press.
Malagola, Carlo. 1888. *Statuti delle università e dei collegi dello studio Bolognese*. Bologna: Zanichelli.
Masi, Michael. 1983. *Boethian Number Theory: A Translation of the* De institutione arithmetica. Amsterdam: Rodopi.
Molland, A. G. 1978. "An Examination of Bradwardine's Geometry," *Archive for History of Exact Sciences* 19, 113–175.
Moody, Ernest A., and Marshall Clagett (eds.) 1960. *The Medieval Science of Weights (Scientia de ponderibus)*. Madison: University of Wisconsin Press.

Mordek, Hubert, Klaus Zechiel-Eckes, and Michael Glatthaar (eds.) 2012. *Die Admonitio generalis Karls des Großen*. Hannover: Hahnsche Buchhandlung.

Mosshammer, Alden. 2008. *The Easter Computus and the Origins of the Christian Era*. Oxford: Oxford University Press.

North, John D. 1976. *Richard of Wallingford. An Edition of His Writings with Introductions, English Translation and Commentary*, 3 vols. Oxford: Clarendon Press.

Peden, A. M. 2003. *Abbo of Fleury and Ramsey: Commentary on the Calculus of Victorius of Aquitaine*. Oxford: Oxford University Press.

Pedersen, Olaf. 1985. "In Quest of Sacrobosco," *Journal for the History of Astronomy* 16, 175–221.

Plofker, Kim. 2009. *Mathematics in India*. Princeton, NJ: Princeton University Press.

Rashdall, Hastings. 1936. *The Universities of Europe in the Middle Ages*. New edition, edited by F. M. Powicke and A. B. Emden, 3 vols. Oxford: Oxford University Press.

Rashed, Roshdi. 2012. *Algèbre et analyse diophantienne: edition, traduction et commentaire*. Berlin: De Gruyter.

Robson, Eleanor, 2007. "Mesopotamian Mathematics," in Victor J. Katz (ed.), *The Mathematics of Egypt, Mesopotamia, China, India, and Islam: A Sourcebook*. Princeton, NJ: Princeton University Press, pp. 57–186.

Sigler, Laurence E. 1987. *Leonardo Pisano Fibonacci, The Book of Squares. An Annotated Translation into Modern English*. Boston: Academic Press.

———. 2002. *Fibonacci's Liber abbaci*. New York: Springer.

Smith, David Eugene. 1958. *History of Mathematics*, Vol. II: *Special Topics of Elementary Mathematics*. New York: Dover.

Smith, David Eugene, and Clara C. Eaton. 1911. "Rithmomachia, the Great Medieval Number Game," *American Mathematical Monthly* 18, 73–80.

Stahl, William Harris, Richard Johnson, and E. L. Burge. 1971. *Martianus Capella and the Seven Liberal Arts. I. The Quadrivium of Martianus Capella*. New York: Columbia University Press.

Strecker, K. 1923. *Poetae Latini aevi Carolini IV, 3*. Berlin: Weidmann.

Taegert, Werner (ed.), 2008. *Zählen, messen, rechnen. 1000 Jahre Mathematik in Handschriften und frühen Drucken. Ausstellung der Staatsbibliothek Bamberg zum Jahr der Mathematik 2008*. Petersberg: Michael Imhof.

Taylor, Jerome. 1961. *The Didascalicon of Hugh of St. Victor, Trans. with Introduction and Notes*. New York: Columbia University Press.

Toomer, G. J. 1998. *Ptolemy's Almagest*. Princeton, NJ: Princeton University Press.

Van Egmond, Warren. 1976. "The Commercial Revolution and the Beginnings of Western Mathematics in Renaissance Florence, 1300–1500." Ph.D. dissertation, Indiana University.

———. 1983. "The Algebra of Master Dardi of Pisa," *Historia Mathematica* 10, 399–421.

———. 2008. "The Study of Higher-Order Equations in Italy before Pacioli," in Joseph W. Dauben et al. (eds.), *Mathematics Celestial and Terrestrial. Festschrift für Menso Folkerts zum 65. Geburtstag*. Halle/Saale: Deutsche Akademie der Naturforscher Leopoldina, pp. 303–320.

Villani, Giovanni. 1991. *Nuova Cronica*, edited by G. Porta. Parma: Fondazione Pietro Bembo/Guanda. http://www.letteraturaitaliana.net/pdf/Volume_2/t48.pdf.

Warntjes, Immo. 2010. "The Argumenta of Dionysius Exiguus and Their Early Recensions," in Immo Warntjes and Dáibhí Ó Cróinín, *Computus and Its Cultural Context in the Latin West, AD300–1200*. Turnhout, Belgium: Brepols.

Woepcke, Franz. 1853. *Extrait du Fakhrī, traité d'algèbre par Aboù Bekr Mohammed ben Alhaçan al-Karkhī*. Paris: L'imprimerie Impériale.

2

Mathematics in Hebrew in Medieval Europe

ROI WAGNER
with contributions by Naomi Aradi, Avinoam Baraness, David Garber, Stela Segev, Shai Simonson, and Ilana Wartenberg

INTRODUCTION

This chapter covers mathematics written in Hebrew between the eleventh and sixteenth centuries in Europe. But the term "Hebrew mathematics" risks conferring an illusory unity on the corpus presented. As we shall see, the mathematical works covered in this section are strongly rooted in the non-Hebrew scientific traditions around which they evolved.

The Hebrew works are linked together by a common thread of canonized Hebrew references and sometimes are concerned with applications to Jewish philosophy or law, but these links are generally weaker than those that bind each work to its specific non-Jewish surroundings. These relatively weaker links, however, should not be ignored. Indeed, they are strong enough to compose some interesting syntheses and sometimes cast Hebrew mathematics in the role of a cultural "go-between."

The twelfth century, which is the starting point of this collection (except for a short extract from Rashi), is not the beginning of Hebrew mathematics. Indeed, one can find some mathematical discussions in the rabbinical literature as early as the *Mishnah* (100–300 CE). But these are few, far between, and do not constitute a systematized body of knowledge. Later on, Jews living in Islamic countries before the twelfth century continued writing mathematical treatises, but they used classical Arabic as their scientific language and addressed their writings to the entire mathematical community, not specifically to Jews.

It was only in the twelfth century that the demand and opportunity arose to produce original and translated Hebrew scientific works. In the first half of that century, the cultural divide between Arabophone Judaism south of the Pyrenees and Hebreophone Judaism to their north led to the composition of several scientific works in Hebrew by Arabophone Jewish scholars. This trend gained momentum the following decades. In the middle of the century, persecutions in Muslim Spain led to the emigration of Arabophone scholarly Jewish families to Christian environments, especially to southern France. In these cultural contexts Jews did not know

Arabic, and the cultural language was Hebrew. Nor did they know Latin, the literary language of the majority culture. Scientific works in the different vernaculars did not yet exist, and Jews did not use them for writing. Thus, it was in these locations that Hebrew emerged as a scientific language.

The first representatives of this transition are also the most famous and most canonized writers of Hebrew mathematics: Abraham ibn Ezra and Abraham bar Ḥiyya (Savasorda). Both are witnesses of the Maghribian/Andalusian branch of Arabic practical mathematics (*muʿāmalāt*)—a branch that did not survive in its original language. Both authors link this branch of mathematics to questions of religious law, astronomy, and number theory, and both had a role in the transmission of this knowledge into Latin. A brief extract of Talmudic exegesis from Rashi, the most renowned commentator on the Bible, shows that the integration of practical measurement and biblical exegesis precedes Ibn Ezra and Bar Ḥiyya; excerpts from Simon ben Ṣemaḥ Duran's *responsa* indicate the lasting presence of this integration.

The next Hebrew mathematical corpus is the thirteenth- and fourteenth-century translations from Arabic to Hebrew executed in Provence by and around the Ibn Tibbon family. They managed to provide Hebrew readers with Hebrew versions of a rather comprehensive sample of canonical Greco-Arabic mathematical texts. Since the present collection excludes literal translations of such identified Arabic sources, we include almost nothing from this important tradition of Hebrew mathematics. Our only representatives of this circle are a short number theoretic excerpt from Qalonymos ben Qalonymos's *Book of Kings* and his translation of a treatise on polyhedra whose source is not identified. But this should not detract from the importance of the translators' work: they introduced into Hebrew an ideal of literal translation, provided an infrastructure for Hebrew science, and created a hitherto nonexistent mathematical vocabulary, coining terms either by allocating new meanings to existing Hebrew words or by introducing loan translations from Arabic.

Next follows the most original scene of Hebrew mathematicians, active in and around the Iberian Peninsula in the fourteenth and fifteenth centuries. This chapter shows the span of their work. We include some arithmetic and geometry by Levi ben Gershon (Gersonides), the most original Hebrew mathematician, whose sources have been difficult to identify. We also include some of the idiosyncratic and original efforts of Abner of Burgos (also known as Alfonso di Valladolid), who followed Archimedean traditions. We further include work by Immanuel Bonfils on circle measurement and decimal fractions (the first documented appearance of the latter on European soil) and a sample of little-known authors of practical mathematics: Jacob Canpanṭon, Aaron ben Isaac, and an anonymous source with a more algebraic flavor.

During the fourteenth century and especially the fifteenth century, numerous Jews had to emigrate from Spain and southern France due to Christian persecution, and their mathematical books and knowledge migrated along with them. Isaac ibn al-Aḥdab, representing the first generation of migrants (he migrated from Spain through the Maghrib to Sicily), carried with him the knowledge produced by commentators on the algebra of Ibn al-Bannāʾ. Later on, the Italian mainland saw Jews working with both older Hebrew versions of Arabic sources and newer Latin and vernacular sources. Our examples include Solomon ben Isaac's treatise on the hyperbola's asymptote and Simon Moṭoṭ's algebra.

The last mathematical scene we sample is the sixteenth century Byzantine Jewish community, which adds Greek sources to the mixture of Hebrew mathematics. Mizraḥi, whose work we represent here, shows his excellent capacity as a mathematical collector and editor—so much so that his work, late and elementary as it was, was still considered worthy of a Latin translation and printing in 1546.

Since the chronological order of these works does not reflect a historical continuity, the order of presentation here follows themes and levels of complexity. We begin with arithmetic and then discuss number theory and combinatorics. We follow this with measurement theory and practical geometry, next include some highbrow scholarly geometry, and finally conclude with algebra. In each section, the presentation begins with the more elementary and proceeds to the more sophisticated.

This chapter is a collaborative effort of several different scholars. I thank Naomi Aradi, Avinoam Baraness, David Garber, Stela Segev, Shai Simonson, and Ilana Wartenberg for making this work possible. Part of the research that made this work possible was supported by the Israel Science Foundation (grant no. 12/10).

NOTE ON TRANSLATION AND TRANSCRIPTION

Since this chapter pools together various existing editions and translations, we could not hope to obtain a uniform translation standard. This should be borne in mind when comparing terms and forms of expression across the different translations. Almost all English translations here are based on a Hebrew manuscript or edition and not on a secondary translation.[1] Some of the translations brought over from English editions were slightly adapted or revised. For biblical quotations we used the Jewish Publication Society Bible.

Where it seemed necessary, we include explanatory footnotes in modern mathematical notation. These should be handled with care. The modern mathematical transcription may not faithfully represent the concepts and procedures reflected in the original sources.

Numbers were either represented as words or according to a key that gave each Hebrew letter a numerical value (the first nine letters represented 1–9, the next nine letters represented 10–90, and the final four letters represented 100–400). The letters were combined additively to form composite numbers, but thousands were sometimes represented apart from lower numbers, because composing too many letters was impractical. Decimal place-value representation was rarely used, and usually only in calculation diagrams or when dealing with large numbers. There, too, Hebrew authors preferred using the first nine Hebrew letters over Arabic numerals. In this volume, numbers written as words are translated as words, and those written as sign combinations are translated into the modern place-value representation with Arabic numerals. However, different manuscripts of the same text are often inconsistent, and some of the translations used here might deviate from this practice.

As for the transliteration of Hebrew names and terms, when they have a standard English transcription (e.g., Reuven, Isaac, Torah), we use that transcription. Other names and terms are transliterated according to the typical phonetic value of the letters, using b/v for hard and soft *bet*, k/kh for hard and soft *kaf*, p/f for hard and soft *pe*, ḥ for *ḥet*, ṭ for *ṭet*, s for both *samekh* and *sin*, ṣ for *ṣadiq*, q for *qof*, sh for *shin*, ʾ for *ʾlef*, and ʿ for *ʿa*yin. When silent or standing for vowels, *ʾlef* and *he* are transliterated accordingly. Long vowels and strong

[1] The exceptions are Levi ben Gershon's *Harmonic Numbers*, which is translated from the Latin edition (the Hebrew original being lost), and his commentary on the parallel postulate, translated by Victor Katz based on a French translation and checked against the Hebrew by Gad Freudenthal.

(doubled) consonants are not distinguished in the transliteration from their short and single parallels (with the exception of names where doubling has become standard, such as in Bar Ḥiyya and Ibn Tibbon).

I. PRACTICAL AND SCHOLARLY ARITHMETIC

This section combines practical and scholarly—as well as earlier and later—Hebrew expositions of arithmetic. From Ibn Ezra's foundational twelfth-century *The Book of Number*, we explore some elementary calculation techniques in decimal numbers and simple and sexagesimal fractions. We follow with a brief discussion of decimal numbers from the unedited arithmetic of a practically unknown Aaron ben Isaac, which sheds some light on the context of practical arithmetic at the time. Immanuel Bonfils (fourteenth century) then shows how to do arithmetic with decimal fractions as well, but does not thereby give up on sexagesimals. The lesser known Jacob Canpanṭon (fourteenth–fifteenth century), whose work has not yet been edited, provides a detailed discussion of irrational root extraction, citing several methods and providing well-reasoned error analysis. Elijah Mizraḥi (sixteenth-century Constantinople) remains at the level of elementary techniques, but provides them with lucid justifications. Finally, Levi ben Gershon sends us back to the early fourteenth century but presents the most scholarly, general, and abstract treatment of the arithmetic of his time, including detailed proofs and covering calculation techniques, proportions, series summation, and typical word problems.

1. ABRAHAM IBN EZRA, *SEFER HAMISPAR* (*THE BOOK OF NUMBER*)

Abraham ben Meir ibn Ezra (ca. 1089–1167)[1] was born in Tudela, ruled by the Emirate of Saragossa. During his lifetime he traveled extensively in North Africa, Spain, Italy, and France. Among his intellectual friends one can find Rabenu Tam and Judah Halevi. In Jewish circles he is a well-known classical poet and writer of biblical and Talmud exegesis, but he also wrote on astrology and ventured into astronomy (including a translation of al-Bīrūnī's commentary on al-Khwārizmī's astronomical tables), calendar studies, mathematics, medicine, grammar, and philosophy.

The Book of Number is the earliest surviving comprehensive Hebrew arithmetic from the Muslim period. It seems to have originated in about 1150 and enjoyed a wide circulation. It made a clear and recognizable impression on several later Hebrew mathematical compositions.

The book opens with methods for applying the four basic arithmetic operations to integers in decimal representation, simple fractions and sexagesimal fractions. It then provides an elementary treatment of arithmetic, geometric and harmonic ratios, and summation formulas for the series of the first n integers and squares. From there the book goes on to apply proportions (the rule of three, but not under this name) to commercial and calendar problems. Next it treats the extraction of square roots, bundled together with some formulas relating sums of squares to squares of differences and sums. The concluding geometry section deals

[1] For a general overview, see [Sela, 2003]; for a catalog of his scholarly writing, see [Sela and Freudenthal, 2006].

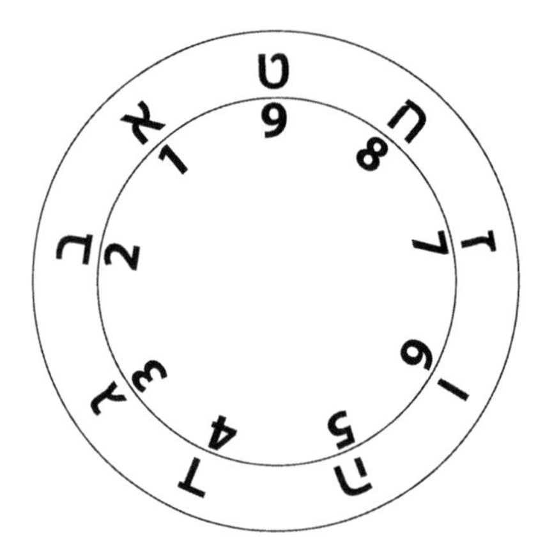

Fig. I-1-1.

with the Pythagorean Theorem and some elementary rectilinear and circle measurement, mentioning Hellenistic and Indian calculations for pi.

The book uses three representations of numbers. The first is fully spelled out number words. The second is the classical Hebrew presentation, related to the Greek and Arabic sources, of expressing numbers by letters according to a fixed key, and combining these letters additively. The third representation is the Hindu-Arabic decimal system, with the nine digits replaced by the first nine Hebrew letters, and zero denoted by a circle. This third representation is used mostly in calculation diagrams, and rarely within the running text. Here the last two representations are rendered by modern numerals. Fractions too are sometimes represented as simple fractions (mostly in arithmetical context) and sometimes as sexagesimal or other *x*-imal fraction systems (mostly in astronomy related contexts).

Numbers and the decimal place-value system (from the Introduction)

In this section Ibn Ezra introduces the Hindu-Arabic decimal system and brings in some Jewish mysticism to complement the presentation. The multiplication sphere is probably Ibn Ezra's original invention.

As God almighty alone created in the upper world nine large spheres surrounding the earth, which is the lower world, and as the author of the Book of Creation [*Sefer Yeṣira*] said that the ways of wisdom are in number, letter and word [*sfar vesefer vesipur*], so the number consists of nine digits which exhaust all numbers. When they are in the first rank, they are called ones. Ten is likened to one and twenty to two. . . . And a hundred is likened to one and to ten, and two hundred is likened to twenty and to two, and so are a thousand and ten thousand the first among [decimal] multiples[2] [*rashey klalim*] for the numbers that follow, which are [marked by the Hebrew letters] א' י' ק' ב' כ' ר'.[3]

This is shown by drawing a circle and writing nine numbers around it [Fig. I-1-1]. Multiply nine by itself. It is a square because its length equals its width; then you see it

[2]Multiplicity (*klal*) is used in opposition to a unity, designating a genuine number, or, more specifically in this text, a power of 10.

[3]That is, א (*alef*) stands for 1, י (*yod*) stands for 10, ק (*qof*) for 100, and ב, כ, ר (*bet*, *kaf*, and *resh*) for 2, 20, and 200, respectively. This is the classical use of the Hebrew alphabet for representing numerals.

as it is. The square is 81, and indeed one, the first among units, is to the left of nine, and 8, which stands in the multiples position for eighty, is to the right of nine. If you multiply nine by eight the result is 72, and indeed 2 is to its left, and 7, standing for 70, is to its right. If you multiply 9 by 7 the result is 63, and indeed 3 is to its left, and 6, standing for 60, is to its right. If you multiply 9 by 6 the result is 54, and indeed 4 is to its left, and 5, standing for 50, is to its right. Since five lies in the middle of the 9 numbers, it is called "round," and goes round into itself, because its square contains five.[4] When you multiply 9 by 5 the units go round to the right and the multiples [of tens] to the left. Indeed, the result is 45 and the 5 is to the right of 9 and the multiples [of tens], which are 4 for 40, to its left. When you multiply 9 by 4 the result is 36 with 3 for thirty. When you multiply 9 by 3 the result is 27 with 2 for twenty. When you multiply 9 by 2 the result is 18 with 1 for ten. Therefore 9 serves to test the multiplication of a number by itself or by another.[5] Thus the Indian scholars based all their numbers on nine and made signs for the 9 numbers 1, 2, 3, 4, 5, 6, 7, 8, 9.[6] I designate them by the [Hebrew letters] א ב ג ד ה ו ז ח ט.[7]

If you ever have a number in the units before the position of multiples of tens, one writes first the number of units and then the number of multiples. And if there is no number in the units, and there is a number in the second rank, which is the tens, one places the image of a wheel O in the first to show that there is no number in the first rank, and writes the number for the tens next. And if its multiples are hundreds and tens, one writes a wheel in the first rank, then the number of tens in the second and the number of hundreds in the third. One then writes the number of thousands, if there are any, in the fourth rank, the number of tens of thousands in the fifth, and the number of hundreds of thousands in the sixth ... and so on without end. If there is a number of units and hundreds but no tens, one writes the number of units in the first rank, and a wheel in the second, and the number of hundreds in the third. Thus one keeps the ranks of the wheel according to the ranks of the number at hand, and places a wheel in the first rank, or two wheels as required at the beginning or the middle. This is the wheel O, for it is "like the whirling [wheel of] dust; as stubble before the wind" [Psalms 83:14], serving only to keep rank. In the foreign tongue [Arabic] it is called *sifra*.

The number one (from Chapter 2)

This treatment of the unit, which is in line with the Hellenistic tradition, attests to the exclusion of one from the sequence of integers.

Know that all numbers are combinations of ones, and one itself is subject to no change, multiplicity or division, but is the ground for all increase, multiplicity and division. One alone is primordial and serves to change all numbers. With its single side it affects what all numbers do with their two sides. Indeed, two precedes three on one side and four follows it on the other, and both sides combined are six which is double three, and the same goes for all numbers; while one has no preceding side, and the side that follows it has two, which is double one.

[4] The unit digit in 5 times 5 is again 5.

[5] This refers to the verification of calculations by "casting nines" (in modern terms, verifying that the results hold when calculated modulo 9).

[6] This is the only place in the text where actual Arabic numerals are used.

[7] The letters are read *alef, bet, gimel, dalet, he, vav, zayin, ḥet, ṭet*.

Multiplication of fractions (from Chapter 5)

This segment discusses various approaches to handling fractions. The use of a common denominator for multiplication of fractions might seem odd today, but the text sheds some light on its context.

To begin with I state the rule that the products of fractions are the opposite of the products of wholes. When one says multiply half by half, it is as if he says, take half of half, and the result is one quarter. We know that the half is taken out of two, whose half is one, and the other half is also one. The product of one by one is one, and the square of the denominator [*more*] is 4, so this last one is a quarter, which is half of half. We do the opposite of our practice with wholes, always taking the value of the product with respect to the square of the denominator. Multiplying a third by a third, the result is a ninth. Multiplying a quarter by a quarter, the result[ing denominator] is 16 and the numerator is one, which is half an eighth. Continue this way until ten and also above, as in one part in 11 multiplied by one part in 11 is one part in 121, which is the square. This way you multiply fractions of some kind by fractions of the same kind, whether equal or one bigger than the other, and then divide by the square of the multiplied denominator.

Example: We want to multiply 3 quarters by 3 quarters. The denominator is 4. For each of the 3 quarters we take 3, and the product is 9. We divide them by 16, which is the square of the denominator, and get its half and half of its eighth. If you wish you may divide 9 by 4, and the result [measured in quarters] is equal because the quarter of a quarter is half of an eighth.

Example: We wish to multiply 3 fifths by 4 fifths. The denominator is 5. For the 3 fifths we take 3, and for the 4 fifths 4. We multiply 4 by 3 yielding 12, which is the product. [Taken with respect to 5 squared] this gives 2 fifths of the square[8] and 2 fifths of a fifth.

And if we are given fractions of two kinds, and are instructed to multiply 2 thirds of one by 3 of its quarters, we seek the denominator for both, multiplying 3 by 4, which is the denominator 12 whose square is 144. So for 2 thirds we take 8 [out of 12], and for 3 quarters 9 [out of 12].[9] We multiply 8 by 9 and get 72, which is half of 144, the square of the denominator. So the result is half of one.

[8]Does the "square" here suggest a figurative model—a square divided into 5 rows and 5 columns? With such a figurative model in mind, it is easy to see that the product of 3 rows by 4 columns equals two full rows (fifths) and two extra subsquares (fifths of fifths).

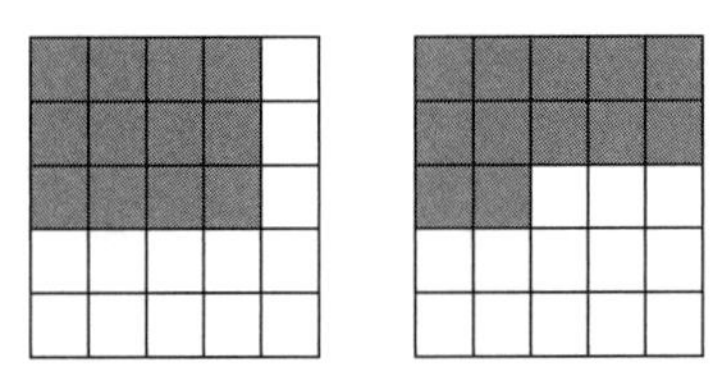

However, the term "square" could simply refer to the squared number, 25. The elliptic formulation would then be odd, but not impossible.

[9]Here Ibn Ezra is finding a common denominator for the two fractions to be multiplied. This might strike us as odd, but may indicate the use of a figurative model as above, where an equal subdivision into 12 rows and 12 columns is preferred over an unequal 3-by-4 subdivision. Alternatively, this preference for a common denominator may be the influence of sexagesimal-like systems, where a multiplication of minutes by minutes (sixtieth parts) is measured in seconds (3600th parts), or may perhaps indicate a general preference for homogeneity in arithmetical operations.

And if you multiply 2 by 3, you also have half the denominator, which is 12. So if you have two [different] denominators it is easier, as there is no need to square the denominator. Consider only the product of one denominator multiplied by the other as if it were the square [in the product of fractions with equal denominators], and divide by it.

Example: We take one denominator 3 for the said thirds and the other denominator 4 for the said quarters. Multiply the one denominator 3 by the other denominator 4 yielding the required 12, with respect to which we take the [following] product: for the 2 thirds we take two (out of the 3), and from 3 quarters we take three (out of 4). We multiply 2 by 3, yielding 6, which is half the product of the denominators.

Commercial problems solved by proportion rules (from Chapter 6)

The following is an example of the usage of proportion in commercial problems (the Rule of Three). Note the use of the wheel symbol for both zero and the position of the unknown number in the diagrams. The first example requires a double use of proportion (Rule of Five), and the second applies the Rule of Three in a false position (making a guess and then rescaling to fit the required result). Note also that the famous "tree" question (which was treated, among others, by Fibonacci) is already referred to as a standard reference problem here.[10]

Question: Reuven hired Simon to carry on his beast of burden 13 measures of wheat over 17 miles for a payment of 19 *pashuṭs*.[11] He carried seven measures over 11 miles. How much shall he be paid? Do thus. You should apply proportions twice, as there is no other way to get it. Suppose he carried the 7 measures according to terms, which is 17 miles. Draw this diagram.

13 7
19 $\underline{0}$[12]

We multiply the extremes, which are 7 and 19, yielding 133. We divide them by 13, yielding 10 wholes and leaving 3 parts of 13 in one *pashuṭ*. But since he carried but the 7 measures for only 11 miles, we should apply another proportion and another diagram thus: 11, 17, wheel, 10 and 3 parts of 13. Here is the diagram:

17 11
3 10 $\underline{0}$

Since we should multiply the extremes and divide the result by 17, and we have in the fourth 3 parts of 13, we should seek one denominator for both. We find it by multiplying 13 by 17, yielding 221, which is the denominator and is one whole. 3 parts of 13 are 51 of 221. We then multiply 11 by 10, yielding 110, and multiply 11 wholes by 51 parts yielding 561. We divide by 221, which is the whole one, yielding 2 wholes, which we join with 110 to make 112, and a remainder of 119 parts of 221. We divide 112 by 17 wholes yielding 6 whole *pashuṭs* and a remainder of 10 whole *pashuṭs*, which we count with respect to

[10] See section II-2-2 on Leonardo of Pisa, *Liber abbaci*, in Chapter 1 of this *Sourcebook*.

[11] Literally, *pashuṭ* means "simple," but here it is used as the denomination of a coin, which is one-twelfth of a *dinar*. In fact, the *pashut* is almost surely the *denier*, a paper-thin lightweight coin of mixed silver and copper, and the standard coin of Europe between the eighth and thirteenth centuries. Both the *pashut* and the *dinar* are also found in the Talmud, referring to small and large denominations of coins. For more information, see [Simonson, 2000b].

[12] The manuscripts use the same symbol for zero and the unknown quantity in the Rule of Three. We underline the latter to prevent confusion.

221 [yielding 2,210], and add the remaining 119 parts of one *pashuṭ*, which is 221 parts, yielding 2,329. We divide by 17, yielding 137. The total is 6 whole *pashuṭ*s and 137 parts of which 221 make a whole.

. . .

[Question:] We subtract from an estate its fifth, seventh and ninth, leaving 10. [To find the estate,] subtract 143 which are the fractions [1/5, 1/7 and 1/9] out of 315, which is the denominator [as calculated earlier], leaving 172. We do the proportion thus.

0 10
315 172

We multiply 10 by 315 yielding 3,150. We divide by 172, yielding 18 wholes and 54 parts of 172. If we take away the fifth, seventh and ninth of this number, we are left with 10 wholes.

And the same method applies to the question of the tree, which has a third in the water and a quarter in the ground, leaving 10 cubits above water. How tall is the entire tree? We seek a number that has a third and a quarter, which is 12. Its third and quarter joined are 7. We subtract them from 12, leaving 5. We do the proportion thus.

0 10
12 5

We multiply the extremes, yielding 120. We divide by 5, yielding 24. This is the height of the entire tree, because its third is eight, and its quarter six, which joined together make 14, and when we subtract them from 24, there remain 10 wholes, no less and no more.

One of the motivations for scholarly presentations of algebra and proto-algebra in the Islamic world was the concern with sharing inheritances according to Muslim law. Here we see this concern reflected from the point of view of Jewish law as well.

Question: Jacob died. His son Reuven produced a deed signed by two valid witnesses that his father Jacob gave him alone his entire estate upon his death. His son Simon also produced a deed that his father ordered to give him half of his estate upon his death. Levi too produced a deed that his father ordered to give him a third of his estate. And Judah also produced a deed that he be given a quarter of his estate. And all carry the same day, time and hour in Jerusalem where time is reckoned.

The scholars of Israel divide it by the claim of each, and the gentile scholars by the proportion of the estate of each. The scholars of arithmetic reckon the estate as one, and when you add its half and third and quarter, the total is two and half a sixth. Consider the one whole as sixty, which has all the mentioned parts, yielding a total of 125. Or we can consider the one whole as 12 and the fractions 13; the final result will be the same either way. Let us calculate Reuven's part according to the proportion of his estate. We will calculate the proportion with respect to sixty, as he requests the entire estate. Say that the estate is 10 dinars which are 120 *pashuṭ*s. This is the proportion rule for Reuven's estate.

0 60
120 125

We multiply the extremes, yielding 7,200, and divide by 125, yielding 57 *pashuṭ*s and 75 parts, which is Reuven's part. [Using this same method, Simon gets 28 *pashuṭ*s and 100 parts, Levi gets 19 *pashuṭ*s and 25 parts, and Judah gets 14 *pashuṭ*s and 50 parts.]. . . .

According to the scholars of Israel, the three elder brothers say to Judah their brother: you contest only 30 *pashuṭs* [a quarter of the estate], and all our claims to them are equal. Take 7 and a half, which is a quarter, and leave us. And each of the three brothers will also take as much. Then Reuven says to Levi: you only contest 40 *pashuṭs*, and have already taken your share of the 30 which all four of us contested. Take a third of the 10, which is [the remaining] quarter of 40, and leave us. And so Levi's share is ten and five sixths. . . . Reuven then says to Simon: you only contest half the estate, which is 60, and the other half is all mine. You've already taken your share of the 40, so you and I contest only 20. Take half of that and leave me. So Simon's share is twenty and five sixths *pashuṭ*, and Reuven's part is eighty and five sixths of one *pashuṭ*.[13]

Extracting square roots (from Chapter 7)

Here we see the technique for extracting square roots. Ibn Ezra first shows how to derive the integer part of the root. Then, after some motivational discussion concerning identities involving squares, he uses rescaling to obtain sexagesimal approximations of irrational roots.

Note that there are 3 squares in the first rank [one-digit numbers], namely 1, 4 and 9. In the second rank [two-digit numbers] there are 6 [squares], namely 16, 25, 36, 49, 64 and 81. And all the ranks that follow these two proceed in the same way: all odd ranks take after the first rank, and all even ranks take after the second rank. Indeed, squares that take after squares in the first rank [namely, one digit squares multiplied by an even power of 10] always have one digit, and those in the second rank have two digits, as do all the numbers that take after them [namely, those multiplied by an even power of 10]. From this analogy you can tell the squares that precede or follow them.

If you know the root of a number in the first or second rank, and want to know the rank of the root of the analogous number [multiplied by a power of 10], do as follows. Know that [the root of] what lies in the first rank is units, in the third rank tens, in the fifth—hundreds, in the seventh—thousands, in the ninth—ten thousands, in the eleventh—hundred thousands, and so on without end skipping from one odd number to the next odd number. And the units are the roots of the second rank, of the fourth rank—tens, of the sixth—hundreds, of the eighth—thousands, of the tenth—ten thousands, of the twelfth—hundred thousands, ever skipping from even to even.[14]

And now I will instruct you on what to do when you know the analogous square and its root. Subtract the square from the desired number [whose root you wish to find], but make sure you take only the square preceding your number. Take the distance between your number and the square, and divide by twice the root of the preceding square. Now do not subtract from your number as much as you can [namely the result of the division times twice the root]—leave enough to fit the square of the result of the division. If the distance to the preceding square equals what comes from the division multiplied by twice

[13] The method of the "scholars of Israel" is reminiscent of Pascal's argument in his letter to Fermat concerning the equitable division of stakes in a game interrupted before its conclusion. See [David, 1998, pp. 85–88].

[14] Ibn Ezra explains that the integer part of the root of a decimal number of the form $\overbrace{x00\ldots0}^{2n}$, where x is a square integer with one or two digits, is of the form $\overbrace{y00\ldots0}^{n}$, where y is a single-digit integer.

the root, joined together with the square of what comes from the division, then the square is true. . . .[15]

Example: We wish to know the square preceding two hundred. This number is of the third rank, which is odd, so we consider the first rank, where the squares are 1, 4 and 9, and their analogous squares are a hundred, 4 hundred and 9 hundred. A hundred is the preceding square, and its root is 10, as we said: the roots which are units for the first rank are tens for the third. Subtract the square [of the root, 100] from our number [200], leaving 100. We already said that the root is 10, so its double is 20. If we divide 100 by 20, and give the result 5, we are left with nothing from which to take the square of the result of the division [after subtracting 5 times 20 from the remaining 100]. So we give 4, which multiplied by 20 makes 80, leaving 20. We then subtract 16 which is the square of what comes from the division, leaving 4. Subtract it from the two hundred, leaving 196, which is the square preceding two hundred. Add 4, which comes from the division, to the first root which was 10. This makes 14, which is the true square root [of 196].

The above calculations obviously relate to what can anachronistically be expressed as $\frac{\sqrt{100a}}{10} = \sqrt{a}$ and $(a+b)^2 - a^2 = 2ab + b^2$. These two identities are easy to understand in hindsight, but it is not clear how they would lead to the root extraction algorithm in a pre-algebraic, intuitive way. The seemingly unremarkable examples in the following paragraph show how a practice of squaring numbers and retracing one's steps can produce such practical intuition. In the first example, sexagesimal-like fractions (here actually septagesimal ones) bring up the rescaling phenomena related to the first identity above, while the second example shows how the intuitive division of a number into an integer and a fractional part makes salient the structure of the second identity.

Example for a calculation that cannot be divided by 60. We multiply 4 sevenths by 4 sevenths. According to the scholars of arithmetic, we multiply 4 by 4 to get 16, divide by 7 [twice] and get 2 sevenths and 2 sevenths of a seventh. According to a way similar to that of the scholars of astrology, make 70 parts, so 4 sevenths are 40 [minute-like parts]. We multiply 40 by 40 to get 1600, divide by 70, and get 22 minutes and 60 seconds, which is the square. Now if he who posed the question turns it around, and asks for the root of this square, so shall we turn the 22 minutes into seconds, and add the 60 seconds that we had in order to get 1600. We consider them as wholes, whose root is 40, and then consider them as minutes, and this is indeed the root.

. . .

Another example: It is said that the square is 11 and a ninth. What is the root? Since it is said to have a ninth, this ninth indicates that there's a third in the root of this square. Subtract the ninth, which is the square [of 1/3] or fraction of fraction, and 11 wholes remain. Their distance from the previous [whole] square [which is 9] is 2 wholes.

[15]The algorithm is as follows. To extract the root of x, take a, an integer approximating the root from below (following the instructions of the previous paragraph). Then consider $x' = x - a^2$. To obtain the next digit of the root, a similar subtraction should be applied to the remainder x'. However, now we should remove from x' not only the square of the chosen integer, b, but also the term $2ab$. So b may be obtained from the quotient $x'/2a$, except that this quotient must be reduced so that b^2 can fit into $x' - 2ab$. The process is reiterated until the integer part of the root of x is obtained.

We turn them into minutes, yielding 120. We divide by twice the preceding root, which is 6, and get 20. So the root is 3 wholes and 20 parts, [which are] a third.

The next paragraph shows how to derive a better approximation of the root of 2 by taking the root of 20,000. This calculation comes after the root of 2 was calculated via the root of 200, yielding the result 1 24′ 52″.

Let's extract this root [of 2] again from 20,000. This number is analogous to units, and 10,000 is analogous to 1, so this is the analogous square. We subtract 10,000 from our number and 10,000 remain. We divide by twice the root, which is 200, but we don't give it as much as we can, leaving enough for the square of the result of the division. So we give it 40, and [after subtracting 40 times 200], two thousands remain. We subtract 1,600, which is the square of the result of the division, leaving 400. Our [cumulative approximation for the] root is 140, and its double is 280. We divide the remainder by it, give it one, and 120 remain. We subtract 1, which is the square of 1, leaving 119, and our [cumulative approximation for the] root is 141. We turn the remainder into minutes, yielding 7,140, and divide by 282, which is double our root, yielding 25 minutes and 19 seconds. [To get the square root of 2], we divide the total of wholes and parts by 100, yielding 1 whole 24′ 51″ 11‴, and this is more precise than the former calculation [derived from the root of 200, which gave 1;24,52].

Mathematics as a natural science (from Chapter 5)

The following statement concludes a short discussion of the evaluation of pi. It likens the discovery of such evaluations to an empirical process rather than a deductive one.

And likewise [i.e., as in the evaluation of pi,] in natural history the scholars found by way of trial and error the true properties of herbs and stones and the parts of the human body, and none of them knows why it is so, except the blessed unfathomable God.

2. AARON BEN ISAAC, *ARITHMETIC*

This section was prepared by Naomi Aradi

The only information known to us today about Aaron and his arithmetical work comes from a single surviving manuscript in Turin. A brief bibliographical reference concerning the manuscript is available in [Steinschneider, 1906, p. 197]. Since the upper part of this manuscript was apparently burned, the first few lines of every folio are illegible. The disruptions in the text indicate that it may be an autograph. The manuscript contains Aaron's arithmetical textbook only. The bibliographers estimate that the manuscript is from fifteenth-century Spain.

The work is divided into three sections. The first section deals with arithmetical operations (addition, multiplication, subtraction, and division) and has four parts: the first three consider respectively operations with integers alone, with fractions alone, and with both integers and fractions, while the fourth part considers roots and progressions.

The second section is devoted to word problems arranged in five parts. The first four parts present addition, multiplication, subtraction, and division problems, respectively, each part in turn discussing cases involving integers alone, fractions alone, and a mixture of both types in an orderly manner. The fifth part concerns the double false position.

The third section is devoted to ratios, including issues related to number theory, such as the discussion on amicable numbers examined in section II-7 below.

Aaron mentions Abraham ibn Ezra when he attributes to him one of his word problems [f. 162a]. Yet this problem does not appear in Ibn Ezra's famous *Book of Number*, which is not mentioned by name in Aaron's work. As he discusses ratios, Aaron notes Greek mathematical terms (in Hebrew letters). This raises the possibility that he knew Greek and perhaps even relied directly on Greek texts, such as the *Introduction to Arithmetic* by Nicomachus.

Included here are some excerpts from the preface, describing Aaron's biographical background, and a mathematical-philosophical discussion of numbers and decimal representations.

I, the aforementioned Aaron, from the day I began to engage in the craft of weaving gold and silk images, although proficient in it with accuracy and precision, realized that I lack some of the arithmetical methods required for the subtlety of the craft. Though I had some knowledge of it, I put my mind and effort to investigate the general practices of number and some of its particular properties, as human knowledge cannot comprehend them all. So I saw fit to write what came within my reach, and put it together into a composition containing the rules of number and most of its many properties, so as not to forget with the passing of events and time what I had learned and to teach them to my sons, God willing.

If a scholar gets hold of this treatise or of *Miqnat Kesef* [acquired property][16] or *ʿIbbur* [intercalation] or . . . *Ṣurat Haʿolam* [the shape of the world], which is *Alfarghani* in another form, and finds that my language is not that of a learned man, he should not judge me harshly, but give me the benefit of the doubt, because I am a craftsman, as I said, and composed these treatises for my son Joseph. Now I add that, by my sins, he is dead, and these compositions are left as they stood, without order and unedited. Indeed, he was to put them in order and edit them, but now he lies deep, and I am at the end of my days, and I never got to teach him. And so, they are as naught.

I say that the kinds of number are two. The first is the number deprived of substance, speech or thought, which is unlimited, because it is a number *in potentia* rather than *in actu*, and has no end. The second kind, which this treatise concerns, is the number initially delimited and bounded by thought. Then, through the rational capacity, one can pronounce it or write it, as one wishes. This number is thus counted and limited. It is divided into two: even and odd, whose foundation is one, because the counted number is a sum of units.

. . .

A necessary comment on the knowledge of letters which are used in arithmetic: I say that since the number deprived of substance is unlimited, the arithmeticians had to use endless figures or letters.

[A few corrupt lines on the finiteness of letters in all languages follow.]

Since there is no letter after ת [the last letter of the Hebrew alphabet, designating 400], you cannot find a letter to express a greater number. Even if you use the letters which indicate large numbers, they will not be sufficient due to the greatness of that number, and if you repeat them several times in writing [using them additively to express higher numbers], much confusion will ensue.

[16] This term is quoted from Exodus 12:44, where it refers to purchased slaves.

The uncounted number is without end, and the letters have an end, and that which has an end cannot count the endless. In other words, the smaller cannot count the larger. So the arithmeticians agreed to set 9 figures with endless ranks. The nine figures are like substance, and in the ranks they are of different form. The first, for example, is one, and can be ten or a hundred or a thousand or larger numbers without end. The same goes for the other 9 figures set by some scholar. These are the figures accompanied by the corresponding Hebrew letters, which you may use to deal with numbers, as I do in this treatise.

0 ט ח ז ו ה ד ג ב א

0 9 8 7 6 5 4 3 2 1

So the ranks are due to the largeness of numbers, and the sparsity of letters is due to the ranks.

3. IMMANUEL BEN JACOB BONFILS, *ON DECIMAL NUMBERS AND FRACTIONS*

Immanuel lived in Orange and Tarascon in Provence (ca. 1300–1377). He made his name as an astronomer, having composed several astronomical treatises and tables, the most famous of which, *Six Wings*, was subsequently translated into Latin and Greek [Solon, 1970]. He also translated from Latin into Hebrew. His mathematical work focuses on the calculation of roots and on circle measurements. His decimal treatment of fractions is the earliest surviving treatment to be recorded in Europe.

This short note on decimals was first identified in a small astronomical codex. Parts of it are available in several manuscripts, but only three of these manuscripts include the decimal treatment of fractions. The note treats multiplication and division of decimal fractions, division of sexagesimal fractions, and rescaling numbers in decimal and sexagesimal systems for the purpose of root extraction. The treatment of decimal fractions and integers is completely homogenized: each decimal place, either integer or fractional, is assigned a positive degree, and these degrees are used to calculate the decimal place of the product or quotient. The exception is units, which are not treated as "zero degree" integers but as a special case. The selection below includes a short fragment on sexagesimal division, which places the decimal treatment in context.

Know that the unit is divided into ten parts which are called prime fractions, and each prime is divided into ten parts which are called seconds, and so on without end. I also want to call to your attention that I am calling the degrees of the tens first wholes, and the hundreds second wholes, and so on without end. The degree of the units, however, I am calling by their name, units, for it is an intermediate between the wholes and the fractions. Therefore, if one multiplies units with units the result is units [which is not the case for any other degree].[17]

Furthermore, I am calling the degrees whose name is greater "greater in name." I mean by that: I am calling the thirds greater in name than the seconds, for the thirds are derived from three, while the seconds from two. Similarly, the fourths are greater in name than the thirds and the fifths than the fourths. This applies to the wholes as well as to the fractions. Furthermore, when I say: add this name to that name, or subtract this name from that name, I mean by that: add the name of the seconds to the name of the

[17] Words in square brackets come from a new manuscript recently reconstructed by Naomi Aradi.

thirds, yielding fifths, or the name of the seconds to the name of the seconds, yielding fourths. Or also, subtract the name of the seconds from the name of the thirds, leaving firsts, or the name of the seconds from the name of the seconds, with nothing left, so it falls into the degree of the units. This applies to the wholes as well as to the fractions. If you subtract a large name from a small name, as when we say: let us subtract the name of fourths from the name of seconds, be it among wholes or among fractions, then it will come into the degree of the seconds on the other side. For instance, when we say: let us subtract the name of the fourths in the fractions from the name of the seconds also in the fractions, then it falls into the degree of the second wholes. Similarly, if we say the same with regard to the wholes, i.e., if we want to subtract the name of the fourth wholes from the name of the second wholes, then it falls into the second fractions.

If you multiply a number by a number, both being wholes or both fractions, add the name of the ranks, and there the product will be among the wholes if both are wholes, and among the fractions if both are fractions. If, however, the one is a whole and the other a fraction, if they have the same name, then the product will fall into the degree of the units. But if the name of one is greater than the other, then subtract the smaller from the greater, and as the number of the remaining name, there the product falls—among the wholes if the name of the wholes is greater, or among the fractions if the name of the fractions is greater. [As when you multiply 3 second wholes by 2 seventh fractions, you subtract 2 from 7, leaving 5. The multiplication yields 6 fifth fractions. For 3 second fractions by 2 seventh wholes, the product will be 6 fifth wholes. Here's a diagram for that.][18]

	Hundreds of thousands	Tens of thousands	Thousands	Hundreds	Tens	Units	Primes	Seconds	Thirds	Fourths	Fifths	Sixths
			2*	5	4	1	3	2	1			
			3	1	2*	5	4	3	2			
						3	1	2	5	4	3	2
					6	2	5	0	8	6	4	
				9	3	7	6	2	9	6		
			3	1	2	5	4	3	2			
	[1]	2	5	0	1	7	2	8				
[1]	5	6	2	7	1	6	0					
[6]	2	5	0	8	6	4						
7	9	4	2*	7	2	5	9	7	5*	6	7	2

[18]The example in the manuscript is 4541.321 times 3135.432. The numbers are originally presented as letters. Naomi Aradi noted that the calculation diagram in the manuscript would make most sense if the first digit of the first multiplicand is rendered as 2 instead of 4 and the third digit of the second multiplicand as 2 instead of 3. Then only a couple of digits at the bottom line need to be amended. The differences between the original and revised versions are marked with asterisks.

If you divide a number by a number, both wholes or both fractions, and the name of their ranks is the same, then the quotient will fall into the degree of the units. For if you subtract the one name from the other name, nothing remains, and it falls into the degree of units. If the dividend is greater in name than the divisor, subtract the name of the divisor from the name of the dividend, and as the number of the name of the remainder, there the quotient will fall on that side, i.e., among the wholes if they were whole, or among the fractions if they were fractions. If, however, the divisor is greater [in name] than the dividend, then subtract the name of the dividend from the name of the divisor, and as the number of the name of the remainder, there the quotient will fall on the opposite side, i.e., among the fractions if both were wholes, or among the wholes if both were fractions. If, however, the one is a whole and the second is a fraction, the name of their ranks being the same or different, then add the name of the ranks, and as the number of the name of the result, there the quotient will fall—among the fractions if the dividend is a fraction, or among the wholes if the dividend is a whole.

. . .

If you want to divide a number of [sexagesimal] degrees, minutes and seconds by another smaller or greater number, then take the number of the lower row, which is the number by which the number of the upper row is divided, and break it all into the kind of the smallest fraction in it. For example, if its smallest fraction is seconds then reduce it all into seconds, and if thirds then reduce it all into thirds, and so on in the same manner. Thereupon take the number of the upper row, which is the number divided by the number of the lower row, and break it into such a kind of fractions, that its distance from the kind of fractions into which you have broken the lower be equal to the distance of the kind of fractions which you want to obtain in the quotient from the degrees.

For example: if you have broken the number of the lower row into the kind of the seconds and you want to obtain in the quotient thirds (or any other kind that you want, but let us suppose in this example that you want thirds), then break the number of the upper row into the kind of fifths, which is removed from the seconds by three degrees, like the distance of the thirds from the end of the degrees, and then you will obtain thirds in the quotient. And if you want to be more accurate and obtain fourths in the quotient, break the number of the upper row into sixths, which is removed from the seconds by four degrees, like the distance of the fourths from the end of the degrees, and then you will obtain fourths in the quotient.

4. JACOB CANPANṬON, *BAR NOTEN ṬAʿAM*

This section was prepared by Naomi Aradi

Rabbi Jacob Canpanṭon lived in Castile in the fourteenth and fifteenth centuries. In Hebrew historiography, he is said to be "one of the rabbis of Spain" and a student of Rabbi Ḥasdai Crescas, who wrote the famous philosophical work *ʾor Hashem*, and the father of Isaac Canpanṭon (1360–1463), *Gaʾon* of Castile. In addition, it is stated that he wrote books on arithmetic, astronomy and the Torah [Hacohen, 1967–1970, vol. 5, p. 94]. According to [Steinschneider, 1893–1901, p. 186], Jacob apparently was already a teacher in 1406 and was no longer alive in 1439. He worked as a mediator in the field of medicine while preparing a Hebrew summary of the Arabic commentary by Solomon ibn Yaʿish on the *Canon* of Avicenna.

Canpanṭon's arithmetical textbook survives in the single manuscript located at the British Library in London. It is titled *Bar Noten Ṭaʿam*, based on a Talmudic saying, indicating the author's intention to explain the reasoning (*ṭeʿamim*, literally, reasons) behind the arithmetic operations, and not merely describe the procedures. The treatise begins with an introduction phrased as a long rhymed poem embedded with biblical and rabbinic expressions describing the circumstances that led the author to write the composition. The treatise was written at the request of one Rabbi Joel ben Daʾud, a close friend of Canpanṭon, who wished to learn mathematics. The book is divided into two sections. The first section is devoted to integers and contains six chapters on addition, subtraction, multiplication, division, proportions, and roots. The second section deals with fractions and discusses the following issues: conversion, fractions of fractions, equalization, addition, subtraction, multiplication, division, ratios, roots, common denominators, completion, and shortcuts.

The passages below are taken from the sixth chapter of the first section dedicated to extraction of roots of integers. In this chapter, Canpanṭon describes iterative root extraction algorithms in his typical lucid and elaborate manner, including a detailed comparative error analysis. As far as is known, the formula $\sqrt{a^2+r} = a + \frac{r}{2a+1}$ was used in Eastern Arab mathematics, but not in the West [Harbili, 2011]. Canpanṭon analyzed when it outperforms the Western formula $\sqrt{a^2+r} = a + \frac{r}{2a}$. Canpanṭon also suggests and iterates the formula $\sqrt{a^2+r} = a + \frac{r}{2a+\frac{r}{2a}}$. I am not aware of earlier precedents for these two contributions.

The discussion starts with a description of the standard algorithm for extracting the integer closest to the square root of a given integer. It then considers further approximations for the roots of non-square integers, where the extracted integer root leaves a remainder.

When there is a remainder after you have completed the extraction of the [integer part of the] root, and you wish to come closer to the truth, consider this remainder. If it is less than the [integer part of the] root, double the root and set it as a denominator to divide the remainder. The result is the addition to the integer [part of the root obtained by the algorithm] in the [new approximate] root. If the remainder is greater or equal to the [integer part of the] root, and you do not intend to come closer to the root except by this step alone, then double the root, add one, and divide the remainder by the sum. The result is the fractions added in the [new approximate] root to the initial integer [obtained by the algorithm].[19]

If you wish to come closer to the truth, even if the truth is invisible to all living beings, as Euclid proved, multiply the integer and fractions by themselves ... and the result will exceed or fall short of the initial number [whose root is being extracted]. Double the root, as we said, and divide the excess or deficit by the result. Subtract the result from the preceding fractions if the number [whose root is being extracted] is less than the square of the root that you extracted in the previous stage; and if the square is less than the number [whose root is being extracted], add the result to the preceding fractions. The

[19]In anachronistic terms, Canpanṭon suggests that when extracting the root of $a^2 + r$, where a and r are integers and $r < a$, then the first approximation for the root should be $a + \frac{r}{2a}$, and if $r \geq a$, then the approximate root should be $a + \frac{r}{2a+1}$.

sum or the remainder will be the fraction added in the root to the initial integer [part of the root obtained by the algorithm].[20]

Here Canpanṭon presents the following example: on extracting the root of 10,375, the result is 101 and the remainder is 174. Since 174 is larger than 101, the new root is 101 and 174/203, which equals 101 and 6/7. Then, since the square of this number is less than 10,375 by $\frac{6}{7} \times \frac{1}{7}$, the next approximation would add $\left(\frac{6}{7} \times \frac{1}{7}\right)/\left(2 \times \left(101\frac{6}{7}\right)\right)$.

You come ever closer to the truth, but you will never attain it. If you look closely, you will see that you can get the [same approximation] with less effort. Indeed, consider the fraction attained. If it is an addition [to the integer part of the root], and it was produced by adding 1 to twice the root [in the denominator], then find the product of this fraction by its complement [with respect to 1]. This product will be the deficit of the square [of the approximate root] with respect to the initial number [whose root is being extracted],[21] and should therefore be divided by double the [approximate] root itself and added to this [approximate] root. This is clearly visible in the previous example (which had [a fraction] added and involved adding one [in the denominator]), where the fraction was six sevenths. Its complement with respect to one is one seventh, and their product is six sevenths of a seventh; this is indeed the deficit we found with respect to the original number [whose root was extracted], and so we instructed to divide it by double the [approximate] root and add [the result] to this root.

But if the additional [fraction] did not involve adding 1 [in the denominator] and falling short [of the initial number whose root is being extracted], then multiply the [additional] fraction by itself, and divide by double the [approximate] root, because this is the excess of the square of the [approximate] root over the original number [whose root is being extracted].[22] We subtract the result from the previous [approximate] root, and so on.

[Do] this if you wish to approach the truth by repeating the procedure, because the more you repeat, the nearer you come to the truth, even if you can never attain it, as we have explained. [If you repeat the procedure], never add 1 to double the root, even if the remainder is very large with respect to the [approximate] root, so as to avoid confusion, for [adding 1] was instructed only for a single [approximation] step. Adding 1 when the remainder is greater or equal to the [approximate] root improves the approximation, as I explained, but if one repeats the procedure, one does not need this addition, because by repeating the procedure one approaches [the truth] very closely even without adding 1. It is better not to add it, so as to maintain a standard form of procedure and prevent confusion.

The reason we say that if we have a remainder smaller than the [approximate] root, then we should divide it by double the root is the following. That which we add to the root

[20]Canpanṭon states the following general principle for iterative root approximation: if a (not necessarily an integer) approximates the root of $a^2 + r$, then the next approximation should be $a + \frac{r}{2a}$, and the same goes for a subtracted r.

[21]Canpanṭon claims that when the root of $a^2 + r$ is approximated by $a + \frac{r}{2a+1}$, then the error (the original number less the squared approximation) is $\frac{r}{2a+1} \times \left(1 - \frac{r}{2a+1}\right) = \frac{r}{2a+1} \times \frac{2a+1-r}{2a+1}$.

[22]Here it is stated that when the root of $a^2 + r$ is approximated by $a + \frac{r}{2a}$, then the error (the squared approximation less the original number) is $\left(\frac{r}{2a}\right)^2$. In the next paragraph Canpanṭon explains that for a reiterated approximation, one should always use this, rather than the previous approximation.

will add to the square its product by twice the previous root and its product by itself, as we explained with regard to integers. But we proceed as if it only adds its product by twice the root. If this were true . . . then we would have this product, which equals the excess of the number [whose root is being extracted] over the square of the integer [received through the root extraction algorithm], and . . . we would reach the required result. . . . [But] that which is added to the root further adds to the square its product by itself. . . . Therefore, when we multiply the root with the addition by itself, the square will exceed the initial number [whose root is being extracted] by the square of the addition.[23]

The explanation is then repeated to obtain the next (subtractive) step of the approximation and its error term, which in anachronistic terms reads: $a + \frac{r}{2a} - \left(\frac{r}{2a}\right)^2 / \left(2\left(a + \frac{r}{2a}\right)\right)$ with the error $\left(\left(\frac{r}{2a}\right)^2/\left(2\left(a + \frac{r}{2a}\right)\right)\right)^2$.

So, when we do not add 1 [to the denominator], and wish to approach the truth, [in the first step] we should only add the fraction of the first step [i.e., the remainder divided by twice the integer approximating the root]. But from there on we must divide the square of the fraction produced at that step by twice the previous [approximate] root. And the result will ever be subtracted from the previous [approximate] root.

Next comes a detailed demonstration of the above reasoning in the extraction of the root of 7. The integer received through the extraction algorithm is 2, and the remainder is 3. The first approximation involving a fraction is therefore $2\frac{3}{4}$. The square of this exceeds 7 by $\left(\frac{3}{4}\right)^2 = \frac{9}{16} = \frac{2}{4} + \frac{1}{4} \times \frac{1}{4}$. Therefore, the next approximation is $2\frac{3}{4} - \frac{\frac{2}{4}+\frac{1}{4}\times\frac{1}{4}}{2\times(2\frac{3}{4})} = 2\frac{3}{4} - \frac{9}{11} \times \frac{1}{2} \times \frac{1}{4} = 2\frac{2}{4} + \frac{1}{2} \times \frac{1}{4} + \frac{2}{11} \times \frac{1}{2} \times \frac{1}{4}$. The square of this then falls short of 7 by $\left(\frac{9}{11} \times \frac{1}{2} \times \frac{1}{4}\right)^2$, and so on.

The reason we say that when the remainder is greater than or equal to the [integer part of the] root, we should divide it by double the root plus one (as long as we do not intend to repeat the procedure so as to further approach the truth and restrict ourselves to this step only), is that if we had not added one, the square of the root consisting of the integer and fraction would exceed the number [whose root is being extracted] by the square of the fraction received in the division. But this would be a quarter or more. For if [the remainder] is equal to the root itself, and we divide it by double the root, the result of the division will be a half. Its square (namely, its product by itself, which is the excess) will then be an entire quarter. And if the remainder is greater than the root, when we divide it by double the root the result will be more than a half, and its square more than a quarter.

Canpanṭon gives the following example: without adding 1, the root of 6 is approximated by $2\frac{1}{2}$, whose square exceeds 6 by $\frac{1}{4}$; if we add 1 to the denominator, we get $2\frac{2}{5}$, whose square is less than 6 by $\frac{2}{5} \times \frac{3}{5}$, which is smaller than $\frac{1}{4}$.

If you divide [the remainder] by double the root plus 1, the square of the root will be less than the sought number by the product of the quotient and its complement with respect to one, which can never in any way reach a quarter. For the product of a portion of a line or a number by its complement never reaches a quarter. Because if you multiply its half by its half the result will be a quarter, and clearly, if you multiply its smaller portion by its larger

[23] To justify the above error estimate, Canpanṭon uses the equality, which in anachronistic terms would be rendered $\left(a + \frac{r}{2a}\right)^2 = a^2 + 2a \times \frac{r}{2a} + \left(\frac{r}{2a}\right)^2$.

[complementary] portion, the product will not be a quarter but smaller than [a quarter] by the square of their distances from half the line or the number.[24]

Canpanṭon then gives examples for the above: $\frac{1}{4} \times \frac{3}{4}$ is smaller than $\frac{1}{2} \times \frac{1}{2}$ by $\frac{1}{4} \times \frac{1}{4}$; 5×7 is smaller than 6×6 by 1×1; 9×3 is smaller than 6×6 by 3×3. In the context of root extraction, the following examples are given: the approximate root of 7 is $2\frac{3}{5}$, and its square is less than 7 by $\frac{3}{5} \times \frac{2}{5} = \frac{1}{5} + \frac{1}{5} \times \frac{1}{5}$; the approximate root of 6 is $2\frac{2}{5}$, and its square is less than 6 by $\frac{2}{5} \times \frac{3}{5} = \frac{1}{5} + \frac{1}{5} \times \frac{1}{5}$.

The reason is that the remainder equals the result of division multiplied by twice the [previous approximate] root plus one. . . . The addition of the result to the previous [approximate] root, however, will add to the square its product by twice the previous root and its product by itself. But its product by itself subtracted from its product by 1 is its product by its complement [with respect to 1].[25] (For example, the product of $\frac{1}{3}$ by 1 equals its product by the parts [of 1], namely by $\frac{1}{3}$, which is itself, and by $\frac{2}{3}$, which is its complement with respect to one; this is clear).

Canpanṭon supplies an example for the reason we do not add one when the remainder is small: if we approximate the root of 29 by $5\frac{4}{10}$, the error is $\frac{16}{100} = \frac{1}{4} - \frac{9}{100}$; if we approximate it by $5\frac{4}{11}$, the error will be $\frac{4}{11} \times \frac{7}{11} = \frac{1}{4} - \frac{2\frac{1}{4}}{11^2}$. The latter is clearly larger than the former. Then Canpanṭon repeats the instruction not to use the $a + \frac{r}{2a+1}$ approximation when conducting a reiterated approximation. Finally, he suggests a different iterable approximation procedure with its own error analysis. This last method is given without justification.

If you wish to approach the truth with little effort, add the remainder [i.e., the difference between the given number and the integer part of the root] to the square of double the [integer part of the] root at hand, and divide by it the product of the remainder and double the [integer part of the] root. Add the result to the [integer part of the] root at hand, and this root will be very near the truth. If you wish to approach the truth further, [divide] the cube of the above remainder by the denominator squared.[26]

Canpanṭon concludes with the calculation of the root of 3 according to this formula. The integer approximating the root is 1, and the error term is 2. The approximation formula yields $1 + \frac{2\times1\times2}{(2\times1)^2+2} = 1\frac{2}{3}$. The error term is indeed $\frac{2^3}{\left((2\times1)^2+2\right)^2} = \frac{2}{9}$. Then the procedure is reiterated with $1\frac{2}{3}$ as the root and $\frac{2}{9}$ as the error. The new approximation is $1\frac{112}{153}$, and the resulting error term is $\frac{2}{153^2}$.

[24] Canpanṭon shows that when r is larger than or equal to a, the error term of the approximation $a + \frac{r}{2a}$ exceeds $\frac{1}{4}$, whereas the error terms of the approximation $a + \frac{r}{2a+1}$ will be smaller than $\frac{1}{4}$. The reasoning goes through a geometric interpretation, and an implicit reference to *Elements* II.5.

[25] Here Canpanṭon provides a proof of the error term formula $\frac{r}{2a+1} \times \frac{2a+1-r}{2a+1}$ for the approximation $a + \frac{r}{2a+1}$. An anachronistic reconstruction of the argument alluded to here would be: $\left(a + \frac{r}{2a+1}\right)^2 = a^2 + 2a \times \frac{r}{2a+1} + \frac{r}{2a+1} \times \frac{r}{2a+1} = a^2 + 2a \times \frac{r}{2a+1} + 1 \times \frac{r}{2a+1} - \left(1 - \frac{r}{2a+1}\right) \times \frac{r}{2a+1} = a^2 + (2a+1) \times \frac{r}{2a+1} - \left(1 - \frac{r}{2a+1}\right) \times \frac{r}{2a+1} = a^2 + r - \left(1 - \frac{r}{2a+1}\right) \times \frac{r}{2a+1}$.

[26] Canpanṭon suggests the approximation $a + \frac{2ar}{(2a)^2+r}$. This may have been obtained from a sort of interpolation between the underestimate a and the overestimate $a + \frac{r}{2a}$ in the form $a + \frac{r}{a+\left(a+\frac{r}{2a}\right)}$. Note also that this is the beginning of the generalized continued fraction representation of the root: $a + \frac{r}{2a+\frac{r}{2a+\dots}}$. The error term is calculated as $\frac{r^3}{\left((2a)^2+r\right)^2}$.

5. ELIJAH MIZRAḤI, *SEFER HAMISPAR* (*THE BOOK OF NUMBER*)

This section was prepared by Stela Segev

Elijah Mizraḥi (ca. 1450–1526), also known by the acronym *ha-reʾem*, was born in Constantinople to a family from the Byzantine Empire, rather than from Spanish exiles. At the time, and especially after the expulsion of the Jews from Spain in 1492, the Jewish community of Constantinople was one of the largest and most important Jewish communities in the world. Mizraḥi became a prominent personality in the community, holding many public positions [Hacker, 2007; Ovadia, 1939].

Mizraḥi was considered by his contemporaries and by later generations to be the most important rabbinical authority in Constantinople and in the whole Ottoman Empire. He wrote several treatises on religious subjects (e.g., a supercommentary on Rashi's commentary on the Torah), and he was also interested in scientific subjects. Besides *The Book of Number*, he wrote a commentary on Ptolemy's *Almagest*.[27]

Mizraḥi's teacher in secular subjects was Mordechai Comtino, whose treatise *On Reckoning and Measurement* was one of the sources for *The Book of Number*. The Karaite Kalev Afendopolo, a colleague, was the author of a commentary on the Hebrew translation of the *Arithmetic* of Nicomachus made from Arabic by Qalonymos ben Qalonymos in 1317.

Mizraḥi's treatise, *The Book of Number* (also known as *Melekhet Hamispar* (*The Number's Craft*)), was widely used during his time as well as later. It can be read in seven extant manuscripts (some of them complete) as well as in the first print edition issued in Constantinople by his son Israel after Mizraḥi's death. The abridged version of the book was partially translated into Latin [Münster, 1546], to be reprinted in 1809.[28]

This essay of approximately 200 pages consists of three articles dealing with arithmetic operations on integer numbers, simple fractions, and sexagesimal fractions. It also includes a chapter on square and cubic roots and on proportions, and a chapter of 99 arithmetic and geometric problems.

For each of his subjects, Mizraḥi first defines and explains the arithmetic operations and related notions, and then describes methods of solution accompanied by examples. He accords special attention to verification techniques. In a separate section at the end of the relevant chapter, Mizraḥi explains and proves his methods. This structure of the book certainly facilitated the publication of the abridged sixteenth-century Latin edition, which contained only the algorithms presented in the first two sections.

Mizraḥi based his work on many sources. He quoted Ibn Ezra, Euclid's *Elements*, and Nicomachus's *Arithmetic*, but it appears that he read also other books, mostly in Hebrew but possibly in other languages too (Greek or Arabic). Evidently, he used Comtino's treatise *On Reckoning and Measurement* and also another arithmetic essay by Isaac ben Moses ʿAli titled *The Art of Number* [Segev, 2010].

Calculating squares by the "thirds method" (from Article I, part 1)

In the section dealing with multiplication of integers, besides algorithms appropriate to numbers written in the positional decimal system, Mizraḥi presents several methods of

[27] Preserved in one manuscript in St. Petersburg, Institute of Oriental Studies of the Russian Academy, C 128.

[28] For an analysis, see [Segev, 2010] and [Wertheim, 1896].

oral calculation. The origin of these methods is not clear. Some of them can be found in earlier Hebrew mathematicians, such as Ibn Ezra's *Sefer Hamispar* (the thirds method) and Comtino's *On Reckoning and Measurement* manuscript (the thirds method and the fifths method). These authors, however, present the methods without justification.[29] Unlike his predecessors, Mizraḥi gives a proof for each method he presents.

The explanations are general and rely on Euclid's geometric propositions from the *Elements*, Book II, interpreted arithmetically. Mizraḥi sometimes quotes Euclid geometrically from Ibn Tibbon's Hebrew translation,[30] but at other times, as here (like Comtino, his teacher), he prefers an arithmetic-algebraic formulation. In this context, it is clear that he interprets Euclid's propositions as dealing with numbers.

Another way to multiply a [two-digit] number by itself: if it has a third, take the third, multiply it by itself, and move the result one [decimal] level higher. Subtract from it the third's square. What remains is the result of multiplying the number by itself [the square of the number].

As an example, if we want to multiply 24 by itself, we take one third of it: 8. Multiply it by itself: 64. Move the result one level higher: 640. Then subtract 64 from it, and what remain are 576. This is the square of 24.

If the number does not have a third, take the closest number which does have a third, either larger or smaller than the given number, and proceed as before. Then, in the case that the number which has a third was less than the given number, add to the result the number which has a third and the given number; in the case that the number which has a third was more than the given number, subtract them from the result.

The example with the number [having a third] which is less than the given number is 25, because the closest number which has a third is 24. We take its third and multiply it by itself as before and we get 576. We add the given number and the number which has a third and we get 625, which is the square of 25. In the example with a number [having a third] which is more than the given number, we choose 23. We take the closest number which has a third, 24, proceed as before and we get 576. Then we subtract from it the given number and the number which has a third, which is 47; the result is 529 which is the square of 23. The method can be used for all numbers.

. . .

[This way of] taking the square of the number's third and moving the result one level higher, then subtracting from it the third's square—the reason for it is known, and is explained by Euclid's *Elements*, Book VIII. For any two square numbers, the ratio of one to the other is equal to the ratio of their sides multiplied by itself.[31] It follows that the ratio of the number's third squared to the number squared is equal to the ratio of the number's third to the number itself, all squared. And the ratio of the number's third to the number is a third, and a third multiplied by itself is a ninth. So, necessarily, the ratio

[29]The "thirds method" can also be found in Gersonides's *Maᶜase Ḥoshev,* but there it appears only for integers that are multiples of 3, and the explanation is based on a numerical example.

[30]Moses ibn Tibbon translated *The Elements* from Arabic into Hebrew in 1270.

[31]*Elements* VIII.5. Euclid's proposition is a more general one, and Mizraḥi infers from it this particular formula.

of the number's third squared to the number squared is a ninth. It follows that the square of the given number will be nine times the number's third squared. Raising the third's square one level, we will get ten times the number's third squared. Subtracting from it one third squared, the remainder will be equal to the square of the given number, since it is equal to nine times the number's third squared, because any two magnitudes whose ratio is one are necessarily equal according to Euclid's *Elements*.[32]

. . .

But if the given number doesn't have a third, we use another number that has a third, which is larger or smaller than the given number by one. The reason for this is also clear. It is well known that the difference between the squares of two numbers is equal to the sum of the two numbers multiplied by their difference.[33] And the product of the two numbers is equal to the product of the smaller number by itself added to the product of the smaller number by the difference between the two given numbers, as is written in Book II of Euclid.[34] And the product of the greater number by itself is equal to the product of the greater number by the smaller one, added to the product of their difference by the greater number, for the same reason.[35] It follows necessarily that the greater number multiplied by itself is equal to the sum of three multiplications: the product of the smaller number by itself, the product of the greater number by the difference, and the product of the smaller number by the difference.[36] The last two products are equal to the product of the difference by the sum of the two given numbers, according to the previous argument. So, necessarily, the greater number multiplied by itself is equal to the smaller number multiplied by itself added to the difference multiplied by the sum of the two numbers.[37]

Therefore, if the required number is greater by one than a number which has a third, it is necessary to add the given number and the number which is smaller than it by one and the square of the number which has a third. The sum is the required square. If the required number is greater by two than a number which has a third, add the given number and the number which is smaller than it by two, multiply the result by two and add to it the square of the number which has a third; this is the required square. If the required number is greater by three, multiply by three the sum of the given number and the number which is smaller than it by three and add it to the square of the number which has a third. And so on in all cases.[38]

But if we work with thirds, we have only numbers which are greater or smaller than the required number by one, for if a number is greater by two than a number which has a third, the same is smaller by one than another number which has a third. And that is the

[32] $10\left(\frac{x}{3}\right)^2 - \left(\frac{x}{3}\right)^2 = 9\left(\frac{x}{3}\right)^2 = x^2$.

[33] *Elements* II.5, $b^2 - a^2 = (b+a)(b-a)$.

[34] *Elements* II.3, $ba = a^2 + a\,(b-a)$.

[35] *Elements* II.2, $b^2 = ba + (b-a)\,b$.

[36] From the previous two results, $b^2 = ba + (b-a)\,b = \left(a^2 + a\,(b-a)\right) + b\,(b-a)$.

[37] From the above, $b^2 = a^2 + (b+a)\,(b-a)$. Note the similarity to *Elements* II.5 above, which is brought up explicitly but is not used in the argument.

[38] $(a+n)^2 = ((a+n)+a)n + a^2$. This could also be obtained directly from *Elements* II.6.

reason the ancients chose to use the thirds ... so there will be no need to multiply the sum of the given number with the number which has a third.

But if we use the fifth (or other parts) there will be cases in which it will be necessary to multiply the sum of the two numbers. I therefore wonder why they chose the fifth part, and not any other, when this method is common to all parts.

Summing the first n integers and cubes (from Article I, part 1)

In the section on multiplication, Mizraḥi presents the sums of several kinds of arithmetic and geometric progressions, as well as of squares and cubes of the first n integers. The sums and their proofs do not use any algebraic terminology or symbolism. Some proofs are based on figurate numbers, but others, like the example below, have a more abstract character, and present original arguments not found in the work of previous authors. Mizraḥi's use of the "recursive" identities like $\frac{1+2+...+n}{n+1} = \frac{1+2+...+(n-1)}{n} + \frac{1}{2}$ or $\frac{1^3+2^3+...+n^3}{1+2+...+n} = \frac{1^3+2^3+...+(n-1)^3}{1+2+...+(n-1)} + n$ clearly illustrates some original pre-inductive reasoning.

But the way the ancients used to sum the natural numbers, that is to multiply the last number by its half added to half, is self-evident. For if you take any number and divide the sum of all its preceding numbers by the number itself you will receive the quotient of the sum of all the numbers preceding the preceding number divided by the preceding number when you add to this quotient a half.[39] So the quotient of one divided by two is half, and the quotient of the sum of one and two divided by three is one integer, and the quotient of the sum of one, two, three divided by four is one and a half, and the quotient of the sum of one, two, three, four divided by five is two, and so on. We always add half.

Therefore, when you want to know the sum of all natural numbers from one to any number, because the quotient of the sum of the preceding numbers divided by the last number will be equal to the sum of the halves added in each step, as we have seen before, we must count the steps from one to the given number. Up to two we have half, and up to three we have one, and up to four we have one and a half, and so on until the last number. And so the quotient of the sum of the preceding numbers divided by the last number will be equal to the former result, and if we multiply this former result with the last number, we receive the sum we wanted.

. . .

But the other kind, the cubes of natural numbers, the ancients knew a way to sum them as well: take the cubic root of the last cube, and follow the former way to sum the natural numbers which are the cubic roots of those cubes, and save the result. Then multiply the saved result by itself; this is the sum of the given cubes.[40]

For example, if you want to know the sum of 1, 8, 27, 64, the cubes of 1, 2, 3, 4, take the cubic root of 64, which is 4, multiply it by half the number of steps [i.e., the given cubes]

[39] $\frac{1+2+\ldots+n}{n+1} = \frac{1+2+\ldots+(n-1)}{n} + \frac{1}{2}.$

[40] $1^3 + 2^3 + \ldots + n^3 = (1+2+\ldots+n)^2 = \left(\frac{n(n+1)}{2}\right)^2.$

added to half, you receive 10. Multiply 10 by itself, the result is 100, and this is the sum of the numbers 1, 8, 27, 64.

. . .

The way they used to add the cubes of natural numbers by multiplying the sum of their cubic roots by itself—its reason is known as well. For if you take the quotient of the sum of the cubes and the sum of their cubic roots, the result is larger than the quotient of the sum of the cubes previous to the last cube and the sum of their cubic roots by the last cubic root.[41] And so on, each quotient is greater than the previous quotient by the last cubic root until we reach the first cube [1^3]. It means that the numbers added at every step are the natural numbers. . . .

Therefore the ratio of the cube of 1 and its cubic root [1] is 1 (namely, they are equal), and the ratio of 1, 8 and the sum of their cubic roots [$1 + 2$] is 3 (namely 3 times, being the former result 1, to which we add 2), and the ratio of 1, 8, 27 and the sum of their cubic roots [$1 + 2 + 3$] is 6 (namely, 6 times, being the former result 1, 2, to which we add 3), and the ratio of 1, 8, 27, 64 and the sum of their cubic roots [$1 + 2 + 3 + 4$] is 10 (namely, 10 times, being the former result 1, 2, 3, to which we add 4), and so on, always in the same way.[42]

So if we want to add the cubes of the [natural] numbers, as many as they are, we take the cubic root of the last cube and we multiply it by its half added to half, and the result is the sum of the cubic roots of the given cubes. Next we want to know the number of steps from the first cube to the last one, and we take 1 for the first step [cube], 2 for the second, 3 for the third, 4 for the fourth and so on until we reach the last cube. Next we add all these numbers and this is the quotient of the sum of cubes and the sum of their cubic roots. We therefore multiply the sum of the cubic roots by the sum of the natural numbers (i.e. the sum of the steps) and we receive the sum of the given cubes.

Simple fractions (from Article I, Parts 1 and 2)

The concept of simple fractions underwent many changes over the centuries before it reached the form we use today. The ancient Egyptians used mostly unit fractions, while the Babylonians used the sexagesimal system not only for integers but also for numbers less than one. Fractions that are written with the numerator above the denominator (but without the

41 $$\frac{1^3+2^3+\ldots+n^3}{1+2+\ldots+n} = \frac{1^3+2^3+\ldots+(n-1)^3}{1+2+\ldots+(n-1)} + n.$$

42 $$\frac{1^3+2^3}{1+2} = \frac{1^3}{1} + 2 = 1 + 2$$

$$\frac{1^3+2^3+3^3}{1+2+3} = \frac{1^3+2^3}{1+2} + 3 = 1 + 2 + 3$$

$$\frac{1^3+2^3+3^3+4^3}{1+2+3+4} = \frac{1^3+2^3+3^3}{1+2+3} + 4 = 1 + 2 + 3 + 4 \ldots$$

horizontal line in between) can be found starting from the ninth century in several Arabic texts, and even earlier in Indian texts.[43] The algorithms used for different calculations were often cumbersome and the definitions inconsistent. In contrast, Mizraḥi presents a well-articulated concept of fractions and sometimes argues with previous authors, as in the following excerpt. The algorithms for calculations presented by Mizraḥi (most of them still in use today) are meticulously explained using propositions from Euclid or Nicomachus. Note, however, that unlike the Greeks, Mizraḥi does not seem to distinguish ratios from fractions and quotients.

In this short excerpt Mizraḥi compares a "relative" notion of fraction (one whole number with respect to another) and an absolute one (parts of the actual unit). He endorses only the former.

I saw some contemporaries who think that the wholes are numbers from one onward, such as one or two or three or any other number, and that fractions are smaller than one, such as half of one or a third or a quarter, or another part of one. Regarding the issue of the parts of one they are right only in one aspect, that is relatively, but they were wrong in thinking of the relative as if it were absolute. That is, they were right when dealing with parts of the whole one, but this "one," you should know that it can be any number we name a whole, because any number can be thought of as a whole in relation to its parts. For example, two with respect to eight is a quarter of eight, and indeed we call the two a quarter of eight when the eight is the whole and the two is the part, the eight being four times two. It [the eight] is the one [whole] whose fractions we calculate, but it [the eight] is not the real one which is indivisible. Now a general definition should include all special cases, whereas if there are two kinds of numbers, wholes and fractions, and if the fractions are parts of the indivisible one, then the definition of the number as the sum of units is contradicted, because the real one [unit] is not a number and neither are its fractions.

In the next excerpt Mizraḥi proves the standard fraction multiplication algorithm.

And I say that the general way to multiply [all kinds of fractions] is to multiply the numerator with the numerator and save the result, then multiply the denominator with the denominator[44] and save the result, and the ratio of the first number you saved to the second one is the fraction or fractions that result from the multiplication.

. . .

The reason for this method, to multiply the numerator with the numerator and the denominator with the denominator, is evident if we know five propositions.

The first of them is that if you take any fraction and multiply its numerator and its denominator by the same arbitrary number, the ratio of the results will be the given fraction relating the original numerator and denominator. The same holds for division. That is, if you divide the numerator and the denominator in arbitrarily many equal parts, and you

[43]The origin of writing fractions vertically probably has to do with the way the division operation was written. The horizontal bar used in the writing of fractions can be found for the first time in the work of al-Ḥaṣṣār from the twelfth century. We can also find this vertical notation in Indian writings from the sixth century [Djebbar, 1992; Mazars, 1992].

[44]Mizraḥi uses the terms "quantity" and "quality" for the numerator and denominator, respectively.

take an equal number[45] of parts of the numerator and the denominator, and you set the ratio of these parts of the numerator to the parts of the denominator, then this new fraction will be equal to the given fraction relating the undivided numerator and denominator. And the reason is in the *Elements* of Euclid, Book V,[46] stating that equimultiplied parts relate to each other as their parts relate to each other. The proof is common to both.

The second proposition is that saying that we multiply one fraction by another is like saying that we take a fraction from another. This is obvious.

The third proposition is that for any number which is multiplied by another, the first number to the product is a fraction derived from the name of the second number.[47] For example, the number two, when multiplied by three, is six. The two will be a third of six, where the name "third" is derived from "three," which multiplies the two. This is explained also in the book of Nicomachus.

The fourth proposition is that if you want to take a part from a fraction, take this part from the numerator of the fraction, and the ratio of this to the denominator of the given fraction is the part taken.[48] For example, if you want to take a third from nine tenths, take a third from nine, which is three, relate them to the tenths, and they make three tenths. This is always so, and is obvious.

The fifth proposition is that if you want to multiply a fraction however many times, take the ratio of the multiplied numerator to the denominator, and the result is the product of the multiplied fraction.[49] For example, if you want to multiply two ninths by three, we multiply two by three, which is six, and relate them to the ninths, yielding six ninths. This is the result of two ninths multiplied by three, and this is also obvious.

Now that you've understood these propositions, the reason for this multiplication is as clear as can be. If we multiply, for example, three quarters by two ninths, it is the same as taking three quarters of two ninths, according to the second proposition. If we could divide the numerator of the two ninths by four to get its quarter, we would take this quarter and relate it to the denominator, and the result would be the quarter of the two ninths according to the fourth proposition. Then we would multiply the numerator by three, and get three quarters of two ninths according to the fifth proposition.

But because the numerator of the two ninths is not divisible by four, we have to multiply it by four (which is the denominator of the three quarters), so that the product of two and four becomes divisible by four. And having done that, we have to multiply also the nine (which is the denominator of the two ninths), by the denominator four, so that the ratio of the products of two and of nine by four equals the two ninths according to the first proposition. We now take the ratio of a quarter of the resulting numerator (which is the product of the numerator by the denominator) and the resulting denominator (which is the product of the denominator by the denominator), and this is a quarter of the two ninths

[45] We read "number" where the text has "parts." According to this reconstruction, Mizraḥi tries to combine the statements about multiplication and division into one.

[46] *Elements* V.15: $\frac{a}{b} = \frac{ac}{bc}$ or $\frac{a}{b} = \frac{a/c}{b/c}$.

[47] If $ab = c$, then $\frac{a}{c} = \frac{1}{b}$.

[48] $\left(\frac{a}{b}\right)/c = \frac{a/c}{b}$.

[49] $\frac{a}{b} \times c = \frac{a \cdot c}{b}$.

according to the fourth proposition. Then we multiply its numerator by three, and this will be three quarters of two ninths, according to the fifth proposition.

Word problems (from Article III, part 1, chapter 1)

In the third article of his book Mizraḥi presents about 100 problems and their solutions. He classifies them in two categories: number problems and geometric problems. The so-called number problems actually deal with a wide range of topics and present various techniques for solving them. Many of them are standard problems that can be found in earlier books, but Mizraḥi's treatment tends to be deep and well reasoned.

The questions selected here are two mathematical riddles about combining and comparing the money of two or three people. Mizraḥi solves one with proportion theory and the other with algebraic-like manipulations (other authors solve them with ad hoc manipulations or double false positioning).

Question [15]: A man told his friend, if you give me one, I will have as much as you. His friend answered: if you give me one, I will have twice as much as you.[50] How much does each one have?[51]

The answer is that this question is also misleading, because it posits that which determines for that which is determined.[52] Here we need the ratio of the two [coins] set apart (one from each man) to the money of both, rather than the ratio of the two [coins] set apart together with the money of the first to the money of the second with one subtracted. However, because the latter ratio determines the ratio between the number two and all the money [of both men], he took one for the other. This is self-evident if we consider the money of both men together as one amount, and set apart two, one from the first man and the other from the second, which are what each said he would give to the other. The money of both men is then divided into three parts: the two set apart by both, the money of the first less one, and the money of the second less one.[53] From this we can determine the answer.

The first man saying to his friend: "if you give me one, I will have as much as you" is the same as if he said: "if I add to my money[54] the two [coins] set apart (the one you want to give me and the one I want to give you), the result will be the same as the amount you have less one. Therefore the money of the first man together with the two set apart, which are two of the three parts composing all the money, will necessarily be half of the money composed of the three parts, because the two former parts together equal the third part.[55]

The second man saying to his friend: "if you give me one, I will have twice as much as you," is the same as if he said: "if I add to my money the two set apart, the sum will be equal to twice the rest of your money." Therefore the money of the second added to the

[50] We omit a variation where the parameters are 2 and 100 instead of 1 and 2.

[51] For a version of this problem by Fibonacci, see Chapter 1, section II-3-2.

[52] This is oddly phrased. In what follows Mizraḥi suggests that to solve the question, one must transform the data so that it fits a standard Rule of Three procedure. If x is the money of the first man and y the money of the second, we are given the ratios $(x+1):(y-1)$ and $(y+1):(x-1)$, whereas we need the ratio $2:(x+y)$.

[53] By considering the parts $a = 2$, $b = x - 1$ and $c = y - 1$, we have as data the ratios $(a+b):c$ and $(a+c):b$, and require the ratio $a:(b+c)$.

[54] From here on the money of the partners is considered as $x - 1$ and $y - 1$, instead of x and y.

[55] $(x-1)+2 = (y-1)$, so $(x-1)+2 = \frac{1}{2}((x-1)+2+(y-1))$.

two set apart will necessarily equal two thirds of the money composed of the three parts, as the two former parts are twice the third part.[56]

So the money of the first man is a third of all the money composed of the three parts, and we have already seen that half of all the money is equal to the money of the first man together with the two set apart. Therefore the difference between the half and the third, which is a sixth [of all the money], equals two, which is what we wanted to find: the ratio between the two set apart and all the money.[57]

So we apply ratios:[58] if one sixth equals two, how much is the whole? This yields twelve. We have already said that the money is composed of three parts: the two set apart, the money of the first man, and the money of the second, and that the sum of the two and the money of the first is half of all the money. So, necessarily, the money of the first man together with the two set apart is six, which are half of twelve. And if we take away the two added to the money of the first, there remain four, the money of the first. So the money of the second will be six. And if we give to each the coin we took, the first will have five and the second will have seven, and this is the money they each have.

Question [35]: Reuven, Simon and Levi went to the fish market and found a fish. Reuven said to his friends: If I gave all my money, and each of you gave half of yours, we could buy the fish. Simon answered and said: If I gave all my money, and each of you gave a third of yours, we could buy the fish. Levi answered and said: If I gave all my money, and each of you gave a quarter of yours, we could buy the fish. What is the ratio between their money, that is, the ratio of each to each?[59]

The answer is that this question is misleading, because the problem omits the ratio between the money of each and the others', stating instead facts that determine them. So if we obtain what is determined by the facts stated in the problem, we will arrive at the answer.

We say that the statements of Reuven and Simon determine that half of Simon's money equals two thirds of Reuven's and one sixth of Levi's. Therefore Simon's entire money equals one and a third of Reuven's and a third of Levi's. That is because all of Reuven's money together with half of the others', as Reuven says, equals all of Simon's money together with a third of the others', as Simon says. Therefore, what Simon added to Reuven's statement—a half of his [Simon's] money—equals what he removed from Reuven's total—two thirds of Reuven's—and to what he removed from half of Levi's money—one sixth of Levi's.[60]

In the same way, according to Reuven's and Levi's statements, it is determined that half of Levi's money equals three quarters of Reuven's and one quarter of Simon's.

[56] $(y-1)+2=2(x-1)$, so $(y-1)+2=\frac{2}{3}((y-1)+2+(x-1))$.

[57] $(x-1)=\frac{1}{3}((x-1)+(y-1)+2)$ and $(x-1)+2=\frac{1}{2}((x-1)+(y-1)+2)$, so $2=\frac{1}{6}(x+y)$.

[58] This is Mizraḥi's term for what is known as the Rule of Three.

[59] Here, if a is the money of Reuven, b is Simon's money, c is Levi's money, and d is the price of the fish, we can write three equations for this problem: $a+\frac{1}{2}b+\frac{1}{2}c=d$, $b+\frac{1}{3}a+\frac{1}{3}c=d$, and $c+\frac{1}{4}a+\frac{1}{4}b=d$. For a version of this problem by Fibonacci, see section II-2-2 in Chapter 1. For Levi ben Gershon's abstract version, see section I-6 in this chapter.

[60] $a+\frac{1}{2}b+\frac{1}{2}c=b+\frac{1}{3}a+\frac{1}{3}c$, so the excess of the right hand over the left hand, which is $\frac{1}{2}b$, equals what the right hand removes from the left hand, namely, $\frac{2}{3}a+\frac{1}{6}c$. We rescale by two and obtain $b=1\frac{1}{3}a+\frac{1}{3}c$.

Fig. I-5-1. The Tel Aviv *Ralbag* street honors Levi ben Gershon. Under his dates, the sign reads, "Acronym for the name of Rabbi Levi ben Gershon: Philosopher, Mathematician, Astronomer, and Commentator on the Bible." Photograph by Phyllis Katz.

Therefore, in the same ratio, a third of Levi's money equals half of Reuven's and a sixth of Simon's.[61]

We already know that Simon's money equals one and a third of Reuven's and a third of Levi's. Therefore Simon's money equals one and five sixths of Reuven's and one sixth of Simon's own. We remove one sixth of Simon's which is common [to both sides], and remain with five sixths of Simon's money being equal to one and five sixths of Reuven's. According to the same ratio, all of Simon's money equals twice Reuven's and its fifth.[62]

Therefore it is determined that if Reuven had five, Simon would have eleven. We already saw that half of Levi's money equals three quarters of Reuven's and one quarter of Simon's. This ratio determines that all of Levi's money will be one and a half of Reuven's and half of Simon's. So it is determined that if Reuven had five and Simon had eleven, as we have already seen, Levi would have thirteen. This determines the price of the fish to be seventeen.

6. LEVI BEN GERSHON, *MAʿASE ḤOSHEV (THE ART OF THE CALCULATOR)*

Levi ben Gershon (1288–1344), who lived his entire life near Orange in Provence, was one of the most prominent medieval Jewish scientists, besides also being a rabbi, philosopher, and biblical commentator. In Jewish circles, he is known by the acronym *Ralbag*, and in Latin circles as Gersonides [Fig. I-6-1]. He wrote numerous commentaries on the Hebrew Bible,

[61] By comparing the second and third identities, we get $\frac{1}{2}c = \frac{3}{4}a + \frac{1}{4}b$. Rescaling the last identity by two-thirds, we obtain $\frac{1}{3}c = \frac{1}{2}a + \frac{1}{6}b$.

[62] By substitution, $b = 1\frac{1}{3}a + \frac{1}{3}c = 1\frac{1}{3}a + \left(\frac{1}{2}a + \frac{1}{6}b\right)$. Rearranging yields $\frac{5}{6}b = 1\frac{5}{6}a$, or $b = 2\frac{1}{5}a$.

a book on logic, four mathematical treatises, and a major philosophical work, *Milḥamot Adonay* (*Wars of the Lord*), which includes a section on trigonometry as part of a longer section on astronomy, in which he criticizes some of the ideas of Ptolemy. During Levi's lifetime, the Jews in Provence (about 15,000 out of a total population of 2,000,000) were under the protection of the pope, then residing in Avignon. Levi was well regarded by the Christian community as a scientist, and some of his works were translated into Latin during his lifetime, although Levi himself probably did not know the language.[63]

Maʿase Ḥoshev[64] is Levi's first book on mathematics and is dated 1321, with a second edition the following year that had minor changes in organization and presentation.[65] The work exists today in twelve manuscripts, nine of the first edition and three of the second. The oldest manuscript is *Parma 2271*, a first edition manuscript, estimated to have been written in Provence late in the fourteenth century. The critical edition and German translation by Gerson Lange of 1909 was based primarily on the Vienna manuscript of 1462, also of the first edition. *Maʿase Ḥoshev* is in two parts followed by a large collection of problems. The first part is theoretical, containing 68 theorems and problems in Euclidean style dealing with arithmetic and combinatorics, with some of the proofs being accomplished by a form of mathematical induction [Rabinovitch, 1970]. Its style is reminiscent of the arithmetic of Jordanus de Nemore (see section II-2-5 of Chapter 1). The second part contains algorithms for calculation and is subdivided into six sections.

Introduction

Signed Levi ben Gershon. In order to acquire a complete ability to do practical crafts, one needs to know the craft—a craft with knowledge of technique, and why to use a particular technique. The practical part of the craft of numbers is one of the practical crafts. So it is clear that it is worthwhile to investigate its theory. Another reason why it is mandatory to investigate this craft and its given theory, is that it is clear that this craft encompasses many different kinds, and each and every kind encompasses many diverse topics, so that you might think they are not all part of the same kind. Because of this, it is clear that you will not complete your acquisition without knowledge of the theory, except with great difficulty. However, with knowledge of the theory, it is possible to complete your acquisition with ease. This is so, because he who knows the theory can, with a single view, understand the practical characteristics of each of the many kinds that the craft encompasses, and he who masters the theory will require just a single view in place of many views for the various topics. Accordingly, we see fit to present the manipulation of numbers and its theory, for our benefit. Along this line of thought, I have divided this book into two sections.

Section One focuses on the principles needed to understand this craft. Section Two focuses on the practical craft of manipulating numbers, one kind at a time, and the

[63] See [Freudenthal, 1992] for more details on Levi's life and work.

[64] The title means, literally, "Thoughtful Application," a biblical reference (Exodus 26:1 and 28:6) to the kind of work used in building the Tabernacle. In the context of that building, this is work that requires thought, planning, and calculation rather than only technical craftsmanship.

[65] The discovery of the second edition is discussed in [Simonson, 2000b].

explanations. And since this book focuses on application and investigation, it is called *Maʿase Ḥoshev*.

However, as far as the instruction in this book is concerned, it is appropriate that he who concerns himself with it should already understand the 7th, 8th, and 9th books of Euclid. It is not our intention to repeat his words in our book, but instead to assume his principles in our development, as they were proved there.

. . .

Arithmetic with proofs (from part I)

This section contains many elementary and less elementary arithmetical facts and problems. Levi endows each fact or problem with a proof in his contemporary adaptation of the Euclidean style, using letters to refer to the quantities discussed but not operating on these letters algebraically. Some proofs depend on recursive calculation and inductive procedures.

1. The product resulting from the multiplication of two numbers one with the other counts every part of the first number as many times as there are ones in the second number.[66]

2. When you have two given numbers, and one is partitioned into some number of parts, then the product of the first number with the second equals the products of each part of the first number with the second number, all added together.[67]

The given numbers are *AB* and *C*, with *AB* divided into the parts *AE*, *ED*, *DB*. I assert that the product of *AB* with *C* is equal to the sum of the products of *AE* with *C*, *ED* with *C*, and *DB* with *C*. In the product of *AE* with *C*, the factor *C* occurs as often as the units in *AE*; also the factor *C* occurs in the product of *DE* with *C* as often as the units in *DE* and in the product of *DB* and *C* as often as the units in *DB*. In *AE*, *ED* and *DB* together, however, are just as many units as in *AB*. Thus in the sum of the products, *C* is a factor as often as the units in *AB*. Also, in the product of *AB* with *C*, *C* occurs as often as the units in *AB*. Thus the product of *AB* with *C* is equal to the sum of these products.

. . .

9. When you multiply one number by a number built[68] from two given numbers and the result is something, then if you multiply a number built from any two of these three numbers by the third number, the result will be the same.

10. When you multiply one number by a number built from three given numbers and the result is some number, then if you multiply any one of these numbers by the number built from the remaining three, the result will be the same.[69]

[66] That is, $ab = a + \ldots + a$, b times.

[67] $(a_1 + a_2 + \ldots + a_n)\, b = a_1 b + a_2 b + \ldots + a_n b$.

[68] The word translated as "built" is *murkav*—defined by Levi as the product or composite of the two numbers. This theorem asserts the associativity and commutativity of multiplication of three numbers. The proof uses theorem 1.

[69] This theorem generalizes theorem 9 to four numbers, and then uses a form of mathematical induction to generalize to n numbers.

Multiply *A* by the number built from *CDE*, giving *FG*. I am saying that if you multiply *D* by the number built from *ACE*, the result will also be *FG*. The proof is that we divide *FG* into parts corresponding to the numbers *CDE* and these parts are *FI*, *IL*, *LG*, then the number of parts is like the number of units in *A*. And, each one of the parts *FI*, *IL*, *LG* counts *D* like the amount of the product *CE*. And this is explained by what came earlier [theorem 9]. Now *FG* counts *D* as much as all its parts together, but all its parts together count *D* like the value of the product *CE* multiplied by *A*. Therefore, *FG* in total counts *D* like the number built from *ACE*. And therefore, the product of *D* with the number built from *ACE* is *FG* also.

And similarly it can be explained that whichever of these numbers is multiplied by the number built from the rest, the result will be *FG*. And like this, the progression can be understood without limit. What I mean to say is that if you multiply a certain number by a number built from four numbers, and let that be some number, then if you multiply any one of these numbers by the number built from the remaining numbers, then the result will be the same. And therefore, the number resulting from the multiplication of any number by the number built from the rest counts the number like the value of the number built from the rest.

. . .

17. If you subtract from a given number a given part or given parts and take from the remainder another given part or given parts and so forth, then the final remainder will be the same and the sum of the pieces taken will be the same, no matter in what order the parts are taken.

Let *A* be the given number, the parts having denominators *B*, *C*, *D*, and let the *B*th part of *A* be subtracted, then *E Cth* parts from the remainder, and *F Dth* parts from the next remainder. I say that the sum of the *B*th part of *A*, *E Hth* parts of the remainder, and *F Dth* parts of this remainder is equal to *F Dth* parts of *A* added to the *B*th part of the remainder and *E Cth* parts of that remainder.[70]

For the proof we set *G* to be the number preceding *B*, and further let the units of the sum of the numbers *E* and *H* equal *C* and the numbers *F* and *I* equal *D*. Let the *B*th part of *A* equal *P*, and the remainder of that from *A* be *J*; let *E Cth* parts of *J* be *K* and let *L* be the next remainder; finally let *F Dth* parts of *L* equal *M*, with *N* being the final remainder. Also, let *F Dth* parts of *A* equal *Q*, with remainder *R*, the *B*th part of *R* equal *S*, with remainder *T*, and *E Cth* parts of *T* equal *U*, with remainder *V*. We claim that *N* equals *V*. The proof: Because *P* is equal to the *B*th part of *A*, therefore *A* has as many parts equal to *P* as *B* has units. But *G* is equal to the number preceding *B*, so *J* has as many parts equal to *P* as *G* has units. Also, we know that *A* is to *J* as *B* is to *G*, since the product of *B* by *P* is *A* and the product of *G* by *P* is *J*. Now set the *C*th part of *J* equal to *W*, so *K* has as many parts equal to *W* as *E* has units, and therefore *L* has as many parts equal to *W* as *H* has units. It is shown similarly that *J* is to *L* as *C* is to *H* and therefore that *L* is to *N* as *D* is to *I*. So it is proved that the ratio of *A* to *N* is composed of the ratios of

[70] $\frac{1}{B}A+\frac{E}{C}\left(A-\frac{1}{B}A\right)+\frac{F}{D}\left[A-\frac{1}{B}A-\frac{E}{C}\left(A-\frac{1}{B}A\right)\right]=\frac{F}{D}A+\frac{1}{B}\left(A-\frac{F}{D}A\right)+\frac{E}{C}\left[A-\frac{F}{D}A-\frac{1}{B}\left(A-\frac{F}{D}A\right)\right].$

the numbers B, C, D to the numbers G, H, I.[71] Similarly, it is proved that the ratio of A to V is composed of the ratios of the numbers D, B, C to the numbers I, G, H;[72] but this ratio is equal to the ratio composed of the numbers B, C, D to the numbers G, H, I, so the ratios of A to N and to V are the same; so N equals V. Therefore, it must also be true that the sum of the numbers P, K, M equals the sum of the numbers Q, S, U. The difference between A and N is equal to the sum of P, K, M, and the difference between A and V is equal to the sum of Q, S, U; but it has been shown that N equals V, so must the sums of P, K, M and Q, S, U be equal.

26. When you add, beginning with one, consecutive numbers, and if the number of these is even, then the sum is equal to the product of half the number with the number following the last one.[73]

Let the numbers be A, B, C, D, E, F,[74] with the number following F being G, and A being one. I say that the sum of A, B, C, D, E, F is equal to the product of half their number with G. For proof, since A is one, the sum of F and A is G. Since the difference between B and one is equal to the difference between F and E, namely one, the sum of B and E is G. Also the difference between C and one is equal to the difference between F and D, so the sum of C and D is also G. Thus in the sum of A, B, C, D, E, F, the factor G occurs as often as half of the number, since the sum of each pair of these numbers is equal to G.

27. When you add consecutive numbers beginning with one, and if the number of these is odd, then the sum is equal to the product of the middle number with the last number.

28. When you begin a sequence of consecutive numbers with one and add together an odd number of terms, multiply half of the last number with the number following it. This product is equal to the sum of the numbers.[75]

30. When you add the sum of the consecutive numbers from one up to a given number to the sum of the numbers up to the number following that number, then the total is equal to the square of the number following the given number.[76]

When you add the sum of the numbers A, B, C, D, E to the sum of the numbers A, B, C, D, E, F, with A being one, then the total is equal to the square of F. To prove this, let

[71]Since $B{:}G = A{:}J$, $C{:}H = J{:}L$, and $D{:}I = L{:}N$, the composition $(B{:}G)(C{:}H)(D{:}I) = (A{:}J)(J{:}L)(L{:}N) = A{:}N$.

[72]Since $D{:}I = A{:}R$, $B{:}G = R{:}T$, and $C{:}H = T{:}V$, the composition $(D{:}I)(B{:}G)(C{:}H) = (A{:}R)(R{:}T)(T{:}V) = A{:}V$.

[73]$1 + 2 + \ldots + n = \frac{n}{2}(n+1)$, where n is even.

[74]Levi proves many of the following results by the method of generalizable example, since he has no way of representing an arbitrary integer. Given that the letters of the Hebrew alphabet can represent numbers, it is not entirely clear whether one should translate the beginning of the proofs by using letters or by using the numbers they might represent. We decided to use letters when Levi specifically notes that A (or aleph) is one, as in theorems 26 and 30, while using numbers when Levi does not so note (as in theorems 41 and 42).

[75]$1 + 2 + \ldots + n = \frac{n}{2}(n+1)$, where n is odd. In theorem 27, Levi shows that the sum of each pair at equal distance from the middle term is equal to twice that term. Then in theorem 28, he notes that twice the middle term is equal to $n+1$, so the product of the middle term with n is equal to the product of twice the middle term with half of n, that is, "half of the last number with the number following it."

[76]$(1 + 2 + \ldots + n) + (1 + 2 + \ldots + n + (n+1)) = (n+1)^2$.

G be the number following F. It has already been proved that the sum of A, B, C, D, E is equal to the product of half of E with F [theorem 28] and that the sum of A, B, C, D, E, F is equal to the product of half of F with G [theorem 26]. But the product of half of F with G is equal to the product of half of G with F, since the factors are in the same proportion. So if you add the sum of A, B, C, D, E to the sum of A, B, C, D, E, F, this is the sum of the product of half of E with F and the product of half of G with F, which is equal to the product of half the sum of E and G with F. But since the sum of E and G is equal to twice F, its half is equal to F. So the sum of the sum of A, B, C, D, E with the sum of A, B, C, D, E, F is equal to the product of F with F, that is, the square of F.

38. When you multiply a given number minus one third of the number that precedes it by the sum of the consecutive numbers from one through the given number, the result equals the sum of the squares of the consecutive numbers from one through the given number.[77]

The following two theorems present a proof in inductive style of the formula for the sum of the cubes of consecutive integers beginning with 1. Notice that the inductive step is proved first.

41. The square of the sum of consecutive numbers from one up to a given number equals the cube of the given number added to the square of the sum of the consecutive numbers from one up to the number before the given number.

Let the numbers be 1, 2, 3, 4, 5. I say that the square of the sum of 1, 2, 3, 4, 5 equals the cube of 5 added to the square of the sum of 1, 2, 3, 4. The cube of 5 is computed by counting 5 once for each unit in the square of 5. But the square of 5 equals the sum of 1, 2, 3, 4 added to the sum of 1, 2, 3, 4, 5 [theorem 30]. So 5 times the sum of 1, 2, 3, 4 added to the sum of 1, 2, 3, 4, 5 equals the cube of 5. But 5 times the sum of 1, 2, 3, 4 added to the sum 1, 2, 3, 4, 5 equals the sum of 5 times 5, that is, the square of 5, and the product of 5 with the sum of 1, 2, 3, 4 and 1, 2, 3, 4, that is, twice 5 times the sum of 1, 2, 3, 4. So the cube of 5 equals the square of 5 plus twice 5 times the sum of 1, 2, 3, 4. Also, the square of the sum of 1, 2, 3, 4, 5 equals the square of 5 plus twice 5 times the sum of 1, 2, 3, 4 plus the square of the sum of 1, 2, 3, 4. Therefore the cube of 5 plus the square of the sum of 1, 2, 3, 4 equals the square of the sum of 1, 2, 3, 4, 5, and this is what we wanted. Finally, we know that one has no number before it. However, its cube equals the square of the sum of the numbers up to it because it is exactly the sum of the number up to it, and hence it is exactly the square of this sum. And this is identical to its cube. This is perfectly trivial.

42. The square of the sum of consecutive numbers from one up to a given number equals the sum of the cubes of the consecutive numbers from one up to the given number.

Let the sum be the sum of 1, 2, 3, 4, 5. I say that the square of the sum of 1, 2, 3, 4, 5 is equal to the sum of the cubes of the numbers 1, 2, 3, 4, and 5. The proof is that the square of the sum of 1, 2, 3, 4, 5 equals the sum of the cube of 5 and the square of the sum of 1, 2, 3, 4 [theorem 41]. But the square of the sum of 1, 2, 3, 4 equals the sum of the cube of 4 and the square of the sum of 1, 2, 3. And the square of the sum

[77] $\left(n - \frac{n-1}{3}\right)(1 + 2 + 3 + \ldots + n) = 1^2 + 2^2 + 3^2 + \ldots + n^2$. We omit the proof, which depends on several earlier theorems.

of 1, 2, 3 equals the sum of the cube of 3 and the square of the sum of 1, 2. And the square of the sum of 1, 2 equals the sum of the cube of 2 and the square of 1, and the square of 1 equals the cube of 1. Therefore, the square of the sum of 1, 2, 3, 4, 5 equals the sum of the cubes of the numbers 1, 2, 3, 4, and 5. And that is what we wanted to prove.

. . .

53. Problem: To find three numbers such that the first, increased by a given part of the sum of the other two, equals the second, increased by another, also given, smaller part of the sum of the other two, and equals the third, increased by a given third part of the sum of both others that is smaller than the other two parts.[78]

. . .

58. Problem: To find three numbers such that the sum of the first and third contains the second as a factor as many times as a given number and such that the sum of the second and third contains the first as a factor as many times as a second given number.[79]

Let the given numbers be *A*, *B*. We denote the number following *A* by *C*; this is the first number. We denote the number following *B* by *D*; this is the second number. We take the product of *A* with *B*, subtract one and designate this by *E*; this is the third number. I say that *C*, *D*, *E* are the three sought numbers. We show that the sum of *C* and *E* is *A* times the factor *D* and the sum of *D* and *E* is *B* times the factor *C*. Since *E* is equal to one less than the product of *A* and *B*, and *C* is equal to the sum of one and *A*, then the sum of *C* and *E* is equal to the sum of *A* with the product of *A*, *B*. But the sum of *A* with the product of *A*, *B* is equal to the product of *A* and *D*. So the sum of *C* and *E* is equal to the product of *A* and *D* and therefore is *A* times the factor *D*. Furthermore, *E* is equal to one less than the product of *A* and *B*, and *D* is equal to the sum of *B* and one, so the sum of *D* and *E* is equal to the sum of *B* with the product of *A*, *B*. But the sum of *B* with the product of *A*, *B* is equal to the product of *C* and *B*, so the sum of *D* and *E* is equal to the product of *C* and *B*. The product of *C* and *B* contains *B* times the factor *C*. So the sum of *D* and *E* contains *B* times the factor *C*, and the sum of *C* and *E* contains *A* times the factor *D*. And this is what we wanted to show.

[78]The problem is to find three numbers G, H, I such that $G + \frac{1}{A}(H+I) = H + \frac{1}{B}(G+I) = I + \frac{1}{C}(G+H)$, where $A < B < C$. Levi gives a general, abstract solution to this problem, which also appears in Diophantus's *Arithmetica*, Book I, #24 (see Appendix 2). However, Levi's solution is very different from that of Diophantus. In a "real-world" form, the problem becomes a recreational problem about three men buying a horse; see Fibonacci in section II-2-2 of Chapter 1 for an example, as well as the work of Mizraḥi above. The problem also occurs in the work of al-Karajī around the year 1000 [Woepcke, 1853, p. 95]. Levi solves this under two separate conditions, first with $A = 2$ and second with $A > 2$. However, the solutions are the same in each case: $G = (A-2)BC + C + B - A$, $H = G + 2(B-A)(C-1)$, $I = H + 2(C-B)(A-1)$. We omit Levi's detailed proof that these are the correct values.

[79]The problem is to find three numbers C, D, E so that $C + E = AD$ and $D + E = BC$, with A, B given numbers. As before, Levi gives a general, abstract solution to the problem. However, this problem appeared earlier as a recreational problem about two men finding a purse; see Fibonacci in section II-2-2 of Chapter 1 for an example. The problem also occurs in the ninth-century work of the Indian mathematician Mahāvīra [Rangācārya, 1912, verse 244]. (See Appendix 3.) Note that the problem as stated is an indeterminate one, but Levi only gives one solution. However, when Levi uses this proposition in problem 18 at the end of the book (see below), he shows how to determine a particular solution when one of the unknowns must have a given value.

Decimal-sexagesimal subtraction (from part II, chapter 1)

We present an example of a subtraction problem with sexagesimal fractions. In the example, "firsts" mean sixtieths, "seconds" mean 3600ths, and so on. But note that Levi writes the whole numbers out in words while using the Hebrew alphabet numerals to write out the fractions. The discussion ends with the possibility of subtracting larger numbers from smaller numbers in circular contexts (e.g., angles, days of the week).

We want to subtract two hundred six wholes, fifty firsts, 37 thirds from 31 thousand and eighty wholes, 46 seconds, 35 thirds, 47 fourths, 53 sixths.

	0	10	7	9		45				
3	1	0	8	0	0	46	35	47	0	53
		2	0	6	50	0	37			
3	0	8	7	3	10	45	58	47	0	53

In the lowest category are the sixths. In that, in the higher row, are 53 sixths. We subtract from that what is in the lower row, but there is nothing, so write 53 in the row of results in the category of sixths. Then we subtract from the 0, that comes next to the 53, what lies below that in the lower row; but there is nothing in the lower row, so write 0 in the row of results in the category of fifths. Then we subtract from the 47 what lies below that in the lower row, but there is nothing, so write 47 in the row of results in the category of fourths. Then we subtract from the 35 in the upper row what lies below in the lower row, which is 37. That cannot be subtracted from 35, so we take one from the category that comes next to 35. So this becomes 60 in this category, which is added to 35, so becomes 95. We subtract 37 from this, so there remain 58. We write that in the row of results in the category of thirds. Now there is left 45 in the following category. We subtract from this what lies below it in the lower row, but there is nothing, so write 45 in the row of results in the category of seconds. There remains now from the 0 to subtract what lies below in the lower row. But 50 is there and we cannot subtract 50 from 0, and also in the category of the whole numbers, that lies next, is nothing that we can bring over. But in the third category from the one in question is a number, namely 8. We take one away from that and put it in the category to the right. We write 7 over the 8 and the 1, that we took away, becomes 10 in the first category. Take one of these into the category of firsts, so there remain 9 in the first category, which we write over the 0. The 1, which was taken, becomes 60 in the category of firsts. From that we subtract 50, so there remain 10, which we write in the row of results in the category of firsts. Now we subtract from the 9 what lies below it in the lower row. That is 6, so there remain 3 which we write in the final row under the units. Furthermore, we subtract from the 7 what lies below in the lower row; there remain 7 because there was 0 there. So we write that in the row of results for the tens. Further we subtract from 0 what lies below in the lower row. That is 2, but we cannot subtract 2 from 0. In the next category is 1, so we take that, which is in this category equal to 10. We write above the 1 a 0, and subtract 2 from the 10. There remain 8, which we write in the row of results in the category of hundreds. Then we subtract from the 0 what lies below in the lower row. But that is 0, so we write 0 in the row of results in the fourth category. Then we subtract from the 3 what lies below in the lower row. That is, however, 0, so we write 3 in the row of results in the fifth category. So the result is thirty thousand eight hundred and seventy three wholes, 10 firsts, 45 seconds, 58 thirds, 47 fourths, 53 sixths. Do the same in similar cases.

Sometimes it can occur in geometric calculations that you must subtract a greater number from a smaller, and indeed that happens in astronomy. You then add the value of the circle circumference, which is equal to 360, to the smaller number, from which you want to subtract, and then you can subtract what you wish, because there is no number in astronomical calculations that is larger than 360. For if it happens that a number will be larger than 360, one takes away that much and uses only the remainder. Such a calculation also appears in the ordinary new moon calculation. If one must subtract a greater number from a smaller number, one adds 7 days to the smaller number, and then you can subtract what you wish, since if the calculator of the new moon obtains a number larger than 7 days, he subtracts off the 7 days and uses only the remainder. Do the same in similar cases.

Mental multiplication technique (from part II, chapter 2)

The basic idea here is a trick for multiplying two digit numbers. It is based on the identity $(a+b)\,(c-b) + ((a+b)-c)\,b = ac$, where b may be an added or a subtracted number. Choosing b such that $a+b$ is a multiple of 10 simplifies the mental calculation.

To make it easier for you, I will give you a number of ways with which to calculate the multiplication of one number by another easily. You already know that multiplying a number of the first rank by a number of the first rank is an easy task, and so is multiplying a "broken" number. By that I mean multiplying a number of the first and second ranks by a number of the first rank.[80] But if you want to multiply one broken number by another broken number, complete one of the numbers to the side that is closest. If you added to this number in order to complete it to the nearest ten, subtract from the other number the amount that you added to the first number, and multiply what is left by the number in your hand, and save the result. If you subtracted from this number in order to complete it, add to the other number the amount you subtracted from the first number, and multiply what remains in your hand by the completed number that is in your hand. After this, look at how much the larger number after the addition or subtraction exceeds the smaller number before the correction. Multiply this excess by the amount you added to one of the numbers, and save the result. This is the second saved value. After this, look at the number that you subtracted. If you subtracted from the big number, then subtract the second saved value from the first saved value, and what remains in your hand is the desired result. If you added to the big number, then add the second saved number to the saved first number, and that is the desired result.

I will give you some examples. We want to multiply 34 by 57. Complete the number 57 to the nearest ten to get 60. Since 60 exceeds 57 by three, subtract three from 34 to get 31. Multiply 31 by 60 to get one thousand eight hundred and sixty, and that is the

[80] By a number of the "first rank," Levi means a number less than 10; a number of the "second rank" is a multiple of 10, while a "broken" number is a two-digit number with neither digit equal to 0.

first saved value. Since 60 exceeds 34 by 26, we multiply 26 by three and that is 78, the second saved value. Since we added to the large number, we add the second saved value to the first saved value and that is one thousand nine hundred and 38, which is the desired result.

In this example of ours, if we lowered 57 to the ten below it, we get 50, and we add 7 to 34 to get 41. We multiply 41 by 50 to get two thousand and fifty, and that is the first saved value. Since 50 exceeds 34 by 16, we multiply 16 by 7 to get 112, and that is the second saved value. Since we subtracted from the big number, we subtract the second saved value from the first saved value, leaving one thousand 9 hundred and 38, which is the desired result.

. . .

Sometimes it will happen that when using this method you will multiply a number by itself, and then this method will make things very easy. For example, you need to multiply 43 by 57. If you complete 43 to 50, you subtract the amount of the completion from 57 to get 50. You will need to multiply 50 by 50 and subtract 7 squared from the result, leaving the desired result. This is very clear from the earlier material at the start of part one of this book. May you understand and discover.

Summing arithmetic and geometric series (from part II, chapter 3)

If you want to add the consecutive numbers from 1 through a given number, take half the square of the given number and add it to half the given number, and that is the desired result. For example, if you want to add one, two, three, four, and so on until ten, including ten, take half the square of ten and half of it to get 55, and that is the desired result. Another way is to multiply this number by half the number that follows it, or half the number by the number that follows it, and that is the desired result. In our example, multiply 10 by half of 11, or half of 10 by 11, to get 55, which is the desired result.[81]

If the numbers follow one another but not [starting with one], that is, if the first is a given number, and the second is twice the given number, and the third is three times that number, and so on until some number [that is, if the numbers are in an arithmetic sequence beginning with the common difference], add up all the numbers until this number in the previous manner, and multiply the result by the first given number, and that is the desired result.[82]

For example, suppose that the first is 7 and the second 14 and the third 21 and the fourth 28, and continue in this way through 9 numbers. You already know that the sum of consecutive numbers from one through nine is 45. Multiply this by 7, which is the first number, to get 315, and that is the desired result.

This is so because the ratio of one to the first is equal to the ratio of two to the second and to the ratio of three to the third and to the ratio of four to the fourth and to the ratio of five to the fifth and to the ratio of six to the sixth and to the ratio of seven to the seventh and to the ratio of eight to the eighth and to the ratio of nine to the ninth. But the ratio of

[81] $1+2+3+\ldots+n = \frac{n^2}{2} + \frac{n}{2} = n \times \frac{n+1}{2} = \frac{n}{2} \times (n+1)$.
[82] $a + 2a + 3a + \ldots + na = (1+2+3+\ldots+n)\,a$.

one to its neighbor is equal to the ratio of all to all. Therefore, the ratio of one to seven is like the ratio of all to all. But seven counts one as many times as there are ones in seven, therefore all of these numbers count 45 as many times as there are ones in seven. Therefore, simply multiply the number 45 by 7 and the result is equal to the sum of these numbers. Ponder this.

. . .

If you want to add the squares of consecutive numbers from one through a given number, take the given number less one third of the number that precedes it, and multiply this by the sum of the consecutive numbers through the given number.[83] For example, if you want to know the sum of the squares of the consecutive numbers through 5, since the number preceding five is four, we subtract one third of four, which is 4 thirds, leaving four less a third. Multiply this by 15, which is the sum of the consecutives through 5, and you get 55, which is the desired result.

. . .

If you want to add the cubes of consecutive numbers from one through a given number, take the square of the sum of the consecutive numbers from one through the given number, and the result is what is desired.[84] For example, if you want to know the sum of the cubes of the numbers from one through six, the sum of the consecutives from one through six is 21, and taking its square gives 441, the desired result.

If the numbers [form an arithmetic sequence beginning with the common difference] and there are a given number for which we want to know the sum of their cubes, find the sum of the cubes of the consecutive numbers from 1 up to the given number and multiply this by the cube of the first number; that is the desired result.[85] The reason for this has already been given earlier. For example, suppose the first number is 4, the second 8, and we follow this law for five numbers. We know that the sum of the third powers of the consecutive numbers up to 5 is equal to 225. We multiply this by 64, the third power of the first number. The result is 14400, which is the desired result.

. . .

Levi next gives rules and examples for summing arithmetic sequences whose first term is not equal to the common difference. He then supplies similar rules for their squares and cubes.

If you want to add a given number of cubes of numbers in an arithmetic sequence whose first term differs from the common difference by a second given number, then if the first number is less than the common difference, take the sum of the squares of these numbers and multiply the result by the triple of the second given number, and save the result. Also, multiply three times the square of the second given number by the sum of these numbers, and save this result. Then, multiply the cube of the second given number

[83] $1^2 + 2^2 + \ldots + n^2 = \left(n - \frac{n-1}{3}\right)(1 + 2 + \ldots + n)$.

[84] $1^3 + 2^3 + \ldots + n^3 = (1 + 2 + \ldots + n)^2$.

[85] $a^3 + (2a)^3 + \ldots + (na)^3 = a^3\left(1^3 + 2^3 + \ldots + n^3\right)$.

by the first number, and add the result to the two earlier saved values, and you have in your hand the adjusted first saved value. After this, take the sum of the cubes of the numbers from 1 through the first given number, and multiply it by the cube of the common difference. Finally, subtract the adjusted saved value from this value, and what remains is the desired result.[86]

For example, if we want to add the cubes of seven numbers, where each number is three greater than its predecessor, and the first is two less than three, then we know that the sum of the squares of these numbers is 952. Multiply this by three times two, which is six, to get 5712, and save this. And also, the sum of these numbers is 70. Multiply this by three times two squared, which is 12, to get 840, and save that. Also, multiply the cube of two, which is eight, by 7, to get 56. Add this to the two saved values to get 6608, which is the adjusted saved value. We calculate the sum of the cubes of the consecutive numbers from one through seven to get 784. Multiply this by 27, which is the cube of three, to get 21168. Subtract from this the adjusted saved value, and what is left equals 14560, and that is the desired result.[87]

If the first number exceeds the common difference by a second given number, take the sum of the given number of consecutive numbers in the arithmetic sequence with the given common difference [and beginning with that common difference]. You already know the manner of doing this from what preceded. Multiply the result by three times the square of the second given number, and save the result. Then take the sum of the given number of squares of consecutive numbers in the [same] arithmetic sequence with the given common difference, and multiply the result by three times the second given number, and save the result. Finally, multiply the cube of the second given number by the first given number and add the result to the two saved values. The result is the first adjusted saved value. After this, take the sum of the given number of cubes of consecutive numbers of the [same] arithmetic sequence with the given common difference, and add it to the first adjusted saved number. This is the desired result.

As a model, consider our previous example, where my intention is that the first number exceeds the common difference by two. We know that the sum of seven consecutive squares beginning with three with a common difference of three is 1260. Multiply 1260 by three times the second given number, which is 6, resulting in 7560, and save this result. And also, the sum of seven consecutive numbers [beginning with three] and with an increase of three is 84. Multiply this by three times the square of the second given number, which is 12, to get 1008. And save this result too. Finally, multiply the cube of the second given number, which is 8, by the first given number, which is 7, to get 56. Add this to the two saved values to get 8624 which is the first adjusted value. Next, the sum of the

[86] If d is the common difference, a the first term, and $t = d - a$, then this algorithm depends on the formula $(dk - t)^3 = [(dk - t) + t]^3 - 3(dk - t)^2 t - 3(dk - t)t^2 - t^3$.

[87] $1^3 + 4^3 + 7^3 + 10^3 + 13^3 + 16^3 + 19^3 = \left(1^3 + 2^3 + 3^3 + 4^3 + 5^3 + 6^3 + 7^3\right) \times 3^3 - \left[\left(1^2 + 4^2 + 7^2 + 10^2 + 13^2 + 16^2 + 19^2\right) \times 3 \times 2 + (1 + 4 + 7 + 10 + 13 + 16 + 19) \times 3 \times 2^2 + 2^3 \times 7\right]$.

cubes of the seven numbers with an increase of three is 21168. Add this to the adjusted first value to get 29792, and that is the desired result.[88]

This is true, because if we subtract 2 from each term, we form an arithmetic sequence with difference 3 [and beginning with 3]. The cube of each term is, however, less than the cube of the original increased term by the triple product of the square of the number with 2, increased by the triple product of the term with the square of 2 and by the cube of 2. If we add these to all the terms, we get what we have asserted and you will find this so. You can find the reason in the other case with a bit of thought.

. . .

If you want to add a given number of terms of a geometric sequence with given ratio, subtract the first from the second and then the ratio of this difference to the first term is equal to the ratio of the difference between the last and first terms to the sum of the entire sequence [up to the term before the last], as is proved at the end of the 9th book of Euclid. As an example, if you wanted to add six proportional numbers with ratio 3 and with first term 4, you already know that the second is 12, and the last is 972. Subtract the first, which is 4, from the second, leaving eight. The ratio of 4 to 8 is one half. Subtract 4 from the last leaving 968. Taking half gives 484, and adding to 972 gives 1456, which is the desired result.

Variations on the Rule of Three (from part II, chapter 6)

You already know that with every four proportional numbers, the product of the first with the fourth equals the product of the second with the third. As this is so, it should be clear to you when given some numbers, and given a second number that is a multiple of one of these numbers, how to extract the rest of the multiples so that the multiples are in the same previous given ratio. It is worth knowing that if you multiply one of the numbers by the given second number, and divide by the number of which it is a multiple, you get the multiple of the number that you multiplied by the second given number.

For example, given the numbers *A*, *B*, *C*, *D*, *E*, and *G*, a multiple of *D*, we want to find multiples of *A*, *B*, *C*, *D*, *E*. Multiply *G* by *A*, and divide by *D*, giving *H*. Because the product of *A* and *G* equals the product of *H* and *D*, so the ratio of *A* to *D* is equal to the ratio of *H* to *G*, and by exchanging, the ratio of *A* to *H* is equal to the ratio of *D* to *G*. Similarly, it is clear that if you multiply *B* by *G* and divide by *D*, the result *I* is the multiple for *B*. So you can multiply *C* by *G* and divide by *D* to get *J*, the multiple for *C*. And, you can multiply *E* by *G* and divide by *D*, to get *K*, the multiple for *E*. So we have found the multiples of *A*, *B*, *C*, *D*, *E*, and they are *H*, *I*, *J*, *G*, *K*. It is clear that *H*, *I*, *J*, *G*, *K* are proportional to *A*, *B*, *C*, *D*, *E*, as was to be proved.

Also, if no number of the proportional numbers was known to us, but we did know the sum of two or three of them, then that is enough to get the numbers. For example, say you know in our last example that the sum of *H*, *G*, *K* equals *M*. We want to get the multiples of the given numbers *A*, *B*, *C*, *D*, *E*. Take the sum of the numbers of which *H*, *G*, *K* are

[88] $5^3+8^3+11^3+14^3+17^3+20^3+23^3=\left(3^3+6^3+9^3+12^3+15^3+18^3+21^3\right)+\left[\left(3^2+6^2+9^2+12^2+15^2+18^2+21^2\right)\times 3\times 2+(3+6+9+12+15+18+21)\times 3\times 2^2+2^3\times 7\right].$

multiples, namely *A*, *D*, *E*, and call this *N*. The ratio of *N* to *M* is equal to the ratio of *A*, *B*, *C*, *D*, *E* to their multiples. Accordingly, multiply *M* by *A* and divide by *N*, to get *H*. And in this way, we find the numbers *I*, *J*, *G*, *K*. I claim that the numbers *H*, *I*, *J*, *G*, *K* are the desired numbers.

. . .

The author writes: this completes the sixth chapter of this work, and with its completion, the book is complete. The praise is to God alone. Its completion was on the first of the month of Nisan in the year 81 of the 6th millennium, when I reached the 33rd year of my years. And bless the Helper. [The year was 5081, i.e., 1321.]

Problem section

Many of the problems that Levi brings are standard problems common to many of his contemporary and predecessor mathematical cultures (Arabic, Indian, Chinese, and Mesopotamian). Note, however, that Levi formulates his problems in abstract terms, and does not shy away from elaborate calculations with fractions.

6. A certain full container has various holes in it. One of the holes lets the contents of the container drain out in a given time; a second hole lets the contents drain in a second given time; and so on for each of the holes. All the holes are opened together. How much time will it take to empty the container?

First, calculate what drains from each hole in one hour and add the values all together. Note the ratio of this to the full container. This ratio equals the ratio of one hour to the time needed to empty the container.

For example, a barrel has various holes: the first hole empties the full barrel in 3 days; the second hole empties the full barrel in 5 days; another hole empties the full barrel in 20 hours; and another empties the full barrel in 12 hours. Therefore, the first hole empties one of 72 parts of the barrel in an hour; the second hole, one of 120 parts; the third hole, one of 20 parts; and the fourth hole, one of 12 parts. When we add them all up, the total that empties from all the holes in an hour is 56 of 360 parts of the full barrel. We divide 360 by 56, to get 6 whole and 25 firsts and 43 seconds. Therefore, the time to empty the barrel is approximately 6 hours, 25 firsts,[89] and 43 seconds. The reason for this is clear.

13. One man hires another to work a given number of days, for a fixed wage. This job requires the hiring of a certain number of men per day, each of whom leads a certain number of animals, each of which carries a given number of measures and walks a given distance. The hired man deviates from some or all of these numbers. How much should his wages be?

Take the product of all the values that were stipulated, and make a note of it. Furthermore, take the product of the actual values that were accomplished; and the ratio of the number you noted to this product equals the ratio of the wages he promised him to the wages he owes him.

[89] Firsts refer here to minutes.

For example, Reuven hired Simon to work 9 days for 10 *litra*. The job stipulated the hiring of 13 men each day, each of whom leads seven animals, each of which carries 15 measures and walks 6 *parsas*.[90] Simon provided 8 days, 17 men, each of whom led 6 animals, each of which carried 11 measures and walked 7 *parsas*. The product of the stipulated numbers, 9, 13, 7, 15, and 6 equals 73 thousand and 710, which is noted. The product of the accomplished numbers, 8, 17, 6, 11, and 7, equals 62 thousand and 832. The ratio of 10 *litra* to what he owes him equals the ratio of the noted value to 62 thousand and 832. If you multiply 10 *litra*, the first number, by the fourth number, which is 62 thousand and 832, and you divide the result by the noted value, you will get the number of *litras* and fractions of a *litra* that he owes him. This is 8 and one half *litra*, and a thousand and 7 hundred and 85 of 73 thousand and 710 parts of a *litra*, which is 5 *pashuts* and 59 thousand and 850 of 73 thousand and 710 parts of a *pashut*.

This is right, because the ratio of what he owes to what he stipulated equals the ratio of what he did to what he agreed to do. And the ratio of what he did to what he agreed to do is composed of the ratios of the numbers that were stipulated to the corresponding numbers that were accomplished. This composed ratio, as we already explained, equals the ratio of the product of the stipulated numbers to the product of the numbers that were accomplished. Use this as a model.

As noted above, problem 18 is an abstract version of the "men finding a purse" problem, a problem that seems to have appeared first in India (see Appendix 3), but is also found in the work of Leonardo of Pisa and other medieval European mathematicians. The original version of the problem was indeterminate, but Levi shows here how to find a definite answer if one is given one additional specific piece of information.

18. We add one number to a second number; and the ratio of the result to a third number is given. When we add the first number to the third number, the ratio of the result to the second number is a second given number. One of the three numbers is known. What is each of the remaining numbers?

You already know how to find three numbers that correctly meet these conditions, so extract them.[91] Since you know one of the numbers corresponding to one of the three, you can extract the other corresponding numbers, and that is what was requested.

For example, when you add the first number to the second number, its ratio to the third equals 3 wholes and 2 fifths and a seventh. When the first is added to the third, its ratio to the second equals 7 wholes and 2 thirds and a fourth. The second number is 30. We want to know: what is the value of each remaining number?

First of all, extract three numbers, using the procedure described in part one of this book. Accordingly, subtract one from the product of 3 wholes and 2 fifths and a seventh with 7 wholes and 2 thirds and a fourth. This leaves 27 wholes and a third of a seventh, which is the first number. Add one to 3 wholes and 2 fifths and a seventh, to get 4 wholes

[90] *Litra* is a coin denomination which contains 20 *dinars*, each dinar containing 12 *pashuṭs*. (See footnote 11.) A *Parsa* is a distance unit.

[91] See part I, problem 58 above.

and 2 fifths and a seventh, which is the second number. Also, add one to 7 wholes and 2 thirds and a fourth, and the result you get is the third number, which is 8 wholes and 2 thirds and a fourth. You already know that the number corresponding to the second number is 30.

First	*Second*	*Third*
27 wholes and a third of a seventh	4 wholes and 2 fifths and a seventh	8 wholes and 2 thirds and a fourth
178 wholes and 98 of 159 parts of one	30	58 wholes and 281 of 318 parts of one

Thus the number corresponding to the first is 178 wholes and 98 of 159 parts of one; and the number corresponding to the third is 58 wholes and 281 of 318 parts of one. These three numbers are what were requested, so investigate and find them.

. . .

We will explain why these corresponding numbers are the desired ones. This is because the ratio of the first of the former numbers to the second of them equals the ratio of the first of the latter numbers to the second of them; and the ratio of the second of the former numbers to the third of them equals the ratio of the second of the latter numbers to the third of them. The compound of these ratios equals the ratio of the first of the former numbers to the third, which equals the ratio of the first of the latter numbers to the third. By adding these together, the ratio of the sum of the first and second former numbers to the third equals the ratio of the sum of the first and second latter numbers to the third. But in our example, the ratio of the sum of the first and second former numbers to the third is 3 wholes and 2 fifths and a seventh. Hence the ratio of the sum of the first and second latter numbers to the third is 3 wholes and 2 fifths and a seventh. Similarly, the ratio of the sum of the first and third former numbers to the second is 7 wholes and 2 thirds and a fourth. Use this as a model.

II. NUMEROLOGY, COMBINATORICS, AND NUMBER THEORY

Numbers were not only treated as mathematical entities—they were also given natural and mystical significance. An excerpt from Ibn Ezra's *Book of One* shows how the different approaches intermingled in scholarship. We follow with two discussions of combinatorics. In the first, ibn Ezra calculates the number of possible conjunctions of a given number of planets from among the seven planets. In the second, Levi ben Gershon engages in an abstract and general discussion of permutations and combinations. We proceed with Levi's elegant treatment of harmonic numbers, showing that different products of powers of two and three cannot have difference one, except for the four known harmonic couples. We end with a discussion of amicable numbers in the Hebrew literature.

1. ABRAHAM IBN EZRA, *SEFER HAʾEḤAD* (*THE BOOK OF ONE*)

This book by Abraham ibn Ezra (see section I-1) summarizes knowledge about numbers from various mathematical disciplines, along with other natural sciences. It also has clear mystical overtones and is written in a terse style that is sometimes hard to fathom.[1] This translation covers the treatment of the number 4.

Four is the first visible square.[2] It is the first even-even [of the form $2n$ with n even], and it is the first composite [non-prime number]. Indeed, the first decimal order [*maʿarekhet*] consists of nine [digits], wherein one is the foundation of counting, leaving four primes: 2, 3, 5, and 7, and the other four composite. Its sum with the numbers that precede it [$1 + 2 + 3 + 4$] is ten, which is the beginning of the analogous multiple [*klal*, the next decimal order].[3] Four is their root [the base of the triangular number 10].

Since numbers unto a prime are like unto an indivisible unit, it is opposite; such are 2 and 5, 3 and 6, 4 and 7, and so on without end.[4] Therefore, the fourth astrological sign is the opposite of the first; for heat and cold are active [properties of the four elements], whereas the remaining two, wet and dry, are passive.[5] Every fourth sign is the opposite of the first in the active [property], but the fiery signs alone are the opposite both in active and passive [properties].[6] For this reason the astrologers say that the quartile aspect [pairs of signs that are three apart] is enmity.

A sextile aspect [pairs of signs that are two apart] is half-friendship, for the third sign is identical in the active property of the element to the first, but opposite in the passive; therefore, they say that it is half-friendship. Trine [pairs of signs that are four apart] is complete friendship, for at the trine aspect is the fifth sign, which has the same element as the first. Therefore, it is the aspect of complete friendship in the active and the passive. And so 1 and 5, since both preserve themselves;[7] likewise 2 and 6, since both are even-odd [of the form $2n$ with n odd]; and so also 3 and 7, since neither are even, and their components are similar.[8] Not so, however, are 1 and 3, 2 and 4, and 3 and 5.[9]

[1] For the various meanings of numbers in Ibn Ezra's work, see [Langermann and Simonson, 2000].

[2] Although one is also a square, it is not so "visibly," since its root is equal to itself.

[3] Compare the opening of *The Book of Number* (section I-1).

[4] This is a possibly corrupt, baffling statement. It might suggest that integers which, like 1 and 4, are three apart, are opposite to each other in that one is prime and the other is not. But this would work only as far as 5 and 8. Another interpretation is that such numbers are opposite in that one is even and the other odd. The next sentences validate this opposition from an astrological point of view.

[5] The signs are divided as follows: Aries, Leo, and Sagittarius are fire signs (hot and dry); Taurus, Virgo, and Capricorn are earth signs (cold and dry); Gemini, Libra, and Aquarius are air signs (hot and wet); and Cancer, Scorpio, and Pisces are water signs (cold and wet).

[6] A pair of signs that are three apart (like $1 =$ Libra and $4 =$ Aries) is always in opposition with respect to the active property of its element (hot/cold). A pair that begins with a fire or air sign is opposite also with respect to the passive property (dry/wet). The text should probably have "fire and air" rather than "fire."

[7] That is to say, 1, 5, and other numbers having either of these as their final digit will preserve their final digit in all higher powers.

[8] According to Comtino, they are similar insofar as both are sums of consecutive odd and even numbers: $3 = 1 + 2$, and $7 = 3 + 4$.

[9] Comtino suggests that this means that we cannot find any full and satisfactory similarity between these pairs, such as were found for the pairs listed above.

Opposition [the aspect of half the orb, where signs are six apart] (and half-wise the quartile aspect [where signs are 3 apart]), are aspects of enmity in the passive property.[10] Do not be puzzled that the seventh house [which is in opposition to the first] indicates one's mate [ʿ*ezer*, woman]; as the author of *Sefer Yeṣira* said: A.M.Sh. for male, and A.Sh.M. for female.[11]

Now the orb, which is one, is divided by its diameter. If the diameter too is cut [in half], the aspect is quartile. If you place a point at the fourth [of the diameter and construct the perpendicular chord], the orb will be divided into three equal sections, forming a triangle. Its half is one-sixth of the orb. No other number can divide it [the circle] except in thought or with fractions.[12]

Four is the beginning of non-equilateral acute triangles. All subsequent consecutive numbers follows its rule.[13] Now in the first obtuse triangle [sides 2, 3, 4] the [square on the] longest side exceeds the [sum of the squares on the other] two by the numerical value of the middle [$4^2 - (2^2 + 3^2) = 3$], but in the acute [triangle] the longest side falls short by the numerical value of the middle [that is, in the triangle 4, 5, 6 we have: $4^2 + 5^2 - 6^2 = 5$]. I will now give you a rule. Given three consecutive numbers, the smallest of which is at least four, and you wish to know by how much the [sum of the squares of] the two [smaller sides] exceeds the [square of] the longest, always subtract four from the middle and multiply what remains by the middle. For example, 10, 11, 12. We subtract 4 from 11, leaving 7; we multiply it by 11, producing 77. It is the deficiency of the square on the longest side [with respect to the squares on the two smaller sides; $10^2 + 11^2 - 12^2 = 77$].

Know that some angles [of triangles whose sides are integer triplets] are very wide; others are close to being right-angles, for example, 4, 8, 9. Likewise there are acute [angles] which are half of a right angle, or a third or as little as one degree. But do not think that you can form a triangle out of any [set of three] number[s] that you want. For an acute [triangle with consecutive side lengths] cannot have [a side] which is less than 4. Nor can one side of a triangle be greater than [the sum] of the two [other sides]. Nor can there be a right triangle in which the numerical values of the sides are distant [from each other]. They are either one set [of consecutive numbers], such as 3, 4, 5 and their multiples; or the two smaller sides are consecutive, for example, 20, 21, 29; or the greater ones are consecutive, as in 5, 12, 13.[14] There are very obtuse triangles, such as 10, 17,

[10]Opposition is the aspect of enmity in the passive, whereas the quartile aspect is an aspect of enmity in the passive only for pairs starting with fire or air signs.

[11]The first house is associated with masculinity, and its opposite, the seventh, with femininity. This is also the case with the permutations of the letters A.M.Sh in *Sefer Yesira*: the first is associated with masculinity, and its opposite among the six possible permutations, which is the fourth permutation A.Sh.M., is associated with femininity. *Sefer Yesira* (the *Book of Creation*; see [Kaplan, 1995]) is a mystical treatise concerning the creative power of letter and number combinations.

[12]The text differs slightly among versions and suggests different readings. In this reading it is claimed that 12 cannot be divided without remainder except by 2, 3, 4, and 6. According to another reading, the point is to show that divisions by four (first of the circle, then of the diameter) can produce the quartile, trine, and sextile aspects.

[13]In the previous sections, the triangles with sides 2, 3, 4 (obtuse) and 3, 4, 5 (right angle) were discussed. The triangle with sides 4, 5, 6 is acute, and so are all the following triangles with consecutive side lengths.

[14]Since there obviously are right triangles where the sides are not multiples of triplets with a consecutive pair (e.g., 12, 35, 37), the meaning of "distant" here and in the final sentence of this paragraph is not quite clear.

26, and slightly [obtuse], such as 10, 23, 26. Look at 10, 24, 26 [which has a right angle]; but 10, 25, 26 is slightly acute. Hence we can form neither a right nor an acute [triangle] in which the sides are distant.

I shall now give you a rule. Know that by reckoning in the first order [i.e., numbers 1–9], there are 84 triangles with no two equal sides, but only 33 of them are true. Likewise, not all [possible] isosceles triangles are true.

Four is the first non-prime [*sheni*, literally, "secondary"].[15] Therefore even numbers are always non-prime, that is, composite, except for the number 2, due to the power of the one, which is its main influence. Every number multiplied by 1 will not increase [1 times *n* is *n*], since it [1] is the essence of every number. Therefore, the first square is 4. Every square multiplied by a square is a square, and divided by a square, is a square; the ratio of a square to a square is also a square. Therefore, the measures of all of the scholars are in squares.[16]

2. ABRAHAM IBN EZRA, *SEFER HAᶜOLAM* (*BOOK OF THE WORLD*)

Sefer Haᶜolam (*Book of the World*) by Ibn Ezra discusses the meaning of celestial conjunctions and aspects. It opens by counting all possible conjunctions of the seven known planets, demonstrating some systematic combinatorial reasoning. To calculate the number of different sets of n elements out of 7 planets, a recursive method is used, taking partial sums of the sequence $1, 2, \ldots, 7$, then taking partial sums of the sequence of these partial sums, and so forth. In this text Ibn Ezra does not mention the symmetry between the combinations of k elements out of 7 and of $7 - k$ elements of 7, but he does note this symmetry and uses it to simplify calculations in his later long commentary to Exodus 33:21 [Ibn Ezra, 1991]. (See [Ginsburg and Smith, 1922] for an earlier translation and analysis.)

If you come across Abu Maᶜshar's[17] *Book on the Conjunctions*[18] *of the Planets*, you should neither like it nor trust it, because he relies on the mean motion for the planetary conjunctions. No scholar concurs with him, because the truth is that the conjunctions should be reckoned with respect to the zodiac. Nor should you trust the planetary conjunctions calculated according to the [astronomical] tables of the Indian scholars, because they are wholly incorrect. Rather, the correct approach is to rely on the tables of the scholars of every generation who rely on experience.

There are 120 conjunctions [of the seven planets].[19] You can calculate their number in the following manner: it is known that you can calculate the number that is the sum [of all the whole numbers] from one to any other number you wish by multiplying this number by [the sum of] half its value plus one-half. As an illustration, [suppose] we want to find the

[15] At the beginning of this text, Ibn Ezra had used the term *murkav,* which we translate as "composite," rather than *sheni* (secondary) used here. He thus transmits both Greek terms, *deuteros* and *sunthetos.*

[16] This is yet another baffling or corrupt sentence. It might refer to the measurements of areas.

[17] Abu Maᶜshar (787–886) is the most prominent astrologer of the Middle Ages. He formulated the standard expression of Arabic astrology in its various branches, creating a synthesis of the Indian, Persian, Greek, and Harranian theories current in his days. Despite the critique presented here, he is also Ibn Ezra's most important Arabic astrological source.

[18] The conjunctions (*Maḥbarot*) are astronomical events where several planets appear close together in the sky. These events are thought to have substantial astrological impact.

[19] The seven planets, in order from the lowest to the highest, are the moon, Mercury, Venus, the sun, Mars, Jupiter, and Saturn.

sum [of all the whole numbers] from 1 to 20. We multiply 20 by [the sum of] half its value, which is 10, plus one-half, and this yields the number 210.

We begin by finding the number of double conjunctions, meaning the combinations of only two planets. It is known that there are seven planets. Thus Saturn has 6 [double] conjunctions with the other planets. [Jupiter has 5 double conjunctions with its lower planets, Mars has 4, and so on. So we need to add the numbers from 1 to 6]. Hence we multiply 6 by [the sum of] half its value plus one-half, and the result is 21, and this is the number of double conjunctions.

We want to find the [number of] triple conjunctions. We begin by taking Jupiter and Saturn, and [then take] any of the other five [planets] with them; the result is the number 5. [Then we move on to conjunctions composed of Saturn, Mars and one of the lower four, then Saturn and the Sun with one of the lower three and so on. Altogether], we multiply it [5] by 3, which is [the sum of] half its value plus one-half, and the result is 15, and those are Saturn's [ternary] conjunctions. Jupiter should have 4 [triple] conjunctions [with Mars and the lower planets; continuing by the same method], we multiply that [4] by [the sum of] 2 plus one-half, and the result is 10. Mars has 3 conjunctions; we multiply them by 2, and the result is 6. The Sun has 2 conjunctions; we multiply them by the sum of 1 plus one-half, and the result is three. Venus has one conjunction with the planets beneath it. So the total is 35, and this is the number of triple conjunctions.[20]

We wish to find out the quadruple conjunctions. We begin with Saturn and Jupiter, and Mars with it. For [these] three [planets] to conjoin [with one of the four lower planets], we start with 4 conjunctions. We multiply them by 2 and one-half and the result is 10, [namely, ten quaternary conjunctions that begin with Saturn and Jupiter]. Then come the [quadruple] conjunctions of Saturn and Jupiter [should be Mars] with the others [lower planets], and we start with 3 [conjunctions]. We multiply that by 2 and the result is 6 [quadruple conjunctions beginning with Saturn and Mars, skipping Jupiter], and [the partial sum] is sixteen. Then we have Saturn with Mars [should be the sun], and there are 2 [quadruple conjunctions]. We multiply them by 1 and one-half and the result is 3. Then comes another conjunction [Saturn, Venus, Mercury and the moon], and [the total] for Saturn is 20 [quadruple] conjunctions. Now we have Jupiter with 3 [conjunctions]. We multiply that by 2 and the result is 6. Then come two [conjunctions]. We multiply that by 1 and one-half and the result is 3. Then comes one conjunction. [The total of] Jupiter's [quadruple] conjunctions is 10. Then we have Mars with two [conjunctions]. We multiply that by 1 and one-half and the result is 3. Then comes one conjunction, making 4 [quadruple] conjunctions [beginning with Mars]. The Sun has one [quadruple] conjunction with the planets beneath it. So the sum total is 35 quadruple conjunctions.[21]

We wish to find the quintuple conjunctions. We find 15 for [those beginning with] Saturn, 5 for Jupiter, and 1 for Mars. So there are 21 quintuple conjunctions. As for sextuple conjunctions, there are 6 for Saturn and 1 for Jupiter, making a total of 7. There is one

[20]In anachronistic terms, Ibn Ezra is calculating that $C_{7,3} = C_{6,2} + C_{5,2} + C_{4,2} + C_{3,2} + C_{2,2}$, where each $C_{k,2}$ is shown to be the sum of integers up to $k - 1$.

[21]In this paragraph, Ibn Ezra calculates quadruple conjunctions according to the anachronistic formula $C_{7,4} = C_{6,3} + C_{5,3} + C_{4,3} + C_{3,3}$, and then uses the above procedure to calculate $C_{k,3}$.

septuple conjunction. So we have obtained 120 conjunctions. All these conjunctions are odd numbers that are divisible by seven [except the last].

3. LEVI BEN GERSHON, *MAʿASE ḤOSHEV (THE ART OF THE CALCULATOR)*

Unlike Ibn Ezra's, Levi's combinatorics from *Maʿase Ḥoshev* (see section I-6) are abstracted from any practical context and include accurate proofs. In part I, Levi proves some of his results on permutations and combinations by using a form of mathematical induction; he presents the inductive step before stating the theorem and writing the initial step.[22] Then, in part II, Levi uses the theoretical results to do calculations.

In what follows, we use the modern words "permutations" and "combinations" for Levi's verbose formulations. Instead of "arrangements of a certain number of distinct terms from another (larger) number of distinct terms that are exchanged either by order or by terms," we write "permutations of n elements out of a set of m elements" or $P_{m,n}$. Instead of "arrangements of a certain number of distinct terms from another (larger) number of distinct terms that are exchanged by terms," we write "combinations of n elements out of a set of m elements" or $C_{m,n}$. And instead of "arrangements of a certain number of distinct terms that are exchanged only by order," we write "permutations of n elements" or P_n.

From part I

63. If the number of permutations of a given number of different elements is equal to a given number, then the number of permutations of a set of different elements containing one more element equals the product of the former number of permutations with the number after the given number.[23]

Let the terms be A, B, C, D, E and their number be G. Let the number following G be H, and let the number of ways to arrange the terms A, B, C, D, E be I. And let the terms A, B, C, D, E, F add one term to the terms A, B, C, D, E, and thus their number is H. We say that the number of permutations of the terms A, B, C, D, E, F is the product of I with H.

The proof is that if you place F first followed by each of the permutations of A, B, C, D, E, the [new] permutations will remain distinct, and therefore the number of permutations where F is the first term equals I. Similarly, since the number of permutations of A, B, C, D, E equals I, therefore the number of permutations of A, B, C, D, F is also I. And when you place E first followed by all these permutations, you are left with distinct permutations, and thus the number of permutations where E is first equals I. And in this way it is explained that when each one of the terms is placed first, the number of permutations is I. If so, then the total of these permutations is I multiplied by the number of terms. However, there are H terms. Therefore, the number of permutations of A, B, C, D, E, F is the product of H with I.

It is clear that among all those permutations counted, there are no two identical ones, because when a certain element is first, there are no identical permutations because the permutations before attaching it were distinct and they remain distinct when it is attached

[22] See [Rabinovitch, 1970] for further discussion of mathematical induction in Levi's work.

[23] $P_{n+1} = (n+1)P_n$. The proof ends with an inductive style point about $n!$.

to them. And there is no doubt about that when the first elements are different. This being the case, it is clear that among those permutations that we have counted, there are no two identical ones.

We also say that there are no permutations besides those. For if there were, let *DFECAB* be such a permutation. But in this case *D* would have been added to the remaining elements *FECAB*, so *DFECAB* is one of the permutations we have counted. And since there are no two identical permutations and there are no permutations except those, it follows that the number of permutations of *A*, *B*, *C*, *D*, *E*, *F* is the product of *H* with *I*, which is what we wanted to prove.

And thus, it can be understood that the number of permutations of a given number of terms is the number built from consecutive numbers starting with one and their number is the number of these terms.[24] The number of permutations of 2 is 2, and that is built from the numbers 1 and 2. And the number of permutations of 3 is the product of 3 with 2, and this equals the number built from 1, 2, 3. And similarly, this can be shown without limit.

64. The number of permutations of two terms from a given number of distinct terms is equal to the product of the given number and the number that precedes it.[25]

65. When you are given a number of terms and the number of permutations of a second given number from these terms is a third given number, then the number of permutations of the number following the second given number from these terms is the product of the given third number by the excess of the first given number over the second number.[26]

Let the terms be *A*, *B*, *C*, *D*, *E*, *F*, *G*, and let their number be *H*. Let *I* be different from *H* and less than *H*. Let the number of permutations of *I* elements from these elements be equal to *L*; let *M* be the number following *I*. Let the difference between *H* and *I* be *N*. I say that the number of permutations of *M* elements from these terms is equal to the product of *L* and *N*. Let one of the permutations of *I* elements be *ABC*; then the remaining elements are *D*, *E*, *F*, *G* and their number is equal to *N*. Putting each one of the elements *D*, *E*, *F*, *G* together with the permutation *ABC*, distinct permutations result and the number of elements in such a permutation is *M*, because one more element has been added. Since the number of *D*, *E*, *F*, *G* is *N*, the new permutations stemming from *ABC* will number *N*, and it is clear that the number of new permutations stemming from each permutation of *I* elements, distinct in their order or in their elements, is *N*. So the total number of these permutations of *M* elements is the number *N* multiplied by the number of permutations of *I* of these elements. But the number of permutations of *I* of these elements is *L*; thus the number of permutations of *M* of these elements is the product of *N* by *L*.

We say that among all these permutations that we have counted, no two are identical permutations. Indeed, to one permutation, distinct elements have been joined, from one time to the next, and from this it follows that these permutations are distinct, and there is no doubt that distinct permutations will not become identical when any element is joined to them. Thus there are no two identical permutations.

[24] $P_n = n!$.

[25] $P_{n,2} = n(n-1)$.

[26] $P_{H,I+1} = P_{H,I}(H-I)$, where, in Levi's lettering, $L = P_{H,I}$.

We say that there are no permutations besides those that we have counted. For if it were possible, let one such permutation be *FDBG*. However, the permutation *DBG* has been joined with each one of the remaining elements at the first place, and one of these elements is *F*. Thus the permutation *FDBG* is one of the permutations that we counted. There being no two identical permutations among those that we counted and there being no permutations besides these, the number of permutations of *M* of these elements is the product of *N* by *L*, which we wanted to prove.

And so it is clear that the permutations of a given number from a second given number of terms is equal to the number built from consecutive numbers. Their number is equal to the first given number, and the last one is the second given number. Let the number of terms be 7, and the consecutive numbers starting from 1 are 1, 2, 3, 4, 5, 6, 7. It is clear that the number of arrangements of two of these is the product of 6 with 7—the number of numbers is 2; they are consecutive, and the last of them is 7. The number of permutations of 3 of them is equal to the product of 5 with the product of 6 and 7, because the excess of 7 over 2 is 5. This equals the product built from 5, 6, 7. Also, these numbers are three numbers; they are consecutive, and the last of them is 7. Similarly, it can be explained with any number you like.[27]

66. When there is a given number of terms, and the number of combinations of a second given number from these terms is a third given number, and the number of permutations of as many terms as this second given number is a fourth given number, then the number of permutations of the second given number from as many terms as the first given number is equal to the product of the third given number by the fourth given number.[28]

Let the elements be *A*, *B*, *C*, *D*, *E*, *F* and let their number be *G*. Let the number of combinations of *H* elements out of *G* be *J* and the number of permutations of *H* elements be *L*. I say that the number of permutations of *H* elements out of *G* elements is equal to the product of *J* and *L*. Let one of the ordered sets from the combinations of *H* elements be *BCD*; then all the permutations of this set produce *L* ordered sets. In the same way, one shows that for each of the ordered sets from combinations of *H* terms out of all the terms one can form *L* ordered sets. So out of the total of all these sets, one obtains ordered sets by multiplying the number by *L*. Since the number of these sets is *J*, the number of the ordered sets is the product of *J* and *L*.

We claim now that among all the chosen ordered sets, no two are the same. For where the elements are different, they are permuted, and the number of permutations is *L*, as we had assumed. Without a doubt, moreover, two ordered sets containing different elements cannot become equal through permutations.

We claim further, that there are no other ordered sets besides the ones we have already counted. For suppose this were possible, say, by the ordered set *FDB*. But all the elements *BDF* are already permuted and one of the permutations is *FDB*. But *FDB* is one of the ones already counted, so there is no further ordered set available. So since, among the counted ordered sets no two are equal and no further ones are available,

[27] In modern notation, Levi has proved that $P_{m,n} = (m-n+1)\ldots(m-2)(m-1)m$.

[28] Using Levi's lettering, this result, in modern notation, is $P_{G,H} = P_H C_{G,H}$.

therefore the number of permutations of H elements out of the elements A, B, C, D, E, F is equal to the product of J and L.

67. When there is a given number of distinct terms, and the number of permutations of a second given number from these terms is a third given number, and the number of permutations of the second given number of terms is a fourth given number, then the number of combinations of the second given number from the given number of terms equals the number of units by which the fourth number counts the third number.[29]

68. When there is a given number of distinct terms, and the number of combinations of a second given number from these terms is a third given number, and the excess of the first given number over the second given number is a fourth given number, then the number of combinations of the fourth given number from those terms is equal to the third given number.[30]

From chapter 4 of part II

If you wish to find the number of permutations of a second given number of elements out of a first given number of distinct elements, you should know that the number of permutations of two elements is equal to the product of the first given number and the previous number and that the number of permutations of three elements has a ratio to that of two elements as the ratio of the difference between the first number and 2 to 1. Furthermore, the number of permutations of four elements has a ratio to that of three elements as the ratio of the difference of the first number and 3 to 1, and so on without end.

Therefore, you proceed as follows: Form the number from the product of as many numbers following each other as the second given number, with the last being equal to the first given number. The result is what one desires. As an example, you want to know the number of permutations of five elements out of eight. Since the second given number is 5, take the product of five consecutive numbers so that the last one is 8, that is, the product of 4, 5, 6, 7, 8. This product is 6720, which is the number of permutations of five elements out of eight. This is so, because the number of permutations of two elements is the product of 7 and 8 and of three elements is the product of 6 by the product of 7 and 8, as was previously shown, and the number of permutations of four elements is the product of 5 by the product of 6, 7, and 8 and of five elements is the product of 4 by the product of 5, 6, 7, and 8, and so forth. This is clear from the above.

If you wish to find the number of combinations of a second given number of elements out of a first given number of different elements, find the number of permutations of the second given number out of the first given number of different elements and note that. Then find the number of permutations of the second number. As many times as that number divides the number first found is what you are seeking. As an example, suppose you want to know the number of combinations of five elements out of eight. Find how often the product of 1, 2, 3, 4, 5 divides the product of 4, 5, 6, 7, 8. The product of 4, 5, 6, 7, 8

[29] $C_{m,n} = P_{m,n}/P_n$.
[30] $C_{m,m-n} = C_{m,n}$.

is 6720 and the product of 1, 2, 3, 4, 5 is 120. Now 6720 contains 120 56 times, so 56 is the desired result. . . .

To make things easy for you, you should also know that the number of combinations of five elements out of eight elements is equal to the number of combinations of three elements out of these elements. In fact, this number is the number of times the product of 6, 7, 8, that is, 336, contains the product of 1, 2, 3, that is, 6. This result is 56, so the number of combinations of three elements out of eight elements is the same as that of five elements. The reason for this was presented earlier.

4. LEVI BEN GERSHON, *ON HARMONIC NUMBERS*

According to Levi ben Gershon himself, *On Harmonic Numbers* was commissioned by Philippe de Vitry, a French composer and music theorist and also an official at the court of Philip VI, to answer a question about numbers formed by powers of 2 and 3. This question was perhaps related to the question of possible ratios giving harmonic tones. Recall that the ratio 2:1 gives an octave, 3:2 a fifth, 4:3 a fourth, and the difference between a fifth and a fourth (the quotient of the ratios) gives 9:8, a single tone. All these ratios are of the form $(n+1):n$, so the question was asked whether there are any other ratios of that form that can come from the basic ratios by composition. Levi's negative answer is based on an ingenious parity analysis of the various numbers involved.

Although Levi wrote the work in Hebrew, it was immediately translated into Latin (perhaps by Petrus of Alexandria), and the Hebrew version has been lost.

In the year 1342 of the incarnation of Christ, our work on mathematics having been completed, I was requested by the noted master of the science of music, Master Philippe de Vitry of the French kingdom, to demonstrate a certain hypothesis postulated in that science: all pairs of harmonic numbers differ by a number, except for 1 and 2, 2 and 3, 3 and 4, [and] 8 and 9.[31]

A harmonic number is described as follows: A number is harmonic if, except for the unit, it is divisible only by 2 or 3, and the factors also are similarly divisible down to unity. Some examples are 1, 2, 4, 8, . . . and 1, 3, 9, 27, . . . , and also 6, 12, 18, 24, ... And I wish to satisfy him and demonstrate this principle in this place in our book. But since all such harmonic numbers are either, first, in continuous proportion with ratio 2, or secondly, in continuous proportion with ratio 3, or thirdly, of the type produced by multiplying a number of the first type by a number of the second, I will demonstrate the theorem on these proportions stated in other terms, as follows:

Of all the numbers successively proportional with ratio 2, and of all the numbers successively proportional with ratio 3, and of their mutual products, any two of these differ from each other by a number, except for: 1 and 2, 2 and 3, 3 and 4, and 8 and 9. If this were not true, there would be another pair of numbers defined as above, consisting of equal numbers or differing just by unity. But this conclusion is false. It appears thus that the initial proposition is true. I will demonstrate in what follows the falseness of the stated conclusion.

To facilitate the comprehension of this demonstration, I propose certain definitions. First, the numbers of the first class will be called "numerical powers of 2"; then the

[31] Note that 1 is not considered to be a number; 2 is the smallest number.

numbers of the second class will be called "numerical powers of 3." Those of the third class will be called "products of a numerical power of 2 by a numerical power of 3." Also, we will call the unit, the "element of the first rank in the first class and the second class," and the first number that follows, we will call the "element of the second rank," and the next number, the "element of the third rank," and so forth, to infinity. Finally, the second, fourth, sixth, eighth ranks and so on, we will call even ranks, the others, odd ranks. This concludes the definitions.

1. Every numerical power of 2 is even, because it is the product of an even number by the unit or by an even number, such as 2, 4, 8, 16, and so on.

2. Every numerical power of 3 is odd, because it is the product of an odd number by the unit or by an odd number, such as 3, 9, 27.

3. Every number that is the product of a numerical power of 2 by a numerical power of 3 is even, because it is the product of an even number by an odd number, such as 6, 12, 18.

4. Every numerical power of 2 is only divisible by another number that is a numerical power of 2, and every number of this same class is divisible by every number of lower rank in this same class and only gives a quotient of this same class. Similarly, every numerical power of 3 is only divisible by a number that is also a numerical power of 3.

This is true after Euclid [IX.11]. It follows that no numerical power of 2 is divisible by an odd number [IX.13].

5. The half of a numerical power of 3 less a unit is equal to the sum of the numerical powers of 3 of rank strictly less including the unit.

This is true by Euclid, IX.38 [IX.35], as one can verify.

6. Every numerical power of 2 and of 3 of odd rank is a square.

This is necessarily true by Euclid [IX. 8].

7. The sum of two numerical powers of 3 is even.

Let the two given numbers be *AB* and *BC*. If we subtract the unit *BD* from the number *BC*, there remains the number *DC* which is even. Since the number *AD* is even, it follows that the number *AC* is even, which was what we wanted to prove. And it has therefore been established that if one adds an even number of numerical powers of three, their sum will be an even number.

8. The sum of an odd number of numerical powers of 3 is odd.

Let the given numbers be *A*, *B*, *C*, *D*. Since the sum of *A*, *B*, *C*, *D* is even according to [7], if we add an odd number *E*, the total will be an odd number, which is what we wanted to prove.

9. The sum of a numerical power of 3 of odd rank and a number of the same class of the immediately following rank will be a square number divisible by 4.

Let *A* be the number of odd rank to which one adds *B*, the number of the following rank. Since *A* is a square by [6], and *B* is triple of *A*, that is, *A* multiplied by 3 is *B*, it follows that $A+B$[32] is four times A, which is a square number. And since 4 is also a square by [6], it follows that the ratio of A to $A+B$ is equal to the ratio of a square number to a square number. But A is a square; thus $A+B$ is a square and since $A+B$ is four times A, it follows that $A+B$ is divisible by 4. Thus the sum of a numerical power of 3 of odd rank and the number of the same class of the immediately following rank is a square number divisible by 4.

From what precedes, it follows that if two successive numerical powers of 3 are added, their sum is divisible by 4, because this sum is equal to four times the smaller of the two numbers. And it follows that if one adds an even number of consecutive numerical powers of 3, the sum is divisible by 4, because the sum of any two of these successive numbers is divisible by 4, after [9], and thus their total is divisible by 4.

10. The sum of any four powers of 3 of consecutive ranks is divisible by 5 and by 8; furthermore, the sum of these same consecutive numbers of the same class is divisible by 4.

Let A, B, C, D be the powers of 3. As 1 is to 9, the third element of this class, so is A to C and B to D. It follows that the ratio of $1+9$ to 1 is the same as the ratio of $A+C$ to A and $B+D$ to B. Therefore, $A+C$ is ten times A and $B+D$ is ten times B. It follows that $A+B+C+D$ is ten times $A+B$. But by [9], $A+B$ is four times A [and $C+D$ is four times C]. Thus $A+B+C+D$ is four times $A+C$ and therefore forty times A. But 40 is divisible by 5 and 8, and so $A+B+C+D$ is divisible by 5 and 8, and it is at the same time the sum of four consecutive powers of 3. And the sum of these same consecutive elements of this class, as we know, is divisible by 4, which is what we wanted to prove.

11. The fourth part of a number divisible by 8 is even.

Since, in fact, it is divisible by 8, it has an eighth part, which part, doubled, is the quarter. But this doubled number may be divided into two equal parts, and such a number is even. It follows that the fourth part of the given number is even, which is what we wanted to prove.

12. No even number is equal to an odd number or to the unit, because it differs by at least a unit from any of these.

13. No even number is equal to another even number, because it differs from any other by at least two.

14. No odd number is equal to another odd number or to the unit, because it differs from any other by at least two.

These things being demonstrated, I will prove that the negation of the stated consequence is true, and therefore, first, that no two numbers of the defined classes are equal.

[32] In the Latin manuscript, the sum of two or more quantities is indicated by placing the letters next to one another, separated by a dot. To make the reading easier, we have used the plus sign in what follows.

In propositions 15 through 20, Levi shows that no two numbers of the same class are equal, that a numerical power of 2 cannot equal a numerical power of 3, nor can a numerical power of 2 or 3 equal a number that is a product of two such numbers.

This decomposition and this mutual comparison of the numbers in question show that the negation of the first conclusion is true, namely that no two harmonic numbers are equal, and that is what we wanted to prove.

It remains to prove the negation of the second conclusion, namely, that no two numbers of the given classes differ by just a unit, with the exception of those that have been indicated above.

In propositions 21 and 22, Levi shows that no power of 2 (except for 1 and 2) can differ by a unit from another such power, that no power of 3 can differ by a unit from another power of 3, and that no product of a power of 2 by a power of 3 can differ by only a unit from another such product, because the parity of all numbers in each class is the same, so any two must differ by at least two.

23. No numerical power of 2 can differ by only a unit from a product of a numerical power of 2 by a numerical power of 3.

Let A be a numerical power of 2 and B the product of a numerical power of 2 by a numerical power of 3. So B is the product of D, a numerical power of 2 and E, a numerical power of 3. If D is greater than or equal to A, then the proposition is true. So suppose that the number A is greater than the number D. It is necessary by [4] that the quotient of A by D is a numerical power of 2. Let us denote that quotient by F. So D times E is B, and D times F is A. But F cannot be equal to E, because they differ by at least a unit according to [12], since one is even and the other is odd. It follows that the number B, which is the product of D and E, is not equal to the number A, which is the product of D and F. Even more, these differ by a number equal to the product of D by the difference between E and F. The resulting number is not less than the number D because the said difference is not less than 1. But the number D is at least equal to 2, since it belongs to the class of powers of 2. Thus the number obtained is at least equal to 2, so the number A differs from the number B by more than a unit, which is what we wanted to prove.

24. No numerical power of 3 can differ by only a unit from a number that is the product of a numerical power of 2 and a numerical power of 3.

The proof is similar to the proof of proposition 23 and is omitted.

25. With the exception of the second term of the class of powers of 2 and the second term of the class of powers of 3, the second term of the class of powers of 3 and the third of the class of powers of 2, and the third of the class of powers of 3 and the fourth of the class of powers of 2, no other numerical power of 2 differs by only a unit from any numerical power of 3.

If this were not true, then either there would be a particular numerical power of 3 that is greater by the unit than a particular numerical power of 2, and therefore when the unit is subtracted from this numerical power of 3 the remainder is equal to the numerical power of 2, but this is false. Or there would be a particular numerical power of 2 that is greater by the unit than a particular numerical power of 3, and therefore when the unit is added to the numerical power of 3, the sum is equal to the numerical power of 2, and this is false. The truth of these negations of the stated propositions will be demonstrated in a clear and evident fashion by the theorems that follow, the first of which is the following.

26. If the unit is subtracted from any numerical power of 3 of even rank, the remainder is not a numerical power of 2.

Let *AB* be such a numerical power of 3, an odd number by [2], from which one subtracts the unit *DB*. There remains the even number *AD*. If this is divided in half at the point *C*, then, by [5], the number *AC* is equal to the sum of all the terms of the class of powers of 3 preceding *AB*. This sum is odd, by [8], because the number of terms preceding *AB* is odd. It follows that the number *AD* is divisible by an odd number. It follows by [4] that *AD* is not a numerical power of 2, and that is what we wanted to prove. We are supposing here that *AC* is not the unit, which will be the case if *AB* is the second term of the class of powers of 3. But this is excluded by the hypothesis, for then *AD* will necessarily be a numerical power of 2, since *AC* will in fact be nothing else than the unit, which thus divides every number and whose double is the second term of the class of powers of 2.

27. If the unit is subtracted from any numerical power of 3 of any odd rank immediately following 4 or any multiple of 4, such as the ranks 5, 9, 13, 17, and so on, the remainder is not a numerical power of 2.

Let *AB* be such a numerical power of 3, which is odd by [2], from which the unit *DB* is subtracted. There remains the even number *AD*. If this is divided in half at the point *C*, it follows by [5] that *AC* is equal to the sum of all the terms of the class of powers of 3 preceding *AB*, and this sum is also equal to the number *CD*. It follows by [10] that this sum is divisible by 5, since the number of terms preceding *AB* is 4 or a multiple of 4. It follows that *AD* is divisible by 5, and the same for *CD*. Thus *AD* is not a numerical power of 2, and that is what we wanted to prove.

28. If the unit is subtracted from any numerical power of 3 of any odd rank that does not follow immediately 4 or any multiple of 4, such as the ranks 7, 11, 15, and so on, the only numbers of the class of powers of 3 of which we now need to speak, the remainder is not a numerical power of 2.

Let *AB* be such a numerical power of 3, an odd number by [2], from which the unit *DB* is subtracted, leaving the remainder *AD*, an even number. This number is divided in half at the point *C*, giving the number *AC*, which is equal to the sum of all the terms of the class of powers of 3 preceding *AB* by [5]. From the number *AC* is subtracted the number *EC* equal to the sum of the first two terms, whose sum is 4. There remains the number *AE*, equal to the sum of all the terms preceding *AB*, the first and second excluded. The number of other terms is 4 or a multiple of 4. It follows that the number *AE* is divisible by 8, since by [10] such a sum is divisible by 8. Let the number *FG* be the fourth part of the number *AE*. To *FG* is added the unit *GH*, which is a quarter of the number *EC*. It follows that the number *FH* is odd and is the fourth part of the number *AC* and also the fourth part of the number *CD*. It follows that the number *AD* is divisible by the number *FH*, which is odd, and therefore by [4] the number *AD* is not a power of 2, and that is what we wanted to prove. Therefore we have demonstrated the truth of the negation of the first proposition, and we know that if a unit is subtracted from any numerical power of 3, except those of the third or the second rank, the remainder is not a numerical power of 2.

It remains to demonstrate now the truth of the negation of the second proposition, to show that if a unit is added to any numerical power of 3, with the first term, the unit, and the second term excepted, the sum is not a numerical power of 2.

29. If the unit is added to any numerical power of 3 of odd rank, with the exception of the first, which is the unit, the sum is not a numerical power of 2.

Let *AB* be any numerical power of 3, of odd rank, to which is added the unit *BD*. From the same number is also subtracted the unit *CB*. There remains the even number *AC*, which is divided into two equal parts at the point *E*. This produces the number *AE*, equal to the sum of all the numerical powers of 3 preceding *AB*, by [5]. This sum is even by [5], since the number of terms preceding *AB* is even. To the number *AE*, which is even, is added the unit. The result is the odd number *EB*. But it is known that this result is half of the number *AD*, because the number *AE* is half of the number *AC* and the unit *CB* is half of the number *CD*. It follows that the number *AD* is divisible by an odd number, and therefore by [4], the number *AD* in question is not a numerical power of 2, and that is what we wished to prove.

30. If the unit is added to any numerical power of three of even rank, with the exception of the second rank, the sum is not a numerical power of 2.

Let *AB* be any numerical power of 3 of even rank, which is odd by [2], to which is added the unit *BD*. Let the number *CE* be the last of all the numbers of any rank preceding *AB*. It follows that the number *AB* is triple the number *CE*. From *CE* is subtracted the unit *FE* and let the number *AB* be decomposed as the sum of *AG*, *GH*, *HI*, such that each of these is equal to the number *CF*, and also the remainder *IB*, which is triple the number *FE*, the unit. It follows that *IB* is 3 and therefore *ID* is 4. Let the number *CF*, which is even, be divided into two equal parts at the point *L*. It follows, by [5], that the number *CL* is equal to the sum of all the numerical powers of 3 preceding the number *CE*, and by [7], this sum is even. It follows by [9] that the number *CL* is divisible by 4. It follows that the number *CF* is divisible by 8, because the ratio of *CL* to 4 is the same as the ratio of *CF* to 8. And since the number *AI* is divisible by *CF*, and the number *CF* is divisible by 8, therefore *AI* is divisible by 8. Therefore, by [11], the fourth part of the number *AI* is even. Let this fourth part be denoted by the even number *MN*. To this is added the unit *NO*, which is the fourth part of the number *ID*. It follows that the number *MO* is odd, and is the fourth part of the number *AD*. It follows that the number *AD* is divisible by the odd number *MO*. It follows by [4] that the number *AD* is not a numerical power of 2, and that is what we wanted to prove.

It appears therefore, in terms of this division of numbers into these three classes and their mutual comparison, that, with the exception of the cases noted above, two arbitrary numbers contained in the classes in question are neither equal to each other nor differ by only a unit. Consequently, any two of these numbers, whatever they are, differ by a number, and that is the principal object of our demonstration.

Thus concludes the treatise of Master Levi ben Gershon on the subject of harmonic numbers.

The next three sections contain ideas on amicable numbers (pairs of integers where the sum of divisors of one equals the other, and vice versa). There are a few discussions of amicable numbers in Hebrew medieval arithmetic. These discussions are undoubtedly based on Thābit ibn Qurra's theorem and proof of the primary calculation procedure enabling one to obtain as many pairs of amicable numbers as one wishes [Berggren, 2005, pp. 560–563]. In an elaborate study, [Lévy, 1996] outlines the references to Thābit's results in Hebrew medieval literature and how the Hebrew adaptations of his theorem circulated in Spain, Provence, and Italy in the fourteenth and fifteenth centuries.

5. QALONYMOS BEN QALONYMOS, *SEFER MELAKHIM* (*BOOK OF KINGS*)

This section was prepared by Naomi Aradi

Qalonymos ben Qalonymos ben Me'ir of Arles (1287–ca. 1329), holder of the title *Nasi* and known in Latin as Maestro Calo, was a prolific translator and an original scholar. His surviving original treatises criticize the ethics of his contemporaries, and his translations cover a wide variety of Arab scholarship. He traveled in the Catalan-Provençal area and worked for a time in Rome at the service of Robert d'Anjou, but even though he was a contemporary of Levi ben Gershon, we have no evidence that they ever met.

The earliest Hebrew treatment of amicable numbers is apparently a passage in a treatise that was identified by [Steinschneider, 1870] as the *Book of Kings* by Qalonymos ben Qalonymos. This composition, currently available in two manuscripts, is a compendium consisting of two sections. The first section is an arithmological summary that enumerates universal properties of the first ten numbers and numerical groupings of beings. In the second section specific properties of the numbers are listed, which are reflected in arithmetical statements and algebraic identities in Euclidean style. The discourse on amicable numbers is a part of a cluster of propositions in the second section, which, according to Lévy, corresponds to excerpts in the booklet of Thābit. In this passage the steps of the algorithm for finding pairs of amicable numbers are outlined briefly.

When we want to find amicable[33] numbers, as many as we wish, we set the numbers proceeding from one in a double proportion, including one. The numbers preceding the last number are summed up, including one $[1 + 2 + ... + 2^{n-1}]$. Then the penultimate number $[2^{n-1}]$ is added to the sum, and the number preceding the penultimate number $[2^{n-2}]$ is subtracted from the sum. The [two] numbers produced by the addition and subtraction [which equal $2^n - 1 + 2^{n-1}$ and $2^n - 1 - 2^{n-2}$] are primes, and neither equals two; if they are not, you proceed until prime numbers come out. Multiply the product of one by the other and by the penultimate number $[(2^n - 1 + 2^{n-1})(2^n - 1 - 2^{n-2})\,2^{n-1}]$, and save the result.

Add to the last number $[2^n]$ the fourth number away [in the list of powers of two, namely, 2^{n-3}] (or one, if 1 is the fourth), and multiply the sum by the last number. Subtract [1] from the product, so that the remainder is prime. Multiply this prime by the penultimate number $[2^{n-1}$, yielding $(2^n(2^n + 2^{n-3}) - 1)\,2^{n-1}]$. The result of the multiplication and the saved number each equal the sum of the parts of the other. The numbers produced in this way are called amicable.

[33] *Ne'ehavim*. Literally, "mutually beloved."

6. DON BENVENISTE BEN LAVI, *ENCYCLOPEDIA*

This section was prepared by Naomi Aradi

Another account of Thābit's theorem is found in a text that is said to be a Hebrew translation from the Arabic of the arithmetical part of an encyclopedia attributed to Abū al-Ṣalt (ca. 1068–1134) prepared by Don Benveniste ben Lavi in Saragossa in 1395. Yet [Lévy, 1996] argues that the main part of the Hebrew text (excluding the opening and ending) is in fact a translation from the Arabic of the arithmetical part of Ibn Sīnā's (Avicenna's) encyclopedia *al-Šifā*. The arithmetic section of the treatise covers virtually the same issues as those discussed in the *Arithmetic* of Nicomachus—types of numbers, ratios, proportions, progressions, and the like. The passage on amicable numbers appears as part of the discourse concerning the even-times-even numbers. First, the definition of the relationship between a pair of amicable numbers is illustrated by the example of the numbers 220 and 284. Then a short rendering of Thābit's general algorithm for finding pairs of amicable numbers is given.

The way of generating [amicable numbers] is by adding the even-times-even numbers together with one [$1 + 2 + \cdots + 2^{n-1}$]. If the sum is prime,[34] and on condition that when the last of them [2^{n-1}] is added, or the one before [2^{n-2}] subtracted, the result after the addition and after the subtraction is prime, then multiplying the result after the addition by the result after the subtraction and then multiplying the product by the last number added [2^{n-1}] is a number that is amicable to another [this number being $(2^n - 1 + 2^{n-1})(2^n - 1 - 2^{n-2})2^{n-1}$].

The number that is amicable to it is the number coming from adding the product of the sum of the above mentioned added and subtracted [numbers $2^n - 1 + 2^{n-1}$ and $2^n - 1 - 2^{n-2}$] by the last number added [2^{n-1}] to the first number that is amicable [$((2^n - 1 + 2^{n-1}) + (2^n - 1 - 2^{n-2}))2^{n-1} + (2^n - 1 + 2^{n-1})(2^n - 1 - 2^{n-2})2^{n-1}$]. These are amicable numbers.

In the margins of the text in [Oxford, Bodleian MS Heb. d. 3/4], a verbose illustration of the general algorithm is given by its application for finding the pair 220 and 284.

A slightly more verbose version of the same procedure appears on another page of the same codex [f. 45r] by the same hand with the following columns of numbers with a derivation of the pair 220 and 284 (compare with Aaron ben Isaac's version below). In anachronistic notation, $a_n = 2^n$, $b_n = 2^{n+1} - 1$, $c_n = b_n + a_n$.

c_n	a_n	b_n
	1	
5	2	3
11	4	7
23	8	15
47	16	31

In the margins of this page, the pair 17,296 and 18,416 is calculated briefly in a different hand as follows: $23 \times 47 = 1081$; $\times 16 = 17296$ and $16 \times 70 = 1120$; $+17296 = 18416$. This applies the algorithm above for $n = 5$.

[34] Note that this condition is redundant. However, the other necessary primality condition is missing.

7. AARON BEN ISAAC, *ARITHMETIC*

This section was prepared by Naomi Aradi

Lévy lists a few additional cases in which a very similar short formulation of Thābit's general algorithm is inserted in other Hebrew manuscripts. We complement Lévy's survey with the arithmetic of Aaron ben Isaac (see section I-2)—an elaborate procedure for finding pairs of amicable numbers.

The deliberation on amicable numbers is included in Aaron's discussion of the types of numbers. Unlike both examples above, Aaron describes the stages of the procedure in a practical manner, assisted by a written calculation table. The practical instructions are demonstrated by detailed examples in which two pairs of amicable numbers are calculated (220 and 284; 17,296 and 18,416). However, using his own guidelines Aaron found another pair of numbers (2024 and 2296) that are not amicable, a fact that he apparently failed to notice. Although Aaron must have drawn the basic algorithm from external sources, perhaps this error could testify to his attempt to implement the algorithm by self-instruction.

[The amicable numbers] are any two numbers, such that the [sum of the] whole parts of each equal the other. You find the first [of each pair] in one manner, and the partners in another. The foundation of both is the doubled numbers called even-times-even, from which we produce primes, and from the primes the amicable numbers.

We first order the doubled numbers, which are even-times-even, starting from one as far as you wish. Then take [the sum of the first two doubles] 1 and 2, which are 3, [and set it] against 2. Take [the sum of the] three doubles from 1 to 4, which are 7, [and set it] against the last double you took, which is 4. Take [the sum of the] 4 doubles from 1 to 8, which are 15, and set it against the last double, which is 8. Take [the sum of the] 5 doubles from 1 to 16, which are 31, [and set it] against 16. [Set] also 63 against 32, 127 against 64, [and 255 against 128.]

A corrupt passage follows, indicating that the third column is the difference between the number in the second column and the number above the corresponding number in the first column. The first column in the diagram below is a reconstruction. In anachronistic terms, $a_n = 2^n$, $b_n = 2^{n+1} - 1$, $c_n = b_n - a_{n-1}$.

a_n	b_n	c_n
2	3	2
4	7	5
8	15	11
16	31	23
32	63	47
64	127	95
128	255	191

From the primes in the third column [c_n] we get the first of each pair of amicable numbers.

The text also notes that 95 will not produce amicable numbers, because it is not prime. The following passage is not quite clear. It instructs the reader to take pairs of successive primes from the third column, but it does not indicate that they are to be multiplied by each other and by the corresponding number in the first column. The examples below, however, fill the gap. The formula is $c_{n+1} \times c_n \times a_n$.

The second [of each pair of amicable numbers] is produced differently. Take the double which follows the double from which you produced the first amicable number [a_3], and add one, which is the first in actu of even-times-evens [a_0]; to the second doubled number of the seconds [a_4] add 2 [a_1]; to the third [a_5], add 4 [a_2], and so on. Then multiply [a_{n+1} by $a_{n+1} + a_{n-2}$], and subtract one from the result. You get a prime. Multiply it by the doubled number preceding it in nature [a_n], and the result is the partner [yielding $(a_{n+1}\,(a_{n+1} + a_{n-2}) - 1)\,a_n$].

I hereby give six examples, three for the first amicable numbers, and three for their partners.

In a corrupt passage, Aaron indicates that the partner of an amicable number can also be derived by summing the parts of the amicable number.

First example: If we want to find the first amicable number of the first pair, we take what remains from 7 and 15 as you see in the third column [of the table], which are the prime numbers 5 and 11 [c_2 and c_3, respectively]. We multiply them by each other, yielding 55. We then multiply 55 by the double number 4 [a_2], yielding 220. This is the first amicable number of the first pair.

By summing the parts of 220, Aaron finds its partner, 284. The next examples for the first numbers of amicable pairs according to this method are $c_4 \times c_3 \times a_3 = 2024$ and $c_5 \times c_4 \times a_4 = 17296$. Aaron fails to note that the first of these does not work, because the construction of the partner involves a non-prime.

We wish to find the partner of the first amicable number we have. Here is how. Take the double number 8 [a_3], and multiply it by itself with 1 added, which is 9, yielding 72. Subtract 1 from 72, there remain 71, which is prime. Multiply it by the doubled number naturally preceding it, which is 4 [a_2], yielding 284. This is the partner of the first amicable number.

Aaron lists the parts of 284 and then produces the next two partners: $(a_4(a_4 + a_1) - 1)a_3 = 2296$ (which fails because the first factor, 287, is not prime) and $(a_5(a_5 + a_2) - 1)a_4 = 18416$.

III. MEASUREMENT AND PRACTICAL GEOMETRY

This section opens with two important treatises that summarize geometric knowledge in a semi-practical style. This does not mean that the authors or readers were ignorant of higher geometry, but that the treatises focused on unmotivated rules or on forms of reasoning somewhat more intuitive than the highbrow Euclidean style. *The Book of Measure* does not provide reasoning at all, but the problems solved are clearly not all practical (one is not likely to know the area of a rectangle and the sum of its side and diagonal without knowing its sides in a practical context). Bar Ḥiyya's *The Treatise on Measuring Areas and Volumes* does provide solid proofs, but it simplifies the Euclidean presentation considerably.

We continue with two discussions of measurement in religious context, which show the relevance of practical geometric reasoning for religious scholars. To conclude, we quote from Levi ben Gershon's discussion of iterative linear interpolation and trigonometry as presented for application in astronomy.

1. ABRAHAM IBN EZRA (?), *SEFER HAMIDOT* (*THE BOOK OF MEASURE*)

This treatise came down to us in Hebrew and in Latin translation. The only surviving Hebrew manuscript attributes it to Ibn Ezra (see section I-1), but this attribution is not considered certain. The treatise opens with an unsystematic collection of notes on arithmetic, which, according to a conjecture in [Lévy and Burnett, 2006], may have been an early draft or collection of notes for Ibn Ezra's *The Book of Number*. It then goes on to treat geometry. The treatise devotes a chapter to measuring triangles, then one to quadrilaterals, another chapter to circles and trigonometry, another to measuring solids, and a final chapter to the use of the astrolabe for determining distances and heights.

The book provides hardly any proofs or arguments, but it summarizes techniques for obtaining geometrical measurements from geometrical givens. It contains no algebraic language and no arsenal of Euclidean building blocks to work with, but the procedures it includes seem to depend on both. [Lévy and Burnett, 2006] conjecture that the book was to be revised and improved, but was then abandoned when Ibn Ezra became acquainted with Bar Ḥiyya's work on geometry (see below).

Preliminary definitions (2.1–2.4)

Because all measurements depend on number, I shall indicate the principles.

The point is in the mind, not in the [drawn] figure. Between two points there is a line; this is "length." Between two lines there is breadth; this is "surface." Likewise, when [one considers] the vertical [dimension], there arises depth, and thus "body" [a solid].

Know that among measurements, there is sometimes a line, at other times a surface, at other times a body.

I shall start now with the measurements of surfaces. Although the [basic] measurement of the area is the square, I shall start by mentioning the triangular figures, because the triangular figure is the principle of all [rectilinear] figures and every figure returns to it.

Triangle measurement (3.1–3.26)

The triangular figure of which the [three] angles are acute and the [three] sides equal [i.e., the equilateral triangle].

Subtract from the square of the side its quarter; its root is the height. The area: multiply the height by half the side, or the side, taken as the base, by half the height. Or take of the square of the height five of its ninths and a fifth of its ninth. Or take of the square of the side its third and add to it its tenth.

Or add the three sides and take the half; observe by how much this half [of the perimeter] exceeds each side, and multiply the excess by itself—this [means] taking its square—and multiply by the first line, which is the excess of the half [of the perimeter], and then one gets a cube; multiply this cube by half the sides [the half-perimeter]; take the root of the product, and in this way [you have] the area.[1]

Problem: The area [of an equilateral triangle] is so much. How much is the side? Take the square of the area; multiply this square by three, take the root of the product and add to it its third; take the root of this sum, and this is the side.

[1] This is Heron's formula applied to an equilateral triangle.

Another: The area is so much. How much is the height? Take a third of the square of the area; take its root and the root of the root, and divide the area by it; and you will find the height.

. . .

The triangular figure of which the angles are acute, the base different [from the two sides], and the two sides equal [i.e., an isosceles triangle]. From the square of one side, subtract the square of half the base; the root of the remainder is the height. The area: multiply the height by half the base, or vice versa.

Problem: The area [of the isosceles triangle] is so much, the height is so much. How much is the base? Divide the area by half the height.

. . .

Another: The base is so much, the height is so much. How much is the side? Add the square of the height to the square of half the base, and take the root of the sum.

. . .

Another: The area is so much, the base exceeds the height by so much. How much is the base and how much is the height? Double the area and take the square of half the excess; add them and take the root of the sum; add to the result half the excess; you will find the base. In the same way, subtract half the excess from the root; you will then find the height.

The triangular figure of which the angles are acute and the sides are different [in length, i.e., a scalene triangle].

Make the base whichever side you wish; take the squares of the other two sides; subtract the smaller from the larger; divide half the remainder by the base; add the result [i.e., the quotient] to half the base. You will then find the larger segment of the fall.[2] If you subtract half the result from half the base, you will then find the smaller segment of the fall. Take the square of the larger segment of the fall; subtract from the square of the longer side [i.e., the longer of the two sides that are not the base]; the root of the remainder is the height [in respect to the base]. Or subtract the root of the smaller segment of the fall from the square of the shorter side; take the root of the number [obtained]; this is [also] the height. It is always the same. The area: by multiplying half the height by the base, or vice versa.

The triangular figure with one right angle and two acute angles.

The area of this figure: multiply half of one short side by the other short side, the two short [sides] being opposite the long side, which is the diagonal.

Problem: [In a right-angled triangle] one of the two sides is so much, the other so much, these two being the short [sides]. How much is the long side? Add the squares of the two [short sides]; the root is the long side.

Another: The area is so much, one of the short [sides] exceeds the other by so much. How much is each side? Double the area; take the square of half the excess; take the

[2]The "segments of the fall" (*mekhona*) are the parts of the base as divided by the perpendicular from the opposite vertex.

root of their sum; add half the excess to the root; you will have one of the [short] sides; and if you subtract from the root [half the excess], you will have the other [short side]. Add the square of this to the square of that; the root of the sum will be the long side [hypotenuse].

. . .

Square and rectangle measurements (4.1–4.25)

The quadrilateral figure of which the length is the same as the breadth and all the angles are right angles [= the square].

Take the square of one side. This is the area.

Problem: We have multiplied the area by so much; the result divided by one side gives so much; what is the length of each side? Divide the result by the number by which the area had been multiplied.

Another: The diagonal is so much. How much is the side? Take the root of half the square of the diagonal.

. . .

Another: We have added the sides and the area; this gives so much. How much is the side? Take the square of half the number of all the sides [= 4] and add it to the sum [of the area plus the four sides]; subtract from the root of this result half the number of the sides [= 2].

Another: We have subtracted the sides from the area. The remainder is so much. How much is the side? Add the square of half [the number] of the sides to the remainder; take its root and add to it half the number of the sides.

. . .

The quadrilateral figure of which the length is greater than the width and the angles are right angles [=the rectangle].

[To obtain] the area, multiply the side of the length by the side of the breadth.

. . .

Another: The area is so much, and the longer side exceeds the shorter by so much. How much is each of the sides? Take half the excess [of the longer side over the shorter]; then take its square; then add this to the area and take the root of the result and add to it half the excess. The longer side is so much. If, in the same way, you subtract [half the excess] you will find the shorter side.

Another: We have added the longer side to the diagonal; this comes to so much; the shorter [side] comes to so much. How much is the longer side and how much the diagonal? Take the square of the sum [of the diagonal and length]; subtract from it the square of the known side; take half the remainder, and divide it by the initial sum; you will find the longer side. If you subtract it [the longer side] from the sum mentioned, you will find the diagonal.

Or take the square of the shorter side, divide it by the initial sum [of the diagonal and length] and add to the result [the quotient] the initial sum; half [this new sum] is the diagonal, from which you will know the [longer] side.

Another: We have added the two sides; they come to so much; the diagonal is so much. How much is each of the sides? Subtract from the square of the sum [of the sides] the square of the diagonal; half of the remainder is the area.[3]

. . .

Another: We have added the two sides and they exceed the diagonal by 4, and the area is 48. How much is the diagonal and how much each side? Take the square of 4 and subtract it from twice the area; divide half the remainder by 4 and you will find the diagonal. One will proceed in the same manner as before when the question is: "The area is so much, the diagonal is so much. How much is each side?"

Another similar [problem]. We have subtracted [all] the sides from the area and the remainder is 20, and one of the sides exceeds the other by 2. How much is each side? Double the 2, add it to the 20; take the number of sides—which is 4—and subtract from this the excess [of the longer side over the shorter]; the remainder is 2; take half of it, then [take] its square, which is one; add it to 24, which makes 25, whose root is 5. Add to it half of 2; so much is the shorter side [= 6].[4]

. . .

Another: We have added [all] the sides and the area, and they come to 76, and one of the sides exceeds the other by 2. How much is each side? Double the 2 and subtract it from the number 76; one knows that the number of the sides is 4; add to them 2, which is the excess of the length over the breadth; this makes 6; take the square of half this amount; add it to 72, and take its root [= 9]; subtract from it 3, which is half 6; thus one has the shorter side.[5]

. . .

Parallelogram measurements (4.44–4.47)

The quadrilateral figure, similar to that which was mentioned above, of which the [opposite] angles are equal, but of which two sides are equal and the other two [equal but] different [from the first two, i.e., a parallelogram that is not a rhombus].

Problem: The longer side is 9, the shorter 5; the longer diagonal is the root of 160 and the shorter one, the root of 52. [What is the area?] Take the square of each of the two sides that differ from each other and add them; this makes 106. Subtract this from the square of the longer diagonal; the difference is 54. Take half of this, which is 27. Divide this by the longer side, which gives 3. Take the square of this, which is 9, and take the square of the shorter side, i.e., 25. Subtract the smaller from the larger. There remains 16,

[3]The text does not contain the completion of the solution, although if one knows the sum of the sides and their product, there is a standard method for finding the two sides.

[4]Although the answer is correct, the method is not at all clear. But the author appears to be reducing the problem to one already solved, namely, determining the sides when the area and the difference of the sides are known. In fact, the instruction to "double the 2 and add it to the 20" converts the problem to the system we would write as $x(y-4) = 24$ and $x = (y-4) + 2$, for which the standard algorithm presented above applies to determine x.

[5]As in the previous problem, the instructions seem to convert the problem to the system $x(y+4) = 72$, $(y+4) = x+6$, to which a standard algorithm applies.

of which the root is 4. If one multiplies this by 9, which is the longer side, it comes to 36. This is the area.[6]

Or take the squares of the two sides and add them. They make 106. Subtract them from the square of the longer diagonal; the remainder is 54. Take half of this, i.e., 27; divide this by 5, i.e., the shorter side [$= 5\frac{2}{5}$]; take the square of the result [$= 29\frac{4}{25}$] and also the square of 9, which we already have [the longer side], and subtract the smaller from the larger [$= 51\frac{21}{25}$]. Take the root of the remainder, i.e., $7\frac{1}{5}$, and multiply this by 5, which is the shorter side. This makes 36, which is, again, the area.

. . .

Circles and circular arcs (5.1–5.27)

The circle.

The area of the circle. You subtract from the square of the diameter its seventh and half of its seventh. Or multiply the square of the diameter by 11 and divide by 14. Or multiply half the diameter by half the circumference. Or a quarter of the diameter by the whole circumference. Or the diameter by a quarter of the circumference.

If you know the diameter, multiply by $3\frac{1}{7}$. You will always find the circumference. Or multiply the diameter by 22 and divide by 7.

If you know the circumference, multiply it by 7 and divide by 22. You will have the diameter.

. . .

Problem: We have subtracted the diameter from the circumference, the remainder is so much. How much is the circumference? And how much is the diameter? Divide the remainder by $2\frac{1}{7}$ and you will find the diameter. Or multiply by 22 and divide by 15 and you will find the circumference. . . .

The arc [or the circular segment].

If one is dealing with a semicircle, its area is like that of half a circle. If it is smaller or larger [than a semicircle], you must know the diameter of the circle from which the circular segment has been cut, and the length of the chord of the arc and of the sagitta.[7] When you know two of these [three] elements, you can determine the third.

Problem: The chord is 8, the diameter, 10. How much is the sagitta? Subtract from the square of half the diameter the square of half the chord; take the root of the remainder, and subtract it from half the diameter; you will find the sagitta [= 2].[8] . . .

[6] To find the area of the parallelogram, one has to find the height. The procedure here and below is similar to that of finding the height and "segment of the fall" in a scalene triangle, described above, except that in this case, the triangle is obtuse.

[7] The sagitta (arrow) is the segment of the diameter perpendicular to the chord lying inside the circular segment.

[8] If d is the diameter, s the sagitta, and c the chord, the Pythagorean Theorem gives $\left(\frac{d}{2}\right)^2 = \left(\frac{d}{2} - s\right)^2 + \left(\frac{c}{2}\right)^2$, which is the identity used here.

Another: The sagitta is 2, the chord, 8. How much is the diameter? Take the square of half the chord; divide it by the sagitta, and add the result to the sagitta; you have the diameter [=10].[9]

. . .

[Table of Sines]

The arc is 90 [degrees] and the Sine 60.[10] If you take the root of half the square of 60, you will find the Sine of the arc of 45 [degrees]. If you subtract the square of half the Sine [of 90 degrees] from its [whole] square, and take the root of the remainder, you will have the Sine of the arc of 60.

The Sine of one degree amounts to one unit and three minutes [1;3].
The Sine of an arc of 5 [degrees] is 5 and 14 minutes [5;14].
The Sine of an arc of 10 is also 10 and 25 minutes [10;25].
The Sine of an arc of 15 is 15 and 32 minutes [15;32].
The Sine of an arc of 20 is also 20 and 31 minutes [20;31].
The Sine of an arc of 25 is also 25 and 21 minutes [25;21].
The Sine of an arc of 30 is also 30 with no subdivisions [30;0].
The Sine of an arc of 35 is 34 and 25 minutes [34;25].
The Sine of an arc of 40 is 38 and 34 minutes [38;34].
The Sine of an arc of 45 is 42 and 25 minutes and 35 seconds and 4 thirds [42;25,35,4].

If you double it and turn the degrees into minutes, the minutes into seconds, the seconds into thirds and the thirds into quarters, you will then find the [square] root of 2 with great precision.[11]

And the Sine of an arc of 50 is 45 and 58 minutes [45;58].[12]
And the Sine of an arc of 60 is 51 degrees and 58 minutes [51;58].
The Sine of an arc of 65 is 54 and 23 minutes [54;23].
The Sine of an arc of 70 is 56 and 32 minutes [56;32].
The Sine of an arc of 75 is 55[13] and 58 minutes [55;58].
The Sine of an arc of 80 is 59 and 8 minutes [59;8].
The Sine of an arc of 85 is also 59 and 46 minutes [59;46].

[9] *Elements* III. 35 states that the product of the two segments of the diameter determined by the chord is equal to the square on half the chord. Thus, if d is the diameter, s the sagitta, and c the chord, we have $s\,(d - s) = \left(\frac{c}{2}\right)^2$.

[10] The Sine of an arc in medieval literature is half the chord of double the angle in a circle of given radius. Thus the Sine of a 90° arc is equal to the radius, which implies that for this table, the radius is 60; this value is often called the "whole Sine." The two calculations in this paragraph are straightforward.

[11] The Sine of a 45° arc is $60\frac{\sqrt{2}}{2}$. Doubling this and changing degrees to minutes, etc., will therefore give $\sqrt{2}$ in sexagesimal form $1;24,51,10,8 = 1.41421356\ldots$ in decimal form.

[12] The value for 55 is missing.

[13] This should read 57.

If you take [an arc of] more than 60 [degrees], consider what is needed to complete an arc of 90 [degrees]. Subtract the square of its Sine from the square of the whole Sine, and take the root of the remainder. You will find what has [already] been indicated.

If you want to know the Sine of an arc that does not appear [in the table], you work it out proportionally: look at what represents [the difference of] 5 [degrees] of arc for the Sine, and consider the proportion that you have to add to the number you already have.[14]

Pyramid measurement (6.11–6.13)

A solid of which the top is square and different from the base, whose breadth is equal to its length [a truncated right pyramid or frustum].

[The side of the square] of the top is 2, that of the base, 4, the height, 10, is the perpendicular. Subtract the top from the base; observe what is the ratio between the remainder and the top—they are equal [2]; one can, then, complete this figure to make it 10, which is like the [original] height [to make a complete pyramid with the height 20]. The volume [of the pyramid thus obtained] in completing the height [of the truncated pyramid] will be $106\frac{2}{3}$ [$= \frac{1}{3} \times 16 \times 20$]. Let us then make the top the base [of a small pyramid]. The volume of this complement will be $13\frac{1}{3}$ [$= \frac{1}{3} \times 4 \times 10$], which one subtracts from that which has been given above. The remainder is $93\frac{1}{3}$. Or let us add the square of the top to the square of the base; let us multiply the side of the base by the side of the top; let us add this result to the squares indicated; and let us multiply the sum of them all by a third of the height; one finds the volume.[15]

Measurements of heights and distances using an astrolabe quadrant (7.1–7.13)

To measure the height of a hill, a tree or a tower.[16]

Set the alidade [lit: "line"] of the astrolabe [lit: "instrument of brass"] [in the quadrant] on which you have graduated the degrees of the sun[17] at 45 degrees [Fig. III-1-1]. Go forwards or backwards until you can see the top [of the object to be measured] in the alignment of the two holes of the alidade. Then measure the distance which separates your feet from the foot of the tree, the hill or the tower. Add to it the height between your eyes and the ground. You obtain the measurement that you were looking for [Fig. III-1-2]. You will measure it in cubits, palms, or any other unit of measure.

Or when you have seen the top [of the object to be measured], turn and, without moving from the place where you are, set the alidade on 45 degrees in the lower quadrant, ... and observe through the hole the point on the ground; then measure the distance that separates that point from the foot of the tree or the tower; you will thus find what you are looking for [Fig. III-1-2].

[14]The procedure suggested here seems to be linear interpolation.

[15]This procedure is in the *Moscow Mathematical Papyrus* (see [Imhausen, 2007, p. 33]).

[16]The descriptions that follow are the standard ways used in the Medieval period to measure unknown heights and distances. See sections I-4-5 and II-5-5 in Chapter 1 for other examples.

[17]Since the astrolabe is typically used to measure the altitude of heavenly bodies, including the sun, in degrees, the author is calling the markings on the circle "the degrees of the sun." In this case, he is just instructing us to set the alidade at 45° so that we obtain an isosceles right triangle.

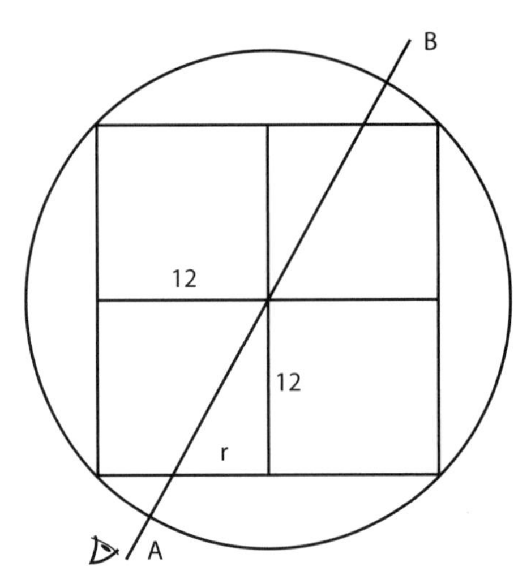

Fig. III-1-1. Diagram of astrolabe. *AB* represents the alidade, and r, measured on a scale of 12, is the gnomonic *umbra recta*.

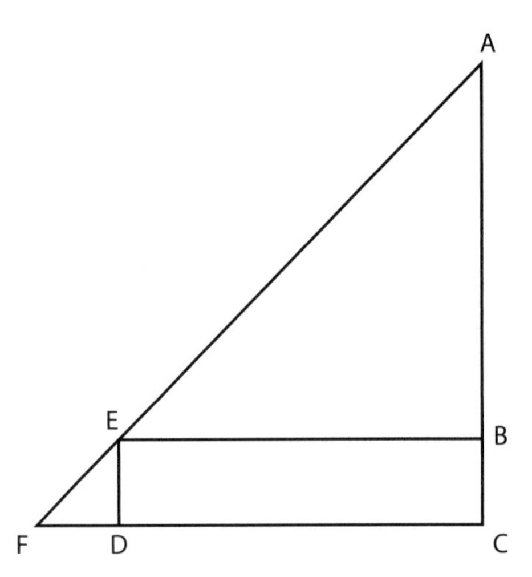

Fig. III-1-2. *A* is the top of the hill, *E* is the eye of the observer, and angle $AEB = 45°$. Thus the height of the hill is $AC = AB + BC = CD + ED$. Alternatively, the astrolabe can be turned around to locate *F* and obtain *AC* as $CD + DF$.

Or suspend the instrument from your right hand and observe how the top [of the object to be measured] looks; then read off [on the astrolabe] the gnomonic *umbra recta*; determine the ratio of 12 to the "number of the umbra" [r]; observe the same ratio between [the distance between the top and the horizontal line issuing from your eyes, h, and] the distance that separates your feet from the place that is sought [d]; add [to the distance obtained by this proportion] the height [m] between your eyes and the ground [Fig. III-1-3].

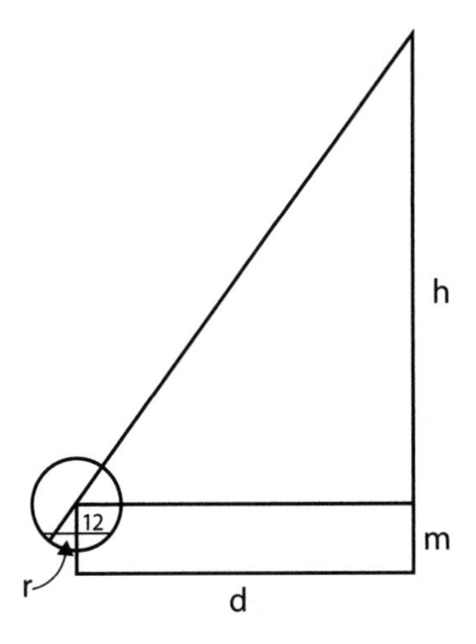

Fig. III-1-3. $12{:}r = h{:}d$, so $h = 12d/r$ and the height of the hill is $h + m$.

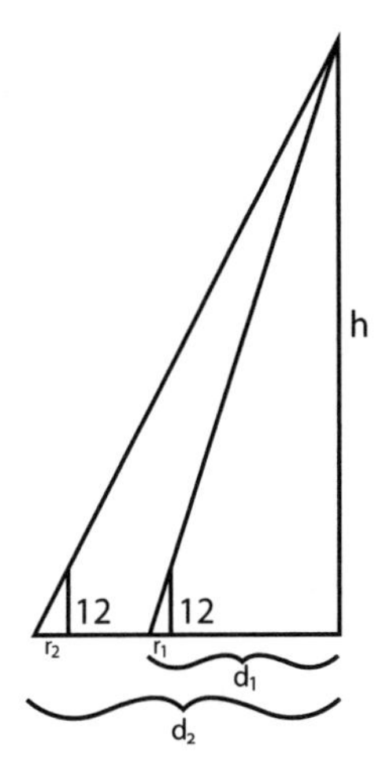

Fig. III-1-4. $h = 12d_1/r_1 = 12d_2/r_2$. If $\Delta = d_2 - d_1$, then $h = 12\Delta/(r_2 - r_1)$, as stated. Also, $d_i = r_i\Delta/(r_2 - r_1)$ for $i = 1, 2$.

If you cannot reach the foot of the tower, suspend the instrument from your right hand and read off the degrees of the height [of the tower] from any place, and know the gnomonic umbra; then, go forwards or backwards and for a second time take the degrees of the top, and know the gnomonic umbra, and know how many graduations separate the two umbras [measured on the astrolabe]. Then multiply by 12 the cubits or palms that you have passed through between the two positions, and divide the product by the difference between the two umbras; add to this the height between your eyes and the ground; such is the measurement that you are looking for [Fig. III-1-4].

If you want to know at what distance you were from the base of the object to be measured in the first position, multiply the cubits that you have passed through in going

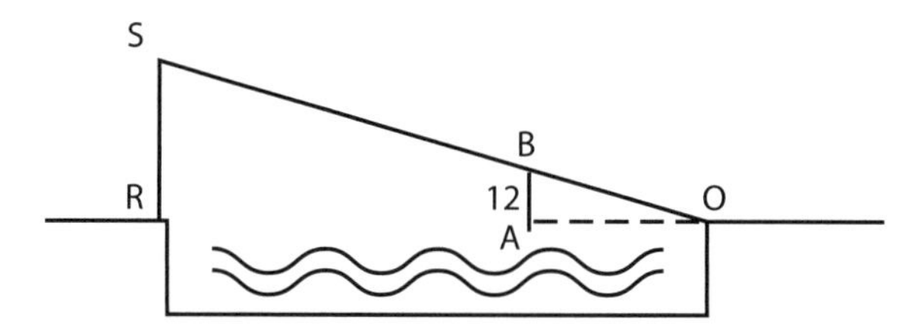

Fig. III-1-5. *OA*:*AB* is the ratio of the gnomonic umbra to 12. This is the same as the ratio *OR*:*SR*, where *SR* is the height of your body.

forwards or backwards by the first umbra, and divide the product by the difference between the two umbras. The product is the distance [of the foot of the tower] to where you were in the first position. Similarly, multiply the cubits by the second umbra and divide the product by the difference; you will find the distance relative to the second position.

. . .

To measure the breadth of a river, stand on the bank and suspend the instrument from your left hand; raise and lower [the astrolabe] so that you can see the other bank of the river; then read off the gnomonic umbra; compare the ratio of the measurement to 12. Having obtained the result, multiply this ratio by the height of your body; such is the measure [of the breadth] of the river [Fig. III-1-5].

. . .

2. ABRAHAM BAR ḤIYYA, *ḤIBUR HAMESHIḤA VEHATISHBORET* (*THE TREATISE ON MEASURING AREAS AND VOLUMES*)

Abraham bar Ḥiyya (ca. 1065–1145) was based in Barcelona, where he worked as a scholar and community leader.[18] His Jewish title was *nasi* (honorary leader), and Arabic title *ṣāḥib ash-shurṭa* (head of the guard, transliterated in Latin as "Savasorda"). He wrote on mathematics, astronomy, astrology, and philosophy, and he is recognized as being the first Jewish scholar in the Arabic-speaking world to write on science in Hebrew. This choice of language was, at least in part, due to the Jewish community's lack of access to the Arab language in Provence, where he visited. His work includes translations from Arabic to Hebrew, and he collaborated with Plato of Tivoli on translations into Latin as well.

Bar Ḥiyya's mathematical work includes *The Foundations of Wisdom and Tower of Faith*, an encyclopedia of which only the introduction and mathematical parts survive,[19] and *The Treatise on Measuring Areas and Volumes*, from which the following selection is taken. This book was partly translated into Latin in 1145 by Plato of Tivoli under the title *Liber Embadorum*, and made an impact on European scholarship.[20]

[18] For general information, see [Steinschneider, 1893–1901; Lévy, 2001, 2008; Langermann, 2007].

[19] A Hebrew-Catalan edition was published in [Millás Vallicrosa, 1952]; subsequent research is available in [Levey 1952, 1954; Rubio, 2000].

[20] See [Curtze, 1902] for an edition of the Latin version with German translation as well as [Millás Vallicrosa, 1931] for an edition in Catalan.

The Treatise on Measuring Areas and Volumes opens with a motivational introduction, pointing out the value of geometry for secular and holy affairs. The first book then gives some basic definitions and theorems that serve as building blocks for the rest of the book. These are mostly theorems from Euclid's *Elements*, Books II and III, along with a basic discussion of similarity and congruence. The treatment is usually more intuitive than Euclid's and is accompanied by arithmetic examples.

Book II is the core of the work. Its first part addresses measurements of squares, rectangles, and rhomboids (deriving their areas, sides, diagonals, etc. from one another), and includes a geometric treatment of quadratic problems. The second part deals with triangles, and the third with general quadrilaterals. Part four deals with circle measurements and includes a trigonometric table for calculating arcs from chords. The fifth part studies the measurement of polygons by triangulation; it ends with some practical notes on measuring sloping and hilly lands.

Book III is a simplified version of Euclid's book on equal division of plane areas. It is one of the few witnesses to this lost work.[21] Book IV provides a brief treatment of solids. The final section deals with practical tips for land measurement, including the use of an instrument shaped as a right angle, and concludes by repeating the warning against simple but false rules of thumb.

This work is not a fully scholarly geometry exposition but a compromise between an introduction to abstract geometry and a measurement manual. It provides a good intuitive introduction to geometrical reasoning and some elementary tips for land measurements (excluding the use of protractors or astrolabes). As the selection below shows, it diverges in several ways from the scholarly Arabic tradition and is probably closer to the popular *muʿāmalāt* tradition, whose Western branch does not survive in Arabic manuscripts.

Motivations for studying geometry (Introduction)

This section discusses the secular and holy motivations for studying geometry, and it warns against rule-of-thumb approximations that may result in unjust distributions of property.

[The scriptures say] "I the Lord am your God, instructing you for your own benefit, guiding you in the way you should go,"[22] that is, instructing you in whatever is useful for you, and guiding you on the way you follow, the way of the Torah. From which you learn that any craft and branch of wisdom that benefit man in worldly and holy matters are worthy of being studied and practiced.

I have seen that arithmetic and geometry are such branches of wisdom, and are useful for many tasks involved in the laws and commandments of the Torah. We found many scriptures that require them, such as "In buying from your neighbor, you shall deduct only for the number of years since the jubilee," and "the more such years, the higher the price you pay; the fewer such years, the lower the price," followed by: "Do not wrong one another, but fear your God."[23] But no man can calculate precisely without falsification unless he learn arithmetic. . . . Moreover, the Torah requires geometry in measuring and

[21] See [Archibald, 1915] for a reconstruction of this treatise and [Hogendijk, 1993] for an edition of the Arabic version.

[22] Isaiah 48:17.

[23] Leviticus 25:15–17.

dividing land, in Sabbath enclosures and other commandments.[24] ... But he who has no knowledge and practice in geometry cannot measure and divide land truly and justly without falsification. ... It suffices to note that the blessed God prides himself in this wisdom, as is written: "He stood, and measured the earth"[25] and "Who measured the waters with the hollow of His hand, and gauged the skies with a span."[26] So you see from these writings that the blessed God created his world in well founded and weighed out measurement and proportion. And a man must be like his creator with all his might to win praise, as all scholars agree, so from all this you see the dignity of these branches of wisdom. He who practices them does not practice something vain, but something useful for worldly and holy matters.

As I see it, Arithmetic, which is useful for worldly matters and crafts as well as for the practice of many commandments, is not difficult to understand, and most people understand it somewhat and practice it, so one does not need to write about it in the holy tongue. Geometry is also as useful for as many matters as arithmetic in worldly matters and commandments from the Torah, but is difficult to understand, and is puzzling to most people, so one has to study and interpret it for land measurement and division between heirs and partners, so much so that no one can measure and divide land rightfully and truthfully unless they depend on this wisdom.

I have seen that most contemporary scholars in *Ṣarfat*[27] are not skillful in measuring land and do not divide it cleverly. They severely belittle these matters, and divide land between heirs and partners by estimate and exaggeration, and are thus guilty of sin. ... Their calculation might mete out a quarter to the owner of a third, and a third to the owner of a quarter, and there is no greater theft and falsification.

If one says that our fathers cared little for calculations and precision, as they say: "If the side of a square is a cubit, its diagonal is one and two fifths,"[28] whereas in such a square, with equal sides and right angles, for each cubit in the sides the diagonal, calculated precisely, is a cubit and two fifths of a cubit and one in seventy parts of a cubit and a small excess [Bar Ḥiyya goes on to quote further Talmudic examples stating that the ratio between the circumference and diameter of a circle is 3, that the excess of the area of a square circumscribing a circle is 1/4, and that the excess of the area of a circle circumscribing a square is 1/3]; so one of our contemporaries, who belittles the division and measurement of land, may wrongfully exclaim: "from the words of our fathers we learn that calculations need not be too precise; we should learn from them, follow them, and calculate by estimate and approximation without study and precision. So you cannot protest and refute us."

I will reply to him and say: "God forbid! Our fathers did not allow us to dismiss calculations, nor steal from heirs, nor give any of them more or less than their fair share,

[24]The text here quotes Numbers 35:5, Deuteronomy 21:2, and Numbers 26:53–54.

[25]Habakkuk 3:6. Here, exceptionally, we use the Bible of King James, as the Jewish Publication Society version interprets the root *mdd* differently, translating the verse as "When He stands, He makes the earth shake," which is not Bar Ḥiyya's reading.

[26]Isaiah 40:12.

[27]The term *Ṣarfat* used here referred at the time to various areas of the Catalan-Occitan region.

[28]Talmud, *Suka* 8a.

as you sinfully do today. Even though they calculated the diagonal of the square and the diameter and circumference of the circle and the area between the circle and the square imprecisely, as you say, nevertheless they warned us and gave us strict orders against stealing and falsifying in measuring land.[29] . . .

All the calculations that you mentioned, where our fathers were imprecise, are completely harmless. Indeed, the approximation of the diagonal of the square is harmless in measuring and dividing land, because you don't need the diagonal when measuring a square. And the calculation of the excess of a square over a circle and vice versa and the diameter and circumference of a circle are required only for Sabbath enclosures and for keeping different kinds of seeds and crops apart. Now anything involving this [Talmudic] approximation renders the commandments obeyed more strictly, rather than more leniently, and does not harm anybody's property. Moreover, in most places where our fathers approximate imprecisely, they make a note, such as "he only gave an approximate figure; and in this case it is in the direction of stringency."[30] . . . This is unlike contemporary errors, which cause harm and great losses to people's property.

Geometric building blocks (from Book I)

The following is a version of Euclid's *Elements* II.4. But note that the diagram is different from that of Euclid, that the proof relies on elementary cut-and-paste rather than on Euclid's bisection of a parallelogram by its diagonal, and that the entire discussion is accompanied by an arithmetical interpretation. This approach represents a line of transmission or pedagogical tradition different from the scholarly Euclidean one, and it characterizes most theorems in the first book of the treatise. The results in this book serve as "building blocks" for the solution of problems in the subsequent books.

I say that in any line divided into two parts, the square[31] of the line by itself equals the square of each part by itself and double the rectangle of one of its parts by the other. I give you an arithmetical example. Let this line be 12 cubits in magnitude, and divide it into two parts, 7 and 5. The square of the whole line by itself is 144. This number equals the square of 7 by itself, which is 49, and the square of 5 by itself, which is 25 (together they make 74), and double the rectangle of 7 by 5, which is 70.

To give an example with a diagram, set the line *AB* of magnitude 12, and divide it at *E* [Fig. III-2-1]. The large part, *AE*, is 7 cubits, and the small part, *EB*, is 5 cubits. We construct the square of the line by itself, making *ABCD*, a square quadrilateral with equal sides and right angles (as all its sides are of length 12) and with area 144 cubit by cubit. We construct the square of the large part by itself, which is the rectangle *AEGH*, and the small by itself, which is *EBIJ*. To complete the large square there remain the rectangle

[29] The text here quotes Leviticus 19:15 and the Talmud, *Baba Meṣiʿa* 107b and 61b.

[30] The text quotes here the Talmud, *Suka* 8a.

[31] The Hebrew uses two terms: *merubaʿ*, which is a general term for a quadrilateral, and *ribuaʿ*, which is the noun for the action of making a *merubaʿ*. Both terms are used here in geometric as well as arithmetic contexts (numerical products and squares). A strict translation would use "quadrilateral" or "the quadrilateral of" throughout. For the benefit of the reader, however, I used "quadrilateral," "rectangle," and "square," according to the context. But note that this choice subtracts from the arithmetic overtones of the book.

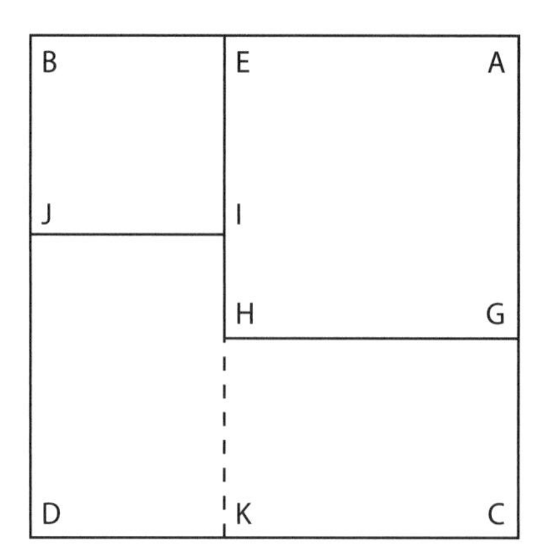

Fig. III-2-1.

GHCK and the rectangle *IJDK*. And we know that both these rectangles are the rectangles of one part by the other, which is 5 times 7. Indeed, the line *AGC* equals the line *AEB*, and the line *AG* equals the line *AE*, which is the large part, leaving the line *GC* that equals the line *EB*. Moreover, the line *CKD* equals the line *AEB*, and the line *CK* equals the line *AE*. So we find that the rectangle *CGHK* is the rectangle of the large part by the small. Further, the line *KD* equals the line *EB*, and the line *JD* equals the line *AE*, which is also the large part by the small. We find that the square of the large part, which is *AEGH*, and the square of the small part, which is *EBIJ*, and twice the rectangle of the large part by the small, which are the rectangles *GHCK* and *IJKD*, complete the large rectangle *ABCD*, as you see in this diagram.

Geometric presentation of quadratic equations (from Book II)

Interpreted algebraically, this section solves the standard compound quadratic equations and justifies the solutions geometrically. One should, however, note its divergence from the Arabic algebraic tradition. First, these problems are not set here as standard problems to which other problems should be reduced. They appear without any distinction among a list of other geometric problems. The choice of 4 (the number of sides) as the coefficient of the linear term might make them appear less generalizable than they actually are. Moreover, the language is entirely geometric, without any special term to designate the unknown side. Finally, the Arabic tradition follows a rather strict normalization of the canonical quadratic problems, where only additive terms are equated; here two problems equate the difference between the area and the sides with a given number. As above, this suggests a line of transmission or pedagogical tradition that diverges from the Arabic scholarly canon.

A square quadrilateral that you take away from the number of its area the number of its four sides, and [you] are left with 21 cubits of its area: what is the area and what is the number of each side of the square?

Answer: Divide the number of the sides, which is four, into two. Multiply the two by itself, which is 4. Add this number to the given number that's left over from the square, and the total is 25. Find the root of 25, which is 5. Add half the sides, which is 2, so the total is 7.

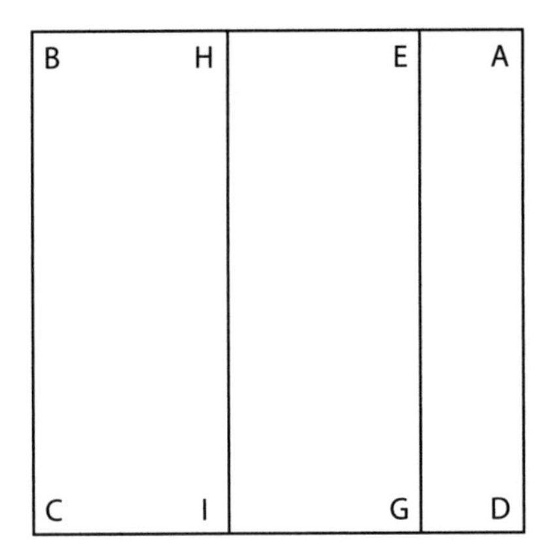

Fig. III-2-2.

This is the side of the square, and its area is 49. He who posed the question subtracted from the area, which is 49, the number of the four sides, each of which is 7 and all four 28, leaving from the square 21, as he told you.

If you want the proof that your answer to him is correct, let this square be *ABCD* with equal sides. It is known that each side has more than four cubits, because he who posed the question said that we subtracted from the square all four sides [considered as a rectangle that has] 4 cubits in width and in length equal to the length of the square, leaving so and so;[32] if the side had no more than 4, he could not have subtracted 4 sides. We therefore subtract from the line *AB* a line whose length has 4, which is the line *BE*, and from the opposite line *CD* a line whose length has 4, which is the line *CG* [Fig. III-2-2]. We make a line from the point *E* to the point *G*, and then divide the line *BE* into two equal parts at the point *H*. The lines *BH* and *HE* are equal, and each has 2 cubits in length. You see in this square that the rectangle marked by *BEGC* is the four sides of the square joined together, as this rectangle is the line *BC* times the line *BE*, which is four, and the line *BC* is the side of the square; we count it 4 times, which is the number of the four sides. So you see that the rectangle *BCEG* is the four sides of the square. When you subtract it from the square, there remains the rectangle *ADEG*, which is known to be 21, as the number left to [him] who posed the question.

Now observe that the line marked *BE* is divided into two equal parts *BH* and *HE*, together with another line, *EA*. But we have already told you[33] that for any line divided into two equal parts and joined with another line, the rectangle of the entire line with the joined line by the joined line, and the square of half the first line, added together, equal the square of half the line with the joined line both together by themselves. Therefore the rectangle of the line *BA* by the line *AE* and the square of the line *EH* by itself together equal the square of the line *HA*. The rectangle of the line *AB* by the line *EA* is the rectangle *ADEG*, because its one side *AD* equals the side of the square and its other side is *EA*, which is the line joined with the line *EB*. This rectangle is 21, and if you add to it the square of the

[32]Note that while the question is set in terms of subtracting numbers, the solution turns the subtracted 4 sides into a rectangle whose sides are 4 and the side of the given square.

[33]Bar Ḥiyya's Book I, §29, equivalent to *Elements* II.6.

line *EH*, which I know to be 4, the total will be 25. Its root is 5, and it equals the square of *HA*. You see that the line *HA* is the root of the square 25. Add *BH*, which is 2 cubits, making a total of 7, which is the line *BA*. And so the rectangle *ABCD* is 49 cubits, as I showed you in this diagram.

If he says: "A square that you add the number of its four sides to the number of its area, making in total 77, how much is this rectangle?"

In this question take half the number of the sides, which is two. Multiply it by itself, making in total 4. Add this to the number that he gave you, 77, making 81. Take the root of this number, which is 9. Subtract from it half the number of sides that you added, and you are left with 7. This is the side of the square, and its area is 49.

[Translation of the proof is omitted.]

A square that you take away its area from the number of its four sides, and are left with three.

In the answer to this question you divide the number of the sides in two. Their half is 2 and their square is 4. Take away the three that you were left with, leaving 1, whose root is 1. Subtract it from half the sides, and you are left with 1, which is the side of the square, or add the root of the 1 you are left with to half the sides, making 3, which are also the side of the square. It can be one and can be 3, as this question has two solutions

[Translation of the proof is omitted.]

Measurements of rectangles (from Book II)

This example shows how quadratic problems are solved by reduction to elementary geometric identities (typically from book II of the *Elements*, which may be interpreted here as a form of "geometric algebra"), rather than to canonical quadratic equations. Note also the terms of the problem: the givens are the difference of two numbers and the root of the sum of their squares, and the numbers are sought. This is then reduced to the case of given products and differences. This way of posing problems echoes pre-Arabic geometric-algebraic traditions.

A non-square rectangle whose diagonal has ten cubits and its length exceeds its width by two: how much is its length, width and area?

Answer: You know that the square of the diagonal is a hundred. Take away the square of the excess of the length over the width, which is 2, and its square 4. You're left with 96 of the hundred. Divide it into two, making 48, which will be the area of the rectangle [this is justified by reference to a previous section]. If you want to know its sides, of which one adds two to the other, divide this excess into two, making one, and its square one. Add this to the area, making 49, whose root is 7. If you add one, which is half the excess, it will be 8, which is the length line. If you subtract one, 6 will be left, which is the width line. 8 times 6 is 48, which is the area.

The proof for this matter: Let the vertices of a non-square rectangle be marked *ABCD* and its diagonal *AD*, which we set as 10 cubits [Fig. III-2-3]. We know that the line *AB*, which is the length, adds to the line *AC*, which is the width, 2 cubits. And we want to know from these two numbers the area of the rectangle and the magnitude of each of its sides. It is known that the rectangle of the line *AB* by the line *AC* is the area, and that the square

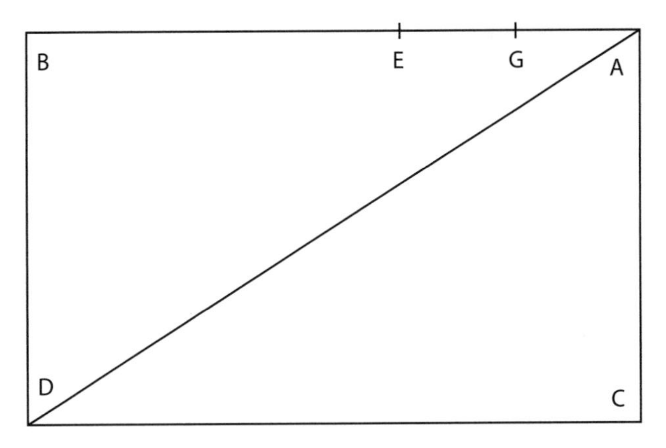

Fig. III-2-3.

of the diagonal equals double the area together with the square of the excess of the length over the width, as I showed you above.[34] Therefore, if you subtract from the square of the diagonal, which is 100, the square of the excess 2, whose square is 4, there remain 96, which is twice the area. Its half is 48, which is the area.

If you want to know the numbers of the sides, you already know that the length adds 2 to the width. Subtract now from the line *AB*, which is the length, a line equal to the line *AC*, which is the width. This line is *BE*, leaving the line *EA*, which is known to be 2, as the excess of the length over the width. If you bisect the line *EA* into two at the point *G*, the line *EG* and the line *GA* will be one cubit each. So *EA* is divided into two equal parts at the point *G*, and you add another line, *BE*. You know that the rectangle of the entire line *AB*, which is the line with the addition, by the line *BE*, which is the added line, together with the square of *GE*, which is the half, equals the square of *GB*, which is the half with the addition.[35] The rectangle of the line *AB* by the line *BE* is the rectangle *ABCD*, which is 48 cubits. Since the line *EB* equals the line *AC*, which is the width, and the square of *EG* is one, added together they are 49 and equal the square of *GB*. Therefore the line *GB* is 7 as the root of the square 49. If you add *AG*, which is one, the entire line *AB* is eight, which is the length of the rectangle. If you subtract from it this line, which is again one, there remain 6 for the line *EB*, and this line equals the width of the rectangle, which is the line *AC*. The rectangle of the length by the width is the area as in this diagram.

In this next example, the solution depends on circle proportion theory, rather than on book II of the *Elements* interpreted as "geometric algebra."

A non-square rectangle whose diagonal together with its side is 18, and its other side 6: how much is its area, and diagonal, and the side added to the diagonal?

He who answers this question will take the square of the known side 6, whose square is 36. He will then divide them by the diagonal and the side, which are 18, making 2. He will add 2 to 18, making 20. It is known that half of 20 is the diagonal, which is 10, and

[34] Bar Ḥiyya's Book I, §30 = *Elements* II.7.
[35] Bar Ḥiyya's Book I, §29 = *Elements* II.6.

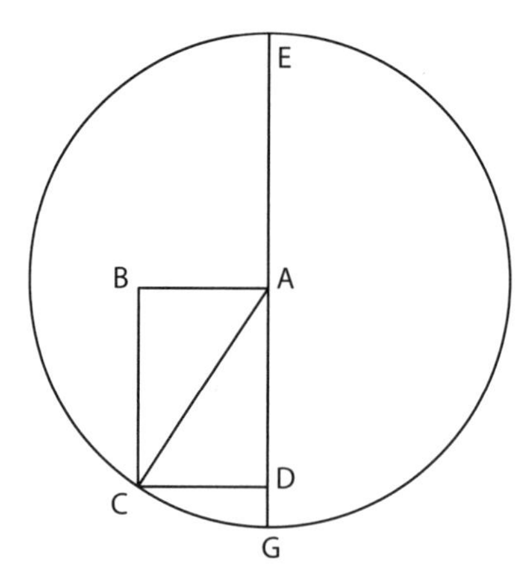

Fig. III-2-4.

what remains from 18 [after subtracting 10] is the side added to the diagonal, which is 8. The rectangle of 8 by 6 is the area, which is 48.

The proof for this answer is this. Let this rectangle be the rectangle *ABCD*, and let its diagonal be *AC* and the unknown side *AD* and the known *CD*. Set the point *A* as a pivot, and make a circle with a compass at distance *AC*. This is the circle marked *ECG*. Extend the line *AD* to the circumference of the circle both ways to the point *E* and the point *G* [Fig. III-2-4]. You have the line *AE* equal to the line AC, which is the diagonal, because they both set out from the pivot called [in Arabic] *markaz* to the circumference. He who posed the question set the line *AC* together with the line *AD* as 18, so the entire line *ED* is also 18 and the entire line *EG* is the diameter of the circle. And it is known, as I expounded among the established reasoning,[36] that the rectangle of the line *ED* by the line *DG*, which complements the diameter, equals the square of the line *DC* by itself, because these are two lines in the same circle dividing each other, and one of them goes through the pivot of the circle. Therefore, if you take the square of the line *DC*, which is six, by itself, and we divide it by the line *ED*, the outcome is the line *DG*, which is 2. The whole line *EG* is 20, and half the line *EG* is the line *AE*, which equals the diagonal. So it is ten, and there remains 8 for the unknown line *AD*, as we answered you, and as seen in this diagram.

Measurements of triangles (from Book II)

Here an extension of the Pythagorean theorem to general scalene triangles is used to determine the height of a triangle from its sides, which in turn would serve to find its area.

If, in the triangle *ABC* which we gave you [with sides 13, 14, and 15], we wish to take the height between the sides *AB* and *AC* onto the base *BC* of length 14 cubits, we should first find the part of the base to one side of the height [Fig. III-2-5]. To extract the longer part of the base, take the square of the longer of the two sides surrounding the top angle of the triangle, between which we take the height. This is the side *AC* of length 15 cubits. We

[36] Bar Ḥiyya's Book I, §33 = *Elements* III.35.

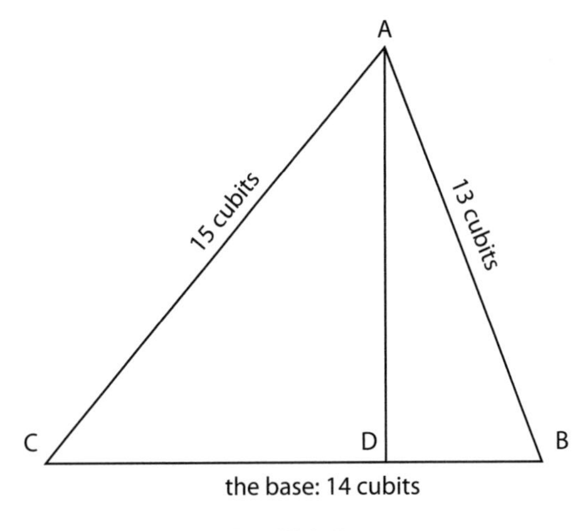

Fig. III-2-5.

add to this square the square of the base. These two squares together are 421. We take away the square of the remaining side *AB*, which is the short side, whose square is 169, leaving you with 252. We divide the remainder in two. Its half is 126. Divide this half by the base, which is 14, and the result is 9, which is the distance along the base from the height to the long side.

. . .

We operate [in] this way for every side for which we want to calculate the height. And once we know the part of the base to one side of the height, we come to know the length of the height thus: we square the side, take away the square of the adjacent part of the base, and take the root of the remainder, which is the length of the height.

. . .

If you ask for the proof for the calculation of the height, look at the diagram which I draw for you now [see Fig. III-2-5]. Know that the square of the side *AB*, which is opposite to an acute angle, as in this triangle, is less than the square of the side *AC* and [the square of] the side *BC*, which is the base; the excess is double the rectangle of *CD*, which is one part, by the entire base *BC*, as taught in geometry.[37] When you divide this excess in two and this half by *BC*, you get *CD*.

Heron's rule for determining the area of a triangle from its sides is brought here as well, without proof (and without reference to Heron).

Although [the calculation of the area of a triangle] is as I said, and all the ways that I showed you are correct and clear to he who understands, and yield a true result, you can find a method for calculating and measuring triangles that does not require the height, namely, the so-called calculation by excesses.

This method here requires that you find half of each of the sides of the triangle, and sum these halves together, and find the excess of the sum over each side, and note down

[37] *Elements* II.13, from Bar Ḥiyya's Book I, §27.

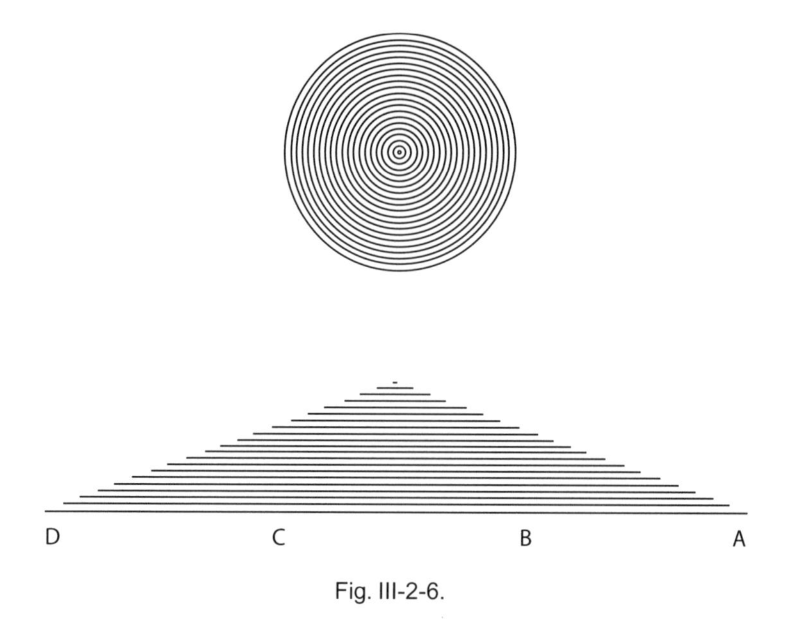

Fig. III-2-6.

these excesses. Then multiply one of them by another, and multiply the product by the third excess, and multiply what you get from this calculation by the sum of the halves that you added together. This number is the square of the area of the triangle. If you take the root of this number, you will find the area.

The author then uses this method to calculate the area of a triangle with sides of length 10, 8, and 6.

This calculation is based on the principles of geometry, and its proof is taught there. We cannot state it here, and we do not have much need for it here, because you saw that this calculation is true by the numbers that I gave you.

Circle and arc measurements (from Book II)

This original presentation of the reduction of the area of a circle to that of a triangle, which is not in Plato of Tivoli's translation, is probably the most famous section of the book. In violation of the classical (Aristotelian) tradition, it depends on decomposing an area into the lines that it contains.[38]

Once we know the circumference and diameter, we know the area of the whole circle, which is half the diameter times half the circumference. The proof for this area: We know that if you open the area of the circle on one side, and straighten all the surrounding lines from the external line [circumference] to the center, the lines surrounding the area of the circle will spread and turn into straight lines, decreasing until they turn into a single point, which is the center point. Such is the line *ABCD* . . . that I have drawn, where the external is the largest, and the next is smaller than the former but larger than the next, and so on to a point, which creates the form of a triangle [Fig. III-2-6]. But we have already taught the area of a triangle, which is as the height times half the base, which is half the diameter times half the circumference.

[38] For an analysis see [Garber and Tsaban, 2001].

This section also presents a table that calculates the arc from the length of a chord. The choice of the full chord, rather than the half chord (Sine) is typical of the Greek, rather than Indian or Arabic traditions. The normalization here is rather idiosyncratic: the diameter is counted as 28 parts, yielding a circumference consisting of 88 such parts.[39] The reconstruction of Table III-2-1 presented below is somewhat conjectural, as the manuscripts are obviously corrupted by scribal errors and often diverge. We also include an example for the use of the table, which depends on rescaling the measurement so that the given diameter fits the table's value of 28.

Suppose you know the diameter of one circle, and you are given a chord, and want to know the length of the arc surrounding that chord. You seek a rule that lets you find the length of the arc from the sagitta or chord, like the rule for finding the length of the diameter from the circumference and the length of the circumference from the diameter. Know that such a rule cannot be given to you, because the ratio between the chord and the arc is not fixed, but changes with the changing arcs and chords. . . . As a result, the reckoning of arcs and chords is difficult for most people. He who reckons it must understand many rules of geometry.

The scholars of astronomy labored on this for their practice, and I copied from this reckoning what I saw fit for this book. I drew for you a table divided lengthwise into 28 parts, as I divided the diameter of the circle into 28 parts. And according to this the circumference is divided into 88 parts. I divided the table along its width into 4 columns, and noted in the first of these columns lengthwise the 28 parts, and in the remaining 3 columns along the width I noted lengthwise the arc fitting each chord from 1 to 28. This table of arcs is divided into 3 columns because I divided each part into 60 seconds [sic], and each of the seconds I divided into 60 thirds, as you see in this table [Table III-2-1]. . . .

An example for this calculation. In a circle whose diameter contains 10 and a half, we take a chord whose length is six and ask to know the length of the round arc over that chord. We multiply the length of the given chord, which is 6, by 28, which is the diameter in the table. We get 168. We divide this number by 10 and a half, which is the diameter of the given circle. The result is 16, which [with respect to the table's diameter] is the chord in the proportion of your given chord to the given diameter. Against it you find in the table 17 large parts and 2 seconds and 16 thirds, which is the arc appropriate for the table's chord. Now multiply again this arc by 6, which is the number of the given chord. The total is 102 large parts, 13 seconds and 36 thirds. Divide this number by 16, which are the table's chord, and it will be 6 large parts, 23 seconds and 21 thirds, which is the measure of the required arc.

Sloping terrain measurement (from Book II)

The area of a parcel of land on a sloping or hilly terrain should be measured, according to this text, as if it were "projected" on an underlying plane. The reason is that crops and structures rise up vertically, and therefore the area must be measured only with respect to the vertical rise it enables. In the case of a hill shaped like a circular arc, the chord-arc table is called for. Although anachronistic, I cannot help but see in the image of a sloping area reckoned with

[39]This value is used because 88 is $3\frac{1}{7}$ times 28. Given the approximation $\pi = 3\frac{1}{7}$, this guarantees that the parts of the diameter and circumference are equal.

TABLE III-2-1

Chord	*Arc*		
	Parts	*Seconds*	*Thirds*
1	1	0	3
2	2	0	8
3	3	0	25
4	4	0	55
5	5	1	44
6	6	2	57
7	7	4	42
8	8	7	1
9	9	9	59
10	10	13	42
11	11	18	33
12	12	24	23
13	13	31	29
14	14	40	0
15	15	50	10
16	17	2	16
17	18	16	36
18	19	33	29
19	20	53	29
20	22	17	10
21	23	45	19
22	25	19	4
23	27	0	1
24	28	50	36
25	30	54	52
26	33	20	55
27	36	29	29
28	44	0	0

respect to a vertical field of crops the modern integral of an area form on a manifold with respect to a vector field. A crucial difference is that the text considers the area globally, rather than breaking it up into small (or infinitesimal) local patches.

All I taught you so far concerns measuring flat lands, where the terrain is spread straight without climbing up or down. But sometimes you come by terrains sloping down from the head of the mountain, or submerged low, or round and curved. The surveyors in these lands are wrong to measure all terrains, high or low, in the same way. You take care, and if you come by a lot [*AB* in Fig. III-2-7] hanging from the head of a mountain, find its height [*AE*], which is the distance from its beginning to its high end, and subtract

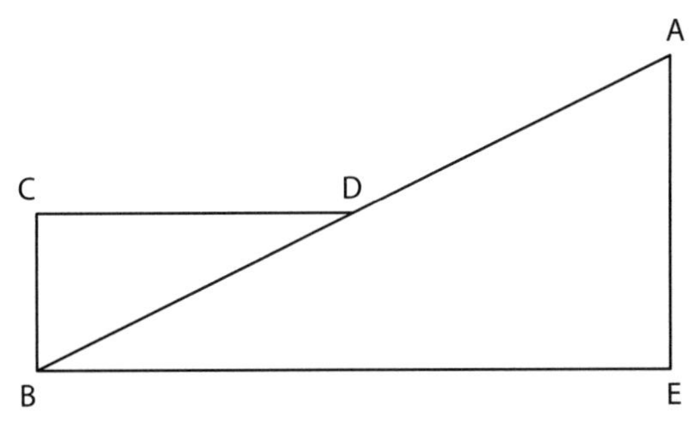

Fig. III-2-7.

the square of the height from the square of the length of the lot; the root of the remaining number times the width of the lot is its area.

If the lot falls low, treat its downward fall in the same way. And for curves on top of the hill, seek and find a way to take its area according to the straight plane on which it sits. Indeed, neither seed nor building rise but according to a straight angle with respect to the straight land, and the excess measured in a high or low land is useless for both seeds and buildings. Therefore you must subtract it and set it against the straight measurement of the plane terrain.

Those clever in measuring land sought to know the height of lands sloping off mountains and hills in order to extract the correct measure of the plane terrain whose area they are to find. They would do thus: They would erect a pole [*BC*] on the lower part of the terrain at a right angle with respect to the plane, and set at the head of that pole another pole [*CD*] at a right angle, and extend this pole set against the other, and lengthen it until it would reach the terrain sloping off the mountain wherever it falls. Then they would measure from the bottom of the pole at the lower part of the terrain to where the other pole reached higher up [*BD*]. This number is always found to exceed that pole which extends from the top of the standing pole to where it touches the ground. They would then use the ratio of the excess to reckon the excess of the area of the sloping terrain with respect to its area if the land were plane. . . .

Suppose the terrain slopes along a round and curved hill, and you wish to find its true area, fitting the flat lot on which the land stands. Erect the first pole [*BC*] at a right angle to the plane land, and set on top of it another pole [*CD*] at a right angle, and extend it until it reaches the curved lot as I showed you for the terrain sloping straight from the head of a mountain [Fig. III-2-8]. You will get the same diagram, except that the line *DB* is a round arc, and so is the entire *AB*, which is the arched terrain, both arcs being part of the same circle. Both arcs are known, as the arc *AB* is the length or width of the terrain you measure, and the arc *DB* is the arc surrounded by the two lines [poles]. You can find the chord of the arc *DB* from the erected triangle *DCB* by squaring the lines *DC* and *CB* surrounding the right angle *C* of the triangle *DCB*, adding the two squares, and taking the root of their sum, which is the length of the chord *DB*.

Now that we know this chord we can reckon to find the sagitta and from that the diameter. It is known that the sagitta divides the arc and the chord into two equal parts. We let *HIG* set out at a right angle from the point *H* at the middle of the chord, dividing the

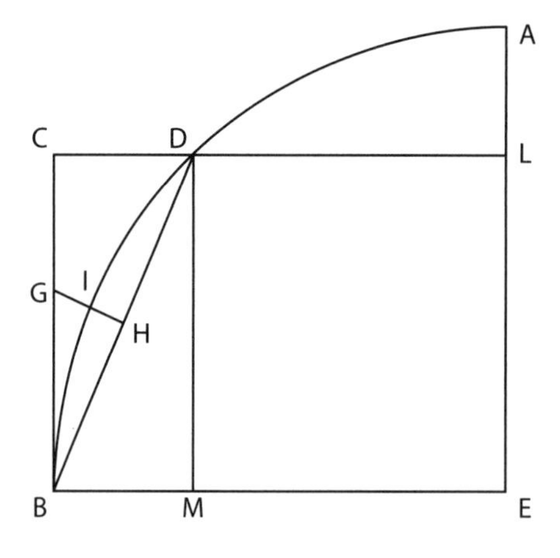

Fig. III-2-8.

arc *DB* in the middle at the point *I*, and extending until it reaches the line *CB* at the point *G*. So we have the triangle *GBH* similar to the triangle *BCD*, and the ratio of *DC* to *CB* as the ratio of *GH* to *HB*. Now *DC*, *CB*, *HB* are all known, and the line *GH* is unknown. If we multiply the line *DC* on one side of the ratio [*muqash*] by the line *BH* on the other side of the ratio [*noqesh*], which are both known, and divide their product by *CB*, which is known, the result is the length of the line *GH*.

From this ratio we come to know the length of the line *HI*, which is the sagitta. Indeed, the ratio of *DC* to *DB* is as the ratio of *GH* to *GB*, and if we multiply *GH* by *DB*, which are both known, and divide the product by the known *CD*, we find the length of *GB*. We measure from the known point *G* the length of *GI*, and subtract from the entire line *GIH*, leaving us with the length of *HI*, which is the sought sagitta. And from this sagitta and the chord *DB* we can find the diameter of the circle as we taught above [using the proportion of the parts of intersecting chords]. Having found the diameter, we can find the chord of double the arc *AD*, as we taught in the table of arcs and chords. Half this chord is as the line *DL*, and if we add to it the known line *DC*, the entire line *CDL* becomes known and is equal to the line *BME*, which should replace the curved line as in this diagram.

Division of areas (from Book III)

Here are two of the more clever divisions of a plane area into halves, borrowed indirectly from Euclid's book *On Divisions*. The first division divides a quadrilateral from an arbitrary point on a side, while the second divides a region partially bounded by a circular arc. The Greek text had disappeared by Bar Ḥiyya's time, but an Arabic abstract of the work by the tenth-century Persian geometer al-Sijzī, including the theorems and a few proofs, may well have been available to him [Hogendijk, 1993]. Compare these to Fibonacci's divisions in section II-4-3 of Chapter 1.

The first construction shows how to divide a quadrilateral in half, where no diagonal divides it in half, starting from a given point on one of the sides.

A different method of division [is effected] by drawing a partition from a point E on the side AB in this irregular quadrilateral shape, as in the diagram I draw [Fig. III-2-9]. First, we divide the quadrilateral [into two equal parts] from the vertex B on the line AB, as you

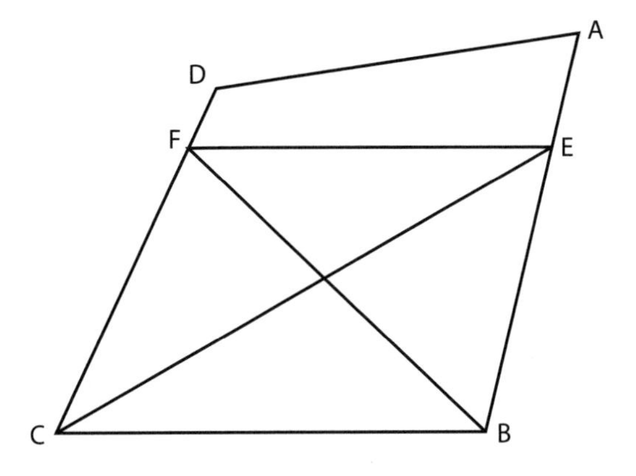

Fig. III-2-9.

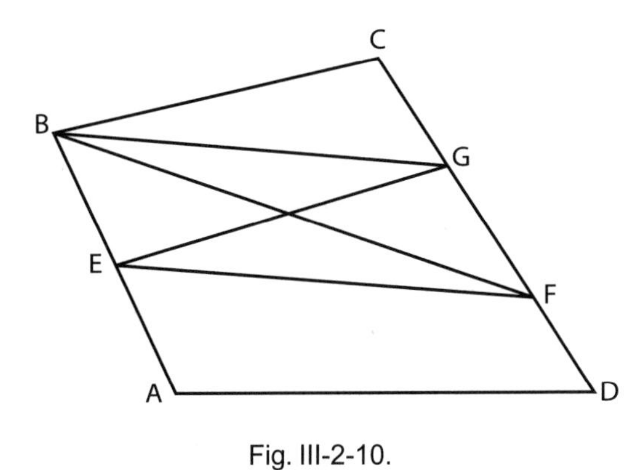

Fig. III-2-10.

were instructed in the previous diagram [not included here]. The parts are the triangle BCF and the quadrilateral ABDF, into which the quadrilateral is divided by the line BF drawn from the point B. We draw a line from point F to point E. If this line is parallel to the line CB, we draw a line from point E to point C, and the quadrilateral is divided in half along the line EC. The quadrilateral AECD equals triangle EBC as in the first diagram for this problem [Fig. III-2-9].

The proof of the division: these are two parallel lines, line EF being parallel to line BC. The entire triangle BCF, which is half of the entire quadrilateral, is equal to triangle BCE, both lying between two parallel lines. Therefore, triangle EBC is half the quadrilateral.

Suppose the line EF is not parallel to the line BC. We draw from point B at the angle of the quadrilateral the line BG parallel to the line EF. It is either interior to or exterior to the quadrilateral. Suppose that it is interior, as in the second diagram [Fig. III-2-10]. We draw a line from point E to point G. It divides the quadrilateral in half, namely, the quadrilaterals AEDG, EBCG. This is known because, as we noted above, triangle FBC is half the quadrilateral, and triangle BGF is equal to triangle BGE, as they lie on the same base between parallel lines. If one adds to each the triangle BGC, then triangle FBC will become

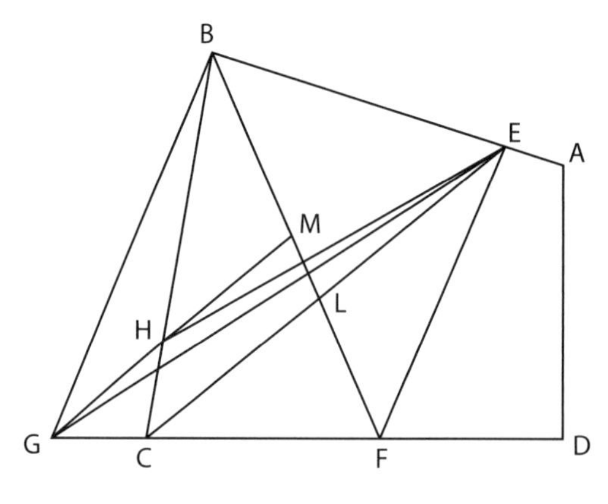

Fig. III-2-11.

equal to quadrilateral EBCG, and therefore quadrilateral EBCG is half of the entire quadrilateral. The other half, as we noted in this second diagram, is quadrilateral AEDG.

Suppose the line falls outside the quadrilateral, as line BG in the third diagram [Fig. III-2-11]. We extend the line CD to the point G, draw lines from E to G and from E to C, draw from G the line GH parallel to the line EC, and then a line from E to H. The quadrilateral is thereby divided into two equal parts, namely the triangle EHB and the pentagon AEHCD.

The proof for this division: Since line EF is parallel to line BG, triangle BFG is equal to triangle BGE. Similarly, triangle GEH will equal triangle CGH, because they lie between the parallel lines CE and GH. If we remove the triangle CGH from triangle BFG, and the triangle GEH from triangle BGE, you will have triangle BGH common to both [remainders]. Therefore remove from these remainders triangle BHG, and there will remain the triangle BCF equal to triangle BHE. The triangle BCF is half of the quadrilateral, and so is triangle BEH as in the third diagram.

You can draw a line from point G parallel to line CE in this third diagram, if you know the ratio of GC to CF, and if you extract, according to the same ratio, LM from the extension of line FL toward B. Draw the line GHM from point G to point M. This line is parallel to line CE, since if you draw the line CL in the triangle FMG, the ratio of CG to CF equals the ratio of LM to LF. Thus the line GH is parallel to the line CE, as we have told you.

In the next construction, the author divides a region partly bounded by a circular arc.

Suppose you have a portion with one round line and the other sides straight, as the portion *ABC* that I draw for you, where the two sides *AB* and *BC* are on straight lines and the side *AC* is somewhat round. This is called a truncated portion. If you wish to cut this shape into two [equal] parts, draw a straight line *AC* and divide it in the middle at the point *E*. From there draw a height on the line *AC* reaching the arc *CA* at the point *G*. Draw a line from *B* to *E* and this truncate is split into two equal parts along the line *BEG*, if it goes along one straight line as in the first diagram [Fig. III-2-12], or along the lines *BE*, *EG* if they are not on a straight line, as in the other diagram [Fig. III-2-13].

You can also divide the second diagram in another way. Draw a line from *B* to *G*, which crosses *AEIC* at the point *I*. Find the ratio of *IE* to *EA*, and divide the line *AB* [at *H*, so the parts are in the same ratio] as the ratio of *EI* to *EA*. We draw a line from *G* to *H*, and the

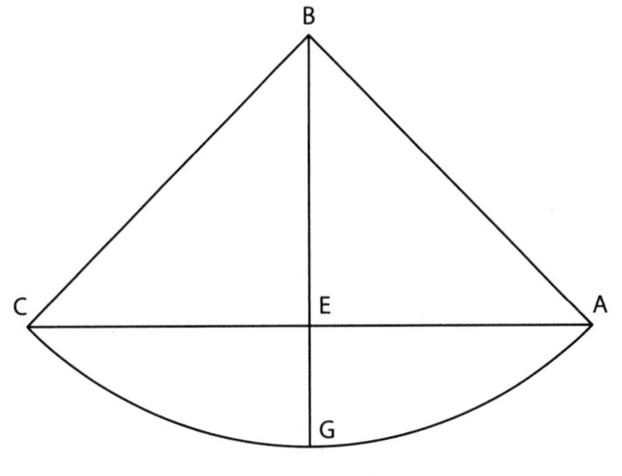

Fig. III-2-12.

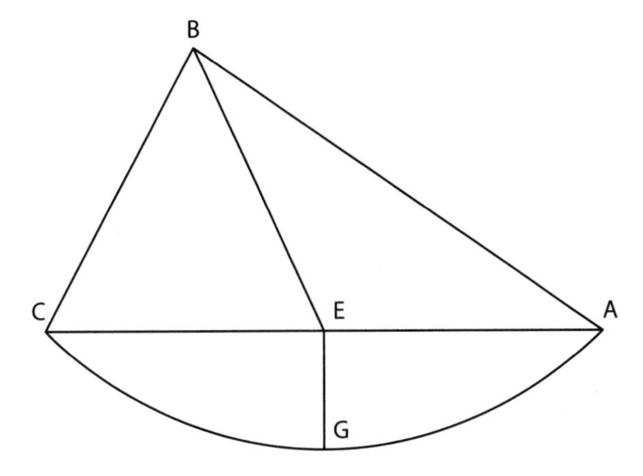

Fig. III-2-13.

truncate splits into 2 equal parts, which are the truncate *GHA* and the irregular *BHCG* as in this diagram [Fig. III-2-14].

The proof for this diagram is that the triangle GHE and the triangle BHE, which are between two parallel lines, are equal. So the truncate *AHG* equals the triangle *ABE* together with the truncate *AEG*, which are both half the truncate *ABC*, as in this diagram.

As we saw in Bar Ḥiyya's introduction, some of the motivation for the Hebrew literature on measurement was religious. Here we include an early example from Rashi, which predates Bar Ḥiyya, and a later example, by Simon ben Ṣemaḥ, which builds on Bar Ḥiyya's work.

3. RABBI SHLOMO IṢḤAQI (RASHI), *ON THE MEASUREMENTS OF THE TABERNACLE COURT*

This section was prepared by David Garber

The following comment is by Rashi (acronym of Rabbi Shlomo Iṣḥaqi, 1040–1105). Rashi spent his life in Troyes, except for a decade of studies in Mainz and Worms. His importance

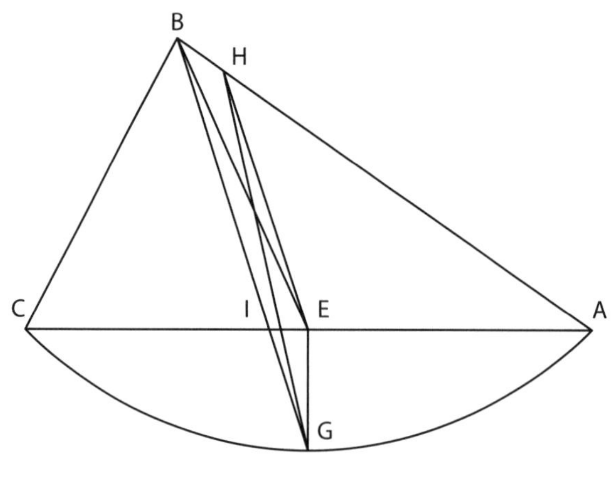

Fig. III-2-14.

among Jewish exegetes cannot be exaggerated, no doubt supported by his clear, succinct, and literal style. His commentary on the Talmud is a standard fixture in Talmud editions.

The commentary here applies elementary geometric considerations to the context of the temple measurements as discussed in the *Mishna*, deploying a cut-and-paste procedure for squaring a rectangle. Albeit elementary, the presentation is clever and elegant.[40]

Rashi commentary to: "The length of the [tabernacle] court shall be a hundred cubits, and the breadth fifty everywhere, the Torah having thus ordained: 'Take away fifty and surround [with them the other] fifty.'" (*Eruvin* 23b)

"Take away fifty," which is the excess of the length over the width, and use them to surround the remaining fifty so as to form [a square of] seventy cubits and 4 palms [*tefahim*]. How so? Split [the remaining fifty by fifty] into five strips of ten cubits breadth and fifty cubits length [Fig. III-3-1]. Set one strip to the east and one to the west. Then the breadth is seventy and the length is fifty. Next set one to the north and one to the south. Then you have seventy by seventy, but the corners are missing ten by ten each with respect to the addition you made. Take from the remaining fifty 4 pieces of 10 by 10, and set in the four corners, so they become complete. There remains one strip of ten by ten cubits left, which is sixty palms by sixty palms. Split them into 30 strips of two palms (then you have 30 strips, each ten cubits in length). Altogether they add up to 3 hundred cubits. Set 70 [cubits, or 7 strips] in each direction, then you have seventy and 4 palms by seventy and 4 palms, but the corners are missing two palms by two palms. You have twenty cubits left. Take eight palms and set them in the corners, so they become complete. You have 18 cubits and 4 palms left of two palms width, that is, but a trifle, which, if you try to split to surround the square, the excess would not reach two thirds of a finger in breadth [Fig. III-3-2].[41]

[40] See [Garber, forthcoming].

[41] Note that the procedure can be continued indefinitely, but that Rashi shows no infinitesimal inclinations here and terminates the calculation when the error becomes small enough for the given context.

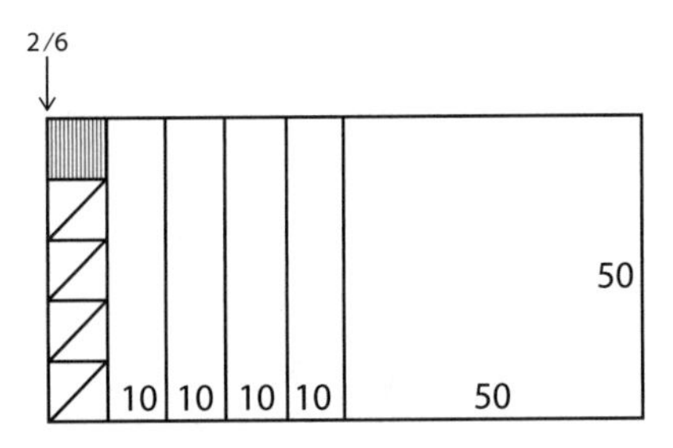

Fig. III-3-1. The rectangular court.

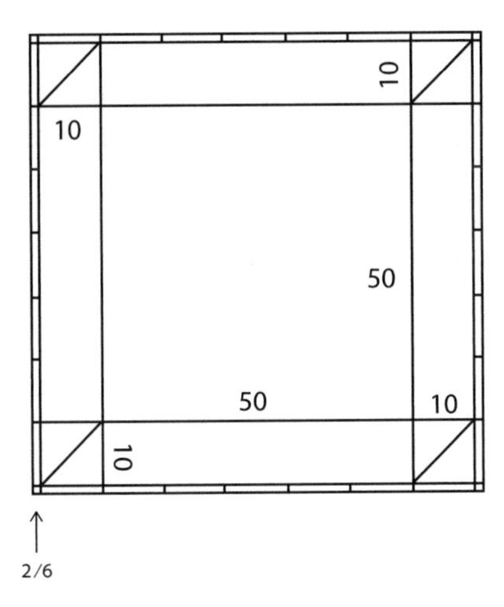

Fig. III-3-2. The squared court.

4. SIMON BEN ṢEMAḤ, *RESPONSA 165 CONCERNING SOLOMON'S SEA*

This section was prepared by David Garber

Rabbi Simon ben Ṣemaḥ Duran (1361 Mallorca–1444 Algiers) was one of the greatest rabbinical authorities of his time. He made his living as a physician in Aragon, but had to flee to Algiers due to the 1391 riots against local Jews. In Algiers he served in the tribunal of Rabbi Isaac ben Sheshet. Ben Ṣemaḥ's responsa book, the *Tashbeṣ* (the acronym for the responsa of Simon ben Ṣemaḥ), deals with about 800 questions and has four parts: three written by himself and one by other rabbis in his family. The book includes several discussions of mathematical aspects of religious law.

This excerpt from question 165 of the first part of the *Tashbeṣ* includes a discussion between Rabbi Simon ben Ṣemaḥ and Enbellshom Efrayim (perhaps the Mallorcan astronomer and brother of Ben Ṣemaḥ's teacher, Vidal Efrayim; "En" is an agglutinative Catalan honorary prefix, equivalent to the Castilian "Don"). The interpretation of the discussion follows [Garber and Tsaban, 1998].

Enbellshom, following Bar Ḥiyya (but without mentioning him as his source; see [Garber and Tsaban, n.d.] for a comparison), claims that religious authorities knew that the value $\pi = 3$, used for religious calculations, was only an approximation, and, moreover, that they were aware of more exact approximations, such as $\pi = 3\frac{1}{7}$, which was known at that time.

According to this view, the Talmudic scholars only used the coarser approximation where it rendered religious rules more strict. This assumption, however, calls for a reinterpretation of various Talmudic debates.

In contrast, Ben Ṣemaḥ, while upholding that religious authorities were no less savvy than Euclid and Archimedes, claims that their calculations sometimes depended on the approximation $\pi = 3$ even if it rendered religious rules more lenient. Nevertheless, he qualifies the injunction to adhere to precise calculations. Ben Ṣemaḥ seems to reconcile the two positions by recalling the instability of measurement units and the authorities' attempts to render religious law accessible to laymen. The underlying concerns of the debate are the place of scientific knowledge in the interpretation of religious texts and the limits of exegetic practices.

In the excerpt presented here, the topic discussed is the measurement of Solomon's Sea, a large ceremonial basin in King Solomon's temple. In Kings I, 7:23, the Sea is stated to be 10 cubits in diameter, 5 cubits high, 30 cubits in circumference, and 2,000 *bat*s in volume (which equal 6,000 *seʾa*s or 450 cubic cubits or 150 kosher ceremonial purification basins (*miqve*)). Since the volume of such a cylinder, assuming $\pi = 3$, should only be 375 cubic cubits, the Talmud suggests that the three bottom cubits were square, and only the top two were circular, yielding the required volume of 450 cubic cubits. Enbellshom quoted Bar Ḥiyya's alternative estimate of the measures of Solomon's Sea (without reference to Bar Ḥiyya, as mentioned above). This revision was rejected by Ben Ṣemaḥ on the grounds stated below.

First we present Bar Ḥiyya's argument, as quoted from Enbellshom by Ben Ṣemaḥ.[42] In this argument, the scriptures are reinterpreted so as to reduce the diameter of the circular part of the Sea to cohere with the prescribed volume, circumference, and $\pi = 3\frac{1}{7}$. Then, assuming that the size of a *miqve* is defined as one part in 150 of Solomon's Sea, it is shown that a kosher *miqve* can be slightly smaller than the standard prescription of 3 cubic cubits. After discussing the supposed true measurements of the Sea, it is explained how a doubt concerning the interpretation of the thickness of the rim of the Sea led the Talmudic scholars to reach a different volume estimate, even though, like him, they supposedly assumed $\pi = 3\frac{1}{7}$.

It is written that [Solomon's Sea] is 10 cubits from side to side, circular in form, 5 cubits high, with circumference 30 cubits round. It seems at the beginning that this neglects the seventh [in $\pi = 3\frac{1}{7}$]. But here things are precise as well, if one looks carefully, for it says that "it was a palm thick, and its brim was made like that of a cup, like the petals of a lily."[43] It appears that the circular part is one palm away from the edge of the square, which makes two palms in diameter out of 5 palms per cubit. ... If you multiply 9 cubits and three fifths [= 10 cubits less 2 palms] by three and a seventh, the external circumference of the Sea will be 30 cubits and a sixth approximately.[44] Its inner circumference should be 30 cubits, so the thickness of the rim, which is like a flower, is about two thirds of a finger.[45]

[42]Levi ben Gershon also considered this issue of the exact dimensions of Solomon's Sea, but there is no evidence that Ben Ṣemaḥ was aware of Levi's work. See [Simonson, 2000a, pp. 7–8].

[43]Kings I, 7:26.

[44]The precise value is $9\frac{3}{5} \times 3\frac{1}{7} = 30\frac{6}{35}$.

[45]This calculation is based on 4 fingers per palm and 6 palms per cubit (as opposed to the 5 palms per cubit above—both possibilities occur in the scriptures). Then, given $\pi = 3\frac{1}{7}$ and the above outer circumference, to obtain

If you multiply half the diameter by half the circumference correctly, and consider the width, you find that the area of the circle is 72 cubits and 2 ninths.[46] This is the volume per one cubit height. For two cubits [in height], it is 144 cubits and 4 ninths. The Sea was 5 cubits high, the first 3 being square and the top two being circular. So the volume of the lower square three cubits being 3 hundred [cubic] cubits, the volume of the entire Sea would be 444 [cubic] cubits and 4 ninths. . . . When you multiply by 4 and a half [*bats* per cubic cubits], it comes to two thousand, in accord with the scriptures: "its capacity was 2000 *bats*."[47]

. . .

The [religious authorities] figured a *miqve* as one part of 150 of the Sea. Since the . . . *bat* is 3 *seʾas*, the Sea was 6000 *seʾas*. Since a *miqve* is 40 *seʾas*, it is one part of 150 of the Sea. The volume of 3 [cubic] cubits is 13 *bats* and a half, which are 40 *seʾas* and a half.[48] According to this calculation, the size of a *miqve* of 40 *seʾas* is a cubit by a cubit at a height of 3 cubits less one part in 27 of a cubit.[49] If [the religious authorities] added to the size of a *miqve* [saying it is one by one by three cubits], they did so for strictness, as was their manner.

Do not think that the Talmudic authorities had figured the square-shaped three cubits [bottom part of the Sea] to be 300[50] [cubic] cubits, and 150 for the circular two cubits [top part of the Sea], with the diameter of the circle ten whole cubits, because then they would have been wrong in two ways, which is not the case, God forbid. First, if the diameter had been ten whole cubits, the volume . . . [of the circular part would have been $2 \times 5^2 \times 3\frac{1}{7} =$] 157 and a seventh. But they figured for the two cubit [high] circular part only 150 cubits [Fig. III-4-1].

Second, if this had been the case, then they would have neglected the missing palm in the diameter, whereas it is explicitly written that it was one palm thick. How could they say something so far from the truth as this? That is impossible. But, it is my opinion that they had a doubt whether the missing palm was on all sides (the diameter being 10 cubits less two palms), or in the diameter (leaving 10 cubits less a palm). The volume per one cubit [height] of the Sea, according to the scholars of geometry [that is, $\pi = 3\frac{1}{7}$], with a diameter of 10 cubits less a palm (based on six palms per cubit, which they agreed upon for strictness),[51] would be 76 [cubic] cubits approximately; the volume per one cubit [height] of the Sea with two palms removed from the diameter, would be 73 [cubic] cubits.[52]

---

an inner circumference of 30 cubits, we require the rim's thickness to be $\frac{6}{35} \times \frac{1}{2 \times 3\frac{1}{7}}$ cubits. Multiplied by 24 fingers per cubit, this makes $\frac{36}{55}$ fingers, which is approximately $\frac{2}{3}$ of a finger.

[46] The actual product is $\frac{30}{2} \times \frac{9\frac{3}{5} - 2 \times \frac{3}{110}}{2} = 72 - \frac{9}{22}$ (where $\frac{3}{110}$ cubits are $\frac{36}{55}$ of a finger, according to the previous calculation). This error is carried over from Bar Ḥiyya.

[47] Kings 1, 7:26.

[48] According to the above, $\frac{2000}{444\frac{4}{9}} = 4\frac{1}{2}$ is the number of *bats* in a cubic cubit, so a *miqve* which consists of 3 cubic cubits has $3 \times 4\frac{1}{2} = 13\frac{1}{2}$ *bats*. A *bat* is 3 *seʾas*, so a *miqve* is $3 \times 13\frac{1}{2} = 40\frac{1}{2}$ *seʾas*.

[49] $1 \times 1 \times 3$ cubits give 40 *seʾas* and a half, so $1 \times 1 \times \left(3 - \frac{1}{27}\right)$ cubits would give precisely 40 *seʾas*.

[50] The text reads 450, but this is clearly an error.

[51] Recall that a cubit sometimes had 5 palms and sometimes 6.

[52] $\frac{3\frac{1}{7}}{4} \times \left(10 - \frac{1}{6}\right)^2 \approx 75.974$; $\frac{3\frac{1}{7}}{4} \times \left(10 - \frac{2}{6}\right)^2 \approx 73.421$.

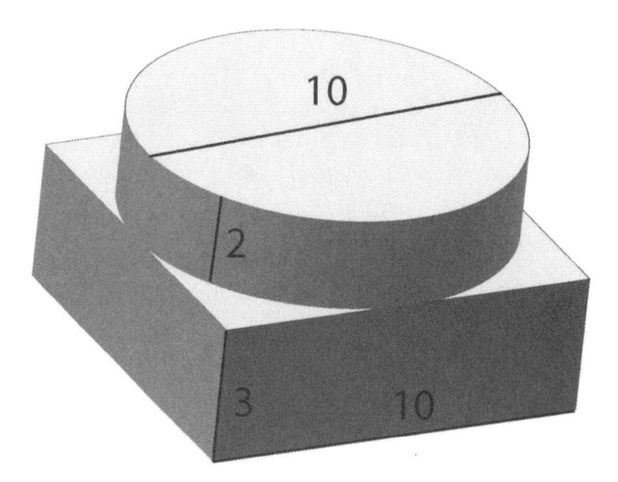

Fig. III-4-1. Solomon's Sea according to the Talmud.

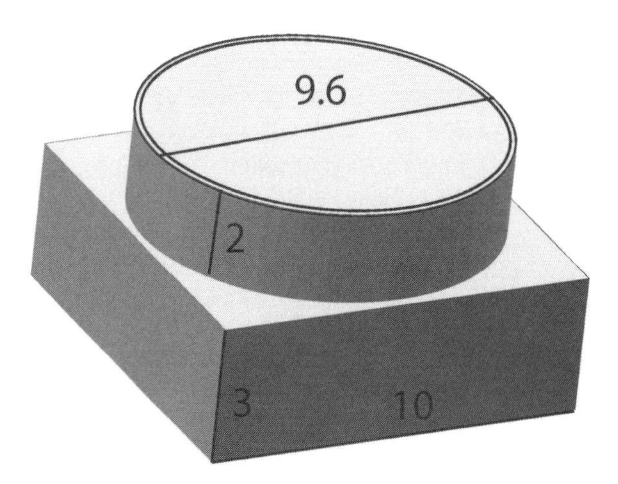

Fig. III-4-2. Solomon's Sea according to Bar Ḥiyya and Enbellshom.

They took an average, and per one cubit [height] of the Sea, they figured 75 [cubic] cubits; per two cubits [height], they figured 150 [cubic] cubits [Fig. III-4-2].

We proceed with several excerpts from Ben Ṣemaḥ's response (leaving out his critique of the calculation error and the wavering between 5 palms and 6 palms per cubit). First, concerning the revised estimate for a *miqve*:

I do not see this as correct. Your error is that you assume that *ḥazal*[53] extracted the volume of a *miqve* from Solomon's Sea, and this is not so. The scriptures do not state that

[53] Hebrew acronym for "the Scholars, Blessed be their Memory." This refers to the religious authorities from the time of the second temple to the closing of the Talmud.

the Sea measured 150 *miqves*. *Ḥazal* first figured that the volume of a *miqve* is cubit by cubit and 3 cubits high exactly, with no approximation. Indeed, they learned from "he shall bathe his body in water"[54] that a person's entire body should be submerged in a *miqve*, and they estimated the size of an average man to be one cubit by one cubit, three cubits tall.

. . .

Then they needed to figure Solomon's Sea (which contained 2000 *bat*s, which are 150 *miqves*) in terms of cubits, given the rate of one cubit by one cubit, 3 cubits high [per *miqve*]. To match this, they stated that only the top two cubits were round, and the 3 bottom cubits were square. If it were all round, it would not have held 2000 *bat*s of liquid.

Second, concerning the attribution of 10 cubits to the width of the square part, rather than the diameter of the circular part:

Your interpretation, that the ten cubits from rim to rim are not the inner measure, but that of the three square lower cubits, as arises from your words, contradicts the scriptures. Indeed, the scriptures say[55] that from rim to rim in the circular part the width was ten cubits. The scriptures do not mention the square part—this is the interpretation of *ḥazal*, in order to fit, as I mentioned, the two thousand *bat*s content. . . . Since the scriptures do not mention the squareness, how can you interpret that the [10 cubits] width concerns that which is not mentioned?

. . .

It seems that you think that the scriptures gave two magnitudes, one for width and one for circumference. And as you see that these two magnitudes do not fit according to the geometers [that is, with $\pi = 3\frac{1}{7}$], and would contradict each other if they concerned the same object, so you interpret them as concerning two [different] objects. You assigned the width to the square part, and the circumference to the circular part, and make up what is missing in the scriptures, namely that the circumference of the square part is 40 cubits and the width of the circular part is 9 cubits and three fifths approximately. This is fitting to your intention, and would have been acceptable by the mind, if the words of *ḥazal* had not forced us to reject it. You turned things around to serve your purpose, as I will show you well in accordance with the scriptures.

Ḥazal, whose words you seek to correct, would not be pleased by them at all, and would reject them. They explicitly said that the width of ten cubits relates to the inside of the circular part, as was stated in *Eruvin* [14a]. . . . Thus your building upon this interpretation is "like a spreading breach that occurs in a lofty wall, whose crash comes sudden and swift."[56]

Third, the attempt to reckon with the thickness of the rim is rejected.

I said that you turned things round to serve your intention. This relates to your attempt at forcing from the scriptures that the width of the circular part was 9 cubits and 3 fifths approximately, saying that the circle was a palm away from the edge of the square, which

[54]Leviticus 16:24.
[55]Kings I, 7:23.
[56]Isaiah 30:13.

makes two palms in diameter, counting five palms per cubit; the two palms, you say, are to be removed. But I cannot find this statement in *ḥazal*, and you too wrote that they were in doubt as to whether the missing palm was in all directions, or referred to the entire diameter, making half a palm on each side. . . . Your vision that the circle is a palm away from the edge of the square is prophetic, because the opinion of *ḥazal* and the literal text of the scriptures is that the width, which is a palm, is so from bottom to top, and does not subtract anything neither at the top or the bottom, since the measurement [of the diameter] is taken on the inside. But close to the rim, this palm dwindles so that at the rim it is [as thin as] a flower. The excess thickness of one palm is on the outside.

Your interpretation also suffers from this, that the ten cubits width in the square part, you measure from the inside, discarding the thickness [of the rim]; but in the circular part you count them with the thickness. . . . Moreover, you add to the thickness two palms as well as the rim, which is like a flower, which contradicts the scriptures.

Finally, here are Ben Ṣemaḥ's alternative explanations for the discrepancy between the Talmudic calculations and the more precise value of π.

We should say, like the *tosafot*, that this is an approximation, and that it neglects the extra seventh [in the value of π], and that the sages also neglected the extra 7 [cubic] cubits and a seventh [that would arise from replacing $\pi = 3$ by $\pi = 3\frac{1}{7}$ in calculating the volume of a cylinder with diameter 10 and height 2, as above], which is only one part in 21 of the circular part, making approximately two *miqve*s and a third.

This is foreign neither to the way people speak, nor to the scriptures, which we find speaking in mundane manner with imprecise statements. . . . Moreover, it is not foreign to *ḥazal* to make things clearer for their disciples, especially as they follow here the literal meaning of the scriptures and rely on them.

If you wish to follow another way, and retain [the measurements stipulated by *ḥazal* as well as the revised value of π] without any difficulties arising from Solomon's Sea, you can say that the cubits measuring the height were smaller and the cubits measuring a *miqve* larger, as is demonstrated in *Eruvin* [3b]. Between the larger and smaller cubits, you can fit the extra 7 [cubic] cubits and a seventh. . . .

I swear that everything you write about the Sea is very precise. I do not dispute it because it is imprecise, but because I see that it does not reflect the view of *ḥazal*.

5. LEVI BEN GERSHON, *ASTRONOMY*

Wars of the Lord was Levi ben Gershon's (see section I-6) major work on religious philosophy, probably completed by 1330. The *Astronomy*, which survives in separate manuscript copies from the rest of the treatise, forms Book V, part 1 of that work. It was translated into Latin in 1342 by Petrus of Alexandria. Meanwhile, chapters 4 through 11 of the *Astronomy* appeared separately as the treatise *On Sines, Chords and Arcs, and the Instrument "the Revealer of Secrets,"* where the latter refers to what is known as the Jacob Staff.

Calculation of Sines and "heuristic reasoning"

In section 3 of chapter four, Levi constructs his table of Sines in a way similar to Ptolemy's construction of his table of chords in the *Almagest*. That is, he first computes the Sines and chords of many angles using basic geometry as well as the sum formula, the difference formula, and the half-angle formula. For example, the chord of 36° is calculated from Euclid's result on the side of a decagon; then he can calculate the Sine of 18°. From the latter and the well-known Sine of 30°, Levi can calculate the Sine of 12° and therefore the Sines of 6°, 3°, $1\frac{1}{2}^\circ$, and 3/4°, as well as the Sines of 15°, $7\frac{1}{2}^\circ$, and $3\frac{3}{4}^\circ$.

He next shows how he will compute the Sine of 1/4°, his desired minimum interval in his table, by a method not strictly "geometrical." This method, "heuristic reasoning" (*heqesh taḥbuli*), is a general one that he uses elsewhere in the *Astronomy*. Levi much preferred a "demonstrative" argument, based on Euclidean precepts, but realized that in certain cases this was impossible. So one alternative was a type of "conditional reasoning," based on making certain assumptions, then checking their consequences, and then repeating until one reached the solution as closely as desired by successive approximation [Mancha, 1998]. For this particular calculation, Levi explains that the tables of Sines with intervals of 1° have errors of as much as 15 minutes of arc when trying to determine an arc corresponding to a given Sine, especially for arcs near 90°. With his method, he believes that "no perceptible error arises from linear interpolation."

We can also easily find the Sine of 1/4° by heuristic reasoning [*heqesh taḥbuli*]. For once the Sine of $8\frac{1}{4}^\circ$ [from the Sine of $(15 + 1\frac{1}{2}^\circ)$] is known, we also know the Sine of $4\frac{1}{8}^\circ$, and, therefore, by successive halvings, we will know the Sine of $\left(\frac{1}{4} + \frac{1}{128}\right)^\circ$. Similarly, from the Sine of $(4 - \frac{1}{4}^\circ)$ [= $3\frac{3}{4}^\circ$] we can proceed until we find the Sine of $\left(\frac{1}{4} - \frac{1}{64}\right)^\circ$. When we investigated this in this way, we found that the ratio of the Sine of $\left(\frac{1}{4} + \frac{1}{128}\right)^\circ$ to the Sine of $\left(\frac{1}{4} - \frac{1}{64}\right)^\circ$ is very nearly equal to the ratio of the first arc to the second one, in such a way that there is no difference between these ratios even to the fourth sexagesimal place, although they differ slightly in the fifth. Therefore, we established [by heuristic reasoning, that is, linear interpolation] that the Sine of 1/4 is 0;15,42,28,32,7. From this amount we can find all the remaining Sines.

In chapter 49, Levi further explains the method of heuristic reasoning. Although the method as described here is the classical method of "double false position,"[57] Levi's main contribution is that one can iterate this method in nonlinear contexts to approximate the desired result. In other words, after calculating a value by means of double false positioning, one can use it as input for a further iteration of double false positioning, and so on. (See [Plofker, 2002] for more details on such iterative approximation in India and Islam.)

It is appropriate to know that it is not possible to do in a quick way a demonstrative research in order to show how our model must be constructed. ... Therefore, the investigations which lead us to the truth necessarily are of the kind of heuristic reasoning, which are made from trial and investigation, which approach step by step to the truth until it is reached. These types of reasoning belong to the category of conditional reasoning, and there are two classes of them, one of which is taken from an excess and a defect; the second one is taken from two investigations in excess or from two investigations in defect.

For illustrating the first class, we say: if when we considered a determined first quantity, it followed an equation greater than that we have by a given second quantity, and if when we supposed a certain third quantity, there followed from it an equation smaller than that

[57] See section V-3 below and section II-1 in Chapter 3 for more on double false position.

we have by a given fourth quantity, it is known, according to the [rules of] proportion, that it is necessary to suppose a mean [quantity] between the first and the third, so that the ratio of the difference between the first and the mean to the difference between the first and the third is equal to the ratio of the second to the sum of the second and the fourth.[58]

To illustrate the second class, we say: if when we supposed a certain first quantity there followed from it an equation greater than that we have by a given second quantity, and if when we supposed a certain third quantity there followed from it an equation greater than that we have by a given fourth quantity, which is smaller than the second one, it is known, according to the proportion, that it is necessary to suppose a fifth quantity so that the third is the mean between the first and the fifth, and the ratio of the difference between the first and the fifth to the difference between the first and the third is equal to the ratio of the second to the difference between the second and the fourth. And one proceeds in a similar way if the second and fourth quantities are derived from equations smaller than those we have.[59]

The conditional reasoning would be simple when we say: if when we suppose a certain first quantity, there followed from it an equation of a given second quantity not equal to the third quantity which we have, it is known, according to the proportion, that it is necessary to suppose a fourth quantity in such a way that the ratio of the fourth to the first one is equal to the ratio of the third to the second.[60]

At this point the Latin manuscript has a long marginal note, not included in the Hebrew original, which clarifies the previous passage with some examples.

Example of excess and defect: the first [quantity] is 10, that will produce 30, which is 6 [units] greater [than what we have], and this is the second; the third is 6, that will produce 15, which is 9 [units] smaller [than what we have], and this is the fourth; the fifth is 8;24, that will produce 24, what we have.[61]

Example of diverse excess: the first [quantity] is 10, that will produce 36, which is 12 [units] greater [than what we have], and this is the second; the third is 8, that will produce 30, which is 6 [units] greater [than what we have], and this is the fourth; the fifth is 6, that will produce 24, what we have.[62]

Example of diverse defects: the first [quantity] is 6, that will produce 24, which is 12 [units] smaller [than what we have], and this is the second; the third is 8, that will produce 30, which is 6 [units] smaller [than what we have], and this is the fourth; the fifth is 10, that will produce 36, [which is] what we have.[63]

The previous cases are [examples of] composite conditional reasoning; the following one is simple. The first [quantity] is 10, which produces 30, which is the second, which is

[58] Suppose you wish to find a such that $f(a) = b$. Suppose further that $f(a_1) - b = b_1$ and $b - f(a_2) = b_2$. Then assuming that f is affine, $\frac{a_1 - a}{a_1 - a_2} = \frac{b_1}{b_1 + b_2}$.

[59] Suppose you wish to find a such that $f(a) = b$. Suppose further that $f(a_1) - b = b_1$, $f(a_2) - b = b_2$, and $b_2 < b_1$. Then assuming that f is affine, $\frac{a_1 - a}{a_1 - a_2} = \frac{b_1}{b_1 - b_2}$. Similarly for $f(a_1), f(a_2) < b$.

[60] This is the simpler case of the Rule of Three, which applies to a linear f.

[61] You wish to find a such that $f(a) = 24$. Suppose that $f(10) - 24 = 6$ and $24 - f(6) = 9$. Then a is 8;24 (sexagesimal) (or 8 2/5).

[62] You wish to find a such that $f(a) = 24$. Suppose that $f(10) - 24 = 12$ and $f(8) - 24 = 6$. Then a is 6.

[63] You wish to find a such that $f(a) = 36$. Suppose that $36 - f(6) = 12$ and $36 - f(8) = 6$. Then a is 10.

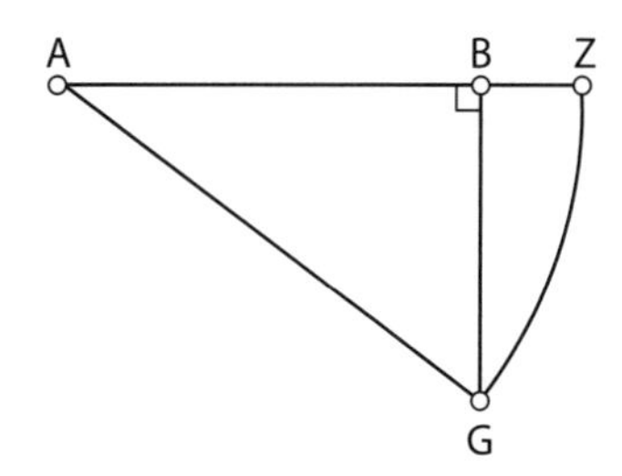

Fig. III-5-1.

not equal to 24, which is what we have, and it is the third. Therefore, 8, which is the fourth and whose ratio to 10 is equal to the ratio of 24 to 30, produces 24.[64]

Solving triangles by means of sine tables

In section 5 of chapter four, Levi treats the solution of triangles. He begins with the procedures for right triangles and then moves on to general plane triangles. Certainly, his methods were not new, as they were available in various Islamic trigonometries. Nevertheless, this was one of the earliest treatments in Europe of the basic methods for solving plane triangles. The methods were then applied later in his work for solving astronomical problems.

The procedure for finding the angles and sides of a triangle when some of them are known.

If two sides of a right triangle are known, the remaining side and angles may be found. Let *ABG* be a right triangle, two sides of which are known; I say that the remaining side and angles are also known. Whichever two sides are known, the third side is known because the square of the side that is the hypotenuse is equal to the sum of the squares of the two remaining sides. Thus, if they are known, it is known; and if it and one of the remaining sides are known, the third side is also known because its square is the difference between the square of the hypotenuse and the square of the other side.

Let us assume that angle *ABG* is a right angle, and that sides *AG* and *BG* are known [Fig. III-5-1]. I say that angle *BAG* is known, for if we consider point *A* as center, draw an arc *GZ* with *AG* as radius, and join line *ABZ*, it follows from the above that line *BG* is the Sine of arc *GZ*. Since lines *AG* and *BG* are known, it follows that line *BG* is known in the measure where line *AG* is the semidiameter of 60. In this measure the arc corresponding to line *BG* considered as a Sine can be looked up in the table of arcs and Sines. When arc *GZ* is found in this way, angle *BAG* is also known, as is clear from Euclid. From it angle *BGA* is also known, because it is the complement in 90°, inasmuch as angles *BAG* and *BGA* together are 90°.

If all sides of any triangle whatever are known, its angles are also known. Let the sides of triangle *ABG* be known; I say that its angles are also known.

We drop perpendicular *BD* from point *B* to line *AG* extended if necessary—in the first figure [Fig. III-5-2] point *D* falls within the triangle, and in the second figure [Fig. III-5-3] it

[64]This relates to the case of a linear f. If $f(10) = 30$, and we wish to get the result 24, the argument of the function should be $10 \times \frac{24}{30} = 8$.

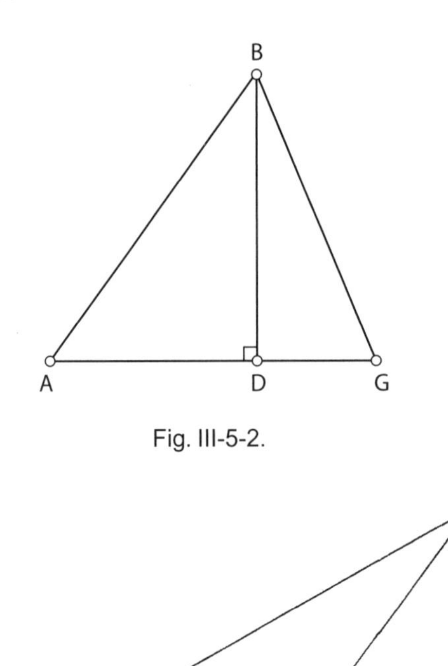

Fig. III-5-2.

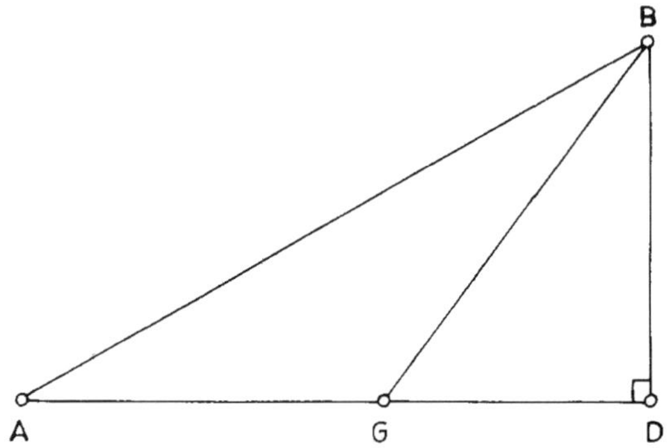

Fig. III-5-3.

falls outside the triangle. I say that the amount of *GD* is known. When we take the excess of the squares of lines *GB* and *GA* over the square of line *AB* in the first figure, or the excess of the square of line *AB* over the squares of lines *GB* and *GA* in the second figure, and divide it by twice line *GA*, the result is equal to line *GD*—as will be clear with a little thought concerning Book II of Euclid [*Elements* II.12, II.13]—and thus the amount of line *GD* is known. Since the square of line *GB*, which is known, is greater than the square of line *BD* by [the amount of the square of line *GD*], the amount of line *BD* is known. Therefore, it follows as before that all the angles of right triangle *BDG* are known. In the first figure this yields angle *BGA* and one part of angle *GBA*, namely angle *GBD*, and in the second figure angle *BGA* is known because angle *BGD*, its supplement in two right angles, is known; also angle *GBD* is known. Moreover, since lines *AG* and *GD* are known, the amount of line *AD* is known. Thus all the sides of right triangle *BDA* are known, and therefore angle *BAG* is known in both figures. The remaining angle *GBA* in the triangle is known because angle *GBD* is known and angle *DBA* is known, from which it follows with a little thought that angle *GBA* is known in both figures. Therefore it is clear that all sides and angles of triangle *ABG* are known, and this is what we sought to demonstrate.

If we know two sides of any triangle whatever and one angle such that one of the known sides subtends it, the other angles and the third side are known.[65]

[65] Although Levi is discussing what is now known as the "ambiguous" case, he is assuming that in any particular problem, one of the unknown angles is assumed to be acute or obtuse, so there is only a single solution.

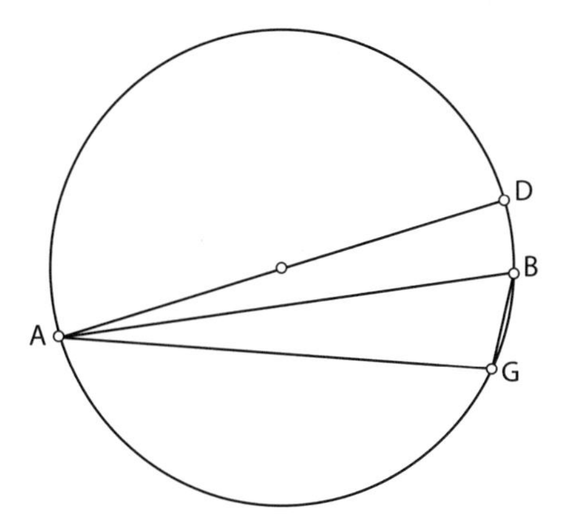

Fig. III-5-4.

Let the two known sides be lines *AB* and *BG* in triangle *ABG*, and let angle *BAG* be known [Fig. III-5-4]. I say that line *AG* is known and that the remaining angles are known. Let us circumscribe circle *BAG* about triangle *BAG*, and let us consider the diameter of the circle to be line *AD*. Since angle *BAG* is known and we consider it as an inscribed angle, . . . arc *BG* is known. Therefore, the amount of the chord of this arc may be found from the table of arcs and Sines in the measure where line *AD* is 120, and so the ratio of line *BG* to line *AD* is known. Since the ratio of line *BG* to line *AB* is also known, the ratio of line *AB* to line *AD* is known. It follows that line *AB* is known in the measure where line *AD* is 120, and thus arc *AB* may be found in the table of arcs and chords. Since both arcs *BG* and *AB* are known, the remaining arc *GA* is known, from which angle *BGA* and line *GA* are known by the aforementioned procedures. Thus it is clear that the sides and angles of triangle *ABG* are all known, and this is what we sought to demonstrate.

You ought to understand from this explanation that if you consider the diameter of the circle to be 60 and the circumference to be 180°, it is not necessary to compute the chords of these arcs in a table of arcs and chords, because the table of arcs and Sines serves this purpose inasmuch as the Sine is half the chord of twice the arc. The ratio of the Sine of any arc to the semi-diameter of the circle is equal to the ratio of the chord of twice that arc to the diameter of the circle. Therefore in this proof we chose the second alternative, so that our remarks would be brief. We mention this here to avoid confusion in our subsequent proofs. From this theorem it follows that in any triangle whose sides are straight lines, the ratio of one side to another is equal to the ratio of the Sines of the angles that they subtend. It also follows with a little thought that if the angles of a triangle with straight sides are known and one side is also known, the remaining sides are known because their ratios to the known side are known.

If two sides of any triangle are known, and the included angle is also known, the remaining angles and sides are known.

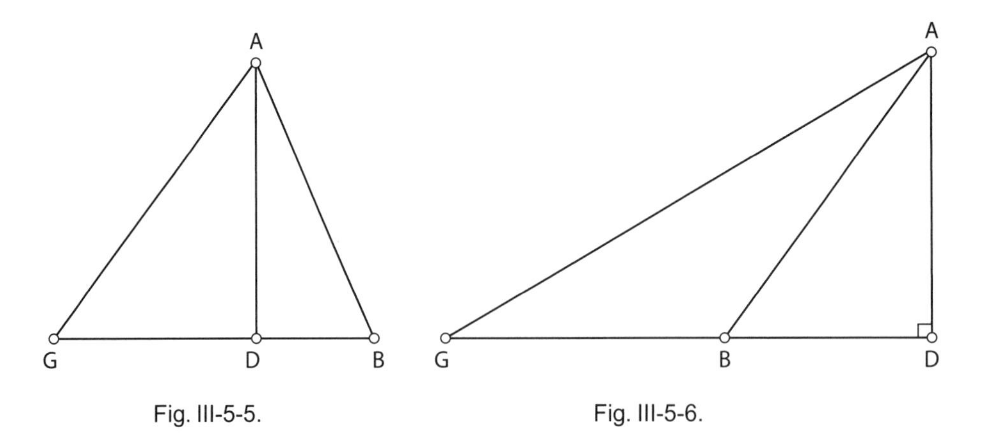

Fig. III-5-5.

Fig. III-5-6.

Consider triangle *ABG* whose sides *AB* and *BG* are known and angle *ABG* is also known; I say that line *AG* is known and that the remaining angles are known. If angle *ABG* is a right angle, this is clear from the preceding. Moreover, if it is acute as in the first figure [Fig. III-5-5] or obtuse as in the second figure [Fig. III-5-6], line *AG* is known. Let us draw perpendicular *AD* from point *A* to line *BG*, extended if necessary. In either case it is clear that angle *ABD* is known because either angle *ABG* or its supplement in two right angles is known. There remains angle *DAB* which is known because it is the complement in a right angle. Therefore all angles and one side of triangle *ABD* are known, and the rest may be found. Moreover, lines *AD* and *DG* are known in both figures. The amount of line *AG* in right triangle *ADG* is known, and thus the angles of triangle *ADG* are known including angle *AGD*. It was already assumed that angle *ABG* is known; there remains angle *BAG* which is known because it is the supplement in two right angles, and this is what we sought to demonstrate.

IV. SCHOLARLY GEOMETRY

This section is dedicated to geometry in the tradition of Euclid, Apollonius, and Archimedes. We start from Levi ben Gershon's attempt to reduce the parallel postulate to a simpler, more intuitive axiom. We continue with Qalonymos ben Qalonymos's discussion of regular polyhedra, Bonfils's circle measurement, and one of the discussions of the asymptote of the hyperbola. To conclude, we bring the work of one of the most enigmatic and unique Hebrew mathematical authors: Abner of Burgos (also known as Alfonso di Valladolid), who, in his treatise on squaring the circle, which survives only in a fragment, made interesting contributions to the quadrature of lunes and the Western reemergence of the conchoid.

1. LEVI BEN GERSHON, *COMMENTARY ON EUCLID'S ELEMENTS*

Levi ben Gershon's (see section I-6) commentary on Books I–V of Euclid's *Elements* was probably written shortly after 1337. We do not know Levi's sources, but there were numerous commentaries on the *Elements* written in Arabic starting in the tenth century, some translated into Hebrew, as well as commentaries written in Latin. The one to which Levi's work seems closest is *The Book Explaining the Elements of Euclid*, originally attributed to Naṣīr al-Dīn al-Ṭūsī (1201–1274), but more recently attributed to his son or one of his students [see Rosenfeld, 1988, pp. 80–85; and Lévy, 1992, pp. 90–91]. Levi's commentary deals with selected definitions, postulates, and theorems from the first five books of the *Elements*.

Here we only present the commentary on the parallel postulate (postulate 5), with Levi's proof of the postulate beginning with his own more "self-evident" postulates. Levi's first postulate is essentially that a straight line can be extended to make it greater than any given straight line. His second postulate is that if the two lines in question (in Euclid's postulate 5) form an acute and a right angle, respectively, with the cutting line, then they approach one another on the side of the acute angle and grow farther apart in the opposite direction. (The latter part is omitted from the formulation but is used in the proof.) This leaves two tasks: (1) to show that when the two intersection angles summing to less than two right angles are obtuse and acute, the lines still approach one another (this is achieved in lemma 9), and (2) that the two approaching straight lines actually intersect (shown in the final proof). (See section IV-5 of Chapter 3 for a "proof" of the parallel postulate by al-Maghribī.)

Euclid says: if a straight line falls on two straight lines, forming [on one side] two interior angles less than two right angles, then, if the lines are prolonged on the same side, they will intersect.

Levi says: This proposition is very profound; it is not easy to validate it. In fact, it is not widely accepted that if the two interior angles are less than two right angles, one being obtuse and the other acute, then the two lines intersect. By the same token, the following statement may not be considered as one of the common notions: if a [straight] line falls on two straight lines, forming on one of the two sides two interior angles less than two right angles, then it follows that any other straight line falling on those [straight lines] forms also on the same side two interior angles less than two right angles. Rather, since in the man of subtle intelligence who immerses himself deeply in this science, this produces a doubt, the more so for any novice, the understanding of this premise is not easy. Furthermore, it has been demonstrated in this science [geometry] that [it is possible] for two lines having between them, at the beginning, a certain distance, that they approach each other as they are extended but never meet, even when prolonged to infinity.[1] This premise can also be subject to doubt, although we acknowledge that it is generally recognized that such lines approach one another.

The premise in question is particularly necessary to this science, as is seen with proposition 29 of the first book of this work [Euclid's *Elements*] and the following—from that premise one derives the properties of parallel lines and the equality of the sum of the three angles of any triangle with two right angles. Thus, if it were to fail, the geometry in its totality, or in its major part, would fail. This is why we thought it appropriate to establish it by a proof. This demonstration takes place after the preceding twenty-eight propositions of the book, since Euclid did not have recourse to this premise for any of these propositions.

We pose as a preamble two well-known premises. The first is that which Euclid has mentioned in the fifth book: in essence, he states that it is possible to multiply any line to

[1] See section IV-5 on the hyperbola and its asymptote.

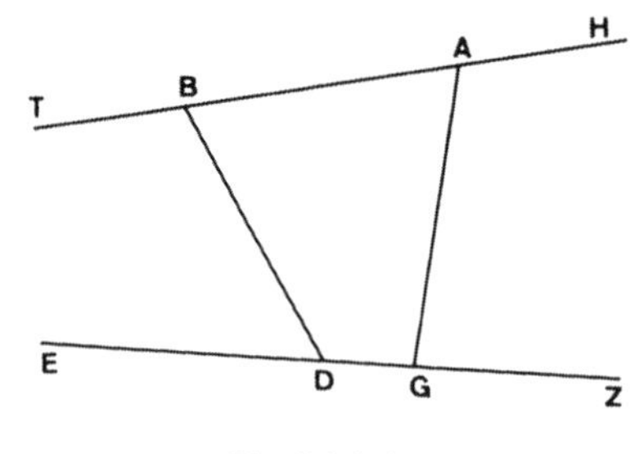

Fig. IV-1-1.

obtain a line greater than a certain given line. This premise is very clear; even more so given all that had been delineated before.

The second premise: the straight line which is inclined [to another straight line] approaches [the second line] on the side where an acute angle is formed [with a line crossing both of these that is a perpendicular from the first line to the second].[2] This is established by considering the sense of the definition; in fact, the notion of inclination means nothing other than the fact that [one line] approaches [the other] in the direction in which they are inclined.

It follows that two straight lines drawn to form two acute angles [with the same straight line] approach one another in the direction [of the acute angles], since each of these is inclined to the other, the notion of acute angle expressing the fact that the line [forming one of the sides of the acute angle] is inclined in the direction [of the line forming one side of the other acute angle]. It follows that on the opposite side they move away, for they approach one another on this side; and also, because on that second side, they depart from the two obtuse angles and each is inclined to the side opposite to that of the other. This is clear and there can be no further doubt as to its truth.

Here ends the commentary on the beginning of [Euclid's] book.

. . .

Here are the propositions that are necessary to demonstrate that: if a straight line falls on two straight lines and forms, on one of the sides, two interior angles less than two right angles, then the lines, if they are prolonged on the same side, will intersect. These propositions have their place after the first 28 propositions of the first book [of Euclid's *Elements*].

[Lemma 1:] There does not exist any quadrilateral figure having all its angles obtuse or having all its angles acute.

Let the quadrilateral be *ABGD*; we claim that it is impossible that all angles be acute or all obtuse. Proof of this impossibility: Suppose that it is possible and assume first that the angles are acute. Prolong the line *AB* in a straight line in two directions to the points *H* and *T*; and also prolong the line *GD* in a straight line in two directions to the points *Z* and *E* [Fig. IV-1-1]. Since each of the two angles *TAG* and *AGE* are acute and thus the angles

[2]This addition to Levi's words helps make sense of Levi's assumption [Lévy, 1992, p. 101]. Tony Lévy justifies it by noting that in Levi's *Treatise on Geometry*, he defines the "approach" of one line to another in terms of a line cutting both. In addition, this addition seems necessary to allow Levi to reach the conclusion in the next paragraph.

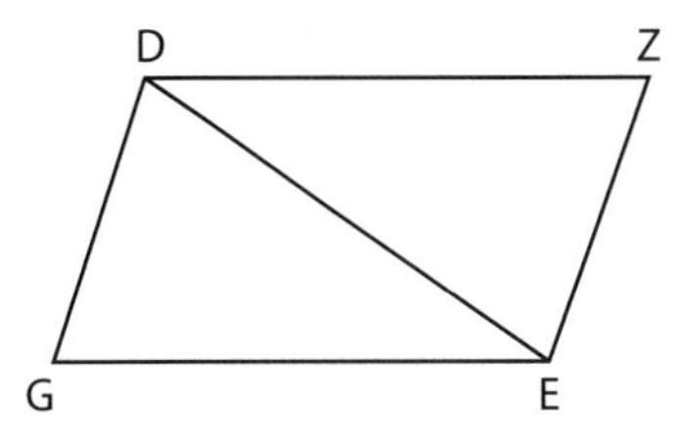

Fig. IV-1-2.

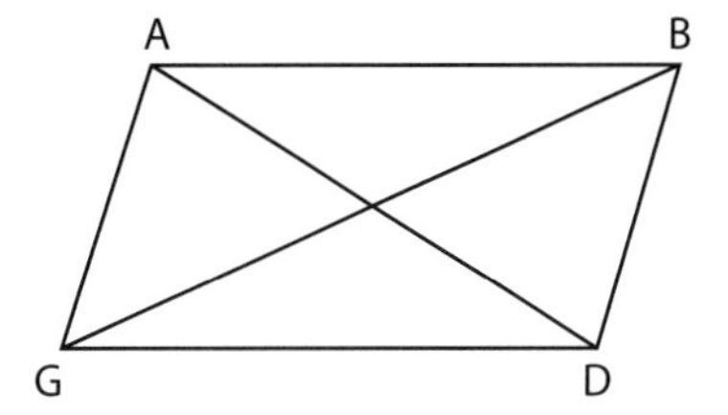

Fig. IV-1-3.

HAG and *AGZ* are obtuse, the two lines *ZE*, *HT* separate from each other in the direction of the two points *H* and *Z* and approach each other in the direction of the two points *T* and *E*. Also, since each of the two angles *ABD* and *BDG* are acute, each of the two angles *TBD* and *BDE* are obtuse, and thus the two lines *HT*, *ZE* separate in the direction of the points *T* and *E*, and approach each other in the direction of the points *H* and *Z*. They therefore separate on the side of the two points *E* and *T*, but also approach on the side of the two points *E* and *T*. This contradiction is impossible. Thus the four angles of the figure *ABGD* cannot all be acute. In the same way, one demonstrates that it is impossible that each of the four angles of the figure *ABGD* is obtuse. As a consequence, there cannot be a quadrilateral figure of which all the angles are obtuse or of which all the angles are acute. QED

[Lemma 2:] We wish to construct a quadrilateral figure for which the opposite sides, taken in pairs, are equal to each other.

Given two straight lines *DG*, *GE* of arbitrary length, meeting at the angle *DGE*, we draw the straight line *DE* [Fig. IV-1-2]. Construct on the line *DE* the triangle *EZD*, where each line is congruent to the corresponding line of triangle *DGE*, that is, the line *EZ* to the line *GD*, and the line *ZD* to the line *GE*. The figure *GDEZ* is thus a quadrilateral of which the opposite sides, taken in pairs, are equal to each other. We have therefore constructed a figure of which the opposite sides, taken in pairs, are equal to each other. QED

[Lemma 3:] Every quadrilateral figure in which the opposite sides are equal to one another also has the opposite angles equal.

Let the quadrilateral *ABGD* have the side *AB* equal to the side *GD* and the side *AG* equal to the side *BD*; I say that the two opposite angles *GAB*, *GDB* are equal and the opposite angles *ABD*, *AGD* are also equal [Fig. IV-1-3]. The proof: Draw the two straight lines *AD*, *BG*. Since the two sides *AB*, *BD* are equal to the two sides *GD*, *AG*, each to its corresponding side [in the triangles *ABD*, *DGA*], and the base *AD* is common, thus the angle *AGD* is equal to the angle *ABD*, to which it is opposite. Similarly, since the lines *AG*, *AB* are equal to the lines *BD*, *DG*, respectively, and the base *BG* is common, thus the angle *GAB* is equal to the angle *GDB*, to which it is opposite. It is thus established that the quadrilateral figure *ABGD* has opposite angles equal. Thus, every quadrilateral figure of which the opposite sides are equal in pairs has the opposite pairs of angles equal. QED

[Lemma 4:] Given an isosceles triangle, if one extends one of the two equal sides in a straight line from their point of intersection by a distance equal to the original length of the

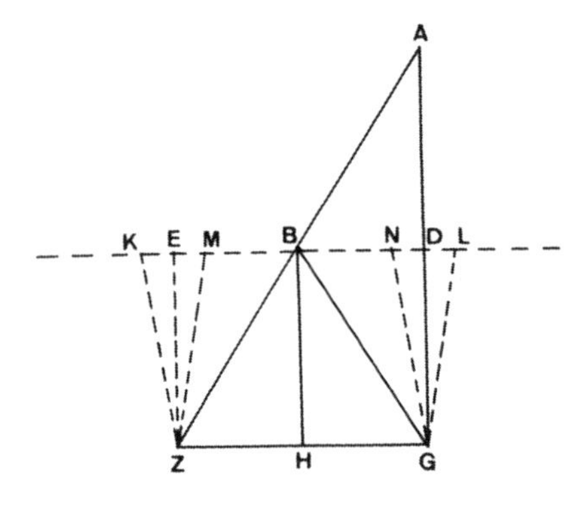

Fig. IV-1-4.

side, and if one draws the base [of the triangle thus formed], this latter forms a right angle with the original base.

For example, let the isosceles triangle be *ABG*, the lines *AB*, *BG* being equal to each other. The straight line *AB* is extended up to *Z* such that *BZ* is equal to one of the two lines *AB*, *BG*; the line *GZ* is drawn [Fig. IV-1-4]. I say that then the angle AGZ is right.

The proof: Divide the line *AG* into two halves at the point *D*, and draw the line *DB*. It results from the eleventh proposition of the first book [of Euclid] that the angle *ADB* is right, and the same for the angle *GDB*. Extend the straight line *DB* to *E*, such that the line *BE* is equal to the line *DB*, and draw the line *EZ*. Since the two lines *EB*, *BZ* are equal, respectively, to the two lines *DB*, *BA*, that is, the line *DB* to the line *BE*, and the line *AB* to the line *BZ*, and since the two vertical angles [at *B*], *DBA*, *EBZ* are equal, then the base *EZ* [of triangle *EBZ*] is equal to the base *AD* [of triangle *ABD*], and the remaining angles [of the first triangle] are equal to the remaining angles [of the second triangle], each to its corresponding one. As a consequence, the angle *BEZ* is equal to the angle *ADB*, and the angle *ADB* is right, so the angle *BEZ* is right. Since the line *EZ* is equal to the line *DA*, which is equal to the line *GD*, [then] the line *EZ* is equal to the line *GD*.

I say that the line *GZ* is [also] equal to the line *DE*. The proof is that it cannot be otherwise; if it were otherwise, then the line *GZ* would either be greater than the line *DE* or would be less than this line. Suppose, first, that the line *GZ* is greater—if this were possible—and divide the line *GZ* into two halves at the point *H*. Since the line *GZ* is greater than the line *DE*, the line *GH*, half of the line *GZ*, is greater than the line *DB*, half of the line *DE*. Also, in the same way, it becomes clear that the line *HZ* is greater than the line *BE*. Extend *BD* in a straight line to *L* so that *BL* is equal to *GH*, and extend *BE* in a straight line to *K* so that the line *BK* is equal to the line *HZ*. Since the two lines *GH*, *HZ* are equal to each other, the two lines *LB*, *BK* are equal to each other. Also, since the line *LB* is equal to the line *GH* and the line *BK* is equal to the line *HZ*, therefore the entire line *LK* is equal to the entire line *GZ*. And since the line *BL* is equal to the line *BK*, and the line *BD* is equal to the line *BE*, it follows that the line *DL* is equal to the line *EK*. The two lines *LD*, *DG* are thus equal to the lines *KE*, *EZ*, each to its corresponding one, that is, the line *LD* to the line *KE*, and the line *DG* to the line *EZ*; and the right angle *LDG* is equal to the right angle *KEZ*; as a consequence, the base *GL* [of triangle *GLD*] is equal to the base *ZK* [of triangle *ZKE*]. Therefore, the figure *LGZK* is a quadrilateral of which the

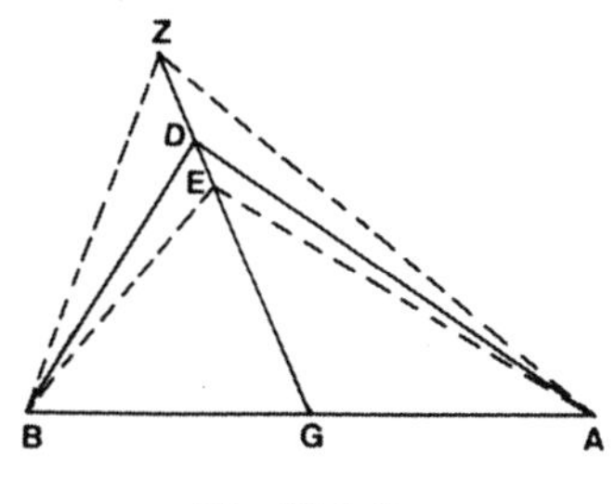

Fig. IV-1-5.

opposite sides are equal in pairs; therefore the opposite angles are equal, the angle *KLG* being equal to the angle *KZG*, and the angle *LKZ* being equal to the angle *LGZ*. As the angle *GDB*, exterior to the triangle *GLD*, is right, the interior angle *GLD* [of the triangle] is less than a right angle; in the same manner, it is shown that the angle *ZKE* is less than a right angle. Since the two angles *GLK*, *LKZ* are acute, the two opposite angles [in the quadrilateral] are [also] acute. Thus, the quadrilateral *GLKZ* has all its angles acute, and this is false. As a consequence, the line *GZ* is not greater than the line *DE*.

The proof that line *GZ* cannot be less than line *DE* is similar.

It has already been demonstrated that [*GZ*] is not greater than [*DE*]. As a consequence, the line *GZ* is equal to the line *DE*. But the line *GD* is also equal to the line *ZE*. The figure *DEZG* is therefore a quadrilateral of which the opposite sides, taken in pairs, are equal; the opposite angles are therefore [also] equal. But the angle *DEZ* is right, and therefore the angle *DGZ* is right. QED

[Lemma 5:] In every right triangle, if the side opposite the right angle is divided into two halves and a straight line is drawn from the point of division to the right angle, then the straight line which results is equal to each of the parts of the divided line.

Let *ABD* be a right triangle with angle *ADB* the right angle. The line *AB* is divided into two halves at the point G and the straight line *GD* is drawn. I say that the line *GD* is equal to each of the two lines *AG*, *GB* [Fig. IV-1-5].

The proof is that it cannot be otherwise. We suppose that the line *GD* were either greater than each of the two lines *AG*, *GB* or that it were smaller. To begin, suppose that it is greater, if that were possible. A line is cut off along the line *GD* equal to each of the two lines *AG*, *GB*, say, *GE*, and the two straight lines *AE*, *EB* are drawn. Since the triangle *AGE* is isosceles, if one of the two sides, suppose the line *AG*, is extended from the intersection of the two sides [*GE*, *GA*] an equal length, to give the line *GB*, and the straight line *EB* is drawn, then the angle *AEB* is right [lemma 4]; but the angle *ADB* is [also] right. One has thus constructed on one of the sides of the triangle *ADB*, that is, the line *AB*, two straight lines drawn from the extremities, meeting in the interior of the triangle and subtending an angle equal to the angle subtended by the two other sides [of triangle *ADB*]—the right angle *ADB* being equal to the right angle *AEB*—and this is false [*Elements* I.21]. By consequence, the line *GD* is not greater than each of the two lines *AG*, *GB*.

The proof that line *GD* is not smaller than the lines *AG*, *GB* is similar.

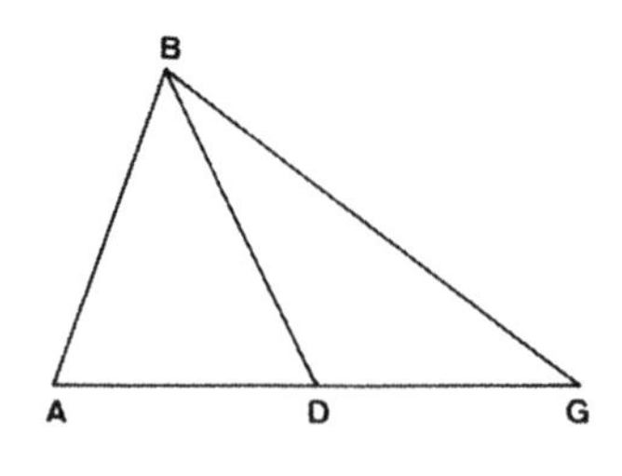

Fig. IV-1-6.

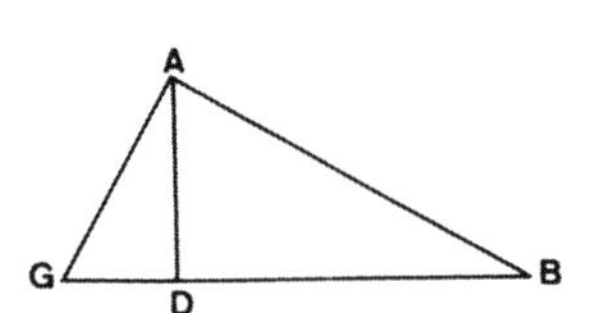

Fig. IV-1-7.

But it has already been demonstrated that [*GD*] is not greater than each of the lines *AG*, *GB*. So the line *GD* is equal to each of the lines *AG*, *GB*. QED

[Lemma 6:] In every right triangle, the non-right angles are together equal to a right angle.

Let the right triangle be *ABG*, of which the angle *ABG* is right. I say that the two angles *BAG*, *BGA*, taken together, are equal to the angle *ABG*, which is right [Fig. IV-1-6].

The proof: *AG* is divided into two halves at the point *D*, and the straight line *DB* is drawn. Then the line *DB* is equal to each of the two lines *DA*, *DG*; the triangle *ADB* is therefore isosceles. Also, the triangle *BDG* is isosceles. Thus, the angle *DBG* is equal to the angle *DGB*, and the angle *DBA* is equal to the angle *DAB*. Therefore, the two angles *DAB*, *DGB*, taken together, are equal to the angle *ABG*, which is right. QED

[Lemma 7:] In every rectilinear triangle, the three angles [together] are equal to two right angles.

The proof: It cannot but be otherwise that either [the triangle] is a right triangle or it is not a right triangle. If it is a right triangle, the property is proved by virtue of the previous proposition. So assume that it is not a right triangle; I claim that the three angles are [also] equal to two right angles.

Example: Let the non-right triangle be *ABG*. The perpendicular is drawn from the point *A* to the base *BG*, supposed unlimited; let the perpendicular be *AD*. If the perpendicular falls on the line *BG* between the points *B* and *G*, which is the case in the first figure [in Fig. IV-1-7], it is clear that the three angles of the triangle *BAG* are [together] equal to two right angles.

The proof: Since the triangle *ADG* is a right triangle, the two angles *DAG*, *DGA* are together equal to a right angle; it is likewise established that the two angles *DAB*, *DBA*

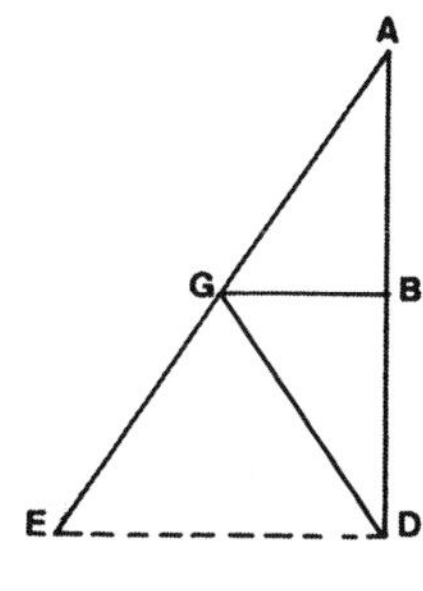

Fig. IV-1-8.

are [together] equal to a right angle. The three angles of the triangle *BAG* together are therefore equal to two right angles.

Similarly, suppose that the perpendicular *AD* falls outside of triangle *ABG*, which is the case in the second figure [in Fig. IV-1-7]. I claim that the three angles of triangle *BGA* are [together] equal to two right angles. The proof: Since the triangle *ADB* is a right triangle, the two angles *DAB*, *DBA* are together equal to a right angle; but the angle *DAB* is itself equal to the [sum of] the two angles *DAG*, *GAB*, and the three angles *DAG*, *GAB*, *DBA* are therefore [together] equal to a right angle. Also, since the triangle *ADG* is a right triangle, the two angles *DAG*, *AGD* are equal to a right angle. The three angles *DAG*, *GAB*, *ABG* being [together] equal to the two angles *DAG*, *AGD*, the common angle *DAG* is subtracted; the angles *GAB*, *ABG* which remain are [together] equal to the remaining angle *AGD*. The angle *AGB* is added [to both]. Then the two angles *AGB*, *AGD* [together] are equal to the three angles *AGB*, *GAB*, *ABG* [together]. But the two angles *AGB*, *AGD* [together] are equal to two right angles; the three angles of triangle *ABG* are therefore also equal to two right angles.

It has thus been demonstrated that the three angles of any rectilinear triangle are equal to two right angles, but this conclusion is not required for the result that we seek.

[Lemma 8:] In every right triangle, if one of the sides containing the right angle is extended by a length equal to itself, and if the side opposite the right angle is extended in the same direction by a length equal to itself, and if a straight line is drawn connecting the extremities of the lines thus obtained, then this line makes a right angle with the extension of the line containing the right angle.

Let *ABG* be the triangle, in which the angle *ABG* is right. The line *AB* is extended to *D*, so that the line *BD* is equal to the line *AB*. The line *AG* is extended in a straight line to *E*, the line *GE* being equal to the line *AG*, and the straight line *DE* is drawn [Fig. IV-1-8]. I say that the angle *ADE* is right.

The proof: Draw the straight line *GD*. Since the right angle *ABG* is equal to the right angle *GBD*, and the line *AB* is equal to the line *BD*, and the line *BG* is common [to the two triangles *ABG*, *GBD*], then the base *GD* is equal to the base *AG*. The triangle *AGD* is therefore isosceles, and one of its equal sides, the line *AG*, [has been extended] a length equal to itself, that is, *GE*. As a consequence [Lemma 4], the angle *ADE* is right. QED

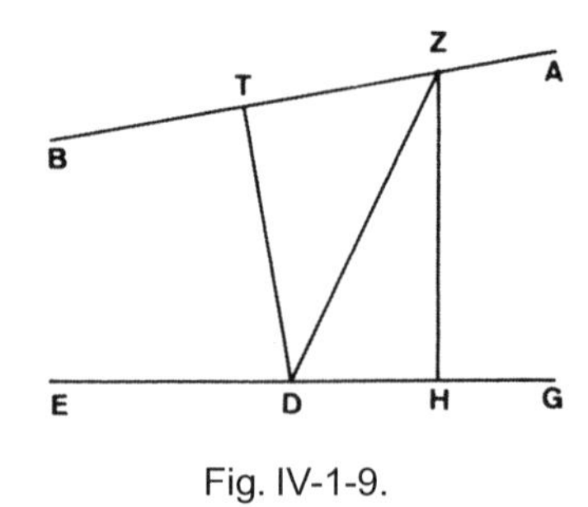

Fig. IV-1-9.

[Lemma 9:] If a straight line falls on two straight lines and forms on one side two interior angles that together are less than two right angles, and if from the vertex of one of the angles a perpendicular is drawn to the second line, then the perpendicular forms, with the other line, an acute angle on the side where the two interior angles are less than two right angles.

Let the two straight lines be supposed unlimited, the lines *AB*, *GE*, and, falling on them, the straight line *ZD* forms [with these] two angles *BZD*, *ZDE*, [together] less than two right angles. I claim that if from point *Z* a perpendicular *ZH* is drawn to the line *GE*, supposed unlimited, the obtained angle *HZB* is acute, that is, the angle that is determined on the side on which the two interior angles are less than two right angles [Fig. IV-1-9]. Similarly, if from point *D* a perpendicular *DT* is drawn to the line *AB*, supposed infinite, the angle *TDE* will be acute.

The proof: Since the three angles of triangle *ZHD* are equal to two right angles, and since the two angles *ZDH*, *ZDE* are also equal to two right angles, the angle *ZDE* is equal to the two angles *ZHD*, *HZD* [taken together]. The angle *BZD* is added [to both]. The two angles *BZD*, *ZDE* are therefore equal to the three angles *ZHD*, *HZD*, *BZD*. But the two angles *ZDE*, *BZD* are [together] less than two right angles; thus the three angles *ZHD*, *HZD*, *BZD* are [together] less than two right angles. But the angle *ZHD* is right. It follows that the two angles *HZD*, *BZD* are [together] less than a right angle, so the angle *HZB* is acute.

Similarly, it is proved that the angle *TDE* is acute. In fact, the angle *BTD*, which is right, is equal to the two angles *TZD*, *TDZ*, since the latter, taken together, form a right angle. The angle *TDE* is added [to both]. The two angles *BTD*, *TDE* are therefore equal to the three angles *TZD*, *ZDT*, *TDE*. But the three angles *TZD*, *ZDT*, *TDE* are less than two right angles. Therefore the two angles *BTD*, *TDE* are less than two right angles. But the angle *BTD* is right; so the remaining angle *TDE* is less than a right angle. QED

[Proof of the postulate of parallels:]
If a straight line falls on two straight lines, forming on one side two interior angles [together] less than two right angles, then the two straight lines, if they are prolonged indefinitely on the same side, will meet.

Let the two straight lines be *AB*, *GD* and, falling on them, the straight line *AE*, forming two angles *BAE*, *AED* [together] less than two right angles. I claim that the two lines *AB*, *GD*, if they are prolonged indefinitely in the direction of *B*, *D*, will meet [Fig. IV-1-10].

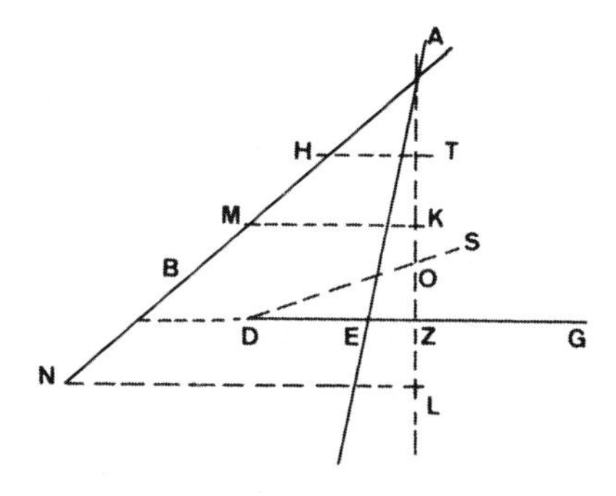

Fig. IV-1-10.

The proof: The perpendicular is constructed from the point *A* to the line *GD*, supposed unlimited; this is the line *AZ*. It is clear from that which precedes that the angle *ZAB* is acute [Lemma 9]. A point, say, the point *H*, is marked anywhere on the straight line *AB*, and from this point the perpendicular to the straight line *AZ* is constructed; this is the line *HT*. Since the angle *ZAB* is acute, it is clear that the point *T* falls on the line *AZ*, extended on the side of *Z*. It has been established that the line *AT*, if multiplied sufficiently many times, will be greater than the line *AZ*. The line *TK* is chosen to be the same length as the line *AT*, and the line *KL* the same length as the line *AK*. When this is done several times, this [line so constructed] will be greater than the line *AZ*.

Suppose that the line *AL* is greater than the line *AZ*. Since the angle *ATH* is right, it is clear that, if the straight line *AH* is prolonged in a straight line an equal length, say, *HM*, and in the same manner the straight line *AT*, giving the line *TK*, then the line that joins the point *K* and the point *M*, that is, the line *KM*, meets the line *AK* in a right angle, as has already been proved [Lemma 8]. Similarly, it is clear that, if the straight line *AM* is prolonged in a straight line an equal length, giving the line *MN*, and the straight line *NL* is drawn, then the angle *ALN* is right.

But since the angle *AZD* is right, the angle *LZD* is also right; the two lines *ZD*, *LN* are therefore parallel on the side [of the triangle *ALN*; *Elements* I.28]. Since the straight line *ZD* is interior to the triangle *ALN*, and the triangle *ALN* is finite, it is therefore possible to prolong the straight line *ZD* continuously in a straight line, such that it exits the triangle *ALN*; but it is not possible that the straight line *ZD*, prolonged indefinitely, exits between the two points *L* and *N*, since the two lines *ZD*, *LN*, which are parallel, would meet, and this is false. I say also that the straight line *ZE* may not exit between the two points *L* and *A*; if this were possible, then it would also be that the line *ZD* would cut the line *AL* at *O*, and in this case the two straight lines *ZO*, *ZDS* would encompass a surface, and this is false. This being so, it remains only that the straight line *ZD*, when it is prolonged indefinitely, must exit the triangle *ALN* between the two points *A* and *N*. As a consequence, the straight line *ZD* meets the line *AN*. QED

2. LEVI BEN GERSHON, *TREATISE ON GEOMETRY*

The *Treatise on Geometry* was probably written shortly after the commentary on the *Elements*. Levi wanted to construct geometry on a stronger foundation than Euclid's.

Unfortunately, the only extant manuscript of this work contains just the first 24 definitions and hypotheses. We present here the twenty-third of these, which distinguishes between the necessary boundedness of actual lines in the world and the indefinite extensibility of mathematical lines.[3]

23. I say that it is possible to prolong a limited straight line in a straight line, continuously, without any determined limit. In fact, the line is always augmentable while still being limited.

This hypothesis has been subjected to doubt. In fact, the Philosopher [Aristotle] contradicts it in his celebrated book *On Physics*. He has explained the impossibility of supposing that a line can be greater than the line that is the greatest that is contained in the universe, that is, the world in its totality. In fact, a line only has existence in a body. Since this premise is very necessary for the geometer, it is appropriate for us to investigate the doubt that befalls it. We say that the aspect through which the Philosopher has contradicted this premise is the necessity for a body to be finite, as has been explained in the place mentioned; but since a line has existence only in a body, it is necessary that the magnitude of the line be limited, that is to say, it is impossible that it can be greater than the straight line contained in the universe, that is, the world in its totality.

This being so, we should investigate from what aspect it is necessary for a body to be finite and limited, that is, not greater than the body of the world. We say that it is manifest that this is a necessity for a body inasmuch as it is a physical body, as has been proven there. However, the geometer supposes that the line is unlimited from the point of view of augmentation, in the sense that one may always add to that which has been added: this does not relate to a line as it is in a physical body, but inasmuch as it is in a mathematical body. From this aspect, the impossibility of the absence of a determined limit in the body has not been demonstrated from it [the body] being always finite. Indeed, the geometer poses this hypothesis from the perspective that it is possible and not from the perspective that it is impossible. As a consequence, in this perspective, no contradiction follows from this premise.

With this premise posited, the geometer recognizes that it is impossible for any magnitude to be infinite, since it exists in actuality and must necessarily be limited. And we claim that this does not contradict the premise that we have mentioned: in fact, what is necessary for a magnitude inasmuch as it has magnitude—and it is this aspect that the geometer considers—is that it not be infinite, but not that it not be greater than the world. Although we acknowledge that the line is always augmentable indefinitely, it does not necessarily follow that the line is infinite in magnitude. In fact, the line, whatever increments it receives, is finite, and that is always so. As for the absence of a limit that we posit with respect to it—it relates to the possibility of augmenting and does not mean that the line is [actually] infinite in magnitude.

[3]The source of the distinction between an unacceptable actual infinity and a permissible indefinite infinity goes back to Aristotle, who is Levi's obvious reference. A possibly related discussion in a mathematical context can be found in Ibn al Haytham's interpretations of Euclid. His concern is the tension between the finite lines that our imagination can contain, and the indefinitely extended lines required by Euclid's postulates [Sude, 1974, 49–55, 88–90].

3. QALONYMOS BEN QALONYMOS, *ON POLYHEDRA*

Qalonymos ben Qalonymos (see section II-5) translated into Hebrew an anonymous Arabic manuscript on polyhedra early in the fourteenth century, a manuscript extending "Book XIV" of Euclid's *Elements*, written by Hypsicles in the second century BCE. Muhyī al-Dīn al-Maghribī used this same manuscript to produce a new Arabic version of the same treatise.

It is not known whether the original from which each borrowed was an Arabic original or a translation from a Greek original. In any case, although the Qalonymos and the al-Maghribī versions are similar, neither is a copy of the other. They have most of their theorems in common, but the ordering and some of the proofs are different. A substantial part of al-Maghribī's version is in section IV-5 in Chapter 3. Here we present two propositions from Qalonymos's version that are not included in that material (although the propositions do occur in the al-Maghribī version). We have chosen to include these in part to demonstrate that Qalonymos believed that there would be a Hebrew-reading audience for these rather advanced geometrical ideas.[4]

[Proposition 18]: We wish to show that the ratio of the [surface] area of the cube to the [surface] area of the icosahedron is as the ratio of the square of the side of the pentagon[5] to [three] and a third times the equilateral triangle whose side equals the root of three times the square on a decagon of the circle that circumscribes the pentagon.

For example, we take AB as the side of the hexagon[6] and divide it in mean and extreme ratio at C. Let D be equal in power [*maḥziq*][7] to AB and AC [i.e., the square on D is equal to the sum of the squares on AB and AC], so D is the side of the pentagon.[8] Let E be equal in power to AB and BC, that is, the root of three times the square of AC, where AC is the side of the decagon.[9] I say that the ratio of the [surface] area of the cube to the [surface] area of the icosahedron equals the ratio of the square of D to three and a third times the equilateral triangle on E. Its proof is: it has already been shown in the preceding proposition that the ratio of the edge of the cube to the edge of the icosahedron is as D is to E, and the ratio of the [square on the] edge of the cube to the square on the edge of the icosahedron is as the square on D to the square on E.[10] Inverting, the ratio of the square [on the edge] of the cube to the square on D is as the ratio of the square on the edge of the icosahedron to the square on E. [For any two lines, the ratio of the square on the first line to twice the triangle on that line is as the ratio of the square on the second line to twice the triangle on that line.] So the square on D is to twice the triangle on D as the square on E is to twice the triangle on E. Therefore, the square [on the edge] of the cube is to the square on D as the square on the edge of the icosahedron is to the square on E. And [the square on the edge of the cube] is to twice the triangle on D, which is twice

[4]For more details on the two manuscripts and on another Hebrew version, see [Langermann, 2014].

[5]The "pentagon" is the pentagon formed by the bases of five triangles of the icosahedron that have a common vertex. Thus the side of the pentagon is an edge of the icosahedron.

[6]The "hexagon" is the hexagon inscribed in the circle that circumscribes the pentagon.

[7]The Hebrew word *maḥziq*, translated here as "equal in power," is a translation of an analogous Arabic word and also of the Greeek *dunamene*.

[8]*Elements*, XIII.9 and XIII.10.

[9]*Elements*, XIII.4.

[10]This is proposition 7 of *Elements* XIV, the work of Hypsicles; see [Montelle, 2014].

the face of the icosahedron, as the square on D, which is the edge of the icosahedron, to twice the triangle on E. The ratio of six times the square of the [edge of the] cube, which is the [surface] area of the cube, to twelve times the face of the icosahedron, which is three fifths of the [surface] area of the icosahedron, is equal to the ratio of the face of the cube to twice the face of the icosahedron, which is the same as the ratio of the square on D to twice the triangle on E. The [ratio of the] surface area of the cube to three-fifths of the surface area of the icosahedron is equal to the ratio [of the square on D] to twice the triangle on E. The ratio of three-fifths the [surface] area of the icosahedron to the surface area of the icosahedron is equal to the ratio of twice the triangle on E to three and one-third the triangle on E. The complete ratio of the surface area of the cube to the surface area of the icosahedron is equal to the ratio of the square on D to three and one-third the equilateral triangle on E.

It is clear from what we have described that the ratio of three-fifths of the surface area of the icosahedron to the surface area of the cube is the same as the ratio of twice the triangle on E to the square on D.

[Proposition 19]: We wish to show that the ratio of the [surface] area of the icosahedron to the [surface] area of the octahedron is equal to the ratio of five times the square on the side of the decagon of the circle to the square on the side of the pentagon.

Let A be the side of the pentagon and B the side of the decagon. I say that the ratio of the [surface] area of the icosahedron to the [surface] area of the octahedron is equal to the ratio of five times the square on B to the square on A. Its proof is, we take C so that its square is three times the square on B. We have already shown that the ratio of three-fifths of the [surface] area of the icosahedron to the [surface] area of the cube is equal to [the ratio of] twice the triangle on C to the square on A. We have already said that the ratio of the [surface] area of the cube to the [surface] area of the octahedron is equal to the ratio of the square on A to twice the triangle on A, for we have already shown in the preceding that the ratio of the [surface] area of the cube to the [surface] area of the octahedron is equal to the ratio of the side of any equilateral triangle to its altitude,[11] which is like the ratio of the square [of its side] to twice its triangle. The ratio of three-fifths of the [surface] area of the icosahedron to the [surface] area of the cube is equal to the ratio of twice the triangle on C to the square on A. The ratio of the surface area of the cube to the [surface] area of the octahedron is equal to the ratio of the square on A to twice the triangle [on A]. In the equality of the ratio, the ratio of three-fifths of the [surface] area of the icosahedron to the [surface] area of the octahedron is equal to the ratio of twice the triangle on C to twice the triangle on A, which in turn is equal to the ratio of the triangle on C to the triangle on A. The ratio of the triangle on C to the triangle on A is equal to [the ratio of the square on C to the square on A and thus to the ratio of three times the square on B to] the square on A. The ratio of three-fifths of the [surface] area of the icosahedron to the [surface] area of the octahedron is as the square on C, which is the same as three times the square on B, to the square on A. The ratio of three-fifths of the [surface] area of the icosahedron to the [surface] area of the entire icosahedron is equal to the ratio of three times the square on B to five times the square on B. In the equality of the ratio, the ratio of

[11] See proposition 12 in section IV-5 of Chapter 3.

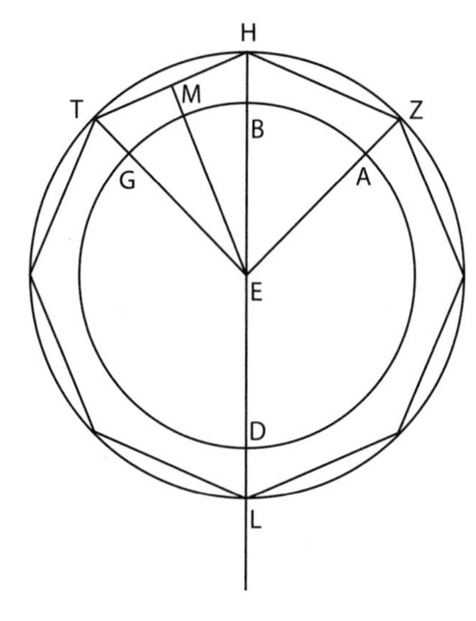

Fig. IV-4-1.

the [surface area of the] icosahedron to the [surface] area of the octahedron is five times the square on B to the square on A.

4. IMMANUEL BEN JACOB BONFILS, *MEASUREMENT OF THE CIRCLE*

This text by Immanuel Bonfils (see section I-3) is a variation on Archimedes's proof for the formula of the area of a circle. It is embedded in a text concerning an evaluation of π (greater than 67,801/21,600 using a 3072-gon), isoperimetric theorems for the circle and sphere, and root extraction. The proof circumvents the construction of a sequence of circumscribed/circumscribing polygons whose areas are arbitrarily close to a given circle; instead it assumes the existence of a circle with a given area and relies on the constructability of a polygon in the ring between any two concentric circles. According to [Lévy, 2012], this proof builds on (and simplifies) the ninth-century proof by *Banū Mūsā*, which relied on constructing a polygon whose circumference is between a given segment and the circumference of a given circle.

Any circle is equal to the product [lit. area] resulting from [the multiplication of] half its diameter by half its circumference. This means that we think of half the circle as if it were a straight line.

Let there be a circle *ABGD* whose center is *E*, and let half its diameter be *EB* and half its circumference arc *BGD*. I say that circle *ABGD* equals the product resulting from the multiplication of line *EB* by arc *BGD* [Fig. IV-4-1].

Proof: If it were not so, let us say that the product resulting from [the multiplication] of line *EB* by arc *BGD* would equal a circle either greater or smaller than the circle *ABGD*.

Let us consider first the greater, that is, circle *ZHTL*, drawn around the center *E*. Inside the circle *ZHTL* let us construct a [regular] polygon with equal angles not touching circle *ABGD*, according to what has been established by [proposition] 13 of [Book] 12 of Euclid [*Elements*], and this figure is *ZHTL*.

Let us draw *EM*, which is the perpendicular to line *HT*. It is evident that the product resulting from [multiplying] *EM* by *MH* is equal to [the area of] triangle *EHT*. This holds for all the triangles constructed in this figure when we draw lines from the center to the vertices of the polygon. Now *MH* is half *HT*. Therefore the product resulting from [multiplying] *EM* by half the perimeter of the polygonal figure equals the entire [area of the] figure *ZHTL*. But *EM* is greater than *EB*, since the [polygonal] figure does not touch the circle [*ABGD*], and half the perimeter of figure *ZHTL* is greater than half the circumference of circle *ABGD*. Therefore the figure *ZHTL* is greater than the product resulting from [multiplying] *EB* by arc *BGD*, which is half the circumference of circle *ABGD*.

We had posited that [the area of the] circle *ZHTL* equals the product resulting from [multiplying] the line *EB* by arc *BGD*. Therefore the figure *ZHTL* is greater than the circle *ZHTL*. This [however] is false, since the circle *ZHTL* circumscribes the figure *ZHTL* and it exceeds the [polygonal] figure by all the [sectors constituted by the] arcs [subtended] by the sides [of the polygon]. Therefore the product resulting from [multiplying] EB by arc BGD is not equal to a circle greater than circle *ABGD*.

The proof of the other alternative is analogous, and we omit it.

5. SOLOMON BEN ISAAC, *ON THE HYPERBOLA AND ITS ASYMPTOTE*

In his *Guide for the Perplexed* I 73, Maimonides claims that some things that are impossible to imagine are nevertheless true. As an example he gives the hyperbola and its asymptote: these lines approach each other indefinitely but never meet.[12] This argument, which is not original to Maimonides, has a long history in the Latin, Arabic, and Hebrew literature [Freudenthal, 1988].

Maimonides's comment inspired several Hebrew investigations of asymptotes. The text quoted here belongs to a Hebrew transmission, which, according to [Lévy, 1989a,b], stems from the Arabic work of al-Ṭūsī. Its proof that a right angle hyperbola approaches its asymptote is clear and concise, even if this manuscript leaves a small gap: it shows that the hyperbola and the asymptote come "much closer" as they extend, but fails to quantify this relation, and therefore does not formally show that they do come *arbitrarily* close. The means to fill the gap, however, are easily accessible to anyone who can follow the text and formulate the gap in a precise manner.

This text is attributed to Solomon ben Isaac, whose identity has not been traced. The manuscript, which seems to be a sixteenth-century Italian copy written during Solomon's life, dates his work to around 1500 [Lévy, 1989a]. Among the other treatments of this problem that belong to the same tradition we should mention the text attributed (falsely, according to [Lévy, 1989b]) to Simon Moṭoṭ [Sacerdote, 1893–1894].

[12] "Know that there are things that a man, if he considers them with his imagination, is unable to represent to himself in any respect, but finds that it is as impossible to imagine them as it is impossible for two contraries to agree. . . . It has been made clear in the second book of the *Conic Sections* that two lines between which there is a certain distance at the outset, may go forth in such a way that the farther they go, this distance diminishes and they come nearer to one another, but without it ever being possible for them to meet even if they are drawn forth to infinity, and even though they come nearer to one another the farther they go. This cannot be imagined and can in no way enter within the net of imagination. Of these two lines, one is straight and the other curved, as has been made clear there in the above-mentioned work" [Maimonides, 1963, vol. I, §73].

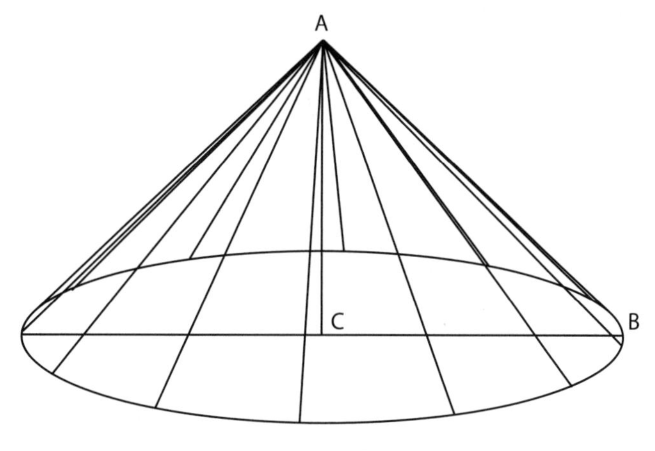

Fig. IV-5-1.

In the text here, we omit the preliminary propositions cited from the *Elements*: the angles of a triangle sum to two right angles; an isosceles triangle has equal base angles and vice versa; an isosceles triangle with a right angle has two half-right angles; in a right angle triangle the height is a mean proportional between the parts of the base as cut by the height; if we divide a line segment, the square on the whole segment equals the squares of the parts and twice the rectangle contained by the parts; and the square built on half a line segment with an addition equals the square on half the line segment and the rectangle contained by the addition and the original segment with the addition.

Given a right angle triangle with two equal sides surrounding that angle, if you set one of those sides erect, and turn the triangle around until it returns to its starting position, then the resulting figure is called a cone with circular base. If you cut this figure by a surface parallel to the base, there will be a smaller circle at the intersection.

An example: a triangle *ABC* of that form, with right angle *ACB* and the side *AC* equal to the side *CB* [Fig. IV-5-1]. When you set the side *AC* erect and turn the triangle *ABC* until the point *B* returns to its place, a figure is formed whose base is a circle, and the entire figure is similar to a round cylinder ending at the top at point *A*. This figure is called a cone, which is what we wanted to explain.

For any cone cut by a plane through its axis (which is the height of the initial triangle that you turned), at the intersection there would be a right angle triangle, whose right angle is the one at the top of the cone, and at its base there would be half a circle with its diameter. If we set this [solid] figure on the intersection, its base will be the aforementioned triangle, the half circle will rise up high, and the diameter of the half circle will be at the bottom, forming the chord of the right angle of the aforementioned triangle. All this is explained in the figure [Fig. IV-5-1], which is what we wanted to explain.

. . .

If one of the sides of the half cone that surround the right angle is extended, and we draw from one of its points a line parallel to the axis of the figure, going into the triangle (which is the base of the half cone), and touching the diameter of the half circle (which is the chord of the right angle of the base triangle), and from the point of the diameter

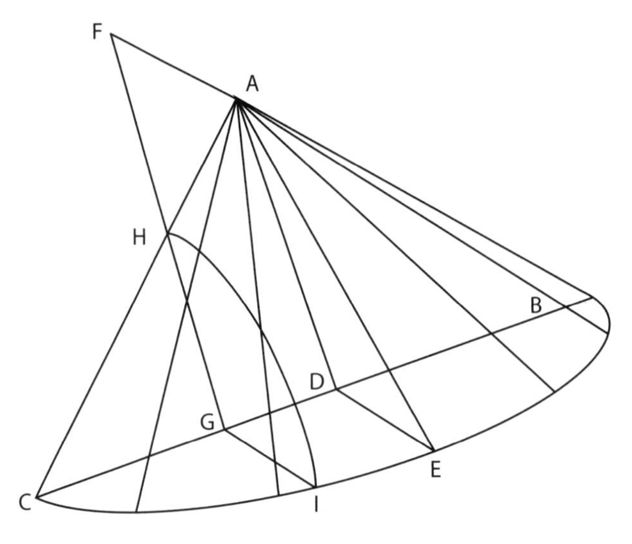

Fig. IV-5-2.

touched by the aforementioned line you draw a height touching the circumference, and you cut the cone by a plane through the aforementioned height and line, then the plane will form a curved line on the surface of the cone.

Let the base of the half cone be the triangle marked *ABC*, and the half circle erected on it *BEIC* with its diameter *BDGC* [Fig. IV-5-2]. Having extended the side *BA* out of the triangle to the point *F*, and drawn from the point *F* a straight line *FHG* parallel to the axis *AD*, and drawn from the point *G*, where the line touches the diameter, a height *GI* touching the circumference, then when we cut the half cone by a plane through the height *GI* and the line *GH*, this plane forms a curved line on the surface of the half cone.

To make things clearer, I form a figure consisting of the entire line *FHG* and the height *GI* together with the plane cutting the figure and the following additions [Fig. IV-5-3]. We bisect the line *FH* at point *J*, and set on the point *H* a height *HK* equal to the line *JH*, which is half the bisected line. We draw a line from *J* to *K* and extend it to *M*. We also extend the height *GI* as far as this line at point *L*.

Concerning the former figure [Fig. IV-5-2], we say that the line *BG* on the diameter equals the line *GF* of the triangle *BGF*. Moreover, the line *GC* that remains on the diameter equals the line *HG* of the triangle *HGC*, and the height *GI* is the mean proportional between the two parts of the diameter, *BG* and *GC*.[13] The two lines *BG* and *GC* equal the two lines *GF* and *GH*, respectively. Therefore the square of the height *GI* is equal to the area surrounded by *GF* and *GH*. In the latter figure [Fig. IV-5-3], the line *FH* has already been bisected at point *J* and added to the line *GH*. Therefore the square of *JG* equals the area surrounded by *GF* and *GH* and the square of *JH*.[14]. Now the height line *GL* equals the line *JG*. Therefore the square of *GL* equals the area surrounded by the

[13]This fact was included in Solomon's list of preliminary Euclidean results.

[14]This fact was included in Solomon's list of preliminary Euclidean results.

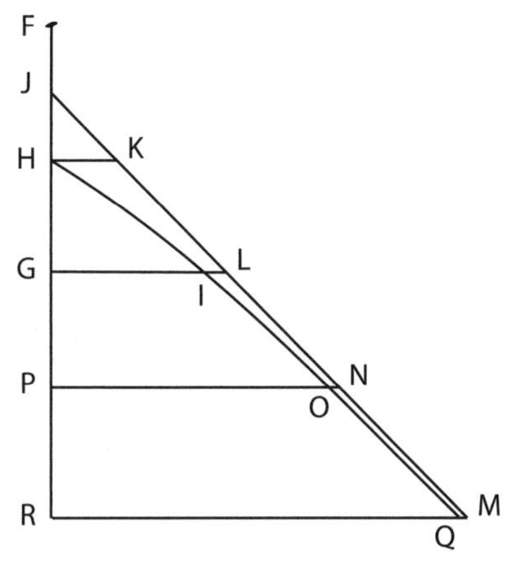

Fig. IV-5-3.

two lines *GF* and *GH* and the square of *JH*. But the square of *GI* has already been said to equal the area surrounded by the two lines *GF* and *GH*. Therefore the square of *GL* exceeds the square of *GI* by the square of *JH*.[15]

Now if you extend the entire diagram so that the line *FJHG* reaches the point *P* and let the height on *P* be *PON*, the former proofs show that the square of *PN* exceeds the square of *PO* by the square of *JH*. And if you extend the entire diagram further, so that *FJHGP* reaches *R* and let the height on *R* be *RQM*, the former proofs also show that the square of *RM* exceeds the square of *RQ* by the square of *JH*. And so ever on, even if you extend the diagram indefinitely, one square will exceed the other by the square of *JH*. Therefore they [the hyperbola and asymptote] cannot meet, because if this were possible, then they [the height to the cone and the height to the plane] would be equal, and the square of the one would not exceed the square of the other. But we have explained that the one exceeds the other by the square of JH. Therefore it is false that they ever meet.

The demonstration that as they [the hyperbola and the asymptote] extend they grow nearer, and that their distance diminishes with respect to the initial distance, which is the height *JH*, is explained by what we have already explained in the previous diagrams before the extension, namely, that the square of *GL* exceeds the square of *GI* by the square of *JH*. Let us complete the square to render this visible.

Let the square of *GI* be the square marked *ABCD* [Fig. IV-5-4]. The three complementing areas [which complete the square of *GI* to the square of *GL*, namely, the three rectangles forming the gnomon *ABGILC*] equal the square of *JH*. Thus *JH* is much greater than *IL*. But *HK* has already been said to equal *JH*. Therefore the line *KH* is much larger than the line *IL*. And so the lines [the hyperbola and asymptote] grow much closer.

Extending the diagram as far as *P* and the height *PON*, we have already explained that the square of *PN* exceeds the square of *PO* by the square of *JH*. When we complete

[15] $GI^2 = GB \cdot GC = GF \cdot GH$; $GL^2 = JG^2 = GF \cdot GH + JH^2$. Therefore, $GL^2 = GI^2 + JH^2$.

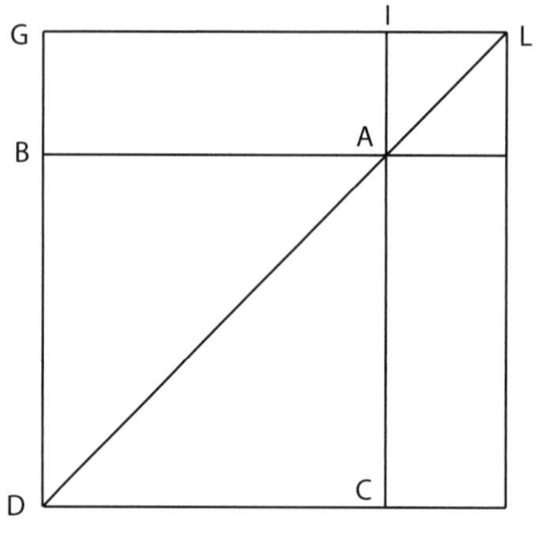

Fig. IV-5-4.

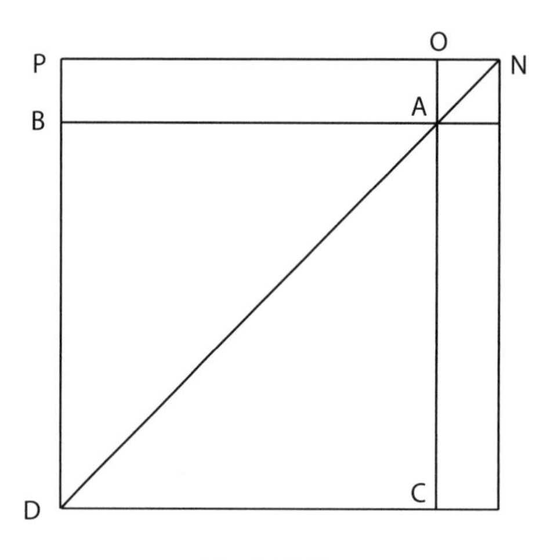

Fig. IV-5-5.

the diagram [to form a square], let the square of *PO* be the square marked *ABCD* [Fig. IV-5-5]. The three complementing areas with respect to the square of *PN* equal the square of *JH*. Since the square of *PN* is much wider and longer than the square of *GL*, and the three respective complementing areas equal the square of *JH*, therefore the line *NO* is much smaller than the line *IL*, and the lines [the hyperbola and asymptote] grow closer. In the same way it is explained that they grow closer as you extend the entire diagram until it reaches the point *R*, and so on indefinitely.

Therefore two lines are drawn, one of which is the straight line *JLNM* and the other is the curved line *HIOQ*. At their inception they had a certain distance, and as they extend they grow nearer. They cannot ever meet even if they are extended indefinitely, which is what we wanted to explain.

6. ABNER OF BURGOS (ALFONSO DI VALLADOLID), *SEFER MEYASHER ʿAQOV* (*BOOK OF THE RECTIFYING OF THE CURVED*)

This section was prepared by Avinoam Baraness.[16]

Abner of Burgos (1270–1348) was a Jewish scholar from Castile, who converted to Christianity and was known after his conversion as Alfonso di Valladolid. After his conversion, Alfonso became engaged in anti-Jewish activity: polemical writings, public disputations, and even inciting the authorities against the Jews.[17] Alfonso was well versed in the Bible and the Talmud, and also in Greek and Arabic philosophy. His philosophical work, *the New Philosophy*, is lost. His extant works are mostly polemical, in Castilian translation. Only one book and a few letters were preserved in the original Hebrew.

The Hebrew mathematical treatise *Sefer Meyasher ʿAqov*[18] by Alfonso is extant in the single manuscript [British Library Add 26984].[19] [Gluskina, 1983] identified the author "Alfonso" as Abner de Burgos—an identification reconfirmed by [Freudenthal, 2005]. *Sefer Meyasher ʿAqov* is very different from the rest of his works known to us. The stated aim of the book is "to inquire whether there possibly exists a rectilinear area equal to a circular area truly, neither by way of approximation as earlier scholars suggested." In the book Alfonso was concerned with the three famous geometrical problems of antiquity, the measurement of curves, plane figures, and solids, and methods for comparing areas circumscribed by mixed straight and curved lines. The treatise contains five chapters, of which the first four prepare the background for the final (and lost) one, where the author was to achieve his goals. The first two chapters offer a historical and philosophical introduction with special attention to the role of motion in geometry. The third chapter consists of 33 geometrical propositions, which are claimed to be "useful for this discipline." Propositions 2–9 of the third chapter are missing, and the text is interrupted at the beginning of the fourth chapter. The original manuscript includes only a few simplistic diagrams for the second chapter.

Whereas Alfonso's philosophical writing sometimes lack clarity and sharpness, the mathematical sections are better organized. The text presented here is from the third chapter, whose content is purely mathematical. Each proposition of this chapter follows the formal Euclidean pattern. It is, however, difficult to point out the organizing principle underlying the chapter as a whole. The propositions deal with a wide range of topics: comparing areas, theorems connecting the length of some segments in polygons (including Ptolemy's theorem on inscribed quadrilaterals), two generalizations of the Pythagorean theorem, characters of compound ratios (in anachronistic terms: products of ratios), two theorems on magnitudes divided into n parts, pre-trigonometric theorems, and some of the classical problems of

[16] Avinoam Baraness expresses his deepest gratitude to Professor Ruth Glasner for significant encouragement and assistance during the writing process. However, he notes that the section on the quadrature of the lune (prop. 23) is due entirely to Tzvi Langermann, as stated in the list of sources at the end of this chapter.

[17] For a detailed documentation of his life and works see [Baer, 1961–1966] and [Sadik, 2012]. Although Abner and Levi ben Gershon were contemporaries, there is no evidence that either was aware of the other.

[18] *The Rectifying of the Curved*, alluding to Isaiah, 40:4 "and the crooked shall be made straight." Clearly, Abner understands the Hebrew in this sense, which is the King James translation, rather than in the sense of the Jewish Publication Society translation.

[19] The manuscript has no colophon and contains many blunders and mistakes. A scientific edition of this manuscript was published by [Gluskina, 1983]. It includes the original text in Hebrew, a Russian translation, and a commentary, along with mathematical remarks by the historian of mathematics B. A. Rosenfeld. Baraness and Glasner are preparing an English translation of the text with mathematical commentary.

antiquity. In this selection, we present Alfonso's quadrature of the lune, his trisection of an angle and his doubling of the cube by means of a conchoid.

The quadrature of the lune (proposition 23)

Since the goal of Alfonso's treatise was the quadrature of the circle, one need not be surprised to find there a treatment of the quadrature of the lune. But Alfonso's introduction makes it clear that this quadrature only gives us reason to believe (rather than a proof) that squaring the circle should be possible. We bring here only the second construction that Alfonso derives from Hippocrates, because its treatment is a little more original. The commentary and translation for this proposition are adapted from [Langermann, 1996].

It is particularly interesting that Alfonso knows quite well that Hippocrates is the author of the quadrature of the lune. [Clagett, 1964–1984, III, pp. 1317–1318] has pointed out that "in none of the many manuscripts of the medieval *Quadratura circuli per lunulas* was Hippocrates named as the author." Nor was his name mentioned by Ibn al-Haytham, the only Arabic authority whose writings on the subject have survived, albeit partially [Suter, 1986].

This construction is essentially the same as Hippocrates's second quadrature as reported by Eudemus [Heath, 1981, 192–193]. However, the construction of Alfonso is simplified considerably. Hippocrates's method has in fact two parts, both of which involve constructing trapezia that are then circumscribed by circles, which form the outer circumference of the lune; the second part in particular has drawn much attention from historians, because it contains one of the earliest known *neusis* constructions. In contrast, Alfonso displays but a single step, and, using Ptolemy's theorem (as Gluskina points out in her commentary), proves that his construction is in fact the trapezium described by Hippocrates. Now we have no evidence that this theorem was known before the time of Ptolemy (second century CE; see [Toomer, 1984, pp. 50–51]), so it seems unlikely that it was part of Hippocrates's own procedure, which was then only summarized by Eudemus; nor is it invoked by Simplicius in his discussion.

Proposition 23: There is yet another figure equal to a rectilinear area. We take straight line *AB* such that when multiplied by itself it will be equal to three times line *BC* multiplied by itself.[20] We take line [*AC*] multiplied by itself to be equal to [the sum of] line *BC* multiplied by itself together with the product of *AB* with *CB*. With these three lines we construct triangle *ABC*, as has been explained. On it we circumscribe [circle] *ABCD* with center *E* [Fig. IV-6-1]. Since line *AC* is greater than chord *CB*, arc *ADC* is greater than arc *CB*. From the latter we mark off arc *AD* equal to arc *BC*. We join lines *AD*, *DC*, *EA*, *ED*, *EC*, *EB*, *BD*.

Now the sides of triangle *ABC* are equal to the sides of triangle *ADB*, respectively. Moreover, *AC* and *BD* are equal to each other. Since *ABCD* is a quadrilateral inscribed within a circle, the product of lines *AC* and *BD*, in other words *AC* multiplied by itself, is equal to the product of *AD* and *CB*—which is [the same as] *CB* multiplied by itself—together with the product of *AB* with either *CB* or *CD*. Therefore, *CB* is equal to *CD*.[21] Accordingly, *AB* multiplied by itself is equal to the sum of chords *AD*, *DC*, and *CB*, when

[20] This construction and the one in the next line can be carried out by the procedure given in *Elements* II.14, "To construct a square equal to a given rectilinear figure."

[21] By Ptolemy's theorem, proved earlier in the text, $AC \cdot BD = AD \cdot BC + AB \cdot CD$. Due to the symmetries of $ABCD$, we have $AC^2 = BC^2 + AB \cdot CD$. But AC was constructed such that $AC^2 = BC^2 + AB \cdot BC$, so $CD = BC$.

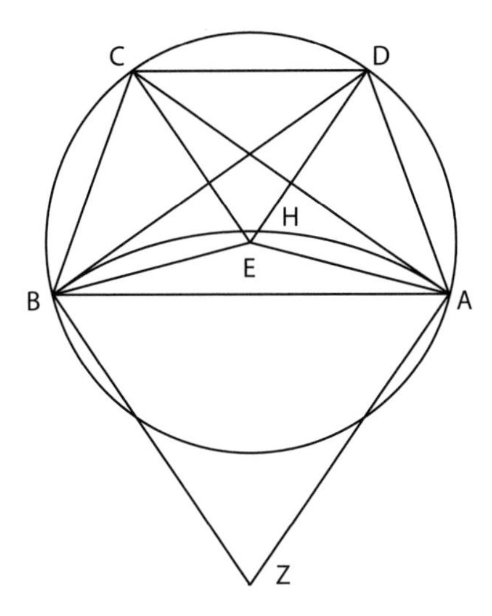

Fig. IV-6-1.

each has been multiplied by itself.[22] We construct on *AB* triangle *ABZ* similar to triangle *AE*[*D*]. With [*Z*] as center we draw sector *ZAHB*. It follows that sector *ZAHB* will be equal to sector *EADB*, and segment *AHB* is equal to the [sum of the] three segments *AD*, *D*[*C*], and *CB*. Therefore, lune *ADCBH* is equal to the rectilinear figure *ADCB*.[23] QED

The conchoid of Nicomedes and its applications (propositions 29–32)

Propositions 29–32, which are presented here, seem to form an independent unit, dealing with the conchoid of Nicomedes and some of its uses. Being located toward the end of the third chapter, it can be considered as one of the book's pinnacles. The quoted text is not long but is representative of the third chapter as well as of Alfonso's general mathematical style. This selection is of both historical and mathematical interest: it attests to the fact that the conchoid was known earlier than hitherto assumed, and that Alfonso's methods seem to be relatively complex and somewhat unique to the author.

Given a straight line (the "ruler" or "canon" *AB*), a point outside it (the "pole" *P*) and a distance (d), the conchoid of Nicomedes is the locus of all points lying at the given distance d from the ruler *AB* along the segment that connects them to the pole *P*.[24]

If *P* is the origin, and *AB* is the line $y = a$, then the curve is defined by the polar equation $r = \frac{a}{\sin\theta} + d$. The curve has two branches on opposite sides of the ruler, to which both are

[22] BC was constructed such that $AB^2 = 3 \cdot BC^2$. The equality of *AD*, *DC*, and *CB* thus establishes Alfonso's claim. Therefore, similar shapes constructed on *AB* and on *BC* (or *AD* or *DC*) will have an area ratio of 3 : 1, as Alfonso claims.

[23] The lune equals the trapezoid *ADCB* less segment *AIIB* plus the sum of the three segments *CB*, *DC*, and *AD*. But these latter three segments equal segment *AHB*, so the lune equals the trapezoid.

[24] The conchoid can be viewed as the trace of a fixed point on a straight line through the pole, moving along the ruler, but it can also be defined in terms of the *neusis* property as the locus of all the points on any line through the pole whose distance from the ruler along the line is constant. It is not clear whether the latter definition, seemingly not involving motion, can be attributed to Nicomedes himself [Sefrin-Weis, 2010, p. 244, fn. 3]. Yet geometric motion was characteristic of Alfonso's approach.

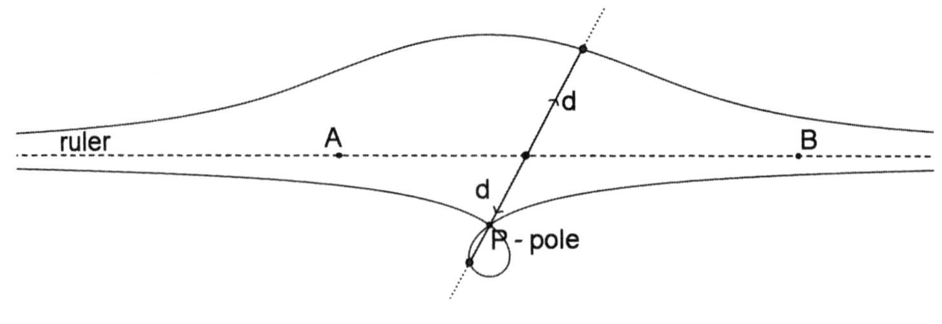

Fig. IV-6-2.

asymptotic. The branch passing on the side of the pole has three different distinct forms, depending on the ratio between a and d: If $a < d$, it has a loop (as in Fig. IV-6-2); if $a = d$, then P is a cusp point; and if $a > d$, the curve is smooth. The other branch does not change topologically.

It is generally accepted that the conchoid, whose name means shell form,[25] was invented and first studied by the Greek mathematician Nicomedes (ca. 280–210 BCE).[26] In his treatise *On Conchoid Lines*, known to us from secondary sources,[27] Nicomedes supposedly described the generation of the curve, its classification into types, some of its properties, and a mechanical device for drawing it. Nicomedes also applied the curve to solve two of the classical problems of antiquity: the trisection of an angle and the doubling of the cube (reduced to the problem of finding two mean proportionals).[28] Both solutions depend on a construction that cannot be implemented with compass and ruler but can be implemented with a conchoid.

It is usually taken for granted that all the applications of the conchoid made in antiquity were developed by Nicomedes himself, and that interest in it was revived in the late sixteenth century [Toomer, 2008]. But Alfonso's text presented below provides rare evidence that the conchoid was known and used in the West in the fourteenth century. In proposition 29, Alfonso constructs the conchoid, and in the three following propositions (30–32) he uses it to trisect an angle, find two mean proportionals, and construct a parallelepiped of the same volume as a given parallelepiped that is also similar to another given parallelepiped. The main questions concern the sources of this knowledge and the manner of its transmission. A close study of the text indicates significant differences between Alfonso's approach to the conchoid and the parallels known from Greek sources.

First, Alfonso stated his aim in constructing the conchoid (proposition 29) as the finding of two asymptotic lines, whereas it is accepted that Nicomedes invented the conchoid to trisect

[25] The curve was called "conchoid" by Proclus, but Pappus called it "cochloid." Heath claims that the latter was evidently its original name [Heath, 1981, p. 238].

[26] Nothing is known of his life; the dating is estimated by references to his work. See [Heath, 1981, p. 238; Toomer, 2008].

[27] Pappus's *Collection*, Eutocius's *Commentary on Archimedes' Sphere and Cylinder*, and Proclus's *Commentary on Euclid* I. See [Toomer, 2008].

[28] In the case where the greater line is double the smaller line. The reduction was established in the fifth century BCE by Hippocrates of Chios. See [Heath, 1981, pp. 244–246].

an angle and duplicate the cube [Heath, 1981, p. 238; Sefrin-Weis, 2010, pp. 126–128].[29] It should be noted, however, that a simpler example of asymptotes was offered by Apollonius of Perga and was well known in the Jewish world through a remark in Maimonides's *Guide for the Perplexed* I.73 [Freudenthal, 1988; see section IV-5 above]. Alfonso probably preferred the more complex example, the conchoid, because he was interested in the uses of this curve as well.

Second, the wording of proposition 29 suggests that *both* branches of the curve are constructed. Indeed, both of them are used in the following propositions: the external one in proposition 30, and the internal one in proposition 31.

Third, the angle trisection (proposition 30) is achieved in a similar manner to that attributed to Nicomedes [Heath, 1981, p. 235; Sefrin-Weis, 2010, pp. 148–149], but the position of the perpendicular (relative to the angle's side) is different, hence the required trisection is constructed outside the given angle, rather than inside. Of the three applications of the conchoid that Alfonso mentions, only this one resembles one of those known from the Greek tradition.

Fourth, and most interesting, proposition 31 is an impressive construction of two mean proportionals that neither resembles nor alludes to any Greek or Arabic solutions known to us.[30] Moreover, proposition 32 seems to be a unique generalization of the problem of doubling the cube, which does not simply rely on the reduction of the problem to that of finding two mean proportionals[31] but shows how to use these proportionals to construct the doubled cube.

We do not know whether Alfonso acquired his knowledge about the conchoid from an unknown Arabic text based on Pappus's *Collection,* or from some oral tradition. Therefore it may not be determined whether he preserves a tradition unknown to us, or whether propositions 30–32 render his own elaboration, based on a fragmentary acquaintance with the known tradition that goes back to Pappus. Like the question of whether and how he "squared" the circle, this question too remains open.

Proposition 29: We wish to find the origin of two lines[32], the one straight and the other curved, so that there is a certain [initial] distance between them, but when produced, the distance between their extremities decreases; one of them approaches to the other, but they do not intersect, even if produced indefinitely.

How? We consider two lines AB, BC enclosing a right angle B, and we pick a point D on the line AB either between A and B or beyond B.[33] Then we move point B on the line BC in the direction of point C, so that the line BD is moved with it in such a

[29] But it was also understood that this curve is interesting in itself: Geminus, in his first classification of lines [Heath, 1956. pp. 160–161] includes the conchoid and the hyperbola in the same subdivision, for having the same subtle distinguishing character: they are "asymptotic." Thus it is possible that the asymptoticity of the conchoid had already been adopted as its main character by some scholars before Alfonso.

[30] See [Clagett 1964–1984, I, pp. 335–345, 658–665, III, pp. 27–30, 849–854, 1163–1179; Heath, 1981, pp. 244–268; Knorr, 1989, pp. 251–319; Rashed, 2011–2014, pp. 60–69, 103–107].

[31] This fact is notably remarkable in light of Heath's comment that since this reduction had been shown, all later mathematicians considered the problem of two mean proportionals rather than the original problem [Heath, 1981, p. 246].

[32] Literally, *yeṣiat shney haqavim.*

[33] The external branch is not mentioned explicitly, but its existence is implied: it is said that point D is picked outside B too. Reading the construction with reference to the letters noted by a prime in the diagram (G' instead of G, etc.) generates this branch.

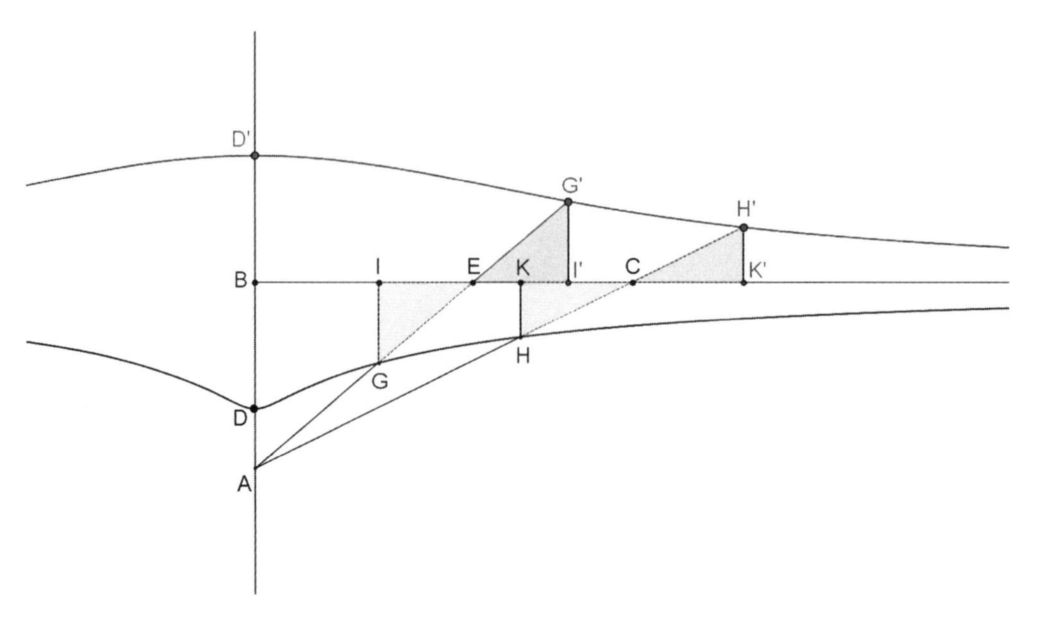

Fig. IV-6-3.

manner that the point *D* is opposite to point *A*, namely, points *A*, *D*, *B* are always collinear [Fig. IV-6-3]. By this motion point *D* describes a segment of a curved line *DGH*, which we call the conchoid.[34] I say that as long as lines *BC*, *DGH* are produced in the direction of *CH*, the distance between their extremities decreases, and they never meet.

The demonstration: We draw the two straight lines *AHC*, *AGE*. The three lines *BD*, *GE*, *HC* are equal to one another. We draw from the points *G*, *H* two perpendiculars *GI*, *HK* onto the line *BC*. Since (i) the square of the line *GE* is greater than the square of *GI*, (ii) the two triangles *GIE*, *KHC* are right-angled, (iii) the hypotenuses *GE*, *HC* are equal, and (iv) the angle *GEI* is greater than the angle *KCH*, it follows that the perpendicular *GI* is greater than perpendicular *HK*. Similarly it can be shown that of the perpendiculars drawn from the conchoid *DGH* to the straight line *BC*, the closer [perpendicular] to the line *AB* is greater than that more distant from it. Hence, as long as the two lines *BC*, *DGH*

[34]The original term is: *haqav haparuṣ*. We are not aware of the use of this word in Hebrew mathematical texts before Alfonso. The root *prṣ* means (1) to expand, to spread out without limits (like the Arabic root *frš*); (2) to crack, to break through (the Arabic root *frḍ* has the related meaning of "to notch," "to make incisions"). In his first classification of lines (given by Proclus; see [Friedlein, 1873, pp. 111; Heath, 1956, pp. 160–162]), Geminus distinguishes between composite and noncomposite lines, and divides the latter class into: (a) those forming a figure (e.g., circle, ellipse, cissoid) and (b) not forming a figure or indeterminate and extending without limit (e.g., straight line, parabola, hyperbola, conchoid). In a second version of the classification [Friedlein, 1873, pp. 176–177; Heath, 1956, p. 160], a fine distinction of the last subdivision is made: "of the lines which extend without limit, some do not form a figure at all, but some *first come together and form a figure, and for the rest, extend without limit*." Following Tannery, Heath concludes almost inevitably that the figure formed is a loop, and that "the curve which has a loop and then proceeds to infinity is a variety of the conchoid itself"—namely, the branch having the loop. These two characteristics of the conchoid may be thought of as associated with the two meanings of *paruṣ* mentioned above (forming a figure—cracked; extending without limit—spreading out).

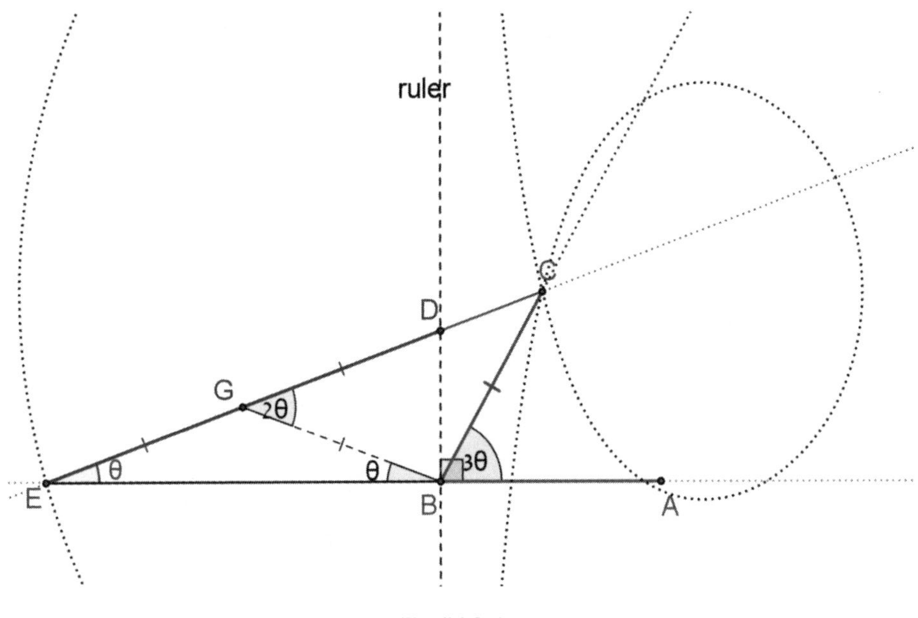

Fig. IV-6-4.

are produced in the direction of *CH*, the distance between their extremities decreases and they never meet. QED

The instrument designed to draw the conchoid is common among craftsmen, and is very useful in this discipline.

Proposition 30: To divide any rectilinear angle into three equal parts.

How? We consider a rectilinear angle *ABC* and erect a perpendicular *BD* upon *AB*, we set the size of *BC* as we wish, and we produce *AB* indefinitely in the direction of *B* [Fig. IV-6-4]. We put the plotting device at the point *C* in such a manner that it meets the two lines *AB*, *BD* at the two points *D*, *E*, such that the line *DE* is twice the line *BC*. This can be done by drawing the conchoid as mentioned above.[35] Then angle *DEB* would be a third of the given angle *ABC*.

The demonstration: Since the line *DE* of triangle *BDE* is the hypotenuse of the right angle *DBE*, and when we cut it at the [mid]point *G* and join *BG*, [the line] *EG* becomes equal to *GB*, which is equal to *DG*,[36] then the angle *BGC*, which is equal to the angle *BCG*, is equal to angle *GBE* together with angle *GEB*. So the angle *DEB* is a third of the two angles *GEB*, *GCB* together, which are equal to the given angle ABC.[37] QED

Proposition 31: We wish to find two straight line segments [that are] mean proportionals between two other straight line segments which are unequal to one another.

[35]To trisect angle *ABC*, Alfonso finds a point *E* on the line *AB* (beyond *B*) so that $ED = 2BC$. This is done by means of a conchoid with pole *C*, ruler *BD* and distance $2BC$. Here Alfonso implicitly assumes that the angle is acute.

[36]Since *BG* is the median to the hypotenuse *ED* of the right-angled triangle *EBD*.

[37]If we set angle $E = \theta$, then angle *BGD* equals 2θ, being an exterior angle to triangle *GEB*. According to the construction, $BC = BG$, so triangle *BGC* is also an isosceles triangle, and the angle *BCD* is 2θ. Therefore the angle *ABC*, being an exterior angle to triangle *BCE*, equals 3θ, and the angle *E* equals one third of the angle *ABC*.

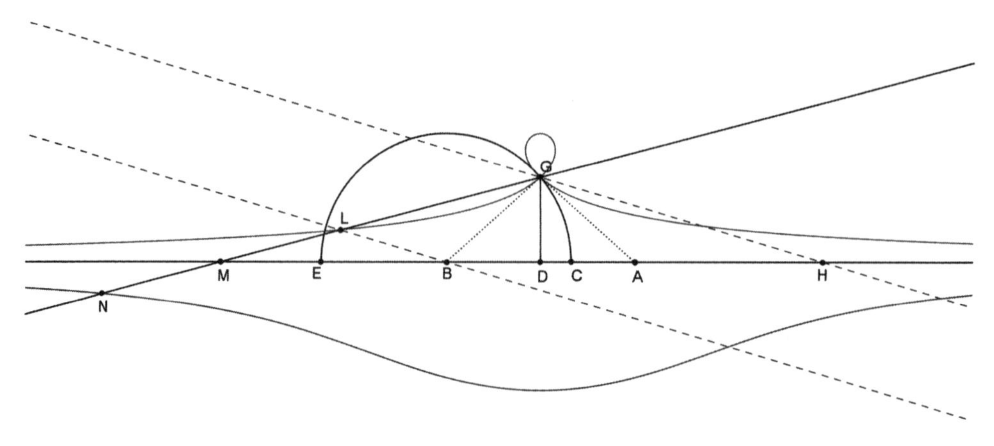

Fig. IV-6-5.

How? We let *AB* be the smaller line and *CE* the larger. We describe a semi-circle whose diameter is *CE*, the larger [segment], and whose center is *B*. *D*, the midpoint of *AB*, falls on the diameter [*CE*]. We draw a perpendicular *DG* on the diameter [and extend it] to the circumference of the circle, and we produce *AD* in the direction of *H* until *AH* equals *AB*. We join *HG*, and draw from the point *B* a line *BL* parallel to *HG* and produce it indefinitely. We draw from point *G* a line *GM* toward the line *CEM*, meeting the line *BL* at the point *L*, in such a manner that the line *LM* is half the diameter [Fig. IV-6-5]. This can be done by drawing the conchoid.[38] I say that the ratio of *AB* to *GL* is as the ratio of *GL* to *BM*, and as the ratio of *BM* to *CE*.

Its demonstration: We produce *GM* until *LN* is equal to the diameter. Since the excess of the product of *MG* by itself over the product of *BG* by itself is as the product of *MB* by itself and by twice *BD*,[39] which is also as the product of *GL* by itself and by twice *LM*,[40] then the product of *AM* by *MB* is as the product of *NG* by *GL*.[41] Moreover, the ratio of *AM* to *NG* is as the ratio of *GL* to *MB*, and the ratio of *AB* to *GL* is as the ratio of half *MB* to *LM*, which is equal as the ratio of *MB* to *LN*.[42] It was already established that the ratio of *AB* to *GL* is as the ratio of *AM* to *NG*, which is equal to the ratio of *LG* to *MB*.[43] Hence, the

[38] Here we use the conchoid with pole G, ruler CE and distance $BE = \frac{1}{2}CE$. By means of this construction one may find a line through G, cutting the parallel through B at point L and line CE at point M, so that $LM = BE$. L is actually the intersection point of the conchoid with the parallel through B.

[39] $MG^2 - BG^2 = (MD^2 + DG^2) - (BD^2 + DG^2) = (MB + BD)^2 - BD^2 = MB^2 + 2MB \cdot BD = MB(MB + 2BD)$.

[40] $MG^2 - BG^2 = GM^2 - LM^2 = (GL + LM)^2 - LM^2 = GL^2 + 2GL \cdot LM = GL(GL + 2LM)$.

[41] Since $MB + 2BD = MA$ and $GL + 2LM = GN$, the last two footnotes yield $MB \cdot MA = GL \cdot GN$, and consequently $GL : MB = MA : GN$.

[42] HB and GL are segments of rays caught between the parallels HG and BL. By triangle proportion theory, their ratio is the same as that of the segments between the parallel and the origin, $MB : ML$. If we divide both ratios in half, we get $AB : GL = \frac{BM}{2} : LM = BM : LN$.

[43] We already have $AB : GL = BM : LN$, which by proportion theory is the same as $(AB + BM) : (GL + LN) = AM : GN$, which is already known to be the same as $LG : MB$.

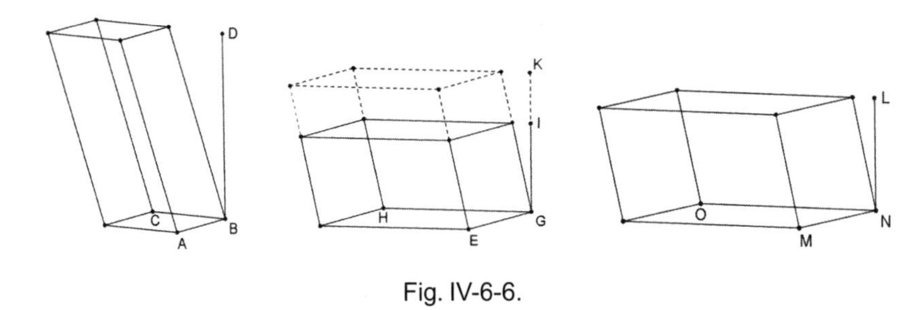

Fig. IV-6-6.

ratio of *AB* to *GL* is as the ratio of *GL* to *MB* and as the ratio of *MB* to *LN*, which is equal to the diameter.

Proposition 32: We wish to construct a polyhedron[44] which is equal [in volume] to a given polyhedron, and which is similar to a second given polyhedron.[45]

How? We set a polyhedron whose base is the area *AB* and whose height is *DB*, and a second polyhedron whose base is the area *EH* and whose height is *IG* [Fig. IV-6-6]. We wish to find a third polyhedron which is equal to the first and similar to the second. We erect upon area *EH* another polyhedron whose height is *GK*, and which is equal to the first polyhedron.[46] We draw between *GK* and *IG* two mean proportionals, *P*, *LN*, so that the ratio of *GK* to *P* is as the ratio of *P* to *LN* and as the ratio of *LN* to *IG*.[47]

We set the ratio of *NM*, which is unknown, to *EG*, which is known, to be as the ratio of *LN*, which is known, to *IG*, which is known. We erect upon the line *NM* an area *MO* which is similar to the area *EH*. It [the area *MO*] would be the base of the required third polyhedron and *LN* would be its height.

Its demonstration: Since the ratio of *GK* to *LN* is as the duplicate ratio of *LN* to *IG*, which equals the duplicate ratio of *NM* to *EG*,[48] which equals the ratio of area *MO* to area *EH*, then the product of *GK* by area *EH* is as the product of *LN* by area *MO*, and the two polyhedra are equal.[49] Therefore the third polyhedron is equal to the first. And since the ratio of *LN* to *IG* is as the ratio of *NM* to *EG*, the third polyhedron is similar to the second.

[44] Though the general word "polyhedron" is used, it seems that Alfonso actually intends to deal here with parallelepipeds or even strictly boxes. However, it is easy to generalize the proposition to prisms.

[45] This proposition appears to be a spatial version of *Elements* VI.25: "to construct a figure similar to one given rectilinear figure and equal to another." It may also be regarded as a generalization of the Delian problem, where the first polyhedron is any parallelepiped with a volume of two cubic units, and the second polyhedron is a cube whose side is one unit. From another point of view, it may also be regarded as a generalization of Euclid XI.27: "to describe a parallelepipedal solid similar and similarly situated to a given parallelepipedal solid on a given straight line."

[46] Alfonso does not give any details about the way this construction should be carried out.

[47] By the previous proposition.

[48] NM is such that $NM : EG = LN : IG$ (*Elements* VI.12).

[49] $GK : LN = (LN : IG)^2 = (NM : EG)^2 = \text{area}(MO) : \text{area}(EH)$ by *Elements* VI.20. Hence $GK \cdot \text{area}\,(EH) = LN \cdot \text{area}\,(MO)$.

V. ALGEBRA

The Hebrew literature does not contain much algebra [Lévy, 2003, 2007]. The only explicitly algebraic treatises known are an anonymous algebra in the tradition of al-Khwārizmī [Lévy, 2002; Aradi, 2013], al-Aḥdab's commentary on Ibn al-Bannā's algebra (see below), Moṭoṭ's algebra written in Italy (see below), and Finzi's translations of the Italian algebra of Maestro Dardi [Wagner, 2013] and of the Arabic algebra of Abu Kāmil's (possibly through a Spanish or Hebrew middleman) [Levey, 1966].

But even before the "official" algebra using the Khwārizmian terms (root/thing, square/property, cube, etc.) and six normal forms of linear and quadratic equations, quadratic problems were treated by methods that go back to Mesopotamia, namely, the reduction of problems to deriving the values of two unknown from their product and sum/difference with an implicit or explicit geometric model. The following selection opens with problems of the latter type (See sections III-1 and III-2 for further examples.). Then we bring extracts from Moṭoṭ and al-Aḥdab expounding the double false position, classical Khwārizmian algebra, and operations with combinations of powers of unknowns—the forerunners of modern polynomials.

1. QUADRATIC WORD PROBLEMS

We begin with some word problems that involve quadratic equations but that do not use explicit algebraic terms or methods. The basic technique is deriving the value of two unknowns with a given product and sum/difference.

a. A quadratic problem from Levi ben Gershon's *Maᶜase Ḥoshev*

Problem 16 of Levi ben Gershon's *Maᶜase Ḥoshev* (see section I-6) is a quadratic problem, whose roots can be traced as far back as ancient Mesopotamia.[1] In Mesopotamia, this was a geometric problem, but it also occurs as a purely arithmetic problem in Diophantus's *Arithmetica*, Book I, #27 [Heath, 1964, p. 140]. Diophantus, however, insists on a rational solution, while Levi does not.

16. We multiply one number by another and get the result. The sum of the two numbers is given. What are each of the numbers?

Take the square of half the sum of the two numbers, and subtract the result from it. Take the square root of what remains, and add it to half the sum of the two numbers, to get the first number. If we subtract it from this half, you get the second number.

For example, the sum of two numbers is 13, and their product is 17. We know that the square of half of 13 is 42 and a quarter. Subtract 17, leaving 25 and a quarter. Extract the square root to get 5 wholes and one first, 29 seconds, 46, 34. Add this to 6 and a half, which is half of 13, to get the first number: 11 wholes, 31 firsts, 29, 46, 34. The second number is: one whole, 28 firsts, 30, 13, 26. The product of one with the other is 17 to a very close approximation.[2]

[1] This problem only occurs in the first edition of the *Maᶜase Ḥoshev*. Levi eliminated it in the second edition.

[2] Levi is using sexagesimal fractions, as was taught in part II of his treatise.

It is impossible to find this number exactly, because 25 and a quarter does not have a true square root, as was explained. This is because the ratio of 25 and a quarter to 25 equals the ratio of one hundred and one to one hundred. But the ratio of one hundred and one to one hundred is not equal to the ratio of a square to a square, since if this were the case, one hundred and one would be a square, because one hundred is a square. But if one hundred and one were a square, then its square root would be a whole number, and that is false.

If the problem was: we multiply a given number by a fixed part of itself; and we add the result to the product of this part with the remaining part of the given number; and the answer is given; what are each of the parts?[3]

Take the square of the whole number, and subtract from it the sum composed of the product of the number with a part of itself and the product of this one part with the other part. Take the square root of what remains, and this is one part. What remains from the number is the fixed part.

For example, the product of ten with a given part of itself, plus the product of this part with the second part, equals eighty. We want to know: what is the given part? The square of ten is one hundred. We subtract eighty from this to get twenty. We extract the square root, which is approximately 4 wholes, 28, 19, 41, 21, and this is one part. What remains is 5 wholes, 31, 40, 18, 39, which is the given part. If you multiply this part by ten and by the leftover, you get eighty to a very close approximation.

b. Quadratic word problems from an anonymous arithmetic

This section was prepared by Naomi Aradi

An anonymous arithmetic textbook, dating from around 1200, which survives only partially in three manuscripts, reveals some calculation methods that are seemingly uncommon in medieval Hebrew arithmetic, such as exceptional root approximations. This treatise does not deal with written calculations but only with mental calculations. It opens with an introduction discussing some special properties of the numbers one to ten and general properties of numbers. In the Geneva manuscript excerpted below, a section containing a discussion of algebraic equations was inserted into the introduction. This section was identified by [Lévy, 2002] as an adaptation of a paragraph from the *Algebra* of al-Khwārizmī. After the introduction come six chapters on addition, subtraction, multiplication, division, ratios, and roots. Apparently two additional chapters appeared before the last chapter (on roots): "Deducing (*hoṣaʾat*) One from Another" and "Converting (*hashavat*) One to the Other." In each of these eight chapters the arithmetic operations are presented with integers, sexagesimal fractions, and simple fractions. A detailed outline of the text can be found in [Aradi, 2013].

The chapter titled "Deducing One from Another," which was preserved only in this Geneva manuscript, is devoted mainly to illustrating ways of solving word problems [Aradi, 2013, pp. 277–292]. These problems are presented with short, unmotivated solutions. They begin with a string of highly standardized commercial problems (applications of the Rule of Three to pricing, salary, and partnership) taken from Abraham Bar Ḥiyya's *Foundations*

[3]The problem can be formulated as follows: given $A = x + y$, and $B = Ax + xy$, find the parts x and y. The solution is based on the observation that $y^2 = A^2 - B$.

of Wisdom. Then, however, follow problems that are exceptional in that they involve non-homogeneous operations, have no commercial application, and lead to quadratic problems, which are infrequent in Hebrew arithmetic writings. Again, the following problems are not solved by reduction to Khwārizmian normal forms as presented in the Moṭoṭ selection below.

A pricing problem

If you give an unknown [quantity of] *kors* [a Talmudic unit of dry measurement] for a price of 60 and you subtract the price of one *kor* from all the *kors*, 4 would remain.[4]

Take half of the 4, and multiply it by itself, which are 4. Add them to the sixty, and take the root, which is 8. Add the 2, these are 10, which is [the number of] the *kors*. Or, if you subtract the 2 from the 8, six remain, which is the cost of a *kor*.[5]

A salary problem

You hired a salaried worker for an unknown [number of] days for an unknown [quantity of] dinars. When you add up the days and dinars they sum up to 40. He worked unknown days so that when multiplied by their salary the total is 12. If you multiply the remaining days by the remaining salary, the total is 192. How much are the unknown days, how much are the dinars, and how much is the [payment for] 10 days?

Divide 192 by 12, yielding 16. Take their root, which is 4. Add one, which is 5. Divide 40 by them, 8 comes out. Take their half, 4, multiply them by themselves, and subtract 12 from the result, 4 remain. Take their root, which is 2, add it to the 4, which is half the 8, and the result, 6, is the unknown [number of] days he worked. Alternatively, find the ratio of the 12 to 192, which is half an eighth. Take the root, which is a quarter, and always add one, which are one and a quarter. Divide 40 by them, the result is 32. Take their half, which are 16, and multiply by themselves, yielding 256. Subtract from these the 192, 64 will remain. Take their root, which is 8, and add it to the 16, these are 24. Add them to the first 6, and the sum, which is 30, is the unknown [number of] days of the job. Subtract them from the 40, and the remainder, which is 10, are the unknown dinars.[6]

A partnership problem

One [gave] 10, the second [gave] 20 and the third [gave] 40. When you multiply the profit of the first and the second by the profit of the third, the result is 48. What was the profit of each one?

[4]Note that we subtract a number designating a price from a number designating volume.

[5]Let a be the number of *kors*, and b the price of a *kor*. ab is 60 and $a - b = 4$. The solution proceeds in the standard manner.

[6]If a and b are the number of days worked and salary received, respectively, and a' and b' are the remaining days and salary, we get $a{:}a' = b{:}b'$. Therefore, $\frac{a'b'}{ab} = \frac{192}{12} = 16$ is the square of $\frac{a'}{a} = \frac{b'}{b} = 4$. Next, given this ratio, $a + b = \frac{a'+b'+a+b}{4+1} = \frac{40}{5} = 8$. We now have $ab = 12$, $a + b = 8$, which is solved in the standard manner. The alternative solution derives $a' + b'$ from $\frac{ab}{a'b'}$ according to the same procedure. Note that the products and sums are not homogeneous quantities.

Add the 10 and the 20, and multiply the result by the 40. The result is a thousand and two hundred. Find the ratio of 48 to the above, which is a fifth of a fifth. Consider them as fractions, and take their root, which is a fifth. Take a fifth as the profit rate for each, so the first is 2, and the second 4 and the third 8.[7]

c. A quadratic problem from Elijah Mizraḥi's *Book of Number*

This section was prepared by Stela Segev

This problem from Elijah Mizraḥi's *Book of Number* (see section I-5) discusses a quadratic equation and is solved by what we now call Viète's formulas. Elsewhere (problems 50–52) Mizraḥi treats the three kinds of quadratic equations with the standard formulas. In both cases, this treatment remains unmotivated.

Question [24]: If you want to know a number which, when its third is multiplied by its quarter (or any other part of it by any other part), the result is the number itself added to seven, for example, or a multiple of the number added to another number, how can one find this number?[8]

The answer is that this question is composed of multiplication, ratio and addition. Therefore we multiply the two fractions without reference to the sought number and apply ratios to the result. We say: if the result equals some multiple of the sought number added to some number according to the question, how much is one whole?[9] We save the resulting multiple [*mnb*]. We then take the resulting number [*mnc*], and look for all the pairs of numbers which, when multiplied, yield that number.[10] We write them down pair by pair, that is, each pair of numbers, whose product is that number, side by side, each with its partner. The number which, when added to the saved multiple, equals its partner—its sum with the saved multiple is the sought number.[11] That holds when the [sought] numbers are whole. But if [the sought number is] a whole number and a fraction, we look for two numbers that bound the sought number between them, and find the sought number easily because it is bounded between the two numbers.

Example with integers: If you ask which number, when its half is multiplied by its quarter, equals the number itself added to 6, you have to multiply the half by the quarter without reference to the sought number, and you get one eighth. Then we apply ratios and say: if the eighth equals the number itself added to 6, how much is one whole? We get 8 [multiples of the sought number] and the number 48.

We save the eight, and look for numbers which, when multiplied, yield forty-eight. These are pairs, each along with its partner: 1 with 48, 2 with 24, 3 with 16, 4 with 12, and 6 with 8—for when any of these pairs is multiplied, they yield 48. Then we check every pair to see which number, when added to the saved [multiple] 8, equals its partner.

[7] Let x be the profit rate. We have: $(10x + 20x)\,40x = 48$, yielding $x^2 = \frac{48}{(10+20)40} = \frac{1}{25}$. Note that the ratio 1:25 has to be "considered as a fraction" for its root to be taken.

[8] If x is a number, we can write the equation as: $\frac{x}{m} \cdot \frac{x}{n} = bx + c$.

[9] $x^2 = (mnb)\,x + (mnc)$.

[10] This refers to all pairs p, q such that $pq = mnc$.

[11] When we find a pair such that $p + mnb = q$, then $p + mnb$ is the sought number. This method depends on what we now call Viète's formulas for quadratic equations.

We find the number 4 which, when added to the saved 8, yields 12. The sum equals its partner, as the partner of 4 is 12. Therefore we assert that the sought number is 12. From this you will be able to find the solution with integers and fractions as well.

2. SIMON MOṬOṬ, *ALGEBRA*

Our knowledge of Simon Moṭoṭ (or, perhaps Miṭoṭ, i.e., "from Ṭoṭ") comes only from the manuscripts of his two mathematical treatises, one on algebra, and the other on the asymptotic property of the hyperbola, but the latter attribution has been questioned by [Lévy, 1989b]. From the dedication of the algebra treatise, we can date his work to the mid-fifteenth century, and place him in contact with the avid copyist and translator of scientific texts, Mordekhai Finzi [Steinschneider, 1893–1901, p. 193; Lévy, 2007].

Moṭoṭ's treatise of algebra belongs to the Italian abbacus tradition—the culture developed around schools where children of merchants were taught arithmetic. Indeed, Moṭoṭ acknowledges that his sources are Christian, and the content fits well. All the problems and proofs that Moṭoṭ presents appear in one way or another already in the algebraic treatises of fourteenth-century abbacus masters [Høyrup, 2007, pp. 147–182]. (See also section III in Chapter 1.) As was often done at the time, Moṭoṭ opens with a brief definition of algebraic terms (the thing, square, cube, and square-square, corresponding to today's x, x^2, x^3, and x^4, respectively), explains how to make arithmetic calculations with roots and with binomials containing roots, then presents the six standard kinds of first and second order equations (in modern transcription: $ax = b$, $ax^2 = b$, $ax^2 = bx$, $ax^2 = bx + c$, $bx = ax^2 + c$, $c = ax^2 + bx$) and culminates with some higher order equations that are reducible to them. He accompanies some of his discussions with geometric proofs and numerical examples.

Moṭoṭ claims that the incompleteness of his sources forced him to make up some contents, but it is not clear what his original ingredients were. Perhaps the specific manuscripts consulted by Moṭoṭ were missing some proofs, which he had to reconstruct from his own memory and ingenuity.

There are, indeed, some idiosyncrasies in Moṭoṭ's work, which may indicate slightly different paths of transmission with respect to the dominant algebraic culture. First, the order of presentation diverges from the order used in all early Italian algebras (Moṭoṭ exchanges the fourth and sixth case; see [Høyrup, 2007, p. 160]). The solutions of the first and third examples also stand out as atypical in abbacus culture. The fact that the fifth case, which may have no solutions, one solution, or two solutions, is treated as if it always had one, suggests that Moṭoṭ's understanding of algebra was probably not state of the art for his time. Here we bring only the introduction and the treatment of the first six cases.

After praising the Lord, whose renown is glorious and illuminates all utterance and action, may his name be blessed and much exalted, I begin and say.

You should know that in the calculus of algebra [lit.: alzibra] the Christians take one part of a question, whose numerical value is unknown, and turn it in their reckoning into a single whole thing, and call it *cosa*. They wish to indicate by this word two things: one whole thing, and a hidden thing that we do not know. I shall follow suit as well in this translation, and call it by the name thing [*davar*]. The product of the thing with itself they call *censo*. I asked the grammarians of their tongue for the meaning of this word, and they said that it designates a definite number, meaning an unknown definite number. And as we found no single word in our tongue for this meaning, and as I did not want to extend my speech by referring to this meaning by two words, nor to invent a new word in our language, I called it a square [*meruba*ʿ], as it is indeed. And the product of the square with itself they call *censo de censo*, and I call it square of the square. And the cube [*me*ʿ*uqav*] number they call *cubo*. And the cube of the cube they call *cubo de cubo*. The units of numbers they called *numeri*, as is their common habit elsewhere.

The discussion of arithmetic with roots is omitted here.

Now, in the name of He who is known among the nations as the creator, I shall begin to discuss the theorems of the calculus of algebra and explain them with my meager intelligence. But before I begin, I present a proposition and its explanation.

I say: you should learn and have in mind that the ratio of the square-square to the cube is as the ratio of the cube to the square and as the ratio of the square to the thing and as the ratio of the thing to the unit. This is because the number of units in the thing is as the number of things in the square and the number of squares in the cube and the number of cubes in the square-square. Remember this proposition, because you will need it in the proofs of the following theorems. Here I begin.

[1] When the things equal units, divide the units by the things and the outcome is the thing. This is self-evident.

Question: I wish to divide the number ten into two parts, such that when one is divided by the other, the quotient is 5.

Practice this method. Say that the part by which one divides is a thing. The part that one divides is necessarily five things, like the outcome of the division.[12] Added together the two parts are six things, and are equal to the number ten. According to the method indicated in this theorem, the number [10] should be divided by 6, which comes to 1 and 2 thirds. Such is the thing.

[2] When the squares equal units, divide the units by the squares and the root of the outcome is the thing.

Question: I wish to find a number such that when its third is subtracted, the square of the remainder is the number 20.

Practice this method. Say that the number, whose two thirds are the root of 20, is one thing. Multiply its 2 thirds by themselves, making 4 ninths the square of the entire number that I wanted to find. According to the method indicated in this theorem, the number 20 should be divided by 4 ninths, and the outcome is 45. Such is the square of the entire number, and its root is what you wanted.

[3] When the squares equal things, divide the things by the squares and the quotient is the thing.

This theorem follows the first theorem because the ratio of the square to the thing is as the ratio of the thing to the unit, as we said in the proposition. Therefore, if one square equals 3 things, for example, then one thing will necessarily equal 3 units.

Question: I wish to find a number such that when a third is subtracted, the remainder is the root of the entire number.

[12] In Italian algebra, the parts would usually be modeled as x and $10 - x$, which yields a slightly more complicated procedure.

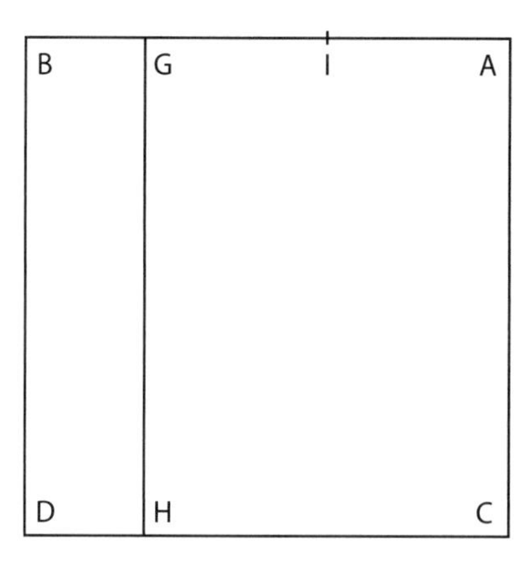

Fig. V-2-1.

Practice this method. Say 2 thirds of this number is one thing.[13] Therefore, the entire number is one thing and a half. So one thing and a half equal one square. According to the indicated method, 1 and a half should be divided by one, yielding 1 and a half. This is the thing, which is two thirds of the number you wish to find. The entire number is therefore 2 and a quarter.

[4] When the things and units equal squares, divide the things and units by the squares. Halve the things coming from the division, and multiply this half by itself. Add the result to the units coming from the division. Take the root of the result, and add to half the things coming from the division. The result is the thing.

To demonstrate this [lit.: to show you this for the eye of the intellect], we draw a diagram and bring a numerical example. Let the line *AB* measure 10, and divide arbitrarily at *G*. Let *AG* measure 8. We set the square *ABCD* on *AB*. From the point *G* we draw a line *GH* parallel to the lines *AC* and *BD* [Fig. V-2-1]. We have the surface *AH*, which is eight things (like the measure of the line *AG* in number, because each unit measure in *AG* holds one thing in the surface *AH*) and the surface *GD*, which measures 20 in area together equal the square *AD*. Now here we face the line *AG* which measures 8 in length, like the number of things. We divide it in half at *I*, and add the line *GB*. It has already been shown in the sixth diagram of Euclid's second book that [the square of *IG*, half the line, and] the rectangle contained by the entire line with the addition [*AB*] and the addition [*GB*] (which equals the surface *GD*, whose area is the number 20 in our example) together, being 36, equal the square of the line composed of half the line [*IG*] and the addition [*GB*], which is the line *IB* in our diagram. Therefore, if you take the root of 36, which is 6, you get the measure of the line composed of half the line and the addition, which is the line *IB*. Add half the things, which is the number 4, as the measure of the line *AI* in number, and 10 will result as the entire line *AB*, the side of the square. This is the thing.

[13] Note the unusual choice of the thing: x is the required number *after* its $\frac{1}{3}$ is subtracted. Therefore, the entire number is $\frac{3}{2}x$, and we have $x = \sqrt{\frac{3}{2}x}$, or $x^2 = \frac{3}{2}x$.

Question: We wish to find a number such that adding to it 28, it will equal two times its square.

Practice this method. Say that this number is one thing. When we've added 28 it becomes one thing and 28 units. These equal two squares. Then, according to the method indicated in this theorem, one thing and 28 units should be divided by 2, the number of squares. The quotient is half a thing and 14 units. Take half of the half thing, which is the quotient. It is a quarter of a thing. Multiply it by itself, it is one part in 16. Add to 14, the number of units in the quotient, yielding 14 and one part in 16. Take its root, which is 3 and 3 quarters. Add it to half the things in the quotient, which is a quarter of a thing, yielding 4. This is the thing.

[5] When the squares and units equal things, divide the things and the units by the squares. Halve the quotient of the things and multiply by itself. Subtract from the result the number which is the quotient of the units. Add the root of the remainder to half the quotient of the things. The result is the thing.

The geometric proof is omitted here.

Question: A merchant went trading with a certain capital, and earned 6. He then returned with that capital and the profit, and made a profit at the same ratio as in the first round, having altogether 27. You wish to know the number of the initial amount.

Practice this method. Say that the initial capital is one thing. He succeeded and made this thing into a thing and 6, and by the same ratio, from one thing and 6 he made 27. The ratio of a thing to a thing and 6 is as the ratio of a thing and 6 to 27 units. We have three magnitudes in proportion. It is known then from proposition 17 of the sixth book of Euclid that multiplying the first by the last equals multiplying the middle by its own image. Now multiply one thing, which is the first, by 27 units, which is the last, yielding 27 things. Then multiply one thing and 6, which is the middle, by itself, yielding one square and 12 things and 36 units. Now subtract the 12 things from these two equal magnitudes, leaving 15 [things] equal to one square and 36 units. According to the method we stated in this theorem, the number of things, 15, and the number of units, 36, should be divided by one, which is the number of the square. The quotient is 15 things and 36 units. Then halve the things, which are 7 and a half, and multiply by themselves. The result is 56 and a quarter. Subtract 36 units, leaving 20 and a quarter. Take its root, which is 4 and a half, and add to half the things, which is 7 and a half, yielding 12. This is the thing which is the initial capital.

[6] When the squares and things equal units, divide the things and the units by the squares. Halve the quotient of the things and multiply the half by itself. Add the result to the quotient of the units. The root of the result less half the things in the quotient is the thing.

The geometric proof is omitted here. This case is not accompanied by a numerical problem.

This is the span of what I sought and found here and there about the calculus of the book of algebra in the books of the Christians. I made up many of these theorems myself. You should know, my dear brother Mordekhai (that he might see offspring and have long life, and that through him the Lord's purpose might prosper),[14] son of our honorable

[14]Isaiah 53:10, quoted in acronym.

master Abraham Finzi (may his memory live in the world to come) that the author of the book of all these theorems brought them in his book without proofs, and none of those who read it knows the methods of this scholar and where he found them. I, your brother, seeing you and my dear friend Rabbi Judah, son of our honorable master Joseph (may God save him and keep him alive), son of our honorable master Avigdor (may his memory live in the world to come), longing to know it, and as he who knows, if we are to call him "one who knows," must know by logical proof—I had to study the proofs and write them for you so as to fulfill your wish.

I was, however, succinct for two reasons. The first is because I trust your good spirit, the divine spirit hovering over all wisdom. The second is the toil and trouble that came upon me to preoccupy my mind and body, and my many dealings in worldly business. But if one of you misses anything due to my brevity and weariness of long proofs, I say that I am willing to further clarify it. One must not be long, except in appealing to God, may He fulfill all thy wishes, let thy springs spread, springs of salvation, Amen. In accordance with your will and the will of your faithful brother, who abides by your command, Simon, son of our honorable master Moses (may God save him and keep him alive), son of our honorable master Simon Moṭoṭ (may his memory live in the world to come).

3. IBN AL-AḤDAB, *IGERET HAMISPAR* (*THE EPISTLE OF THE NUMBER*)

This section was prepared by Ilana Wartenberg

Isaac ben Solomon ibn al-Aḥdab (Castile, ca. 1350–Sicily, ca. 1430) was a Jewish polymath. His prolific writings covered a wide range of fields: astronomy, reckoning of the Jewish calendar, mathematics, exegesis, and poetry. After leaving Castile, he studied with Muslim scholars in North Africa. Then, on the way to the Holy Land, he was shipwrecked in Syracuse, Sicily. At the request of the local Jewish community there, Isaac composed *The Epistle of the Number*.[15]

The Epistle of the Number is the first and only known Hebrew version of the succinct Arabic mathematical tract *Talkhīṣ aʿmāl al-ḥisāb* (*A Summary Account of the Operations of Computation*) written by the famous Moroccan mathematician Aḥmad ibn al-Bannāʾ (1256–1321) (see parts I-1 and II-1 of Chapter 3). *The Epistle of the Number* is not only the first Hebrew text we know of that contains explicit algebraic materials but it also includes numerous arithmetical themes.

The Epistle of the Number presents a perfect translation of *Talkhīṣ aʿmāl al-ḥisāb* as well as detailed mathematical explanations accompanied by a multitude of numerical examples and philological and philosophical discussions. (See section II-1 in Chapter 3 for a direct translation of this part of *Talkhīṣ* from the original Arabic.) The Hebrew text follows the structure of its main Arabic source: the first part is dedicated to arithmetic, that is, arithmetical operations on three types of known quantities (numbers): integers, fractions, and roots. The second part of the book presents three methods to determine the value of the unknown: the Rule of Three, the rule of false position, and algebra. The part on algebra includes lengthy discussions of algebraic elementary entities (numbers, roots, and squares)

[15] *The Epistle of the Number* survives in a fragmentary unicum [Cambridge University Library, Heb. Add. 492.1, ff. 1b–38b]. For various analyses of the text and its context, the author's life, and a critical edition and English translation, see [Lévy, 2003; Wartenberg, 2007, 2008a,b,c, 2013, 2014].

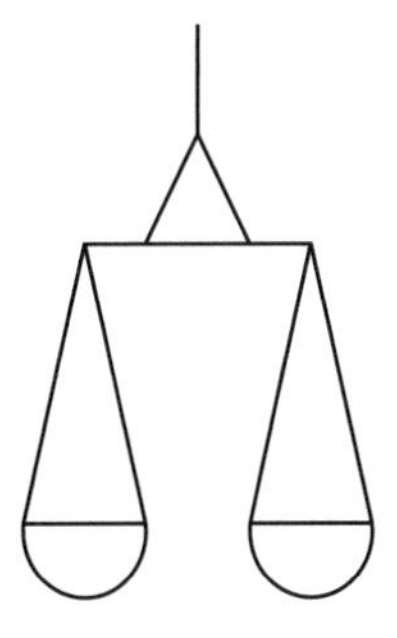

Fig. V-3-1.

and algebraic operations (restoration, opposition, and equation). It also presents a detailed analysis of algebraic expressions, the "ancestors" of modern polynomials. At the very end of the truncated unicum we find a series of various problems (e.g., charity distribution, time-velocity-distance calculations) that are solved by algebraic methods.

Double false position

The method of scales is usually referred to in modern mathematics as the rule of false position. This method was widely known in the Arabic mathematical tradition.[16] It is an arithmetical procedure that enables one to find the value of the unknown in what could be anachronistically termed "linear" problems by balancing out the errors of wrong guesses. The standard method of double false position requires two different guesses, and it can be used in problems with two unknowns. However, where there is only a single unknown, the method can be adapted to a single guess. Both possibilities are treated below.

The basic tool used in this method are scales, where one writes one's guesses, the results of substituting these guesses into the problem, and the resulting errors. Since the errors are considered as absolute numbers (rather than signed numbers), the algorithm depends on the direction of errors. Ibn al-Aḥdab treats all such possibilities, but here we only include an example where the errors are positive.

He [the author of the *Talkhīṣ*] says: **The method of scales is part of the art of mathematics**[17] **and their shape is to be drawn as follows** [Fig. V-3-1]:

Place the given known number on the fulcrum. Take one of the pans, and then take any number you wish and carry out the procedures which were given, whether addition or subtraction or another procedure. Then compare the result with the number on the fulcrum. If you find that it is the same value, then the value you have chosen is the unknown number.

[16]According to [Suter, 1901, p. 31], the method of scales was already known in Baghdad in the ninth century at the time of al-Khwārizmī; it was also widely used in North Africa and in the Middle East. See [Youschkevitch, 1976, pp. 45–48]. The method of double false position appears earlier in the Chinese work *Nine Chapters of the Mathematical Art* (first century BCE) [Dauben, 2007, pp. 269–274]. See section II-3-2 in Chapter 1 for its appearance in Fibonacci's *Liber abbaci*.

[17]The Hebrew term *hamelakhot halimudiyot* usually refers to astronomy or mathematics, the latter being most fitting in our context. Here it is used to translate the Arabic term صناعة الهندسية, which literally means "the art of geometry." See Section II-1 in Chapter 3.

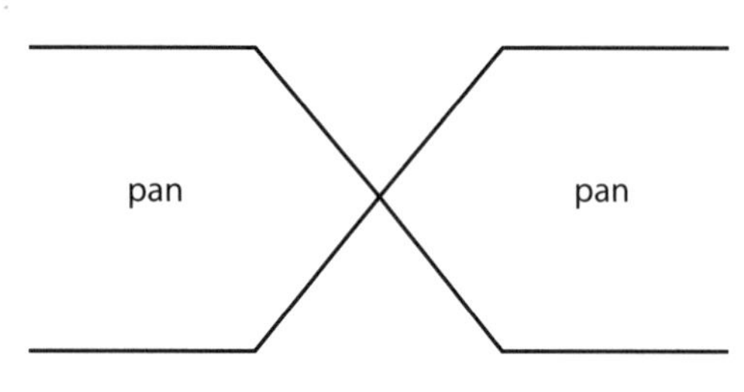

Fig. V-3-2.

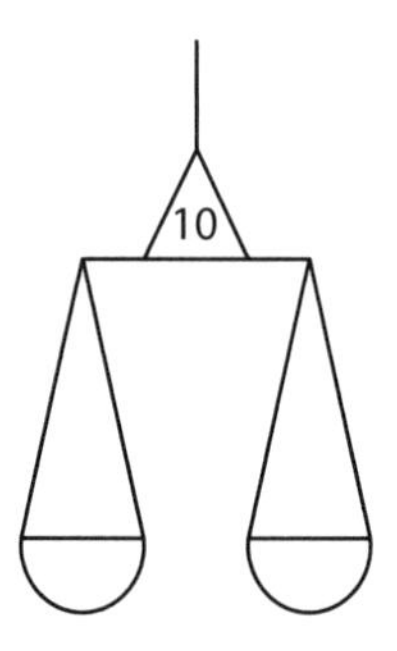

Fig. V-3-3.

He says that this is part of the science of mathematics because at times he adds, at times he subtracts, and takes the intermediate value and this is found in mathematics such as the mean, minimal and maximal distances found in [astronomical] tables, and so on.

You already recognize from the form of the fulcrum that it is the upper part of the scales, which is called the *level*, and the pans are hung by threads. Draw it as follows [Fig. V-3-2]: or in any form you wish.

We shall give the example we have given in the part on the proportional numbers,[18] i.e., a wealth [that is, a certain amount of money], from which one has subtracted both a third and a quarter, and ten is left. How large is the wealth?[19] Draw the scales as follows [Fig. V-3-3], and write ten on the fulcrum, which is the given known in the problem. Then take whichever number you wish, write it in one of the pans, and proceed according to what is mentioned in the problem, by subtracting its third and its quarter and taking the rest, comparing it with the ten on the fulcrum, i.e., inspect whether it is the same value, less or more.

. . .

[18]This refers to the Rule of Three, which was explained right before the method of scales.

[19]If we use anachronistic notation and denote the unknown wealth by x, the problem is: $x - \frac{1}{3}x - \frac{1}{4}x = 10$. We will use the notation: $f(x) = x - \frac{1}{3}x - \frac{1}{4}x$. Note that here the operation is linear. In general, the method works also for affine functions, where $f(x) = Ax + B$.

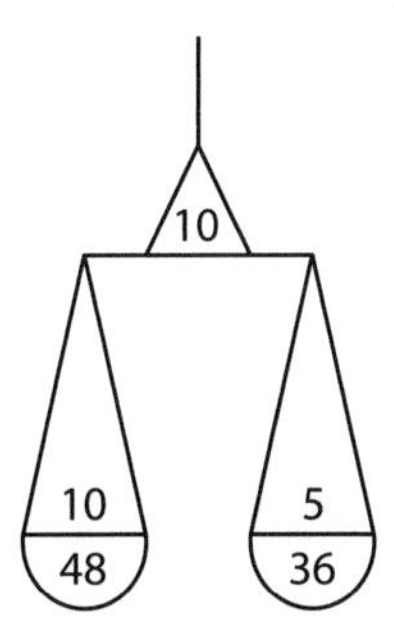

Fig. V-3-4.

He says: **If you err then write the error above the pan if it is superfluous, or underneath, if it is deficient. Then place in the other pan any number you wish, except for the first one you have chosen; follow the same procedure as you have done in the first case. Then multiply the one error by the other integer. Inspect whether the errors are superfluous or deficient. Subtract the smaller error from the larger one and subtract the smaller multiplication from the larger one. Divide the error between the multiplications by the remainder of the errors.**

Commentary: if it did not occur that you have chosen the right number, but a different one, which is the case most of the time—the other case [choosing the right number] is rare—in any case, after you subtract one third and one quarter, the error in relation to the ten on the fulcrum will be either larger or smaller than it; this is what he meant by "if you err."

For example, you draw the scales as follows; take, for example, the number 36, write it in one pan, subtract its third, 12, and its quarter, 9, altogether 21, and the remainder is 15. Compare it to the ten and there is an error, because there are five superfluous ones, i.e., units. Write the five, which is the error, above the pan, because it is superfluous. If the remainder were 8, being less than 10, you would write it beneath the pan. Then place any number you wish on the other pan, except for the 36 that you have already chosen, e.g., 48. Write it in the other pan, subtract its third, 16, and its quarter, 12, altogether 28, the remainder is 20. Compare it to the ten on the fulcrum, there is an error here too, because there are ten superfluous ones. Write it above the pan because it is also superfluous [Fig. V-3-4].[20]

. . .

Then multiply each error in each pan by the integer in the other, i.e., multiply the 5 above the 36, which is the error in the first pan, by the integer in the other pan, 48, the result is 240. Also, multiply the ten above the second pan, which is its error, by the other, it becomes 360. 240 and 360 are called *multiplications*. This is the first procedure for the case when the errors are above the pans.[21]

[20] $f(36) = 15$ and $f(48) = 20$; $f(36) - 10 = 5$ and $f(48) - 10 = 10$.
[21] $48 \times (f(36) - 10) = 48 \times 5 = 240$; $36 \times (f(48) - 10) = 36 \times 10 = 360$.

. . .

He said: **Then inspect whether both errors are larger** than the ten in the pan, i.e., such as in the first example, **or deficient**, i.e., like in the second example.[22] **Then subtract the smaller from the larger**, i.e., the smaller of errors from the larger one, i.e., subtract the smaller error, which is 5 in the first example, from ten, which is above the second pan, and 5 remains. The **smaller of multiplications** refers to subtracting the smaller multiplication, such as 240 in the first example, **from the larger one**, which is 360, and the remainder is 120. **Divide the remaining part of the multiplication**, which equals 120, **by** 5, **the remainder of the errors.** The result, 24, is the unknown sought number.[23]

. . .

He said: **If one of them is superfluous and the other is deficient, divide the sum of the multiplications by the sum of the errors.**[24]

. . .

He says: **If you wish, place in the second pan the first number or another one, and subtract from it the part which is compared to what is above the fulcrum, then multiply it by the integer in the first pan and multiply the error of the first by the integer in the second pan. Then, if the error of the first pan is deficient, add the multiplications. If it is larger, take the difference between them and divide it by the part, i.e. the number, in the second pan.**

Commentary: This is another method in the procedure of the scales in which you take whichever number you wish and write in one pan, do with it as you have in the first procedure, write this error above if it is superfluous, or below, if it is deficient.

We demonstrate it with an example: let there be ten on the fulcrum and place 36 in the first pan. We follow the same procedure as before; the error, 5, will be above the pan.

. . .

We then turn to the second pan and write 48 in it as we have done before. . . . If we take 48 and subtract its third and its quarter, twenty is left and it is that term with which we compare. In this procedure, we write the error between this number and the ten above the fulcrum neither above nor below the pan, but we write the integer either above or below as we wish. . . . Multiply this part by the other, which is 20, when the number written in the pan is 48. If the first pan had 36, we would multiply 20 by 36, becoming 720. We multiply 5, which is the error above the first pan, by 48, and it becomes 240 [Fig. V-3-5]. Since the error is above the pan, which is superfluous, we take the increment between the two multiplications, i.e. we subtract 240 from 720, and the remainder is 480. We divide it by

[22] This second example is omitted here.

[23] The operation here is: $\frac{36\times(f(48)-10)-48\times(f(36)-10)}{(f(48)-10)-(f(36)-10)}$. This number solves the problem $f(x)=10$. Indeed, the ratio $(48-x):(48-36)$ should be the same as the ratio $(f(48)-f(x)):(f(48)-f(36))=(f(48)-10):((f(48)-10)-(f(36)-10))$. The value of x can now be derived by means of proportion theory.

[24] Since the author considers the absolute values of the errors $f(a)-10$, the difference in the denominator has to be replaced by a sum when one error is negative and the other is positive.

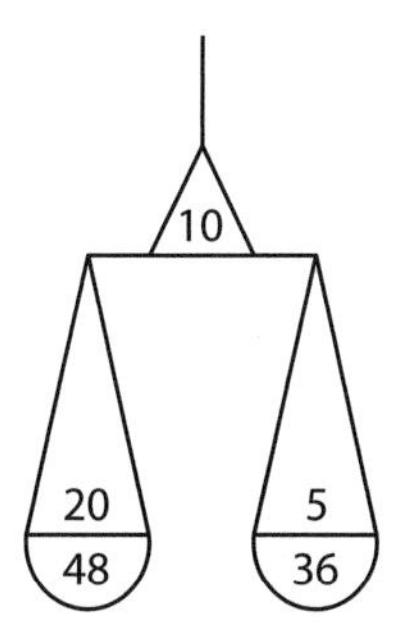

Fig. V-3-5.

20, which is called *division by what is above the second pan*, and the result is 24; it is the desired solution[25] and [Fig. V-3-5] is its form.

The elements of algebra and their operations

Here are presented the algebraic terms (root or thing, square or estate, etc.). These terms are presented as analogous to numbers and decimal ranks, but unknown algebraic terms are carefully distinguished from known numbers, which could be roots and squares as well. The discussion refers to the possibility of unknown algebraic terms that are not determined with respect to each other (a root which is not the root of the given square, that is, referring to a different unknown).

He said: **The application of restoration[26] is upon three species: numbers, things, and estates. The things are the roots. The squares are the result of the root multiplied by itself.**

Commentary: As previously explained, the number has three ranks: units, tens, and hundreds and all other [numbers] are composed by them. Also, in this science, ranks are set which are numbers, roots and squares and the rest are formed by them. This is what he meant when he said: the application of restoration is upon three species. Even though there are many species, cubes, and others, here, the main aim of this science is to reduce them all to these three species.

The three species are numbers, things, and estates. He said that the things are the roots previously mentioned in the book, and in this science one names them things. The estates are the result of the root multiplied by itself. In other words, when one multiplies the root by itself, this multiplication is named an estate in this science, and it is the square that was mentioned in the book.

The commentary of these two: he set [aside] the number because these two are also named in this science by a different name, but the number is not. A complete commentary about these three species is as follows: the numbers could be any number, whether units,

[25] Here we have the solution $\frac{36\times f(48)-48\times(f(36)-10)}{f(48)}$, which applies when f is linear ($f(x) = Ax$). This solution retains its meaning even if we only have one sampling of f, in which case it reduces to $\frac{48\times 10}{f(48)}$, namely, to a simple application of the Rule of Three in a single false position.

[26] "Restoration" is one of the elementary algebraic operations (*al-jabr*), that of adding a subtracted term to both sides of an equation.

tens, or hundreds, other ranks or their composition, either large or small, such as 5, 9, 11, and 120, and in general, every number, either large or small. For this reason this rank is named the number.

The roots are the roots of the squares. It is known that every number can be a root of a square. . . . That number is a root to that square and for every number [it is the] same. However, the number is called a number in itself and it is not called a root unless related to a square. Therefore, the number given here is a number as such. The given root here is a number which is a root extracted from a certain square and it does not have a determined value. That is why it is called a thing, i.e., a certain thing among the numbers that is a root to a certain square.

. . . There is no interest in a square [per se] in the Science of the Number but [rather] in the result of the multiplication of a number by itself. For this reason, the given square here is an unknown square and that's why they named it an estate.

Also, the roots, that are called things, and the squares, that are called estates, are always set in the problems as unknown. [As for] the numbers, there is no way to set them as unknown, but only as known, since the number has no relation to others, such as the root has with the square or the square has with the root, because the number, as I said, is a number in itself, the root is a root to a square, and the square is a square to a root, in the way of all [things which are] related.

This is the explanation of the three species, on which restoration acts.

The hundreds are formed by tens and units and the thousands are formed by hundreds, tens are formed by units, and all the other ranks are formed by previous ranks. Also, in this science, the squares are joined by the multiplication of the root by the root and are called estates. The cube [is formed] by the multiplication of the root by an estate. If one multiplies the root by a cube, it becomes of a different rank, called the estate of an estate, it is the multiplication of an estate by an estate. If one multiplies an estate by a cube, the result becomes of a different rank, called an estate-cube. A cube [multiplied] by a cube becomes a cube-cube, and so on with the rest. The repetitions of estates and cubes are called an estate-estate and a cube-cube-cube or their composition, an estate-estate-cube, an estate-cube-cube and similarly [with] the rest. As numbers have ranks, so do these [objects].

Just as numbers can be added, subtracted, multiplied, and divided, so can numbers, roots, and squares. The other ranks are also added together, subtracted from one another, multiplied and divided by each other. He says: the number, which is 5 and 7 added together, equals what 8 and 4 add up to. He also says that roots are equal to a number, or roots are equal to an estate, or roots and a square [equal to] a number and so on. . . .

There are 4 types of the number,[27] the first of which is an isolated number, such as 5, 30 or others, and it is called integer. Also in this science, we find a root by itself or an estate and the term in which one of these is found is called integer.

[27]The following categorization does not appear in the *Talkhīṣ*. The four types are: whole terms (e.g., a number, a root, or a square); connected terms (sums whose parts are not related, that is, are not of the same kind and cannot form a ratio—e.g., numbers counting different kinds of objects, or squares and roots of different unknowns); added terms (e.g., sums of numbers of the same kind, roots and squares of the same unknown); and subtracted terms (differences of numbers or algebraic terms).

The second type refers to having two numbers or more, where each stands by itself, such as 4 and 5 or 8 and 12, etc., such that one number is not related to the other, and it is called connected. Similarly, in this science it is said that a term has, for example, a root, a square and a cube, and each one stands by itself; i.e., the roots are not of the squares and the squares are not of the cubes. Rather, each stands by itself and is added to the other and is also called a connected term. This is scarcely encountered in this science, unless within numbers. . . .

The third: when he says, for example, one and a quarter and one and a half, where the quarter is related to one. So is one half, because one quarter is one quarter of the one and this is called the additive, because the part is added to the integer. Also, in this science, it is said that a term has squares and roots, i.e., an estate and things, or a cube and an estate which are related, meaning that the things are things of the estate and the estate is an estate of the cube and it is called additive. Also, an estate plus a number and a thing plus a number, are called additive even though they are not related or connected.

The fourth: as he says, in "number ten minus 2" or "30 minus 5" or "one minus one third," where it is called deficient or subtractive, so it is in this science, it is said that a term has an estate less a thing, or a cube less a square, which is called deficient or subtractive.

Multiplication and division of algebraic terms

The fragment below comes after a systematic treatment of adding and subtracting compound algebraic terms and a discussion of multiplying binomial sums and differences of algebraic terms. Each algebraic term is assigned a degree (1 for root, 2 for square, etc.), and the degree of the product is the sum of the degrees of the multiplied terms. A notation that places the degree above the coefficient (e.g., 2 with superscript 3 for "2 cubes") is introduced but is not actually put to use. These operations are presented as arithmetical operations in their own right, but their discussion is interspersed with applications to simplifying equations (adding and subtracting terms and reducing the degree of an equation through division by the term of least degree).

Whenever we wish to multiply ranks by ranks, the custom is to write them in two lines, one beneath the other. The degrees of the superior line are above and the degrees of the inferior line are beneath. For example, you wish to multiply 3 cubes and 7 estates and ten things by 9 cubes and six estates and 5 things. Write them in this form:

1	2	3
10 5	7 6	3 9
1	2	3

Then, we actually write three lines. In the middle one write the degrees by order, i.e., start by the larger and end by the smaller. In the lower line, write the outcome of the multiplication of the additives, and in the upper [line, write] the outcome of the multiplication of the subtractives, each below the degree which corresponds to it or above it in this form:

0	1	2	3	4	5	6
		50	35 60	15 42 90	18 63	27
		50	95	147	81	27

Start with the degree 6, as it is the biggest in this multiplication. At the [left] end, we write a zero, as this is appropriate when there is a number in the multiplication. Then, multiply 3 by 9, and the outcome is 27. Add the degrees, 6, and this is why we write the 27 under the 6. Then, multiply 3 by 6, and it becomes 18. Its degree is 5, and that is why we write it under the 5. Then, multiply 3 by 5, it becomes 15, and its degree is 4. That is why we write it under the 4. All multiplications of the 3 are hereby ended.

We return to the 7. Multiply it by 9, it becomes 63 and adding its degrees it becomes 5. That is why we write it under the 5. Then multiply 7 by 6. It is 42 and their added degrees are 4. That is why we write it under the 4. Then, multiply 7 by 5, it is 35 and its degree is 3. Write it under the 3 and the multiplication of the 7 is [hereby] ended.

Return to the 10 and multiply it by 9. It is 90 and its degree is 4. That is why we write the 90 under the 4. Multiply 10 by 6, it is 60 and its degree is 3. Write the 60 under the 3, then, multiply 10 by 5. It becomes 50 and its degree is 2. Write the 50 under the 2. Then, add each species with each other and the result is 27 square-square-estates plus 81 square-cubes plus 147 square-estates plus 95 cubes plus 50 estates. Write them each under its own species.

Another example: we wish to multiply 2 estates less 3 things by 3 cubes less 4 estates in this form:

2		1
2 3	Less Less	3 4
3		2

Multiply 2 by 3. They are [the multiplication of] an additive by an additive, or say an integer by an integer. It is 6 and the addition of their degrees is 5. That is why we write it underneath the 5 [see table below]. Then, multiply 2 by less 4, and it becomes subtractive 8. It is subtractive so we write it above the 4, as the addition of their degrees results in 4. Then, multiply less 3 by 3, it becomes subtractive 9 and its degree is 4. That is why we write it above the 4 as well. . . . Then, multiply less 3 by less 4 as if the word less were not there. It is 12 and its degree is 3. That is why we write it under the 3. Then, add each species with its own and the outcome is 6 cube-estates less 17 square-estates and 12 additive cubes. We write them below each other in correspondence to their species.

	9 8	
3	4	5
12		6
12	Less 17	6

. . .

He said: **when one divides a species of these species by its inferior, subtract the degree of the divisor from the degree of the divided term and the rest is the degree of the outcome of the division.**

Commentary: it is already known that the lower of the species that has a degree is the root. Next, the greater [species] is the square, i.e., the estate, and after it the cube, and so on. One knows in the division of integers that the division has two forms: first, the division of a large number by a small number, or express it as "a superior number [divided] by an integer number," and this is called simple division. The second [division] is the division of a small number by a large number and [the former] is called by the denomination [of the latter].

The second [division] is not commonly found in these chapters. One knows this from the condition that the writer set here about the knowledge of the outcome of the division, by subtracting from the degree of the divided term the degree of the divisor. This necessitates that the degree of the divisor is smaller than the degree of the divided term. Also, the author tells us next not to divide the lower of the species by the superior. His instructions are that when dividing integers, it is a number by a number. However, the divided term is [of] superior [degree] and the divisor [is of] inferior [degree].

For example, one wishes to divide ten estates by 5 roots. Divide ten by 5, as with integers, and the outcome of the division is 2. In order to know [of] what [species] this 2 is, subtract the degree of the divisor, roots with degree of 1, from the degree of the divided term, which is estates of degree 2. 1 remains, and it is the degree of roots. Here, the 2, the result of the division, is 2 roots. Then, if you wish, [you can] test the division from what you already know, by multiplying the result by the divisor, and the outcome is the divided term. Also, in this example, multiply the outcome, 2 roots, by 5 roots, the divisor, and it becomes ten estates with a degree of 2.

. . .

He said: **when dividing one species by the same one, the outcome is a number.**

Commentary: when dividing roots by roots and subtracting the degree of the divisor from the degree of the divided term, nothing remains. Thus, the outcome of the division is a number without a degree. For example, when ten roots are divided by five roots, the outcome of the division is 2 and they are *zuzim*,[28] because the degree of the divisor is 1, and so is the degree of the degree of the divided, a root. When subtracting 1 from 1, nothing remains. . . .

He said: **when dividing one of these species by a number, the outcome is of the same species.**

Commentary: as the number has no degree, when dividing any of these species by it, the divisor, which is the number, has no degree to subtract from the degree of the divided term. Thus, the degree of the divided term remains in its place. For example, when 12 cubes are divided by 4 *zuzim*, the outcome is 3 cubes.

He said: **if the divided term contains subtractive terms, divide each term, in both the subtractive and the subtrahend, by the divisor, and then subtract the outcomes.**

[28] A Hebrew coin denomination.

Commentary: shortly the author mentions that no species containing subtractives can divide. However, the divided term, even though it is subtractive, can be divided. He gives the method here, [i.e.,] dividing the subtractive alone by the divisor and dividing the term subtracted from as well. Then, subtract the outcome of the first from the outcome of the second. The remainder is the answer being sought.

An example here: you wish to divide 4 estates less 6 roots by two roots. Divide the subtractive, 6 roots, by the divisor, 2 roots, and the outcome is 3 *zuzim*. Then, divide the part from which the subtractives are subtracted, 4 estates, by the divisor, which is 2 roots[, producing 2 roots]. One subtracts the first outcome, 3 *zuzim*, from the second outcome, 2 roots, and the final outcome is 2 roots less 3 *zuzim*, which is the result of the division by its divisor, which is 2 roots. . . .

He said: **one shall not divide by a subtractive.**

Commentary: the divided is not subtractive. The reason being is that you already know that the definition of division, according to what the author wrote in the chapter on division, is the decomposition of the divided term into equal parts, and their number is reflected by the divisor in units. I have already explained this definition there. It follows from it that the knowledge of the units of the divisor and all these species in these chapters are unknown. When the species is subtractive, we do not know what is left of its subtracted term for us to divide by this remainder.

For example, if the divisor is a square less 2 roots, when we subtract two roots from the square, then we will not know how many roots are left in a square to divide by it. Similarly in all species, and this is explained.

He said: **This completes what we wished to know. Thank God, may His name be magnified and blessed.**

Commentary: He bestowed praise upon the Almighty, may His name be magnified, who helped him complete his mission as the erudite authors rejoice upon the completion of their composition and they thank the Almighty, may His name be blessed, who favored them and helped them make their name respected. We praise the exalted God of Israel by every blessing and praising, who helped us with its explanation. He will help us with His mercy, with everything of the honor of His name. He will save us. He will tell about our sins for his great and fearful name, may His name be blessed and magnified. Amen.

. . .

The commentator said: I have also seen [it appropriate] to write in this chapter a supplement [written] by their [i.e., Arab] scholars in this science, as I have seen it to be obligatory and useful in all species of restoration.

One already knows what was mentioned about division in [the part about] restoration, [i.e.,] that one does not divide by a subtractive [term] and also, that a lower species does not divide an upper species. It is known that there are instances in the problems where it is required that the lower divide the upper and, also, divide by a subtractive.

Therefore, when this is given in the problem, one has to apply this from the known rule, because when one multiplies the outcome of the division by the divisor, the outcome of the multiplication is the divided term. Therefore, multiply what was given from the result of the division by its divisor. The outcome of the multiplication is equal to the divided term.

The example mentioned in the third problem [that is, split 10 into two numbers, such that one divided by the other is 4], in which he said: when one divides the one part of the ten by the other, the outcome of the division is 4, because, if we let the one part [be] a thing and the other [be] ten less a thing, then this is a division by a subtractive ... and it does not divide. Indeed, he states that the outcome of the division is 4. Therefore, when multiplying the 4 by ten less a thing, the divisor, the outcome is 40 less 4 things and it equals the divided term. When one completes and opposes,[29] it becomes 40 equals 5 things. When one restores, 8 is obtained, and it is the one part of ten and the other is equal to two.[30] Use it in analogy.

Sometimes, there are instances in such problems, where you are not able to determine the solution by this rule. Therefore, you need to know a different rule. In the example here, in which one asks: Ten, partition it into two parts, divide each one by the other and add the outcome of the two divisions. The sum is 2 and one sixth.[31]

If one lets the one part [be] a thing and the other be ten less a thing, then there is no way to divide by each other, because the one is division of a lower by upper and the second is a division by a subtractive, as mentioned. There is no way to solve it by the mentioned rule because the given in the problem is the outcome of the two divisions, added together. We do not know what will come out of each division in order to multiply it by its divisor, to give the divided term.

Therefore, you need here a different rule. The rule is that for all the numbers [in the problem], divided by one another, i.e., by each other, add the outcomes of the two divisions and keep them. Then, multiply each of those numbers by itself, add the outcomes, and then it becomes the multiplication of one of the numbers by the other and the kept outcome.

An example of this rule: 4 and 6. Divide the six by the 4, and from the division we obtain one and a half. Divide the 4 by the six, the outcome is 4 sixths. Add the latter with the one and half, it becomes 2 and one sixth, which is the outcome of the divisions. Keep it and multiply 4 by itself, and it becomes 16. [Multiply] six by itself, and it becomes 36. Add 16 to 36, and it becomes 52. Also, when multiplying 4 by six, it becomes 24. Multiply it by the kept 2 and one sixth, this is 52. And this equals the sum of the two multiplications, because it is also 52.[32]

After knowing this rule, return to the problem that he set. He says that when one divides the one part of the ten, which is a thing, by the other part, which is ten minus a thing, divide the ten less a thing by the thing, add the outcomes of the two divisions, and the outcome is 2 and one sixth. Do as mentioned in the rule of multiplication. Multiply the thing by itself, and it becomes an estate, multiply the ten less a thing by itself, the result is 100 minus 20 things and an estate. Add it with the estate which results from the multiplication of

[29]Opposition is the algebraic operation of subtracting equal terms present in both sides of an equation (*al-muqābala*).

[30]The problem is to split 10 into two parts, x and $10 - x$, such that their quotient is 4. Instead of the inadmissible division $\frac{x}{10-x} = 4$, the commentator introduces the equation $x = 4\,(10 - x)$ to obtain $x = 8$.

[31]The problem is: $\frac{x}{10-x} + \frac{10-x}{x} = 2\frac{1}{6}$, which includes inadmissible divisions. To solve it, rearrange it as: $x^2 + (10 - x)^2 = 2\frac{1}{6}x\,(10 - x)$.

[32]$\frac{4}{6} + \frac{6}{4} = 2\frac{1}{6}$. This is equivalent to $4 \times 4 + 6 \times 6 = 4 \times 6 \times 2\frac{1}{6}$.

the thing by itself, and it becomes 100 less 20 things plus 2 estates. Then, multiply the ten less a thing by a thing, and obtain 10 things less an estate. Multiply it by 2 and one sixth, as [previously] mentioned to be the outcome of the two divisions. The outcome of the multiplication is 21 things and two thirds of a thing less 2 estates plus one sixth [of an estate] equals the 100 less 20 things and 2 estates.

When one proceeds by restoration and opposition, one obtains an estate and 24 *zuzim* equal 10 things [we omit the derivation of the last equality from the previous one]. This is the fifth type of the six types of restoration. Do as is written there, by taking half the number of roots, 5, squaring it, and it becomes 25. Subtract from it the number of *zuzim*, 24 and 1 is left. Take its root, which is also 1, and subtract it from the number of half the roots, 5. The remainder is four, and it is the one part of the ten. The other one is six. Also, if one adds the one to half the number of estates it becomes 6; it is one part of the ten and the other is 4.

SOURCES

I. Practical and Scholarly Arithmetic

1. Abraham ibn Ezra, *Sefer Hamispar*, translated by Roi Wagner from the Hebrew and German from Abraham ibn Esra. 1895. *Sefer ha-Mispar, Das Buch der Zahl*, edited and translated by Moritz Silberberg. Frankfurt am Main: J. Kauffmann, pp. 1–4, 12, 32–33, 45, 51–52, 56–58, 62–63, 70–71, 73–74,
2. Aaron ben Isaac, *Arithmetic*, translated by Naomi Aradi and Roi Wagner from the manuscript, Turin, Biblioteca Nazionale Universitaria A V 15 (IMHM: f 41584), fifteenth century, 18b–21a.
3. Immanuel ben Jacob Bonfils, *On Decimal Numbers and Fractions*, translated by Roi Wagner based on Solomon Gandz. 1936. "The Invention of the Decimal Fractions and the Application of the Exponential Calculus by Immanuel Bonfils of Tarascon (c. 1350)," *Isis* 25(1), 16–45, and Tony Lévy. 2003. "Immanuel Bonfils (XIVe s.): Fractions décimales, puissances de 10 et opérations arithmétiques," *Centaurus* 45, 284–304, with the assistance of a somewhat deviant third manuscript edited by Naomi Aradi.
4. Jacob Canpanṭon, *Bar Noten Taʿam*, translated by Naomi Aradi and Roi Wagner from the manuscript London, British Library Or. 1053 (IMHM: f 5932), ff. 1b–65b (cat. Margo. 1012, 1), fifteenth century, pp. 34b–40b.
5. Elijah Mizraḥi, *Sefer Hamispar*, translated by Stela Segev from the edition in Stela Segev. 2010. "*The Book of the Number* by Elijah Mizraḥi: A Mathematical Textbook from the 15th Century (Hebrew)." Ph.D. thesis, Hebrew University of Jerusalem, and from the printed edition, Elijah Mizraḥi. 1534. *Sefer Hamispar* (Hebrew). Constantinople. Article I, part 1, pp. II-8, 43, 48, 56–57 65–68, 78, part 2, pp. 29a, 31a–b, Article III, part 1, chapter 1.
6. Levi ben Gershon, *Maʿase Ḥoshev*, translated by Shai Simonson from the Hebrew edition in Gerson Lange. 1909. *Sefer Maassei Chosheb—Die Praxis des Rechners, Ein hebraeisch arithmetisches Werk des Levi ben Gerschom aus dem Jahre 1321*. Frankfurt am Main: Louis Golde, after consulting the other available manuscripts. The problems at the end were already included in Shai Simonson. 2000b. "The Missing Problems of Gersonides: A Critical Edition," *Historia Mathematica* 27, 243–302, 384–431.

II. Numerology, Combinatorics, and Number Theory

1. Abraham ibn Ezra, *Sefer Haʾeḥad*, translation revised by Roi Wagner from Tzvi Y. Langermann. 2001. "Studies in Medieval Hebrew Pythagoreanism: Translations and Notes to Nicomachus' Arithmological Texts," *Micrologus* 9, 219–236, based on the editions in Abraham ibn Esra. 1921. *Buch der Einheit*. Edited and translated by E. Müller. Berlin: Welt Verlag; Abraham ibn Ezra. 1856. "Sepher Haeḥad," *Jeschurun* 1,

3–10 (Hebrew); and Abraham ibn Ezra. 1867. *Sepher Ha-Echad*, edited and commented by S. Pinsker and M. A. Goldhart. Odessa: L. Nitzsche (Hebrew).

2. Abraham ibn Ezra. *Sefer Haʿolam*, based on Shlomo Sela. 2010. *Abraham ibn Ezra: The Book of the World*. Leiden: Brill.
3. Levi ben Gershon, *Maʿase Ḥoshev*, translated by Shai Simonson from the Hebrew edition in Gerson Lange. 1909. *Sefer Maassei Chosheb—Die Praxis des Rechners, Ein hebraeisch arithmetisches Werk des Levi ben Gerschom aus dem Jahre 1321*. Frankfurt am Main: Louis Golde.
4. Levi ben Gershon, *On Harmonic Numbers*, translated by Victor Katz from the Latin edition and French translation in Christian Meyer and Jean-François Wicker. 2000. "Musique et Mathématique au XIVe Siècle: Le *De Numeris harmonicis* de Leo Hebraeus," *Archives internationales d'histoire des sciences* 50, 30–67.
5. Qalonymos ben Qalonymos, *Sefer Melakhim*, adapted and translated by Naomi Aradi and Roi Wagner from the Hebrew and French in Tony Lévy. 1996. "L'histoire des nombres amiables. Le témoignage des textes hébreux médiévaux," *Arabic Sciences and Philosophy* 6, 63–87.
6. Don Benveniste ben Lavi, *Encyclopedia*, adapted and translated by Naomi Aradi and Roi Wagner from the Hebrew and French in Tony Lévy. 1996. "L'histoire des nombres amiables. Le témoignage des textes hébreux médiévaux," *Arabic Sciences and Philosophy* 6, 63–87.
7. Aaron ben Isaac, *Arithmetic*, translated by Naomi Aradi and Roi Wagner from the manuscript, Turin, Biblioteca Nazionale Universitaria A V 15 (IMHM: f 41584), fifteenth century, 206b–208a.

III. Measurement and Practical Geometry

1. Abraham ibn Ezra (?), *Sefer Hamidot*, adapted with minor changes from the English translation in Tony Lévy and Charles Burnett. 2006. "'Sefer ha-Middot': A Mid-Twelfth-Century Text on Arithmetic and Geometry Attributed to Abraham ibn Ezra," *Aleph* 6, 57–238.
2. Abraham bar Ḥiyya, *Ḥibur Hameshiḥa Vehatishboret*, translated by Roi Wagner from the Hebrew-German edition, Abraham Bar Chija. 1913. *Chibbur ha-Meschicha weha-Tischboret: Lehrbuch der Geometrie*, edited and translated by M. Guttman. Berlin: Vereins Mekize Nirdamim.
3. Rabbi Shlomo Iṣhaqi (Rashi), *On the Measurements of the Tabernacle Court*, translated by David Garber and Roi Wagner from *Talmud Bavli, Eruvin* 23b.
4. Simon ben Ṣemaḥ, *Responsa on Solomon's Sea*, translated by David Garber and Roi Wagner from Simon Ben Ṣemaḥ Duran. 1998. *Sefer haTashbets* (Hebrew). Jerusalem: Shlomo Aumann Institute and Jerusalem Institute.
5. Levi ben Gershon, *Astronomy*. The translations from chapter four are taken from Bernard L. Goldstein. 1985. *The Astronomy of Levi ben Gerson (1288–1344)*. New York: Springer, pp. 41, 43–47. The translations from chapter 49 are taken from J. L. Mancha. 1998. "Heuristic Reasoning: Approximation Procedures in Levi ben Gerson's Astronomy," *Archive for History of Exact Sciences* 52, 13–50, pp. 15–16.

IV. Scholarly Geometry

1. Levi ben Gershon, *Commentary on Euclid's Elements*, translated by Victor Katz from the French in Tony Lévy. 1992. "Gersonide, Commentateur d'Euclide: Traduction annoteé de ses gloses sur les *Éléments*" in Gad Freudenthal (ed.), *Studies on Gersonides: A Fourteenth-Century Jewish Philosopher-Scientist*. Leiden: E. J. Brill, 83–147, pp. 100–102, 104–115. Translation checked against the Hebrew original by Gad Freudenthal.
2. Levi ben Gershon, *Treatise on Geometry*, translated by Victor Katz from the French in Tony Lévy. 1992. "Gersonide, Commentateur d'Euclide: Traduction annoteé de ses gloses sur les *Éléments*" in Gad Freudenthal (ed.), *Studies on Gersonides: A Fourteenth-Century Jewish Philosopher-Scientist*. Leiden: E. J. Brill, 83–147, pp. 146–147. Translation checked against the Hebrew original by Gad Freudenthal.
3. Qalonymos ben Qalonymos, *On polyhedra*, translation adapted from Tzvi Langermann. 2014. "Hebrew Texts on the Regular Polyhedra," in Nathan Sidoli and Glen Van Brummelen (eds.), *From Alexandria, through Baghdad: Surveys and Studies in the Ancient Greek and Medieval Islamic Mathematical Sciences in Honor of J. L. Berggren*. New York: Springer, 411–469, pp. 440–441.

4. Immanuel ben Jacob Bonfils, *On Measurement of the Circle*, translation adapted from Tony Lévy. 2012. "Immanuel ben Jacob of Tarascon (Fourteenth Century) and Archimedean Geometry: An Alternative Proof for the Area of a Circle," *Aleph* 12, 135–158.
5. Solomon ben Isaac, *On the Hyperbola and Its Asymptote*, translated by Roi Wagner from the Hebrew and French in Tony Lévy. 1989a. "Le chapitre I,73 du Guide des Egarés et la tradition mathématique Hébreaïque au moyen age. Un commentaire inédit de Salomon b. Isaac," *Revue des études juives* 98(3–4), 307–336.
6. Abner of Burgos (Alfonso di Valladolid), *Sefer Meyasher ᶜAqov*. Proposition 23 translated from the Hebrew by Tzvi Langermann in "Medieval Hebrew Texts on the Quadrature of the Lune," *Historia Mathematica* 23, 31–53; propositions 29–32 translated by Avinoam Baraness from the Hebrew edition in Gluskina, Gitta (ed). 1983. *Meyasher ᶜAqob*. Moscow: Izdatelstvo Nauka and the manuscript London, British Library Add. 26984 (IMHM: f 5656; cat. Neub. 1002), fifteenth century.

V. Algebra

1. Quadratic word problems: a. Levi ben Gerson, Maᶜase Ḥoshev, translated by Shai Simonson from the Hebrew edition in Gerson Lange. 1909. *Sefer Maassei Chosheb—Die Praxis des Rechners, Ein hebraeisch arithmetisches Werk des Levi ben Gerschom aus dem Jahre 1321*. Frankfurt am Main: Louis Golde. b. Anonymous arithmetic, translated by Naomi Aradi from the manuscript Bibliothèque de Genève, MS héb. 10/1 (IMHM f 2320), fols.2a–38a, fourteenth–fifteenth century. This translation also appears in Naomi Aradi. 2013. "An Unknown Medieval Hebrew Anonymous Treatise on Arithmetic," Aleph 13, 235–309. The particular texts are from fols. 25b, 26b, 29a. c. Elijah Mizraḥi, Sefer Hamispar, see I, 5. These problems are on pp. 92, 94, 96.
2. Simon Moṭoṭ, *Algebra*, translated by Roi Wagner from Naomi Aradi's Hebrew transcript of the manuscript Berlin Staatsbibliothek Or. Oct. 244/14 (IMHM: f 1996), ff. 113a–120a, fifteenth century with help from the French translation in Gustavo Sacerdote. 1893–1894. "Le livre de l'algèbre et le problème des asymptotes de Simon Moṭoṭ," *Revue des études juives* 27, 91–105; 28, 228–246; 29, 111–126.
3. Ibn al-Aḥdab, *The Epistle of the Number*, translated by Ilana Wartenberg in Ilana Wartenberg. 2007. "The *Epistle of the Number* by Isaac ben Solomon ibn al-Aḥdab (Sicily, 14th century): An Episode of Hebrew Algebra." Ph.D. thesis, Université de Paris 7, Tel Aviv University, pp. 443–459, 462–468, 539–560.

REFERENCES

Aradi, Naomi. 2013. "An Unknown Medieval Hebrew Anonymous Treatise on Arithmetic," *Aleph* 13, 235–309.

Archibald, Raymond C. 1915. *Euclid's Book on Division of Figures*. Cambridge: Cambridge University Press.

Baer, Yitzhak. 1961–1966. *A History of the Jews in Christian Spain*. Philadelphia: Jewish Publication Society of America.

Berggren, J. Lennart. 2005. "Mathematics in Medieval Islam," in Victor J. Katz (ed.), *The Mathematics of Egypt, Mesopotamia, China, India, and Islam: A Sourcebook*. Princeton, NJ: Princeton University Press, pp. 515–675.

Clagett, Marshall. 1964–1984. *Archimedes in the Middle Ages*. Madison: University of Wisconsin Press and Philadelphia: American Philosophical Society.

Curtze, Maximilian. 1902. "Der 'Liber Embadorum' des Savasorda in der Übersetzung des Plato von Tivoli," *Abhandlungen zur Geschichte der Mathematischen Wissenschaften* 12, 1–183.

Dauben, Joseph. 2007. "Chinese Mathematics," in Victor J. Katz (ed.), *The Mathematics of Egypt, Mesopotamia, China, India, and Islam: A Sourcebook*. Princeton, NJ: Princeton University Press, pp. 187–384.

David, F. N. 1998. *Games, Gods and Gambling: A History of Probability and Statistical Ideas.* New York: Dover.

Djebbar, Ahmed. 1992. "Le traitement des fractions dans la tradition mathématique arabe du Maghreb," in P. Benoit, K. Chemla, and J. Ritter (eds.), *Histoire de fractions, fractions d'histoire.* Basel: Birkhäuser, pp. 223–245.

Freudenthal, Gad. 1988. "Maimonides' *Guide for the Perplexed* and the Transmission of the Mathematical Tract 'On Two Asymptotic Lines' in the Arabic, Latin and Hebrew Medieval Traditions," *Vivarium* 26(2), 113–140.

———. (ed.), 1992. *Studies on Gersonides: A Fourteenth-Century Jewish Philosopher-Scientist.* Leiden: E. J. Brill.

———. 2005. "Two Notes on *Sefer Meyasher ʿAqov* by Alfonso, alias Abner of Burgos," in G. Freudenthal, *Science in the Medieval Hebrew and Arabic Traditions.* Farnham, UK: Ashgate/Variorum Collected Studies, part IX.

Friedlein, Godofredi. 1873. *Procli Diadochi In primum Euclidis Elementorum librum commentarii.* Lipsiae: B. G. Teubneri.

Garber, David. Forthcoming. "The Approximation to the Square Root of 2 in Rabbinical Writings," (Hebrew).

Garber, David, and Boaz Tsaban. 1998. "Mathematical Appendix to the TASHBETS Responsa" (Hebrew), in Simon ben Ṣemaḥ Duran (ed.), 1998. *Sefer ha Tashbets* book (Hebrew). Jerusalem: Shlomo Aumann Institute and Jerusalem Institute, pp. 423–446. Available at http://u.cs.biu.ac.il/~tsaban/Pdf/tashbets.pdf.

———. n.d. "An Early Source for a Mathematical Method Appearing in the TASHBETS Responsa Book" (Hebrew). Available at http://u.cs.biu.ac.il/~tsaban/Pdf/rt.pdf.

———. 2001. "A Mechanical Derivation of the Area of the Sphere," *American Mathematical Monthly* 108, 10–15.

Ginsburg, Jekuthiel, and David E. Smith. 1922. "Rabbi Ben Ezra and Combinations," *Mathematics Teacher* 15(6), 347–356.

Gluskina, Gitta (ed). 1983. *Meyasher Aqob.* Moscow: Izdatelstvo Nauka.

Hacker, Joseph. 2007. "Mizraḥi, Elijah," in M. Barenbaum and F. Skolnik (eds.), *Encyclopedia Judaica,* second edition. Gale Virtual reference library, vol. 14, 393–395.

Hacohen, Naftali Yaacov. 1967–1970. *Otzar Hagedolim Alufey Yaakov* (Hebrew). Haifa.

Harbili, Anissa. 2011. "Les Procédés d'Approximation dans les Ouvrages Mathématiques de l'Occident Musulman," *Llull* 34, 39–60.

Heath, T. L. 1956. *Euclid, The Thirteen Books of the Elements,* vol. I. Mineola, NY: Dover.

———. 1964. *Diophantus of Alexandria: A Study in the History of Greek Algebra.* Mineola, NY: Dover.

———. 1981. *A History of Greek Mathematics,* vol. I. Mineola, NY: Dover.

Hogendijk, J. P. 1993. "The Arabic Version of Euclid's *On Division,*" in M. Folkerts and J. P. Hogendijk (eds.), *Vestigia Mathematica: Studies in Medieval and Early Modern Mathematics in Honour of H. L. L. Busard.* Amsterdam: Rodopi, pp. 143–162.

Høyrup, Jens. 2007. *Jacopo da Firenze's Tractatus algorismi and Early Italian Abbacus Culture.* Basel: Birkhäuser.

Ibn Ezra, Abraham. 1991. "Commentaries to Exodus," in *Torat Hayim.* Jerusalem: Mosad Harav Kook.

Imhausen, Annette. 2007. "Egyptian Mathematics," in Victor J. Katz (ed.), *The Mathematics of Egypt, Mesopotamia, China, India, and Islam: A Sourcebook.* Princeton, NJ: Princeton University Press, pp. 7–56.

Kaplan, A. 1995. *Sefer Yetzirah, The Book of Creation.* Lanham, MD: Jason Aronson.

Knorr, Wilbur. 1989. *Textual Studies in Ancient and Medieval Geometry.* Boston: Birkhäuser.

Lange, Gerson. 1909. *Sefer Maassei Chosheb—Die Praxis des Rechners, Ein hebraeisch arithmetisches Werk des Levi ben Gerschom aus dem Jahre 1321.* Frankfurt am Main: Louis Golde.

Langermann, Tzvi. 1996. "Medieval Hebrew Texts on the Quadrature of the Lune," *Historia Mathematica* 23, 31–53.

———. 2001. "Abraham bar Ḥiyya Savasorda," in T. Hockey (ed.), *The Bibliographical Encyclopaedia of Astronomers.* New York: Springer, pp. 95–96.

———. 2014. "Hebrew Texts on the Regular Polyhedra," in Nathan Sidoli and Glen Van Brummelen (eds.), *From Alexandria, through Baghdad: Surveys and Studies in the Ancient Greek and Medieval Islamic Mathematical Sciences in Honor of J. L. Berggren*, New York: Springer, pp. 411–469.

Langermann, Tzvi, and Shai Simonson. 2000. "The Hebrew Mathematical Tradition," in Helaine Seline (ed.),*Mathematics across Cultures*. Dordrecht: Kluwer, pp. 167–188.

Levey, Martin. 1952. "The Encyclopedia of Abraham Savasorda: A Departure in Mathematical Methodology," *Isis* 43(3), 257–264.

———. 1954. "Abraham Savasorda and His Algorism: A Study in Early European Logistic." *Osiris* 11, 50–64.

———. 1966. *The Algebra of Abū Kāmil: Kitāb fi al-Jābr waʾl Muqābala in a Commentary by Mordecai Finzi*. Madison: University of Wisconsin Press.

Lévy, Tony. 1989a. "Le chapitre I,73 du Guide des Egarés et la tradition mathématique Hébraïque au moyen age. Un commentaire inédit de Salomon b. Isaac," *Revue des études juives* 98(3–4), 307–336.

———. 1989b. "L'étude des sections coniques dans la tradition medieval Hébraïque. Ses relations avec les traditions Arabe et Latine," *Revue d'histoire des sciences* 42(3), 193–239.

———. 1992. "Gersonide, Commentateur d'Euclide: Traduction annoteé de ses gloses sur les *Éléments*," in Gad Freudenthal (ed.), *Studies on Gersonides: A Fourteenth-Century Jewish Philosopher-Scientist*. Leiden: E. J. Brill, pp. 83–147.

———. 1996. "L'histoire des nombres amiables. Le témoignage des textes hébreux médiévaux," *Arabic Sciences and Philosophy* 6, 63–87.

———. 2001. "Les Débuts de la Literature Mathématique Hébraïque: La Géometrie d'Abraham bar Ḥiyya," *Micrologus* 9, 35–64.

———. 2002. "A Newly-Discovered Partial Hebrew Version of al-Khwarizmi's Algebra," *Aleph* 2, 225–234.

———. 2003. "L'algèbre arabe dans les textes hébraïques (I), Un Ouvrage Inédit D'Isaac ben Solomon al-Aḥdab (XIV siècle)," *Arabic Sciences and Philosophy* 13, 269–301.

———. 2007. "L'algèbre arabe dans les textes hébraïques (II). Dans l'Italie des XVe et XVIe siècles, sources arabes et sources vernaculaires," *Arabic Sciences and Philosophy* 17, 82–87.

———. 2008. "Abraham bar Ḥiyya," In H. Selin (ed.), *Encyclopaedia of the History of Science, Technology and Medicine in Non-Western Cultures*. New York: Springer, p. 5.

———. 2012. "Immanuel ben Jacob of Tarascon (Fourteenth Century) and Archimedean Geometry: An Alternative Proof for the Area of a Circle," *Aleph* 12, 135–158.

Lévy, Tony, and Charles Burnett. 2006. "'Sefer ha-Middot': A Mid-Twelfth-Century Text on Arithmetic and Geometry Attributed to Abraham ibn Ezra," *Aleph* 6, 57–238.

Maimonides, Moses. 1963. *The Guide for the Perplexed*. Chicago: University of Chicago Press.

Mancha, J. L. 1998. "Heuristic Reasoning: Approximation Procedures in Levi ben Gerson's Astronomy," *Archive for History of Exact Sciences* 52, 13–50.

Mazars, G., 1992. "Les fractions dans l'Inde ancienne de la civilisation de l'Indus à Mahavira (IXe siècle)," in P. Benoit, K. Chemla, and J. Ritter (eds.), *Histoire de fractions, fractions d'histoire*. Basel: Birkhäuser, pp. 209–218.

Millás Vallicrosa, Jose M. 1931. *Abraam bar Hiia, Libre de Geometria. Hibbur Hameixihà vehatixboret*. Barcelona: Alpha.

———. 1952. *La obra enciclopedica Yesode ha-Tebuna u-Migdal ha-Emuna de R. Abraham bar Ḥiyya Ha-Bargeloni*, edición critica con traducción, prólogo y notas. Madrid and Barcelona: Casa Provincial de Caridad de Barcelona.

Montelle, Clemency. 2014. "Hypsicles of Alexandria and the Fourteenth Book of Euclid's *Elements*." (in press)

Münster, Sebastian. 1546. *Sphaera mundi, autore Rabbi Abrahamo Hispano filio R. Haijae. Arithmetica secundum omnes species suas autore Rabbi Elija Orientali*. Basel: H. Petrum.

Ovadia, A. 1939. "Rabi Elija Mizraḥi," *Sinai* 5: 397–413, *Sinai* 6: 73–80.

Plofker, Kim. 2002. "Use and Transmission of Iterative Approximations in India and the Islamic World," in Y. Dold-Samplonius et al. (eds.), *From China to Paris: 2000 Years Transmission of Mathematical Ideas*. Stuttgart: Franz Steiner, pp. 167–186.

Rabinovitch, Nachum. 1970. "Rabbi Levi ben Gershon and the Origins of Mathematical Induction," *Archive for History of Exact Sciences* 6, 237–248.

Rangācārya, M. (ed.) 1912. *Mahāvīra's Ganitāsarasangraha*. Madras: Government Press.

Rashed, Roshdi. 2011–2014. *A History of Arabic Sciences and Mathematics*, edited by Nader El-Bizr. London and New York: Routledge and Beirut: Centre for Arab Unity Studies.

Rosenfeld, B. A. 1988. *A History of Non-Euclidean Geometry: Evolution of the Concept of a Geometric Space*. Translated from the Russian by Abe Shenitzer. New York: Springer.

Rubio, Mercedes. 2000. "The First Hebrew Encyclopedia of Science: Abraham bar Ḥiyya's *Yesodei ha-Tevunah u-Migdal ha-Emunah*," in S. Harvey (ed.), *The Medieval Hebrew Encyclopedias of Science and Philosophy*. Dordrecht, Boston, and London: Kluwer, pp. 140–153.

Sacerdote, Gustavo. 1893–1894. "Le livre de l'algèbre et le problème des asymptotes de Simon Moṭoṭ," *Revue des études juives* 27, 91–105; 28, 228–246; 29, 111–126.

Sadik, Shalom. 2012. "Abner of Burgos," in Edward N. Zalta (ed.), *The Stanford Encyclopedia of Philosophy*, Fall 2012 edition. Available at http://plato.stanford.edu/archives/fall2012/entries/abner-burgos.

Sefrin-Weis, H. (ed.). 2010. *Pappus of Alexandria: Book 4 of the Collection*. New York: Springer.

Segev, Stela. 2010. "*The Book of the Number* by Elijah Mizraḥi: A Mathematical Textbook from the 15th Century" (Hebrew). Ph.D. thesis, Hebrew University of Jerusalem.

Sela, Shlomo. 2003. *Abraham ibn Ezra and the Rise of Medieval Hebrew Science*. Leiden: Brill.

Sela, Shlomo, and Gad Freudenthal. 2006. "Abraham ibn Ezra's Scholarly Writings: A Chronological Listing," *Aleph* 6, 13–55.

Simonson, Shai. 2000a. "The Mathematics of Levi ben Gershon, the Ralbag," *Bekhol Derakhekha Daehu* (Bar-Ilan University) 10, 5–21.

———. 2000b. "The Missing Problems of Gersonides: A Critical Edition," *Historia Mathematica* 27, 243–302, 384–431.

Solon, Peter. 1970. "The *Six Wings* on Immanuel Bonfils and Michael Chrysokokkes," *Centaurus* 15(1), 1–20.

Steinschneider, Moritz. 1870. "Das Königsbuch des Kalonymos," *Jüdische Zeitschrift für Wissenschaft und Leben* 8: 118–122.

———. 1893–1901. *Mathematik bei den Juden*. Berlin, Leipzig, Frankfurt: Kaufmann; reprint. Hildesheim: G. Olms, 1964 and 2001.

———. 1906. "Mathematik bei den Juden,1551–1840," *Monatsschrift für die Geschichte und Wissenschaft des Judenthums* 2, 195–214.

Sude, Barbara Hooper. 1974. "Ibn al Haytham's Commentary on Euclid's *Elements* (*Sharh Musadarat Kitab Uqlidis Fi-Al Usul*), Books I–VI." Ph.D. thesis, Princeton University.

Suter, Heinrich. 1901. "Das Rechenbuch des Abû Zakarijâ al-Ḥassâr," *Bibliotheca Mathematica* 2, 12–40.

———. 1986. "Die Kreisquadratur des Ibn al-Haitam," in *Beiträge zur Geschichte der Mathematik und Astronomie im Islam*, Vol. 2. Frankfurt am Main: Institut für Geschichte der arabisch-islamischen Wissenschaften, pp. 76–90.

Toomer, G. J. 1984. *Ptolemy's* Almagest. New York: Springer.

———. 2008. "Nicomedes," in *Complete Dictionary of Scientific Biography*. Retrieved September 15, 2012 from Encyclopedia.com: http://www.encyclopedia.com/doc/1G2-2830903168.html.

Wagner, Roy. 2013. "Mordekhai Finzi's Translation of Maestro Dardi's Italian Algebra," in A. Fidora, H. Hames, and Y. Schwartz (eds.), *Latin-into-Hebrew: Texts in Contexts*. Leiden: Brill, pp. 195–221, 437–501.

Wartenberg, Ilana. 2007. "*The Epistle of the Number* by Isaac ben Solomon ibn al-Aḥdab (Sicily, 14th Century): An Episode of Hebrew Algebra." Ph.D. thesis, Université de Paris 7, Tel Aviv University,

———. 2008a. "*The Epistle of the Number*: An Episode of Algebra in Hebrew," *Zutot: Perspective on Jewish Culture* 5, 95–101.

———. 2008b. "*Iggeret ha-Mispar*: by Isaac ben Solomon ibn al-Aḥdab (Sicily, 14th century) (Part I: The Author)," *Judaica* 64(1), 18–36.

———. 2008c. "*Iggeret ha-Mispar*: by Isaac ben Solomon ibn al-Aḥdab (Sicily, 14th century) (Part II: The Text)," *Judaica* 64(2/3), 149–161.

———. 2013. "*The Epistle of the Number*: The Diffusion of Arabic Mathematics in Medieval Europe," in G. Federici-Vescovini and A. Hasnaoui (eds.), *Circolazione Dei Saperi Nel Mediterraneo: Filosofia E Scienze (Secoli IX–XVII): atti del VII Colloquio internazionale della Société internationale d'histoire des sciences et de la philosophie arabes et islamiques* (Florence, February 16–28 , 2006). Florence: Cadmo, pp. 101–110.

———. 2014. "The Naissance of the Medieval Hebrew Mathematical Language as Manifest in Ibn al-Aḥdab's *Iggeret ha-Mispar*," in N. Vidro, I. Zwiep, and J. Olszowy-Schlanger (eds.), *A Universal Art: Hebrew Grammar across Disciplines and Faiths*. Leiden: Brill, pp. 117–131.

Wertheim, Gustav. 1896. *Die Arithmetik des Elia Misrachi. Ein Betrag zur Geschichte der Mathematik, Zweite verbesserte Auflage*. Frankfurt am Main: Braunschweig, Friedich Vieweg und Sohn.

Woepcke, F. 1853. *Extrait du Fakhrī, traité d'algèbre par Abou Bekr Mohammed ben Alhaçan al-Karkhī précédé d'un mémoire sur l'algèbre indeterminée chez les Arabes*. Paris: L'imprimerie Impériale.

Youschkevitch, Adolf P. 1976. *Les mathématiques arabes: VIIIe–XVe siècles*. Paris: Vrin.

3

Mathematics in the Islamic World in Medieval Spain and North Africa

J. LENNART BERGGREN

INTRODUCTION

The mathematics of medieval Islam includes the mathematical theories and practices that grew, and often flourished, in that part of the world where the dominant religious and cultural influence was the religion of Islam. The historical period involved is roughly the 700 years from 750 CE to 1450 CE, although the earliest mathematical works date from around 825. Geographically, medieval Islam extended from the Iberian Peninsula through North Africa and the Middle East, to the central Asian republics of the former Soviet Union, Afghanistan, Iran, and even parts of India. Although the mathematical treatises were written in a number of languages, including Persian and Turkish, the principal language used was Arabic, and for that reason it is occasionally called Arabic mathematics. This name, however, too easily becomes "Arab mathematics," with the implication that most of its practitioners were Arabs, even though many of its practitioners were, for example, Iranians, Egyptians, and Moroccans. So, I will generally designate this mathematics as "mathematics in the Islamic world."[1]

In this chapter, we consider the mathematics that was written in Spain and North Africa. So we begin with a brief history of the Muslim West, a history that is not so well known as that of the East and is, in many respects, quite different.[2] At the beginning of the eighth century, Spain was controlled by a Germanic tribe known as the Visigoths, with Toledo as their capital. Originally followers of a Christian heresy known as Arianism, they had converted to Catholicism in 589. North Africa had been controlled by the Vandals, who had replaced Rome

[1]Islam is a religion, but it created an Islamic culture that included significant, and often thriving, communities of other religious groups—such as Christians and Jews. These groups played an important role in the history of Islamic mathematics, especially during the early period. Yet Islamic mathematics was mathematics created by a people who were mainly Muslims in a culture whose dominant element was the religion of Islam.

[2]This account relies heavily on [Lévi-Provençal, 1986], [Ṣāʻid al-Andalusī, 1991], and [Djebbar, 1995].

as rulers of the area, and they were, in turn, replaced by the Byzantines with the victory of the Emperor Justinian in 534.

Byzantine rule was not long lived, however, and by the end of the seventh century North Africa had become part of the Umayyad Empire, ruled by the Caliph in Damascus. Muslim expansion in the West did not stop there, however, and in 711 CE the Berber commander, Tariq ibn-Ziyad, led a small reconnaissance force from North Africa which landed at Gibralter and then, joined by other Muslim troops, fought a seven-year war of conquest. That war ended by bringing most of Spain under the rule of the Umayyad Caliph. The Emirate of the Umayyads, whose capital was Cordova, was known in Arabic as al-Andalus, and Muslim immigrants to the Emirate came from two ethnic groups, Berbers, who were indigenous to North Africa, and Arabs. (On a number of occasions, open warfare broke out between the two groups.)

In 750 the situation changed drastically when the ʿAbbasids, with strong ties to Persia, overthrew the Umayyad caliphate in Damascus and transferred the capital to Baghdad. Most of the prominent Umayyads were killed, but some escaped the slaughter and fled, eventually to al-Andalus. In the spring of 755 one of these immigrants, who became known as ʿAbd al-Raḥmān I, marched up the Gualdalquivir River to Cordova and took the city. He ruled al-Andalus for more than 33 years (until 788) as the "Immigrant Emir" and lavished money and attention on the various arts. And it was during his reign that the study of the sciences began in al-Andalus.[3]

His great-great grandson ʿAbd al-Raḥmān II reigned from 822 to 852 and was, in the words of Lévi-Provençal, "renowned as an organizer and builder and a patron of arts and letters."[4] According to [Djebbar, 1995, p. 800], "it is from this period that the bibliographers have given us names [of scientists], and sometimes information about their [scientific] profile and activities."

The most famous ruler of this transplanted Umayyad Dynasty was ʿAbd al-Raḥmān III, who proclaimed himself Caliph, in direct opposition to the Egyptian dynasty of the Fatimids, who had ruled much of North Africa. He ruled for a half century, from 912 to 961, and his reign was known as "the golden age" of al-Andalus.[5] His son and successor, al-Ḥakam II, who reigned from 961 to 977, was a firm supporter of the sciences, and Ṣāʾid says of him that even during the time his father was living "he began his effort to support the sciences and befriend scientists. He brought from Baghdad, Egypt, and other eastern countries the best of their scientific works." It was just after the end of al-Ḥakam's reign that Ibn al-Samḥ (979–1035) was born, who studied and worked in three different cities of al-Andalus and whose work on the ellipse is excerpted in this chapter.

Al-Ḥakam's son, Hishām, was very young when he inherited the throne on the death of his father, and he was deposed by a coup led by his chamberlain. This man's first act on seizing power was to "take hold of the libraries" and to order all books dealing with scientific subjects (other than medicine and mathematics) to be destroyed.[6] This reign of intellectual

[3]See [Ṣāʿid, 1991, p. 61]. (Ṣāʿid wrote his history of science of all nations up to his time in 1068.)

[4]See [Lévi-Provençal, 1986, p. 82[b]].

[5]According to [Lévi-Provençal, 1986, p. 84[a]], "[out of the civil and military chaos that preceded his reign] he contrived to make a pacified, prosperous and immensely rich State. ... Cordova far surpassed the other capitals of Western Europe and enjoyed in the Mediterranean world a reputation and prestige comparable to that of Constantinople."

[6]According to [Ṣāʿid, 1991, p. 61].

terror lasted until the end of the Umayyad Caliphate in 1031, at which time al-Andalus broke up into a number of small kingdoms. During this period, again according to Ṣāʾid, the study of the various sciences began to recover and, writing of his own time in the mid-eleventh century, he says, "The present state, thanks to Allah, the Highest, is better than what al-Andalus has experienced in the past; there is freedom for acquiring and cultivating the ancient sciences and all past restrictions have been removed" [Saʾid, 1991, p. 62]. It was at this time that ʿAbd al-Raḥmān ibn Sayyid of Valencia (ca. 1070) lived, and his study of figured numbers is cited by the thirteenth-century mathematician Ibn Munʿim in his own treatise (excerpted in this chapter) on the subject. Most intriguing, however, is ibn Bājja's mention of ibn Sayyid's study of twisted cubic curves.[7] He studied them as intersections of surfaces and used them to solve two problems leading to equations of degree higher than three. Unfortunately, no surviving copies of his works are known. Also, it was in this period, in 1085, that al-Muʾtaman ibn Hūd, the ruler of Saragossa, wrote his *Book of Perfection* (*Kitab al-Istikmāl*), which showed not only his own mathematical abilities but also the rich legacy of ancient mathematics and great works from Eastern Islam that was available in Islamic Spain in the late eleventh century.

However, the dissolution of a powerful empire into (often quarreling) principalities also led to weakness, and very shortly after the time of al-Muʾtaman, al-Andalus was taken over by two successive dynasties, both from North Africa and both of Berber origin: the Almoravids (in 1086) and the Almohads (in 1145). The first-named actively opposed the scientific investigations that had previously been supported. The second group had to withdraw after their defeat in 1212 when most of Spain fell to the Reconquista (Latin Europe's "reconquest" of what, prior to the Muslim conquest, had been a Christian land). Just before this time, Seville was the political and cultural capital of al-Andalus, and it was here that the Moroccan mathematician, Ibn al-Yāsamīn (d. 1204), of whom we shall say more below, studied. Andalusian contributions to mathematics continued into the thirteenth century. Indeed, the noted mathematician Ibn Munʿim (d. 1228) came from the same general area of al-Andalus as Ibn Sayyid, although he worked most of his life in Marrakesh.

The Reconquista did not attempt to retake North Africa, however, and it is to that part of Western Islam, known as the Maghrib, that we now turn. The Maghrib is presently understood as being that part of North Africa from the western borders[8] of Egypt to the Atlantic Ocean. (The central and eastern Maghrib were also known to the Arabs as Ifriqiyya.) In medieval times, however, things were not so cut and dried. The tenth-century geographer, al-Muqaddasī, called the whole of the western part of the Muslim world beginning with Sicily the Maghrib. And Ibn Khaldūn writes in his *Kitāb al-ʿibar*[9] that some Arab geographers considered even Egypt to be part of the Maghrib. When choosing selections for this book, we have generally taken them from Andalusia and the Maghrib as the latter is understood today. We have, however, occasionally taken a wider view of the region and have included interesting reatises of al-Kishnāwī, who came from what is modern-day Nigeria; al-Maghribī, because he was born in the Maghrib; and Ibn al-Majdī, who lived in Cairo and wrote a commentary on a major Moroccan author.

[7] These are the so-called gauche curves, for example, the curve given parametrically by $x = t$, $y = t^2$, $z = t^3$.

[8] The Arabic word for "the west" is *al-maghrib*.

[9] [De Slane, 1925, p. 193].

It is in and around Qairawān, in modern Tunisia, where one finds, from the end of the eighth century, the first evidence of mathematical activity—notably in the fields of commercial arithmetic (*muʾāmlāt*) and the calculation of the shares of an estate due to the various legal heirs (*ʿilm al-farāʾiḍ*). Some of these writings were, four centuries later, still studied in Bougie,[10] a scientific center in the Maghrib. Qairawān was, in the ninth century, a center of theological debates, and it may have been its status as an intellectual center that attracted such scholars as Abū Sahl of Qairawān, whose parents emigrated from Baghdad and who wrote the first mathematical book in the Maghrib whose title is known: *The Book on Hindu Calculation*. (The text of this work has not been found.)

Part of the attraction of the Maghrib for scholars from the East was likely the fact that the reigning dynasty in Ifriqiyya, the Aghlabids, ruled the region in the name of the ʿAbbasids in Baghdad and modeled its style of government and civil institutions on those of Eastern Islam. For example, the Aghlabid ruler Ibrāhīm II (875–902) even founded a House of Wisdom, whose name and activities (such as translation of scientific works) were modeled on those of the House of Wisdom founded by the caliph al-Maʾmūn in the early ninth century in Baghdad. However, Ibrāhīm was the last important Aghlabid ruler, and the Fatimids of Egypt brought the dynasty to an end within eight years of his abdication in 902.

We have little information about mathematical work in the Maghrib during the centuries immediately following the end of the Aghlabid rule, and it is only from the eleventh century that we have the names of a number of astronomers/astrologers. One indication of how scholars moved from place to place in that century is the life of Abū al-Ṣalt, who wrote on geometry, astronomy, and logic. Originally from al-Andalus, he spent most of his life in Egypt and then went to Ifriqiyya, where he died in 1134. His book on geometry has not yet been found, but some of his astronomical works have survived. (Two of these appear to have been written when he was imprisoned—in a library!—for the failure of an engineering project that went awry due to poor quality cords.)

About half a century after Abū al-Ṣalt, at the beginning of the rule of the Almohads, the mathematician Abū al-Qāsim al-Qurashī lived. Like Abū al-Ṣalt, he was originally from al-Andalus (more precisely, Seville) but later lived in Bougie, where he died in 1184.[11] His commentary on the *Algebra* of the Egyptian, Abū Kāmil, was highly regarded long after his death and was still studied and taught in the Maghrib as late as the fourteenth century. Despite the interest it generated, the commentary has not yet been found, but Djebbar has suggested that it may have been the study of Abū al-Qāsim's work that led Ibn al-Bannāʾ, some decades later, to write his work on the fundamentals of algebra, the *Kitāb al-ʾuṣūl*, excerpted in this chapter. Al-Qurashī was also a specialist in division of estates, and he advocated the use of prime factorization of numbers to reduce fractions to a common denominator. The method was adopted by mathematicians in succeeding centuries, such as Ibn al-Jayyāb of Seville (thirteenth century) and al-Qalaṣādī (d. 1486), whose work also appears in this chapter; this is the method most of us learned in school. But the legal scholars in the Maghrib continued to use the older methods—as they do even today.[12]

[10]The modern Béjaïa, a port in Algeria. It was here that Fibonacci, working in his father's business, learned mathematics (see section II-2-2 in Chapter 1).

[11]Fibonacci would have been about fourteen when Abū al-Qāsim died.

[12]According to [Djebbar, 1995, p. 12].

Other mathematicians of the period of Almohad rule were Abū Bakr Muḥammad al-Ḥaṣṣār,[13] Ibn al-Yāsamīn, and Ibn Munʿim, whose work is excerpted here. Al-Ḥaṣṣār wrote at least two mathematical works. One, the *Book of Demonstration and Recollection*, dealt with calculation and is the first book known using the horizontal bar to separate the numerator from the denominator of fractions. This practice spread rapidly in mathematical teaching in the Maghrib, where Fibonacci (who died sometime after 1240) learned of it and used it in his famous *Liber abbaci*. The book was also translated into Hebrew in the thirteenth century by Moses ibn Tibbon. The other, the *Complete Book on the Art of Number*, of which only the first part has been found, dealt with some of the topics in his first book but also explained the exact calculation of a cube root of a whole number as well as perfect and amicable numbers.[14]

Ibn al-Yāsamīn, whose mother was a black African and whose father was a Berber, is best known for his highly popular *Poem of Algebra,* but his most important work was his *Fertilization of Spirits with the Dust Numerals*, a large work treating both techniques of calculation by itself and its application to geometry. His use of symbolism, identical with that of such later mathematicians as Ibn Qunfudh and then al-Qalaṣādī, permits him to write and solve equations and manipulate polynomials in an abstract manner. Because of the quality of the work, it is surprising that no later mathematician refers to it.[15]

As for the third mathematician, Ibn Munʿim, his life is discussed in the introduction to the two sections appearing here from his great work, the *Fiqh al-ḥisāb*, written in the early years of the thirteenth century. It is thanks to references in this work that historians are as well informed as they are about the mathematicians of his epoch and their works.

Ibn al-Bannāʾ was born in 1256—after the collapse of the Almohad dynasty—in Marrakesh, which from 1062 to 1248 had been the capital of the whole Maghrib. For more details, see the introduction to the selections from his works (sections I-1, II-1, and III-2) and, of course, the selections themselves! Bear in mind that when reading the works of Ibn al-Bannāʾ, we are reading the work of "the last Maghribi mathematician who had an active research career, in the sense that he attacked new problems for his epoch and produced original solutions or advanced new ideas" [Djebbar, 1991, p. 20].

I. ARITHMETIC

1. IBN AL-BANNĀʾ, ARITHMETIC

Aḥmad Ibn al-Bannāʾ was born in Marrakesh, Morocco, in 1256, and he died, it seems, in July 1321 in the same city. He was regarded as a learned and pious person of good character. The only record of travels in his life indicates that several times he went to the capital, Fes, at the invitation of the reigning sultan.

Three of his works are excerpted in this chapter, the first of which is *A Summary Account of the Operations of Computation.* The second is his commentary on that work, *Raising the Veil*

[13]Since he refers to writers of the eleventh century, and Ibn Munʿim in the thirteenth century refers to him as deceased, he most likely worked during the twelfth century.

[14]We have not been able to obtain a copy of this work.

[15]It is the subject of an unpublished master's thesis [Zemouli, 1993].

on the Various Procedures of Calculation. The third, which we shall excerpt in section II-1 on algebra, is his *Fundamentals and Preliminaries for Algebra*.[1] All three of these works would have fit well with his role as a teacher of mathematics at the Jāmiʿ mosque in Marrakesh. Other works include *The Abridgement of Rūmī Arithmetic*[2] and writings on astronomy and the calendar.

Along with many in his time, Ibn al-Bannāʾ believed in the mystical significance of numbers and divination, and he belonged to the mystical order known as the Ṣūfīs. During his lifetime he acquired a reputation as one who could use his mathematical powers to perform magic.

The selections here are from the first two of the three works, starting with the introduction to *Raising the Veil*. Excerpts from this work are in bold, while excerpts from the *Summary Account* are not. In the selections from the latter we have included examples from the footnotes of M. Souissi's French translation of the work, which he took from the commentary of ʿAlī b. Muḥammad al-Qalaṣādī on the *Summary Account*, namely, his work *Taqrīb al-aqsā min masāʾil Ibn al-Bannāʾ* (*Easing Difficulties in the Problems of Ibn al-Bannāʾ*).

From the *Summary Account* and *Raising the Veil*

Glory to God, the dazzling Truth, He who has given us reason, by the aid of which we apprehend and understand. He who has made of calculation [an instrument permitting us] to get to the quantities of his creatures and the marvels of his production, an instrument of which a number of questions relate to religion and which He has glorified in attributing to himself, for the All High says, "We suffice as Calculator."[3] May prayers be on his illiterate Prophet, Muḥammad, who has attained the purpose of everything (even if he is neither a calculator nor a writer[4]), on his family and his Companions, noble and good, and may salvation be theirs.

Having said this, my book—which I have written as a resume of the operations of calculation, to render its ideas accessible and to specify its rules and its structure—has gathered together the practical art of number with its two parts: the known and the unknown. I thus wanted to show clearly what it contained as a science, to explain that which the non-initiate believes to be beyond comprehension, and to show its foundations, rules and structures. I have organized my book on this according to sections, so that it will be easy to study. I have entitled it *Raising the Veil on the Various Procedures of Calculation*, and I ask God for aid and success.

Calculation is the manipulation of numbers in two ways: union and separation. The art of number is in two parts, and these are the two subjects shown in the work [i.e., the *Summary Account*]. It [calculation] is used in inheritance, transactions

[1] The Arabic titles of these three works are, respectively, *Al-talkhīṣ aʿmāl al-ḥisāb*, *Rafʿal-ḥijāb ʿan wujūh aʿmāl al-ḥisāb*, and *Al-ʾuṣūl wa al-muqaddimāt fī al-jabr wa al-muqābala*.

[2] According to the *Encyclopædia of Islam* [Bearman et al., 1986, Vol. VIII, p. 601], *Rūm* occurs in Arabic literature with reference to the Romans, the Byzantines, and the Christian Melkites interchangeably. "*Rūmī* arithmetic" refers to a decimal, nonpositional system of arithmetic using 27 symbols. It was used in the Maghrib for administrative and commercial purposes. Because it originated in Greece, it was called *Rūmī*.

[3] Quoted from the *Qur'ān*, chapter 14, verse 47.

[4] Muslims regard Muḥammad's lack of any formal education as important, because it is evidence that the beautiful prose of the *Qur'ān* must have originated from Allah (via the angel Gabriel), just as Muḥammad said.

and other domains, because inheritance has foundations established by law, and transactions have foundations established by city dwellers, foundations which the art of number treats. And it is this treatment that is called calculation.

Part I: On Whole Numbers

Chapter 1: The different categories of number and the orders

A number is a collection of units. According to its origin it falls into one of two categories, whole number or fraction.

The commentary explains and broadens this remark.

They [numbers] are considered from the point of view of expressing their names and they are then called "wholes." [And] they are considered from the point of view of their ratios, the one to the other, and they are called "fractions." And they are considered from the point of view of their similarity with lines and surfaces, and those are called "roots."

Whole numbers are of two kinds: even and odd. There are three kinds of even numbers: evenly-even, evenly-odd, and evenly-even odd,[5] and there are two kinds of odd numbers, odd primes and oddly odd.[6]

Section [i]

Because numbers increase without limit, one has given them three orders (*martaba*), also called indices, to which all of them are referred. Each order admits of nine numbers.

The first order goes from one to nine and carries the name "units."

The second order goes from ten to ninety and carries the name "tens."

The third order goes from one hundred to nine hundred and carries the name "hundreds."

Twelve simple names serve for the numbers. Combined together they form all the names of numbers. The nine first [names] appertain to the order of units, the tenth to that of tens, the eleventh to that of hundreds, and the twelfth is that of a thousand. The last-named [again] takes the place of units, and the cycle begins again from here.

Each number is known by its index of position[7] and by its name. The index is equivalent to[8] the order of the number: the index of units is one, that of tens is two, that of hundreds is three, and so on. The name designates the number that occupies a given position. The name of one is "units," two is "tens," three is "hundreds."

[5]Euclid, *Elements* VII, Def. 8, defines an "evenly-even" number as one that is "measured by an even number according to an even number," such as 24, whereas Nicomachus, in his *Arithmetica,* defines it as a number that, unlike the example of 24, can be expressed only as the product of two even numbers, i.e., a number of the form 2^n. Ibn al-Bannāʾ's discussion of perfect numbers [see below] shows that he is following the latter definition. Euclid defines an "evenly-odd" number as one that is "measured by an even number according to an odd number" (Def. 9), but does not mention the last of the three "even" types.

[6]Euclid, *Elements* VII, Def. 10, defines this as a number that is "measured by an odd number according to an odd number." (For example, 63 measured by 3 is 21.)

[7]I chose this translation of *ʿussihi* on the basis of the options in M. Souissi's dictionary of mathematical Arabic, since I am using "order," another possible translation, to translate *martaba.*

[8]Arabic = *ʿibāra ʿan.*

Know that the description of "number" given there [i.e., in the *Talkhīṣ*] is just a reminder of that which is in the soul, which is [made of] accident and difference. Certain [people] have believed that it would be definable, and that its definition is "a multiplicity composed of ones (or units)."[9] [But] their belief is not exact because "multiplicity" is number itself, not a genus for number. The reality of multiplicity is that it is composed of units, When someone says "[Number is] a multiplicity composed of units," it is as if one were to say "a multiplicity is a multiplicity," because "multiplicity" is a name for that which is composed of units. And saying "[composed] of ones or of units" is a redundancy, and one cannot understand the sense of this term, nor know it, other than by multiplicity.

He who defines number as being a discrete, ordered quantity[10] [gives] a true definition—except that the representation of quantity in the soul needs to be defined by part, or by division, or by equality. But one can only represent a part, or division, by multiplicity. And, as for equality, quantity is better known to the intellect than it, because equality is an accident belonging to quantity, which should be included in its definition.[11]

One will say, thus: equality is a union of that which is in quantity. As for "order," which has been used to define number, one can only understand it when one has understood number.

So, these descriptions are nothing but reminders of that which is in the soul, similar to reminders by examples and synonyms, and number is one of the notions that one conceives by themselves, and if one evokes them by these things [examples and synonyms] it is solely to indicate or distinguish them.

One cannot, then, argue against anything that has been given as a definition of number, other than to say that some definitions are clearer, nearer, or more adequate than others. As when one says "that which is composed of units" is more adequate than "units composed," because units do not constitute a genre for "composed" or "decomposed," for composition is a general accident for "units" and things other than units, and not a difference. In this case one has taken that which should be a genre in place of a difference and that which should be a difference in the place of a genre.

Section [ii]—On knowing the index of position of a repeated (*mukarrar*) number.[12] You multiply the number of repetitions by three and you add to the result the kind of that number. It is what is sought. Or, inversely, if you have [a number with] a certain number of places[13] and you want to name it, you divide the number of positions by three, so you

[9]This is the definition found in Euclid's *Elements* VII, Def. 2.

[10]"Quantity" was one of Aristotle's ten basic categories around which we can organize our experience of the world. Within this category two subdivisions were "continuous" and "discrete."

[11]The author here is referring to the ancient dictum that one should not define a concept by one that is less well known than that being defined.

[12]That is to say, numbers in whose name the word "thousand" is repeated, for example the number name "ten thousand-thousand" has two occurrences of the word "thousand." According to the rule given here, then, its index is $2 + 2 \times 3 = 8$, the first "two" being the index of 10.

[13]All of "[a number with] a certain number of places" is in Arabic *manāzil*, i.e., simply "places."

will have a remainder less than or equal to three. The quotient is then the number of repetitions of the number, which one identifies by means of the remainder.

This which is said to find the index, and the name is clear because after the first three orders they [the units] repeat three other orders, which are the thousands, their tens and their hundreds. And after them the third group of three orders repeats, which are: the thousand of thousand, their tens and their hundreds, and so on. After each three orders three other orders repeat, except for the first three orders, which are not repetitions.

Chapter 2: On addition

Addition consists of reuniting numbers, one to another, in order to express them by a single name. It is divided into five kinds:

1) addition of numbers having no particular relation to each other,
2) addition of numbers having a known difference,
3) addition of numbers in their natural order, of their squares and of their cubes,
4) addition of the sequence of odd numbers, their squares and their cubes,
5) addition of the sequence of even numbers, their squares and their cubes.

Addition of numbers having no special relation between themselves consists of adding one number of multiple digits to another number of multiple digits. One should place one of the numbers on one line and the second below in such a way that each unit of one order should be below that of the corresponding order. Then one adds each digit of one order of one of the terms [to be added] to the corresponding digit of the other. In the case where the latter order is empty [i.e., its digit is zero] the first digit plays the role of the sum of this order and the corresponding order. The result obtained is the answer.[14]

One begins adding either with the first or last digit, but it is better to begin with the first[15] because it is more orderly.

The most one can obtain by adding is only one order.[16] The check of addition is that if you subtract one line from the answer, you obtain the other line.[17]

The result of addition is the gain of at most one place and the most subtraction can lose is one place. This is because if one adds two positions the result can only grow and be placed towards the following position, and none other. Thus it gains [at most] one position. And if you subtract one position from one position it is not possible that you subtract it from this position and from that following it, but it is

[14]Note the brevity of Ibn al-Bannāʾ's exposition. Al-Qalaṣādī helps the student in his commentary: "There are three possibilities [when one adds two corresponding digits]—the sum may consist entirely of units [i.e., is at most 9], or entirely of tens or of both units and tens. In the first case, put the result above the units. In the second, above the tens, while placing zero above the units. And in the third case, place the units above the units and the tens above the tens."

[15]Since Arabic is written from right to left and the units digit is written first, that is considered the first digit.

[16]For example, one can add units to units and get tens but not hundreds.

[17]Souissi notes that Ibn al-Bannāʾ has not yet discussed subtraction, but this offhand reference to a later topic is not unusual with Ibn al-Bannāʾ.

possible that nothing remains in the position following it. In this case you lose one position, which is the opposite of addition.

[Introduction of arithmetic and geometric series]

Be aware that known differences [in a sequence of numbers] are of two kinds. [One is] difference in quality, when the numbers are in geometric ratio, increasing by different numbers, and are qualitatively similar when one takes the ratio of the one to the other, such as the ratio of a half, or of a third, or some other.

. . .

Ratio is the measure of one of two numbers by the other, and it is an application.[18] Proportionality is the similarity of ratios, and is therefore a quality. Numbers in proportion are those in which there exists a proportionality. When we say that they increase in quality and are similar in quality it is in taking account of the two extremities of the ratio.

For example, in the case of numbers that follow one another in the ratio of one-half, the two extremities of the second ratio are divided into two ratios, each of them being equal to the first from the point of view of quantity of their extremities and of their quality. As for the extremities of the third ratio they are divided in four ratios, each of them equal to the first. Each of the ratios is a multiple and the ratio of the sum of these differences that there are between them is similar, from the point of view of its quality, to the first ratio.

It is called geometric ratio because it is specific to magnitudes studied in geometry.

The second kind of difference is difference in quantity, which is when the numbers are in a numeric ratio, such as the sequence of whole numbers, for which the difference is one, and the sequence of odd numbers, for which the difference is two, and similar examples. The numbers increase by equal numbers but differ in quality when they are put in ratio to one another, and it is called "numeric ratio" because it is specific to number. This will appear later in the summation of families of numbers.

The second kind indicated in the work [the *Talkhīṣ*] appertains to the category of differences in quality and to that of differences in quantity. It is the addition according to a known ratio, geometric or arithmetic, but when one uses the term "ratio" it often concerns geometric ratio, and this is why we have said that the second kind is addition according to a known difference and not according to a known ratio. Know it.

There are two kinds of geometric difference, for the ratio is either that of one-half or something else, and each of these two is subdivided into four kinds, since the number of elements in a progression is either evenly-even, or something else, and it either begins with one or with something else. These are the eight different kinds,

[18] Perhaps what is meant here is that, for example, one thinks of the ratio 6:2 as representing the application of a length of two units along a length of six units, which can be done three times before the larger length is measured.

each kind posing two problems: determining one of the proportional numbers of the progression or their sum. So there are sixteen problems, which are all solved with the aid of the procedure indicated in the work.

In adding numbers with known differences, for example in the case of the [problem of the] chess board and analogous problems, the condition is that the value of the first term equals one, and one progresses by doubling the values successively until one reaches the last term.[19] Then you double the number of the corresponding square after the last term and subtract one from the result obtained. The remainder is the number sought [i.e., the sum of the series].

In saying [in our book] "like the cells of the chess board," we mean that the number of terms of the sequence is evenly-even and that the terms are among themselves in the ratio of one-half.[20] **In order to solve problems about cells of the chess board one must know that the position of each term of the sequence is that of the order of the exponents of the numbers whose increase is geometric and this is the case where the terms begin with one.**

The peculiarity of this sequence is that the number in each cell exceeds by one the sum of those that precede it, as we shall demonstrate later. We have drawn attention to this peculiarity by saying in the book, "You obtain the sum of that which is contained in the second cell and in that which precedes plus one," and we have not said "and this is the content of the third cell," because this is known by the product and the index, that which is not known. It is a fact that it exceeds by one the sum of all that precedes it.

If the first term is different from one, multiply the remaining by the first term. You obtain the answer sought. If the numbers are in a different ratio [than 1/2] then multiply the smallest term by the amount by which the largest exceeds it. Then divide [the product] by the difference between the smallest term and the one immediately following it. Add the quotient to the largest. That is the answer.

If [instead] the numbers are related by a known difference, but other than by multiplication, then multiply the difference by the number of terms less one. To this product add the first term. The sum is equal to the last term. Add this term to the first, multiply the result by the half of the number of terms and you obtain the answer.

To sum the sequence of [whole] numbers you calculate half the product of the last term by the sum of that term and one.

[19]In what follows I have rewritten the passage to make it say what it should say, which is that $1 + 2 + 2^2 + \ldots + 2^k = 2^{k+1} - 1$. The passage says, literally, "You add one to the unity, which is found in the first square [of the chessboard], to have what is put in the second. Next you square the number obtained and the result is then the sum of the contents of the second square and of that which precedes it plus one. Then you continue to square the result, to double the number of the corresponding square, up until the last term and to subtract one from the result obtained. The remainder is the number sought." The first two sentences are true but irrelevant to the problem Ibn al-Bannā᾽ has set himself—summing a geometric series. The rest of the passage is not true.

[20]Mention of the cells of a chessboard refers to an Arabic tale of the legendary inventor of the game who asked that as a reward he be given the number of grains of wheat obtained by putting one grain on the first cell of a chessboard, two on the second, four on the third, eight on the fourth, etc. So many grains of wheat (the number being on the order of billions of trillions) would be a handsome reward indeed.

The sum of the squares [of the first n whole numbers] you obtain by multiplying the sum [of the sequence of whole numbers] by two-thirds of the last term augmented by a third of unity.

The sum of the cubes is obtained by squaring the sum [of the sequence of whole numbers].

To sum the sequence of odd numbers you square half the sum of the last term plus one.

The sum of the squares of the terms of this sequence is obtained by multiplying the sixth of the last term[21] by the product of the next two terms [in the sequence of whole numbers].

The sum of the cubes of the terms of this sequence is obtained by multiplying the sum of [of the odd numbers] by one less than its double.

In adding the sequence of even numbers you always add two to the last term of the sequence and you multiply half of that sum by half of the last term.[22]

The sum of the squares [of the terms of the sequence of even numbers] is obtained by multiplying two-thirds of the last term augmented by two-thirds of one by the sum,[23] or by multiplying the sixth of the last term by the product of the two whole numbers that immediately follow it.

The sum of the cubes [of the terms of this series] is obtained by multiplying the sum[24] by its double.

[Another application of addition]

. . . we will derive perfect, abundant and deficient numbers:[25] If the sum of successive evenly-even numbers beginning with one is a prime number (which, thus, cannot be decomposed into prime factors) its product by the last term of the sequence is a perfect number. Its product by the following term is an abundant number, and its product by the term that precedes the last is a deficient number.

The double of any perfect number is an abundant number, and the half of any perfect number is a deficient number.[26]

"Perfect" signifies that the number, such as six, is equal to the sum of its proper divisors. Abundant is that, such as twelve, the sum of whose divisors is larger than

[21]By "the last term," he means the last number whose square is being summed. So for the sum of the squares of the first four odd numbers the last term refers to 7.

[22]Here "the last term" refers to the last number being summed, so in the case of summing the first four even numbers the last term is 8.

[23]Here "the last term" means the last even number being squared, and "the sum" means the sum of the first n even numbers.

[24]Here "the sum" refers to the same thing as in the previous footnote.

[25]So far as is known, "perfect numbers" appear first in Book VII of Euclid's *Elements* (Def. 22), where Euclid defines them as "a number equal to the sum of its parts [proper divisors]," the first such being $6 (= 3 + 2 + 1)$. Abundant ("over-perfect") numbers are those for which the sum of the proper divisors exceeds the number, and deficient numbers are those for which the sum of the proper divisors is less than the number.

[26]The rule for generating perfect numbers that Ibn al-Bannāʾ gives is found in Euclid's *Elements*, IX, Prop. 36. The rule, of course, generates only even perfect numbers, and whether Ibn al-Bannāʾ thought all perfect numbers were even or he was just writing somewhat loosely, in the context of his discussion of the particular perfect numbers formed by the rule he had just given, is unclear. In any case, it is still not known whether there are any odd perfect numbers. (In 1991 it was shown that if there is one, it must be larger than 10^{300}.)

it, and deficient is that, such as fourteen, the sum of whose divisors is smaller than it.

He goes on to give a number of other rules and then states the following.

The reason these procedures are valid can be understood with the aid of three premises. One of them is the peculiarity of the geometric progression, that the product of the two extremes is equal to the product of any two numbers of the progression that are equidistant from the two extremes (and equal to the square of the middle term if the number of terms in the progression is odd), as we shall show in our following discussion of ratios, with the help of God's power.

The second premise is that the sum of the two extreme terms [in an arithmetic progression] is equal to the sum of any two numbers of the progression that are equidistant from the extremes (and equal to double the middle term if the number of terms in the progression is odd), because the sum of two numbers is equal to their sum after having added to the one what has been subtracted from the other.

The third premise is that the meaning of "product" is the addition of the one of these numbers as many times as there are units in the second.

Chapter 3: On subtraction

Subtraction consists of searching for the remainder obtained by removing one of the numbers from the other. It is of two kinds:

1) the operation where one subtracts the smaller number from the larger once, and
2) that in which one subtracts the smaller from the larger more than once until the latter is exhausted or until one obtains a remainder less than the smaller.

He explains the procedure of subtraction and then continues.

You begin subtraction either from the first order or the last. It is better, contrary to addition, to begin with the last.

Subtraction can reduce the order of a number by at most one unit. To check subtraction, add the remainder to the number subtracted. The sum is equal to the number from which one subtracted. Or you subtract the remainder from the number from which you subtracted and the difference will be equal to the number subtracted.

[Subtraction as a check of arithmetic operations]

The second kind [of subtraction] admits of three sorts of subtractions. They are those used frequently in checking operations. The first sort is subtraction of nines, the second is the subtraction of eights, and the third is the subtraction of sevens.

In the subtraction of nine, there remains from the unit of each order one. You then take the number with the [digits of the] different orders, as if these latter all belonged to the order of units, and you subtract the nines.

In subtraction by eight, from each ten there remains two and from each hundred four. An even number of hundreds and the units of a higher order are [all] divisible by eight, and so an odd number of hundreds give four as a remainder. And [so] you multiply the [number of] tens by two and you add that to the four [in the case of an odd number of hundreds] and to the units. And you reduce it eight at a time.

In subtraction by seven, from each ten there remains three, from each hundred two, from each unit of a thousand six, from each ten thousand four, from each hundred thousand five, from each thousand thousand one. And from then on the cycle repeats. You check by seven by means of the letters that make up the phrase "I require his friendship"[27] repeated under the units of the corresponding orders. You multiply each digit of each order by the number represented by the letter placed below it. You subtract sevens and you place the remainder above. Then you add the remainders found in each order, which are taken as units, and you [again] subtract sevens.

You may also, if you wish, operate thus: Multiply the last [i.e.,leftmost] digit by three, reduce it by sevens, and add the result to the following digit. Multiply the sum by three and place the remainder above. Then add the remainders of each order, making them play the role of units. Then subtract sevens. Add the remainder to the following order. If the following order is empty multiply the remainder obtained by three, subtract sevens, and work in this way until you arrive at units.

He gives one more method as well, similar to the one just above, and then devotes a short section to using the material he has just given to checking the four arithmetic operations.

Chapter 4: Multiplication and its remarkable properties

Ibn al-Bannāʾ begins by defining multiplication using the same idea as Euclid does in Book VII (Def. 15) of his *Elements*: "Multiplication consists in repeating one of the two numbers [being multiplied] as many times as there are units in the second." He then explains a common method of multiplication, one he calls "multiplication by shifting," which was explained by Kūshyār ibn Labbān (among many others) and is set out in [Berggren, 1986]. He next explains a special method for multiplying a number by itself. His third and last method he explains as follows:

The third kind is multiplication without shifting, which has a number of variants. We mention [among them] multiplication by means of a grid, described as follows: You construct a rectangle which you divide both in length and width into as many strips [of equal width] as there are orders in each of the two factors. You draw diagonals from the lower right side of each of the resulting squares to the upper left side.

You place the multiplicand above the rectangle so that each of its digits corresponds to one column. And you place the multiplier either at the left or the right of the rectangle, descending[28] along its side and, again, making each digit correspond to one strip.

[27]The phrase is a mnemonic, based on using an alphabetic system of numeration (called the *abjad* system) to change the consonants in the mnemonic words to a sequence of numerals. In this case the words are the Arabic version of "I require his friendship," which require only six letters in Arabic. This gives six numerals 1, 3, 2, 6, 4, and 5 as the multipliers for checking by sevens. Souissi gives the example of reducing 3,405,131 by 7. One writes under 3405131 the sequence A – G – B – W – D – E – A, so that A goes under "1" on the right, "G" under the "3" next to it, etc. One multiplies the numerical value of each letter of the mnemonic by the digit above it, mentally reduces each product by 7, and adds the results. One gets $1 + 2 + 2 + 2 + 0 + 6 + 3$, which reduces by 7 to 2, the remainder on dividing 3,405,131 by 7.

[28]Because the Arabs wrote the numbers beginning with the units digit, the reader would have understood that the units digit of the multiplier would face the topmost strip, the ten's digit the one below that, etc.

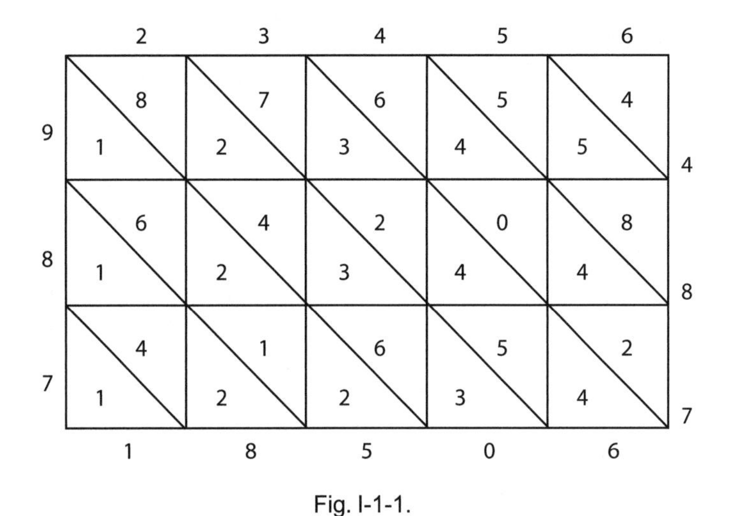

Fig. I-1-1.

Now you multiply the multiplicand, digit by digit, by all the digits of the multiplier. You put each of these partial products in the square where the corresponding [horizontal and vertical] strips cross, placing the units above the diagonal and the tens below.

Then, beginning at the top right corner [of the rectangle] you add the numbers contained between the diagonals without canceling [anything] and you carry each number to the next diagonal row to make the sum with numbers contained therein. The total you get is the answer.

In his *Talkhīṣ*, Ibn al-Bannāʾ gives the example of multiplying 23,456 by 789 shown in Fig. I-1-1.

The multiplicand, 23,456, is written digit by digit along the top of the grid, and the multiplier, 789, is written along the left side (beginning with the 7 next to the bottom left square). In each cell one places the product of the number heading the column by that at the left of the row, with the units digit being written above the diagonal of the cell and the tens digit below. Adding up the numbers diagonal by diagonal and beginning with the 4 in the upper right corner, then the 5, 5 and 8 in the diagonal below that (and carrying the tens digit as a unit digit to the diagonal next on the left as necessary), and so forth. Ibn al-Bannāʾ obtains the result 18,506,784.

Ibn al-Bannāʾ then gives a number of shortcuts for special cases, such as multiplying by a number ending in one or more zeros. We quote two of them, one for multiplying a number by another consisting entirely of nines and one for multiplying any two numbers.

The method [of multiplying by a number consisting only of nines] consists of adding to the orders of the multiplicand as many zeros as there are nines in the multiplier, then to subtract the multiplicand from the number thus formed.

Another category is known as "quadrature." Here it is: Take half the sum of the two factors and raise it to the second power. Then subtract from the result the square of half the difference. What remains is the result of the multiplication.[29]

[29]Notice the application of the identity $a \cdot b = [(a+b)/2]^2 - [(a-b)/2]^2$, which could be useful if one had access to tables of squares. However, the author is unaware that any such tables were produced. In section IV-1 in this

Chapter 5: On division

Division consists of decomposing the dividend into as many equal parts as there are units in the divisor. One understands by division the ratio of one of the two numbers to the other.

The majority of mathematicians mean by division in an absolute sense the knowledge of that part of the dividend that belongs to each of the units in the divisor.

We omit Ibn al-Bannāʾ's explanations of various ways to divide and pass on to two following parts of this chapter. The first part sets out a solution to the problem of dividing a given sum, S, into three quantities x, y and z that are proportional to given quantities a, b, and c. In modern notation, the problem is to solve the equations $x{:}y = a{:}b$ and $y{:}z = b{:}c$, given that $S = x + y + z$, and the method Ibn al-Bannāʾ explains became known in the West as the "method of proportional parts." Ibn al-Bannāʾ calls the process "allotments" and explains the solution as follows.

There is a sort of division distinguished by the name "allotments." The method in this case is to add the parts of the allotments [i.e., a, b, and c] and take the sum as denominator. Multiply each part of the allotment by what is to be divided [S], and divide the product obtained by the denominator. The quotient is the number sought.

If all the parts of the allotments have fractions in them multiply them by the least number divisible by all the denominators [of the fractions]. If all the parts have a common divisor, suppress this divisor and replace each part by the result.[30]

[Factoring numbers]

To factor numbers there is a preamble that you must memorize: Every number not ending in a units digit [i.e., which ends in one or more zeros] is divisible by ten, five and two which [divisibility by two] is characteristic of even numbers. If it ends in five it is divisible by five. If it ends in an even units digit it is even and one exhausts it by means of three subtractions [usually]. If it is exhausted by subtracting nines it is divisible by nine, six and three, but if three or six remain then it is divisible by three or six.

He then goes through a number of other cases that lead to his final rule.

You continue to divide the number you want to factor in terms of primes until you find one by which it is divisible, or until you arrive at a number whose square is larger than the given number, or until the quotient is less than or equal to the divisor and the division leaves a remainder. Then you know that the [given] number is [itself] prime.

He then goes on the explain the method he calls "the sieve" to find the prime numbers in a sequence of consecutive whole numbers. He does not mention the discoverer of this method, Eratosthenes of Alexandria, in this regard.[31]

---

chapter, on Ibn ʿAbdūn, the reader will find the identity underlying the method of quadrature used to solve a problem asking to find the dimensions of a rectangle given its area and diagonal.

[30] Al-Qalaṣādī gives an example of the procedure Ibn al-Bannāʾ outlines, which is included in the selection from al-Qalaṣādī in section I-2.

[31] We know that Eratosthenes discovered this method, which he called a "sieve," because of Nicomachus's description of it in his *Introduction to Arithmetic*.

Part II: On Fractions

A fraction is the ratio between two numbers when it is one or more parts. The ratio between a part and the number having the same name is called a fraction. In the framework of our study six chapters are devoted to fractions.

In the [Arabic] language, "fraction" means the "breaking" of something, but in numbers it is not used in the sense of where one says of a terrain that it is broken, i.e., that it has mountains and valleys. The application of the [term] fraction to number is made in suppressing the preposition ["to"].[32] Thus what is intended is a number to fractions, and one has said this of it because of the diversity of its relations, as the diversity of the land which has mountains and descents. This resembles their comparison of numbers with "plane," "linear" or "solid."

Again some linguistic remarks are made, and then he continues as follows.

According to this, if one were to ask, for example, what is the ratio of two to three, one would say two-thirds, and if one were to ask what is two of three, one would say "two parts" or "one part." A number in relation to a number is one or more parts, and the ratio between them is a fraction whose name is a half or one of the other names for fractions.

Chapter 1: Nomenclature of fractions and finding their numerators.

The fractions have ten simple names, the first of which is one-half, the largest fraction,[33] then the third, the fourth, the fifth, the sixth, the seventh, the eighth, the ninth, the tenth and then the part.[34] Fractions can be written as duals or as plurals.[35] Addition of each of these fractions ceases when the numerator is one less than the denominator.[36]

The simple fractions may be joined, the one to the other,[37] which then forms a name composed of [the names of] two or more of them.

Finding the numerator consists in reducing all the fractions given in a problem to the smallest of the fractions. The numerator differs according to the kind of fraction, of which there are five: *simple* fractions [common fractions a/b, where $a < b$], fractions *related* [*to others*], written $\frac{a\,c}{b\,d}$, which is interpreted as $\frac{c}{d} + \frac{a}{b} \cdot \frac{1}{d}$, *different* fractions $[\frac{a}{b} + \frac{c}{d}]$, *partitioned* fractions $[\frac{a}{b} \cdot \frac{c}{d}]$, and fractions *separated by a sign of subtraction*.

[Finding the numerators of these five types of fractions]

The numerator of a simple fraction is the number on top of it.[38]

[32] Instead of the language of ratios, where one says, for example, "ratio of three *to* five" one says, in the context of fractions, "three fifths."

[33] That is, the largest of the ten with "simple names."

[34] Any fraction we would write as $1/n$, $n > 10$.

[35] Arabic nouns have three numbers: singular, dual, and plural. For example, 1/6 is *suds* and 2/6 (the dual) is *sudsān*.

[36] The Arabs only recognized fractions less than 1. The word for denominator is *samiyy*, which comes from the Arabic word for "name," just as our "denominator" comes from the Latin word for it.

[37] The Arabic term here comes from the name for the standard way of forming the possessive case in Arabic, hence "one-third of one-seventh" instead of "one twenty-first."

[38] Initially the numerator was simply written above the denominator, as in Kūshyār's example of 317/511 quoted in [Berggren, 2007, p. 536]. According to [Djebbar, 1992], it was the twelfth-century Maghribi mathematician

The numerator of a related fraction is what is above the first denominator multiplied by the following denominator and added [to the number placed above that], and so on to the end of the line, or the number written on top of the first denominator multiplied by the following denominators and that which is above the second denominator [multiplied] by the denominators following, and so on until the end of the line, and then one adds up all the products obtained.[39]

For the numerator of different fractions one multiplies the numerator of each part by the other denominators, and then adds the results.

The numerator of partitioned [fractions] is obtained when one multiplies the numbers written above the line[40] by each other.

For the numerator of subtracted fractions, one operates, for that which is discontinuous, as for different fractions, and one subtracts the smaller from the larger result. But for those that are continuous, one multiplies the numerator of that [fraction] from which one subtracts by that of the [fraction] subtracted, and one also multiplies it by the denominators and one subtracts the smallest result from the largest.

If, in a problem, a [whole] number precedes fractions, one multiplies it by the denominators, and one adds the product to the numerator. But if it is found after them [the fractions] one multiplies it by the numerator. And, if it is found in the middle of them [then] if it is related to the fractions before it, it is considered as being at the end, and if it is related to the fractions following it, it is considered as being at the beginning. You calculate its numerator according to that one of the two ways in which it is related. And, with the remaining [fractions] it will be like the different [fraction]: [in the situation where it is] at the beginning or the end, you multiply it by the numerator of the remaining fractions.

2. ʿALĪ B. MUḤAMMAD AL-QALAṢĀDĪ, *REMOVING THE VEIL FROM THE SCIENCE OF CALCULATION*

The Andalusian religious scholar and mathematician ʿAlī b. Muḥammad al-Qalaṣādī was born in the early fifteenth century in what he describes in his autobiography as "a small town," Beza, northeast of Granada, and, after working in a number of cities in Spain and North Africa, he died in Béja, Tunisia, either late in the fifteenth or early in the sixteenth century. He wrote not only on religious law, Arabic grammar, and composing verse, but also on mathematics, including a commentary on Ibn al-Bannāʾ's *Talkhīṣ*. Along with other Maghribi mathematicians of his time, al-Qalaṣādī made use of mathematical symbolism in some of his works, the symbols being a letter from the appropriate Arabic word.[41]

---

al-Ḥaṣṣār who first used a horizontal bar in a mathematical work to separate the numerator from the denominator, a practice that later mathematicians in that region, including Ibn al-Bannāʾ, followed. Prior to that, however, jurists and specialists in the division of estates had used the fraction bar. See [Djebbar and Guergour, 2013].

[39] Al-Qalaṣādī gives the following example to fill in details: Four-ninths, five-eighths of a ninth, and one-half of an eighth of a ninth is written as $\frac{1\,5\,4}{2\,8\,9}$. The numerator of the corresponding isolated fraction is $4 \cdot 8 \cdot 2 + 5 \cdot 2 + 1 = 75$.

[40] This is a clear reference to the fraction bar mentioned in footnote 38.

[41] The symbolism can, in fact, be found in the works of earlier fourteenth-century writers, such as Ibn Qunfudh.

For example, the letters ل ج ك م ش[42] stood for "thing" (i.e., an unknown), "square," "cube," "root," and "equal," respectively. Souissi notes[43] that al-Qalaṣādī would also simply put the three dots, seen in the first of the Arabic letters, above a numeral to indicate the corresponding multiple of the unknown.

Some of his works were widely influential in the Maghrib as well as in the West. One of these works, *Easing Difficulties in the Problems of Ibn al-Bannāʾ* was, as the title suggests, a commentary on the *Summary Account*. Souissi translates examples from al-Qalaṣādī's commentary in his French translation of the *Summary Account*, some of which were included in the previous section. The following extracts from al-Qalaṣādī are, however, from another work of his, *Removing the Veil from the Science of Calculation*.

Preface

Praise to God. In the name of God, the Merciful and Compassionate, may God's blessings and favor be on our lord and master, Muḥammad, his family, and his companions. ... Praise to God, Who is prompt in his accounts in the book of the Day of Judgment, who abundantly spreads his benefits, who opens doors. ...

Now, to come to the matter at hand, this is an abridgement, sufficiently extended and rich in content, equally far from insufficiency and prolixity, which I have extracted from my work titled *Raising the Vestments of the Science of Calculation*. The aim of this abridgement is, on the one hand, to offer ample provision to students and, as well, to serve as a manual for those endowed with superior intelligence. ... This treatise consists of an introduction, four parts, and a conclusion. Each part contains eight chapters.

Part I

In part I, after chapter 1 on addition, the next chapter covers subtraction. After discussing two examples where it is not necessary to "borrow" from one place to the lower place, al-Qalaṣādī turns to the procedure for subtracting, say, 386 from 725. Notice that he does not borrow but adds the appropriate power of 10 to the top and then to the bottom.

Chapter 2: On Subtraction

Put it down it as follows:

$$\begin{array}{l} \underline{3\,3\,9\quad} \\ 7\,2\,5 \\ 3\,8\,6 \end{array}$$

Now subtract the six from the five. This is not possible, so add ten to the five and subtract the six. Nine remains. Put it above the line. Then add the ten, as a one, to the eight, so nine results. Subtract it from two, which is not possible. So add ten to the two, which results in twelve. Subtract nine and three remains. Put it above the line. Then add the ten, as a one, to the three, so four results. Subtract it from seven, giving three. Then what remains will be three hundred thirty-nine, thus 339.

[42] From the left the letters are shīn, mīm, kāf, jīm, and lām.

[43] See [Ibn al-Bannāʾ, 1969, p. 91, n. 1].

And if someone says to you, take away three thousand nine hundred seventy-eight from five thousand seven hundred two, arrange it this way:

1724
———
5702
3978

Next, take away the eight from the two. This is not possible, so add ten to the two, resulting in 12. Now subtract the eight. Four remains and put it above the line. After this, add a unity to the 7 [in the lower row] to get eight. Take it away from zero, but this is not possible. So add ten to the zero and subtract the eight. Two remains. Put it above the line. Then add unity to the nine. The result is ten. Subtract it from the seven. It is not possible, so add ten to the seven, resulting in seventeen. Subtract [the ten] from the sum and seven remains. Put it above the line. Next add unity to the three, resulting in four. Subtract it from five and one remains. Put it above the line. The remaining will, then, be one thousand seven hundred twenty-four. Thus, 1724.

In the following discussion of multiplication, al-Qalaṣādī does not, as Ibn al-Bannāʾ does, explain it in terms of addition. Rather he relates it to the idea of finding unknowns from knowns.

Chapter 3: On Multiplication

Multiplication is the act of finding an unknown number from two known numbers. It may be performed in different ways:

Slantwise multiplication: The practice of this operation consists in placing the multiplier on a line and, below it, the multiplicand, in such a way that the first place of the multiplicand is found below the last place of the multiplier.[44]

For example, if someone says to you, multiply seventy-three by fifty-two, put it down this way,

5 2
7 3

with a broken line above these numbers. Then multiply the seven by the five, resulting in thirty-five. Put the five above the seven and the three to the left of it. Next, also multiply the three by the five, yielding fifteen. Put the five above the two numbers multiplied the one by the other [i.e., above the five and the three] and the unity before, above the five [that you just put down]. Next, move the three [of the 73] to the right below the units [of the 52] and the seven in the tens place. Multiply the seven by the two, resulting in fourteen. Put the four above the multiplicand and the unity before it. Then multiply the three by the two, which gives six. Put it above the two numbers multiplied the one by the other. Then, draw a line above the results and add them [the results] above the line. It will be three thousand seven hundred ninety-six.

[44] Because Arabic is written from right to left, "the first place" refers to what we would call "the lowest" or "units" place, and "the last place" is what we would call "the highest place." But, for the convenience of the reader, we will, in what follows, tacitly change the text to reflect our direction of writing.

We reproduce here the explanatory figure from Woepcke's note to the text:

```
 3 7 9 6
---------
       6
   1 4
   1 5
 3 5   ____
     | 5 2
_____|
    7 3
      7 3
```

In the following section of his *Arithmetic*, al-Qalaṣādī gives rules for multiplication that appear, in part, in Abū al-Wafā'’s treatise on the "arithmetic of the hand" and, in other parts, need pen and paper to perform.

Shortcuts to Multiplication

We also add to this chapter a number of fundamental rules, which are sufficient in some cases:

Any number multiplied by zero produces zero.

Any number multiplied by the unit produces the same number.

To multiply a number by two add it to itself.

To multiply a number by three, add it to its double.

To multiply a number by four, double it twice.

To multiply a number by five put a zero to the right of it and take its half. For example, if someone says to you: Multiply sixteen by five, put a zero to its right, it will be one hundred sixty. Take the half of that, eighty, the number sought.

. . .

To multiply a number by six, add it to half of its product by ten.

. . .

To multiply a number by seven, put a zero to its right and subtract its triple from its product by ten.

. . .

To multiply a number by eight, put a zero to its right and subtract its double.

. . .

To multiply a number by nine, put a zero to its right and, again, subtract it from the result. It will be the number you sought.

. . .

To multiply a number by ninety-nine put two zeros to its right and, again, subtract the number from the result.

. . .

To multiply a number by ten put a zero to its right or, by a hundred, put two zeros to its right.

. . .

To multiply a number by eleven add it to itself, with the placement of the digits changed [i.e., shifted one place to the left].

His example for 352×11 is below. One then adds the two lines.

```
 3 5 2
3 5 2
```

To multiply a number by twelve, place below the number the same number so that the places match up. Then, place it again below the two first, but so that the units of the lowest line match up with the tens of the two lines above. Add all of this and the result will be the number sought.

...

To multiply a number by fifteen, if it is even, add it to its half and put a zero to its right. If it is odd, subtract one, add it to half of what remains [after subtracting], and put a five to the right.

...

And [so], if someone says to you: Multiply twenty-nine by fifteen, add fourteen to the multiplicand and then put a five to the right of the sum. Thus, 435.

To multiply a number by a number formed by two equal digits, multiply the number by that digit and add the result to itself with the places shifted one to the left. For example, if someone says to you: Multiply five hundred thirty-four by eighty-eight, multiply the multiplicand by one of the eights and add the result to itself, changing one place. The number sought will result, namely forty-six thousand, nine hundred ninety-two, thus 46992.

Chapter 4: On division

His explanation of division is different in kind from that of multiplication.

Division is the decomposition of the dividend into parts equal to the divisor. The unit is to the result of the division as the divisor is to the dividend.

...

Chapter 5: Decomposing numbers into the factors composing them

One who studies this science must know thoroughly [the results in] this chapter. This is because all the operations rest on it, which is like the axis around which they turn, or the sun that makes them clear.

The practice of this operation consists in reducing[45] the number, if it is even,[46] by nine. If the number reduces [in this way] then it has a ninth, a sixth and a third, such as thirty-six. But, if three or six remain it has [only] a third or a sixth, such as forty-eight and seventy-eight. But if none of these occur, reduce it by eight. If it reduces it has an eighth and a fourth. But if four remains it has a fourth. If it does not reduce and four does not remain, reduce it by seven. If it reduces it has a seventh, like ninety-eight. If it does not reduce [by seven] it has only a half,[47] such as forty-six. Then see if the half has other parts, of which the first is eleven.

[45]That is, "dividing."

[46]The parity of the number, of course, has nothing to do with casting out nines. Al-Qalaṣādī simply means that he is first going to discuss the reduction of even numbers.

[47]If it reduced by five, it would, since it is even, end in a zero and so would obviously be a multiple of ten.

If the number is odd, one reduces it by nine. If it reduces by nine it has a ninth and a third, such as sixty-three.

Having finished his discussion of reducing odd numbers, al-Qalaṣādī then adds a general remark.

If the number ends in a five it has a fifth, and if it ends in a zero it has a tenth, a fifth, and a half.

A Practical Method for Performing Reductions

As for reduction by nine, you add the parts [i.e., digits] of the number, one by one, as if they were units, and you reduce the sum by nine.

He then gives the example of 234, whose digits sum to 9, so that 234 has, as he puts it, "has a ninth and [being even] a sixth." He continues as follows.

And if someone says to you, "Reduce three hundred eighteen," put it thus, 318. Operate as before [with nine] and three remains. You then say that the number has a third and [since it is even] a sixth.

He points out that if you reduce 1827 by nine, you find that it has a ninth and a third but no sixth, since it is odd.

As for reducing by eight, ignore the hundreds, if they are even, because they are [in this case] reducible. Multiply the number in the tens place by two and add the result to the number in the units place, and reduce the sum. If it reduces by eight, the number has an eighth and a fourth, but if four remains, then it has [only] a fourth. ... And, if there is an odd number of hundreds their remainder [on reducing by eight] is four. Add the four to the units and to what comes from [reducing] the tens [by eight].

He gives the example of 512, where one adds the four remaining from the reduction of the odd number of hundreds to the two units, and that sum to the two that comes from reducing the single ten. The result is 8, so the number is reducible.

As for reducing by seven,[48] think of the leftmost digit as tens and add it to the digit to its right, considered as being units. Reduce the sum by seven. Then add the remainder [after this reduction], thought of, again, as tens, to the next digit to the right, and continue the reduction in this way.

He gives the example of 5236, where one takes the 5 as fifty and adds to it the 2, considered as units, resulting in 52. Reducing it by 7 leaves a remainder of 3. Now take it as 30 and add it to the next digit down, 3, to get 33. Reducing this by seven leaves 5. Take this as 50 and add the final 6 to get 56. This is reducible by 7, and so 5236 is reducible by 7.

Chapter 6: Denomination

The meaning of this term is the division of a small number by a large number. The way to do it is to decompose the number after which one names [i.e., the denominator] into the factors of which it is composed, to place them under a line and to then divide the number which you want to denominate by the factors, one after the other. You then obtain the desired result.

[48]The process he describes here is simply the process of long division, which is quite different from the process Ibn al-Bannāʾ describes in chapter 3 of part I ("On Subtraction") of his *Talkhīs* (see section I-1). Interestingly enough, he uses reduction by 7 in his chapter on checking arithmetic operations.

For example, if someone says to you, “Denominate nineteen according to thirty-five,” decompose the denominator into seven and five and place a line above these numbers. Next, divide the numerator by five, the result being three, with a remainder of four. Put the remainder [on the line] above the five, and the result [the quotient] above the seven. Because these numbers are smaller than the others, you will have the desired result, i.e. three sevenths and four fifths of one seventh. Thus:

$$\frac{4\ 3}{5\ 7}$$

And if someone says to you, “Denominate seventy-five according to one hundred forty-four,” you decompose the denominator into nine, eight and two and divide the numerator first by two, obtaining thirty-seven with a remainder of one, which you put above the two. Then divide the quotient by eight to get four [with a reminder of five, which you put above the eight].[49] Put the four above the nine.[50] The result will be four ninths and five eighths of a ninth and a half of an eighth of a ninth. Write it as follows:

$$\frac{1\ 5\ 4}{2\ 8\ 9}$$

Al-Qalaṣādī gives one more example, to denominate 196 according to 385, which follows exactly the pattern of the preceding example to obtain

$$\frac{1\ 4\ 5}{5\ 7\ 11}.$$[51]

Chapter 7: Division of shares

The practice of this operation consists in adding all the parts, to decompose that which comes in [i.e. the total capital] into the factors of which it is composed, and to place those in reserve in the third column. Then set the quantity that is to be divided [i.e., the profit] in the second column, which comes after the column of the sum of the portions. After this multiply each person’s portion by the quantity that is to be divided, and divide the result by the factors placed in reserve. You will obtain the result sought.

For example, if someone says to you, “One of three men has twenty-two dinars, another nineteen, and the third seven. They go into business together and they gain twelve dinars.” Add up their portions and you will have forty-eight, which is composed of six and eight. Put these numbers after [to the left of] the column of property, i.e. in the column of gain. Next multiply the portion of each by the gain, namely twelve. Then divide the result first by six and then the quotient by eight.[52] The first will receive five and

[49] At some point in the history of this text, the bracketed phrase I have inserted, clearly essential, was left out.

[50] Note that in both examples, al-Qalaṣādī proceeds through the factors of the denominator in increasing order.

[51] Al-Qalaṣādī’s algorithm for converting fractions written in this way to common fractions is in the section quoted below from the introduction of part II of his work.

[52] This is one example of the use of the rules Ibn al-Bannāʾ gives in chapter 3 (On Subtraction) for (what amounts to) dividing by small integers. Notice, too, that—as in the preceding chapter—al-Qalaṣādī proceeds through the factors of the divisor in increasing order.

four-eighths dirhams, the second four and six-eighths and the third one and six-eighths. It [the sum of the fractional parts] comes out to two wholes. Put them in the column of twelve, thus:

6 8	12	48	
0 4	5	22	Zaid
0 6	4	19	Umar
0 6	1	7	Baqr

If you notice that all of the parts have a common factor, suppress it and reduce each portion to the corresponding part in the factor column. Then multiply by the property.

For example, if someone says to you, "Of three men one has sixty-three [dinars], the other thirty-five and the third twenty-one. They go into business and gain fifty-one dinars." Every person's share has one-seventh [as a factor], so reduce each person's portion to its seventh. Then the first will have nine, the second five, and the third three. The sum is seventeen, and this is the factor.[53] Multiply each of the reduced parts by the property and divide the result by the factor, i.e., seventeen. The first gets twenty-seven dinars, the second fifteen and the third nine, thus:

17	51	17	119	
	27	9	63	Zaid
	15	5	35	Umar
	9	3	21	Baqr

Al-Qalaṣādī then discusses the case when the shares of all or some of the partners contain fractions. He says that, in this case, one should do the following.

Find the smallest number that contains all the denominators [as factors]. Multiply the numerator of each [partner's] portion[54] by this number and divide the result by the denominator [of the person's share when expressed as a fraction]. Then you will have the value of this portion.

We omit the example.

Chapter 8: On Checking

For addition, the operation [of checking] consists of reducing each of the two numbers added,[55] adding the two remainders, and reducing the sum [of the remainders] in the same way. That which remains is the response. Then you reduce the result [of the addition you are trying to check]. It will [if you have added correctly] be the same as the response.

For example, if someone tells you, "Add thirty-four to fifty-three," set them down thus:

5 3
3 4

[53] Since it is prime, it has only itself as a factor.

[54] That is to say, "Express each contribution as a fraction with the aforesaid number as denominator."

[55] This was discussed in chapter 3 ("On subtraction") in the selections above from Ibn al-Bannāʾ's *Summary Account* (section I-1 above).

Add, following the rules previously given. You get the sum, eighty-seven, thus: 87. If you reduce by seven the number added the result is six, and the remainder of the addend is four. The sum of the two remainders is ten its remainder is three. And the same is the remainder to the result.

Part II: On Fractions

The introduction and chapters 1–4 of this part list the five kinds of fractions that Ibn al-Bannā᾽ discusses in chapter 1 of part II of his text and deal with the four arithmetic operations on fractions, which al-Qalaṣādī writes with the numerator above the denominator. Here we present al-Qalaṣādī's discussion of one of the five types, namely, "relative fractions," which he has already used in chapter 6 (On Denomination) of part I.

From the Introduction

As for the numerator of a relative fraction, the way to find it is to multiply that which is found above the first factor [of the denominator] by that which comes after the corresponding factor, to add to the result that which is found above this [latter factor], and to multiply similarly by the third factor and the others.

For example, if someone says to you, "Convert four-fifths and three-sevenths of a fifth and five-eighths of one-seventh of a fifth," put it down thus:

$$\frac{5\;3\;4}{8\;7\;5}$$

Then multiply the four by the seven. The result is twenty-eight. Add the three to it, which will give thirty-one. Multiply this by the eight and add the five and you will have two hundred fifty-three, which will be the numerator of the problem, thus 253.

Chapter 8: On Transforming Fractions

Transforming is passing from a fraction with one name to [one with] another name. The practice of this operation consists in multiplying the total numerator of the fraction to be transformed by the factors [of the denominator] of the fraction to which one is transforming, and to divide, first of all, that which comes out by the factors [of the denominator] of [the fraction] to which one is transforming, and then the result by those of the [fraction] into which one transforms.

For example, if someone says to you, "Five sevenths and a half of one seventh, how many thirds of an eighth are they?" Put it down this way:

$$\frac{1\;5}{2\;7}\text{, how many}$$

$$\frac{\quad}{3\;8}\text{?}$$

Then, multiply the total numerator of the fraction to be transformed, which is eleven, by the factors of the denominator of the fraction into which one transforms. The result is two hundred sixty-four, so 264. Divide this result first by the factors of the denominator of the fraction being transformed and then by those of the fraction into which you are transforming. You will have for the result six eighths and six sevenths of a third of an eighth, so:

$$\frac{0\ 6\ 0\ 6}{2\ 7\ 3\ 8}.$$

3. MUḤAMMAD IBN MUḤAMMAD AL-FULLĀNĪ AL-KISHNĀWĪ, ON MAGIC SQUARES

The Nigerian author Muḥammad al-Fullānī al-Kishnāwī, who died in Cairo in 1741, wrote a substantial treatise on constructing magic squares, known in Arabic as "harmonious arrangements." He is included here because his work is of mathematical interest and, because of his birthplace, he seems to fit as well with the Maghrib as with the eastern part of the Islamic world, which was the focus in the previous sourcebook [Berggren, 2007]. In this treatise[56] he relies heavily on earlier authors, including M. Shabrāmallisī (written ca. 1600). However, on reading his treatise, it is obvious that al-Kishnāwī had a gift for explaining complex procedures clearly, and it is regrettable that only the part of his work dealing with squares of odd order[57] has survived.

Here is al-Kishnāwī's second method of constructing simple magic squares of odd order followed by a method he explains for constructing bordered magic squares[58] of odd order.

One begins in any one of the cells of the square whatsoever. One moves in regular steps: by one cell in either of the two principal directions [i.e., up/down or right/left] from file[59] to file and by two cells in the other [principal direction].[60] One proceeds in this way until one has entered as many numbers as there are divisions on a side. With this the first cycle [of placement] ends.

One then moves to begin the second cycle. For this displacement, and the following ones, there are two methods. The one consists of moving from the cell reached (staying within the file determined by the displacement from file to file) by a number of consecutive cells [counting the cell you depart from as "one"] equal to the number of divisions of the side and in the direction in which you moved two cells. The other consists of moving from the cell reached (staying within the file determined by the displacement by two files) by three cells in the direction of the displacement from file to file. The cell you reach by either of these two methods will be the point of departure for the second cycle.

The author gives the examples here [Fig. I-3-1]. The 5×5 square on the left he generated by the first method and that on the right by the second method. Bear in mind that in the square on the left, the cell with "5" was reached by moving one to the left of 4 and then two down,

[56]I became aware of al-Kishnawī's work in reading [Sesiano, 1994], from which comes the information concerning al-Kishnawī.

[57]The order of a magic square is the number of cells on its side.

[58]These are squares that remain "magic" when the outer border of cells is removed, when the one inside that is removed, etc.—down to and including the innermost (three-by-three) square.

[59]Here this word denotes either a row or a column.

[60]The author describes this as a knight's move (in the game of chess). Note that when one reaches the edge of the square, one continues as if the first row were below the fifth and the rightmost column were to the left of the leftmost column.

18	10	22	14	1
12	4	16	8	25
6	23	15	2	19
5	17	9	21	13
24	11	3	20	7

13	25	7	19	1
17	4	11	23	10
21	8	20	2	14
5	12	24	6	18
9	16	3	15	22

Fig. I-3-1.

so "the file determined by the displacement from file to file" is the leftmost, whereas in the second case, the second file from the bottom was reached by a displacement of two files, so one stays in the second file from the bottom to get to the cell for "6."

The author then describes a very different method for generating magic squares, one that Ibn al-Haytham described around 1000 CE; it can be found in [Berggren, 2007, pp. 658–659].

Later in the treatise the author describes a number of well-known methods for constructing various kinds of bordered magic squares. He then writes that one finds

> a few methods and ways of constructing [them] among specialists, in which some are instructions resting on trial-and-error which are arduous for those not practiced in the area but the others are valuable precepts generally.

One of these latter he describes as follows.

> One always places the first of the numbers in the middle of the right side. Counting the cells of the side after the middle cell down to the lower right corner, beginning with 1, one puts in each cell its number until one arrives at the corner indicated, which one leaves empty but places the number one has reached in the lower left corner. Then one advances with the consecutive numbers that follow it in the cells of the lower file, one after the other, following the corner one has filled in until one comes to the middle square. One does not put here the number one has reached, but in the opposite cell of the topmost file, i.e., its middle. One then imagines that the number one has placed in the middle of the topmost file was placed in the middle cell of the left file, and one then commences counting with this number [along the left file] as if it were in the middle, advancing through the consecutive numbers following this number in the cells of this file, until one comes to the upper left corner. In each cell, including that in the upper left corner, one puts the number reached [in the counting].
>
> One now imagines that the number in the upper left corner occupies the middle of the top file, and one counts the cells of the topmost file from there to the right, beginning with the number one has imagined occupies the middle of the side in question. One places the number attained [in the counting] in each cell, but not in the corner cell when one reaches it. When this is done, the numbers will appear distributed in the border in the way characterized above.

16	115	114	113	112	11	17	18	19	20	116
15	33	96	95	94	29	34	35	36	97	107
14	32	46	81	80	43	47	48	82	90	108
13	31	45	55	70	53	56	71	77	91	109
12	30	44	54	60	65	58	68	78	92	110
121	101	85	73	59	61	63	49	37	21	1
120	100	84	72	64	57	62	50	38	22	2
119	99	83	51	52	69	66	67	39	23	3
118	98	40	41	42	79	75	74	76	24	4
117	25	26	27	28	93	88	87	86	89	5
6	7	8	9	10	111	105	104	103	102	106

Fig. I-3-2.

Now one fills in the second [inner] border in the same way, beginning with the last of the numbers attained in the outer border. Then one places the numbers in the third border according to what has been explained, beginning with the number following the last of the numbers placed in the second border.

Continuing always in this way, one fills one border after another until one reaches, in the middle, a three-by-three square.[61] One fills in this three-by-three square and then fills in the empty cells as one knows how, the "equalizer"[62] used for the boundaries of each magic square being the equalizer of the large magic square that contains them.[63]

Fig. I-3-2 shows al-Kishnāwī's example of a bordered magic square of order 11.

[61] The author calls this a "square of three."

[62] This refers to the number $n^2 + 1$, where n is the order of the magic square. It occurs in the discussion of the trial-and-error methods. In the case of the 11×11 square, the equalizer for each of the borders would be 122.

[63] In his discussion of the trial-and-error method, the author explains that if a cell in a given border is occupied by the number c, then the one opposite it contains the equalizer minus c, i.e., $n^2 + 1 - c$. (In the case of interior cells in a border "opposite" means vertically or horizontally, and for corner cells it means diagonally opposite.)

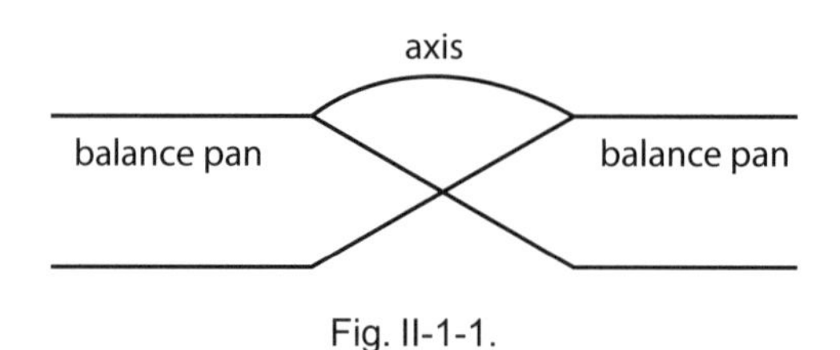

Fig. II-1-1.

II. ALGEBRA

1. AḤMAD IBN AL-BANNĀʾ, ALGEBRA

In [Sabra, 1986, p. 1139], A. I. Sabra remarked "in accordance with the general conception of *ḥisāb* being concerned with the determination of unknown numerical quantities from known ones, books on *ḥisāb* usually include sections on algebraic problems. Indeed, a number of treatises purporting to be on *ḥisāb* are almost entirely on algebra." Thus it is no surprise to see a section in Ibn al-Bannāʾ's *Summary Account* on algebra, from which the first selection on algebra is taken.

From the *Summary Account*, Book Two

Rules for obtaining an unknown beginning with quantities supposed to be known.

This book is divided into two sections, the first on the use of ratio and the second on *al-jabr* and *al-muqābala*.[1]

Section One: Operations on ratios

These operations are of two sorts: by four numbers in proportion and by the method of balance pans (*kaffāt*).[2]

Ibn al-Bannāʾ explains the various ways to use the Rule of Three for solving proportions and then turns to the method of balance pans.

The method of balance pans forms part of the art of geometry,[3] and its diagram is that you draw a balance, as in [Fig. II-1-1].

You place the number supposed to be known on the axis and place on one of the two balance pans any number you want. You then perform on it the operations indicated in the problem—addition, reduction, whatever—and you compare the result to the number placed on the axis. If you find it exactly then the number on the balance pan is the unknown number.

But, if the result is in error, then write the error above the balance pan if it is in excess and below the balance pan if it is less. Then, on the other balance pan put any number, other than the one you picked first, and operate just as you did with that first number. Next multiply the error of each balance pan by the (supposed) exact number of the other. Now look at the result. If the errors are both greater or [both] less, subtract the smaller from the

[1] "*Al-jabr* and *al-muqābala*" was the medieval Arabic term for what became "algebra." For Ibn al-Bannāʾ's explanation of what the phrase meant, see the excerpt "[Definition of Algebra]" below.

[2] As the reader will see, this latter method is the one known today as the method of double false position. See section II-3-2 in Chapter 1 for this method in Fibonacci's work and section V-3 in Chapter 2 for a commentary on Ibn al-Bannāʾ's work.

[3] Ibn al-Bannāʾ regarded it as "part of geometry," because it involved diagrams.

larger, take away the smaller of the two products from the other, and divide the difference of the two products by the difference of the two errors. But, if one of the two errors is more and the other less, you divide the sum of the two products by that of the two errors.

Or, if you wish, you can take for the second pan the first number, or another, to make the part corresponding to it that you compare with the number on the axis. Multiply this part by the [supposedly] exact number of the first pan, multiply the error of the first by the [supposedly] exact number of the second. Then, if the error of the first is by deficiency, you add the two products. But if it is by excess you take the difference. You divide the result by the part of the second pan and you obtain the number sought.

Souissi (See Ibn al-Bannā'. 1969) gives two examples from al-Qalaṣādī's commentary of Ibn al-Bannā''s work.[4]

Find a number whose products by 2 and 3 sum to 10.

Try 4 and you get 20, 10 more than what you wanted. You place this 10 above the first pan. Try 5 and you get 25, 15 more than what you wanted. You place this 15 above the second pan.

So the diagram looks like [Fig. II-1-2].

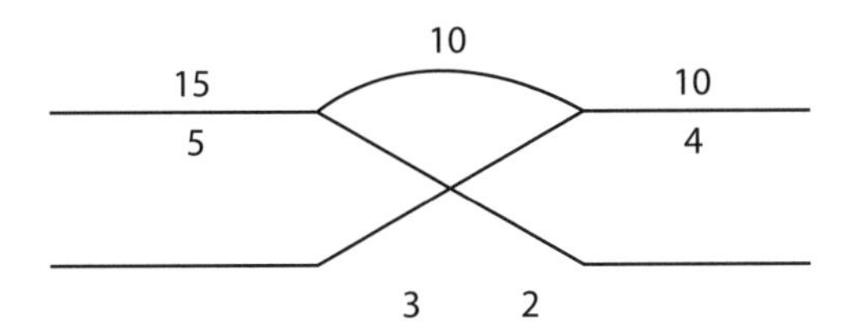

Fig. II-1-2.

Then $4 \times 15 - 5 \times 10 = 10$, and $15 - 10 = 5$.

Since $10{:}5 = 2$ that is the answer.

The sum of the products of a number by 6 and by 7 is equal to 25. What is the number?

Choose 6 first and this gives 78, whose excess over 25 equals 53. This goes above the pan. Now suppose it is equal to 1, so the sum of the products is 13. The difference of 13 and 25 is 12, so this goes below the pan.

The diagram looks like [Fig. II-1-3].

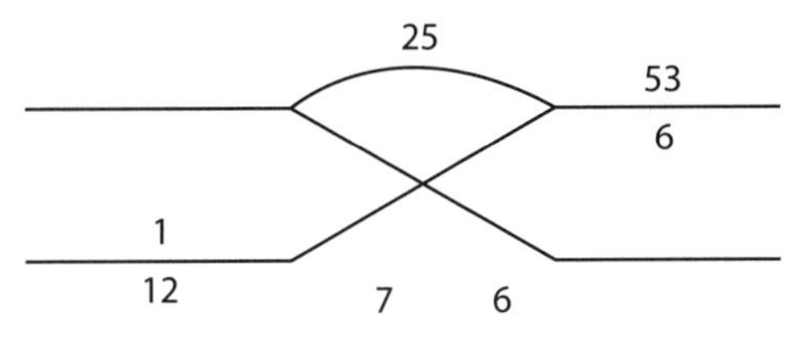

Fig. II-1-3.

[4]Note that the two problems, both of the same type, are nothing more than solving the equation $ax + bx = c$. They are included here because al-Qalaṣādī does not solve them as we would today.

Then $6 \times 12 + 1 \times 53 = 125$ and $53 + 12 = 65$.
Finally, 125:65 = 25:13 = 1 12/13, which is the correct answer.

Section Two: *Al-jabr* and *al-muqābala*

[Definition of Algebra]

Chapter 1—The Meaning of *al-jabr* and *al-muqābala*

Al-jabr signifies "restoring," as we have indicated in Book One of this work.[5] And "*al-muqābala*" means "subtracting each species from the one like it [on the other side of the equation] in such a way that [in the end] the equation does not contain the same species on both sides."

An equation means restoring the term subtracted to make it added, to subtract among like unknowns [on either side of the equation] the positive from the positive and the negative from the negative.

Algebra turns around three species: number, unknowns and squares.

The unknowns are roots. The square is the number obtained by multiplying the root by itself. The known number is that which is referred neither to a root nor to a square. These three species may be equated to each other, either singly or in combinations. The result is six cases: three simple and three combined.

The first type of simple equations according to contemporary usage is "squares equal roots," the second type is "squares are equal to a number," and the third type is "roots are equal to a number."

Among the three combined cases the first, which is the fourth [of the six cases], is that the number is isolated.[6] In the fifth it is the root that is isolated and in the sixth it is the square that is isolated.

Chapter 2—Solving the Six Cases

Among the three types of simple equations you divide by the [number of the] squares that of the quantity to which it is equal. But if there are no squares, you divide by that of the roots. The quotient in the first and third cases is the root, and in the second case it is the square.

When one knows the root, one has the square by multiplying the root by itself. When one knows the square, one deduces the root.

For the fourth case,[7] the method consists in taking half the number of the roots, squaring it, adding the result to the number, and then taking the square root of the sum. You then subtract half of this number from this root, and what is left is the root [i.e., the solution].

The sixth case is like the fourth as far as the solution goes, but in the last operation one adds the half to the square root of the sum to obtain the root.

[5] That is to say, adding to both sides of an equation quantities that have been subtracted from one side or the other.

[6] That is, the equation is of the form $x^2 + bx = c$.

[7] In his initial discussion of the solutions of the last three cases, ibn al-Bannāʾ assumes that the number of squares is one, i.e., that the coefficient of the square term is unity.

For the fifth case, you subtract the number from the square of half the number of the roots, and you take the square root of the remainder. If you add this square root to the half [of the number of roots], you obtain the larger root. When you subtract it, you obtain the smaller root. But, when the square of the half is equal to the number, the half is the root and its square is the known number.

In the three combined cases of equations whenever one has a number of squares larger than unity, one reduces it to unity and reduces the other coefficients in the same amount. But, if one finds in one of the three cases [that] the number of squares is less than unity, then restore it in such a way that it becomes equal to unity and transform by the same amount the other coefficients.

And this is the way that reduction and restoration are done, but if you wish you can also divide all the numbers by the number of the squares and the problem reduces[8] to the quotient. You make each new term correspond to the original term.

According to the preceding discussion, Ibn al-Bannāʾ's fourth type of equation of degree at most two is "numbers equal a square plus things" (i.e., $c = x^2 + bx$), and he gives the rule that "the thing" (x) is equal to $\sqrt{\left(\frac{b}{2}\right)^2 + c} - \frac{b}{2}$. In *Raising the Veil*, he gives a justification of the procedure.

[Proof of the validity of the solution for the fourth type of equation]

The justification of the procedure in composite types [of equations] is derived from [the rule for finding] the product by means of quadrature, which we have given in that place [where we discussed multiplication].[9] You always take, as the two numbers multiplied [in applying the rule of quadrature], the square [x^2] and the number [$c = x^2+bx$]. Their difference will be [equal to] things [bx] in the fourth type. You multiply one of these by the other [and this will be] some squares [namely, cx^2], to which you add the square of half their difference [$((b/2)x)^2$]. This gives [according to the rule of quadrature] the square of half of their sum. You take its root, and this will be half of their sum, and this will be some things [namely, $\sqrt{\{(b/2)^2+c\}}x$], which you keep. Then you consider half of their sum, which will be a square plus half the things with it [$x^2+(b/2)x$], because the number is equal to a square plus things, and if you add a square, this will be two squares plus the things, and half of this is a square plus half of the things. You compare this with the result you kept, and the result is that things are equal to a square, which is the first type [of equation, in this case, $x^2 = [\sqrt{\{(b/2)^2+c\}} - (b/2)]x$].

Chapter 3—On Addition and Subtraction [in algebra]

Addition of terms of different types is done by the conjunction "and." The taking away of terms of different types is done without actually doing the subtraction, but for terms of the

[8] The Arabic word here is "returns," which is borrowed from the science of dividing inheritances, in which one problem is replaced by ("returns to") an equivalent one.

[9] The technique of quadrature uses the identity $\{(m+n)/2\}^2 - \{(m-n)/2\}^2 = m \cdot n$ to find the product of two numbers m and n.

same type it is done by subtracting the smaller term from the larger. And the subtraction of different types of terms is done by the particle of exception.[10]

An exception is either from the two sides or from one of them, and it is either one kind or two different kinds. The particle of exception may figure in both terms of the difference or in only one of the two. It may also belong to only one kind or to two different kinds. The method consists in adding the number excepted to each of the two terms of the difference, and so you effect the subtraction.

Chapter 4—On Multiplication and on the Recognition of Exponents[11] and the Name of the Corresponding Term

Be aware that the exponent of the unknown is one, that of the squares two, and that of the cubes three. And the word corresponding to exponent one is "unknown," that corresponding to exponent two is "square" and to three it is "cube." Beyond this, one has three for each cube [repeated] and two for each square. When you multiply different types [of terms], add the exponents of the multiplier and the multiplicand, and the sum will be the exponent of the product. And when you multiply a known number by a given species, the exponent of the product will be that of the given species.

If you have an equation between squares and squares; cubes and squares; or even between cubes, squares and unknowns; or analogous equalities not having any known number term, then subtract the smallest exponent from the exponent of each term and, with what remains, form an equation on the model of the initial equation.

The product of two positive or two negative terms is positive, and the product of positive by negative is negative.

Chapter 5—On Division

If you divide one of the species by a species of lower degree,[12] subtract from the exponent of the dividend that of the divisor; the remainder is the exponent of the quotient of the division.

If you divide one of the species by a species of the same degree, the quotient is a number.

If you divide one of the species by a number, the quotient is of the same species.

If the dividend contains a difference, divide each term of the difference by the divisor, and subtract the quotient of the subtracted term from that of the term from which one subtracts; the result is the quotient.

A species of lower degree is not divisible by one of higher degree.

One cannot divide by an expression containing a subtraction.

[10]This is the term *illā* in Arabic. As we would say, one writes a minus sign.

[11]By translating the Arabic work *'uss* as "exponent," I am following the suggestion in [Souissi, 1968, p. 78]. Souissi notes that "One meets this term for the time in the work of Ibn al-Bannā' and his commentators." In any case, it is quite different from the phrase "the number of its rank," which al-Kāshī (who lived on the other side of the Islamic world 150 years later) used for the same concept in his *Reckoners' Key* [Berggren, 2007, p. 537].

[12]The Arabic here and in the following sentence, where we write "same degree," reads, literally, "by a lower species" and "by the same species," respectively.

From the *Book on Fundamentals and Preliminaries for Algebra*

This work was one of the three that Ibn al-Bannā᾿ wrote dealing in part or in whole with algebra, the other two being his *Summary* and *Raising the Veil.* He wrote it in Morocco early in his career, sometime in the last third of the thirteenth century, and a number of writers attacked him for borrowing, sometimes word-for-word, from an earlier writer, al-Qurashī. Since al-Qurashī's work has not been found, we cannot judge the extent to which, and how, Ibn al-Bannā᾿ used the other's work. But, whatever may be the case, Ahmed Djebbar (a noted authority in the history of arithmetic and algebra) has remarked that in his section on problems, it is clear that Ibn al-Bannā᾿ made heavy use of the *Algebra* of the tenth-century Egyptian mathematician Abū Kāmil, although regrouping the problems according to subject. Indeed, Djebbar describes Ibn al-Bannā᾿'s work as "the last great work of the Muslim West to belong entirely to the tradition of Abū Kāmil" [Djebbar, 2005, p. 86].

As for the influence of this work, despite its clear pedagogical intent, we do not know to what extent it was used in teaching. According to a private communication from Djebbar, there are eight known manuscripts in libraries from Teheran to Morocco. So it received a certain amount of study, but more than three times that number of copies have been found of Ibn al-Bannā᾿'s *Summary.*

We begin with Ibn al-Bannā᾿'s introduction and some rules for algebraic operations from part one of the book.

Introduction

Praise to God, who alone is worshipped, the Eternal who has created all things and perfected their qualities, who has estimated then and has measured their quantity. . . .

Having said this, I have composed this book on *al-jabr* and *al-muqābala* and I have done it in two parts, one on the foundations and preliminaries, which are the basis for the operations of algebra, and one part on problems of algebra on which the student may practice and by which he may become attentive to the way of solving any problem put to him.

Even though countless works have been composed for this purpose this work contains their foundations and includes their branches and their sections. It is a small volume but contains much science. Because it is so short it resembles those abridgements [one sees], but because of the richness of its science it makes the long books unnecessary. I have emphasized those kinds of procedures that the mind can easily comprehend, which are immediately evident to the intelligence. I have not given demonstrations because these depend on preliminary propositions, for many of which one needs the book of Euclid.[13]

I ask God to aid me in attaining what I hope for and for protection from error and fault.

As Ibn al-Bannā᾿ says in his introduction to the work, it consists of two parts: part one on the basics of algebra and part two on problems. Part One begins with a brief preliminary discussion of numbers, after which Ibn al-Bannā᾿ deals with (in this order) multiplication, division, root extraction, addition, and subtraction. Each of these chapters contains problems illustrating the rules. He then continues with a section on unknowns

[13] He refers, of course, to Euclid's *Elements.*

and algebraic manipulation before concluding Part One with a discussion of the solution of equations. Since the procedures in this last section are very similar to the rules in the previous selection, they are not included here.

Part One: On Fundamentals

[From the chapter on multiplication]

The following rule states the distributive law for multiplication and its consequence when both multiplier and multiplicand are expressed as a sum of parts.

Know that the result of the product of a number by a number is exactly the result of the product of one of them by all the parts[14] of the second, one by one, then the sum of this. It is also the product of each of the parts of one of them by each of the parts of the other and then all added together.

The square of a number plus the square of one of its two parts is equal to the square of the other part plus the product of the first part by the double of the number. It is also equal to the half of the square of the other part, to which one adds twice the product of the first part plus the half of the second part by itself.

The product of a number by one of its parts is equal to the square of that part plus the product of the two parts.

The following rules deal with associativity of products.

The product of a first number by a second, then of a third number by a fourth, then the product of the result of the one by the result of the other is equivalent to the product of the first number by the second number, the result by the third number, [and] the result [of that] by the fourth, however they have been permuted in the multiplication.

One knows, beginning from there, that to multiply the product of two numbers by a third number is equivalent to multiplying one of the numbers by the product of the other two. And that to multiply one number by the square of a second is equivalent to multiplying the second by the product of the two.

He now states the rule of signs for multiplication of positive and negative numbers.

The product of two terms added or two subtracted is added and the product of a term added or subtracted by its opposite [i.e. a term subtracted or added] is subtracted.

[From the chapter on division]

The division of one number [by another] is equivalent to the division after having multiplied each of the two by another number. One will know, beginning from there, that, any two numbers being divided—the one by the other—the multiplication of the result by their product is equal to the square of the one of the two [numbers] that is the dividend, and that the division of the square of the dividend by their product is then equal to the division of the dividend by the divisor and to the division of their product by the square of the divisor.

Any two numbers being divided, each by the other, the product of the two results is always equal to one.

[14]That is, however one may express the second number as a sum.

The following rule will be applied in Ibn al-Bannāʾ's solution to a problem excerpted below about the division of ten into two parts satisfying a certain condition.

[For] Any two numbers being divided, each by the other, the sum of their squares is equal to the multiplication of their product by the sum of the two results of the division.

[Problems on division]

If someone says: Divide the root of ten by two plus the root of seven, the procedure consists of multiplying two plus the root of seven by its *apotome*,[15] which is the root of seven less two. The result is the product, which is three. One will divide by this the dividend, which is the root of ten, and one will then multiply the result of the division by the *apotome* of the divisor, which is the root of seven less two. What comes out of this will be the result of the division, which one wants to know: the root of seven plus seven-ninths less the root of four and four-ninths.

[From the chapter on extraction of roots]

If one says, eight plus the root of sixty, what is its root? The procedure for this ... is to subtract one-fourth of the square of the smaller of the two expressions from one-fourth of the square of the larger and take the root of the difference—which is one [$\surd\{(1/4)8^2 - (1/4)(\surd 60)^2\} = 1$]. Then you add this one to half the larger of the two numbers and you take the root of this sum, which will be the root of five. Then you subtract the one from half of the greater of the two numbers, and you take the root of the remainder, which will be the root of three. The sum of these two square roots is the root of five plus the root of three, and this is the root requested [$\surd\{8 + \surd 60\} = \surd 5 + \surd 3$].[16]

[From the section on unknowns and powers]

Know that each unknown, whose ratio to unity you do not know, is called "thing." This is because that is what it is when you do not know its value, and when [there is] another unknown, which does not carry this name at the same time as it, but whose ratio to it you are able to know, in this case you give to it [the latter unknown] the measure of this ratio. If the [second] unknown were two times the object designated as "thing" one would say that it is "two things." And if it were its half, one would say it is "half a thing."

And if one calls the unknown "thing," it is to distinguish between it and all the [other] unknowns, and this name is to all the things what the unity is to the numbers, which is to say that they are composed of it.

The thing is also called "root," because every number is the root of its square, even if one does not know its value. "The thing" and "the root," then, both designate the same object.

Any unknown that is the product of the thing by itself is called *māl*. And one does not call any unknown other than it by this name, unless one knows its ratio to it. In such a case, its name will, equally, correspond to the value of this ratio—as has been done

[15]The term is Greek and is found in, among other works, Book X of Euclid's *Elements*.

[16]The general rule is $\sqrt{a+\sqrt{b}} = \sqrt{\frac{a}{2} + \sqrt{\frac{a^2}{4} - \frac{b}{4}}} + \sqrt{\frac{a}{2} - \sqrt{\frac{a^2}{4} - \frac{b}{4}}}$.

[above] for the thing. The name by which this unknown is called [i.e., *māl*] is a convention to distinguish the square of the thing from the square of objects other than it.

If one multiplies the thing by its square, which is *māl*, the result of the product is called cube to distinguish this unknown from all the [other] unknowns. One gives it this name because it is a cube, even if one does not know its value. In the same way, if the thing is multiplied by the cube the result is called *māl-māl*, because it is equal to the product of the *māl* by itself. And if the thing is multiplied by the *māl-māl*, the result is called *māl*-cube, because it is equal to the product of the *māl* by the cube.

And if one does not assign exponent to a number and one [instead] puts the thing in the first place, the square in the second ... the cube-cube in the sixth [place] and so on indefinitely, the positions are distinguished from one another by their distance from the position of the number, i.e. three for each cube, two for the square, [etc.].[17]

If one wants to multiply one of these types, whatever one it is, by another, the position of the result of the product will be distant from the position of one of the unknowns of the product by the distance of the position of the second [unknown] from the number. So the position of the product is equal to the sum of the positions of the two factors.

Part Two: Problems

[1][18] Divide ten into two parts such that if you divide each of the two by the other, the sum [of the ratios] is equal to four and a fourth.[19]

Your set one of the parts equal to a thing, and the other part is then ten minus a thing. Then you square each of these and sum the two squares. This will be equal to one hundred plus two *māl* minus twenty things. You save this [result]. Then you multiply the product of the two parts by the result of the ratios, which is four and a quarter. This will be forty-two things and a half of a thing less four *māl* and a quarter of a *māl*. This is equal to the sum of the two squares that has been kept.

You restore and you simplify and you get the fifth type [of equation].[20] You determine the thing by defect, which will be two, and it is one of the two parts. The thing, by excess, will be eight, and it is the other part.

The principle behind this solution has been set out in the chapter on division in the first part, namely: For every pair of numbers of which one divides each by the other, the sum of their squares is equal to the multiplication of their product by the sum of the two results [of the divisions].

Ibn al-Bannāʾ gives also a variant of this method and then gives another approach, in which he introduces an auxiliary unknown, the "dinar," in daily life a unit of currency.

You may solve it in another way, which is to suppose that one of the two parts is the thing, the other, then, being ten less a thing. Then, you divide ten less a thing by a thing,

[17]He refers here to the way in which polynomials were expressed by two rows of cells, those of the top row labeled "unit," "thing," *māl*, etc., with the respective coefficients of these terms being written underneath them. See the examples and further explanation in [Berggren, 1986].

[18]The numbering of the problems is mine. Ibn al-Bannāʾ does not number them.

[19]Compare this to the similar problem in the work of Fibonacci (section II-3-2 in Chapter 1 of this sourcebook).

[20]For the types of equation see the selection from the *Summary*, "Chapter 1—The meaning of *al-jabr* and *al-muqābala*."

and you suppose the result equal to some unknown among the many possible unknowns, say, a dinar. The product of a dinar by the thing will then be ten less the thing, and because of this the result of the division of a thing by ten less a thing will be equal to four and a quarter less one dinar. Then multiply it by the divisor, which is ten less a thing, and you then obtain fifty-two and a half less five things and a quarter of a thing less ten dinars. You equate the result to the thing, which is what was divided, then you restore and you simplify and you obtain fifty-two and a half equal to ten dinars and six things and a quarter of a thing. Take away, from both sides,[21] six things and a quarter of a thing. The equation becomes ten dinars equal fifty-two and a half less six things and a quarter of a thing. The dinar is thus equal to five and a quarter less five-eighths of a thing. But we supposed that the product of the dinar by the thing is [equal to] ten less a thing. You multiply, in place of the dinar, that to which it is equal, which is five and a quarter less five-eighths of a thing, by the thing. The result of the product is five things and a quarter of a thing less five-eighths of a *māl*, which is equal to ten less a thing. Your restore and simplify, and you come down to the fifth type. So understand the method of the dinar and the ingenious procedure that is [used], and it serves in a number of problems.

[2] One divides two hundred twenty dirhams among ten men in such a way that the second takes double the first, the third triple the first, the fourth quadruple, and the multiples increasing in this way up to the part of the tenth, which is ten times that of the first.

[Solution] You take the share of the first equal to "thing" or some good,[22] the part of the second two things or two goods—it is double the first—the part of the third three times that of the first, and in this way the multiples follow, up to the part of the tenth, which is ten times that of the first. Then you add all of these, which gives fifty-five things or goods. You equate this to two hundred twenty and find that the part of the first is four and that of the tenth is forty. Learn this and solve all similar problems in the same way.

In what is labeled as Problem 1, Ibn al-Bannāʾ introduced an auxiliary unknown, which he called a "dinar." In this next problem he introduces two auxiliary unknowns, called "dinar" and "dirham," also a unit of currency, although he lets the third unknown be an arbitrary number. This problem of "buying an animal" is very common in medieval problem collections and occurs in both Chapters 1 and 2 of this sourcebook (section II-2-2 in Chapter 1 and section I-4 in Chapter 2).

[3] [Three men want to buy an animal together.] If the first says to the second and third, "If one takes half of what you have and adds it to what I have, I will have the price of the animal," and if the second says to the first and the third, "If one takes a third of what you have and adds it to what I have, I will have the price of the animal," and if the third says to the first and the second, "If one takes a fourth of what you have and adds it to what I have, I will have the price of the animal" [how much does each have?].

[Assume that the amount of the first man is a thing, that of the second is a dinar, and that of the third is equal to any number you wish, say, three dirhams.] You take, then, the half of what the second and third have, you add it to what the first has and you will have ... that the price of the animal is a thing plus half a dinar plus a dirham and a half. Then

[21] Literally, "from the two members."

[22] Depending on context, the Arabic term used here, *māl*, can mean what we would write as x^2, or a certain sum of money, or simply something of value, as in a merchant's "goods."

you take a third of what the first and third have and add it to the second. The price of the animal is then a dinar plus one-third of a thing plus a dirham. [So] this is equal to the thing plus a half dinar plus a dirham and a half, the first [expression for the] price. You simplify and you have: a half dinar is equal to two-thirds of a thing plus a half dirham. So the dinar is [equal] to a thing and a third of a thing plus one dirham, and this is what the second man has. So the price of the animal is [equal] to a thing and two-thirds of the thing plus two dirhams. Then you take the fourth of what the first and the second have and you add it to what the third has. This will be three dirhams and a fourth of a dirham and three-sixths plus a half of a sixth of a thing, and this is the price of the animal. It is equal to a thing and two-thirds of a thing plus two dirhams, and this is the first [expression for the] price. You simplify, and you come to the third type [of equation]. The thing will be equal to a dirham and two-thirteenths of a dirham, and this is what the first has. What the second has is [equal] to two dirhams and seven-thirteenths of a dirham, and that of the third [is] three [dirhams]. The price of the animal is three dirhams plus twelve-thirteenths [of a dirham].

If you wish to get rid of the fractions, multiply the price of the animal and what each one has by any number whatever having a thirteenth and you will have whole number [solutions]. This problem, just like the preceding one, is indeterminate. Think about it and take it as a model.

[4] A man goes to the market and he finds that a pound of honey costs four dirhams, a pound of oil two dirhams and vinegar costs one dirham for five and one-third pounds. From all of these he takes forty-five pounds for forty-five dirhams. How much has he taken of each kind?

Say that the amount of honey he takes is one thing and the amount of oil he has taken is one thing. So the amount of vinegar he has taken is forty-five pounds less two things. It is clear that the price of honey he has taken is four things and the price of oil he has taken is two things and the price of vinegar eight dirhams and three-eighths and a half of an eighth of a dirham less three-eighths of a thing.[23] You add up all these prices, and set it equal to forty-five dirhams. You arrive at thirty-six dirhams and four-eighths and a half of an eighth of a dirham equal to five and five-eighths things. So you have ended up with the third type. So the thing will be six and a half, and it is the number of pounds of honey he has taken. And their price is twenty-six dirhams. He has also taken six and a half pounds of oil, and their price is thirteen dirhams. Finally he has taken thirty-two pounds of vinegar, and their price is six dirhams.

That was a good method, by which one obtained equal quantities of two [or more] kinds of commodities. If you wish, set the one of them [equal to] a thing and the other [equal to] that which you wish as things, and the [amounts of different] kinds [of commodities] will then be different.

[5] A good has a root,[24] and if you add [to the good] ten of its roots plus twenty dirhams that also has a root.

[23] Ibn al-Bannāʾ does not go into the details of converting $45 - 2x$ pounds of vinegar into dirhams when 5 1/3 pounds of it sell for one dirham, but the arithmetic involved assumes the reader has mastered the arithmetic of fractions.

[24] That is to say, the value represented by the "good" is the square of a rational number.

Solution: A good plus ten of its roots and twenty dirhams therefore has a root, so its root is necessarily greater than the root of the good. Since twenty is less than the square of half the number of roots [i.e., 25] one knows that the excess of the [second] root over the thing [i.e., the root of the good] is less than five. (We know this by what we have explained about the fourth among the [six] types of equations relative to the completion of the square that we have used in its solution.) So one must set the square root of the good plus ten roots plus twenty dirhams equal to a thing, in order to eliminate the good in the comparison (of the two members of the equation), and one adds to it [the thing] a number that must be less than half the roots and whose square is greater than twenty.

. . .

In our problem you can set the root of the good plus ten roots plus twenty dirhams [equal to] a thing plus four and a half. You take its square and it will be: a square plus twenty dirhams and a quarter dirham plus nine roots is equal to a good plus ten roots plus twenty dirhams. You simplify it, and the thing will be equal to a fourth and the good will be equal to half of an eighth.

If you set the root of the good plus ten roots plus twenty dirhams equal to a number, its square will be greater than twenty, and if you subtract the thing, it will be a good procedure. Think about this.

You can [also] solve it in another way, which is to set a good plus ten roots plus twenty dirhams equal to a square. Clearly, the square [that is] equal [to the first member of the equation] is greater than twenty. You subtract twenty [from the two members of the equation], and there remains a good and ten roots equal to a square less twenty. This is the fourth type of equation. You raise half the number of roots to the square, which will be twenty-five. You add it to the number [of the second member]. It will be a square plus five, which should have a [rational] root to the end that one can subtract half the roots so that the thing remains.

This leads to the problem of finding a square [such that], if you add five to it, it will still be a square. We have already shown how to do this when we said, "A good has a root [and] if you add ten to it, it still has a root."[25]

So you look for this, and you know that the square is more than twenty, supposing that the number that is added to the thing for finding the root is one or more fractions. When you have found it, you equate it to a good plus ten roots plus twenty dirhams, and you end up with the fourth type. This problem is also indeterminate.

[6] Determine three different goods that have the same ratio and are such that the sum of the squares of the smallest and middle is equal to the square of the largest.

[Solution] Since they are proportional, the product of the smallest by the largest is equal to the square of the middle. Set the smallest equal to any number you wish, say, "one," and you set the middle equal to "a thing." Then you divide the square of the middle by the smallest, and there results a good. Then you add the square of the smallest to the square of the middle, and you equate that to the square of the largest. So a *māl-māl* is equal to

[25] This was done in an earlier problem, not presented here.

a *māl* plus one. This is an example of the sixth type [of equation] because the *māl* is the root of the *māl-māl*. It is as if one had said, "A square is equal to its root plus one." You solve it as before. [The root] will be [equal to] a half plus the root of one and a fourth, and it is the square of the middle. The middle will thus be the root of [the whole quantity] one half plus the root of one and a fourth. The smallest is one and the largest is one-half plus the root of one and a fourth. These are the three sums of money satisfying the conditions given.

2. MUḤAMMAD IBN BADR, *AN ABRIDGEMENT OF ALGEBRA*

Virtually nothing is known about the life of the writer Muḥammad ibn Badr other than that he was the author of the book, *An Abridgement of Algebra*. Because he wrote on mathematics and his name was Ibn Badr, A. Djebbar has suggested[26] that he may have been part of the family whose members included the tenth-century Cordovan mathematician ʿAbd al-Raḥman ibn Badr. (The Arabic text attributes the *Abridgement* to the latter!)

On the Definition of Algebra (*al-jabr* and *al-muqābala*)

Al-jabr [literally, "restoration"] is adding in anything taken away [in an equation] so that it does not lack, and *al-muqābala* is canceling any kind from its corresponding [kind] so that the two sides [of an equation] do not have a quantity of the same kind.

For example, if someone were to say to you, "One hundred dirhams less ten things is equal to seventy," you would have to use *al-jabr* on the hundred with the ten things taken away, and it will be [simply] one hundred dirhams. And you add to the seventy dirhams the ten things with which you restored the hundred. [Now] you compare the seventy against the hundred, so you take them away from the hundred. Thirty dirhams will remain equal to ten things, and so the thing is three.

And so if we take away ten of its equals from the hundred, there will remain seventy. And it is equal to the hundred less ten likenesses of the thing, which was three, as we said. And we do a similar thing in everything in which there occur the less and the more. So understand.

Now this completes the preliminaries for algebra, which [we shall explain] in seventeen sections, six of which are [devoted to] the six known problems.[27] The second six are chapters about roots. [Then] section thirteen is a preliminary to the products of things, squares and cubes. And the fourteenth is their addition and subtractions. The fifteenth is the division of one of them by the other. The sixteenth is knowing the product of the deficient by the deficient, and the deficient by the excess, and the excess by the excess. And the seventeenth is knowing what *al-jabr* and *al-muqābala* are, which we have just explained.

[26]See [Djebbar, 2005, p. 133]. Some sources suggest that Ibn Badr lived in the thirteenth century, but the evidence so far is rather weak.

[27]Ibn Badr's word that is translated here as "problems" is *masāʾil*. He is referring to the six possible equations of degree at most two that one gets when constrained to write equations using only positive quantities, such as $bx = c$, $ax^2 = bx$, $ax^2 = c$, $ax^2 + bx = c$, etc. Despite the apparent reference here to the numbering of sections of the book, the only numeration is the occasional reference to "the second (or third)" problem of a certain type.

On the multiplication of things, *māl*, and cubes with numbers, one by another

Know that the multiplication of things by numbers produces things, the multiplication of *māl* by numbers [produces] *māl*, and the multiplication of cubes by numbers [produces] cubes. Similarly, when you multiply a number by any one of these types it is that very same type. And know that the product of things by things is *māls*, and the product of things by *māls* is cubes, and the product of *māls* by *māls* is *māl-māls* and the product of *māls* by cubes is *māl*-cubes and, also, the product of cubes by cubes is cube-cubes.

. . .

Know that the product of the positive by the positive is positive, and the product of negative by negative is positive and the product of positive by negative is negative and the product of negative by positive is negative.

All that precedes concerning multiplication [of roots, *māl*, etc.] becomes intelligible by this reasoning which I am going to mention, namely, make the "thing" a "degree,"[28] and for the *māl* two degrees, and the cube three degrees.

And so, when we multiply some of these unknowns[29] and what is composed of them, one by another, the result of each one of them [is found by] adding the degree of the multiplicand to the degree of the multiplier, and the sum of these degrees is applied to each of the three preceding cases. If the result is one degree, it corresponds to things; if two degrees, to *māls*; if three degrees, to cubes. Therefore, if you multiply cubes by cubes, you will have to add the degrees of cubes with the degrees of cubes, and there results six. This six is the degree of "cube-cube." In the same way, if you multiply *māl* by *māl* you add the degree of *māl* to the degree of *māl*, and the result will be four, which is the degree of *māl-māl*.

Also, if the result of the addition is nine this will be the degree of cube-cube-cube. And if the result is ten this would have to be the degree of *māl-māl-māl-māl-mal*, five times in this case, because the degree of each *māl* is two. Or you can say also that it is the degree of cube-cube-cube multiplied by thing.

Example of this: If you say, multiply twenty cubes by twenty cubes. You will get [from multiplying the numbers] four hundred. Then you will add the degrees of the cubes, which are three, to the degrees of the other cubes, which are three. You will get six. So you will say the category (*bāb*) of six is cube-cube, and if you want you say *māl-māl-māl* because each *māl* has two degrees according to what we explained earlier, and each cube has three degrees. Thus it is that you will say the result is four hundred cube-cubes or four hundred *māl-māl-māl*.

In the same manner, if it is said to you: multiply twenty cube-cubes by twenty cube-cubes, you will multiply twenty by twenty and sum the degrees that correspond to the cubes, which are [all together] twelve degrees, because the degree of each cube is three, and so it is the category of cubes repeated four times, or the category of *māls* repeated

[28]The Arabic term is *daraja*, the same term used for "degree" in geometry.

[29]Literally, "questions" or "problems" (*masāʾil*).

six times. Thus you say the result is four hundred cube-cube-cube of the cubes, or if you want you may say four hundred *māl-māl*, ... repeated six times.

Thus you operate in all the other cases of this type in which the degree is multiple. Understand!

[Problem]: If it is said to you "multiply ten dirhams and a thing by ten dirhams and a thing," then the operation in that case is that you multiply the ten dirhams by the ten dirhams, which will give you one hundred dirhams. Then you multiply the ten dirhams by the thing and the other ten dirhams by the thing. You obtain twenty things. Then you multiply thing by thing and you obtain *māl*, so there will be with you one hundred dirhams and *māl* and twenty things.

And if it is said to you "multiply ten dirhams and a thing by ten dirhams less a thing," then you operate according to what was just said and there results to you one hundred dirhams plus ten things and minus ten things minus *māl*. And the plus ten things disappear with the minus ten things. There remain with you one hundred dirhams minus *māl*.

And if someone says to you "multiply ten dirhams and a thing and *māl* and cube by ten dirhams and thing and *māl* and three cubes," do it according to the places (*manāzil*) of the numbers.

On the sum and difference of things, *māl*s, and cubes with each another

Know that the sum of things [added] to things, and *māls* to *māls*, and cubes to cubes is something of the same type. It does not sum to a type unless one of that type is equal to one of the other. For example, if you say "add a thing to four things," then, if the first "thing" is equal to the "thing" from the four things, you say that the sum is five things. But if it is not equal to it, then you say that the sum is equal to four things and a thing. Similarly, if it [is] said to you "add ten cubes to ten cubes," then if each one of the ten cubes is equal to each one of the other ten cubes then you say in adding them [that] the union is twenty cubes. But if that is not so, then you say that the union of them is ten cubes and ten cubes. Similarly for *māl*s, also. If the *māl* is equal to the *māl,* then you combine them. But if it is not equal to it, then you must necessarily keep them separate.

Ibn Badr then makes corresponding statements for the case of subtraction. The following problem is one of a group involving squares and subtraction. Ibn Badr clearly looks on this problem as being about finding a *māl*, given information about it, whereas we would think the point of the problem is to find the "thing."

Problem: If it is said to you a *māl*: its third is subtracted, then what remains is multiplied by three roots of the original *māl* and the original *māl* is returned.

Its rule is that you make your *māl* a *māl*, then you take away its third. There remains for you two-thirds of the *māl*. So multiply that by three roots of the original *māl*, and that is three things. This amounts for you to two cubes. So they equal the original *māl*, which is a *māl*. So divide the two cubes by the *māl*. It results in two things equal a dirham, since the two things multiplied by the *māl* yielded a *māl* that is the same as the two cubes. So you say: What number, multiplied by a *māl*, amounts to a *māl*? You find that it is a dirham. So the one thing is half a dirham. So the *māl* is a quarter of a dirham, so the desired *māl* is a quarter of a dirham. . . .

Various problems

[On mail.] If it is said to you that a courier leaves a city and he is ordered to travel twenty *farsangs*[30] per day. He goes five days, and then another courier is sent after him with the order to travel thirty *farsangs* per day. In how many days will he overtake him?

The rule for that: if you make the days that the second travels a "thing," then the days that the first travels will be "a thing and five." Then you multiply the count of days of the first by the number of *farsangs* he travels per day and the days of the second by the number of *farsangs* that he travels in a day. And one of the two numbers [resulting] is equated to the second. And you will have a hundred of number and twenty things equal to thirty things. And so remove the twenty from the thirty and divide the hundred by what remains [i.e., ten]. Your result is ten, and it is the value of the thing, namely, the number of days of the second courier, and the first travels the same [number of days] and five [more], i.e., fifteen days.

But if the question is how many days the first will go, then we make them the thing and the days of the second will be a thing less five. Then we multiply the days of each one of them by what he travels. We equate one of the two [resulting] numbers to the second. Your result is that the value of the thing is fifteen, and that is what the first travels. And the second is what the first travels, less five days, and that is ten.

[On troops.] Some troops go raiding, and the first of them [the raiders] gains as booty two dirhams, and they [the amounts each successive soldier gains] increase by threes. And the total of their booty is one hundred twenty-six [dirhams].

The rule for this is that you make the number of troops the thing. You subtract one from it, and there remains a thing less one. And multiply that by the increase, so there results three things less three dirhams. Then add to that the booty from the first, and you get three things less one [unit]. And that is what the [one who had the] greatest [booty] gained. And so add what the first gained to what the highest gained. It comes to three things and one [dirham]. Now multiply it by half the number of the troops, and you have a square and half a square and half a thing. And this is equal to what the troops gained as booty, i.e., one hundred twenty-six [dirhams].

Take two-thirds of each of these terms, and you have a square and one-third a thing is equal to eighty-four [dirhams]. Halve the number of things and multiply that by its equal, and add the result to the number of dirhams. You take the root of what results and subtract from it half the number of things. What you have left is nine, and it is the number of troops.

And if you wish to check this, multiply the number of troops less one by the excess, and add to the result what the first took, which is two. You get twenty-six, which is the greatest booty obtained. Now add the booty of the first to the greatest booty and multiply that by half the number of troops, and you obtain one hundred twenty-six, which is the total booty we mentioned at the outset.

And in this problem, and the remaining [problems] about troops, as well as the problems of the mail, only one procedure is used: the steady increase of the messengers.

[30] An ancient measure of length. It varies according to time and place, but it was (very roughly) about six kilometers.

Ten messengers. The distance traveled by the first is two, the gradual increase between the ten is three. What distance does the last of them travel, and how much is the total distance traveled by all of them?

The operation for this is that you multiply the number of messengers less one by the gradual increase. You add to the product the distance that the first travels, which gives to you the distance the last travels. Then you add the distance the first travels with the distance the last travels. You multiply the result by half the number [of messengers], and you obtain the total traveled by all messengers. Understand it!

[On wheat and barley.] You buy three units of wheat for ten dirhams and you buy, you know not how many, units of barley for twelve dirhams. You sell each unit of wheat for the value of [i.e., the value you paid for it] each unit of barley and each unit of barley for the value of each unit of wheat. And you have an excess of four dirhams.

The rule is that you take the [number of units of] barley as the thing and then divide the twelve dirhams by the thing, resulting in twelve divided by the thing, which will be the price of each unit of barley. Then divide the ten by three, the number of units of wheat, and you get three and a third, which is the price for which you buy each unit of wheat. Multiply the number of the barley, the thing, by the price for which you have bought each unit of wheat. You get three things plus a third of a thing, which is the value of the barley. Then multiply the units of wheat by the price you paid for each unit of barley, which is twelve divided by the thing, and you get thirty-six divided by the thing. Add all of this and the result will be the value of the wheat plus the value of the barley obtained in the sale. The result of the addition is three things plus a third of the thing plus thirty-six dirhams divided by the thing. This is equal to the value of the wheat, ten, plus the value of the barley, which is twelve, plus the four dirhams which was the excess, the sum total being twenty-six. Then multiply everything you obtained by the thing, the product of the thing by thirty-six divided by the thing being thirty-six. The result of all this is three *māl* plus a third of a *māl* plus thirty-six units (no longer a quotient) is equal to twenty-six things. Then take three-tenths of each term, and when you have done this, you obtain a *māl* plus ten units plus four-fifths of a unit is equal to seven things plus four-fifths of a thing.[31] Divide the [number of] things by two, and the result will be three and nine-tenths. Multiply this by itself, and it will be fifteen and a fifth and a tenth of a tenth. From this subtract the ten units and the four-fifths of a unit, leaving you with four units plus two-fifths plus a tenth of a tenth. Take its root, two and a tenth. If you add this to half the number of roots, you get six, which is the unknown quantity of barley.

[On dowries.] If it is said to you: a woman has married three husbands. The first has fixed for her as dowry an unknown amount. The second has fixed three times that of the first, and the third has fixed four times that of the second. The total is eighty. What amount has each of them fixed?

The rule for this is that you make what the first fixed [as dowry] for her the thing. [Then] what the second fixed for her is three things and what the third fixed for her is twelve. And so sixteen things are equal to eighty. And so, divide eighty by sixteen things and what the

[31] Note that the author has now reduced the problem to one of the canonical forms of quadratic equations.

first fixed for her is five, and the second fifteen, and the third sixty. Add up all of that and it is eighty, as required.

Fourth similar problem. If it is said to you: a woman has married three husbands. The first fixed as her dowry an unknown amount. The second fixed as dowry the root of that which she has given to the first, plus one dirham. The third fixed three times the amount of the second, plus four dirhams. The total is forty.

The rule for this is that you suppose the amount the first husband fixed is *māl.* Then what the second husband fixed as dowry is the root of the square plus one, i.e., a thing plus one dirham. And what the third husband fixed is three times that of the second plus four dirhams, i.e., three things plus seven dirhams. Then you have that a square plus four unknowns plus eight dirhams is equal to forty.

Apply *al-jabr* and *al-muqābala* and operate as for the fourth of the sixth problems. The result is that the first [husband] fixed sixteen as her dowry, the second (the root of that which the first fixed plus one) five, and the third (three times that which the second fixed plus four dirhams) nineteen dirhams. And if we sum what each of them fixed as her dowry, the result is forty, as was required, so understand [it].

[On problems of commerce.] If someone says to you: a man has a certain amount of capital. He trades with it, and he doubles it. He gives one dirham to charity. He trades with the rest, and he doubles it. He gives two dirhams to charity. Then he trades with the rest, and he doubles it. He gives three dirhams to charity. Nothing remains of his capital.

The rule for this is to make the capital the thing. You double it, as was said, and you give one dirham to charity. You have two things less one dirham. You do business with the rest and it doubles, so it will be four things less two dirhams. Your give to charity two dirhams, so you have four things less four dirhams. You do business with it and it doubles, so you have eight things less eight dirhams. From this you give three dirhams to charity and you have eight things less eleven dirhams. Nothing is left. Apply *al-jabr* and *al-muqābala* and the result is that the capital is one and three-eighths, which is what you wanted to know.

III. COMBINATORICS

1. AḤMAD IBN MUNᶜIM, *FIQH AL-ḤISĀB* (*ON THE SCIENCE OF CALCULATION*)

Aḥmad ibn Munᶜim al-ᶜAbdari (d. 1228) came from Denia, on the Spanish coast near Valencia, but spent most of his life in Marrakesh, probably at the Almohade court during the reign of Muḥammad al-Nāṣir (1199–1213). He wrote works on geometry and magic squares, but none of these is extant. The following two excerpts are from his major work *Fiqh al-ḥisāb* (*On the Science of Calculation*), the first on figured numbers and second on combinatorics.

Figured numbers have a long history, going back to the Pythagoreans in ancient Greece. They refer to arrangements of objects (e.g., pebbles or other elements) on a plane surface in the shape of polygons—triangles, squares, pentagons, and the like (Fig. III-1-1).

In the following selection on figured numbers ("numerical figures" in Ibn Munᶜim's phrase), Ibn Munᶜim draws attention to a number of properties of these numbers, stating how they can be generated recursively and giving a rule for calculating the sum of consecutive

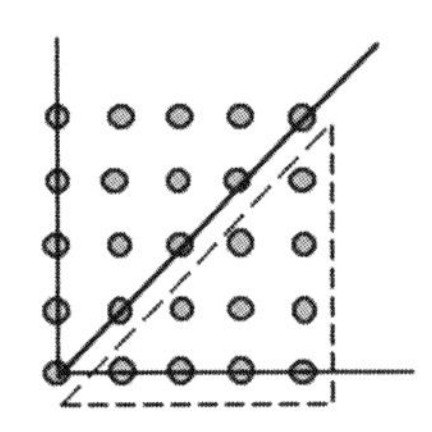

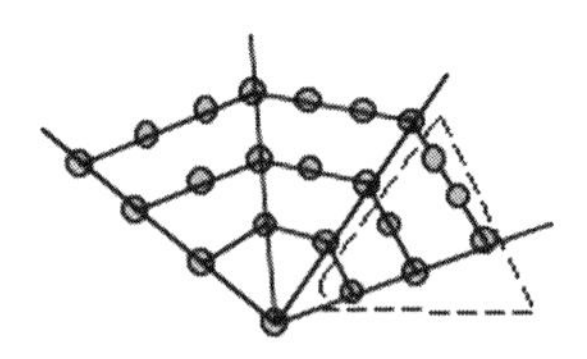

Fig. III-1-1. The figure shows, on the left, a square with five objects on each side and, on the right, a pentagon with four objects on each side. The significance of the triangles enclosed in dashed lines is explained in the text.

figured numbers of a given type. Since one such sum is the sum of square numbers, this result sets what appears to be an interesting, but isolated, result in the context of a wider set of results—something that has always been of interest to mathematicians.

Ibn Munᶜim's phrasing is sometimes a bit vague, and I have occasionally added words in square brackets to specify the meaning intended. However, this is done only once in each translation: the reader should then supply the same explanatory phrase each time it is needed.[1]

In general, Ibn Munᶜim calls all figured numbers having the same number of sides a "species" and, within a species (say, the hexagonal numbers) those having a different number of elements on a side are different "figures."[2]

[From the introduction to the section on figured numbers]

When I wanted to write, for my book, this part on numerical figures, I studied the works of the arithmeticians, and I saw that their books dealt deeply with this subject. . . .

When I had just finished studying a short treatise by Ibn Sayyid,[3] I noticed that he had inserted a problem about which he was mistaken. As a matter of fact, he indicated a method to calculate the sum of numerical figures of odd and even sides, and with even or odd values, although this method [of his] only allows calculating the sum of the squares of successive even natural numbers or the sum of successive odd natural numbers (. . .) I thought about it, and some excellent method appeared to me to determine, with geometrical proofs, the sums of numerical figures according to their natural divisions . . . I am going to present what Ibn Sayyid said about it and, in addition, what I have introduced.

Ibn Munᶜim next introduces a table (reproduced here as Table III-1-1).

[Section 9]

And, when the natural numbers are added in order, what is composed by them are the triangles in sequence, and from the triangles are composed the squares, and from the triangles with the squares are composed the pentagons,[4] and from the triangles with

[1] I thank both Professors Djebbar and Lamrabet for answering a number of questions about this work.

[2] The Arabic for "species" is *nau*ᶜ (a word Ibn Munᶜim also uses to denote the different sections of his book). The Arabic for what I have translated as "figure" is *ṣūra*, whose general meaning is "form" or "shape."

[3] 'Abd al-Raḥmān ibn Sayyid of Valencia (late 11th c.)

[4] This is illustrated by the triangles in dashed lines seen in Fig. III-1-1.

TABLE III-1-1

Row of sides	1	2	3	4	5	6	7	8	9	10
Row of triangles	1	3	6	10	15	21	28	36	45	55
Row of squares	1	4	9	16	25	36	49	64	81	100
Row of pentagons	1	5	12	22	35	51	70	92	117	145
Row of hexagons	1	6	15	28	45	66	91	120	153	190
Row of heptagons	1	7	18	34	55	81	112	148	189	235
Row of octagons	1	8	21	40	65	96	133	176	225	280
Row of enneagons	1	9	24	46	75	111	154	204	261	325
Row of decagons	1	10	27	52	85	126	175	232	297	370

the pentagons are composed the hexagons and from the triangles with the hexagons are composed the heptagons, and so on without end. And each of these figures is equilateral.

. . .

And, if we want to see the arrangement of their composition and decomposition to know their differences in length and width by the triangles and the figures, we set down [in the table] rows beginning [in the first row] from the unit according to the sequence of number: one, two, three, setting thus to work. Then you add the one with the two, [obtaining] three, which you place [in the second row] under the two in the [first] row, and before the three [in the second row] you place one, since the unit is, potentially, every figure. And three is in fact the first of the triangles. Then to three you add the number next to two, which is three, so six. And it is in fact the next triangle after three. Then you add the four to the six, so ten, and it is in fact the triangle next to the six. And so on, without end. And so it will be:

[Proposition] The triangles, according to their arrangements, will differ by the natural numbers, which are their sides. Each triangle differs from the one before it by its side. Then you compose the row of the squares, so, you add one [at the beginning of the second row] with the first triangle. In fact it is four, and so it is in fact the first square. Then you add the triangle that is three with the triangle that is next to it, six, so it is the square that is next to the four, and it is nine. And in that way you add each triangle with the triangle that is next to it and [in this way] the squares arise in sequence. Then the pentagons arise by adding the first potential triangle [1] to the first square [4] and it [5] will in fact be the first pentagon. Then [you add] the triangle that is next to it [i.e. next to the potential triangle], and that is three, to nine, and it will be twelve, the second pentagon. And it is thus that each pentagon arises. Then, according to this arrangement, arise the hexagons and the figures whatever their number of sides.

And I have drawn a table in which it appears how these figures arise in order. And so the first row is the natural numbers in a row. And the second row is the triangles that arise naturally [from it]. And the third row is the squares according to their nature according to what arises from the triangles. And the fourth row in it is the pentagons, which arise from the triangles with the squares. And, in the same way, the hexagons, heptagons and octagons. And so the rows arise, and this [table] is its figure.

Proposition: Every column passes through figures of different species, and [each entry in a column] differs [from the one above] by a triangle whose side is one less than its [the given figure's] side.

Example of heptagon, four on a side [thirty-four], therefore subtract from it the triangle whose side is one less than [a triangle of side] four, which is six. There remains twenty-eight, which is the hexagon of side four. We [then] subtract from it the triangle, which is six, and twenty-two remains, which is the pentagon of side four. We subtract from it six, and sixteen remains, which is a square of side four, from which we subtract six and there remains ten, which is triangle of side four.

And so you have learned this special property. [Now,] if any particular side in the column of natural numbers is specified to us, how do we determine the size of the equilateral figure of any species we want? And, conversely, if we are given the size of any particular species how is its side determined in the row of natural numbers.

For example: The side is four, how much is its octagon? Subtract the three from the eight[5] and there remain five, and so there are five differences and they come from the triangles whose side is one less than four, namely three, which is the triangle [with] six [elements]. And because there are five differences and each difference is six then the sum of the amounts of the differences is thirty. And we add to it the triangle [of side] four, which is ten, and that [the sum, 40] is [the number of elements in] an octagon [of side] four.

And the converse of that is: An octagon of amount forty [what is its side?] Make the unknown side the "thing."[6] And add up from one to the thing according to the sequence of [natural] numbers and let us denote its triangle[7] according to what we premised as "half *māl* and half thing."[8] And this is the triangle to which is added the differences that add up to the octagon. Then the triangle whose side is less by one is known, and so one is subtracted from the thing, to get "thing less one." So the triangle is "thing less one."

And according to what we premised, that is "half *māl* less half thing," which is the difference. Then the amount of the difference from the octagon to the triangle is [also] known, and it is five [of these] differences. You multiply it [the 5] by half *māl* less half thing and it will be two *māl* and a half *māl* less two things and half a thing. And if you add [to this] the triangle that is half *māl* and half thing it will be "three *māl* less two things is equal to the octagon," which is forty. So the thing turns out to be four, which is the side of the octagon.

And if you want to learn the number of the sides for the figure, in our example, [by] geometry:

[As before, we know that] three of the squares of the side of the octagon equals forty and twice the side of the octagon. And so the square of the side of the octagon is equal to thirteen and a third and two-thirds of the side of the octagon.

In the following, segment AB is taken to represent the unknown side of the octagon.

[5]There are five species after the third (the triangle) up to and including the octagon.

[6]Ibn Munᶜim is now using an algebraic approach to find the number of elements on the side of an octagon having 40 elements. The "thing" denotes what we would call the "unknown."

[7]That is to say, "the number of elements in the triangle whose side is the unknown number, x."

[8]Earlier in the treatise Ibn Munᶜim gave the rule for finding the sum $1 + 2 + \ldots + x$, namely, $x^2/2 + x/2$.

[Let us assume that] AG is two-thirds of one from the [point of view of] number. And the square of AB will be equal to the product of AB by AG (which is two thirds the side of the octagon) together with the product of AB by BG. And so the product of AB by BG is thirteen and one-third. And we divide AG into two halves at point K, and the product of AB by BG together with KG^2, is thirteen and four-ninths and that will be equal to KB^2 [by *Elements* II.6]. And so KB is three and two thirds; add AK, which is one third of a unit, to it. It adds up to AB, and that is four, the side of the octagon. [Problem ends.]

We now do what we are able to do of the properties of the table, and we say that these figures differ between species. The triangles differ from one another by the natural numbers, and these numbers differ by ones. So the triangles differ from one another by numbers that differ by ones. And the square areas differ by numbers that differ by two. And the pentagonal numbers differ by numbers that differ by threes. And the row of the hexagons differs among themselves by fours. And these increases are, in the triangles, the successive sides, and in all the other figures characteristics by which each figure differs over the one of the same type immediately preceding it in the row.

And other properties appear in this table; the more you investigate its nature [the more] its properties appear. For example, the triangle that is six, when the square which follows it in the [same] column is added to it and from the sum one side, three, is taken away, there remains the pentagon that follows it in the column, which is twelve. If the triangle six, is added to this it is eighteen, [and, again, if] the side, [three], is subtracted from that, there remains fifteen, which is the hexagon that follows it in the column, etc.

Proposition: Whenever you add a triangle and subtract from the sum its side these results the figure of a different species that is below it in the column. And the same holds for all other figures in a given column.

And when we added the units in the first of the matter there came about the numbers measuring lengths, which are similar to lines. Then we add the numbers [measuring lengths] in succession and there arise the planes in succession, and they are a collection of numbers similar to plane numbers. And so we now begin to want to add the planes in succession, and there result the numbers similar to solids.

And the following appears to us from the matters found in this table: If the first triangle is divided by its side, one, the result is one. Then if we divide [the first] two triangles by [the sum of] their sides, the result is one and one-third [$= (1 + 3)/(1 + 2)$] and if we divide three triangles, in order, by their sides, in order, these results one and two-thirds. And there results from the division of the first of them one and each differs from the previous by one third.

Then you investigate the squares in this way [and] you find that the results differ from one another by two-thirds. Then you look into the pentagons and you find that it differs by three-thirds. Then the matter is like that in [the remaining] figures, and so we know from this property found in it [the table] the way to proceed in all the figures.

And if we want to add up from triangle one to triangle five in sequence we subtract one, which is the first one, from five and four remains and this four is thirds, which is one and one-third. You add to it the first one and it is two and one-third. Then you multiply by it the sum of the numbers one through five (what is between the one and the five in the succession of the numbers) and what results is the sum of the triangles one to five.

And also, since the results for the squares differed by two-thirds we subtract from their number one and multiply the remainder by two-thirds. What results we add to it the one. Then we multiply the sum by the total number of sides [that is however many squares we wanted to add up] and it becomes [the sum of the square numbers].

The general method of working in all the figures is that you note the number of figures from the figure [in question] to the triangle[9] and for each figure, you take one-third. And the sum of the thirds is the number of thirds that is the result of dividing [the sum of] those figures [of a certain type] by the sum of their sides. Then you subtract the first one from the number of the figures and you multiply the remainder by the sum of those thirds and to the result you add the one and multiply the sum by the total of the sides. It results in the total of all the figures, from the one to the figure of whatever number [you chose].

And we multiply like that when [e.g.,] we want to add the hexagons from one to five according to the sequence of hexagons. And let us count from the hexagon to the triangle, and it is four figures. And we take one-third for each figure, so it will be one and one-third, and you recognize that the results of the divisions of [the sum of] the hexagons by [the sum of] their sides differ by one and one-third. Then you take away one from five and the remainder, four, is the number of sides that remain, which you multiply by the difference, which is one and one-third, and you get five and one-third. You add to it the one that you subtracted, which gives six and one-third. Then you add the natural sides from one to five, which is fifteen. You multiply that by six and one-third, which is ninety-five, which is the sum from hexagon one to hexagon five, in the sequence of the hexagons. And it is like that in all of this area of work.

Ibn Munᶜim follows up on the last part of this selection with more material on summing polygonal numbers of a given species. Specifically, for a given species of figured number, with n elements on each of its sides, he gives (for figures of types 3 to 10) a table of the coefficients of the cubic polynomial that expresses the sum of the figured numbers of that type with sides $1, \ldots, n$ [Fig. III-1-2].

Some of these coefficients turn out to be negative. For example, for the hexagonal numbers, row five, this table shows that the sum of these from one to seven points on a side is $(2/3)7^3 + (1/2)7^2 - (1/6)7$, that is, 252. In the table the negative sign is signaled by prefixing the word *illā* to the fractional coefficient, which (either in that form or simply as *lā*) was a common way of expressing the negative sign in the Maghrib and al-Andalus.

The four columns of the table [Fig. III-1-2] are headed (reading from right to left) "column of the figures," "taken of its [the side's] cube," "taken of its square," and "taken of it." In the rightmost column the rows are labeled, reading down from the first, "the sides," 'triangles," "squares,"..., "decagons." Note that the Arabic expression for –1/6 appears in the leftmost column of the fifth row. (The forms of the numerals, with the exception of 4 and 5, are the same as ours—since it is from the western part of the Islamic world that our numerals derive.)[10]

[9]The count includes both the triangle and the figure in question, so for the triangle, the count would be "one," for the squares "two," etc.

[10]The numeral "4" appears in the numerator of the fraction "–4/6" in the last column of row 8, and that for "5" appears in the numerator of "–5/6," directly below –4/6. Note the horizontal bars separating numerators and denominators of the fractions and, for example, in the last row of the second column, apart from the direction of writing, the fraction 1 1/3 is written exactly as we would write it today. However, the fraction 1 3/6 in the entry directly above that one is incorrect and should be 1 1/6.

جدول النسبة اشكال نوع نوع من الكبو أضلاعها

نسبة ماخوذة من القلع الاكبر

أسماء الاشكال	[illegible]	[illegible]	[illegible]
الاضلاع		١/٢	١/٢
المثلثات	١/٦	١/٢	١/٣
المربعات	١/٣	١/٢	١/٦
المخمسات	١/٢	١/٢	
المسدسات	٢/٣	١/٢	١/٦ V/
المسبعات	٤/٦	١/٢	٢/٦ V/
المثمنات	١	١/٢	٣/٦ V/
المتسعات	١ ٣/٦	١/٢	٤/٦ V/
المعشرات	١ ١/٣	١/٢	٥/٦ V/

Fig. III-1-2. From a manuscript provided by Prof. D. Lamrabet, with his kind permission.

TABLE III-1-2

Sum	The example written in a table										
1	1	Line of pom-poms of ten colors									
10	9	1	Line of pom-poms of nine colors								
45	36	8	1	Line of pom-poms of eight colors							
120	84	28	7	1	Line of pom-poms of seven colors						
210	126	56	21	6	1	Line of pom-poms of six colors					
252	126	70	35	15	5	1	Line of pom-poms of five colors				
210	84	56	35	20	10	4	1	Four colors			
120	36	28	21	15	10	6	3	1	Three colors		
45	9	8	7	6	5	4	3	2	1	Two colors	
10	1	1	1	1	1	1	1	1	1	1	One color
all	10th	9th	8th	7th	6th	5th	4th	3rd	2nd	1st	

In [Berggren, 2007], I translated Section 11 of the *Fiqh*, titled "On Denumerating the Words Such That a Person Cannot Express Himself without One of Them," up through Problem 5. The section deals with the problem of counting the number of possible words that can be formed from the Arabic alphabet with its 28 letters that obey three rules of Arabic orthography, rules that Ibn Munʿim states in his introduction. Here, I recapitulate the first five problems and then continue with Problem 6.

Ibn Munʿim began by posing a problem whose connection with that of counting possible words is not immediately obvious.

[Section 11]

[Problem 1] Ten colors of silk being given, with which we want to make tassels[11] of [respectively] one, two, three etc. colors, we want to know what is the number of tassels of each type, the colors of each tassel being known. Or, what is the number of all the tassels taken together, account being taken of the different numbers of colors of the tassels. We arrange the colors, one after another, in a line, according to the size of the page, as in the example [shown in Table III-1-2].

The bottom row lists the colors from right to left as first, second, and so forth. Above that the row labeled "one color" records the fact that if the tassel only has one color, there is only

[11] In the earlier translation, I translated the French *houppes* as "pom-poms." This translation uses "tassels" instead.

one way to use the first, or the second, or the third, and so forth. And in the last column (headed "sum") the sum of these possibilities, ten, is recorded. The next row up, labeled "two colors" has nothing above "first," since one needs two colors. Above "second" it has 1, because there is only one way to make a tassel of two colors from only two colors. Moving to the next column, since the third color could be combined with either the first or second color one enters, above "third," the number 2. And similarly for the other entries in this row, the sum total of all entries, 45, being recorded in the leftmost column.

Continuing up, row by row, the "line of tassels of eight colors" has 1 above "eighth." Then if an additional (ninth) color is given, that color can obviously replace any one of the eight colors counted in the previous cell, which gives eight more possibilities. There are now nine $(8 + 1)$ possible tassels with only 8 colors chosen from the first 9 colors. Going now to ten colors, the possible choices of eight colors that include the tenth color is exactly the number of tassels using 7 of the first 9 colors. That is, obviously, the sum of the number in the first three cells of the row for 7 colors, namely, $28 + 7 + 1$ (i.e., 36). It is thus apparent how the entire chart is generated, working upward and from right to left, from the first row of all 1s.

[Problem 2] We want to know a canonical procedure to determine the number of permutations of the letters of a word of a known number of letters in which no letter is repeated.

Ibn Muʾnim works out, case by case, the solution for words having at most 5 letters and obtains for each case the expected factorial. Problem 3 then shows how to solve the same question as in Problem 2, but allowing letters to be repeated a known number of times.

For the next problem one must know that all 28 letters of the Arabic alphabet are consonants, and words are built on triliteral roots.[12] So, for example, the triliteral root k-t-b carries the basic meaning of "writing." One of three vowels (a, i, u), not considered as being letters and indicated by marks above or below the letters, is attached to the consonants to build words from a root. Thus *kataba* means "he wrote" and *kutiba* means "it was written." If a vowel is not attached to a consonant, then a sign (a small circle, called *sukūn*) is written above the consonant. For example, the quadriliteral word *barhana* (to prove/demonstrate) has a *sukūn* over the "r." However, a word cannot have two consecutive *sukūn*s, nor can it begin with a *sukūn*.

[Problem 4] Find the number of configurations of a word of which the number of letters is known, taking account of the vowels and *sukūn*s that succeed one another on the consonants.

Table III-1-3 records the results of Ibn Munʿim's solution to this problem for words having anywhere from one to ten letters.

TABLE III-1-3
Number of configurations of a word by the succession of vowels and *sukūn*s but not according to permutations of its letters.

10 Letters	9	8	7	6	5	4	3	2	1
507,627	133,893	35,316	9315	2457	648	171	45	12	3

[12] In relatively few cases, a word may be built on more or fewer than three consonants.

[Problem 5] Having shown this, we go back to our problem: To know the number of words composed of letters of the alphabet with the smallest having three letters and the largest ten.

We describe the procedure, to start, when in the word no letter is repeated. The method here consists of considering the number of letters of the alphabet as being colors of silk and asking: "How many tassels do they make so that the number of colors of each tassel is, for example, three?" You obtain a result, which you keep. Since you have supposed that the number of colors of a tassel is three you put, on a line, the sequence of whole numbers from one to three, thus: 1, 2, 3. Then you begin by multiplying one by two, then the product by three, which follows on the line. And if the following were another [number], you would multiply the product by it. Then you multiply the result by the number of configurations of the word, according to the vowels and the *sukūns*. Then you keep the result and you multiply it by the number of tassels of silk that are, each, of three colors (as we have supposed in this example). The result is the number of words of the alphabet, each of which has three letters. You proceed in the same manner for more or fewer than three letters. Then you sum the number of groups of letters of the whole,[13] and it will be the number of words composed of letters of the alphabet, having [at least] three letters and not more than ten, and in which no letter is repeated.

[Problem 6] A preliminary proposition for that which we plan to establish: being given a collection of silk threads, of known colors, we want to make tassels so that in each one there will be a given number of threads of a given number of colors.

For example, given a collection of silk threads of ten colors, we want to make tassels so that each one has three threads of two colors, of which one is of one color and the other two are of another color but no two different tassels have the same number of filaments of the same color.

If you think about the problem you find that the number of tassels is twice the number [obtained] if you did not mention threads and said: "We want to make tassels so that there are two colors in each one [tassel]." It is thus because the two threads may be of the first color in one tassel and of the second color in another one.

Speaking generally, whether the number of threads of [each] color be larger or smaller, the number of tassels made by the colors of one tassel is equal to the number of configurations of letters of a word of which the number of letters is equal to the number of colors of the given tassels and of which the number of repetitions of each repeated letter is equal to the number of colors having the same number of threads.[14]

For example, given a tassel with eight threads of five colors of which two are one color each and six are, two-by-two, of one and the same color. How many tassels of eight threads of five colors are there so that two of the threads are each of a different color and six threads fall into pairs, each pair a different color?

If you want to solve this problem you arrange the threads in front of you according to [Table III-1-4].

[13]That is to say, that one sums over the different possible numbers of letters in the words (from three to ten in this case).

[14]That is, to each group of threads having a given color one assigns a letter, and all groups having the same number of threads are assigned the same letter.

TABLE III-1-4

Thread of first color	*Thread of second color*	*Thread of third color*	*Thread of fourth color*	*Thread of fifth color*
g	g	k	k	k

Then you think about the problem and you note that the two first colors have the numbers of their threads equal. You write, then, under each of these, the same letter, say, "g, g." Then you note that the three other colors have the numbers of their threads equal. You write, equally, under each of these, the same letter, say, "k, k, k." One is, thus, led by this figure to the combination of five letters, of which three repeat [one letter] and two repeat [another letter]. The [number] of permutations [resulting from this combination] is then equal to the number of tassels composed of the colors of one and the same tassel, i.e., ten.

Proof of this: This comes from the fact that for each permutation of the letters you can find a combination of colored threads, and conversely. In fact, corresponding to a permutation in which the first and third letters are the same, there is a tassel in which the number of threads of the third color is equal to that of the first color [in the given tassel].

The demonstration is analogous for the rest of the colors and it will be thus, however many there may be.

Section:

[Definition]: The "combination," in our convention here, is a collection of letters.[15]

[Problem 7] Explicitly: Given a nine-letter word, made up, to begin with, of five [distinct] letters, of which two do not repeat, two repeat twice, and one repeats three times, how many combinations result from these letters?[16]

You write them [the nine letters] in this way:

a b c e i
c e i
i

and below columns having the same number [of letters] you write identical letters.

a a d d r

One is then taken back to a combination of five letters, of which two are repeated twice. The number of permutations of the letters of this figure will be, then, the number of combinations of nine letters of which two are not repeated, two repeated twice, and one repeated thrice—thirty combinations.

[15] As the following arguments show, the combination may contain repeated letters, and two combinations are equal if each has the same letters appearing the same number of times. So that {a, a, a, g, g} and {g, g, a, a, a}, for example, represent the same combination.

[16] That is to say, given five distinct letters, say, a, b, c, e, and i, how many words can you make in which one of the letters repeats three times, two repeat twice each, and two do not repeat?

[Problem 8] We return to our problem. We say that the words of which the letters repeat and which are composed of the letters of the alphabet are, taking account of repetitions [of their letters], either words of ten letters or of less than ten (and this down to two). . . .

One therefore has:

Nine distinct letters and the tenth repeats one of the nine, or
Eight are distinct, the ninth and the tenth repeat just one letter among the eight or two, or
Seven are distinct and the three remaining letters repeat one letter of the seven, or two, or three, or
Six are distinct and four repeat just one letter, or even two. And if they repeat two letters, this will either be once for one of them and thrice for the second, or twice for the one and twice for the other. Or [the four] repeat three letters and this will be twice for one and once for each of the other two [of the three repeated] letters. [Or, they repeat four letters, once each.]
Five are distinct and the five others repeat either one or two of the five letters. And if they repeat two, this will be either once for the one and four times for the second, or twice for the one and three times for the second. Or the five letters repeat three letters, once for one and twice for each of the other two, or three times for the one and once for each of the two others; or the five letters repeat four letters, twice for one of them and once each for the remaining three. Or the five repeat five letters, once each.
Or there are four distinct . . .
Or there are three distinct . . .
Or there are two distinct and eight repeat one and the same letter, or two—in such a way that one of them is repeated once and the second is repeated seven times, or, . . . or one of them is repeated four times and the other is also repeated [four times].
And if nine letters repeat one letter then the ten are simply the repetitions of a single letter.

Ibn Munʿim now begins a case-by-case treatment of the above.

If the tenth letter repeats one of the nine others, there will be, for the ten letters, nine combinations, each [having] ten letters of which one letter is repeated twice.[17]

If the ninth and the tenth repeat a single letter of the eight letters, then there will be, for the ten letters, eight combinations, each having ten letters and each having one letter repeated three times.

But, if the ninth and tenth repeat two of the eight letters, there will be, for the ten letters, twenty-eight combinations—each of ten letters and having two letters repeated, each twice. The way to obtain these twenty-eight combinations is to say: A tassel with ten threads of eight colors of silk, with six threads each of one color and four, in pairs of the same color. How does one obtain tassels, each of ten threads of eight colors, with six, each of one color and four made up of pairs of the same color? One obtains this as in the preceding [discussion].

[17]This makes it clear that a "combination" is not a word but simply a list of letters that may be written down in a number of ways to make up a word.

In this way one determines the number of the species of combinations, the procedure having been already explained. The way to determine this is to consider the distinct letters as colors and those that are repeated as threads [of a given color], the calculation being done as previously.

[Problem 9] This having been shown, we return to our remarks, saying: We want to know the number of words which are such that a human being cannot say anything but one of them, knowing that the smallest word must have one letter and the largest ten, the ten letters being, perhaps, all repeated or, perhaps, all distinct, or, perhaps, being in part repeated and in part distinct—however the person may pronounce them.

As for when all the letters are distinct, we have already shown how to enumerate them.

Now we speak of those of which the letters are repeated, and begin with those which, taking account of repetitions, have ten letters. Then we move on to those that are less in number until we arrive at those which, taking account of repetitions, have two letters. For those which, taking account of repetitions, have ten letters, we have subdivided [the calculation thus].

We begin with those we treated first, the word of ten letters, of which nine are distinct and the tenth repeats one of the nine. We say: Given twenty-eight[18] colors of silk, we want to make tassels so that each tassel has nine colors, which correspond to the number of distinct letters [in the word]. There follows a result which will be "the first" and which you must keep.

Then you say: Given ten letters, of which nine are distinct and one of which is repeated twice, how many combinations are possible for the ten letters? One obtains, as previously, nine combinations of ten letters of which nine are distinct and one of which is repeated twice. You keep this and it will be "the second kept."

Next, you determine, as previously, the number of permutations of letters of a word of ten letters of which one is repeated twice. You obtain a result which you keep and will be "the third kept."

Next, you determine, as previously, the number of configurations of a word of ten letters according to the vowels and *sukūns*. You will keep the result, and it will be "the fourth kept," that is to say, the number by which, taking account of the vowels and *sukūns,* the number is multiplied.

Then you multiply the first kept result by the second, the product by the third, and [that] product by the fourth. The result is then the number of words such that one cannot pronounce a word of ten letters, of which one is repeated twice, without it being one of them. . . .

Equally, we determine the number of words composed of letters of the alphabet, each having ten letters, two of which are each repeated twice. We say: Given twenty-eight colors of silk we want to make tassels so that each tassel has eight colors, which correspond to the number of distinct letters [in the word]. You will obtain a result which you will keep and will be "the first kept."

Then you say: Given ten letters, of which eight are distinct and the ninth and tenth repeat twice [one letter] among them [the eight]—how many combinations issue from

[18]These represent the 28 letters of the Arabic alphabet. Ibn Munʿim often uses this phrase.

these ten letters? You obtain, as previously, twenty-eight combinations, each of ten letters of which eight are distinct and two repeat two letters among the eight. You keep the number and it will be "the second kept."

Next you determine the number of permutations of a word of ten letters, of which eight are distinct and two repeat twice among them. You will obtain a result which you will keep and will be "the third kept."

Next, you determine the number of configurations of the letters of a word of ten letters according to the vowels and *sukūns*. You will keep this, and it will be "the fourth kept," the best way being to keep the result from the preceding question, so that you do not get tired by calculating [this "fourth kept"] each time.

Then, you multiply the first kept result by the second, the product by the third, and [that] product by the fourth. The result is then the number of words such that one cannot, in any way, pronounce a word of ten letters, of which eight are distinct and two repeat twice, without it being one of them.

You determine, in the same way, the species of all the combinations of ten letters containing repetitions. And you will proceed in the same way for [words of] nine letters with repetitions, for those of eight with repetitions, and thus up to two letters. Then you sum everything and to it you add that which you have obtained previously as the number of combinations without repetition of words of ten letters. That which you will obtain will be the number of words such that a human being cannot speak without pronouncing one of them.

The eleventh section is finished. May God shed his benefits on our lord, Muḥammad, on his family, and his companions, and may he grant them peace.

[Examples of using the tables]

Since the science of calculation is the privileged domain for passing from general rules to particular cases and because we have treated general procedures in the eleventh section of the first chapter of this book, leaving aside examples because of lack of time, we have composed, when we had some more time (which happened after the work had been recopied and was in the hands of students), the examples that follow and we have tackled the particular aspects of the section, in the form of a complement that we have joined to the examples at the end of the section. May God grant us His assistance, there being no other God than He.

[How to Use Table III-1-5]

Locate in the table, on the "Line of Successive Integers," the number of letters in your word and enter by the column corresponding to this number. Then, you locate in the table, on the Column of Successive Integers, the number of letters of your language, which is twenty-eight for the Arabic language, and you enter by the line corresponding to that number. Then, the number contained in the cell where this line meets the column where you have already entered is the number of combinations of words made up of distinct letters of your language in the number which you have supposed for that word.

TABLE III-1-5
For determining the number of combinations, one-by-one, etc. . . . , up to ten, of letters of an alphabet with twenty-eight letters.[18]

1	2	3	4	5	6	7	8	9	10
1									
2	1								
3	3	1							
4	6	4	1						
5	10	10	5	1					
6	15	20	15	6	1				
7	21	35	35	21	7	1			
8	28	56	70	56	28	8	1		
9	36	84	126	126	84	36	9	1	
10	45	120	210	252	210	120	45	10	1
11	55	165	330	462	462	330	165	55	11
12	66	220	495	792	924	792	495	220	66
13	78	286	715	1287	1716	1716	1287	715	286
14	91	364	1001	2002	3003	3432	3003	2002	1001
15	105	455	1365	3003	5005	6435	6435	5005	3003
16	120	560	1820	4368	8008	11,440	12,870	11,440	8008
17	136	680	2380	6188	12,376	19,448	24,310	24,310	19,448
18	153	816	3060	8568	18,564	31,824	43,758	48,620	43,758
19	171	969	3876	11,628	27,132	50,388	75,582	92,378	92,378
20	190	1140	4845	15,504	38,760	77,520	125,970	167,960	184,756
21	210	1330	5985	20,349	54,264	116,280	203,490	293,930	352,716
22	231	1540	7315	26,334	74,613	170,544	...	...	...
23	253	1771	8855	33,649	100,947	...	...	...	...
24	276	2024	10,626	42,504	134,596	...	...	...	...
25	300	2300	12,650	53,130	177,100	...	...	...	...
26	325	2600	14,950	65,780	...	...	...	...	...
27	351	2925	17,550	80,730	...	...	...	...	...
28	378	3276	20,475	98,280	...	...	...	...	...
1 letter	2 letters	3 letters	4 letters	5 letters	6 letters	7 letters	8 letters	9 letters	10

The 252, for example, in the first row of Table III-1-7 below arises from the fact that if each of two distinct letters are repeated five times then the number of possible 10-letter words that can be formed from these two letters is simply the number of ways one can choose five of the ten places where one of the two letters will be repeated, which is $C(10, 5)$, that is, 252.

[18]This table represents $C(n, p)$, $1 \leq n \leq 28$, $1 \leq p \leq 10$. For example, the entry 126 in the (9, 4) position means that 126 four-letter "words" can be made from a nine-letter alphabet. Note the binomial coefficients appearing in the first few lines of the table.

TABLE III-1-6
Configurations of words in which all the letters are different or in which only one is repeated.[19]

	1	2	3	4	5	6	7	8	9	10
10 letters	3,628,800	1,814,400	604,800	151,200	30,240	5,040	720	90	10	1
9	362,880	181,440	60,480	15,120	3,024	504	72	9	1	
8	40,320	20,160	6,720	1,680	336	56	8	1		
7	5,040	2,520	840	210	42	7	1			
6	720	360	120	30	6	1				
5	120	60	20	5	1					
4	24	12	4	1						
3	6	3	1							
2	2	1								
1	1									

TABLE III-1-7
Number of permutations of letters of a word [of n letters, $4 \leq n \leq 10$], with only two letters repeating [respectively, k and p times, $2 \leq k, p \leq 8$; $4 \leq k + p \leq n$].

Column of the number of repetitions of each of the two repeated letters in the word.	10 letters	9	8	7	6	5	4
5,5	252						
6,4	210						
5,4	1,260	126					
4,4	6,300	630	70				
7,3	120						
6,3	840	84					
5,3	5,040	504	56				
4,3	25,200	2,520	280	35			
3,3	100,800	10,080	1,120	140	20		
8,2	45						
7,2	360	36					
6,2	2,520	252	28				
5,2	15,120	1,512	168	21			
4,2	75,600	7,560	840	105	15		
3,2	302,400	30,240	3,360	420	60	10	
2,2	907,200	90,720	10,080	1,260	180	30	6

[19] The entry in row n, column k is $n!/k!$.

TABLE III-1-8
Configurations of letters of words in which four letters are repeated.

Line of the repetitions of each letter in the word				
3 3 2 2	4 2 2 2	3 2 2 2	2 2 2 2	
			2,520[20]	8 letters
		7,560	22,680	9 letters
25,200	18,900	75,600	226,800	10 letters

[Using Table III-1-11 below]
A method for determining the number of combinations coming from a word of which the number of letters is known and in which some letters repeat a known number of times.

You locate in the column of numbers the letters repeated and the number of repetitions of each of these letters. Then you enter by the corresponding line. You enter equally by the column of words. Then the number in the cell where the line and the column meet is equal to the number of combinations coming from the word that you have considered and in which some letters repeat. This is what we wanted to show [i.e., explain].

[Application] Method for determining the number of words made up of letters of the Arabic alphabet, of which the largest (taking account of affixes and repetitions) has ten letters and the smallest has only one.

First we deal with words composed of ten distinct letters. You enter as indicated in the table that has been prepared for this [Table III-1-5] and you obtain 13,123,110, which you keep first. This is the set of words formed from combinations of letters of which each is made up of ten distinct letters.

Secondly, from the table of permutations of a word [Table III-1-6], you take the number of permutations of a word of 10 distinct letters. 3,628,800, which is the second number you keep.

Then, from the table of vowels and *sukūns* that succeed one another on the letters [Table III-1-3], you take that which corresponds [to the number of configurations] of a word of ten letters. You obtain 507,627, which you keep, and which will be the third kept.

Then you multiply the first result by the second, and then the product by the third, and you obtain the number of words of ten letters, all distinct, formed starting with twenty-eight letters [of the alphabet] and it is the result which we seek.

Ibn Munʿim then lists all the other cases (nine distinct, . . . , two distinct, one only), then he begins to list the cases for ten letters with one repeated, then two, etc, then the cases for nine, etc. He then continues:

We take the example of a word of nine letters of which two are not repeated, two are repeated twice, and one is repeated three times. So it has five distinct letters. You calculate the number of combinations of five elements formed of distinct letters of the [Arabic] alphabet, taking it from [Table III-1-5], and it is 98,280 [= $C(28,5)$], which you keep first.

[20]For example, this value is $8!/(2!2!2!2!)$.

TABLE III-1-9
Permutations of letters of words of at most ten letters, and at least six, with repetition of three of those letters.

Column of the number of repetitions of each of the three letters in the word	10	9	8	7	6
3 3 4	4,200				
3 3 3	16,800	1,680			
2 4 4	3,150				
2 3 5	2,520				
2 3 4	12,600	1,260			
2 3 3	50,400	5,040	560		
2 2 6	1,260				
2 2 5	7,560	756			
2 2 4	37,800	3,780	420		
2 2 3	151,200	15,120	1,680	210	
2 2 2	453,600	45,360	5,040	630	90

TABLE III-1-10
Permutations of letters of a word with ten letters, with repetition of five letters.

	2 2 2 2 2
10 letters	113,400

Then [from Table III-1-9] you calculate the number of permutations of your word of 9 letters which repeat according to your hypothesis. You obtain 15,120, which you keep second.

Then you take from the table [Table III-1-3] of vowels and *sukūns* which follow one after the other on the letters [the number] that corresponds to your word of nine letters. You obtain 133,893, which you keep third.

Then you extract from the table of combinations [Table III-1-11, row 2 from the bottom, column 2, in the fourth subtable] that which it needs for your word as combinations. You obtain 30, which is the fourth result.

Then you multiply the first result by the second, the product by the third, and this last product by the fourth. You obtain 5,968,924,232,544,000 [a number just short of 6 quadrillion!]. That is the number of words of nine letters etc. And this is what we wanted to explain.

A Problem by Way of Example:

We want to know how many words of nine letters one can form with five distinct given letters of which two are not repeated, two are each repeated two times, and one is repeated three times.[21]

[21] In this part of the example, the letters are given in advance, but only the pattern of repetitions concerns us. The specific letters repeated two or three times are of no concern.

TABLE III-1-11

Number of combinations derived from a word in which some letters are repeated [an even number of times].

Line of the number of letters repeated in the word.	Words of 10	9	8	7	6	5	4	3	2 letters
2 2 2 2 2	1								
2 2 3 3	6								
2 2 2 4	4								
2 2 2 3	20	4							
2 2 2 2	15	5	1						
2 4 4	3								
3 3 4	3								
2 3 5	6								
2 2 6	3								
2 2 5	12	3							
3 3 3	4	1							
2 3 4	24	6							
2 2 4	30	12	3						
2 3 3	30	12	3						
2 2 3	60	30	12	3					
2 2 2	35	20	10	4	1				
5 5	1								
4 6	2								
3 7	2								
2 8	2								
4 5	6	2							
3 6	6	2							
2 7	6	2							
4 4	6	3	1						
3 5	12	6	2						
2 6	12	6	2						
3 4	20	12	6	2					
2 5	20	12	6	2					
3 3	15	20	6	3	1				
2 4	30	20	12	6	2				
2 3	42	30	20	12	6	2			
2 2	28	21	15	10	6	3	1		
10	1								
9	2	1							
8	3	2	1						
7	4	3	2	1					
6	5	4	3	2	1				
5	6	5	4	3	2	1			
4	7	6	5	4	3	2	1		
3	8	7	6	5	4	3	2	1	
2	9	8	7	6	5	4	3	2	1

You multiply the number of permutations of these letters, 15,120,[22] by that which corresponds to it in the table of vowels and *sukūn*s, and which is 133,893.[23] The result will be [2,024,462,160].

If, among the letters that are given, those that are repeated are also given,[24] that is to say the given letters being, for example, c, d, e, g, i, the non-repeated letters are c and d, those which are repeated twice are e and g, and the one repeated three times is i, the letters being set out as follows–c, d, e, e, g, g, i, i, i–then the result of the [preceding] product is your sought result. But, if the letters repeated are not given, you multiply the preceding result by the number of combinations coming from these repeated letters, which is 30. Then you obtain [60,733,864,800], which is the result demanded. And this is what we wanted to explain.

The end. May the benediction of God be on our lord, Muḥammad, on his family and his companions and may he grant them his salvation. There is no intercession or power without God, the most high and omnipotent.

2. IBN AL-BANNĀʾ ON COMBINATORICS, *RAISING THE VEIL*

In this selection from his commentary on the section on addition of his *Talkhīs*, Ibn al-Bannāʾ gives rules for computing the number of possible k-letter words it is possible to form from an alphabet of n letters. (No letters are allowed to be repeated.) He begins by counting the number of combinations of n letters taken k at a time and then achieves his goal by multiplying these numbers by the number of permutations of k distinct letters. (His opening example, of counting three-letter words, would be of special relevance to the Arabic language, since the vast majority of words in Arabic grow out of a three-letter "root," or "radical," as it is usually called. However, two letters of a given radical may be the same.)

It is of interest that he introduces the topic as being an application of the rule for computing the sum of consecutive squares, and he phrases the rules in terms of summing triangular numbers—which he refers to simply as "triangles." In fact, Section I-1 presented Ibn al-Bannāʾ's rules for summing squares, and earlier in the section excerpted here, he notes that the sum of successive triangles from side 1 to side n is equal to the sum of successive odd squares from 1 to n^2 when n is odd and is equal to the sum of successive even squares from 1 to n^2 when n is even.

The summation of squares is used in the determination of three-letter words for counting the contents of languages and for similar things. For example: How many three-letter words are there made up of letters of the alphabet in one figure without permuting them [i.e., without counting permutations of a given choice of three letters as being different words]? In fact, the number of three-letter words is equal to the sum of the triangles of which the side of the last is the number of the letters of the alphabet less two; and the sum of the triangles is obtained by multiplying the side of the last by the product of the two integers which follow it, and then taking one-sixth of the result, as one

[22] As in the problem above.

[23] Again, as in the problem above.

[24] Here, unlike the first part, one chooses which of the five letters are not repeated and which are repeated two or three times.

does for the summations of the squares of the odd numbers and the squares of the even ones.

This is because the number of two-letter words is obtained by multiplying the given number of letters [in the alphabet] by half of the second number before that one, beginning with the given one.[25] The number of three-letter words is obtained by multiplying the number of two-letter words by a third of the third number before the given number, beginning with the given one. The number of four-letter words is obtained by multiplying the number of three-letter words by a quarter of the fourth number before the given number, beginning with the given one. And thus, in general, you multiply the number of combinations preceding the desired combination by the number preceding the given number whose distance from it is equal to the number of combinations sought, and of that result, one takes the part given by the number of the sought combinations. The reason for this follows clearly from this chapter.

As for the number of two-letter [combinations], it is equal to the sum of the consecutive integers from one up to the number preceding the given number. As to the number of three-letter [combinations], to each of the two-letter combinations is associated one of the elements of the remaining set of letters. The number of three-letter combinations thus obtained is equal to the product of the number of two-letter [combinations] by the given number of elements less two, which is the third number before the given number. Since the three-letter combinations come from distinct two-letter [combinations], it follows necessarily that there are three repetitions of the same three-letter combination, itself and two permutations. For example, if one combines a and b with g, this is the same as combining a and g with b or combining b and g with a. These three three-letter combinations are actually the same combination, and their number is three because of the disposition of pairs of letters. It is therefore necessary to take a third of the number of two-letter [combinations] and multiply it by the number of given elements, or to multiply the number of two-letter [combinations] by a third of the number of elements.

As to a four-letter [combination], it contains four of the three-letter combinations, since the two-letter [combinations] are six in number, and we take the product of this with a third of the third number before four [i.e., two]. There results four, which is, as we have indicated, the number of three-letter combinations contained in the four-letter combination. From each three-letter combination, combined with one of the remaining letters of the set of letters, there results four similar figures, which do not differ except in order. Thus, it is necessary to take one-quarter of the result. The same, for the five-letter [combinations]—one necessarily gets five repeated figures because each five-letter combination contains five four-letter [combinations]. Since the combinations are such that for each, one letter is suppressed, thus the number of distinct combinations corresponds to the number of letters of the word.

It follows from this, that if a set of numbers is given, and we wish to know the number of combinations from these having a given number of elements, we take the product of the numbers of the sequence of integers starting from the greatest term which is the number of elements of this set, and take the number of terms equal to the number of

[25]Thus, since it begins the counting with the given number of letters of the alphabet in question (say n) the "second number before that" would be $n - 1$.

elements of the combination. Further, we put as a divisor of this product the successive numbers of which the greatest term is the given number [of letters in each combination] and which begins with one and with two. Then we cancel the common terms between the first numbers and the second; and when we do that, all the second numbers are cancelled; then you multiply the remaining first numbers together and you get the combinations from this set of numbers.

An example of this: if one is given twenty eight letters, how many five-letter combinations can one find? Proceed by what we have said. The result will be, in this example, the product of five by twenty four, multiplied by thirteen, then by nine, then by seven.[26] We deduce necessarily from this, that being given two arbitrary successive integers, the result of the product of one of them by half of the other is the number of two-letter combinations contained in the greatest of the two, and this is the triangle of the smaller of the two, after that which we have said. Being given also three successive integers, the result of the product of one of them by the half of the second and by the third of the third is equal to the number of three-letter combinations in the largest of these three. It is also the sum of the successive triangles up to the triangle of the smallest of the three numbers, and it is equal to the sum of the squares of successive odd numbers beginning with one up to the smallest of the three numbers if it is odd, or also equal to the sum of the squares of successive even numbers beginning with two up to the smallest of the three numbers, if it is even, as is apparent by induction. This is why what we have said in this work about the method relative to the summation of odd squares and of even squares proves necessary.

As to the combinations one obtains by permutation from one figure having a given number of elements, you take consecutive integers beginning with two or with one, and such that the last term is equal to the given number of elements, then one takes the product, and the result is the number of permutations coming from this single combination of elements. In fact, two letters contain two figures, a first figure and a permuted one. If one then adds a third letter, it will be in each of the two figures, either as the first, the middle, or the last, and that gives six figures. If one then adds a fourth letter, it will be in each of these six figures, either as first, as second, as third, or as fourth; so we have twenty four figures coming from one four-letter combination.

Thus, the two-letter permutations are two, the three-letter ones are the result of the product of two by three, the four-letter ones are the result of the product of two by three by four, and so on for the following numbers. It appears clear from this, that the product of two consecutive integers is equal to the number of combinations two by two, with their permutations, contained in the greatest of the two integers, and that the product of three consecutive integers is equal to the number of combinations three by three, with their permutations, contained in the greatest of the three integers, and that the product of four consecutive integers is equal to the number of combinations with their permutations contained in the greatest of the four integers. And this is true for all numbers after this.

[26] $C(28,5) = \frac{28 \cdot 27 \cdot 26 \cdot 25 \cdot 24}{1 \cdot 2 \cdot 3 \cdot 4 \cdot 5} = 5 \cdot 24 \cdot 13 \cdot 9 \cdot 7.$

3. SHIHĀB AL-DĪN IBN AL-MAJDĪ, ON ENUMERATING POLYNOMIAL EQUATIONS

Shihāb al-Dīn ibn al-Majdī, who died in 1447, was an astronomer who was responsible for regulating the times of the five daily prayers at al-Azhar mosque in Cairo.[27] He was a prolific author whose works were still studied in Egypt in the nineteenth century. He was also a competent mathematician who, among other treatises, wrote a commentary on Ibn al-Bannāʾ's *Talkhīs*, whose section on enumeration appears earlier (see section III-2).

The section from Ibn al-Majdī's work excerpted here deals with the problem of enumerating all possible polynomials in one variable of a given degree. One already finds, in al-Khwārizmī's *Algebra* in the early ninth century, a list of the six possible equations of degree two or fewer—three of them having only two terms and another three having three terms.[28] Omar Khayyam, who died in 1131, listed the twenty-five possible equations of degree at most three in his *Algebra*, and very likely it had been done before him. Ibn al-Majdī's work continues the enumeration to higher powers of the unknown, and he uses the word "species" to refer to the monomial summands. So "numbers" make up one species, "things" a second species, "squares" a third species, "cubes" a fourth, and so on.

Ibn al-Majdī's goal is to show that the number of polynomial equations is unlimited. He begins with a quick listing of how the numbers of possible equations increase as one goes from two to three to four species. He then shows how to enumerate possible equations by breaking the enumeration down into subcases, the polynomials in each subcase having some property in common. For example, to take the case of equations involving at most four species, all equations in which the sum of two species is equal to the sum of the two other species would be one subcase. The other would be those in which one species is equal to the sum of the three remaining species.[29] (This approach links al-Majdī's problem with the problem of partitioning a given positive integer into two or more positive summands, although al-Majdī makes no reference to the latter problem.)

Know that the problems that calculation by *al-jabr* and *al-muqābala* reverts to are not restricted to the six mentioned.[30] Rather, they are restricted to them only in cases of equations between the three species: numbers, things and squares. But, if you look at species higher than them, you obtain an unrestricted number of equations, since the number of species is unbounded.

[For example,] if you look at cubes together with what comes before them, the equations are restricted to twenty-five forms: six of them simple [i.e., one species equals] one species; twelve of them trinomials [i.e., one species equal to] two species; four of them are quadrinomials (with 1 equal to 3), and three of them are quadrinomials (with 2 equal to 2). And if you look at five species, their forms are 90: ten binomials, 30 trinomials, 20 quadrinomials (with 1 equal to 3) and 15 quadrinomials (with 2 equal to 2), and 10 with five species (2 equal to 3) and also five with five species (1 equal to 4).

[27] I thank Prof. A. Djebbar and Dr. M. Bagheri for aid in translating the text in this section.

[28] Since $ax^2 + bx + c = 0$, with all coefficients positive, has no positive solutions, equations of this type were not considered.

[29] Ibn al-Majdī refers to these two subcases as "two equal two" and "one equals three." He uses the Arabic numerals, rather than words, in such expressions and also uses them (in most cases) to express the results of his counting.

[30] He is referring to the six standard forms of first and second degree equations in one unknown.

And the way to enumerate each case of these species, however many [species] there may be, is that we multiply the number of those species by half the number preceding it. And that is the number of binomials,[31] which expresses the number of simple [equations].

Then the number of simple equations is multiplied by the third number before it,[32] which expresses the number of trinomials in it. And your work is finished if the number [of species] is three. But, if not, then you know the quadrinomials are of two kinds, 1 and 3, or 2 and 2.[33] So, in the first [case] you multiply one by three to get three, which you keep. Then you multiply the trinomials by the fourth number [$n - 3$], you divide the result by what you kept, and there results the first kind of quadrinomials. And, as for the second [kind], multiply two by two, [which gives] four, and keep it. Then you multiply as before[34] and divide by what you kept, and the second kind of quadrinomials results. And your work is finished if the number is four.

But, if it is five, then the composition [of equations involving at most four of the five species] is as was premised for the quadrinomials, and what was before it. And, as for [equations with five species using all] five terms, there are two kinds: 1 equals 4 and the second, 2 equals 3. And so, multiply the number of the first type of quadrinomials [i.e., 1 equals 3] by the fifth number before it [$n - 4$] and divide the result by four. There results the first kind for five species. And multiply the number of the second kind of their quadrinomials[35] by the number five before [$n - 4$] and what you obtain you will multiply by two, and you will divide the result by 3, i.e., [you multiply by] what results from the division of 2 by 3, as though they were numerator and denominator [of a fraction]. The result is the second type of the five species.

But when the numerator of the first is one, the multiplication has no effect.[36] And know that if we speak here of three species or more we do not, here, want thereby what is above number only, like roots and squares and what is above that. Rather the intent is more general than it, either from the direction of freeing it from number or line, from it [?] or both of them, so know that.

[Case of 3 species]

An example in three species is that we multiply 3 by half of two, which [the "two"] precedes it, and 3 results, and it is the [number of] binomials. Then we multiply the binomials by the third number [before the number of species], which is 1. And 3 results also. And it is the trinomials, and the case finishes, and so the totality of their forms is only six.

[31] *Al-thanāʾī.*

[32] The count to "the number before it" starts with the number in question. So, for example, 2 is the third number before 4. He usually writes, when discussing, say, "the third number before it," simply "the third number."

[33] He now indicates the various subcases by this shorter expression.

[34] That is to say, multiply the number of trinomials by $n - 3$.

[35] That is, those using any four of the five species.

[36] That is, when the number of species available is equal to the number of monomials used, the number $n - p$ becomes $n - (n - 1)$, i.e., 1.

[Case of 4 species]

And if there are four species, then multiply four by one and a half.[37] Six results, the binomials. Then [multiply] this six by the third number, which is two, so twelve, and it is the trinomials. And then multiply that by the fourth before it, which is one, and so twelve, and divide by the denominator of a third [i.e. three] multiplying by its numerator. 4 results, and it is the quadrinomials that are in it with 1 equals 3. Then divide the twelve [obtained just above] by the denominator of a fourth multiplied by its numerator. The result is three, and this is what is in it from the quadrinomials in which two are equal to two. So their composition is finished.

And if there are five species, then multiply the number by 2, which yields 10, and these are binomials. Then multiply that ten by the third number, and 30 is obtained, and these are the trinomials. Then multiply that thirty by the fourth number, and 60 results. Multiply that by one and divide the result by 3, as we premised in the corresponding case earlier. 20 results and it is the first of the quadrinomials. And if the sixty is divided by 4, 15 results, which are their second. Then multiply the first from the quadrinomials, 20, by the fifth of the number, and divide the result by 4. There results 5, the first of the pentanomials. Then multiply the second of the quadrinomials by 2, and divide the result by 3. The result is 10, which is the second [summand] of the five species. So their composition is finished.

[Concluding remarks]

And this is a vast chapter; its figures of its composition are unbounded[38] because of the multiplicity of species. And it is not our purpose to describe the listing of the forms of each composition. Rather, the goal is to demonstrate that the equations are not limited to the six well-known [equations], except with respect to their being only three species with their combination, because the species on the two sides of the number[39] are not bounded.

And it appeared from that that the well-known six are a very small part of the totality that occurs in equations, although most of the problems are centered on them. And, for that reason they [people] think there is a limit to what is sought, and they confine themselves to it, and from that they derive all their operations. And on top of this, when some equations occurred other than one of the six well-known [equations], its solution was difficult by those methods that have been mentioned, except in some of its problems that one could reduce to them [the six].

[37] He has combined the two steps for computing the binomials, namely, multiplying the number of species by one less than itself and then dividing by 2.

[38] That is to say, all the different equations are unbounded.

[39] Here the author may be referring to what we would call the positive and negative powers of the unknown, since one finds in Arabic manuscripts polynomials represented by two rows of cells, one row above the other. The cells in the top row are labeled "number" and, on either side of that, the names of the powers of x relevant to the problem. Below each of these cells the coefficient of the power named above is written. Hence the positive powers of x are on one side of the "number" and the negative powers of x on the other side.

IV. GEOMETRY

1. ABŪ ʿABD ALLAH MUḤAMMAD IBN ʿABDŪN, *ON MEASUREMENT*

Abū ʿAbd Allah Muḥammad ibn ʿAbdūn (923–976 (?)) was a mathematician who was born in Cordova and taught mathematics there. He then went on to become a physician as a result of his studies in the East. Later he returned to Cordova as the physician of Caliph al-Ḥakam II. His only known mathematical work is *On Measurement*, of which only one copy survives. Many of the methods in this treatise can be found in texts written in ancient Babylon, and the editor of the text, Ahmed Djebbar, points out that Ibn ʿAbdūn's treatise marks the extension of a pre-algebraic tradition of measuring surfaces from the Eastern Islamic lands to al-Andalus and subsequently to the Maghrib.

At the beginning of the manuscript of this treatise, Ibn ʿAbdūn is referred to as *muhandis* and *faraḍī*. The first denotes someone involved with measuring (theoretical or practical, e.g., surveying), and the second denotes a specialist in the arithmetical procedures necessary to calculate the legal heirs' shares of an inheritance according to Islamic law. I have selected problems[1] dealing with the measurement of squares, rectangles, triangles, circles, and segments of circles among the plane figures and truncated pyramids among solid figures. One interesting feature of the treatise—one that takes it above the simple recitation of procedures for measuring figures—is that the author, for each type of figure, identifies the key components that can be measured and then, in a series of problems, discusses how each component can be derived from the others. Similar problems are found in Chapter 1, section II-4-2 and II-4-3, and in Chapter 2, sections III-1 and III-2.

6. If you are told, "We add the sides and the area and it is one hundred forty, what are the sides?" The calculation is that you add up the number of the sides, which is four, and take its half, two. Multiply it by itself, it is four. Add it [the product] to one hundred forty, which is one hundred forty-four. Then take the root of that, twelve, and take away from it half of the four and the remainder is equal to each of its sides.

The following sequence of problems is interesting in that after the first of the three has been solved, all the values are known. So the point of the remaining two problems is less finding the answer than learning mathematical methods.

17. If you are told that the diagonal is ten, and one side exceeds the other by two, what are its two sides? The way to calculate this is that you multiply the diameter by itself, which is one hundred, and you multiply the two by itself, which is four, and you subtract it from one hundred. The remainder after that is ninety-six. You take half of that, which is forty-eight, which is the area. Now it is as if you are told, "A rectangle, whose length exceeds its width by two, and its area is forty-eight. What is each of the sides?" So you work as we described to you [earlier], and you will hit the mark, Allah willing.

He then continues with the assumption that the diagonal is ten.

18. And, if you like, when you have arrived at the knowledge of the area, which is forty-eight, multiply the diagonal by itself, which is one hundred. Then double the area, which

[1] Prof. Djebbar has supplied (in brackets) numeration for the problems in his edition of the text, and I have followed his numeration.

is ninety-six. Add it to the hundred, which is one hundred ninety-six. Take its root, which is fourteen. Take half of that, which is seven, and multiply it by itself, forty-nine. Subtract the area from it, so one remains. If you add it to the seven it is the longer side, and if you subtract it from the seven, what remains is the shorter side.

And, be aware that if you multiply the diagonal by itself, and double the area, and if what comes out as a result of doubling the area is greater than the product of the diagonal by itself, then the problem is impossible.

In problem 20 Ibn ʿAbdūn returns to the situation discussed in problems 17 and 18.

20. If the question is as above, except that he says, "We add the longer side together with the diagonal and it is eighteen, and the shorter side is six, how much is the diagonal and how much is the longer side?"

The way to calculate it is you multiply the eighteen by itself, which is three hundred twenty-four. Then you multiply the six by itself, thirty-six, and you subtract it from the three hundred twenty-four, after which the remainder is two hundred eighty-eight. Take half of that, one hundred forty-four, and divide it by eighteen. Out of the division comes eight. And it is the [longer] side. Subtract it from eighteen, remainder ten, and that is the diagonal.

After a discussion of various sorts of quadrilaterals, Ibn ʿAbdūn turns to triangles.

Be aware that triangles are of three kinds, and they are acute, right and obtuse.

59. And as for the acute angle triangle, it is that in which if you multiply each of the sides containing the acute angle by itself and add the results, the sum is greater than the product of the side subtending the acute angle by itself by twice the product of what is between the foot of the perpendicular and the acute angle by the line the perpendicular falls on.[2]

60. For example: We make a triangle in which each side is ten. How much is its area? So we do not find the area without knowing its perpendicular. But we are able to measure it without a perpendicular, which is that we add its three sides, then take its [the sum's] half, and then note what is between the half [of the sum] and each of the sides.[3] And we keep it in mind. Then we multiply them [each of the three differences] by each other. And that would be as if you found that it [the semiperimeter] exceeded each side by five, and there are three sides. And so it is as if you have five, five, and five and if you had multiplied each of them together you would have one hundred twenty-five. Then you will multiply this one hundred twenty-five by half of the sides that you added [i.e., half the sum of the sides]. And take the root of what you get, and that is the area.

61. And so if you want to find the area of a triangle prior to [knowing] its perpendicular, then multiply its side by itself, 100, and take away from it half the base multiplied by itself and take the root of what comes out. And that is the perpendicular. And multiply it by half the base. It is [very nearly] forty-three and a third. It is the area. . . .

Problem 62 addresses the case of an equilateral triangle whose altitude is $\sqrt{15}$ and asks for its side.

[2]That is to say, if ABC is a triangle with an acute angle at B, pick one of the other angles, say, A, and drop a perpendicular from A onto BC, say, AD. Then the sum of the squares on AB and BC exceeds the square on AC by twice the product of DB by BC. This result in found in Euclid's *Elements* II.13.

[3]That is, we subtract each side from half the sum of the three sides.

63. And if the question is as above [i.e., the triangle is equilateral] except that he says, "The area of a triangle is such and such, what is each of its sides?" The calculation of that is that you always multiply the area of the triangle by itself and you multiply the result by three. And what it is you take its root. And you add to what results its third. And you take the root of the result. And what results is each of its sides.

. . .

71. As for a scalene triangle, one of whose sides is thirteen, the second fifteen and the base is fourteen, knowledge of its area is by deriving its altitude. And knowledge of its altitude is by making the foot of its altitude known. And if you want to know that you multiply the one side, i.e., thirteen, by itself, and you multiply fifteen by itself, and you subtract the smaller from the larger. You take half of the remainder, which is twenty-eight. Divide it by the base, and you get, from the division, two. If you add it to half the base, you have the longer section created by the foot of the perpendicular on the base, which is nine. And if you subtract it from half the base, it is the shorter, which is five.

And if you want to know the altitude, then you multiply the one segment [it creates on the base] by itself and the side next to it [the segment] by itself. And you subtract the smaller from the larger. And then take the root of the remainder, and that is the altitude.

Ibn ʿAbdūn now concludes Problem 71 by applying Problem 61 to find the area.

72. And there is another method to know the segments of the base. That is that if you want to know the shorter segment, then you multiply the base by itself and the short side by itself, and from what results from adding [these two], you subtract the product of the longer side by itself. You take half of the remainder, and you divide it by the base. What you obtain is the shorter segment.

He gives a similar rule for finding the longer segment of the base, in general terms, and then he works through that procedure in terms of the numerical example of a scalene triangle. He concludes by saying that if we have obtained the longer segment, we can subtract it from the base to find the shorter segment, "and you have obtained knowledge of the perpendicular in the way I explained to you, Allah willing."

The series of problems 106–109 is part of a section dealing with the relations between various dimensions of a truncated square pyramid whose base has side 4, whose top has side 2, whose height (from the center of the bottom to that of the top) is 10, and whose volume is $93\frac{1}{3}$. In problem 106, Ibn ʿAbdūn assumes one is given the side of the top, the height, and the volume, and the problem is to find the side of the base. In problem 107, the task is to find the altitude when one is given all other data. Problem 108 simply shows another method for solving problem 107.

109. And for it there is another way,[4] which is that you multiply the four, which is the base, by itself to obtain sixteen, and [you multiply] the two, which is the top, by itself to get four. Then you multiply the [side of the] top by [that of] the base, which is eight. And you add that to the sum of the product of the base by itself and the top by itself to obtain twenty-eight. Take a third of that, which is nine and a third. Multiply that by the perpendicular, which is ten, and it is ninety-three and a third.

[4]He explained one method at the very beginning of the section. The method of problem 109 is the same method used in the ancient Egyptian *Moscow Mathematical Papyrus*. [See Imhausen, 2007, p. 33.]

With problem 114, Ibn ʿAbdūn begins his treatment of circular figures.

114. The width of a circular figure is its diameter, ten. What is its area? Its calculation is that you multiply its diameter by itself and what the multiplication arrives at, you will subtract from it its seventh and a half of its seventh, and what remains is its area.

115. And knowledge of its circumference is that you multiply the diameter by three and a seventh, and what the multiplication attains is its circumference, and that is thirty one and three-sevenths.

116. And, if you wish, multiply the diameter by itself and then, always, by ten,[5] and you take the root of that, and it will be, approximately, the circumference.

The last section, on circular segments, comes after problem 121, which explains how to calculate the area from the circumference.

Circular segments:

Know that there are three kinds of circular segments: the semicircle, the segment less than the semicircle and the segment more than the semicircle.

As for what is half a circle, it is when its arrow[6] is equal to half its chord. And what is less than half a circle is that whose arrow is less than half its chord. And what is more than half a circle is that whose arrow is more than half its chord.

Problem 122 deals with measurement of various parts of half a circle and introduces nothing that would not be clear to a person who had understood the section on measuring circles. Ibn ʿAbdūn then proceeds as follows.

123. As for segments [of circles] less than half a circle, [say,] one whose arrow is two and whose chord is eight, knowing its area is prior to its arc and knowing its arc is by knowing the diameter of the circle of which it is a segment. So you halve its chord [to get] four. And so you multiply it by its like, which gives sixteen, and you divide it by its arrow, so that you have eight. And then you add to it its arrow [which will give] ten, and that will be the diameter of the circle whose segment is that arc [Fig. IV-1-1]. And if you want to know the arc, add to the arrow its seventh and then you add the result to the diameter of the circle whose segment is that arc. From the result you subtract what is between the arrow and half the diameter of the circle. And what remains after the addition and subtraction is the arc.[7] That less than half a circle is that whose arrow is less than half its chord, and that more than half a circle is that whose arrow is greater than half its chord.

[5]The word "always" means that the reader is to understand that this "ten" is a constant, to be used even when the diameter is not ten.

[6]The arrow of a segment is the length of the straight line joining the center of its arc to the center of its chord. It was a standard number in medieval measurements involving the circle, including trigonometry. For a circle of radius one, it corresponds to the modern versed sine of half the arc. The chord of the semicircle is, of course, the diameter of the corresponding circle.

[7]In the numerical example chosen by Ibn ʿAbdūn, when the arrow is 2 and the chord is 8, the diameter is 10 (which is exact), and the arc length is 9 2/7, close to the actual value of 9.273. The general formula is $s = a + \frac{1}{7}a + 2r - (r - a) = \frac{15}{7}a + r$, where s is the arc length, a the arrow, and r the radius. This is exact (assuming $\pi = \frac{22}{7}$) if s is half the circumference, for then $a = r$ and $s = \frac{15}{7}r + r = \frac{22}{7}r$.

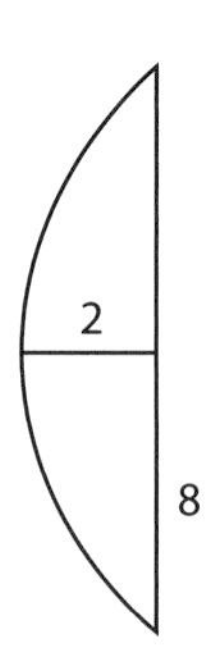

Fig. IV-1-1.

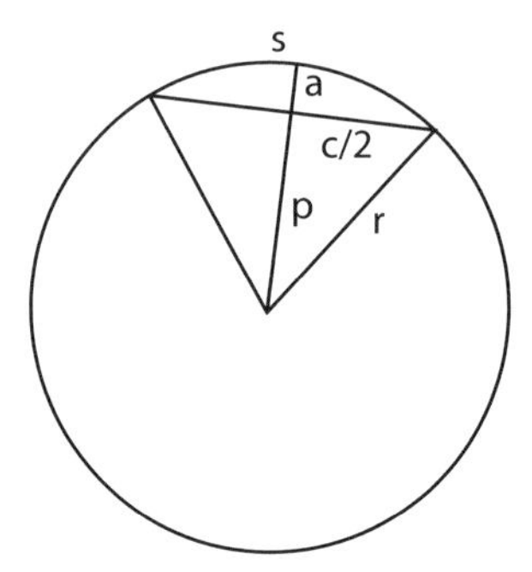

Fig. IV-1-2.

And there is another method for knowing the arc, which is not easy for a beginner in the science of calculating.

124. If you want to know the area of the arc,[8] multiply half of the arc [$s/2$] by half the diameter of the circle [r], and remember what the result of the multiplication is. Then you see what is between the arrow and the half diameter [$p = r - a$]. Whatever that is, multiply it by half the chord [$c/2$]. Then you subtract what is obtained by the multiplication from what you kept in memory, since the segment is less than half a circle. (And if it is more than half a circle, you would add it to what you kept in memory.) And what remains after the subtraction of the number from the number you remembered is the area of the arc.[9] And this is the diagram [Fig. IV-1-2].

2. ABŪ AL-QĀSIM IBN AL-SAMḤ, *THE PLANE SECTIONS OF A CYLINDER AND THE DETERMINATION OF THEIR AREAS*

Abū al-Qāsim ibn al-Samḥ (984–1035) was a student of the famous astronomer Maslama al-Majrīṭī (Maslama of Madrid). For a time he lived in Cordova before moving to Granada,

[8] At the beginning and end of this problem, "arc" (Arabic *al-qaws*) means the segment contained by the arc, but elsewhere it means the arc itself.

[9] The general formula is $A = \frac{rs}{2} - \frac{pc}{2}$, that is, the area of the sector less the area of the triangle.

where he wrote on astronomy, astrology, and mathematics. His writings include works on the astrolabe and the earliest known work on the equatorium,[10] as well as treatises on arithmetic and geometry.

In this selection from his work, *The Plane Sections of a Cylinder and the Determination of Their Areas*,[11] Ibn al-Samḥ's aims are two. In the first part, he introduces a figure constructed by what he calls a "triangle of movement" and then an oblique section of a right circular cylinder, which he knows is an ellipse. He shows that they share the same properties and therefore are the same figures. In the second part, he determines the area of an ellipse, in an approach following Archimedes's lead in *Measurement of the Circle.*

Ibn al-Samḥ's "triangle of movement" is constructed by fixing one side [Line *NT* in Fig. IV-2-1] of a triangle [Triangle *NET* in Fig. IV-2-1] and moving the intersection of the other two sides in such a way that their sum is always equal, although the lengths of each will vary as their intersection moves.[12]

In the second part, Ibn al-Samḥ finds the area of the ellipse by relating its area to that of its inscribed and circumscribed circles. To do this, however, he proves that a series of ratios involving the ellipse and its inscribed circle are equal to the ratio of the major axis to the minor axis. First, he treats the ratios of the semi-chords in these two figures (propositions 6 and 7) and then of the trapezoids determined by them. Next he establishes the ratios of the inscribed polygons made up of those trapezoids (proposition 12), and then finally that of the ellipse and its inscribed circle (proposition 13).

Propositions 13, 14, and 15 prove, respectively:

If E is an ellipse with inscribed circle C_i, then $C_i{:}E =$ (minor axis):(major axis).
If E is an ellipse and O an arbitrary circle with diameter d, and if a is a line segment satisfying $a{:}d = d{:}$(minor axis), then $E{:}O =$ (major axis):a.
If E is an ellipse, then (inscribed circle):$E = E{:}$(circumscribed circle).

Proposition 16 is phrased in a way that echoes proposition 1 of Archimedes's *Measurement of the Circle*, each expressing the area of a curved figure (an ellipse in the one case, a circle in the other) in terms of a certain right triangle.

We begin with a part of Ibn al-Samḥ's introduction.

When one cuts a right cylinder of which the bases are two circles by a plane that is not parallel to the base, the section that results may be generated by fixing a side of a triangle and making each of the two remaining sides rotate [each around its endpoint on the base] in the plane of the triangle [so that they remain joined and their sum does not change], until they have returned to the original position. We formulate a proof of that in what follows.

[10]This was an instrument for graphically finding the positions of the sun, moon, and planets.

[11]This work survives only in part, and that part in a Hebrew translation made by Qalonymos ben Qalonymos (see sections II-5 and IV-3 in Chapter 2).

[12]This method is the basis for a practical method of tracing out an ellipse on the ground, a method sometimes referred to as the "gardeners" method.

As will appear below, Ibn al-Samh assumes that the section of the cylinder is an ellipse.[13] The proof strategy is to establish similar properties for the two figures, then to superimpose the two figures and, using the properties, show that the two curves are completely identical. Ibn al-Samḥ begins by giving some definitions for the curve generated by rotating the triangle.

We say that the figure obtained by the motion of the triangle is called *elongated circular*, a name derived from the configuration. In fact, it is a circular contour which is elongated. Neither the circularity nor the extension of the length [alone] characterizes this in a unique fashion. And this designation is required with a view to the way this figure is generated, since the process for constructing it combines circular motion and rectilinear motion, rectilinear motion being the extension in length. The movement of the common extremity of the two sides that turn generates what is called the *contour of the elongated circular figure*. [Curve *BGAD* in Fig. IV-2-1] The fixed side of the triangle is called the *central side* [or *central line*, line *NT* in Fig. IV-2-1]. Similarly, the two remaining sides, those that rotate, are called the *moving sides*. The triangle itself is called the *triangle of movement*.

After what we have indicated about this construction, the two moving sides come to coincide, in the course of their rotation, with the central side, making with the latter a unique straight line. The rectilinear and circular extension is thus at its maximum, and the excess of the one of the two [sides] over the other is at its maximum, the difference between the two equaling exactly the central side. It is also clear that, in rotating, one of the two [sides] grows at the same time that the other diminishes. ... That one of the two grows while the other diminishes, means that they become equal in certain positions. This equality is true in two positions, on one side and the other of the central line. In this case, they have the name *equal moving sides*. The perpendicular to the central side drawn from their common extremity then cuts the central side into two halves. The [cutting] point is *the center of the figure*. It is the center of two circles, one of which passes through the other extremity of the given perpendicular—which represents its semi-diameter—the circle being tangent to the figure. As to the other circle, the extremities of its diameter are situated where the two sides of the triangle coincide in making a single straight line, the distance to the center being the greatest at any point. It appears therefore that the diameter is equal to the sum of the two moving sides. ... The great circle described here is tangent to the figure. ... The great circle is called *circumscribed*, and the small circle, *inscribed*.

All the lines passing through the center and cutting the figure are divided in half by the center. One calls them the *diameters*. The greatest of these is the diameter common to the figure and the circumscribed circle; the smallest of these is the diameter common to the figure and the inscribed circle. Every line cutting the curve without passing through the center is called a *chord*. Among the chords, those which are cut into two halves by one or the other of the greatest and smallest diameters, make a right angle with that line. And if one of the two diameters cuts a chord at a right angle, it also cuts it into two halves. The straight line making a right angle [with the greatest diameter] at an extremity of the central

[13] In proposition 9 of his *Conoids and Spheroids*, Archimedes shows that he knew this. Much later, Serenus of Antinoë (probably late fourth century CE) stated explicitly that any section of a right circular cylinder by a plane not parallel to its base is an ellipse. And in the ninth century, Thābit ibn Qurra proved this in an Arabic treatise.

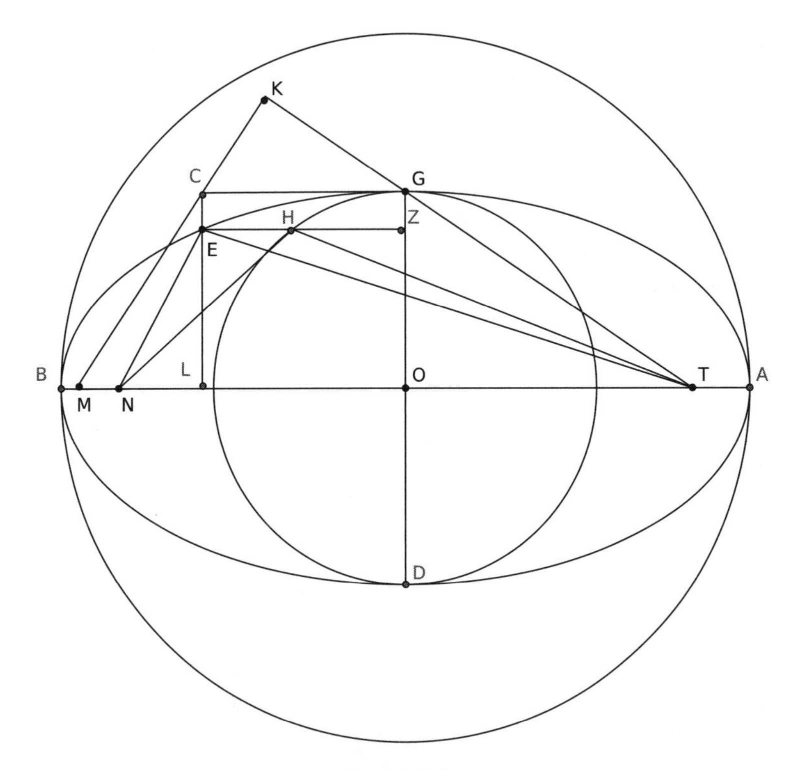

Fig. IV-2-1.

line, and cutting the great circle, is called an *invariant line*. The part of this line inside the elongated circular figure is called the *separated line*.

Ibn al-Samḥ now proves six results about the elongated circle. The statement and proof of the sixth result is given here, along with analytic proofs of two of the earlier results ibn al-Samḥ uses in that proof.

Proposition 6: If in any elongated circular figure, one chooses an arbitrary point on the circumference and draws from this a line that is perpendicular to the smallest diameter, then the ratio of this perpendicular to the part which is contained in the inscribed circle is equal to the ratio of the greatest diameter to the smallest diameter.

Let *ABCD* be the given figure with inscribed circle *GHD* [Fig. IV-2-1]. Let the point *E* be chosen on the circumference of the elongated circular figure, through which one draws a perpendicular *EHZ* to the smallest diameter. I claim that the ratio of *EZ* to *HZ* is equal to the ratio of the greatest diameter to the smallest diameter, and is also equal to the ratio of half the greatest diameter to half the smallest diameter.

Proof: Let *TN* be the central line, *O* the center of the circle. Connect *TG*. It has been shown that *TG* is equal to half the greatest diameter.[14] Prolong [*TG*] to *K* so that *KT*

[14]This follows from the construction of the elongated circular figure, since *G* is the point at which the two moving sides are equal.

is equal to *TE*. It is clear that *GK* is equal to the excess of *AO* over *EN*.[15] Drop a perpendicular *EL* to *AB*, and choose *M* on *AB* such that the ratio of *ML* to *OG* is equal to the ratio of *OG* to *OT*, so *ML* then forms a continuous proportion with *OG* and *OT*. Then prolong *LE* in a straight line up to *C* [on KM] and join *CG*.

Ibn al-Samḥ has already proved that under these conditions we have $ET{:}TM = TO{:}TG$ (proposition 3) and $ZE^2 = ZH^2 + GK^2$ (proposition 5). We give these proofs analytically using the standard notation for ellipses. In fact, our standard equation for an ellipse comes directly out of Ibn al-Samḥ's definition, which is effectively the two-focus definition of an ellipse. Set the equation of the figure to be $b^2x^2 + a^2y^2 = a^2b^2$, with $AO = a$, $GO = b$, $OT = ON = c$, and $a^2 - b^2 = c^2$. Let $E = (x, y)$, $M = (x + b^2/c, 0)$, and note that $ET^2 = x^2 + y^2 + c^2 + 2cx$ and $EN^2 = x^2 + y^2 + c^2 - 2cx$. It follows that $ET^2 - EN^2 = 4cx$. But $ET + EN = 2a$. So $ET - EN = 2cx/a$. Thus $ET = a + cx/a$ and $EN = a - cx/a$. Also, $TM = x + b^2/c + c$. Thus $ET{:}TM = c/a = TO{:}TG$, as desired (proposition 3). Furthermore, since $ZH^2 = b^2 - y^2$ and $GK^2 = (AO - EN)^2 = (cx/a)^2$, it follows that the sum of those two squares is equal to $x^2 = ZE^2$ and proposition 5 is proved.

Thus, after what we have established, it is known that the ratio of *KT*—which is equal to *ET*—to *TM* is equal to the ratio of *TO*—which is half of the central line—to *TG*—which is equal to half of the greatest diameter. These lines form the same angle so the triangle *GTO* is similar to the triangle *TKM*; therefore angle *K* is equal to angle *O*. But angle *O* is a right angle, so angle *K* is a right angle. There remains the angle *TGO*, which is equal to the angle at *M*. The angle *L* is right, as is the angle *O*. The triangle *CML* is therefore similar to the triangle *GTO*. Thus the ratio of *TO* to *OG* is equal to the ratio of *CL* to *LM*. But the ratio of *TO* to *OG* is equal to the ratio of *GO* to *LM*; thus the ratio of *GO* to *LM* is equal to the ratio of *CL* to *LM*. As a consequence, *OG* is equal to *CL*, and they are parallel. The line *GC* is therefore equal to the line *OL*, and they are parallel. Further, *OL* is equal to *ZE*, and therefore *GC* is equal to *ZE*, and it has been proved that the square of *ZE* is equal to the squares of *ZH* and *GK* [taken together]. But the square of *GC* is equal to the squares of *GK* and *KC* [taken together], and the angle *K* is right. Thus the squares of *GK* and *KC* [taken together] are equal to the squares of *GK* and *ZH* [taken together]. Subtract the square of *GK*, which is common. So the square of *KC* is equal to the square of *ZH*. But *GC* is parallel to *OL*, and angle *KGC* is equal to angle *T*. Further, the angle *O* is right as is the angle *K*. The triangle *KGC* is therefore similar to the triangle *GOT*. Therefore, the ratio of *GC* to *KC* is equal to the ratio of *TG* to *GO*. But *KC* is equal to *ZH*; *ZE* is equal to *GC*; and *GT* is equal to *AO*. Consequently, the ratio of *EZ* to *ZH* is equal to the ratio of *AO* to *OG*. That is what we wished to prove.

Ibn al-Samḥ next establishes the same property as in proposition 6 for the curve that is a section of right circular cylinder by a plane not parallel to a base. He calls this curve an ellipse, so he knew, and evidently assumed the reader would know as well,[16] that this is the same curve as Apollonius had defined using a section of a cone.

Having established, in the introduction, that which is necessary about the curve obtained by the rotation of [two edges of] a triangle, we now introduce what is necessary

[15]The Levy/Rashed translation supplies the following argument: $ET + EN = 2AO$, and $KT = TE$. Hence, $KT + EN = 2AO$. But $KT = TG + GK$, and $TG = AO$, so $GK = AO - EN$.

[16]See footnote 13.

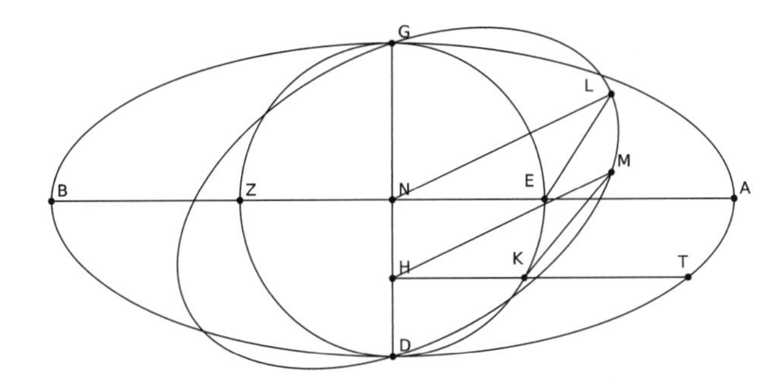

Fig. IV-2-2.

about the section of a cylinder. If one cuts a right circular cylinder by a plane that is not parallel to the base, the point where the plane cuts the axis of the cylinder is called *the center of the figure*, and the straight lines cutting the ellipse and passing through the center are *the diameters*. If one cuts the cylinder by a plane parallel to the base and passing through the center of the figure previously obtained, the section will be a circle,[17] in particular a circle inscribed in the ellipse. In fact, if one rotates the ellipse about the common line—which is in the plane of the circle—the circle reappears in the interior of the ellipse; and the line common to the two sections[18] is a diameter of the ellipse, the smallest of all the diameters [called the minor axis]. The diameter that cuts this in a right angle is the greatest of all the diameters [called the major axis]. The diameters that are nearer to the smallest diameter are smaller than the diameters that are further away and approach the greatest diameter.

The circle inscribed in an ellipse, whose diameter is the smallest diameter, is equal to the circle of the base of a cylinder of which the ellipse is a section. In fact, the base of the cylinder is the same as the section passing through the center of the ellipse parallel to the given base. And the circle cutting the cylinder and passing through the center of the ellipse is inscribed in the ellipse, since it has, in common with the latter, the smallest diameter [the minor axis].

Proposition 7: Given an ellipse with an inscribed circle. Through an arbitrary point on the minor axis, different from the center, draw a line parallel to the major axis until it reaches the ellipse. Then the ratio of this line contained in the ellipse, to its part which is contained in the circle, is equal to the ratio of the major axis to the minor axis.

Let the ellipse be *ABGD*, the inscribed circle be *GEZ*, and the point *N* the center of the circle and the center of the ellipse [Fig. IV-2-2]. Let *AENZB* be the major axis and *DNG* the minor axis, which is also a diameter of the circle. Let *H* be an arbitrary point on *DG* and through it, draw the line *HKT* parallel to the line *AB*, which is thus perpendicular to the diameter *DG*. I say that the ratio of *HT* to *HK* is equal to the ratio of *AB* to *GD*.

Proof: Imagine a cylinder whose base is the circle *GEDZ*; imagine that the line *DG* is fixed as the circle pivots around it; imagine the movement of the ellipse *DAG* during

[17]This would, of course, be true regardless of whether the plane passed through the center.

[18]That is, common to the ellipse and the circle containing its center and parallel to the base of the cylinder.

the pivot until it reaches the surface of the cylinder, thus becoming the curve *DLMG*, with the [semi]major axis *NA* becoming the line *NL* and *HT* becoming *HM*. Join *E* and *L*, *K* and *M*. The angle *KHG* being right, it will be the same for the angle *MHG* since its initial configuration does not change. The line *GN* is perpendicular to the plane *ENL*, and every plane passing through the line *NG* is perpendicular to the plane *ENL*, as has been shown by Euclid [*Elements* XI, definition 4]. Similarly, every plane passing through the line *HG* is perpendicular to the plane *KHM*.

But the plane of the circle passes through the line *NG*; so each of the planes *LEN* and *HKM* is perpendicular to the plane of the circle. But the lateral surface of the cylinder makes a right angle to the plane of the circle, so the common sections [of the planes and the cylinder], that is, *LE* and *MK*, are perpendicular to the plane of the circle. Two perpendiculars to the same plane are parallel to each other, so *LE* is parallel to *MK*. Since the angle *LNG* is right, and the angle *MHG* is also right, the line *LN* is parallel to the line *MH*. Similarly, the line *EN* is parallel to the line *KH*, and these lines are in the plane of the ellipse. Therefore, the triangle *LNE* has its sides parallel to the sides of the triangle *MKH*, and the angles of the two triangles are [therefore] equal.[19] In fact, the two lines *LN*, *NE* contain the angle *ENL*; the two lines *MH*, *HK* contain the angle *KHM*, and the lines are not in the same plane. The two angles *ENL*, *KHM* are therefore equal. Thus it has been established that the angles of the triangle *LNE* are equal to the angles of the triangle *MKH*. The two triangles are therefore similar.

The ratio of *LN* to *NE* is therefore equal to the ratio of *MH* to *KH*. But *LN* is equal to *AN*, and *MH* is equal to *HT*. It follows that the ratio of *AN* to *NE* is equal to the ratio of *TH* to *HK*, and that is what we wanted to prove.

To show that the elongated circle and the section of the cylinder are identical, Ibn al-Samḥ proves proposition 9 by superposition and by using the analogous results of propositions 6 and 7. We omit the details of the proof.

Proposition 9: We wish to show that the elongated circular figure, formed by the rotation of a triangle, is equal to the oblique section of a cylinder, of which the greatest diameter is equal to the greatest diameter of the circular elongated figure formed by the rotation of the triangle, and where the smallest diameter is equal to the other smallest diameter. The one figure coincides with the other, point by point, and they are identical.

In the remaining propositions given here, Ibn al-Samḥ determines how to calculate the area of an ellipse.

Proposition 12:[20] Suppose one inscribes in a quarter of an ellipse, beginning at the extremity of the major axis and ending at [the extremity of] the minor axis, consecutive chords, and one draws from the extremities of the chords perpendiculars to the minor axis, which intersect the quarter of the circle inscribed in the ellipse. Then one draws the associated chords in the arcs of the circle thus determined. Thus one forms two polygonal areas, one inscribed in the ellipse and the other in the circle, such that the ratio of the area inscribed in the ellipse to the area inscribed in the circle is equal to the ratio of the major axis to the minor axis.

[19] *Elements* XI.10.

[20] This is a lemma for proposition 13.

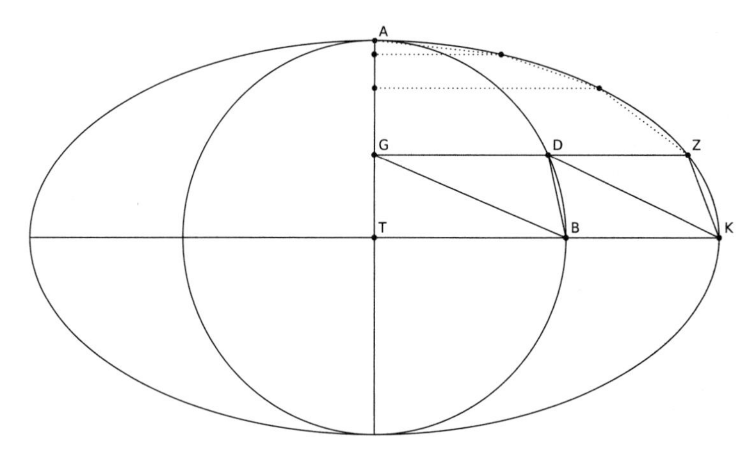

Fig. IV-2-3.

Let *ATK* be the quarter of the ellipse defined by two semi-axes—the semi-major axis being *KT* and the semi-minor axis being *AT* [Fig. IV-2-3]. Let *ADBT* be the quarter of the circle inscribed in the quarter of the ellipse, *T* being the center. One draws chords in the ellipse, *KZ* being one of them. One drops from *Z* a perpendicular to *AT*, the line *ZDG*, with *D* being on the circumference of the circle. In the circle, one draws the chord *DB*. One thus obtains two polygonal regions inscribed respectively in the two figures, *KZGT* in the ellipse and *BDGT* in the circle. I say that the ratio of the polygonal region inscribed in the ellipse—that of which *KZ* is a side—to the polygonal region in the circle—that of which *BD* is a side—is equal to the ratio of *TK* to *AT*.

Proof: Draw the lines *KD* and *BG*. These divide each of the two quadrilaterals into two triangles. The two triangles *KZD* and *BDG* have equal altitudes. The ratio of the area of one to the area of the other is therefore equal to the ratio of the base *ZD* to the base *DG*. Similarly, the ratio of triangle *KDB* to triangle *BGT* is equal to the ratio of *KB* to *BT*. But the ratio of *KB* to *BT* is equal to the ratio of *ZD* to *DG* as has been previously established.[21] As a consequence, the ratio of the four triangles, taken two by two, is the same; and remains so when they are joined. Thus the ratio of quadrilateral *KD* to quadrilateral *BG* is the same ratio, that is, the ratio of *KB* to *BT*.

One proceeds in the same fashion for the remainder of the regions bounded by the chords and the perpendiculars, of which the ratios—the one to the other—will be the same. But the proportional magnitudes remain proportional when one combines them. Thus the ratio of the set of regions inscribed in the quarter of the ellipse *KTA* to the set of regions inscribed in the quarter of the circle is equal to the ratio of *KT* to *BT*.

That which has been done in the two quadrants can also be done in the other quadrants that complete the figures. Thus, the ratio of the surface inscribed in the ellipse, contained by the chords taken on the arc of the semi-ellipse and by the major axis, to the region inscribed in the semi-circle, bounded by the chords taken on the semi-circumference and by the diameter of the semi-circle, is equal to the ratio of the major axis to the minor axis. And it is the same for the other half of the ellipse—that which completes the figure in its

[21] In proposition 6.

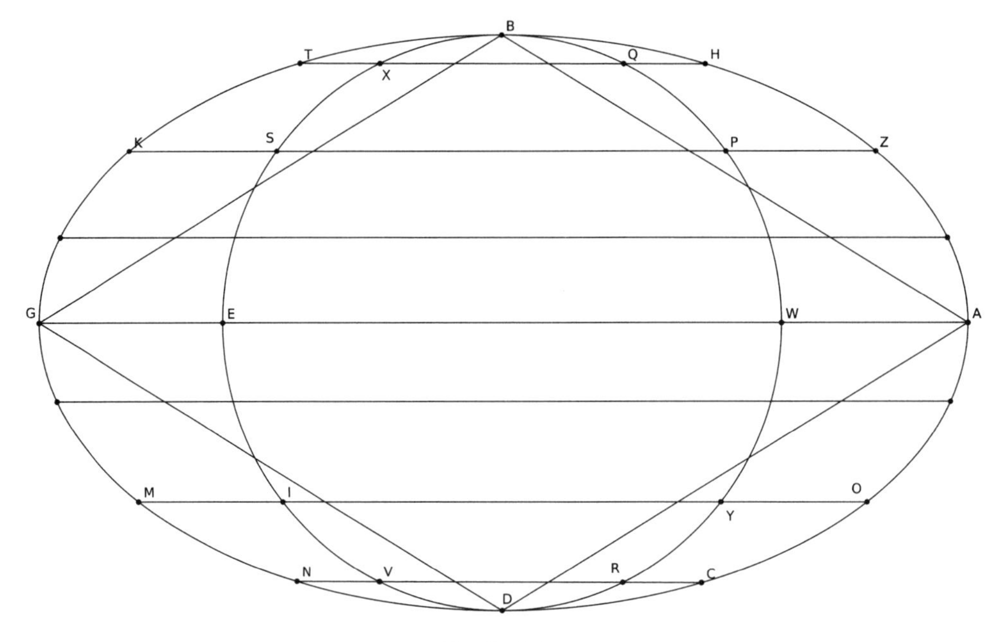

Fig. IV-2-4.

totality—and for the remaining semi-circle, which is inscribed and which completes the circle. The method is the same. The ratio of the entire figure inscribed in the ellipse to the entire figure inscribed in the circle is thus equal to the ratio of the major axis to the minor axis. That is what we wished to prove.

Proposition 13: We want to show that the ratio of the surface of the small circle, that which is inscribed in the ellipse, to the surface of the ellipse, is equal to the ratio of the minor axis to the major axis.

Let the ellipse be *ABGD* with major axis *AG* and minor axis *BD* [Fig. IV-2-4]. Let the small circle, that which is inscribed in the ellipse, be *BWDE*, with diameter *WE*. I claim that the ratio of the surface *ABGD* to the circle *EBWD* is equal to the ratio of *AG* to *BD*.

Proof: The ratio of *WE*, the minor axis, to *AG*, the major axis, is equal to the ratio of circle *EBWD* to ellipse *ABGD*, and it cannot be otherwise. If this were possible, then the ratio would be equal to the ratio of the circle to a magnitude either smaller or larger than the ellipse. To begin, we assume that the magnitude is smaller than the ellipse, and that this magnitude is the surface *L*.

The surface *L* is therefore less than the ellipse, the difference being the magnitude of the surface *U*. Join *AD*, *DG*, *GB*, *BA*. These lines in the ellipse form, in the direction of the center, a region greater than half of the ellipse, the parallelogram *ABGD*. Now cut each of the arcs cut off by these lines into two parts, and draw the chords. These chords cut off, in the direction of the center, a region greater than half the elliptical segments defined by the arcs. If one continues in this manner, one eventually has a remainder in the ellipse smaller than the surface *U*. The elliptical segments defined by the arcs *AZ*, *ZH*, *HB*, *BT*, *TK*, *KG*, *GM*, *MN*, *ND*, *DC*, *CO*, *OA*, form the region smaller than the surface *U*. It follows that the polygonal surface formed by the lines and inscribed in the ellipse is greater than the surface *L*.

We now draw, between the extremities of the arcs found by the division, lines parallel to the major axis. The circumference of the circle is thus divided into arcs at the points *P*, *Q*, *X*, *S*, *I*, *V*, *R*, *Y*. One now connects these points by chords as before. Then the ratio of the [polygonal] figure inscribed in the circle, passing through W, *P*, *Q*, *B*, *X*, *S*, *E*, *I*, *V*, *D*, *R*, *Y*, to the polygonal figure inscribed in the ellipse, passing through *Z*, *H*, *B*, *T*, *K*, *G*, *M*, *N*, *D*, *C*, *O*, *A*, is equal to the ratio of *WE* to *AG*,[22] which ratio is equal to the ratio of the circle to the surface *L*. Thus, the ratio of the figure inscribed in the circle to the figure inscribed in the ellipse is equal to the ratio of the circle to the surface *L*. But the figure inscribed in the circle is smaller than the circle, and the figure inscribed in the ellipse is greater than the surface *L*. Under these conditions, the ratio of a smaller to a greater would be equal to the ratio of a greater to a smaller, and this is impossible. It is therefore impossible that the ratio of *WE* to *AG* is equal to the ratio of the circle to a magnitude smaller than the ellipse.

We omit the proof that the magnitude is greater than the ellipse.

Proposition 14: We want to show that the ratio of an ellipse to any circle is equal to the ratio of the major axis to a line such that its ratio to the diameter of the circle is equal to the ratio of the diameter to the minor axis of the ellipse.[23]

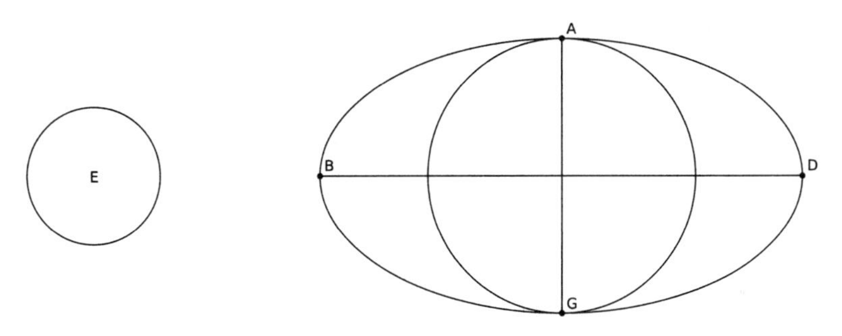

Fig. IV-2-5.

Let the ellipse be *ABGD* with the inscribed circle *AG* [Fig. IV-2-5]. Let *E* be another circle. Let *Z* be a line such that the ratio of *Z* to the diameter of *E* is equal to the ratio of the diameter to *AG*. I claim that the ratio of ellipse *ABGD* to circle *E* is equal to the ratio of diameter *BD* to the line *Z*.

Proof: The ratio of circle *AG* to circle *E* is equal to the ratio of the square of *AG* to the square of the diameter of *E*. But the ratio of the square of *AG* to the square of the diameter of *E* is equal to the ratio of the line *AG* to the line *Z*, since the three lines are proportional. The ratio of circle *AG* to circle *E* is therefore equal to the ratio of *AG* to *Z*, and the ratio of ellipse *ABGD* to circle *AG* is equal to the ratio of *BD* to *AG*. Under these

[22] By proposition 12.

[23] This is equivalent to proposition 5 of Archimedes's *Conoids and Spheroids*, which states: The ratio of the area of any ellipse to that of any circle is the same as the ratio of the rectangle contained by the major and minor axes of the ellipse to the square of the diameter of the circle.

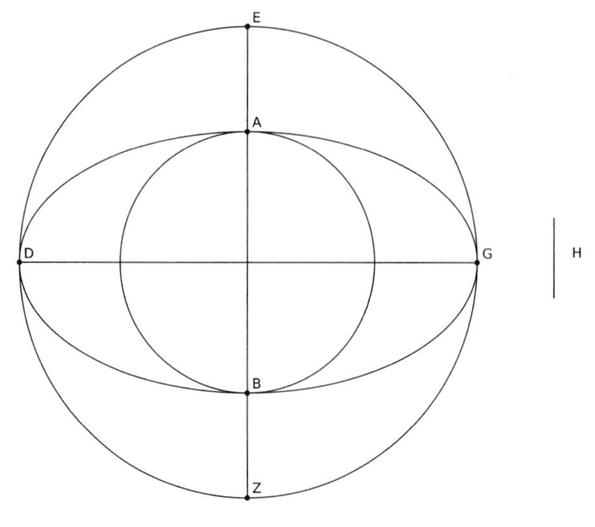

Fig. IV-2-6.

conditions, if one considers the ratio of equality,[24] the ratio of ellipse *ABGD* to circle *E* is equal to the ratio of *BD* to *Z*.

Proposition 15: We wish to show that the ratio of the small circle to the ellipse is equal to the ratio of the ellipse to the great circle.

Let *EGZD* be the great circle, *ABGD* the ellipse, and *AB* the small circle [Fig. IV-2-6]. I say that the ratio of circle *AB* to ellipse *ABGD* is equal to the ratio of ellipse *ABGD* to circle *EZGD*.

Proof: We fix a line *H* such that the ratio of the minor axis *AB* to the major axis *GD* is equal to the ratio of *GD* to *H*. We have established that the ratio of *AB* to *H* is equal to the ratio of the square of *AB* to the square of *GD*. But the ratio of the square of *AB* to the square of *GD* is equal to the ratio of circle *AB* to circle *EZ*. It follows that the ratio of circle *AB* to circle *EZ* is equal to the ratio of *AB* to *H*.

The ratio of the ellipse *ABGD* to a certain surface *T* is equal to the ratio of *GD* to *H*. But we have established that the ratio of circle *AB* to ellipse *ABGD* is equal to the ratio of *AB* to *GD*. If one considers the ratio of equality, it follows that the ratio of circle *AB* to surface *T* is equal to the ratio of *AB* to *H*. But we have shown that the ratio of *AB* to *H* is equal to the ratio of circle *AB* to circle *EZ*. It follows that the ratio of circle *AB* to circle *EZ* and its ratio to the surface *T* are the same. The surface *T* is therefore equal to circle *EZ*. But given our hypotheses, the ratio of circle *AB* to ellipse *ABGD* is equal to the ratio of the ellipse to the surface *T* [prop. 13]. The surface *T* being equal to the circle *EZ*, it follows that the ratio of circle *AB* to ellipse *ABGD* is equal to the ratio of ellipse *ABGD* to circle *EZ*. That is what we wanted to prove.

[24]The phrase "ratio of equality" refers to *Elements* V, definition 17 and proposition 22, which says—in a particular case—that if one is given magnitudes a, b, c and other magnitudes d, e, f, and if $a{:}b = d{:}e$ and if $b{:}c = e{:}f$, then $a{:}c = d{:}f$.

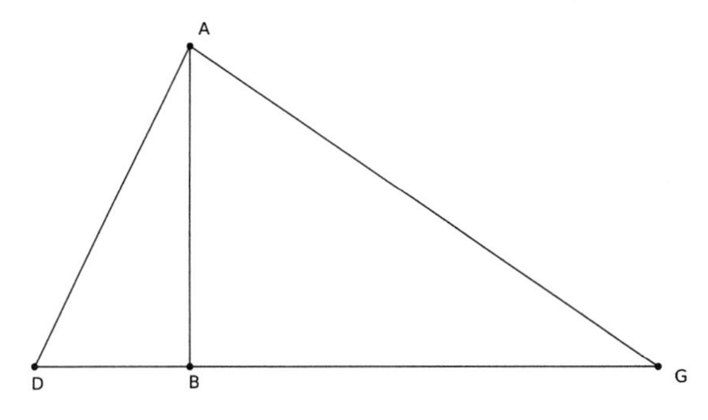

Fig. IV-2-7.

Corollary: It follows that the ratio of the small circle to the great one is equal to the duplicate ratio of that of the small circle to the ellipse[25] and that the ratio of the great circle to the small circle is equal to the duplicate ratio of that of the great circle to the ellipse.

Proposition 16: Every ellipse is equal to the right triangle of which one of the sides containing the right angle is equal to the circumference of the inscribed circle and of which the second side is equal to half of the greatest diameter.

Let the circumference of the inscribed circle be *AB*; let half of the greatest diameter of the given ellipse be *BG* with the angle *ABG* being right [Fig. IV-2-7]. Join *A* and *G*. I claim that the triangle *ABG* is equal to the aforementioned ellipse.

Proof: Prolong *GB* in a straight line, such that *BD* is equal to half of the smallest diameter. In virtue of what has been established by Archimedes, triangle *ABD* is equal to the small circle. In virtue of what we have ourselves established [prop. 14], the ratio of the ellipse to triangle *ABD* is equal to the ratio of triangle *ABG* to triangle *ABD*. The ellipse is therefore equal to triangle *ABG*. That is what we wished to prove.

Corollary: It results from what we have established that if we take five-sevenths and one half of one-seventh of the smallest diameter, and multiply this by the greatest diameter, we obtain the area of the ellipse. In fact, the area of triangle *ABG* is obtained by multiplying half of *AB* by *GB*. But half of *AB* is three times and one-seventh of *BD*. Consequently, the quarter of the half of *AB* is five-sevenths and one-half of one-seventh of *BD*. It follows that the product of five-sevenths and one-half of one-seventh of twice *BD* by twice *BG* measures the triangle *ABG*. That is what we wished to prove.

Proposition 17: Every ellipse is equal to the circle of which the diameter is the mean proportional between the two axes of the given ellipse.

Let the minor axis be *A* and the major axis *G* [Fig. IV-2-8].[26] One takes the line that is the mean proportional between these two, *B*. The ratio of *A* to *B* is equal to the ratio of *B* to *G*. And when three lines are proportional, the circles of which they are the diameters

[25]The duplicate ratio of *a* to *b* is the ratio of *a* to *b* compounded with itself. We would say that this is the ratio of *a* squared to *b* squared.

[26]In the proof of proposition 17, the text alternates between referring to *A*, *B*, *G* sometimes as circles and sometimes as lines representing their diameters. Prof. Levy informs me (private communication) that the Hebrew manuscript shows circles, so I have followed the manuscript in the diagram.

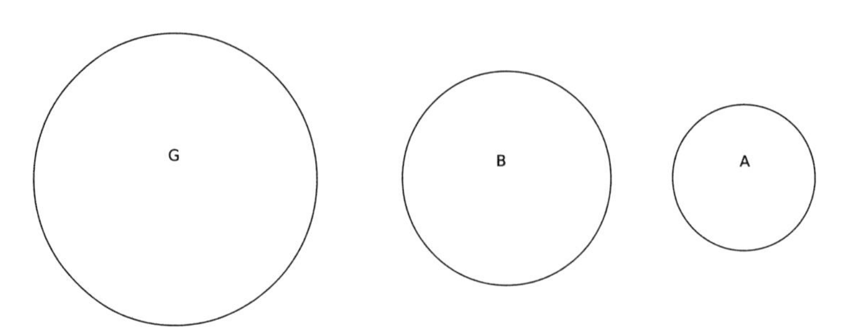

Fig. IV-2-8.

are also proportional. The ratio of the circle [of diameter] *A* to the circle [of diameter] *B* is equal to the ratio of the circle [of diameter] *B* to the circle [of diameter] *G*; and the ratio of the circle [of diameter] *A* to the circle [of diameter] *G* is equal to the square of the ratio of the circle [of diameter] *A* to the circle [of diameter *B*].

But the circle [of diameter] *A* is that which is inscribed in the ellipse, and the circle [of diameter] *G* is that which is circumscribed about the ellipse. We have established that the ratio of the circle inscribed in the ellipse to the circle circumscribed about the ellipse is equal to the square of the ratio of the circle inscribed in the ellipse to the ellipse itself. It follows that the square of the ratio of the circle [of diameter] *A* to the circle [of diameter] *B* is equal to the square of the ratio of the circle [of diameter] *A* to the ellipse. As a consequence, the ellipse is equal to the circle [of diameter] *B*.

3. ABŪ ʿABD ALLAH MUḤAMMAD IBN MUʿĀDH AL-JĀYYĀNĪ, ON RATIOS

The little that we presume to know of the life and career of Ibn Muʿādh has been reconstructed from a scattering of clues gleaned from his works as well as from a few secondary references. For instance, his name, Abū ʿAbd Allah Muḥammad ibn Muʿādh al-Jayyānī, implies that he was from Jaén in Andalusia. He is known to have been a *qādī* (religious judge) and in fact came from a family whose members included a number of such learned officials. Recent scholarship suggests that he was born in the early eleventh century, so, since his *On the Total Solar Eclipse* describes an eclipse that he observed on July 1, 1079, we assume that, given his advanced age at that time, he probably died not long after. Beyond this meager and tentative biographical sketch, there is little or nothing that we can reasonably venture.

Ibn Muʿādh composed astronomical tables that Gerard of Cremona translated into Latin. These tables became known as the *Tabulae Jahen* (*Jaen Tables*). The following extensive extract from a treatise he wrote defending Euclid's view of ratio and proportion,[27] one of a number of medieval publications on the topic, shows that he understood the theory well. He argues persuasively for it acceptance.

We intend to explain what may not be clear in the fifth book of Euclid's writing[28] to such as are not satisfied with it, even though I am convinced, as to Euclid's reasoning, that his intentions and his aims are, to any true and fair examiner, more evident and more clearly

[27] *Elements* V.
[28] Throughout this excerpt, "his book" refers to the *Elements*.

subdivided than much by means of which one has tried to clarify it and hoped to illustrate it. Only some things were left aside because his reasoning and the arrangement of his writing were sufficient to make them understood and realized.

Concerning that which met with opposition and people thought incomplete or unclear, so that they made it complete or clear according to their own thinking, well, their reasoning on this point is more fit for explanation and more in want of proof than Euclid's writing. So, for instance, is our opinion on the writing ascribed to Syndus,[29] for he who considers his proofs of that which Euclid gives without proof, as: "If two straight lines are drawn over less than two right angles, they will inevitably meet" and "the diameter bisects the circle" and "two straight lines do not enclose a surface" will see weakness and debility in them and the necessity of assuming things which are more extensive than those Euclid claims to assume as self-evident.

Now somebody may remark: "If Euclid's reasoning in his writing is as you mention, and nobody makes it clearer than Euclid did, why then do you desire to explain his reasoning in the fifth book and elucidate what is obscure in it?" Then I say: By this writing we only mean to explain that Euclid's reasoning in this book is very sound and his method very clear. For many think that Euclid approaches the explanation of ratio by a door other than its proper door, and introduces it in a wrong way by his definition of it by taking multiples, and separating from its definition concerning its essence that which is understood by the very conception of ratio. And they judge that there is no obvious connection between ratio and taking multiples.

But upon my life, with nothing is ratio more closely connected than with taking multiples of the two magnitudes being compared. On the whole I say that the fifth book contains some obscurity, and that the study of it may be tiring, but all strivings are measured by the nobility of the object to be achieved, and he who aspires to a beauty's hand must pay the price. . . .

I say, and God be my help, that the things that come under the [concept] "quantity" used in the art of geometry are five, viz., number (which is the first and simplest of them), line, surface, angle and solid. Let us use the term magnitude in general to denote any of these things as a name for the species comprising all of them. When some magnitudes all belong to one of these five species, let them be called "homogeneous."[30] These have the property that by frequent multiplication, the lesser can equal the greater or exceed it. Also by frequent division, the greater can become equal to the less or fall short of it. When there are two homogeneous magnitudes and the lesser measures the greater, then the lesser is called a part of the greater and the greater is called the multiple. It is also called the superior to that part. When the lesser of the two homogeneous magnitudes measures the greater, or when there is a third magnitude of the same species that measures each of the two, then they are called commensurable. And when the lesser does not measure the greater, and no third magnitude is found that measures them, then they are called incommensurable.

Ratio is the size of a magnitude as compared with another magnitude of the same species, viz., a comparison is made between the two magnitudes so that the size of one

[29]No writer of this name is known.

[30]In using the terms "magnitude" and "homogeneous" to include numbers, Ibn Mu 'adh was deviating from what had been common mathematical usage since the time of Plato and Euclid in the fourth century BCE.

of them as compared with the other may be known.[31] Proportion is equality of ratios.[32] When a magnitude is related to another magnitude, the latter is called consequent and the former the antecedent, and either of antecedent and consequent is called the companion of the other.

The whole is greater than the part, and likewise it is usual to admit that the part is less than the whole. Now from the whole only such parts are taken that do not amount to the whole; for if it exceeds the whole, it is called greater than it, or a multiple of it, and the term "taking parts" is not applicable to it. We however will introduce in this writing of ours the practice that when we take parts of some magnitude, we will pay no attention to the question [of] whether such parts do equal the magnitude they are taken from or fall short of it or exceed it. To do so would only add the difficulty of repeating our reasoning and prolixity in what we intend to explain hereafter, if God is willing. For this will neither damage the proof in any way, nor will it deviate from any assumed base or be at variance with it. It only serves to facilitate the reasoning and to shorten it. So we absolve ourselves from the obligation to indicate in every case which magnitude is the greater in order to take parts from it.

If the magnitudes are commensurable, then one of them contains exact parts completely filling up the other. For if the third magnitude, which measures both, is produced, this third is a part of each of them, and when it is taken a number of times from one of the two magnitudes and exhausts that magnitude in numbering, then evidently the measured magnitude is "parts" of the other. If however the magnitudes are incommensurable, none of them contains exact parts of the other. As to the terms of the proportion, the first of them contains as many parts of the second as the third contains such parts of the fourth. This is evident, no veil is over it, and it needs no proof. Euclid deals with it in the same way in the seventh book,[33] but it is only effective if the first magnitude is commensurable with the second and the third commensurable with the fourth, so that each contains a whole number of parts of its companion.

When however the companions are incommensurable and it is impossible to express the size of each antecedent as compared with its consequent and to reduce it to a partition and to express it by means of parts, then it is evident, although the well-known approach of expressing [magnitudes] by parts does not exist, ... that it is impossible for the ratio of the first to the second to be like the ratio of the third to the fourth when the first contains more or fewer parts of the second than the third of the fourth. This needs no proof, because it appeals immediately to the mind. For when a magnitude contains more parts of a second magnitude than a third magnitude contains such parts of a fourth magnitude, then the size of the first as compared with the second is not like the size of the third as compared with the fourth. Over this there is no veil, nor is it made clearer by prolixity in the reasoning, since, in this case as well, things that are clear and evident to the mind without need of proof are not made clearer by prolix explanation, because there is no method to make clear what is already clear in it. This preamble, however, detracts nothing from the fact that in this case, there is no expressing [the magnitudes] by means of parts that cover

[31]This is substantially the definition of ratio in Euclid's *Elements* V, Def. 3

[32]Euclid (*Elements* V, Def. 6) considers proportion not in terms of equality of ratios but as a relationship between magnitudes ("Magnitudes which have the same ratio are called proportional.")

[33]This is the book of his *Elements* in which Euclid deals with ratios of whole numbers.

each of the two antecedents and that it [the relationship] can only be expressed by less and more.

However what is less than a thing is not determined, neither is what is more. Equality only is determined. But we often find this situation in this art [geometry]. For instance, when we want to know the length of the circumference of a circle and its size as compared with the diameter, we learn this by less and more; for likeness and equality fail in it, because equality never occurs between a curved line and a straight one, for they are not of the same kind.[34]

And so too are the commensurable and the incommensurable among the magnitudes, these being not of the same kind. For expressing by parts is impossible among incommensurables, just as it is impossible with one special measure among circumference and diameter. About the circumference it is only said that it is greater than the sides of every figure inscribed in the circle and less than the sides of every figure circumscribed about the circle. And likewise, when we relate some magnitude to a magnitude incommensurable with it and are obliged to express this [relationship] or to approximate it by some expressible ratio for some design or other, [in such a case] we say that it is greater than so-and-so many parts [of the other] and less than some other number of parts [of that other].[35]

So too it is with irrational roots. If we are obliged to realize them in parts, we say that the root of a given number which is not a square is greater than the root of each square less than the given number and less than the root of each square greater than the given number; and the omission and impossibility of expressing do not intrude upon the purport of it and neither affect the sound base nor transform it.

When now it is all true what we have mentioned about this, we only expatiate on it and bring it forward and emphasize it in order that it may be a basis to rely upon in what follows, if God is willing. This is that when the ratio of the first to the second is as the ratio of the third to the fourth, it cannot happen that, of any parts of the second, the first [magnitude] should contain anything more than the third contains of such parts taken from the fourth—and no less either. However no indefiniteness must be seen in our words "any parts of the second ... anything," because if the names of the parts of the second contained in the first differ from the names of the parts of the fourth contained in the third, the ratios are different and not equal, irrespective what parts are concerned.[36]

On this point we will give some elucidation by means of a particular case, because there is some help in such an example, which nobody will refuse, although we have fastened down already the general idea of what applies to all particular cases in a sufficient way. We will give it however for the sake of greater clarity [Fig. IV-3-1].

[34]This view, that the straight and the curved are of different genera and so the one cannot be measured by the other, goes back to Aristotle. Abū Isḥāq al-Ṣābī criticized Abū Sahl al-Kūhī, a distinguished geometer of the late tenth century, for ignoring this principle in one of his results.

[35]One possibility is that Ibn Mu'ādh had in mind here Archimedes's result, which was well known to Muslim scholars, that the circumference of any circle exceeds three times its diameter by less than one-seventh part of that diameter but by more than ten seventy-first parts of the diameter.

[36]Ibn Muʿādh is saying here that, for example, if the first contains three seventh-parts of the second and the third contains three fifth-parts of the fourth magnitude then, even though the first and third each contain the same number of parts of the second and fourth, they are different parts and so there is no proportion in this case.

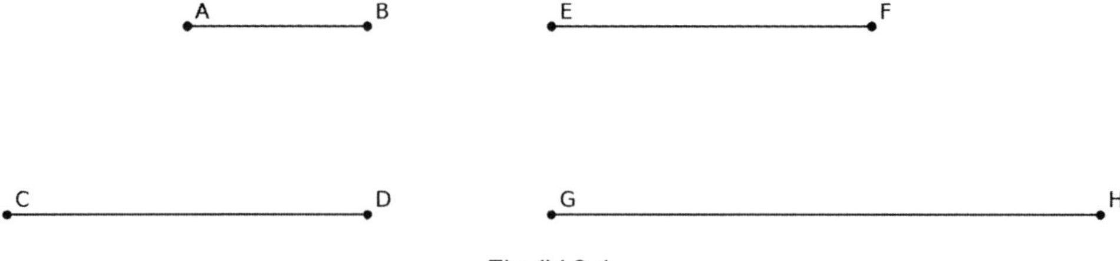

Fig. IV-3-1.

So we say: When somebody supposes the ratio of a magnitude AB to a magnitude CD to be as the ratio of EF to GH, then the reasoning about this is clear if AB and CD are commensurable; I mean that AB contains as many parts of CD as EF contains such parts of GH. But if they are incommensurable, then, although we are unable to express how many parts AB contains of CD, anyone who supposes these magnitudes to be proportional is not allowed to say, for instance, that AB is more than two thirds of CD, and EF less than two thirds of GH or equal to two thirds of GH. Also he may not say that AB is two thirds of CD, and EF more than two thirds of GH nor less either; nor that AB is less than two thirds of CD, and EF more than two thirds of GH or equal to two thirds. These things are essentially absurd. Equally, when we examine the other parts and cite them, for what is an absurd view in respect of the parts we mentioned is also absurd in respect of the other parts, be they ever so small.

Therefore if somebody thinks these magnitudes to be proportional and says that AB is, for instance, more than three hundred and sixty-seven ten-thousandth parts of CD and that EF is less than such parts taken from GH or equal to them, there is no difference between this and the preceding. This we adduced, although our reasoning needs no more of the kind in order that it might be clear that the theorem is quite general, so that the condition applies to all parts, small or large, no parts being distinguished above others. This is that when of proportional magnitudes, the ratio of the first to the second is as the ratio of the third to the fourth, it is impossible to find parts of the second and fourth equal in number and denomination, whatever parts they may be, so that the parts of the second are found to exceed the first magnitude, unless the parts of the fourth, too, exceed the third magnitude; or that the parts of the second are found to be equal to the first magnitude, unless the parts of the fourth magnitude, too, are equal to the third magnitude; or that the parts of the second are found to fall short of the first magnitude, unless the parts of the fourth, too, fall short of the third magnitude. For if parts of the second are found exceeding the first and parts of the fourth being equal to the third or falling short of it, the first contains fewer parts of the second than the third of the fourth, and of this we proved already the absurdity. And likewise, when parts of the second are found being equal to the first magnitude and parts of the fourth exceeding or falling short of the third magnitude, the first contains fewer or more parts of the second than the third of the fourth; and likewise, when parts of the second are found falling short of the first magnitude and parts of the fourth being equal to the third magnitude or exceeding it, the first contains more parts of the second than the third of the fourth.

All this is absurd, and what remains for us in this preamble is to show whether it is convertible or not. We then say that it is convertible, *viz.*, when there are four magnitudes while of all parts equal in number and denomination taken from the second and the fourth

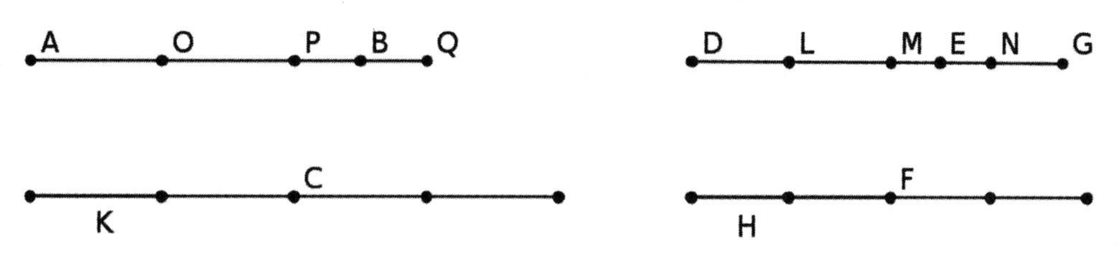

Fig. IV-3-2.

the parts of the second are not found to exceed the first magnitude, unless the parts of the fourth too exceed the third magnitude, nor are the parts of the second found to be equal to the first magnitude, unless the parts of the fourth, too, be equal to the third magnitude, nor are the parts of the second found to fall short of the first magnitude, unless the parts of the fourth too fall short of the third magnitude, then the ratio of the first to the second is as the ratio of the third to the fourth.

If, for instance, we take for the four magnitudes AB, C, DE, F, and suppose that of all the same parts found of C and F, the parts of C are not found to exceed AB, unless the parts of F too exceed DE, and the parts of C are not found at be equal to AB, unless the parts of F too be equal to DE, and that the parts of C are not found to fall short of AB, unless the parts of F too fall short of DE, then I say that the ratio of AB to C is as the ratio of DE to F [Fig. IV-3-2].

The proof of this is that it cannot be otherwise. For if that could be so, let the ratio of AB to C be as the ratio of DG to F.[37] If GE is less than F, we say: we divide F again and again, first taking the half, then which follows in the division, *viz.*, the third and so on, until we come to the first part of F that is less than GE and call that part H. This is less than GE. From C we take the same part; let it be K. Then we subtract from DE pieces equal to H, beginning to subtract from D, until it comes to the first mark near E not reaching up to G, for GE is greater than H; and let these pieces [i.e., the ones equal to H] be DL, LM, MN. Then we subtract from AB pieces equal to K; it will contain K as many times as DE contains H. The result passes B and exceeds AB, just as DN exceeds DE, because we have supposed so about the magnitudes AB, C, DE, F. Let these parts be AO, OP, PQ. Consequently AQ is parts of C, and DN is the same kind of parts of F, in the same number. But AQ is greater than AB, and DN is less than DG; therefore AB contains fewer parts of C than DG of F. And yet the ratio of AB to C should be as the ratio of DG to F. This is a contradiction; it is impossible, and consequently the ratio of AB to C is unlike the ratio of something greater than DE to F. Likewise it is proved to be unlike the ratio of something less than DE to F. Consequently the ratio of AB to C is only as the ratio of DE to F, and this is what we wanted to prove.

Since what we have previously said and emphasized is now clear, and we have gotten a sound idea of proportion in the case of commensurable magnitudes (where expressing

[37]The existence of a fourth proportional for arbitrary magnitudes, represented in this case as a line segment *DG*, was a common assumption in Greek and medieval Islamic mathematics. In the case when the three given magnitudes are line segments, the existence of a fourth proportional is proved in *Elements* VI.12.

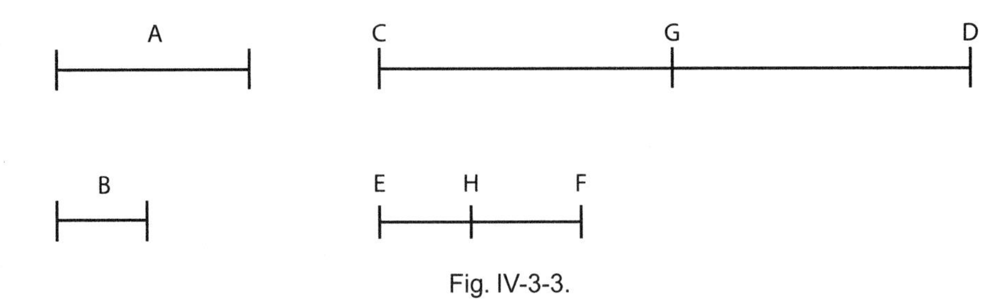

Fig. IV-3-3.

by means of parts is possible in an exact way) and in the case of incommensurable magnitudes, although expressing by means of parts in the strict sense is not attainable here, what we have mentioned may suffice, if God, who is lofty, is willing.

Now, let us say what is our intention in this writing. This is the view that brought Euclid to his definition of ratio by means of taking multiples with restriction to that, without completing all that is accidental and complementary to ratio and its increasing and decreasing; and to the neglect of all this being content with taking multiples. Therefore I say, and God be my help for right conduct, that, as [the concept of] proportion arises, as we mentioned, from the consideration of parts and their comparison, all that occurs with the parts we take occurs likewise with the multiples, whether it be exceeding, falling short or being equal, and what is unattainable with parts is also unattainable with multiples, because multiplication and partition are of one category. For they both originate from unity, only that multiplication originates from it by increase and enlargement, and partition inversely by decrease and diminution. The same principle applies to each of both. The two magnitudes indeed, when it is possible to find parts of one of them that are equal to some parts of the other, then it is also possible to take a multiple of one of them that is equal to some multiple of the other; which will occur in the case of commensurable magnitudes. But when it is entirely impossible to find parts of one of them that are equal to parts of the other, then it is also impossible that there is some multiple of one of them equal to some multiple of the other; which will occur in the case of incommensurable magnitudes.

And likewise, when to two unequal magnitudes a third magnitude is found less than these two, measuring them and being a part of each of them, then also a common multiple of them will be found, so that the conformity of the conditions of the multiples to the state of affairs with the parts is yet more evident.

I will now present something of which an example is required, *viz.*, that if A is some multiple of B, and the same multiple is taken of A and B, *viz.*, CD and EF, then I say that CD is the same multiple of EF as A is of B [Fig. IV-3-3].

Proof: We divide CD in its parts equal to A, which may be CG and GD, and divide EF in its parts equal to B, which may be EH, HF, then CG is equal to A and EH equal to B, and likewise CG is the same multiple of EH as A is of B, and likewise GD of HF. Therefore the whole, CD, is the same multiple of EF as A is of B, and the proof is complete.

Now we say that when two different magnitudes are taken, and parts are taken of one of them, and these parts are compared with the other magnitude, then their condition as to exceeding, falling short and being equal is the same as the condition of a certain multiple taken of the partitioned magnitude in respect of a certain multiple taken of the other magnitude: if [the number of] the multiple of the partitioned magnitude is equal to

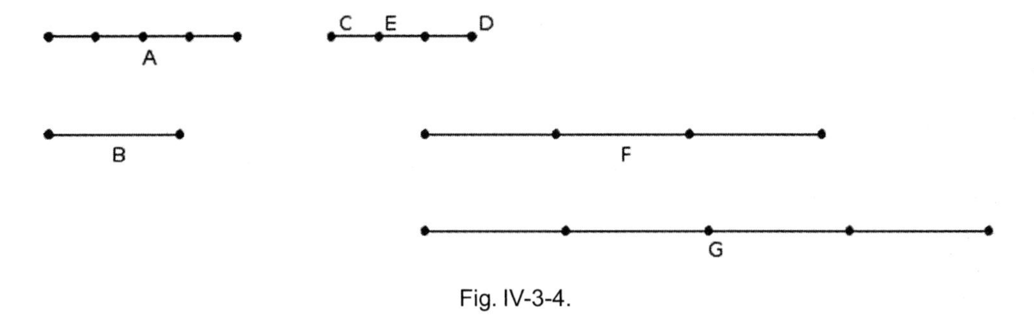

Fig. IV-3-4.

the number of the parts, and likewise [the number of] the multiple of the other magnitude is equal to the denominator[38] of the parts.

An example of this is that *A* and *B* are different magnitudes [Fig. IV-3-4], and a number of parts is taken of *A*, which may be *CD*, one of the parts being *CE*. And, of the magnitude *A*, a multiple is taken, corresponding to the number of parts *CE* contained in *CD*, which may be *F*. And, of *B*, a multiple is taken corresponding to the multiple that *A* is of its part *CE*, which may be *G*. Then I say that the condition of *CD* in respect of *B*, as to being equal, exceeding, and falling short, is the same as the condition of *F* in respect of *G*. So that *F*, which is a multiple of *A*, corresponds in condition with *CD*, which is parts of *A*, and *G* corresponds in condition with *B*, of which it is a multiple: if *CD* is equal to *B*, then *F* is equal to *G*; and if *CD* is greater than *B*, then *F* is greater than *G*; and, if *CD* is less than *B*, then *F* is less than *G*.

The proof of this is, that *G* is the same multiple of *B* as *A* is of *CE*; and also, that *F* is the same multiple of *A* as *CD* is of *CE*. Therefore *A* is a multiple of *CE*. Now the same multiple is taken of *A* and *CE*, which are *F* and *CD*, so that *F* is the same multiple of *CD* as *A* is of *CE*. And *A* is the same multiple of *CE* as *G* is of *B*, therefore *F* is the same multiple of *CD* as *G* is of *B*. Consequently, if *CD* is equal to *B*, then also its multiple, which is *F*, is equal to *G*, which is the same multiple of *B*; and if *CD* is less than *B*, then also *F* is less than *G*; and if *CD* is greater than *B*, then also *F* is greater than *G*. This is what we intended to demonstrate.

Now that this is clear we say: When we intend to compare some magnitude with another magnitude by means of its parts, and substitute a multiple of each of them for the parts in the way we have illustrated in this diagram, then whatever was clear to us about the parts, as to being equal and exceeding and falling short, all is clear to us about the multiples. And likewise, when we take the converse of this proposition, we say: When two magnitudes are different, and when a different multiple is taken of each of them, then the condition of the multiple of one magnitude in respect of the multiple of the other magnitude is [the condition] of being equal, exceeding, or falling short in accordance with the condition of certain parts of each of the two magnitudes in respect of the other magnitude itself; when the denominator of the parts is the number of the multiple taken of the magnitude that is not partitioned, and the number of the parts is equal to the number of the multiple taken of the partitioned magnitude. Therefore it is sound what we have

[38] If *A* is divided into *n* parts, then *n* is the "number that names" (i.e., the "denominator of") one of those parts.

said and mentioned before, *viz.*, if multiples are substituted for parts, they do not give up anything of them nor do they clash with it.

Now Euclid, instead of taking parts of the second and the fourth and comparing each of them with its companion, I mean comparing the parts of the second with the first magnitude and comparing the parts of the fourth magnitude with the third magnitude, only takes a same multiple of the second and the fourth and substitutes the multiples of each of them for its parts, and takes the same multiple of the first and the third and substitutes the multiple of each of them for itself. And all that is clear in the case of the multiples and happens with them in respect of the multiple of the other is clear in the case of the parts and happens with them in respect of the other magnitudes.

We will illustrate this further with an example and say: When there are four magnitudes, and when, whatever parts equal in number and denomination be taken of the second and the fourth,[39] the parts of the second are not found to exceed the first magnitude, unless the parts of the fourth are found to exceed the third magnitude, and the parts of the second are not found to fall short of the first magnitude, unless the parts of the fourth are found to fall short of the third magnitude, and the parts of the second are not found to be equal to the first magnitude, unless the parts of the fourth are found to be equal to the third magnitude; then, all same multiples being taken of the first and the third and all same multiples being taken of the second and the fourth, if the multiple of the first exceed the multiple of the second, the multiple of the third too will exceed the multiple of the fourth, and if the multiple of the first is equal to the multiple of the second, the multiple of the third is equal to the multiple of the fourth, and if the multiple of the first falls short of the multiple of the second, the multiple of the third falls short of the multiple of the fourth. Also I say: When the multiples are in this condition, then the parts are in the first-mentioned condition. And if any one of these two conditions[40] is found to be satisfied so is the other too, and the proportionality of the magnitudes in order is assured, and the ratio of the first to the second is as the ratio of the third to the fourth.

For instance, let the magnitudes be A, B, C, D [Fig. IV-3-5], and let the aforesaid condition as to the parts be found [to hold]; I mean that, whatever equivalent parts are taken of B and D, the parts of B are not found to exceed A, unless the parts of D too exceed C, and the parts of B are not found to be equal to A, unless the parts of D too be equal to C, and the parts of B are not found to fall short of A, unless the parts of D too fall short of C. Then I say that whatever same multiple be taken of the first and the third, and whatever same multiple be taken of the second and the fourth, the multiple of the first will not be found to exceed the multiple of the second, unless the multiple of the third too be found to exceed the multiple of the fourth; and the multiple of the first will not be found to fall short of the multiple of the second, unless the multiple of the third too be found to fall short of the multiple of the fourth.

The proof is that it cannot be otherwise. For if such were possible, let equimultiples be taken of A and C, say, E [and] F, as well as equimultiples of B and D, say, G and H. And

[39] To say that parts of the second and fourth are of the same denomination is to say that, e.g., they are both third parts, or both fourth parts.

[40] The phrase "two conditions" refers to the statement regarding exceeding, falling short, or equaling applied to parts or multiples.

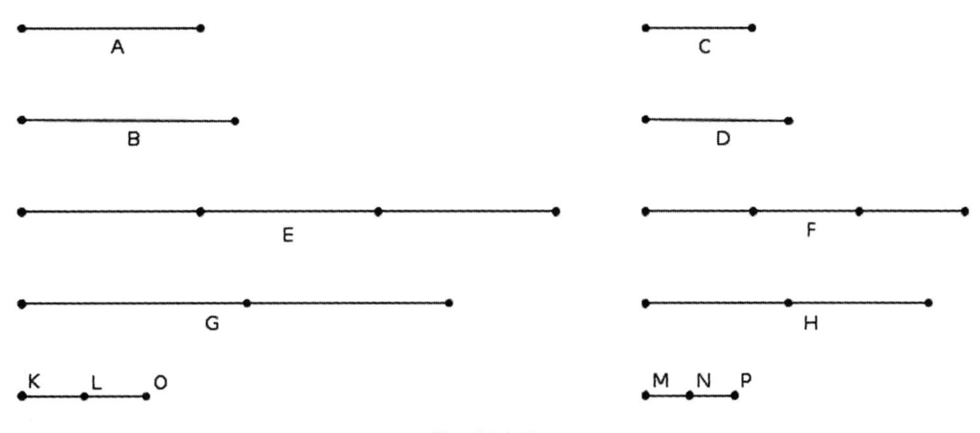

Fig. IV-3-5.

suppose that the condition of *E* in respect of *G* be unlike the condition of *F* in respect of *H* as to being equal, exceeding or falling short, namely, *E* greater than *G* and *F* not greater than *H*, if such were possible.

Let us now take of *B* the part denominated after the number of the multiple that *E* is of *A* and *F* is of *C*, for these are the same. This may be *KL*. And of *D* the same part, which may be *MN*. Then we make *KO*, which must be the same multiple of *KL* as *G* is of *B* and *H* is of *D*, for these are the same. Likewise we make *MP*, which must be the same multiple of *MN* as these multiples were. Now *KO* is parts of *B*, and *MP* is the same parts of *D*. Therefore *A* and *B* are two different magnitudes, and of one of them, *B*, parts are taken, *KO*. And of *B* also a multiple is taken, *G*, the number of which is the number of the parts in *KO*. And of *A* a multiple is taken, viz., *E*, the number of which is the denominator of the parts, in respect of which it has been demonstrated in the foregoing proposition that the condition of *E* in respect of *G* is like the condition of *A* in respect of *KO*, as to being equal, exceeding, and falling short.

Now we supposed *E* to be greater than *G*, and therefore *A* is greater than *KO*. The magnitudes *C* [and] *D* behave in the same way, both their multiples and their parts, so that the condition of *F* in respect of *H* is like that of *C* in respect of *MP*, as to being equal or exceeding or falling short. But *F* was not greater than *H*, so *C* is not greater than *MP*. And *A* is greater than *KO*, and *KO* and *MP* are parts, equal in number and denomination, of *B* and *D*. Therefore parts are found of *B*, viz., *KO*, less than *A*, and parts are found of *D*, viz., *MP*, not less than *C*. This is a contradiction. It is impossible, because we have supposed about the magnitudes that no parts of them could be found other than satisfying the condition we have mentioned as to being equal or exceeding or falling short. Hence, it is clear that when the said condition of the parts is found, the condition of the multiples is found too, and the proportionality is found. And this is what we wanted to prove.

By the same procedure it is made clear to us, if we suppose about these magnitudes that the condition of the multiples is found, that the condition of the parts is found too and that the proportionality is found.[41]

[41] If $(m/n)B$ is less than, or equal to, or greater than A implies $(m/n)D$ is less than, or equal to, or greater than C, respectively, then mB is less than, or equal to, or greater than nA implies mD is less than, or equal to, or greater than nC, respectively, and conversely.

As it is clear now conclusively that the definition of ratio by means of multiples is sound, and that all that is found with parts is found likewise with multiples, and that these two methods of reaching the truth and to ascertain the soundness [of a statement about proportion] are equal, since one of them is not opposed to the other nor does one of them imply what the other declares impossible or declare impossible what the other implies. So we say that the searching of proofs about ratio and what is additionally or accidentally connected with it, and the investigation of this with multiples is more elegant than with parts and is more convenient for many reasons. One of these is that to take multiples of each of the two magnitudes is not difficult, be it a small or a large multiple; but to take parts of some magnitude in such a way that these parts exceed that magnitude, and to call what is greater than the thing parts of it, that is difficult, since it arises from an unusual conception.

For in parts of a thing is only expressed what is less than it. This we mentioned already in the preface of this writing. If therefore we avoid this view and strive to take parts of the greater of the two (companions) only, be it the antecedent or the consequent, then we are not allowed nor able to do so before we know, in each case of a ratio, whether the antecedent or consequent is greater. With multiples we do not need [to know] anything of the kind. The facility of this [method of multiples] as compared with the difficulty [of the method of parts] only he knows who has tried the exposition of the definition of ratio by means of parts as well as by means of multiples, for there is between them a large difference in difficulty of research and awkwardness of exposition. Moreover if we would carry out the definition of ratio by means of parts, we would have to alter the necessary propositions that are advanced in the fifth book [of Euclid's *Elements*], and we would have to transpose them from multiples into parts. But taking parts gives more work and difficulty in research, because in the case of multiples, the magnitude is only multiplied by some number, whereas in the case of parts, one needs the number of the parts and of the denominators of the parts and besides [one needs] the whole. Truly the knowledge of the large in arithmetic is easier than the knowledge of the small.

4. AL-MU'TAMAN IBN HŪD, *KITĀB AL-ISTIKMĀL* (*BOOK OF PERFECTION*)

Al-Mu'taman ibn Hūd (d. 1085) is an important figure in the history of mathematics in Andalusia. Until recently his works were thought to have been lost, but in the late 1980s Professors Ahmed Djebbar and Jan Hogendijk discovered manuscripts of his extensive survey of the mathematics of his time, his *Kitāb al-Istikmāl* (*Book of Perfection*). Ibn Hūd had planned for the book to have two "genera," but he had only finished the first when he became king of Saragossa, Spain, in 1081 and evidently had no time to write the second before he died four years later. Ibn Hūd had an elaborate division of his "genera" into species, subspecies, and sections. The work, definitely not intended for beginners, sheds unexpected light on the mathematics of Ibn Hūd's time and is a fascinating blend of mathematics from Greek and Arabic sources, as well as what appear to be some original contributions of Ibn Hūd himself. The section numbers of each result from [Hogendijk, 1991] are given as well as the subject matter.

312.14: Heron's Theorem

In this selection from his *Istikmāl*, Ibn Hūd gives what appears to be a new proof of Heron's Theorem, stated by Heron of Alexandria in *Metrika* I,8: If ABG is any triangle

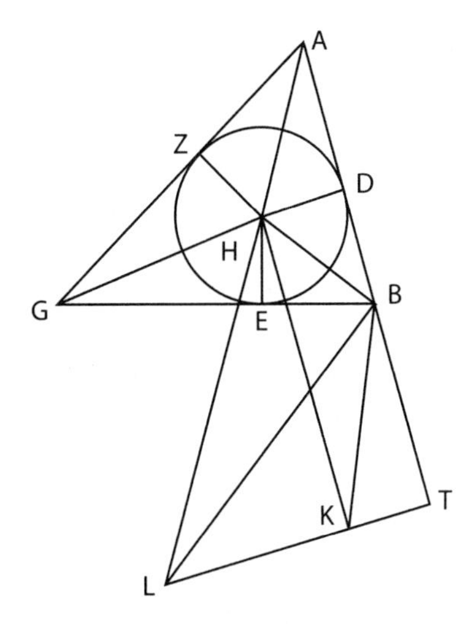

Fig. IV-4-1.

with sides a, b and g, and if s denotes half of its perimeter, then, in modern notation, $A = \sqrt{s \cdot (s-a)}\sqrt{(s-b) \cdot (s-g)}$.[42] As the reader will see from the translation of the Arabic text, the author phrased this equation as proportion: the product $s(s-a)$ is to the area of triangle ABG as that area is to the product $(s-b) \cdot (s-g)$.

Professor Hogendijk, who has studied the geometrical portions of the treatise extensively, suggests that Ibn Hūd's account here relies on an Arabic translation of a Greek treatise. It is clear, in any case, that the author of the original assumed that the reader was familiar with the basic facts of the incircle of a triangle. Since these are by no means commonly known today, the notes for this section add some reminders of these facts to aid the reader.

[For] each triangle the ratio of the surface that is made of half the sum of its sides by the excess of that half over one of the sides to the surface of the triangle is as the ratio of the surface of the triangle to the surface that is made from the excess of half the sum of the sides over one of the two remaining sides by [the excess over] the other.

Example: In triangle *ABG*, the line *AB* is produced to *T* and *ABT* is made equal to half of *AB* and *BG* and *GA*, and *BT* is the excess of *AT* over *AB* [Fig. IV-4-1]. And let *BD* be the excess of *AT* over *AG*, and [so] *AD* is the excess of *AT* over *BG*.[43] Then I say that the ratio of *TA* by *AD* to the surface of triangle *ABG* is as the ratio of the surface of triangle *ABG* to the surface of *DB* by *BT*.

Proof: We mark point *H*, the center of the circle made in the triangle *ABG*,[44] and we produce perpendiculars *HD*, *HE*, and *HZ* onto *AB*, *BG*, and *GA* [respectively].[45] And we produce from point *H* the line *HK* parallel to line *DT* and line *TK* parallel to the

[42]See section II-4-1 of Chapter 1 for another proof of this theorem.

[43]This deduction is correct, but Ibn Hūd does not justify it. It seems he felt the reader could supply the necessary argument.

[44]That is, what is now called the "incircle" of triangle *ABG*, whose construction is explained in *Elements* IV.4.

[45]That the incircle would be tangent to *AB* at *D*, which has already been defined, is something that Ibn Hūd assumes the reader knows. He also assumes his readers know that point *H* is the intersection of the three angle bisectors, which is shown in *Elements* IV.4.

perpendicular *HD*. And we produce from point *B* the perpendicular *BL* to line *HB* which meets *TK* at the point *L*, and we join *LH*.

Then since the two angles *HBL*, *HKL* are right angles and subtend line *HL*, they [i.e., their vertices] are on a semicircle [with *HL* as diameter], and so angle *HLB* is equal to angle *HKB*, and angle *HKB* is equal to angle *KBT*. And angle *KBT* is equal to angle *HGZ*, because line *KT* is equal to *HD*, and *HD* is equal to *HZ*, and *TB* was shown to be equal to *ZG*. And the two angles that are at the two points *Z*, *B* are right angles, so by subtraction, angle *LHB* is equal to angle *GHZ*. And the angles *GHZ*, *BHD*, *DHA* are equal to two right angles, because the six angles at point *H* are equal in pairs. So the angles *BHL*, *BHD*, *DHA* are equal to two right angles. And so the line *AHL* is continuous in a straight line.

And angle *HBL* is right, and so the two angles *HBD*, *LBT* are equal to a right angle. And so angle *BLT* is equal to angle *HBD*. Thus the two triangles *BHD* and *BTL* are similar. Hence the ratio of *TL* to *BD* is as the ratio of *BT* to *DH*, and so the surface of *TL* by *DH* is equal to the surface of *TB* by *BD*.

And so, if we make line *AT* a common altitude, the ratio of *AD* to *DH* is as the ratio of *DA* by *AT* to *DH* by *AT*, the consequent [the latter term of the ratio] being the area of triangle *ABG*. And the ratio of *AD* to *DH* is [also] equal to the ratio of *AT* to *TL*.[46] And this equals the ratio of the surface of *AT* by *DH* to the surface of *TL* by *DH*, the consequent being equal to *TB* in *BD*.

Hence the ratio of surface *TA* in *AD* to the surface *HD* in *AT*, whose consequent is equal to the surface of the triangle, is equal to the ratio of the surface of *HD* in *AT* to the surface of *DB* in *BT*.

312.16: Cross Ratio

In Book VII of his *Mathematical Collection* the Greek geometer, Pappus of Alexandria, provides a commentary on Euclid's lost work, *The Porisms*, a work that left traces in Arabic writings. Pappus's work gives us information about the contents of Euclid's *Porisms* and contains a number of lemmas to *The Porisms* proving the invariance of cross ratios,[47] a powerful tool in geometry with versions true in projective geometry and non-Euclidean geometry as well.

Pappus proves, with reference to Fig. IV-4-2, that the ratio of the rectangle whose sides are *BD* and *EG* to the rectangle whose sides are *BG* and *DE* is as the ratio of the rectangle whose sides are *BZ* and *HT* to the rectangle whose sides are *BT* and *ZH*. Ibn Hūd proves the equivalent statement, in which the antecedents and consequents in each of these two ratios are interchanged. He then proves a form of the general theorem on cross ratios according to which (again with reference to Fig. IV-4-2, but this time using line *CNMT* instead of line *BZHT*) the ratio of the rectangle whose sides are *BG* and *DE* to the rectangle whose sides are *BD* and *EG* is as the ratio of the rectangle whose sides are *CT* and *NM* to the rectangle whose sides are *CN* and *MT*.[48]

[46]By the similarity of the two triangles, *ADH* and *ATL*.

[47]Neither Greek nor Arabic had any special term for this ratio. It was not named until the nineteenth century, when Lazare Carnot called it *le rapport anharmonique*. Much later in that century, the English mathematician W. K. Clifford introduced the term "cross ratio."

[48]Ancient and medieval geometers moved freely between the ratio of two rectangles whose sides are line segments a, b and c, d and the composition of ratios, $a : c$ and $b : d$. Ibn Hūd refers to the operation of composition as "doubling."

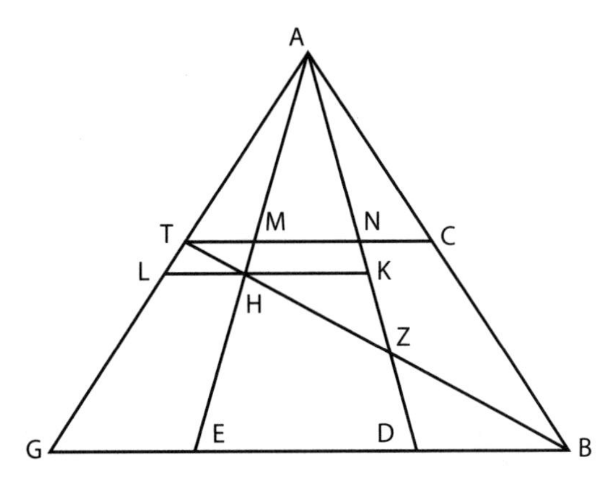

Fig. IV-4-2.

In every triangle, if two points are marked on its base and they are joined to the apex of the triangle, and any line cuts the two sides of the triangle, then it cuts the two lines mentioned in ratios corresponding to the ratio of [the corresponding parts of] the base. That is to say, if the ratio of the base to either of the two segments at its extremities is as the ratio of the other extreme [segment] to the remaining segment, then the ratio of the segment of the cutting line between the two sides of the triangle to the part corresponding to the one that the base is antecedent to is as the ratio of the other part corresponding to the antecedent from the base to the consequent.

And if the ratio of the base to the segment to which it is the antecedent is greater, [then] the ratio of the [segment cut off from the] cutting line to the other segment corresponding to it [i.e., to the consequent of the ratio whose antecedent is the base] is greater than the ratio of the segment that is the antecedent for the corresponding consequent. And if the ratio of the base to the segment to which it is the antecedent is less, [then] also the ratio of the [cutting] line to the other segment corresponding to it is less than the ratio of the segment that is the antecedent for the corresponding consequent.

And, in general, if the ratio made up of the ratio of the base to one of the two extreme parts composed with the ratio of the part [of the base] that is near it [the extreme part chosen] to the remaining part, [then] it [the doubled ratio] is as the ratio of the cutting line to the part corresponding to the part to which the base is related composed with the ratio of the part that is near it to the remaining part of it.

For example: On the base *BG* of triangle *ABG* [Fig. IV-4-2] two points *D* and *E* are marked, and both of them are joined to *A*. The two sides *AB* and *AG* are cut by a line *BZHT*. Then I say that if the ratio of *BG* to *GE* is as the ratio of *BD* to *DE*, then the ratio of *BT* to *TH* is as the ratio of *BZ* to *ZH*.

And if the ratio of *BG* to *GE* is greater than the ratio of *BD* to *DE*, then the ratio of *BT* to *TH* is greater than the ratio of *BZ* to *ZH*. And if the ratio of the line *BG* to *GE* is smaller than the ratio of *BD* to *DE*, the ratio of *BT* to *TH* is smaller than the ratio of *BZ* to *ZH*.

Proof: From point *H* we produce a line parallel to line *BG* which meets the two lines *AD* and *AG* at the two points *K* and *L*. And if we make line *HL* a mean in the ratio between the two lines *BG* and *GE*, then the ratio of *BG* to *GE* is as the ratio of *BG* to *HL* composed with the ratio of *HL* to *GE*. But the ratio of *BG* to *HL* is as the ratio of *BT* to *TH*, and the ratio of *HL* to *EG* is as the ratio of *HA* to *AE*. Therefore, the ratio of *BG* to *GE* is as the ratio of *BT* to *TH* composed with the ratio of *HA* to *AE*.[49]

And as for the ratio of *ED* to *DB*, if we make the line *HK* a mean in the ratio between them, the ratio of *ED* to *HK* composed with the ratio of *HK* to *DB* is as the ratio of *ED* to *DB*. But the ratio of *DE* to *HK* is as the ratio of *EA* to *AH*, and the ratio of *HK* to *DB* is as the ratio of *HZ* to *ZB*. And so the ratio of *BG* to *GE* composed with the ratio of *ED* to *DB* is as the ratio of *BT* to *TH* composed with the ratio of *HA* to *AE* composed with the ratio of *EA* to *AH* composed with the ratio of *HZ* to *ZB*.

But the ratio of *HA* to *AE* composed with the ratio of *EA* to *AH* is the ratio of similarity, so there remains that the ratio of *BT* to *TH* composed with the ratio of *HZ* to *ZB* is as the ratio of the line *BG* to *GE* composed with the ratio of *ED* to *DB*.

But the ratio of *BG* to *GE* composed with the ratio of *ED* to *DB* is as the ratio of the surface *BG* in *DE* to the surface of *GE* in *BD*. And so the ratio of the surface *BG* in *DE* to the surface *GE* in *DB* is as the ratio of the surface *BT* in *ZH* to the surface of *TH* in *ZB*.

And so if the ratio of *BG* to *GE* is as the ratio of *BD* to *DE*, the surface *BG* in *DE* is equal to the surface *GE* in *DB*. And so the surface *BT* in *ZH* is equal to the surface *TH* in *ZB*. And so the ratio *BT* to *TH* is as the ratio of *BZ* to *ZH*.

And if the ratio of *BG* to *GE* is greater than the ratio of *BD* to *DE*, the surface *BG* in *DE* is greater than the surface *GE* in *DB*. And for that [reason] the surface *BT* in *ZH* is greater than the surface *TH* in *ZB*. And so the ratio of *BT* to *TH* is greater than the ratio of *BZ* to *ZH*.

And if the ratio of *BG* to *GE* is less than the ratio of *BD* to *DE*.[50] ...

And if the line cutting the two sides, *AB* and *AG*, does not pass through the point *B* but cuts *AB* like *TMNC*,[51] we join the point *B* with the point *T*. And we have proved that the ratio of *TB* to *BZ* composed with the ratio of *ZH* to *HT* is as the ratio of *TC* to *CN* composed with the ratio of *NM* to *MT*. But the ratio of *BT* to *BZ* composed with the ratio of *ZH* to *TH* is as the ratio of *BG* to *BD* composed with the ratio of *DE* to *EG*. And so the ratio of *TC* to *CN* composed with the ratio of *NM* to *MT* is as the ratio of *GB* to *BD* composed with the ratio of *DE* to *EG*. And that is what we wanted to prove.

312.18: Ceva's Theorem

Until the recent investigation of Ibn Hūd's *Book of Perfection*, scholars thought that the Italian geometer Giovanni Ceva, in his 1678 work *De lineis rectis,* was the first to state a theorem now named for him. However, we now know that Ibn Hūd stated and proved the theorem some 600 years before Ceva. Ibn Hūd's proof, in turn, was based on a theorem that had been known about 900 years before his time by the Greek geometer, Menelaus. Menelaus's Theorem states that if sides *GA* and *GB* of triangle *ABG* [Fig. IV-4-3] are cut internally by a transversal *ZE*, which also cuts the side *AB* extended at *D*, then $GZ{:}ZA = (GE{:}EB) \cdot (BD{:}DA)$.[52]

[49] The argument here comes down to a proof of the case of Menelaus's Theorem for plane triangles applied to the configuration *BEGTAHB*. The case is similar for the argument in the next paragraph.

[50] Since the idea here is the same as in the case of "greater than," we have omitted the proof.

[51] Line *TMNC* appears to be parallel to *BG*, but the text does not mention it and the proof does not require it.

[52] In ancient texts this operation on ratios, written here as multiplication, is called "composition" or, as Ibn Hūd called it, "doubling." There are a number of other cases of Menelaus's Theorem, but we have stated here the one that Ibn Hūd used.

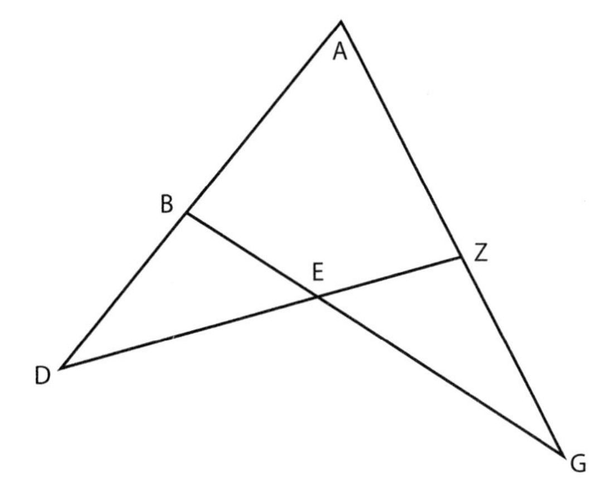

Fig. IV-4-3.

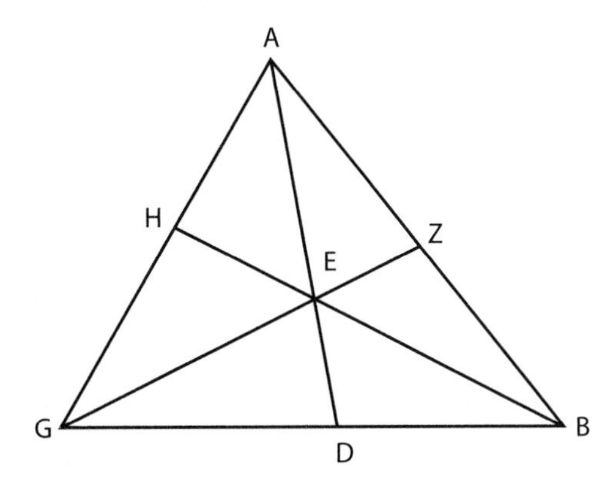

Fig. IV-4-4.

[In] every triangle in which from each of its angles a line issues to intersect the opposite side, such that the three lines meet inside the triangle at one point, the ratio of one of the parts of a side of the triangle to the other [part], doubled with the ratio of the part [of the side] adjacent to the second term [of the first ratio] to the other part of that side is as the ratio of the two parts of the remaining side of the triangle, if [this last] ratio is inverted, and conversely.

Example of this: (In) triangle *ABG* from point *A* line *AD* has been drawn to meet side *BG* at point *D*, and from point *B* line *BH* has been drawn to meet line *AD* at point *E* and side *AG* at point *H*, and points *G*, *E* are joined, until line *GE* meets side *AB* at point *Z* [Fig. IV-4-4]. I say that the ratio of *BZ* to *ZA* doubled with the ratio of *AH* to *HG* is as the ratio of *BD* to *DG*.

Proof of this: The ratio of *BZ* to *ZA* is as the ratio of *BE* to *EH* doubled with the ratio of *HG* to *GA*. But the ratio of *AH* to *HG* doubled with the ratio of *HG* to *GA* is as the ratio of *HA* to *AG*. Thus the ratio of *BZ* to *ZA* doubled with the ratio of *AH* to *HG* is as the ratio of *BE* to *EH* doubled with the ratio of *HA* to *AG*, which (ratio) is as the ratio of *BD* to *DG*.

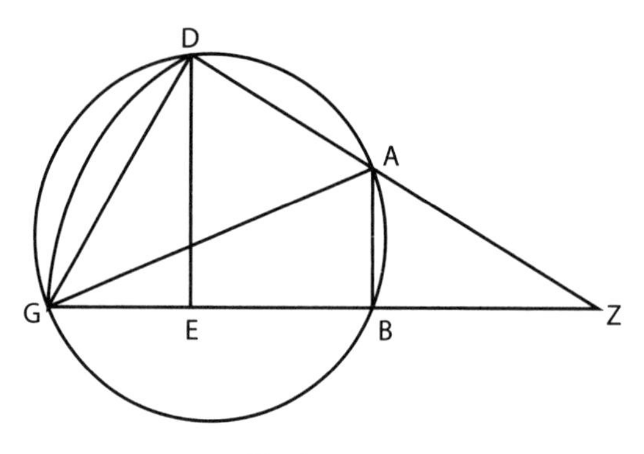

Fig. IV-4-5.

Thus the ratio of *BZ* to *ZA* doubled with the ratio of *AH* to *HG* is as the ratio of *BD* to *DG*. That is what we wanted to demonstrate.

Here it has become clear that if the ratio of *BZ* to *ZA* doubled with the ratio of *AH* to *HG* is as the ratio of *BD* to *DG*, and *BH*, *GZ,* and *AE* are joined, it [*AE*] passes through point *D*. For if it would not pass through it, line *BG* would be divided at two different points in the same ratio on the same side [i.e., internally], which is a contradiction. Thus if line *AE* is extended, it passes through point *D*. That is what we wanted to demonstrate.

511.1: Continued Proportion

[Berggren, 2007] gives geometrical constructions from some of Islam's outstanding geometers, such as the problem of inscribing an equilateral triangle in a square and constructing a regular nine-sided polygon in a circle. Another important problem, one that the mathematicians of both ancient Greece and medieval Islam solved in a number of ways, is the following. Given line segments a and d, construct segments b and g that are in continued proportion, that is, $a{:}b = b{:}g = g{:}d$.[53] A geometric solution to this problem requires the use of conic sections, and Ibn Hūd's construction, which follows, uses a parabola.[54]

We can also find this by means of a parabola and a circle. Let the two lines be lines *AB*, *BG* [Fig. IV-4-5], and let them contain a right angle. Let us draw through points *A*, *B*, *G* a circle. Let *BG* be the greater of the two lines, and let us make [it] the parameter and the axis of a parabola with vertex at point *G*. Let it meet the circle at point *D*. We draw from point *D* an ordinate *DE*. I say that lines *DE, EG* are the two mean proportionals.

Proof of this: We join *AD,* and we extend it rectilinearly to meet line *BG* at point *Z*. We join *DG*. Then, since the angle at point *D* is right, the ratio of *ZE* to *ED* is equal to the ratio of *ED* to *EG*. Since *BG* is a parameter of [conic] section *GD,* the rectangle *BG, GE* is equal to the square of *DE*, so the ratio of *BG* to *DE* is equal to the ratio of *DE* to *EG*. Thus line *BG* is equal to line *ZE.*

[53]Euclid explains the construction of a single, mean proportional between a and g in *Elements* VI, proposition 13. In the case when two means are required and $d = 2a$, the problem is equivalent to the ancient problem of doubling the cube.

[54]See section IV-6 of chapter 2 for a different construction of two mean proportionals using the conchoids of Nicomedes.

If we subtract the common [part] *BE, ZB* is equal to line *EG*. Thus the ratio of *ZE to ED* is equal to the ratio of line *ED* to line EG, and the ratio of line *EZ* to line *ED* is equal to the ratio of line *BZ* to line *BA.* Thus the ratio of line *BG* to line *ED* is equal to the ratio of line *ED* to line *EG,* and is equal to the ratio of line *EG* to line *AB.*

511.2-7: Alhazen's problem

A famous problem in geometrical optics is a problem generally referred to as "Alhazen's problem,"[55] which concerns reflection in mirrors whose surfaces are curved. The problem requires that one make the following construction. Suppose one is given a spherical or conical mirror, concave or convex, and an object (thought of as being a point) visible in the mirror to an observer (represented by another point). The question is: At what point on the mirror will the observer see the object? (The limiting case would be an object such as the sun, which is, effectively, infinitely far from the mirror.)

As part of his solution to this problem, Ibn al-Haytham gave six difficult geometrical lemmas, which were adapted by al-Muʾtaman ibn Hūd in his *Istikmāl.* In some cases Ibn Hūd followed Ibn al-Haytham's ideas, but in a number of cases he introduced new techniques, which simplify and shorten Ibn al-Haytham's proofs. We present these lemmas and their demonstrations here and conclude with a description of how the lemmas relate to the solution of Alhazen's problem.

Ibn al-Haytham's statements and proofs of the lemmas were translated from Arabic and published in [Sabra, 1982]. The translation here, of Ibn Hūd's adaptation of the lemmas, is that of Prof. Jan Hogendijk, who published the Arabic text as well. The footnotes and diagrams are, apart from minor modifications, based on those in Prof. Hogendijk's paper. For the reader who wishes to consult that paper, we have retained the numbering of the propositions in the article, which follows the numbering in Ibn Hūd's work. Hence, the numbering here runs from 2 to 7 rather than from 1 to 6.

Ibn al-Haytham uses lemmas 2–4 in the proof of the three remaining lemmas, which are used to solve various cases of Alhazen's problem. For details of the applications the reader may consult Hogendijk's paper. The six lemmas, taken as a whole, are ample evidence of the sophistication of Andalusian geometry in the eleventh century.

Al-Muʾtaman presents both the analysis and synthesis of the construction in 2 below, and since Ibn al-Haytham does not present an analysis, it seems reasonable to assume that al-Muʾtaman is the source for the analysis, which is reproduced here.

2. If two lines which are known in position intersect, and if there is an assumed point between them, we want to show how we draw from the assumed point a line which meets them, such that the part cut off from the extended line by the two lines which are known in position is equal to a known line [Fig. IV-4-6].[56] Thus let the two intersecting lines which

[55] So-called after Ibn al-Haytham of Basra (d. 1048), who discussed the problem in his famous book, *The Optics.* This large work, written in Egypt sometime between 1030 and 1040, was translated into Latin in the twelfth century. It set Western optics on a new course and formed the basis for research in optics during the sixteenth and seventeenth centuries.

[56] In a note Hogendijk points out that this problem was stated and solved by Pappus of Alexandria in the early fourth century CE, and then in the ninth century by Aḥmad ibn Mūsā and Thābit ibn Qurra.

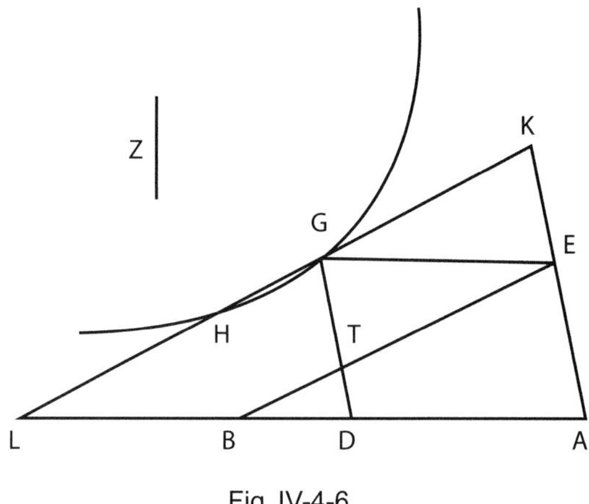

Fig. IV-4-6.

are known in position be lines *AB*, *GD*[57] and (let) the assumed point (be) point *E*, and (let) the assumed line (be) line *Z*. We want to draw from point *E* a line which meets lines *GD*, *AB* in such a way that the part of it which is cut off by them is equal to line *Z*.

Analysis: We assume that this (line) already exists, let it be line *ETB*, which meets *GD* at point *T* and line *AB* at point *B*, and (we assume) that line *TB* is equal to line *Z*. Then, since point *E* is assumed, if there are drawn from it lines *EA*, *EG* parallel to the assumed lines *GD*, *AB*, then points *A*, *G* are assumed. If there is drawn from point *G* a line parallel to line *EB*, namely, line *GL*, it is equal to line *EB*. But line *AD* is equal to line *BL*. If *LG* is extended to meet line *AE* at *K*, line *GK* is equal to line *ET*. Thus, if we subtract line *LH*, which is equal to line *GK*,[58] [from *KH*], line *GH* is equal to line *TB*. Since lines *AE*, *AB* are assumed, and point *G* is assumed, therefore if we draw through point *G* the hyperbola with asymptotes lines *EA*, *AB* it passes through point *H* [converse to *Conics* II-8[59]].

Synthesis: We draw from point *E* line *EA* parallel to line *GD* and line *EG* parallel to line *AB*. We construct on point *G* a hyperbola with asymptotes lines *EA*, *AB*, namely, section *GH*. We let fall on it line *GH* equal to line *Z*. We draw from point *E* line *EB* parallel to line *GH* to meet line *GD* at point *T*. I say that line *TB* is equal to line *Z*.

The proof of the synthesis is clear from *Conics* II-8.

3. We want to show how we draw from an assumed point on an assumed circle a line such that the part of it that is cut off by the circumference of the circle and by an assumed chord in it (the circle) is a line equal to an assumed line [Fig. IV-4-7].[60] Thus let the circle be circle *ABG*, (let) the chord (be) *BG*, (let) the assumed point (be) point *A*, and (let) the assumed line (be) *D*. We want to draw from point *A* a line which meets chord *BG* in such

[57] Point *B* itself is not "assumed." In Greek geometry, a known line may be indicated by two points, which may be unknown.

[58] This defines the position of point *H*.

[59] If a straight line meets a hyperbola in two points, produced both ways it will meet the asymptotes, and the straight lines cut off on it by the section from the asymptotes will be equal.

[60] This proposition is a simplification and a generalization of Ibn al-Haytham, *Optics*, Y, 33–34 in [Risner, 1572], lemmas I and II in [Sabra, 1982, pp. 315–320].

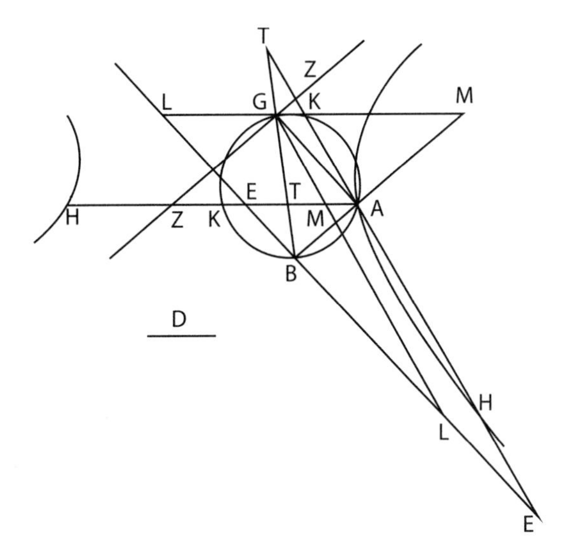

Fig. IV-4-7.

a way that the part of it which is cut off by the circumference of the circle and chord *BG* is equal to line *D*.

Thus let us join *AB*, *AG*. We draw from points *B*, *G* two lines parallel to lines *AG*, *AB*, namely, lines *BE*, *GZ*. We construct on point *A* a hyperbola with asymptotes lines *BE*, *GZ* [*Conics*, II-4]. Let [*H* be chosen so that] the ratio of *AH* to *BG* is equal to the ratio of *BG* to *D*. Then, if we want the line which is cut off by the circumference of the circle and the chord to meet the chord outside the circle, we let line *AH* fall on the (conic) section. If we want (it) to meet the chord inside the circle, we let line *AH* meet the opposite section[61] at point *H*, if line *AH* is not less than the minimum (straight) line[62] which issues from point *A* to the opposite section. But if it [*AH*] is less than it (the minimum straight line), there cannot be drawn from point *A* a line in such a way that the part of it which is cut off by the chord *BG* and the circumference of the circle inside the circle is a line equal to line *D*. Thus let line *AH* meet the (conic) section or the opposite section at point *H*, the asymptotes at points *Z*, *E*, the chord *BG* at point *T*, and the circle at point *K*.[63] I say that line *KT* is equal to line *D*.

Proof of this: we draw from point *G* line *GML* parallel to line *AH*, to meet *AB* at point *M*, and *BE* at *L*. Then, since the rectangle *BT*, *TG* is equal to the rectangle *AT*, *TK*,[64] the ratio of *BT* to *TA*, which is equal to the ratio of *BG* to *AZ*,[65] is equal to the ratio of *KT* to *TG*. Thus. *alternando*,[66] the ratio of *BG* to *KT* is equal to the ratio of *AZ* to *GT*, which is

[61]That is, the opposite branch of the hyperbola.

[62]That is, the shortest straight line. The Arabic is simply "the small line," and the terminology is found in *Conics*, V.

[63]Points *K*, *H*, *E*, *Z*, *T*, *L* and *M*, each occur twice, once for the case where the line cut off meets the chord outside the circle and once where it meets inside.

[64]*Elements* III.35, 37.

[65]These two ratios are equal because *BA* is parallel to *GZ*.

[66]The term is a technical term, coming from *Elements* V.16, which states (in abbreviated form) if $a{:}b = c{:}d$, then $a{:}c = b{:}d$.

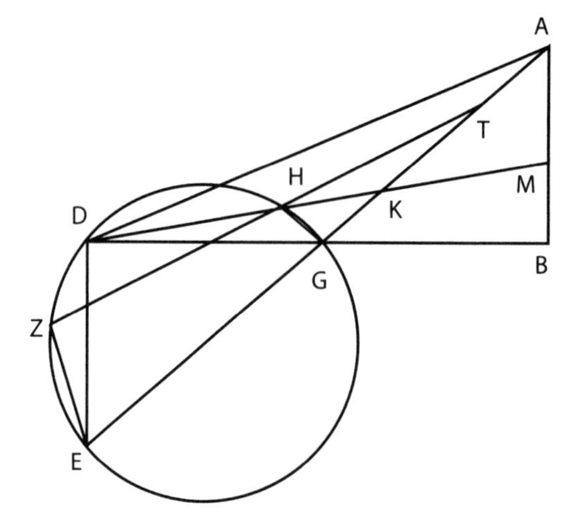

Fig. IV-4-8.

equal to the ratio of *LM* to *BG*, because of the similarity of the two triangles.[67] Since line *AZ*, which is equal to line *MG*, is equal to line *EH*, and line *AE* is equal to line *GL*, (thus) line *ML* is equal to line *AH*. Thus the ratio of *LM* to *BG* is equal to the ratio of *BG* to line *D*.[68] It has been proved that the ratio of line *LM* to line *BG* is equal to the ratio of line *BG* to line *KT*. Thus line *KT* is equal to line *D*. That is what we wanted to prove.

4. If there is an assumed right-angled triangle, and [if there is] an assumed point on one of the sides containing the right angle or on the rectilinear extension of it, how do we draw from it a line such that the ratio of [the part] of it that is cut off by the two other sides to what is cut off from the side containing the point from the two remaining sides[69] is equal to an assumed ratio?

Thus let the triangle be triangle *ABG*, angle *B* in it be right, and point *D* (be) on line *BG* [Fig. IV-4-8].[70] We want to draw from it a line [i.e., *DM*] such that the ratio of the part of it which is cut off by lines *AB*, *AG* [i.e., *KM*] to the part which falls between its intersection with line *AG* and point *G* [i.e., *KG*] is equal to an assumed ratio.

Thus let us draw from point *D* line *DE* parallel to line *AB*, to meet line *AG* at point *E*. We circumscribe a circle about triangle *GDE* [*Elements* IV.5], and we join line *DA*. We make on diameter *GE* at point *E* angle *GEZ* equal to angle *DAB* [*Elements* I.23]. We draw from point *Z* a line that meets diameter *EG* in such a way that the part of it that falls outside the circle is equal to the line, the ratio of *AD* to which is [equal to] the assumed ratio [from 3 above]. Let it be line *ZHT*. We join *DH*. Let *DH* meet diameter *EG* at point *K* and line *AB*

[67]The proof comes from the similarity of the two pairs of triangles *ATG*, *LGB* and *ZTG*, *MGB* and *Elements* V.18.

[68]By hypothesis, $AH:BG = BG:D$.

[69]The text here is mathematically incorrect, but no plausible emendation is apparent. However, the text of the following paragraph makes it clear what is meant.

[70]The text contains three figures, illustrating the three different possibilities for the position of *D* relative to the segment *BG*. We reproduce the diagram for the case when *D* falls on the extension of *BG* to the left of *G*.

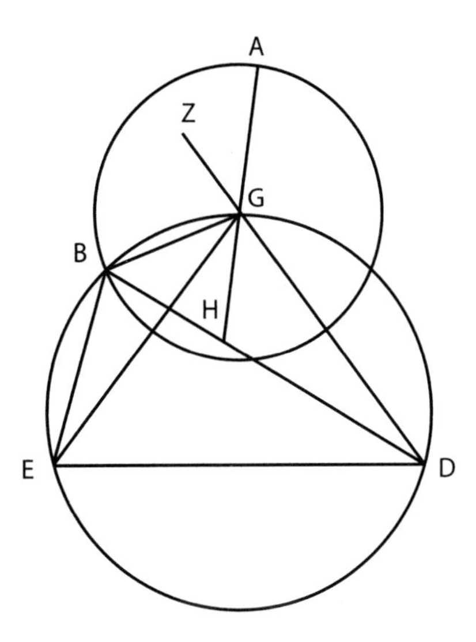

Fig. IV-4-9a.

at point *M*. I say that the ratio of *MK* to *KG* is equal to the ratio of *AD* to *HT*, which is the assumed [ratio].

Proof of this: We draw line *GH*. Because angle GHK is equal to angle *GED* [*Elements* III.22],[71] which is equal to angle *GAM*, and angle *MAD* is equal to angle GEZ, and angle *GEZ* is equal to angle *GHT* [*Elements* III.22], therefore angle KHT is equal to angle KAD [by subtraction]. But the angles at point K are equal. Thus triangles *KAD*, KHT are similar, and in the same way, triangles *KAM*, *KHG* [are similar]. Therefore, the assumed ratio of *AD* to *HT* is equal to the ratio of *MK* to *KG*. This is what we wanted to prove.

5. We want to show, if there are an assumed circle and two assumed points such that both of them [are] or one of them is outside it (the circle), how we draw from them two lines which meet at the circumference of the circle such that the tangent at their point of intersection bisects their angle of intersection.

Thus let the circle be circle *AB*, let its center be point *G*, and the assumed points are points *D*, *E*. Their distances to the center are either the same or they are not the same. Let their distances to the center first be the same. We join *GD*, *GE*, *DE*, and we circumscribe about triangle *GDE* a circle, which meets circle *AB* at point *B*. We join *DB*, *BE* [Fig. IV-4-9a]. I say that point *B* is as we wanted.

Proof of this: We extend line *DG* towards *Z*. Then, since angle *GBD* is equal to each of the angles *GDE*, *GED* [*Elements* III.21], it is half of the exterior angle *EGZ*. But angle *DBE* is equal to angle *DGE*. Thus half of angle *DBE* plus angle *GBD* is [a] right [angle]. If one draws a perpendicular to line *GB* from point *B*, it [the perpendicular] is tangent to the circle and it bisects angle *DBE*.

Again, let the distances of the two points to the center not be the same, and let them be like points *E*, *H*. We join *HG*, and we extend it to meet the circle at point *A*. We join

[71] The opposite angles of quadrilaterals inscribed in circles are equal to two right angles.

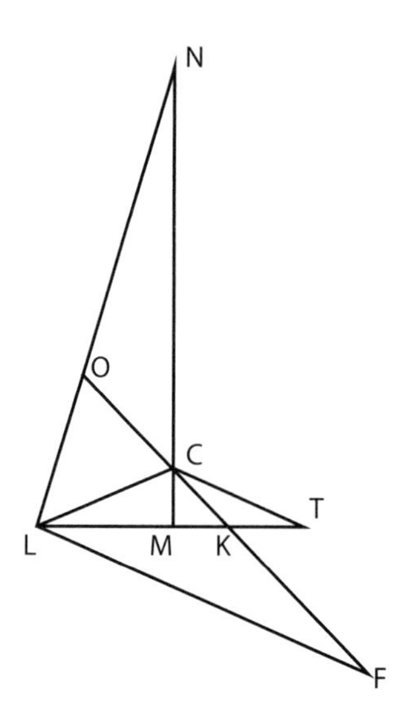

Fig. IV-4-9b.

GE. We assume line *TL* [Fig. IV-4-9b], and we divide it at point *K* such that the ratio of *TK* to *KL* is equal to the ratio of *HG* to *GE*. We bisect it at point *M*, and we draw from point *M* a perpendicular *MN*. We make on point *L* angle *MLN* equal to half of angle *EGA*. We draw from point *K* line *KCO* in such a way that the ratio of *CO* to *OL* is equal to the ratio of *HG* to *GA*.[72] We make on line *HG* at point *G* angle *HGB* equal to angle *COL*. Let line *GB* meet the circle at point *B*. We join *HB*, *BE*. I say that the tangent [to circle *AB*] at point *B* bisects angle *HBE*.[73]

Proof of this: We join *LC*, *CT*, and we draw from point *L* line *LF* parallel to line *CT*, to meet line *CK* at point *F*. We make on line *GH* angle *HGD* equal to angle *CLF*. Let line *GD* meet line *BH* at point *D*. We join *DE*. Then, since the ratio of *HG* to *GB*, which is equal to *GA*, is equal to the ratio of *CO* to *OL*, and the angle at *G* is equal to the angle at *O*, triangle *GHB* is similar to triangle *COL*. Thus angle *GHD* is equal to angle *LCF*. But angle *HGD* is equal to angle *CLF*. Therefore the ratio of *GH* to *GE* is equal to the ratio of *TK* to *KL*, which is equal to the ratio of *CK* to *KF*, and equal to the ratio of *CL* to *LF*, which is equal to the ratio of *HG* to *GD*.[74] Thus the ratio of *HG* to *GD* is equal to the ratio of *HG* to *GE*. Therefore line *GE* is equal to line *DG*. We extend line *DG* toward *Z*. Then angle *AGZ* is twice angle *CLK*, but the whole angle *AGE* was assumed to be twice angle *NLM*. Therefore angle *ZGE* is twice angle *OLC*, which is equal to angle *GBH*. Since angle

[72]A marginal remark in the manuscript states, correctly, "It (the construction of *KCO*) has been shown in the previous proposition."

[73]The construction is essentially that of Ibn al-Haytham in *Optics* V, but the proofs are not the same.

[74]$GH:GE = TK:KL$ according to the definition of *K*. And $TK:KL = CK:KF$, because *CT* and *LF* are parallel. Since $LM = MT$ and *CM* is perpendicular to *LT*, we have angle $CLM =$ angle *CTM*. Because *CT* and *LF* are parallel, angle $CTM =$ angle *MLF*. Hence angle $CLK =$ angle *KLF*, so $CK:KF = CL:LF$, by *Elements* VI.3. Moreover, $CL:LF = HG:GD$, because triangle *CLF* is similar to triangle *HGD*, which is a consequence of angle $GHD =$ angle *LCF* and angle $HGD =$ angle *CLF*.

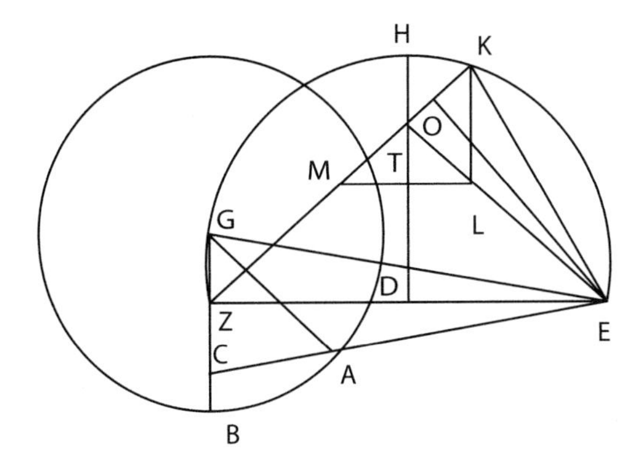

Fig. IV-4-10.

ZGE is [an] exterior [angle] of the isosceles triangle *GED*, it is twice angle *GED*. Therefore angle *GBD* is equal to angle *GED*, so they are on the same arc.[75] Thus the tangent at point *B* bisects angle *DBE*, as above. That is what we wanted to prove.[76]

6. We want to show, if there are an assumed circle, an assumed diameter in it, and an assumed point outside it, how we draw from it (the point) a line which ends at the assumed diameter, in such a way that the part of it cut off by the circumference of the circle and the assumed diameter is equal to the part of the assumed diameter between the center and the end of the line which is drawn from the assumed point [Fig. IV-4-10]. Thus let the assumed circle be circle *AB*, its center [be] point *G*, the assumed diameter [be] *GB*,[77] and the assumed point [be] point *E*. [We want to draw a line *EC* such that $AC = GC$.]

We draw from point *E* a perpendicular *EZ* to line *GB*, and we join *EG*. We circumscribe a circle about triangle *EZG*,[78] and we bisect *EZ* at point *D*. We draw perpendicular *DH*. We draw from point *Z* line *ZK* to meet line *DH* at point *T* and the circumference at point *K* in such a way that line *TK* is equal to half of *GB*.[79] We join *ET*, and we draw from point *K* line *KL* parallel to line *HD*, to meet line *ET* at point *L*. We draw from point *L* line *LM* parallel to line *EZ* to meet *ZT* at point *M*. We join *KE*. Then angle *EKZ* is equal to angle *EGZ* [*Elements* III.21]. Thus we cut off from angle *EGZ* angle *ZGA* equal to angle *LKM*, and we join *EA* to meet diameter *BG* at point *C*. I say that line *CA* is equal to line *CG*.

Proof of this: We draw from point *E* to line *TK* a perpendicular *EO*. Then, since angle *LMK* is equal to angle *EZK*, and angle *KLM* is right, triangles *KLM*, *ZOE* are similar.

[75]That is, a circle passes through *G*, *B*, *D*, and *E* (by *Elements* III.21).

[76]The proof is much shorter than that of Ibn al-Haytham. Note, however, that there is no figure for the case when the two points are outside the circle.

[77]For some reason the diagram shows *GB* as a radius.

[78]Only a part of the circle is shown in the figure.

[79]For the construction, see proposition 3 above. Neither al-Muʾtaman nor Ibn al-Haytham discusses the question of for which points *E* the construction is possible. The problem can be seen as a limiting case of the "problem of Alhazen," if we suppose that *E* is the eye, and that the object is the sun, which is supposed to be at infinite distance, so that its rays are parallel to *GB*. If $AC = CG$, point *A* is the point of reflection. Thus the problem of finding *A* can be solved for all points *E* such that the perpendicular *EZ* intersects the diameter *BG* between points *B* and *G*.

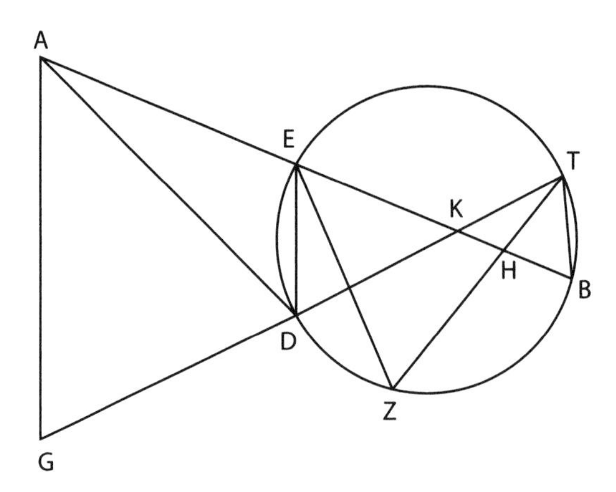

Fig. IV-4-11.

Triangle *ETZ* is isosceles, so line *LT* is equal to each of the lines *TM*,*TK*.[80] Thus line *KM* is twice line *KT*. But line *KT* was assumed to be equal to half of line *GB*, which is equal to line *GA*. Thus line *AG* is equal to line *KM*. Thus[81] the ratio of *KM* to *EZ* is equal to the ratio of *KL* to *EO*. But the ratio of *EZ* to *EG* is equal to the ratio of *OE* to *EK*,[82] and *KM* is equal to *AG*. Therefore, *ex aequali*,[83] the ratio of *AG* to *GE* is equal to the ratio of *LK* to *KE*. But angle *LKE* is equal to angle *AGE*. Thus triangle *AGE* is similar to triangle *LKE*. By subtraction,[84] *GAC* is equal to angle *KLT*, and angle *AGC* is equal to angle *LKT*, and triangle *LKT* is isosceles. Therefore triangle *ACG* is isosceles, so line *AC* is equal to line *CG*. That is what we wanted to prove.

7. We want to show, if there are two lines assumed in position, containing an angle, and if one of them is assumed in magnitude, and if one marks a point between them, how we let pass through it (the point) a line such that the ratio of it to the part cut off by its extension[85] from the assumed line is equal to a given ratio. Thus [Fig. IV-4-11] let the two lines which are assumed in position be lines *AB*, *AG*, and let *AB* be assumed in magnitude. We mark point *D* between them. We want to draw from point *D* a line that ends at *AG* and at line *AB* such that the ratio of it to the part which is cut off by it from *AB* on the side of *B* is equal to an assumed ratio.

Thus let us join *AD*. We draw from point *D* line *DE* parallel to line *AG*, to meet *AB* at *E*. We circumscribe a circle about points *B*, *D*, *E*. We make on line *DE* angle *DEZ* equal to angle *DAB*, to meet the circumference at point *Z*. We draw from *Z* line *ZHT*, to meet *AB* at point *H* and the circumference at point *T*, in such a way that the ratio of line *AD* to line *HT* is equal to the assumed ratio [see 3]. We join *TD*; let it meet *AG* at point *G* and *AB* at point *K*. I say that the ratio of *GK* to *KB* is the assumed ratio.

[80]Because $ED = DZ$ and *DH* is perpendicular to *EZ*, we have $ET = TZ$, whence by similar triangles $LT = TM$. Because *KL* is perpendicular to *LM*, it follows that $LT = TM = TK$.

[81]That is to say, because of the similarity of triangles *KLM* and *EOZ*.

[82]Triangles *EZG* and *EOK* are similar because *Z* and *O* are right angles, and angle *ZGE* is equal to angle *ZKE*.

[83]*Ex aequali* (*bi l-musāwāt*) indicates reasoning in the following form: if $a{:}b = d{:}e$ and $b{:}c = e{:}f$ then $a{:}c = d{:}f$. See *Elements* V, definition 17 and proposition 22.

[84]That is to say, if we take the supplements of the equal angles *GAE*, *KLE*.

[85]In Fig. IV-4-11 one does not need to extend line *GK*. The "assumed line" is the line assumed in magnitude (in the figure it is *AB*).

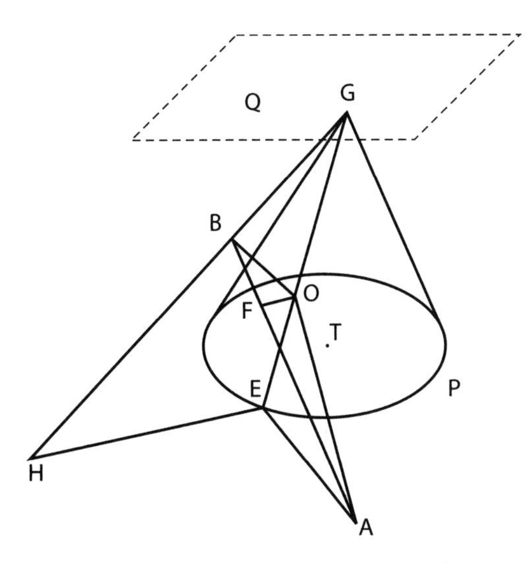

Fig. IV-4-12.

Proof of this: We join *TB*. Then, since angle *DEB*, which is equal to angle *BAG*, is equal to angle *DTB* and angle *DEZ*, which is equal to angle *DAB*, is equal to angle *DTZ*, and since the two angles at points *K* are equal, triangles *GAK*, *DAK* are similar to triangles *KTB*, *KTH*. Thus the ratio of *AD* to *TH*, which is equal to the assumed ratio, is equal to the ratio of *AK* to *KT*, which is equal to the ratio of *GK* to *KB*. That is what we wanted to prove.

To give an example of how these results are used to solve Alhazen's problem, consider reflection on the convex surface of a right circular cone with vertex G. In Fig. IV-4-12, let *P* be the plane containing the base of the cone and *Q* the plane through the vertex parallel to *P*. Let the point *A* represent the viewer's eye, the point *B* the object to be seen in the mirror and *O* the point of reflection. Points *B*, *O*, *A* lie in a plane that contains a line *a* passing through *O* and tangent to the surface of the cone. By the Law of Reflection, the incidence angle of line *BO* with line *a* equals the angle of reflection of *OA* with *a*.

Ibn al-Haytham considers three cases depending on the positions of *A* and *B* relative to the plane *Q*.[86] In the case shown in Fig. IV-4-12, where *A* and *B* are on the same side of the plane *Q*, he constructs line *GBH*, where *H* is the intersection of *GB* with plane *P*, and he uses a geometrical construction to reduce the case to that in which the object is at *H*, which reflected to *A* at a point *E* in the "mirror" formed by the base of the cone.

In the second case [Fig. IV-4-13], in which he uses lemma 6, he assumes the point *B* is on plane *Q*. As Hogendijk remarks, this is the case when *H* is a "point at infinity" (as in the case of the sun). Ibn al-Haytham draws (in *Optics* V:57) in plane *P* the diameter *TM* parallel to *GB*,[87] and he now wants to construct in the plane *P* point *E* on the circle in such a way that a ray from *A* is reflected by point *E* in a direction parallel to *TM*. Suppose that *E* is the

[86]In the following exposition we follow Hogendijk very closely, with only very minor editorial changes. However, footnotes have been added to help the reader follow Ibn al-Haytham's solution.

[87]If one is working in the plane *BGT* this is just an application of *Elements* I:31: to draw a straight line parallel to a given straight line through a point, *T* (not on the given line).

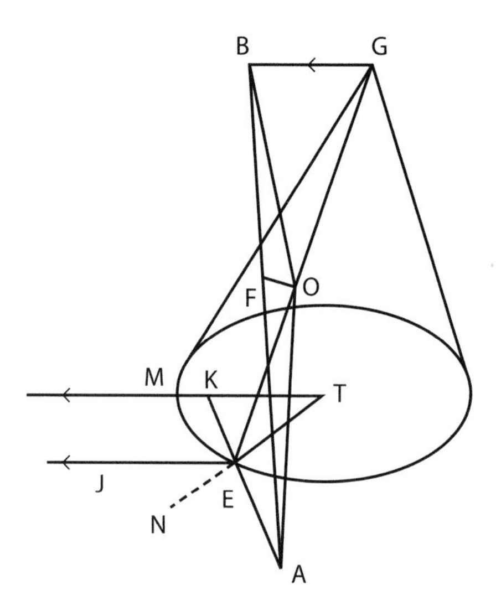

Fig. IV-4-13.

required point,[88] and extend *AE* to meet diameter *TM* at point *K*. Then angle *TEK* = angle *KTE*.[89] Thus point *E* can be found by means of lemma 6.[90] Ibn al-Haytham now uses the same construction as in the first case to find point *O* such that the ray from *A* to *O* is reflected toward *B*. Lemma 6 can be called Alhazen's problem for the case where the object is at infinity.

However, if point *B* is on the other side of plane *Q* Ibn al-Haytham draws (in *Optics* V:59) *BG* and extends it to meet *P* at point *H*.[91] Then he finds, by lemma 7, *E* as the point on the circle in such a way that the tangent at *E* bisects angle *HEA*. In the same way as above, he constructs the desired reflection point *O*.

5. MUḤYĪ AL-DĪN IBN ABĪ AL-SHUKR AL-MAGHRIBĪ, *RECENSION OF EUCLID'S ELEMENTS*

The following selections are from the *Recension of Euclid's Elements* by Muḥyī al-Dīn al-Maghribī. He was born in the Islamic West and, at some point, went to Aleppo, where he worked at the court of Sultan Al-Nasir (1237–1260). After Aleppo fell to the Mongols, he went to Maragha in Persia, where he worked at the observatory of Nāṣir al-Dīn al-Ṭūsī. He died in 1290.

[88] At this point Ibn al-Haytham begins an analysis of the problem. So he supposes he has done the construction.

[89] Extend *TE* toward *N* and let *EJ* be the reflected ray. Then *EN* is the normal to the circle at *E*, so angle *AEN* = angle *NEJ* because the angle of incidence is equal to the angle of reflection. We also have angle *AEN* = angle *KET*, and, because *EJ* is parallel to *TM*, angle *NEJ* = angle *KTE*.

[90] In the notation of this paragraph and the terminology of lemma 6, line *TKM* contains part of the assumed diameter (i.e., a radius) of the assumed circle of radius *TE*. Lemma 6 describes the construction of a line *AEK* such that *EK* = *KT*. But this means triangle *KTE* is isosceles with angle *TEK* = angle *KTE*, which is the construction required here.

[91] We omit the figure Hogendijk supplies. The reader may easily picture the situation in terms of the figure for the previous case.

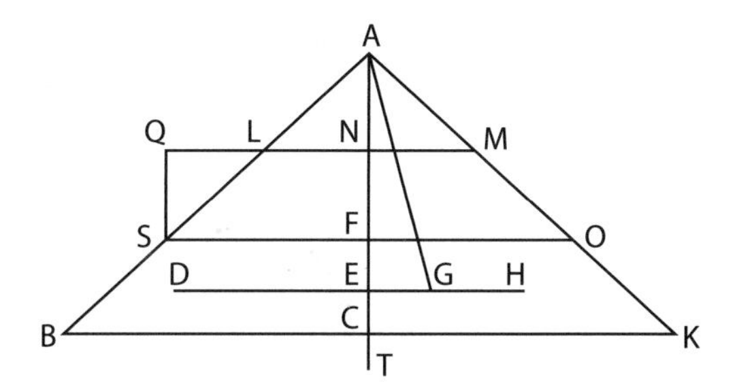

Fig. IV-5-1.

[On the Parallel Postulate]

In his *Recension of Euclid's Elements*, al-Maghribī endeavored to correct the faults of the work itself as well as those of the three previous recensions of the *Elements* that he had seen. One of the features he felt needed correction was the basis for Euclid's theory of parallels, namely, the fifth postulate of Book I, which a number of ancient and medieval authors tried either to prove or to replace with one that seemed more obvious. (See section IV-1 in Chapter 2 for a "proof" of the parallel postulate by Levi ben Gerson.)

For his part, Muḥyī al-Dīn, to avoid using the parallel postulate, introduced the following definition of parallel lines: parallel straight lines are those that lie in the same plane surface and are such that if a straight line falls on any two of them at random it makes the two angles on one side equal to two right angles.

Euclid proved the second half of this statement in *Elements* I.29 by using postulate 5. And on the basis of I.29, Euclid proved I.34, whose conclusion includes the statement that in a parallelogram opposite sides and angles are equal to each other.

Muḥyī al-Dīn proceeded differently. His definition of parallel lines amounts to saying that they are lines satisfying the third part of the conclusion of *Elements* I.29. But it is the first part of the conclusion of I.29 that is required in the proof (in I.34) of the equality of opposite sides of a parallelogram. And it is this that Muḥyī al-Dīn needs to prove to show that *QS* and *NF* (in Fig. IV-5-1) are equal and, then, on the basis of that, he can develop the theory of parallels. As Sabra points out in his introduction to Muḥyī al-Dīn's piece, most medieval Arabic treatments of the problem proved the parallel postulate from a postulate equivalent to it rather than—as G. Saccheri (1667–1733) did—try to prove it by contradiction.

Lemma: If a line falling on two straight lines makes the two angles on one of the two sides less than two right angles, then the two lines, if produced indefinitely on that side, meet.

Example: The line *AG* has fallen on the two lines *AB*, *GD*, making the two angles *BAG*, *DGA* less than two right angles [Fig. IV-5-1]. I say that if the two lines are indefinitely produced, they meet.

Demonstration: If one of the two angles is right, we complete the demonstration as will be shown. But if not, we produce *DG* indefinitely and on it let fall the perpendicular *AE* (12),[92] which we then produce indefinitely on the side of *T*. Then, on the point *A* of the

[92]Numbers in parentheses refer to the numeration of theorems in al-Maghribī's book. In the section quoted here these coincide with the numbering of propositions in T. L. Heath's translation of Book I of the *Elements*.

line *AE*, we construct the angle *KAE* equal to angle *BAE* (23). And we indefinitely produce the lines *AB*, *AK* in the directions of *B*, *K* (Post.[93]) [i.e., Euclid's postulate 2 = Maghribi's postulate 2]. We mark on *AB* the point *L* and cut off *AM* equal to *AL* (3) and join *ML*. Then, because the angles *KAE*, *BAE* are acute (Post.[94]), the line *ML* cuts *AT* in *N*; and since the two sides *AL*, *AM* are equal, and the side *AN* is common and the two angles at *A* are equal, then the two angles *N* in the two triangles *ANL*, *ANM* are right (4 and Post.[95]).

Now if point *N* lies between the points *E*, *T*, we complete the construction; otherwise, we cut off the lines *MO*, *LS* equal to *AM*, *AL* (3), and join *SO*, which cuts *AT* in *F*. We then prove as before that the two angles *F* in the two triangles *AFO*, *AFS* are right angles (4).

Now if point *F* falls between the points *E*, *T*, that will be sufficient for us; otherwise, we produce the line *NL* indefinitely and from point *S* draw *SQ* perpendicular to it (12).

Then, since the angle *ANL* is right, and the angle *LQS* is right, and the angles *ALN*, *SLQ* are equal (15), and the lines *AL*, *LS* are equal, then the lines *AN*, *QS* are equal (26). Further, since the angle *QNF* is right, and the angle *NFS* is also right, as we showed, and the angle *Q* is right, the surface *FQ* is a parallelogram, and, therefore, the lines *QS*, *NF* are equal (34).

We continue with this construction, cutting off from *AT* multiples of the line *AN* until we reach a multiple greater than *AE*;[96] let it [this multiple] be *AC* and let the line which has cut off *AC* from *AT* be the line *KC*. Then, the angles at *C* are right, as we showed for the angles *ANL*, *ANM*. And the angles at *E* are also right. Therefore, *ED* does not meet *CB* (28), nor does it meet *AE* (Post.). So, it necessarily meets *AB*.

It is evident from this and from [*Elements* I.] 28 that every two lines in a plane surface either meet each other or are parallel[97] for if a straight line falls on them, making the two angles on one of the two sides less than two right angles, they meet. But if it makes (the angles) equal to two right angles, then the two lines are parallel. Thus the doubt concerning this question has been removed by virtue of what we have posited as an emendation of the preliminaries[98] [to Book I of the *Elements*].

[93] Sabra's "Post(ulate)" translates "al-Maghribī's word *muṣādara*, a standard term for "postulate." Since al-Maghribī' does not number his postulates, Sabra puts in square brackets, following the reference, an explanation of what he believes the former was referring to.

[94] Although he designates this as a "postulate," [Sabra, 1969, p. 16, n. 78] points out that Muḥyī al-Dīn's defines an acute angle as an angle less than a right angle.

[95] As [Sabra, 1969, p. 16, note 79] points out, this is in fact al-Maghribī's definition: "If a straight line standing on a straight line makes the two angles on both sides equal to each other, each of the angles is a right angle."

[96] Here al-Maghribī applies what has become known as the axiom of Eudoxus, but (unlike Euclid, who uses it tacitly) he states it explicitly as a "common notion": For any two magnitudes of which one is greater than the other, if the smaller is multiplied indefinitely it will reach, by multiplication, a magnitude greater than the given, greater magnitude.

[97] This is not the truism it sounds. What al-Maghribī means is that it is impossible that two straight lines should continuously approach each other in a certain direction without eventually meeting.

[98] The Arabic is *al-muṣadara*, which Sabra translates as "postulate." However, [Sabra, 1969, pp. 1–2] points out that the word is also used to refer to other mathematical preliminaries (e.g., definitions) as well, and in this case he suggests "the premises" as an alternate. We have elected to use an even more general term, since his treatise is based on the emendation of Euclid's definition of parallels.

[Book XV]

Euclid's *Elements* originally contained 13 books, but at some point in the history of that work a treatise written, perhaps in the mid-second century BCE, by a Greek author named Hypsicles (of Alexandria) was added to it as a fourteenth book. Hypsicles's treatise, on the comparison of the icosahedron and dodecahedron, continued the investigation of the regular solids found in Book XIII of the *Elements*.[99] (According to his preface, he was inspired to do so by what he considered to be an unsatisfactory treatment of the topic by Apollonius of Perga.) At some later time, an unknown author (or authors) wrote an inferior fifteenth book, still on the theme of the regular solids.

Al-Maghribī included parts of this fifteenth book as part of Book XIV of his own *Recension*, along with parts of *Elements* XIII itself. But in his own Book XV, which is excerpted here, he extended Hypsicles's treatment to the remaining three regular solids: the cube, tetrahedron, and octahedron. This work was not original to him but was adapted from an unknown Arabic work that Hogendijk denotes by the letter A. A Hebrew translation of A, by the fourteenth-century Jewish scholar from Arles, Qalonymos ben Qalonymos, has recently been published (see section IV-3 of Chapter 2).

A reference in the following translation of the form X:Y refers to proposition Y of Book X of al-Maghribī's *Recension of the Elements*. A number in brackets following such a reference refers to the number in Heiberg's edition of the Greek text of the *Elements* (and in Heath's English translation). We have made some minor changes to mathematical terminology in the translation, and not all footnotes from Hogendijk's paper have been retained.

The Fifteenth book, on the Relation between the Five Solids

1. The square of the altitude of every equilateral triangle is three-fourths the square of its side. Thus let triangle *ABG* be equilateral and its altitude (be) *AD*. I say that the square of *AD* is three-fourths the square of *AB*.

Proof: Angles *ABD* and *AGB* are equal, the two angles *D* are right (angles), and line *AD* is common to triangles *ADB*, *ADG* [Fig. IV-5-2]. Thus lines *BD* and *DG* are equal. Hence the square of *BD* is one-fourth of the square of *BG*, that is, *AB*, II:4. The (sum of the) squares of *AD*, *DB* is equal to the square of *AB*, I:47. But the square of *DB* is one-fourth of the square of *AB*. Hence, by subtraction, the square of *AD* is three-fourths of the square of *AB*. That is what we wanted to prove.

2. The side of the octahedron is equal to the altitude of the triangle of the tetrahedron inscribed in the same sphere.

Proof: The square of the diameter of the sphere is one and a half times the square of the side of the tetrahedron, XIV:1 [= *Elements* XIII.13]. The square of the diameter of the sphere is also twice the square of the side of the octahedron, XIV:5 [= *Elements* XIII.14]. Thus one and a half times the square of the side of the tetrahedron is equal to twice the square of the side of the octahedron, so three-fourths of the square of the side of the tetrahedron is equal to the square of the side of the octahedron. But three-fourths of the square of the side of the tetrahedron is equal to the square of the altitude of its triangle.[100]

[99] Hypsicles's treatise is found in some Arabic editions of the *Elements* as Book XIV and was translated into Latin by Gerard of Cremona. See [Montelle, 2014] for a new English translation.

[100] Proved in the preceding theorem of Maghribī's *Recension*, (hereafter, "M") XV:1.

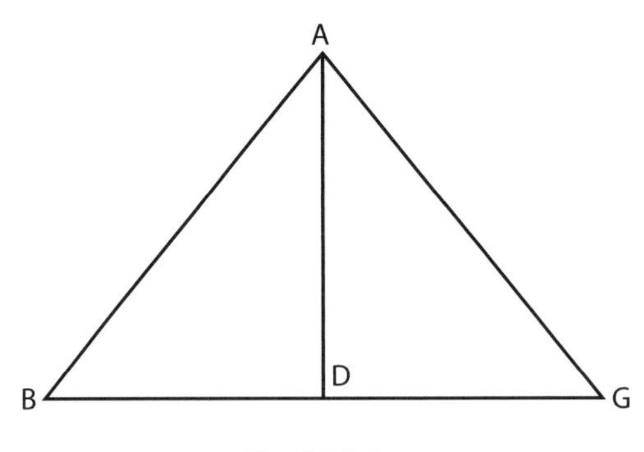

Fig. IV-5-2.

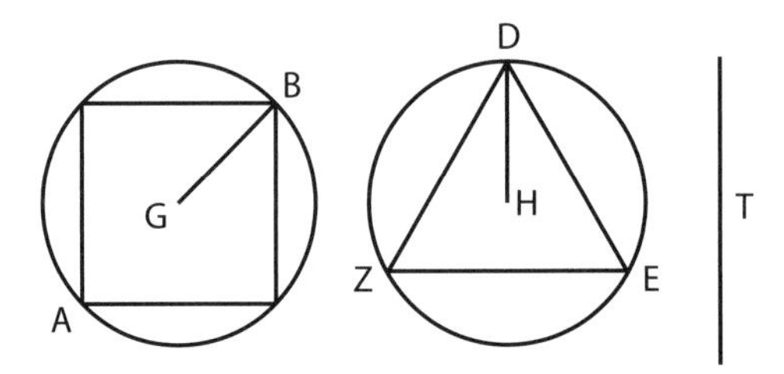

Fig. IV-5-3.

Thus the square of the side of the octahedron is equal to the square of the altitude of the triangle of the tetrahedron. Thus the side of the octahedron is equal to the altitude of the triangle of the tetrahedron. That is what we wanted to prove.

It is clear from this that the triangle of the octahedron is three-fourths of the triangle of the tetrahedron, since the ratio of the triangle of the octahedron to the triangle of the tetrahedron is equal to the ratio of the squares of the sides, and the square of the side of the octahedron is three-fourths of the square of the side of the tetrahedron.

3. The square of the side of the cube and the triangle of the side of the octahedron constructed in the same sphere are circumscribed by the same circle.

Thus let the square of the side of the cube be *AB*, (let) the radius of its circumscribed circle (be) *GB*, (let) the triangle of the octahedron (be) *DEZ* and (let) the radius of its circumscribed circle (be) *HD* [Fig. IV-5-3]. Let the diameter of the sphere be line *T*. Then the square of *T* is three times the square *AB*, XIV:3 [= *Elements* XIII.15], but the square *AB* is twice the square of *GB*, I:47, so the square of *T* is six times the square of *GB*. The square of *T* is twice the square of *DE*, XIV:5 [= *Elements* XIII.14], and the square of *DE* is three times the square of *HD*, XIII:12 [= *Elements* XIII.12], so the square of *T* is six times the square of *HD*. Thus the square of *GB* is equal to the square of *HD*, so *GB* is equal to *HD*. Thus the circle which circumscribes the square of the side of the cube is equal to the circle which circumscribes the triangle of the octahedron. That is what we wanted to prove.

It has become clear from this that the perpendicular falling from the center of the sphere on the square of the side of the cube is equal to the perpendicular falling from its center

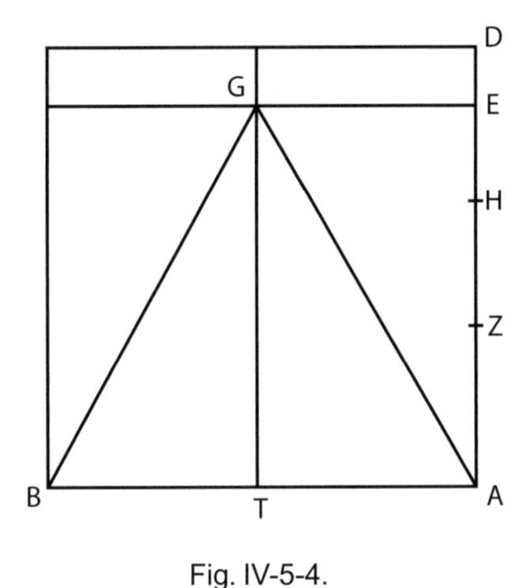

Fig. IV-5-4.

on the triangle of the octahedron, because the perpendiculars falling from the center of the sphere on equal circles drawn on the surface of the sphere are equal.[101]

It has also become clear that the square of the side of the octahedron is one and a half times the square of the side of the cube, since the square of the side of the octahedron is three times the square of the radius of the circumscribing circle, that is (the circle) which circumscribes the square of the side of the cube, and (since) the square of the side of the cube is twice the square of the radius of this circle. Thus the square of the side of the octahedron is one and a half times the square of the side of the cube.

8. The ratio of the triangle of the octahedron to the square of the side of the cube is equal to the ratio of half the altitude of the triangle to two-thirds of its side.

Proof: We make the triangle of the octahedron *ABG*. We construct on side *AB* square *DB*, and we draw from *G* a line parallel to *AB*, which meets *DA* at *E*. We bisect *AE* at *Z* [Fig. IV-5-4]. Then *AZ* is half of the altitude of triangle *AGB*. We make *AH* two-thirds of *AB*. Then the ratio of *AZ* to *AD* is equal to the ratio of half of rectangle *EB*, that is triangle *AGB*, I:41, to the square *DB*. But the ratio of *AD* to *AH* is equal to the ratio of the square *DB* to the square of the side of the cube, because the square of the side of the octahedron is one and a half times the square of the side of the cube, XV:3, and (because) *AD* is one and a half times *AH*. Thus, *ex aequali*, the ratio of *AZ*, that is, half the altitude of triangle *ABG*, to *AH*, two-thirds of its side, is equal to the ratio of triangle *ABG*, that is, the triangle of the octahedron, to the square of the side of the cube. That is what we wanted to prove.

12. The ratio of the surface of the cube to the surface of the octahedron is equal to the ratio of the side of an equilateral triangle to its altitude.

Proof: The ratio of the square of the side of the cube to the triangle of the octahedron is equal to the ratio of two-thirds of the side of the equilateral triangle to half of its altitude, XV:8. Thus the ratio of six times the square of the side of the cube, that is the surface of the cube, to six times the triangle of the octahedron, that is three-fourths of the surface

[101] This is proved in a lemma in the end of M:XIII, which is a special case of proposition 6 of Book I in Theodosius's *Spherics*.

of the octahedron, is equal to the ratio of two-thirds of the side of the equilateral triangle to half of its altitude, V:14 [= *Elements* V.15]. Thus the ratio of the surface of the cube to eight times the [triangle of the] octahedron, that is the surface of the octahedron, is equal to the ratio of two-thirds of the side of the equilateral triangle to two-thirds of its altitude, that is to say, equal to the ratio of the side of the equilateral triangle to its altitude.

It has become clear from this that the ratio of the surface of the cube to the surface of the octahedron is equal to the ratio of the square of a line to twice the equilateral triangle constructed on it, because the ratio of every line to the altitude of its triangle is equal to the ratio of the square of that line to the product of it and the altitude of its triangle according to the lemma,[102] and (because) the product of the altitude and the base is twice that triangle. By the "triangle of any line" we always mean the equilateral triangle constructed on it. Similarly the square of it, and the pentagon of it, and so on. And that is what we wanted.

17. The surface of the octahedron is one and a half times the surface of the tetrahedron inscribed in the same sphere.

Proof: Every two triangles of the octahedron are one-fourth of its surface, and every triangle of the tetrahedron is one-fourth of its surface. Thus the ratio of every two triangles of the octahedron to a triangle of the tetrahedron is equal to the ratio of the surface of the octahedron to the surface of the tetrahedron, V:14 [= *Elements* V.15]. But every two triangles of the octahedron are one and a half times every triangle of the tetrahedron, by prop.[103] XV:2. Thus the surface of the octahedron is one and a half times the surface of the tetrahedron inscribed in the same sphere. That is what we wanted to prove.

18. Every parallelepipedal solid with base equal to twice the face of a tetrahedron and height equal to the diameter of the sphere circumscribing the tetrahedron is nine times the tetrahedron.

Proof: The parallelepipedal solid with base equal to the face of the tetrahedron and height equal to its height is three times it [the tetrahedron], XII:6 [= *Elements* XII.7]. Thus the parallelepipedal solid with base twice the face of the tetrahedron and height equal to its height is six times it [the tetrahedron]. Thus the parallelepipedal solid with base twice the face of the tetrahedron and height one and a half times its height is nine times it [the tetrahedron]. But one and a half times the height of the tetrahedron is equal to the diameter of the circumscribing sphere, because the height of the tetrahedron is two-thirds of the diameter of the sphere, prop. XIV:1 [= *El*ements XIII.13]. Thus the parallelepipedal solid with base twice the face of the tetrahedron and height equal to the diameter of the circumscribing sphere is nine times the tetrahedron. It follows from this that the parallelepipedal solid with base twice the face of the tetrahedron and height one-ninth of the diameter of the [circumscribing] sphere is equal to the tetrahedron.

19. Every parallelepipedal solid with base equal to the square of the side of the octahedron and height equal to the diameter of the (circumscribing) sphere is three times the octahedron.

[102] It is not clear what lemma al-Maghribī is referring to here. It must have been a special case of *Elements* VI.1.

[103] The abbreviation "prop." for "proposition" here and further on in the text is intended to reflect the abbreviation in the text "*sh*" for "*shakl*" (meaning "proposition" in this context).

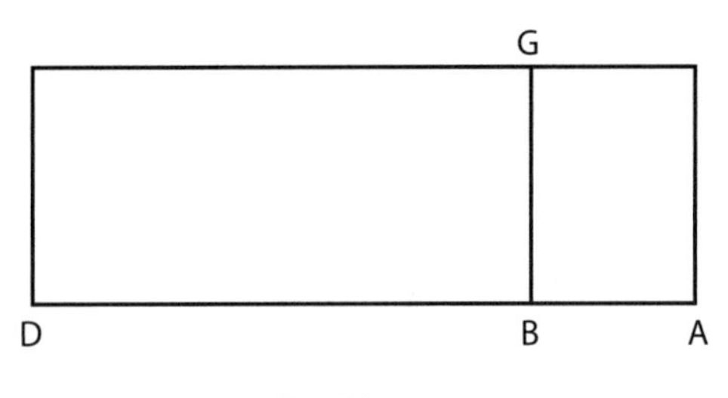

Fig. IV-5-5.

Proof: The parallelepipedal solid with base the square of the side of the octahedron and height equal to the radius of the sphere circumscribing the octahedron is three times the pyramid with base the square of the side of the octahedron and height equal to the radius of the sphere, that is to say, half of the octahedral solid. Thus the parallelepipedal solid with base the square of the side of the octahedron and height equal to the diameter of the sphere is six times half of the octahedral solid. Thus it [the parallelepiped] is three times it (the octahedron). It follows from this that the [parallelepipedal] solid with base the square of the side of the octahedron and height one-third of the diameter of the [circumscribing] sphere is equal to the octahedral solid. It has become clear from this that the parallelepipedal solid with base three times the square of the side of the octahedron and height equal to one-ninth of the diameter of the sphere is equal to the octahedral solid.

20. The ratio of the tetrahedral solid to the octahedral solid is equal to the ratio of the side of the (i.e. any) equilateral triangle to three times its altitude.

Proof: We make *AB* the side of the tetrahedral solid. We draw from point *B* the perpendicular *BG*, and we cut it off equal to the altitude of the triangle of the tetrahedron [Fig. IV-5-5]. Then it [*BG*] is equal to the side of the octahedron, XV:2. We extend *AB* in a straight line, and we cut off from it *BD* [equal to] three times *BG*. We complete the rectangles *AG*, *GD*. Then rectangle *AG* is twice the triangle of the tetrahedron, and *BD* is three times the side of the octahedron. Thus the parallelepipedal solid with base *AG* and height one-ninth of the diameter of the [circumscribing] sphere is equal to the tetrahedron, XV:18 and the parallelepipedal solid with base *GD* and height one-ninth of the diameter of the sphere is equal to the octahedral solid, XV:19. The ratio of the parallelepipedal solid with base *AG* and height one-ninth of the diameter of the sphere to the parallelepipedal solid with base *GD* and height one-ninth of the diameter of the sphere is equal to the ratio of *AG* to *GD*, that is to say, *AB* to *BD*. But *BD* is three times the side of the octahedron. Thus the ratio of the tetrahedral solid to the octahedral solid is equal to the ratio of the side of the triangle of the tetrahedron to three times its altitude.

It has become clear from this that the ratio of the tetrahedral solid to the octahedral solid is equal to the ratio of the side of the tetrahedron to three times the side of the octahedron.

21. For every equilateral and equiangular solid circumscribed by a sphere, if its solid angles and the center of the sphere are joined by straight lines, [the solid] is divided into equal and similar pyramids, the number of which is equal to the number of faces of the solid.

[Proof omitted]

23. The ratio of the solid of the cube to the octahedral solid is equal to the ratio of the side of the (i.e., any) equilateral triangle to its altitude.

Proof: The ratio of the solid of the cube, that is six times the pyramid with base the face of the cube and vertex the center of the sphere circumscribing the cube, XV:21, to six times the pyramid with base the face of the octahedron and vertex the center of the circumscribing sphere, that is three-fourths of the solid of the octahedron, XV:21, is equal to the ratio of the pyramid with base the face of the cube and vertex the center of the sphere, to the pyramid with base the triangle of the octahedron and vertex the center of the sphere. [But this ratio is equal to the ratio of the face of the cube to the face of the octahedron,][104] because the height of the pyramid with base the face of the cube is equal to the height of the pyramid with base the face of the octahedron, XV:3. But the ratio of the face of the cube to the face of the octahedron is equal to the ratio of two-thirds of the side of the equilateral triangle to half of its altitude, XV:8. Thus the ratio of the cube to the octahedron is equal to the ratio of two-thirds of the side of the equilateral triangle to two-thirds of its altitude, which is the ratio of the side of the equilateral triangle to its altitude. That is what we wanted to prove.

26. The ratio of the radius of the sphere circumscribing the octahedron to the perpendicular of its solid is equal to the ratio of the altitude of an equilateral triangle to half its side.

Proof: The perpendicular of the octahedron is equal to the perpendicular of the cube, XV:3. But the square of the radius of the sphere is three times the square of the perpendicular of the cube, XIV:3[105] Therefore it is three times the square of the perpendicular of the octahedron. But the square of the altitude of the equilateral triangle is three times the square of half of its side, XV:1. Therefore the ratio of the square of the radius of the sphere to the square of the perpendicular of the octahedron is equal to the ratio of the square of the altitude of the (equilateral) triangle to the square of half of its side. Therefore the ratio of the radius of the sphere to the perpendicular of the octahedron is equal to the ratio of the altitude of the triangle to half of the side of the triangle. That is what we wanted.

V. TRIGONOMETRY

1. ABU ʿABD ALLAH MUḤAMMAD IBN MUʿĀDH AL-JAYYĀNĪ, *BOOK OF UNKNOWNS OF ARCS OF THE SPHERE*

To the biographical material on Ibn Muʾādh in the introduction to section IV-3, we add here that Ibn Muʾādh, in addition to his astronomical handbook, the *Jaen Tables*, wrote a mathematical treatment of some problems in astrology (titled *Projection of the Rays*),

[104]Hogendijk added the passage in square brackets to make mathematical sense.

[105]Euclid proves in *Elements* XIII.15 that the square of the diameter of the circumscribing sphere is three times the square of the side of the cube. Al-Maghribī concludes in a corollary to M XIV:3 (= *Elements* XIII.15) that the square of the radius is three times the square of the perpendicular from the center to any face of the cube.

the treatise *On Twilight* excerpted in section V-2, and a treatise on a solar eclipse that he had observed late in life. He also composed the first known work on spherical astronomy independent of any applications to astronomy. That work, excerpted here, is titled *Book of Unknowns of Arcs of the Sphere.*[1] Most scholars conclude that Ibn Muʾādh probably learned of this material from sources originating in the East; however, nothing in the work points to any particular Eastern source. And, at this time, all treatments known to us from the East were composed in the context of astronomical problems, whereas Ibn Muʾādh gives an entirely mathematical treatment of the subject.

[Introduction]

In this book we want to find the magnitudes of arcs falling on the surface of the sphere and the angles of great arcs occurring on it as exactly as possible, in order to derive from it the greatest benefit towards understanding the science of celestial motions and towards the calculation of the phenomena in the cosmos resulting from the varying positions of celestial bodies.

So we present something whose value and usefulness in regard to understanding this [subject] are great. As for premises that were derived by scholars who preceded us, we give just the statements, without proof, so that we may arrive at acknowledgement of their proof. We will mention them since the knowledge of their proofs and attached things takes in itself its demonstration. Sometimes we have presented a matter without any proof, content to fix the attention of the reader on that. We have written our book for those who are already advanced in geometry, rather than for beginners.

And our first premise from that is the Transversal Theorem, an example of which is: Two arcs of great circles, *AB* and *AG*, meet at point *A*, and each of them is less than a semicircle [Fig. V-1-1]. Between them two arcs of great circles, *BD* and *EG*, intersect at point *W*. Then I say that the ratio of the Sine[2] of arc *BA* to the Sine of arc *EA* is composed of the ratio of the Sine of arc *BD* to the Sine of arc *WD* and of the ratio of the Sine of arc *GW* to the Sine of arc *GE*. And similarly, it will happen if you begin with the ratio from the side of the arc *AG*.

Ibn Muʾādh states the analogous proportion for the ratio of the Sine of *GA* to the Sine of *DA*, and then the theorem for the ratio of the Sine of *BE* to the Sine of *EA*, namely, $\mathrm{Sin}(BE)/\mathrm{Sin}(EA) = (\mathrm{Sin}(BW)/\mathrm{Sin}(WD))\cdot(\mathrm{Sin}(GD)/\mathrm{Sin}(GA))$ and the analogous relation for $\mathrm{Sin}(GD)/\mathrm{Sin}(DA)$. These are the four basic cases of the oldest theorem in spherical trigonometry.[3] The theorem is first encountered in a book, *Spherics*, by the Greek astronomer/mathematician, Menelaus of Alexandria, written in the first century of our era. We find it again in the *Almagest* of Claudius Ptolemy, written in the second century. The theorem was much elaborated by mathematicians

[1]Because titles in the manuscript sources vary, this work is also referred to as *Determining the Magnitudes of Arcs on the Surface of the Sphere*. I have chosen the title used here because it is the title in the edition of the Arabic text I have consulted.

[2]Here and throughout these translations, the capital letters S and C, are used to remind the reader that the medieval Sine and Cosine of an arc were line segments. The Sine of an arc was equal to one-half the chord of double the arc, and the Cosine was the Sine of the complement of the arc.

[3]It was originally, and often thereafter, stated for chords of arcs rather than for Sines. (As Ibn Muʾādh points out in the remarks below, any statement involving ratios of two chords has an analogue for ratios of Sines.)

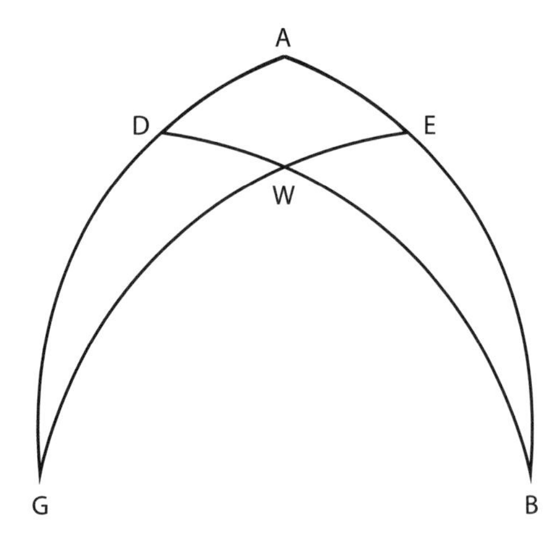

Fig. V-1-1.

in the Islamic world and in the latter part of the ninth century. Thābit ibn Qurra wrote a treatise listing and proving eighteen different cases of the theorem.

We say that there are two kinds of things found in a triangle, sides and angles. There are three sides and three angles, but there is no way to know the triangle completely, i.e. [all] its sides and its angles, by knowing only two of the six. Rather, from knowing only two things, be they two sides or two angles or one side and one angle, it [the triangle] is unspecified. For it is possible that there are a number of triangles, each of which has those [same] two known things, and so one must know three things connected with it [the triangle] to obtain knowledge of the rest. Thus it is impossible to attain all of it knowing less than three members: three sides, three angles, two sides and an angle or two angles and a side.

These selections from the introductory material end here, and we pass on to his statements and proofs of two theorems concerning two circular arcs of which the ratios of their chords or Sines are known as well as either the difference between the larger and smaller arc or the sum of the two arcs. The following selection deals with the case when the difference of the two arcs is known.

[Determining arcs when Sine ratios and sums or differences are known]

We shall mention what we must, and we say that if the ratio of the chord of an arc to the chord of another arc is a known ratio [and] one of the arcs exceeds the other arc by a known arc then it is not possible to find two arcs other than the two first arcs of which one exceeds the other by an excess equal to that of the first two such that the ratio of the chord of the longer of the two of them to that of the shorter of the two is as the ratio of the chord of the longer of the first two to the chord of the shorter.[4]

[4]This theorem amounts to saying that if α and β are unequal arcs in a circle, with $\alpha > \beta$, such that both the ratio, Crd α/Crd β, and the difference, $\alpha - \beta$, are known then α and β are uniquely determined. For clarity he proves that if α and β are unequal arcs in a circle, with $\alpha > \beta$ such that both the ratio Crd α /Crd β and the difference $\alpha - \beta$ are

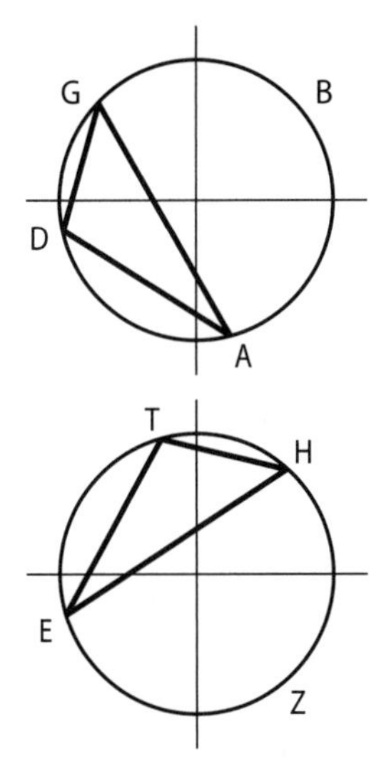

Fig. V-1-2.

For example, arc *ABGD* has chord *AD*, and arc *ABG* has chord *AG*, and the excess of arc *ABGD* over arc *ABG* is [the known] arc *GD*. And the ratio of chord *AD* to the chord *AG* is a known ratio [Fig. V-1-2]. Then, I say that it is not possible that there be in a circle equal to *ABGD* two other arcs not equal to these two arcs one of which exceeds the other by an arc equal to *GD* and [such that] the ratio of the chord of the longer of the two of them to the chord of the shorter is as the ratio of the chord of *AD* to the chord of *AG*.

Its proof is that we construct a circle equal to circle *ABGD*, namely circle *EZHT* and, if possible, we cut it into two arcs not equal to the two arcs *ABGD* and *ABG*, satisfying the conditions we mentioned. Let arc *EZHT* be the larger of the two and *EZH* the smaller. And suppose *HT*, the excess, is equal to arc *GD*, the first excess, and that the ratio of the chord of *ET* to the chord of *EH* is as the ratio of the chord of *AD* to the chord of *AG*, but that arc *EZH* is not equal to arc *ABG*, if that were possible.

And we draw the two chords, *GD* and *HT*, which are equal because they are the chords of the two equal arcs. And angles *A* and *E* are equal since they are on the circumference of equal circles and on equal arcs. And the ratio of *ET* to *EH* is as the ratio of *AD* to *AG*. And so the two triangles are similar. And for that [reason] the ratio of *ET* to *AD* is also as the ratio of *TH* to *DG*. And chord *TH* is equal to chord *DG*. And so chord *ET* is equal to the chord *AD*, and similarly chord *EH* is equal to chord *AG*. And so the angles of the triangle are like the angles of the other triangle. And angle *D* is equal to angle *T*, and so the base [of arc] *ABG* is equal to the base [of arc] *EZH*. And similarly the longer arc, *ABGD* is equal to the longer arc, *EZHT*. But these two were not equal to each other, which is a contradiction. And so the two arcs, *EZHT* and *EZH* are equal to the two arcs *ABGD* and *ABG*. So two arcs are not found in a circle equal to circle *ABGD* satisfying the conditions we mentioned unless they are equal to the two arcs *ABGD* and *ABG*, which is what we wanted to show.

known and if α' and β' are unequal arcs in a another circle, equal to the first, and if Crd α'/ Crd β' and $\alpha' - \beta'$ are equal to Crd α/ Crd β and $\alpha - \beta$, respectively, then $\alpha = \alpha'$ and $\beta = \beta'$

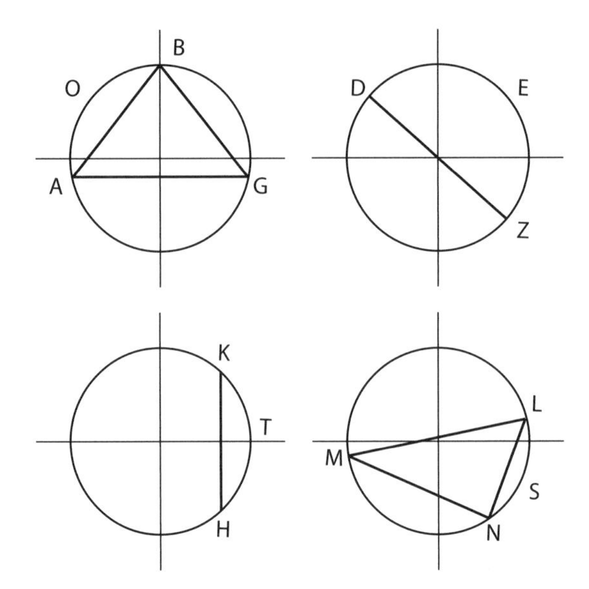

Fig. V-1-3.

And similarly we will also prove that, in two unequal circles, if the arc that is the excess in one of them is similar to the arc that is the excess in the other, and the ratio between the two chords in the one circle is as the ratio between the two chords in the other circle then the two arcs of the one circle are similar to the two arcs of the other [circle], each to the corresponding one.[5]

Ibn Muʾādh now deals with the case in which the sum of the two arcs is known.

Similarly, I also say that if there are two arcs from one circle, or from two equal circles, such that the two, when added together, do not make the circumference of a circle but either something more or something less,[6] and their two chords have a certain ratio, then there are no two arcs, for example of that circle, other than [equals of] the first two arcs, whose sum is equal to the sum of the first two and the ratio of the chord of one of them to the chord of the other is as the ratio between the chords of the first two.[7]

Its example is that there are two arcs, *ABG* and *DEZ* of two equal circles that [together] do not contain a circle exactly, rather exceed it or fall short of it. And the ratio of the chord of *AG* to the chord of *DZ* is a known ratio. Then I say that one cannot find two other than [ones equal to] *ABG* and *DEZ* whose sum is equal to that of arcs *ABG* and *DEZ* and the ratio of the chord of one of them to the other is equal to the ratio of the chord of *ABG* to the chord of *DEZ* [Fig. V-1-3].

Its proof is that if it [what is claimed] is false then it is possible [that two other such arcs exist], and so let us suppose there are two circles equal to circles *ABG* and *DEZ* that have two arcs satisfying the conditions mentioned (which we omit for brevity). Say these two circles are *HTK* and *LMN* and the two arcs are arcs *HTK* and *LMN* and the ratio of the

[5]This theorem extends the previous one to the case not discussed, namely, when α and β are in a circle of one size and α' and β' are in an unequal circle.

[6]Ibn Muʾādh tacitly assumes that the sum of the two arcs is known.

[7]Given two arcs α, β such that $\text{crd}(\alpha)$:$\text{crd}(\beta) = r$, then if arcs γ and δ have the property that $\gamma + \delta = \alpha + \beta$ and $\text{crd}(\gamma)$:$\text{crd}(\delta) = r$, we then have $\alpha = \gamma$ and $\beta = \delta$.

chord of arc *HK* to the chord of arc *LN* is as the ratio of the chord of *AG* to the chord of *DZ*. And, if possible, the arc we have premised, *HTK*, is not equal to premised arc *ABG*, and, similarly, the following arc, *LMN*, is not equal to the following arc, *DEZ*.

Because arc *HTK* is not equal to arc *ABG* and so one of them will be greater than the other, and we suppose [*A*]*BG* is greater. We cut from it the equal of *HTK*, say, *AOB*, and because the sum of the two arcs, *HTK* and *LMN*, is equal to the sum of the two arcs, *ABG* and *DEZ*, and the excess of *ABG* over *HTK* is arc *BG*, then the excess of arc *LMN* over arc *DEZ* is equal to *BG*. And so we cut from arc *LMN* the equal of arc *DEZ*, say arc *LM*. And we divide *MN* like arc *BG* and we make the chords *AB*, *BG*, *LM* and *MN*. [And] since we postulated that the ratio of *HK* to *LN* is as the ratio of *AG* to *DZ*, and the chord of *HK* is equal to the chord of *AB*, and the chord of *DZ* is equal to the chord of *LM*, the ratio of *AB* to *LN* is as the ratio of *AG* to *LM*, and angle *A* is equal to angle *L*. And, because of that, triangle *ABG* is similar to triangle *LMN*, and angle *B* is equal to angle *N* and angle *G* is equal to angle *M* and the bases of these angles[8] will be equal, and so arc *AOB* will be equal to arc *LSN*. But arc *AOB* was equal to arc *HTK*, and so arc *HTK* is equal to arc *LSN*. But the sum of arc *LSN* and arc *LMN* is exactly the circumference of the circle. And so the sum of the two arcs, *HTK* and *LMN* is exactly the circumference of a circle. But we supposed that that was not the case [and] this is a contradiction, impossible. And so their ratio cannot be other than that of the first two, which is what we wanted to prove.

And the affair will be like that in two unequal circles as regards similarity [rather than equality] of the arcs, satisfying these conditions analogous to what we stated in the theorem preceding this, God willing. And so:

Given that this has been clearly proven and we have two arcs of which the ratio of the Sine of the one to that of the other is known and we except those arcs whose Sines are used from being a semicircle. Rather it is less. And you will remember it, so we do not have to repeat it—unless we especially need to. Neither of the two arcs is known but one of them exceeds the other by a known arc less than one hundred eighty degrees. Then we are able to produce two arcs satisfying this property. And no one can say that two other arcs than these two can be found satisfying this property.

Similarly, if we have two arcs such that the ratio of the Sine of one to the Sine of the other is known, but each one of them individually is unknown, however the sum of the two is known and it is not a semicircle, then we can produce two arcs satisfying this property. And it is not possible that two other arcs, different from these two but satisfying this property, can be produced. This is because everything that holds for the whole chords belonging to arcs hold for the Sines of their [the arcs'] halves. And everything that fails for the wholes fails for the halves—after we have required, in the case when Sines of arcs are used, that each arc be less than one hundred eighty degrees. For we established the proof with arcs having whole chords and each arc was completely smaller than the circumference of a circle. And the Sine belongs to half the arc, and half of what does not make up a full circle must be less than half a circle. And we proved in the two preceding theorems the impossibility of finding these conditions in variable chords and arcs and, like that, it is forbidden in arcs and their Sines.

[8] This phrase must refer to the chords that the equal angles subtend.

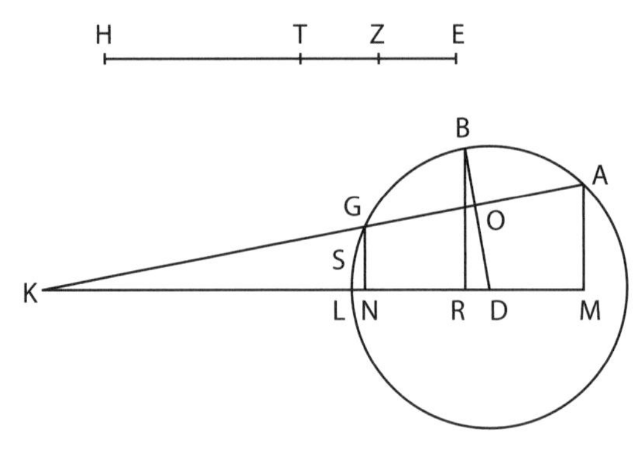

Fig. V-1-4.

In the second theorem[9] we require that the sum of the two arcs whose chords are in the proportion is not the circumference of a circle. For, if it is, the problem is "indeterminate," because every two arcs whose sum is the circumference of a circle have equal chords, and so the ratio of their chords is the ratio of equality.[10]

Ibn Mu ͨādh continues in this vein and then writes the following.

And having mentioned all of this, let us mention the way to deduce the unknowns of the two theorems. We begin with the first:[11] We draw a circle *ABG*, whose center is *D*, and two known, unequal magnitudes, *EZ* and *ZH* [Fig. V-1-4]. And we make the known arc, *ABG*, less than a semicircle (according to the condition we premised). We say that in circle *ABGD* are two unknown arcs, assuming each of them is less than one hundred eighty degrees and that one of them exceeds the other by a known arc *ABG*, and the ratio of the Sine of one of them to the Sine of the other is like a certain ratio, *EZ* to *ZH*. And so, if we want to derive those two arcs, I say that the way to do this is to draw the chords and arcs according to the conditions these premises require, not by cutting from the circle a conjectural segment matching the theorem in a practical way of knowing. But, what we do is to make the chord of the arc *ABG* the line *AG*. Then we cut from the longer of the two magnitudes, *EZ* and *ZH*, the equal of the shorter, *ZT* which is equal to *ZE*.[12] And we make the ratio of *HT* to *TZ* equal to the ratio of the line *AG* to its extension, the line *GK*. And we join *K* with *D*, the center of the circle, by line *KLD*. Then I say that arc *LBA* is the larger of the two arcs we are seeking and arc *LSG* is the smaller, and they satisfy the preceding condition.

Its proof is that we produce from the two points, *A* and *G*, perpendiculars to the line *DL*, and they are lines *AM* and *GN*, which are parallel. And so the ratio of *AK* to *GK* is as the ratio of *AM* to *GN*. And because we postulated that the ratio of *HT* to *TZ* is as the ratio of *AG* to *GK*, when we invert, the ratio of *TZ* to *TH* is as the ratio of *GK* to *AG*. And if we compose, the ratio of *ZT* to *ZH* is as the ratio of *KG* to *KA*. And the ratio of *KG* to KA is as

[9]The reference is to the theorem quoted above on the sum of arcs.

[10]In modern terms, the ratio is 1.

[11]The reference would seem to be the analogue of the first theorem quoted from Ibn Muͨādh for Sines rather than for chords.

[12]The author takes the longer segment to be *ZH*.

the ratio of *GN* to *AM*. And so the ratio of *GN* to *AM* is as the ratio of *ZT* to *ZH*. And *ZT* is equal to *EZ*. And each of the two lines *AM* and *GN* is the Sine of each of the two arcs *ABGL* and *GSL*.

And as for knowing the amount, that is easy if we take into consideration their conditions. We divide the chord *AG* into halves at *O* and we produce line *DOB*. And so *OG* is known since it is half the chord of arc *ABG*, which is the Sine BG.[13] And *DO* is known since it is the Sine of the complement of arc *BG*, and it is also the remainder of the completion of *BO* to the center. And, because the ratio of *OG* to *AG* is as the ratio of *TZ* to *TE*,[14] and the ratio of *AG* to *GK* is as the ratio of *HT* to *TZ*, and by the ratio of equality in the extremes, the ratio of *OG* to *GK* is as the ratio of *HT* to *TE*. So, because of that, *GK* is known.[15] And all of line *OK* is known. And so you multiply it by its equal, and line *OD* by its equal, and add the two of them and take the root of the sum. That is the line *DK*, so it is known. And we produce from *B* a perpendicular onto *DK*, which is *BR*, and it is the Sine of arc *BGL*. And the ratio of line *DK* to line *KO* is as the ratio of line *DB* to line *BR*, since the triangles are similar. And so we multiply the magnitude of line *KO* by sixty, which is the magnitude of line *DB*, and we divide the result by *DK* and what comes out is *BR*. And we take the arcSine of that[16] which is *BGSL*. And so if we add to it the arc *AB*, which is half the arc *ABG*, the sum is arc *ABGSL*, the larger of the two we are seeking. And if we subtract from it the arc *BG*, which is also half of arc *ABG*, arc *GSL*, the smaller of the two we are seeking, remains—which is what we wanted to demonstrate.

Behold we have premised this, so let us mention things that pertain to the figures [i.e., various sorts of spherical triangles] and are characteristic of them, i.e., if one of them is necessary the other is equally so.

[Possible sides of a right-angle triangle according to the remaining angles.]

We premise: In a triangle with a right angle, if one of the other angles is less than a right [angle] then the side subtending that angle is smaller than ninety degrees. If there is in it an angle greater than a right then the side subtending it is more than ninety degrees, and if it is also right, then the side subtending it is equally ninety degrees.

In the same way also, conversely, if there is in it [the triangle] a side greater than ninety (and by 'side' I mean one of the two sides containing the right angle, not the one that subtends it) then the angle that subtends[17] it is greater than a right. And if it is less than ninety then the angle is less than a right. And if it is, equally, ninety then the angle that subtends it is right. Also, it is necessary to know that in any triangle, if the three angles are added together then they are greater than two right angles.[18]

And in a right angle triangle with two acute angles, the side subtending the right angle is smaller than a quadrant. And if its two remaining angles are obtuse then the side subtending the right angle is less than a quadrant. And if, of its two remaining angles

[13] *B* is the midpoint of arc *AG* because it was specified above as the point where the radius to the midpoint of the chord of arc *AG* intersects that arc (this follows easily from *Elements* III.3, 30).

[14] Since the consequent of each of the two ratios is double the antecedent.

[15] Because the remaining three terms in the preceding proportion are known.

[16] Literally, "we arc that, as arcing of Sines."

[17] In Arabic usage, not only can a side "subtend" an angle, but an angle can also "subtend" a side.

[18] Doubtless, astronomers from at least the time of Menelaus in the first century of our era had, for one reason or another, summed the angles in specific spherical triangles, but here Ibn Muʿādh is making the general statement about all spherical triangles. This illustrates his interest in the mathematics of spherical arcs.

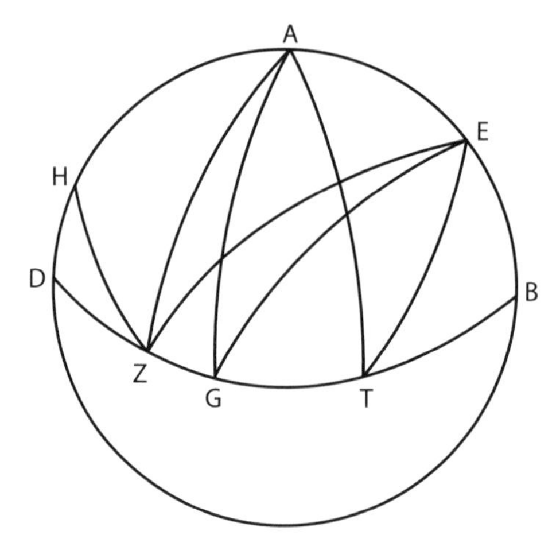

Fig. V-1-5.

[i.e., the two not initially assumed to be right], one is acute and the other is obtuse then the side subtending the right angle is larger than a quadrant. And if one of the two remaining angles is right as well, then the side subtending the first right angle is ninety. And, similarly, the side subtending the remaining right angle is also ninety, whatever the third angle might be.

And, conversely, i.e., if each of the two sides bounding a right [angle] is less than ninety then the side subtending the right is smaller than ninety, and—similarly—if each of the two sides bounding a right is more than ninety then the side that subtends the right is less than ninety also. And if one of the two sides bounding a right angle is more than ninety and the other less, then the side subtending the right is more than ninety.

And we will prove all of that together in one figure [Fig. V-1-5]. For this we make a right angle, *ABG*, with two great circle arcs. Then we extend arcs *BA* and *BG* until they meet at the point opposite *B*, point *D*. And, of the figures arising and the varying theorems in the right angle triangle [of the hypothesis] none of their types or forms go beyond the two arcs *BAD* and *BGD*.[19]

Now let point *A* be the pole of circle *BGD* and point *G* the pole of arc *BAD*, and so *B* is the right angle[20] in the triangle. So, when we make one of the sides of the triangle, say, *BE*, less than a quadrant then I say that, whatever the angle subtending the side *BE*, it is, indeed, less than a right angle. The proof of that is that we take any point whatever on arc *BGD*, say, *Z*, and we join *E* with *Z*. Then I say that angle *EZB* of triangle *EZB* is less than a right since when we join *A* with *Z*, however we place it [*Z*], angle *AZB* is right, since *A* is a pole [of arc *BGD*]. And so angle *EZB* remains smaller than a right. And if we make side *BH* bigger than a quadrant then the angle that side *BH* subtends is bigger than a right angle, since, wherever the angle falls on arc *BGD*, if we join that place with point *A*, the resulting angle is right. And the angle that side *BH* subtends, like angle *HZB*, is bigger than right.

[19]That is, they will all be contained in the sector *BGDAEB* in Fig. V-1-5.

[20]If *C* is a great circle on a sphere and it passes through a pole of a great circle *C**, then it will intersect *C** at right angles.

In the following section Ibn Muʿādh lays out the properties of spherical triangles in which at least one side is a quadrant. He begins by listing possibilities for the other sides and the consequences of these for angles of the triangle.

[Properties of triangles with one side a quadrant]

And let us prove the properties of a triangle one of whose sides is ninety [degrees]. So I say that if the magnitude of one of a triangle's sides is a quadrant then, if there is only one other side in it that is a quadrant, each of the two angles subtending these two sides are right. And if all the sides are ninety then all of its angles are right.

And if no other side [than the first mentioned] is a quadrant of a circle then, if one side of the triangle is less than a quadrant, the angle subtending it is less than a right angle. And, if the side is greater [than a quadrant] then the angle subtending it is obtuse. And if it is ninety we have previously mentioned this, namely, that the angle subtending it is right.

In the following, one side is still a quadrant, but Ibn Muʾādh now lists possibilities for the two angles at the endpoints of the quadrant and the consequences of these for the sides of the triangle.

And conversely[21] also, namely, if one of the two angles that are on the extremities of the quadrant is acute then the side that subtends it is less than ninety. And if it is obtuse, then that which subtends it is greater than ninety. And if it is right we have already stated it, namely, that the [angle] subtending it is ninety.

Also, if each of the angles on the two extremities of the quadrant is acute, then the angle subtending the quadrant is obtuse. And, in the same way, if each of the angles at the two extremities of the quadrant is obtuse, then the angle that subtends this quadrant is also [a reference to the previous result] obtuse. And if one of the two angles at the extremities of the quadrant is acute and the other is obtuse then the angle subtending the quadrant is acute,

And conversely[22]—and before [going on] it is necessary to know that the sides of any triangle total less than three hundred sixty degrees—if each of the remaining two sides [i.e., the two not assumed to be quadrants] is smaller than ninety, the angle that subtends the quadrant is obtuse. And, similarly, if each of the two remaining sides is greater than ninety, then the angle subtending the quadrant is obtuse. And, if one of the remaining two sides is more than ninety and the other smaller than ninety, then the angle that subtends the quadrant is smaller than a right [angle].

And also, when the angle that subtends a quadrant is obtuse, then the two sides containing it are both greater than ninety or both less than ninety and if it is acute then without doubt one of the two sides containing it is more than ninety and the other is less than ninety.

And the demonstration of these properties will be clear from the previous proposition to him who ponders it [Fig. V-1-6]. That is, if we make the side *AB* a quadrant and make each of the angles *BAG* and *ABG* right then each of the two sides, *AG* and *BG* are quadrants. And if we make the angle *ABD* obtuse the side *AD* is, without doubt, more than ninety,

[21] This does not mean that the following statement is the converse of the preceding statement; rather it means that he is now changing focus, as described in the preceding comment.

[22] Again, the term signals not a logical converse of the preceding statement but a change of focus.

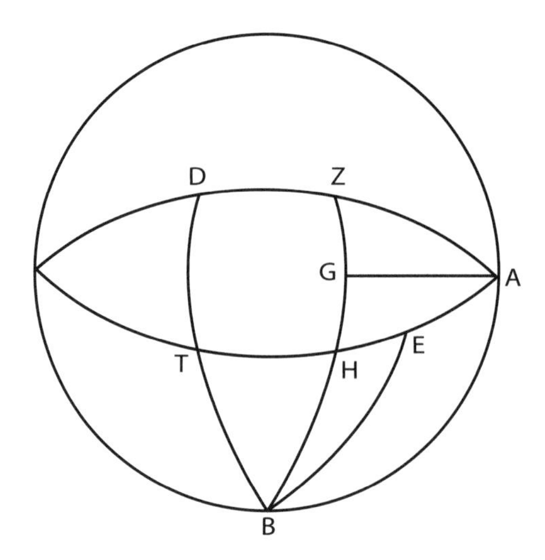

Fig. V-1-6.

and if we make angle *BAE* the side *BE* is, without doubt, less than ninety. And if we make angle *BAD* obtuse then angle *D* is obtuse because triangle *BZD* has angle *Z* right while the side *BZ* is greater than a quadrant. And similarly, if we make angle *ABE* less than a right angle, angle *AEB* is obtuse since in triangle *BEH* angle *H* is right, and side *HB* is less than ninety and so angle *BEH* is acute and angle *BEA* is obtuse.

And if we make angle *BAT* acute and angle *ABT* is obtuse, angle *ATB* is acute, since triangle *BHT* has a right angle at *H* and side *HB* is less than ninety. And all the rest of what we mentioned is evident.

And I also say that about the triangle whose angle is acute. And it was premised before that the angles of each triangle are greater than two right angles, but its proof is not in this book.

[The Sine theorem]

We also premise a theorem of great usefulness and abundant benefit in general. [In] any triangle whose sides are arcs of great circles [lit., "great arcs"], the ratio of the Sine of each of its sides to the Sine of the opposite angle [lit., "the angle subtending it"] is a single ratio [i.e., one and the same for each side].

For example, let *ABG* be an arbitrary triangle. I say that the ratio of the Sine of the side *AB* to the Sine of angle *AGB* is as the ratio of the Sine of side *AG* to the Sine of angle *ABG*, and it is also as the ratio of the Sine of side *BG* to the Sine of angle *BAG* [Fig. V-1-7].

Proof: We start with any side we wish and we find its pole that is in the direction of the angle [of the given triangle] opposite it. So we take the side *AB* and we find its pole that is in the direction of *G*, namely, *D*. We join *G* and *D* and we extend [the arc joining them] beyond *G* [lit., "in the direction of *G*"] so that it meets arc *AB* at *E*. We make the side *A*[*B*]*Z* a quarter circle, we draw *DZ*, and we produce *AG* to *H* [on *DZ*]. By what we demonstrated earlier, the two angles *H* [i.e., *DHG*, *GHZ*] are two right angles and arc *HA*

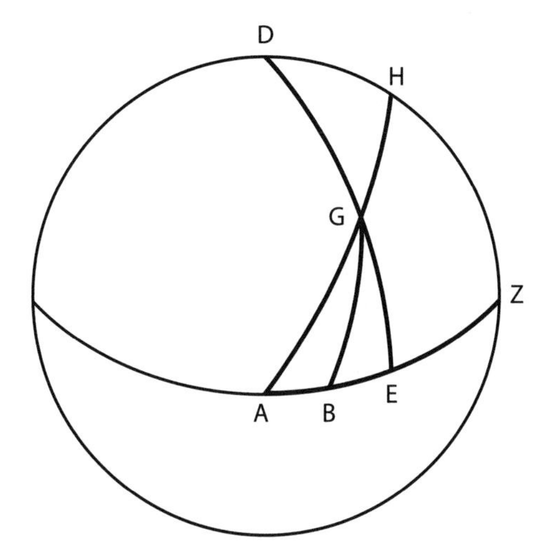

Fig. V-1-7.

is a quarter circle. Arc *HZ* is the magnitude of angle *HAZ*. And the ratio of the Sine of arc *DZ* to the Sine of arc *HZ* is composed of the ratio of the Sine of arc *DE* to the Sine of arc *GE* and the ratio of the Sine of arc *AG* to the Sine of arc *AH*.[23] If we convert this ratio of magnitudes to four magnitudes (some of the magnitudes being equal and, so, cancelling)[24] there remains the ratio of the Sine of arc *AG* to the Sine of arc *GE* being as the ratio of the Sine of arc *AH* to the Sine of arc *HZ*. But arc *AH* is ninety degrees so its Sine is 60.[25] And arc *HZ* is the magnitude of angle *HAZ*. So the ratio of the Sine of arc *AG* to the Sine of arc *GE* as the ratio of 60 to the Sine of angle *GAE*. And the product of the Sine of arc *AG* by the Sine of arc *GAE* is equal to the product of 60 by the Sine of arc *GE*.

Similarly, it may be demonstrated that in triangle *GBE*, the Sine of arc *BG* is to the Sine of arc *GE* is as the ratio of 60 to the Sine of angle *GBE*. So the product of the Sine of arc *BG* by the Sine of angle *GBE* is equal to the product of 60 by the Sine of arc *GE*.

But the product of 60 by the Sine of arc *GE* is equal to the product of the Sine of arc *AG* by the Sine of angle *GAE*. So the product of the Sine of arc *AG* by the Sine of angle *GAE* is equal to the product of the Sine of *BG* by the Sine of angle *GBE*. Hence, the ratio of the Sine of arc *AG* to the Sine of angle *GBE* is as the ratio of the Sine of arc *BG* to the Sine of angle *GAB*. And the Sine of angle *GBE* is equal to the Sine of angle *GBA*. So we have proved that the Sine of arc *AG* to the Sine of angle *ABG* is equal to the ratio of the Sine of arc *BG* to the Sine of angle *BAG*.

And, in the same way, if we make the side *GB* the base and find its pole, as we did above in the case of arc *AB*, it becomes clear that the Sine of arc *AG* is to the Sine of

[23]This follows from the transversal theorem applied to the figure *DGAEZH*.

[24]Note that arcs *DZ* and *DE* are both quarter circles. This translation is a bit loose, but it captures the essentials of Ibn Mu'adh's thought.

[25]Ibn Mu'adh is assuming that 60 is the radius of the circle on which Sines are calculated. This was the standard value in Islamic trigonometry.

angle *ABG* as the Sine of arc *AB* is to the Sine of angle *AGB*. And that is what we wanted to prove.

And from that it will be clear that [in particular] for any right triangle, the ratio of the Sine of any of its angles to the Sine of the opposite side is as the ratio of sixty to the Sine of the [side] subtending the right angle.

And, so that it is not necessary to repeat the preceding words later, we make this premise: The Sine does not exceed sixty, which is the Sine of the quadrant. But the arcs that are sides for a triangle can exceed a quadrant.

For that reason if some line is produced as known then it is a Sine for two arcs. For example, if a line is thirty then it is the Sine for an arc whose magnitude is 30 [degrees], and this is the arc of this Sine, the arc named by it, but it is also the Sine of an arc whose magnitude is one hundred fifty. For that reason, it is always necessary to know whether the arcs are greater than a quadrant or less. For, if we know the Sine, which is what always results in the operations, we are forced to find its arc. And it became clear to us that for each Sine there are two arcs, [so] we need to recognize which of the two arcs is the desired one.

When we want the arc that is less than a quadrant we need say nothing more than that it is from a Sine. ... As for the other [arc], it is the supplement (relative to one hundred eighty) of the first. So we say, if we want the first, "speaking simply" and, if we want the second, we say that we take the large arc, or (speaking concisely and summarily) we take the large arcSine.

In the next section Ibn Muʿādh applies the sine law to a right triangle, *ABG*, in which the angle *A* is a right angle [Fig. V-1-8]. Denoting its angles by *A*, *B*, *G* and the arcs opposite these as *a*, *b*, *g*, respectively, then Ibn Muʿādh states the following rules:

$\mathrm{Sin}(g):\mathrm{Sin}(a) = \mathrm{Cos}(B):\mathrm{Cos}(b)$, and

$\mathrm{Sin}(A):\mathrm{Sin}(G) = \mathrm{Cos}(b):\mathrm{Cos}(B)$, which, he points out, can also be written as
$\mathrm{Sin}(A)\cdot \mathrm{Cos}(B) = \mathrm{Sin}(G)\cdot\mathrm{Cos}(b)$

He does not offer any proofs of these rules.[26] (As he said at the beginning, he is not writing for beginners in geometry!) His statements are as follows:

And, with respect to the figure of the right triangle, such as triangle *ABG*, if we suppose angle *A* is right then what we mentioned from the preceding proof shows also that the ratio of the Sine of side *AB* to the Sine of side *BG* is as the ratio of the Sine of the complement of angle *B* to the Sine of the complement of side *AG* subtending it. And, if we want, we say that the ratio of the Sine of the right angle *A* to the Sine of any angle we choose of the triangle is as the ratio of the Sine of the complement of the side on which the two corresponding angles fall to the Sine of the complement of the third angle. And, if we want, we say that the product of the Sine of the right angle (which is always sixty) by the Sine of the complement of either of the two remaining angles is equal to the product of the Sine of the third angle by the Sine of the complement of the side that subtends the first.

[26]The first theorem follows from the transversal theorem, while the second one is equivalent, given the law of sines.

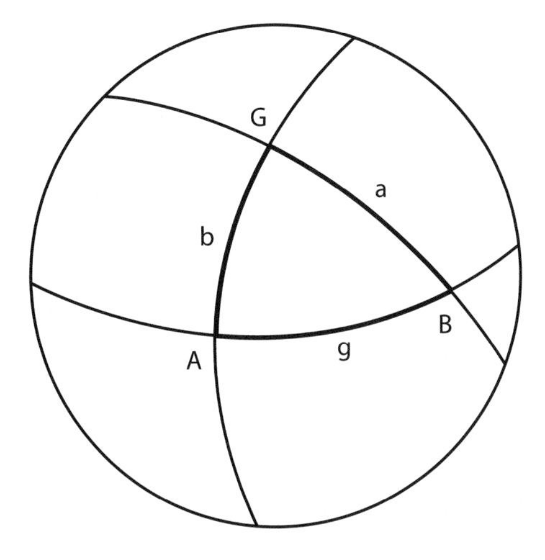

Fig. V-1-8.

This selection from Ibn Muʿādh concludes with his discussion of how to solve two types of triangles. The first is the case when an angle and the two sides bounding it are known, and the second is the case when all sides of the triangle are known. The first case is exactly the situation one has in determining the direction of Mecca from a given locality, a direction known in Arabic as the *qibla*, where the vertices of the spherical triangle are the North Pole, Mecca, and the given locality. The two known sides of this triangle are the two arcs joining the North Pole to Mecca and to the locality in question, and the known angle is the difference in longitude between Mecca and the locality in question. However, before quoting Ibn Muʾādh's explanation of how to solve these two cases, it is necessary to show his solution to two of the sixteen auxiliary problems that precede the solutions of different cases of triangles.

[Solving triangles]

Now we have mentioned what we think are sufficient preliminaries, so let us turn to the different types of problems. The first of these is the right triangle. So let the triangle be *ABG*, with angle *A* right, and angle *B*, or *G*, known. It makes no difference which, so let it be *B*. And let side *BG*, which subtends the right [angle], be known. Then I say that the whole triangle—its sides and its angles—is known. And from what we premised: the ratio of the Sine of side *BG* to the Sine of angle *A* (which is 60) is as the ratio of the Sine of side *AG*, which is unknown, to the Sine of angle *B*. Hence the Sine of *AG* is known by multiplying the Sine of the known side, *BG*, by the Sine of angle *B* and dividing the result by sixty. The result is the Sine of side *AG*, and its arcSine is the side *AG*, if the angle *B* is acute. But if it is obtuse the resulting arc of the Sine is the large one [and] the resulting arc will be the side *AG*. And this problem is the first, resulting from the relation of only four quantities.

And, if you want to know the side *AB* [in the above right-triangle], this problem we make the ninth.[27] And from what we premised the product of the Sine of *BG* and the

[27]This problem is more complicated because the angle opposite it is not known, so the Sine Law is not directly applicable, as it was in the case of side *AG*.

Cosine of *AB*, the unknown, and the result multiplied by the Cosine of angle *B*, is equal to the product of the Sine of *BA*, the unknown, times the Cosine of *BG* and the result multiplied by sixty [the Sine of angle *A*].[28] And because of that, the ratio of the Sine of *BA* by[29] the Cosine of *BA* is as the ratio of the product of the Sine of *BG* by the Cosine of angle *B* to the product of the Cosine of *BG* by sixty. And this ratio is known, I mean the ratio between the Sine of *BA* to its Cosine.[30] Since these four Sines are known, we deduce *AB* as we premised. And it is necessary that you know in advance if *AB* is less than ninety, since it and its complement together will then be ninety. And then we derive it by using what we proved about the derivation of the two unknown arcs of known ratio,[31] their sum being known.

But if it is greater than ninety, then its supplement [that is to say, its difference with 180°] is what is added to ninety [to obtain the arc *AB*]. And so we deduce it by the calculation of two unknown arcs of known ratio, the excess of one of them over the other being known.[32]

And as for knowing if the side *AB* is greater or less than a quadrant then we consider the side *BG* and angle *B*. And if each of them is less than ninety then we make the calculation according to that. But if they differ, I mean if the side *BG* is greater than ninety and angle *B* is less than ninety, or angle *B* is greater than right and the side *BG* is less than a quadrant, then *AB* will be greater than a quadrant and we make the calculation according to that. And thus the two unknown sides, i.e., side *BA* and side *AG* are calculated from knowing side *BG*. And the easier way to calculate side *AB* is that we first calculate the side *AG* and then we calculate the side *AB* from knowledge of the sides *AG* and *BG*, and it is exactly problem four.

We now pass from samples of the sixteen problems to the application of the techniques shown in these problems to the solution of spherical triangles in which three of the six possible things are known.

And from these sixteen problems are deduced the unknowns of all types of triangles from their knowns. Then, after the demonstration of right triangles we take other kinds of triangles. The first of these is that we make triangle *ABG* and we make *AB* known and *AG* known and angle *BAG* known (but not right), and we want to know all the other sides and angles of the triangle [Fig. V-1-9].

We take the shorter of the two known sides, say, *AB*, and from *B* we produce an arc perpendicular to *AG*, the perpendicular *BD*. We now have triangle *ABD* with its angle *D* right and angle *A* known and side *AB* known. And this is the first of the six cases from which we deduced the sixteen problems. We will first derive the side *BD*.

[28]This result, which can be expressed as $\text{Sin } a \text{ Cos } g \text{ Cos } B = \text{Sin } g \text{ Cos } a \text{ Sin } A$, follows from the earlier result $\text{Sin } a \text{ Cos } B = \text{Sin } g \text{ Cos } b$. For then $\text{Sin } a \text{ Cos } g \text{ Cos } B = \text{Sin } g \text{ Cos } b \text{ Cos } g$. But another consequence of the transversal law is that $\text{Sin } A \text{ Cos } a = \text{Cos } b \text{ Cos } g$, so the right side becomes Sin *g* Cos a Sin *A*, as desired. See section V-4 for proofs of these results by Jābir ibn Aflaḥ, and see [Van Brummelen, 2009, chapter 4] for more details.

[29]The text uses *min* where one would have expected *ilā*.

[30]Ibn Muʿādh includes a table of tangents, in steps of one degree, from 1° to 89° in his book. (For the last degree, he also tabulates it for 15, 30, 45, and 59 minutes!) Again, showing his mathematical orientation, he refers to it as a "Table Composed of the Division of the Sine by the Cosine," rather than by the traditional phrase "the reversed shadow," which originated in the theory of sundials.

[31]Literally, "two arcs unknown as to quantity (*al-qawsain al- majhūlatain al-kammiyyat*)." And he means that the ratio of their Sines is known—not the ratio of the arcs themselves.

[32]The two arcs in question are *AB* and *AB* – 90°. Their difference is 90°, a known angle, and the ratio of their Sines, Sin(*AB*)/Sin(*AB* – 90) (=Sin(*AB*)/Cos(*AB*)), is also known.

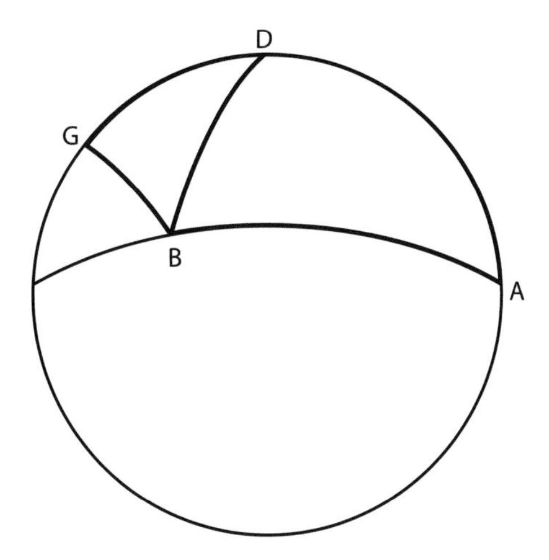

Fig. V-1-9.

To do this we will derive the side *AD*, which is the ninth problem. And so, if *AD* is smaller than what we postulated *AG* to be, then the perpendicular [BD] falls between *A* and *G*. And, if it is larger, then the perpendicular falls on arc *AG* beyond *G*. And we proved them together, wherever they fall.

The triangle *BDG* is right at *D*. And *GD* will be known. For *AD* is known and *AG* was known, and so it remains that *GD* is known. And *DB* is known, and they [the two sides just mentioned] contain the right angle. So all sides and angles of triangle *BDG* are known. And so, *BG* is known, as one of the sides of the original triangle. And if angle *AGB* is wanted, then it will be derived from the right angle triangle *BDG*. And hence angle *ABG* will be known [as the sum of angles *ABD* and *DBG*]. And if you want to know angle *ABG* from the first triangle you derive in the one right angle triangle [ABD] angle *ABD*, and, from the other right angle triangle, *BDG*, [you derive] angle *GBD*. And when you will know these two angles you will know angle *ABG*. And this is all we want to say about the triangle in which an angle and the two sides containing it are known.

The fourth case is that of a triangle in which all sides are known. We shall not give the proof, but pass on to case 5, in which Ibn Muʿādh uses what is known as the polar triangle of a given spherical triangle to find the sides when all three angles are known. (The polar triangle of a given triangle *ABG* is a triangle *DEZ* whose vertices are the poles of the three sides of the given triangle. We have put in bold type the three steps by which Ibn Muʿādh constructs the polar triangle of the given triangle, *ABG*.) A key fact in the proof, although Ibn Muʿādh does not state it, is that if *XY* and *YZ* are intersecting arcs of distinct great circles on a sphere, and *P*, *Q* are their respective poles, then the great circle arc *PQ* is a measure of the angle *XYZ* subtending it. Within Ibn Muʿādh's demonstration are the facts needed to show that the polar triangle of the polar triangle of *ABG* is triangle *ABG* itself.[33]

[33]In modern terms, a spherical triangle and its polar triangle are dual figures. However, there is no evidence that Ibn Muʿādh was interested in a theoretical study of the relation between a spherical triangle and its polar, to which he attaches no special name.

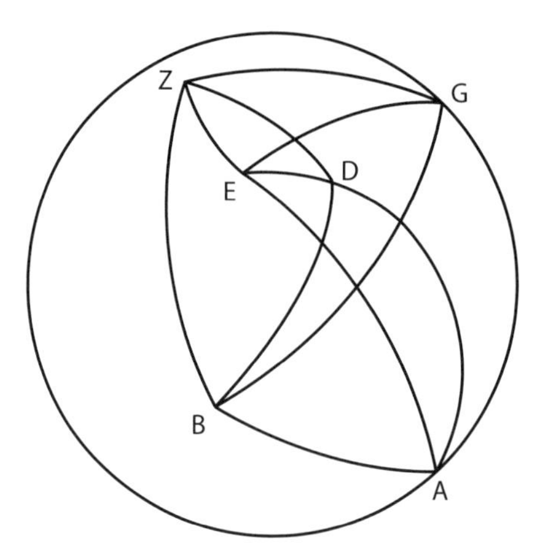

Fig. V-1-10.

The fifth case:

Let there be a triangle *ABG* in which all angles are known and we want to know the sides. This is a difficult, obscure case to deal with. Its solution would be easy with care and reflection, if studied closely. . . .

If you want to know the side *AB*, for example, then at point *A* on arc *AB* erect a right angle in the direction of *G*, say, angle *BAD* [which will then be] known. *AD* falls either outside or inside the triangle. And if it [the angle at *A*] is acute, it [*AD*] falls outside and so let it be obtuse, as in the first figure [Fig. V-1-10]. We make arc *AD* ninety, so that point ***D* is the pole of arc *AB***. Also, we join *D* with *B* so that angle *DBA* is right[34] as well.

And if angle *GBA* is acute, as in the first figure, then arc *BG* is inside triangle *ADB* as in the first figure. And so the first figure is the case when one of the two angles *A* and *B* of the original triangle is obtuse and the second is acute. And the second figure [Fig. V-1-11] is for the case when angles *A* and *B* are both acute. And we make the third figure [Fig. V-1-12] for the case when both are obtuse.

In all cases angle *GAD* will be acute. And we increase it so that it will be perpendicular to arc *GA*. And it is angle *GAE*, and so it will be equal to angle *BAD* [which was already shown to be right]. And we make arc *AE* a quadrant so that point ***E* will be the pole of arc *AG***. And we join *D* with *B* so the angle *DAB* is right. And the angle *BAG* is assumed known and angle *BAD* is right, so the remaining angle *DAG* is known and it is acute. And angle *EAG* is right, and so it remains that angle *DAE* is always a known acute angle. And it, when angle *BAG* is acute, is equal to it. And when angle *BAG* is obtuse, angle *DAE* is joined to it. I mean that if it were joined to it so that the sum of them, joined, would be two right angles. Arc *DE* is the magnitude of angle *DAE*, since each of the two arcs, *AE* and *AD* is ninety, and so arc *DE* is known. Also, if we make at point *B* what we made at point *A*, i.e., we complete the known acute angle *GBD* so that it makes a right angle with arc *GB*, namely, *GBZ*, then it is equal to angle *ABD*. And we make arc *BZ* a quadrant of a

[34]This is because *D* is a pole of arc *BA*.

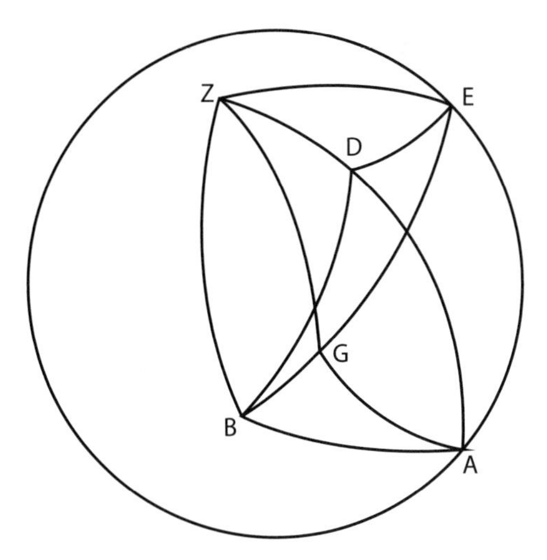

Fig. V-1-11.

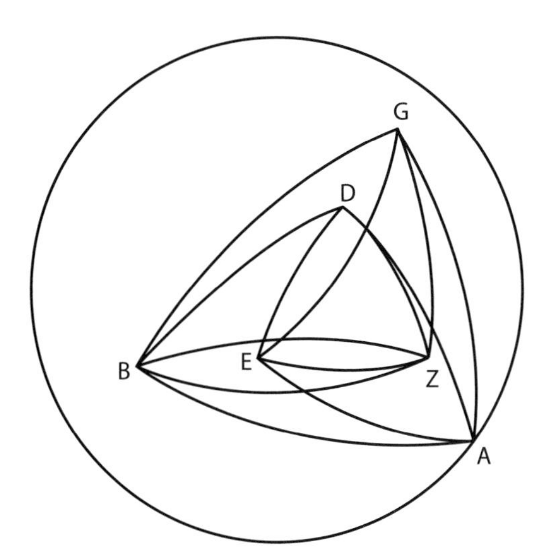

Fig. V-1-12.

circle so that the point ***Z* is the pole of arc *GB***. And we join *D* with *Z* so that the arc *DZ* is the measure of the known acute angle *DBZ*, according to what we mentioned earlier about the arc *DE* [which was shown to be the measure of angle *DAE*]. And so arc *DZ* is known.

And we join *G* with *E* and with *Z*, and because *E* is the pole of arc *AG*, *EG* is a quadrant. And because *Z* is the pole of arc *BG*, arc *GZ* is a quadrant as well.

We join *Z* with *E* and because angle *EGA* is right [since *E* is a pole of arc *AG*] and angle *BGA* is known, angle *EGB* is known. And angle *ZGB* is right [since *Z* is a pole of arc *GB*] and so angle *EGZ* is known. And it is the magnitude of arc *EZ*. And so the three sides of triangle *DEZ* are known.

And we showed how to derive an angle from the fourth type [where all sides are known]. And so angle *ZDE* is known. And angle *ZDB* is right. So angle *BDE* is known. And angle *ADE* is right and so angle *ADB* is known, and it is the magnitude of side *AB*, since point *D* is the pole of arc *AB*. And these are the three figures. And the explanation [given here] applies generally to all of them [the figures].[35]

Ibn Muʿādh concludes this case with a summary of the argument.

2. ABŪ ʿABD ALLAH MUḤAMMAD IBN MUʿĀDH AL-JAYYĀNĪ, *ON TWILIGHT AND THE RISING OF CLOUDS*

Ibn Muʿādh's authorship of *On Twilight* has only recently come to light, in great part no doubt because the work was known only in its Latin form (translated by Gerard of Cremona in the twelfth century), under which guise it was consistently misattributed to Ibn al-Haytham (965–1041) from the late Middle Ages on.[36] Only with A. I. Sabra's recognition of a fourteenth-century Hebrew version of the treatise, translated by Samuel ben Yehuda and explicitly ascribed to Ibn Muʿādh, has this error been definitively rectified.

Of Ibn Muʿādh's two basic aims in *On Twilight*, the first is to establish precisely why the sky lights up before sunrise and remains lit well after sunset. He starts by noting that dawn and dusk are essentially alike insofar as they share the same primary cause (i.e., the sun). Having thus isolated the primary or immediate cause, Ibn Muʿādh then broaches the subject of secondary or mediate cause by describing the formation of night in terms of the conical shadow cast in the night sky by the earth as it blocks the sun. Everything outside this nocturnal cone is, of course, continuously open to solar illumination, yet despite such constant illumination, the heavens appear dark throughout the night. The reason, Ibn Muʿādh concludes, is that, except for where it concentrates to form stars and planets, the ether pervading the celestial sphere is too rare to absorb light passing through it, and the same holds for the pure air surrounding the terrestrial globe.

We are therefore bound to concur with Ibn Muʿādh that the secondary or mediate cause of twilight is something denser and more receptive to illumination than pure air or ether. Perhaps, he opines, we can look to the earth itself for this denser medium. Why not suppose, then, that what we see at twilight is merely solar illumination of the eastern or western fringes of the horizon? Because, asserts Ibn Muʿādh, if this were the case, the time interval between first light at dawn and sunrise would be far shorter than it actually is. Therefore, the earth itself cannot be the sought-after medium. But if it is neither earth, nor air, nor ether, then we are left with only one option: the requisite agent must be the moist vapors that rise continuously into the upper atmosphere from the earth's surface. Lying well above that surface, such vapors must be illuminated by the sun's rays to give us the first intimations of morning light long before the sun itself peeps over the horizon.

With the immediate and mediate causes of twilight pinpointed, Ibn Muʿādh has achieved the first goal of this treatise. His second goal is to determine just how high the vapors he has adduced reach into the atmosphere. Four parameters are needed: the circumference of the earth (24,000 miles), the ratio of the solar to the terrestrial radius (5.5:1), the mean radius of the solar orbit in terms of earth radii (1110), and the depression of the sun below the horizon

[35] Of course, once side *AB* is known, one can use the law of sines to determine the other two sides.

[36] The introduction to this selection is adapted from [Smith, 1992].

at the first instant of twilight (19°). Along with these parameters, Ibn Muʿādh supplies three preliminary theorems that he deems useful as background to the determination itself. In the first he demonstrates that, if an illuminating sphere (e.g., the sun) is the same size as the sphere it illuminates, it will illuminate precisely half. In the second he shows that, if the illuminating sphere is larger than the illuminated one (as is the case with the sun and the earth), it will illuminate more than half. And, finally, he undertakes to prove that, when any two spheres are connected by tangents, the two face-to-face segments delineated by those tangents are completely open to one another, while the remaining two are completely occluded from one another.

Ibn Muʿādh is now prepared to calculate the maximum altitude that the vapors can attain. The first step in the process is to compute how great an arc on the earth's surface is illuminated by the sun on the basis of two of the previously given parameters: the relative size of the sun and earth in terms of terrestrial radii (5.5:1) and the mean distance between the sun and earth, also in terms of terrestrial radii (1110). Ibn Muʿādh calculates the resulting arc to be 180°27′52″ The second step is to define the highest point to which the rising vapors can reach. As Ibn Muʿādh shows geometrically, this point lies at the intersection of the plane of the horizon and the leading edge of the nocturnal cone when the sun is exactly 19° below the horizon. The third step, therefore, consists of determining exactly how high this point lies above the earth's surface. To this end, Ibn Muʿādh offers the following construction.

Let *EFN* in the accompanying figure (Fig. V-2-1) represent a cross-section of the earth through the viewer's zenith, *H* being the center and *E* being the given viewpoint upon its surface. Line *KE*, which is tangent to the earth at point *E*, thus represents the horizon-line or line of sight from *E*. Let line *TL* represent the leading edge of the nocturnal cone, so that it is tangent to the earth at *L* and to the sun in the direction of *T* (the sun therefore rises clockwise in the plane of the diagram). Thus *Q*, where horizon-line *KE* and leading edge *TL* of the nocturnal cone intersect, represents the highest point to which the vapors that are illuminated by the sun can ascend. Let line *IN* represent the trailing edge of that cone, so that it is tangent to the earth at *N* and to the sun in the direction of *I*. Thus, both *TL* and *IN* will eventually meet beyond *L* and *N*. Let line *GFH* connect the centers of sun and earth. Let line *BHD* pass through *H* parallel to horizon-line *KE*, and assume that both lines effectively coincide at such a great distance as that between sun and earth. Finally, connect lines *LH*, *QZH*, and *EH*.

The problem, then, is to calculate the length of *QZ* in terms of terrestrial radius *ZH*, and the result depends upon our knowing the following: (1) angles *EHB*, *QEH*, and *QLH* are right angles; (2) angle LQH = angle EQH, so that angle QHE = angle QHL; (3) $LH = ZH = EH$; (4) arc $LFN = 180°27'52''$, so that its half (arc LF) $= 90°13'56''$; (5) angle $BHF = 19°$, so that angle $LHB = 71°13'56''$; (6) angle $LHE = 18°46'4''$, so that its half (angle QHE) $= 9°23'2''$, whose complement (angle HQE) $= 80°36'58''$; (7) *ZH*, which represents the earth's radius, is just over 3818 miles. Given these data, it is an easy matter to calculate *QZ*. By Ibn Muʿādh's reckoning, it turns out to be just under 52 miles.[37]

[37] [Smith, 1992, p. 89, n. 16] points out in a footnote, "There is no evidence that anyone before Ibn Muʿādh attempted to derive the height of the atmosphere on the basis of a rigorous geometrical method, and we know that his determination of 52 miles, or thereabouts, remained standard until the 17th century."

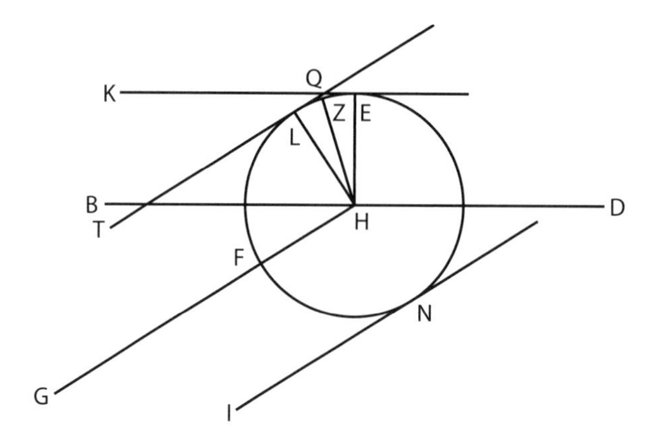

Fig. V-2-1.

In this treatise I wish to show what twilight is and what necessary cause governs its appearance, and from there I will proceed to determine the highest distance above the earth's surface that the subtle vapors arising from it can reach. I therefore maintain that morning and evening twilight are of a similar form, for indeed the one depends on the waxing of the sun's light while the other depends upon its waning. Yet the colors of both [kinds of twilight] are distinguished by the different horizons at which the sun appears, because whenever the sun lies not very far above the eastern horizon, its color there is different from the color it presents to our eyes when it lies the same distance above the western horizon. And likewise, even when the sun itself is hidden [below the horizon], its radiation, as well as whatever in the ether appears [to take on] its illumination, differs according to whether the sun lies in either [an eastward or westward] direction. For in the east the color of its illumination is white and clear, whereas when it lies to the west its color inclines somewhat toward red. But what sort of thing might this illumination be, how might it appear where it does, and what cause must be responsible for it? In response to this question we will set forth some introductory points in order to provide an explanatory basis for what we wish to demonstrate later on.

To start with, the entire heavenly sphere is brightened and illuminated by the major [celestial] light source (i.e., the sun), except for the shadow cast by the earth in the shape of a cone, a shadow which forms night. . . . Now [authorities] assert that it is neither the earth itself nor the outermost regions of it distinguishable by us at this time [of first light) which absorb the sun's radiation, for when a viewer stands on a level plane, his vision extends only about three miles in any direction. Furthermore, if the viewer happens to stand on the highest possible mountain—and that does not even reach 8 miles according to what authorities concerned with such matters claim—then his vision will extend no more than about 250 miles. And this point is surely obvious, given what necessarily follows from the earth's shape and given the height of the viewer's standpoint above its surface.[38]

[38]Ibn Mu'ādh's aim in this paragraph is to establish that it is not the horizon itself that we see illuminated at the moment of dawn—in short, that the necessary intermediate cause is not the earth or its outer fringes. [Goldstein, 1977, pp. 111–112] explains how Ibn Mu'ādh would have arrived at the stated parameters of 3 and 250 miles, assuming that an average man is about 6 feet tall. According to Goldstein, Ibn Mu'ādh's claim that 8 miles is the limit for the tallest possible mountain cannot be traced to any known source.

But the [sun in its daily] orbit passes over the aforementioned space [of 250 miles] in a quarter of an hour. It therefore follows that the sun should rise soon after dawn, by a quarter of an hour or less. In fact, the time between the appearance of dawn and the appearance of the sun is a good deal more than that.[39] What we have just said, however, is only a rough approximation for the person who has not mastered geometry. Actually, the line of sight does not reach any point on the earth already illuminated by the sun without also reaching and perceiving the edge of the sun itself, for two lines touching any given point on a circle from opposite directions join together to form a single straight line.

Hence, when [dawn first appears] to us, the source of illumination is not the earth itself, according to what we have said, nor is it the air that fills the entire [enveloping] sphere, according to the previously established point that the sun's radiation falls continuously, day and night, on all, or on the greater part of the air without any of it showing on account of the air's rarity. But the only object above the earth that is denser than air consists of vapors rising from the earth, and there is never any lack of such vapors which might be illuminated by the sun. Then, when the cone of [nocturnal] shadow passes out of that part of the sphere of vapors enveloping the earth to which our vision reaches, the body of the sun is open to those vapors, its rays strike them, its radiation is absorbed by them, they shine forth that absorbed light to us, our vision reaches them, and the light in them appears to us, just as we see it appear in clouds from the coloring of rising vapors, and just as in dew colors appear in the shape of an arc of a circle, and in other ways.

Therefore, when we want to know the maximum elevation of those vapors above the earth's surface, four things, no fewer and no more, are necessary and sufficient for determining that amount. And this, surely, is how need is [properly fulfilled]—according as there is no making do with less than, nor requirement for more than [accomplishes a given end].[40] And [in our case] the four things [we need to know] are: the size of the earth, the size of the sun, the distance of the sun's center from the center of the earth everywhere [along its annual orbit], and the amount below the horizon the sun lies when dawn appears. But the earth's size provides a standard for the rest of these measurements, and, according to what authorities have said and have shown by indisputable propositions, the size of any great circle encompassing it is 24,000 miles. Those authorities have also claimed that, taking the radius of the earth as one unit, the sun's radius is 5.5 units, while

[39]The argument to this point can be summarized as follows: (1) the sun orbits the earth in 24 hours; (2) the earth's circumference is 24,000 miles, which constitutes the projection of the sun's daily orbit upon the earth's surface at the equator (Ibn Muʿādh in fact cites this figure later on in the treatise; see footnote 48 below); (3) thus, since the sun's motion projects over 1000 miles of surface in an hour at the equator, it must project over only 1/4 of that (i.e., 250 miles) in 15 minutes (these distances will of course decrease with an increase in latitude either north or south). The guiding assumption for the rest of the argument seems to be that, by the time the sun's light has pervaded the entire 250 miles of space between horizon and elevated viewer, the sun itself will begin to peep over the horizon. Thus, from first light at the horizon to the total pervasion of the intervening space (and, thus, the first appearance of the sun over the horizon) fifteen minutes will elapse. By the same reasoning, presumably, the lower the viewer's elevation, the shorter will be the interval between first light, total pervasion of the intervening space, and first appearance of the sun—hence the author's qualification of "a quarter of an hour or less." Here we have a clear indication that Ibn Muʿādh ignores, or at least discounts, the effect of atmospheric refraction on horizon phenomena, since he assumes that, at the very instant the sun appears over the horizon, its cusp and the viewpoint will be connected by a straight line tangent to the horizon.

[40]This is a fairly clear articulation of the principle of logical economy commonly described as Ockham's Razor: *frustra fit per plura quod per pauciora fieri potest.*

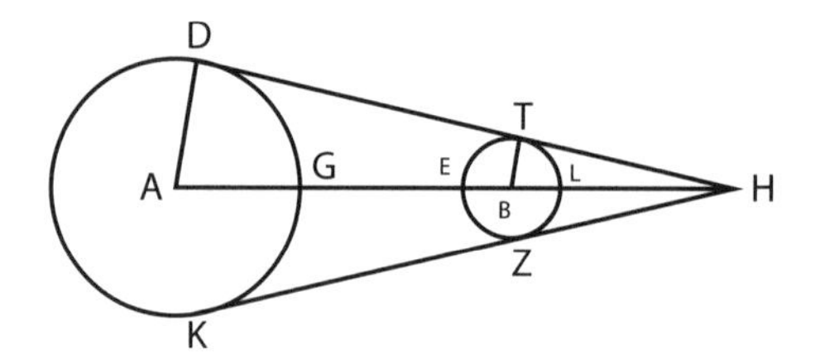

Fig. V-2-2.

the distance between the sun's center and the earth's center (although not at every point [in the sun's annual orbit]) is around 1110 units; and they have claimed that the sun's distance below the horizon when dawn first appears is 18 degrees. But now it is found to be 19 degrees, and I will carry out our calculation accordingly. In order, then, to fulfill what we plan [in this work], I will set forth some very helpful propositions which I have at hand. [I do so on the understanding] that he who presents a matter with elaboration is worthy of having his word accepted as long as no one else contradicts it, since he who presents the elaboration knows what no one else knows and achieves what no one else has achieved, and so he who presents something that he has grasped before anyone else has grasped it deserves to be followed when he is not judged to be [intellectually] suspect.

[Proposition 1] Accordingly, I maintain that, if any two spheres between which no other body lies to block one from the other are of equal size, then half of either one of them is face-to-face with half of the other (and by [saying that] one is face-to-face with the other), I mean that, if one of them is luminous and the other receives its light, then half of the one receiving that light is illuminated and shines.[41]

Ibn Muᶜādh's proof of this obvious proposition and the accompanying figure are omitted.

[Proposition 2] If one of the spheres is larger than the other, however, the portion of the smaller one that is face-to-face with the larger one comprises more than half, while the portion of the larger one that is face-to-face with the smaller one comprises less than half.

For example, let there be two spheres *A* and *B* [Fig. V-2-2], and let sphere *A* be the larger. Then I will draw a plane surface passing through the center of both. It will therefore cut each sphere in half along the two circles *AGD* and *BEZ*. Next, I will extend *AB* and draw it rectilinearly toward [and beyond] *B*, and I will assume that the ratio of the radius of circle *AG* to the radius of circle *BE* is as *AH* is to *BH* (an explanation of this is near to hand in the sixth and fifth books of Euclid's *Elements*).[42] Finally, from point *H* I will draw line *HTD* tangent to circle *AGD*. I say, then, that this same line is also tangent to circle *BEZ*.

Accordingly, I will connect *A* with *D* by line *AD*. Therefore, it is perpendicular to line *HD*. I will also draw line *BT* from point *B* orthogonal to line *HTD*, and because the two lines *BT* and *AD* are perpendicular to line *HD*; they are parallel. Furthermore, since line

[41] As [Goldstein, 1977, p. 113] points out, this theorem and the one following it are similar to the first two theorems of Aristarchus of Samos's *On the Sizes of the Sun and the Moon* [Heath, 1913, pp. 355–361]. This treatise would have been readily available to Ibn Muᶜādh through Qusṭā ibn Lūqā's tenth-century Arabic version.

[42] Particularly apposite in this case is *Elements* VI.12, where Euclid explains how to derive a fourth proportional to any three given straight lines.

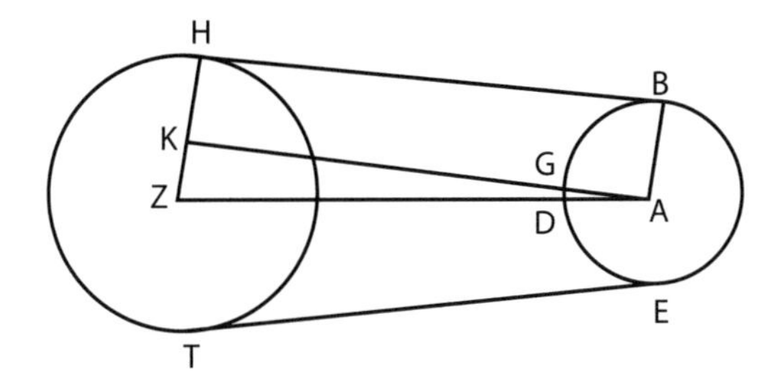

Fig. V-2-3.

BT is parallel to the base *AD* of triangle *HAD*, the ratio of *AD* to *BT* will be the same as that of *AH* to *HB*. But we have already established [by construction] that the ratio of *AH* to *HB* is as the ratio of the radius of circle *AD* to the radius of circle *BE*. Thus, line *BT* is a radius of circle *BE*. Thus, too, point *T* lies upon the circumference of [that] circle, and we have constructed the two angles [*DTB* and *HTB* at point] *T* as right angles. Line *HTD* is therefore tangent to the smaller circle. But we have already drawn it tangent to the large one; so the same line is tangent to both at the same time.

Now, as in the previous case, I will draw line *HZK* from point *H* tangent to both circles at the same time on side *Z*. Hence, section *DGK* of the larger circle *A* is face-to-face with the smaller circle *B*, and [that section] is smaller than a semicircle, because angle *DAH* is smaller than a right angle, since that angle lies in the same triangle *DAH* as right angle *ADH*. Consequently, section *DG* is smaller than a quarter of a circle, and, likewise, section *GK* is equal to it. Therefore, section *DGK* is smaller than a semicircle. In addition, because line *TB* is parallel to line *DA*, angle *TBH* will be equal to angle *DAH*. Therefore, section *TL* will be similar to section *DG*, and the whole section *TLZ* will be similar to section *DGK*. Hence, each of them is smaller than a semicircle; section *TEZ* therefore remains greater than a semicircle, and that is the section of the smaller circle which is face-to-face with the larger one. And that is what was claimed.

Therefore, *TEZ* and *DGK* are the two sections of the two circles that stand face-to-face (provided of course, I emphasize, that no part of one is hidden from the other circle), and section *TEZ* is larger than a semicircle, while section *DGZ* is smaller than a semicircle. And that is what we wanted [to prove].

[Proposition 3] I also maintain that, when there are two circles and two lines are drawn, each of them tangent to both circles at once, according to the previous model, then in neither of the two sections that are face-to-face with one another is there a spot [in] which anything is hidden from the other circle; nor in the two remaining sections of the two circles—the sections not face-to-face—is there any spot which is visible from the other circle; nor can what lies on the hidden part of one of the circles be directly linked to the other circle without being blocked by the part of its own circle that hides it.

Ibn Muʿādh's proof of this obvious proposition is omitted.

In order to follow the plan we have laid out for ourselves, we must now find out how large an arc on the earth is illuminated by the sun, an arc, we have already established, that comprehends more than half the earth's circumference. Accordingly, I will suppose that the two circles *ABGDE* and *ZHT* [Fig. V-2-3] represent [great] circles of the earth and sun cut out from each by some plane surface. Circle *A*, then, represents the earth, and circle *Z* the sun.

Next, I will draw the two lines *BH* and *ET* tangent to each of the circles in the manner discussed previously. Section *BGDE* of the earth is thus illuminated by the sun, as we have already shown, and it is larger than a semicircle. If we wanted to determine the size of this section, then we would connect *A* with *B*, as well as with *Z*, and we would connect *Z* with *H*. Thus, *BA* and *HZ* are parallel, because both of them are perpendicular to line *BH*, which is tangent to the two circles. Now I will cut from line *HZ* a segment *HK* that is equal to line *BA*, and I will connect *A* with *K*. Consequently, *AK* is perpendicular to *HZ*, for it is parallel to *BH*, since it connects the endpoints of two equal and parallel lines *BA* and *HK* [whose other endpoints are connected by *BH*]. Each angle [*AKH* and *AKZ*] at *K* is therefore a right angle. Moreover, because line *HZ* is 5.5 times as long as *AB*, line *KZ* remains 4.5 times as long as *AB*, while the mean distance *AZ* [between earth and sun] is found to be 1110 times as long as *AB*. Therefore, line *AZ* consists of 1110 units, of which line *KZ* contains 4.5. So also,[43] in the measure that line *AZ* subtended by right angle [*AKZ*] is 60;00 parts, line *KZ* is 00;14 + 3/5 parts. Thus, angle *KAZ* is 14 – 1/3 x 1/5 [i.e., 13 + 14/15] minutes, according to the 90 degrees in a right angle, and arc *GD* is the same size. But arc *BG* is 90 degrees, because angle *BAG* is right. Therefore, arc *BD* is 90 degrees and 13 + 14/15 minutes, and arc *DE* is equal to arc BD. As a result, the entire illuminated arc BGDE is around 180 degrees and 27 + 4/5 + 1/3 x 1/5 [i.e., 27 + 13/15] minutes, and that is what we wanted to determine.

Let us now broach the question we raised about the cause of the dawn's appearance, the form of its appearance to us, and its configuration in the eastern horizon. Accordingly, I will suppose that circle *AB* [Fig. V-2-4], with *A* as the viewpoint, is marked out on the earth's sphere [having been] cut from the earth by a plane surface passing through the zenith and through the two centers of the earth and sun.

Then I will have a line pass through point *A* tangent to the circle and will extend it in both directions toward endpoints *G* and *D*. It is therefore obvious that the [line of] sight does not fall upon anything whatever that lies below line *GAD* on the side of *B*, for the earth blocks it from us, since the line of sight follows only a straight line (and Euclid has indeed demonstrated that from a point of tangency no line can be drawn between a circle and the line tangent to it).[44] Hence, the line of sight does not fall upon anything that lies below line GAD but does fall upon whatever lies above it.

Next, I will assume that *L*, *E*, *U*, and *Z* define the cone of [nocturnal] shadow cast by the earth shortly before dawn, when the sun lies more than 19 degrees below the horizon, by a minute or so, for instance. Thus, everything within the designated cone, whose vertex is *L* and whose base is [arc *UZ* on] the earth, is hidden from the sun, being neither open

[43]In this, as in subsequent calculations, Ibn Muʿādh is working within the sexagesimal (base-60) system on the basis, essentially, of Ptolemy's analysis of chords in *Almagest* I, 11. Accordingly, he is operating under the assumption that the diameter of a circle is subdivided into 120 parts of which the radius thus contains 60. Given these assumptions, the calculations in this passage are as follows: (1) *AZ* ~1110 units, and *KZ* ~4.5. (2) Sine *KAZ* ~*ZK*/*AZ* ~4.5/1110 ~.004054 ~00;14 + 3/5. Hence, angle *KAZ* ~13′56" ~13 +14/15′. (4) Therefore, 2× angle *KAZ* ~27 + 13/15 minutes ~27′52". (5) Therefore, arc *BGDE* ~180°27′52". According to [Doncel, 1982], Ibn Muʿādh derived his values for trigonometric functions from al-Khwārizmī-Maslama's tables. See also [Villuendas, 1979].

[44]*Elements* III.16.

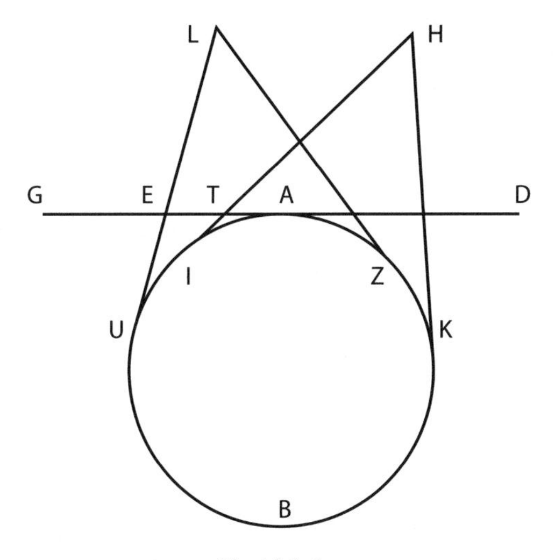

Fig. V-2-4.

to it nor illuminated by it, so that [everything within that cone] is truly dark. Meanwhile, anything that lies outside that cone is open to the sun, and so its rays and its light fall upon that thing. Still, any such body, if it is quite tenuous, does not hold forth to our sight what it takes on of that radiation. For that reason, then, the subtle air within the cone [of nocturnal shadow] is the same to our eyes as that outside the cone, and the entire [sphere of] ether appears uniform in regard to light and darkness. Of course we know that, not being open to the sun, the air that surrounds us, as well as whatever [within it] is near us, is dark to us. Moreover, we know that whatever enters [the lighted portion of ether] either above, to the right or left, or behind or in front [of the cone of shadow] is illuminated, being open to the sun; and yet what we see in both cases is the same to our visual comprehension.

It is therefore evident that nothing appears to our eyes in the state of shining or illumination except through the convergence of three conditions, one of which is that the object not lie below line *GAD*, for if it does, then the earth's sphere intervenes between it and the eye so that the sight does not grasp it whether it is luminous or dark. The second condition is that the object not lie within the cone of [nocturnal] shadow, for if it does, then it is dark, since it is blocked from the face of the sun and, accordingly, from its illumination. The third condition is that the object be denser than the subtle air filling the [surrounding] sphere, for we have already seen that the upper air outside the cone lies above line *GAD* and that, notwithstanding this fact, none of the light within it appears to us on account of its tenuousness and subtlety. Furthermore, because we are looking just before dawn in an area of the entire [heavenly] sphere that is not illuminated, and because no part of it differs from any other part, we know that in that [area] there is neither a point nor a single place in which those three conditions converge. Now at point *E*, where the boundary of the cone [of nocturnal shadow] intersects line *GAD*, we already find two conditions met, since that point lies neither below line *GAD* nor within the cone. Therefore, the sun's radiation falls upon it, and for that reason it must appear in darkness to our eyes only because of

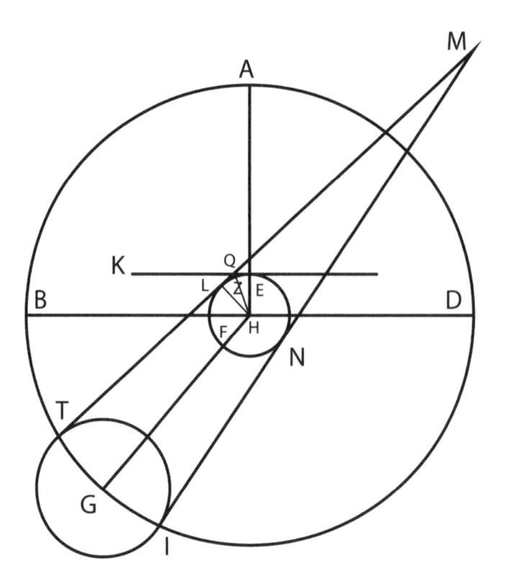

Fig. V-2-5.

its failure to meet the third condition, which is density. So it is now established that the air at point *E* is rare [and that] the dense vapors rising from the earth, which are denser than the air, do not reach it.

Then, very soon afterwards, the sun rises, and its depression below the horizon becomes precisely 19 degrees. Let the cone [of nocturnal shadow at that time] be represented by the figure described by *H*, *T*, *I*, and *K*. Then, where nothing luminous existed before, something luminous appears on the horizon. We know that this is the first instance and the first place in which the three aforementioned conditions converge, for just before, by an immeasurably tiny amount of time, nothing was illuminated in that place. Moreover, point *T* is the first place in which it happens simultaneously that [the object] does not lie below line *GAD* and does not fall within the cone of [nocturnal] shadow. Thus, point *T* is the first place in which the third condition—the one involving density—is met. Point *T* is thus the last and highest place to which the vapors rise; they do not fall short of it, nor do they surpass it. If, however, they were to fall short of it, then point *T* would lie in the subtle air, and nothing of the light in it would appear to us, just as light does not appear in that [part of the air] that lies beyond it toward *E*. Yet if those vapors were to surpass point *T*, then [the region toward] point *E* would have been illuminated for us before [the region around point *T*], since we have assumed that no sensible object [i.e., other than the vapor] lies anywhere between the two positions at *T* and *E*. Therefore, point *T* is the ultimate place to which the vapors can attain as they ascend, and it is the intersection of line *GAD*, tangent to the earth's sphere, with line *HTI*, which forms the edge of the cone [of nocturnal shadow] cast [by the earth] when the sun lies 19 degrees below the horizon.

Since, then, we want to determine that point's distance above the earth's surface, we shall mark off circle *ABGD* [Fig. V-2-5], which passes through the center of the sun when its distance below the horizon is 19 degrees and dawn is just beginning to appear.

That circle will therefore cut the earth's sphere along [great] circle *EZN*. Then, let line *AEH* pass through the zenith [at *A*], as well as through the earth's center [*H*] and through line *BHD*. As a result, line *BHD* cuts the earth in two equal portions, [one defining] the area of visibility and [the other defining] the area of invisibility.[45] The area of visibility is therefore that area which lies on the side of *A* above [line *BHD* cutting] the earth, whereas the area of invisibility is that area which lies [below line *BHD*] on the side of *G*. But what we say here is not quite accurate; actually, the area of visibility only lies above the extension of line *EQK* tangent to the [terrestrial] sphere at the viewpoint [*E*]. However, the size of the earth is nugatory compared to the size of this sphere [*ABGD*, so *EQK* and *BHD* can be treated as if they coincided].[46]

Now I will suppose arc *BG* to be 19 degrees, which was the distance of the sun below the horizon at the very inception of dawn. The center of the sun thus lies on point *G*. So I will draw the sun's circle *TI* at that point, with its radius 5.5 times the length of line *EH*, so that circle *ABGD* cuts the sun through *G*. Next, I will extend line *HG*. Then I will draw the two lines *TLM* and *INM* tangent to the two circles of the earth and sun and encompassing the portion of the earth illuminated by the sun, which lines touch the earth at the two points *L* and *N* and form the boundary of the cone of [nocturnal] shadow. Line *TLM* thus intersects line *EK* at point *Q*. Hence, according to what we showed in the preceding figure, point *Q* is the place of [first] illumination at the inception of dawn, and it is also the highest point to which vapors can rise.

If, then, we wanted to determine the distance of that point from the earth's surface, we would connect *H* with *Q* by line *HZQ* and would also connect *H* with *L*. Consequently, section *LFN* is illuminated because it is face-to-face with the sun. But we have already shown that this section is 180 degrees, 27 minutes, 52 seconds, while arc *FL* is half that: i.e., 90 degrees, 13 minutes, 56 seconds, and that is the size of angle *LHF*. Angle *BHF* was already [assumed to be] 19 degrees, since that is the distance of the sun below the horizon. Therefore, angle *LHB* remains 71 degrees, 13 [minutes], 56 [seconds]. But angle *EHB* is 90 [degrees], since it is a right angle. Thus angle *EHL* remains 18 [degrees], 46 [minutes], 4 [seconds], and line *QH* bisects it, which is obvious.[47] Therefore, angle *QHE* is 9 [degrees], 23 [minutes], 2 [seconds]. Angle *HQE* is thus the complement of a right angle, which makes it 80 [degrees], 36 [minutes], 58 [seconds]. Hence, its Sine, which is line *HE*, is 59;11,48 parts according as line *QH* comprises 60;00 parts. However, according as line *HE* comprises 60;00 parts, *QZH* comprises 60;48 + 5/6 parts, but on that basis line *HZ* comprises 60;00 parts. Thus, *ZQ* remains 00;48,50 parts. In miles, of which the earth's circumference contains 24,000, that will amount to 51;47,34 + 6/11.

[45] Although Ibn Muᶜādh fails to stipulate it explicitly, line *BHD* must be constructed perpendicular to line *AEH* for the specified conditions to hold.

[46] To treat *EQK* and *BHD* as if they actually coincided is legitimate within the framework of the celestial sphere to which, according to Ptolemy. *Almagest* I,6, the earth bears the ratio of a point. Within the framework of the solar orbit (represented by *ABGD*), however, the ratio is 1:1110 according to the parameters offered by Ibn Muᶜādh. Although this ratio is not large, it is nonetheless significant enough to force him to admit that "what we say here is not quite accurate."

[47] That line *QH* bisects angle *EHL* depends upon proving the equality of sections *LQ* and *EQ* of intersecting tangents *TLM* and *EQK*. The basis for such a proof is *Elements* III.17, whence it is an easy matter to demonstrate that any two tangents dropped from a single point (e.g., *LQ* and *EQ* dropped from point *Q*) are equal.

And that represents the highest distance that the vapors rising from the earth can reach. And that is what we wanted [to determine].[48]

Here is the end of what the author meant to accomplish in this letter. Nevertheless, in the Arabic version there follow certain things which I have omitted, since there is nothing worthwhile in them. For nothing is contained in them beyond some lines in which the author praises God in the Saracen fashion and rebukes those who have questioned what use there might be in what he has said in this letter. And he has claimed that those who do not understand insensible things along with sensible things must be confuted. But since there is nothing worthwhile in what he says there, I have omitted it.[49]

3. ABŪ ʿABD ALLAH MUḤAMMAD IBN MUʿĀDH AL-JAYYĀNĪ, ON THE *QIBLA*

In the brief account of the career of Ibn Muʿādh, which introduced the excerpt from his defense of Euclid's definition of proportion (section IV-3), the *Jaen Tables* are mentioned. In chapter 18 of this work appears the first exposition in al-Andalus and the Maghrib of a scientifically correct method for finding the direction of prayer in Islam (i.e., the direction of Mecca relative to the local horizon), which is known in Arabic as the *qibla*. Although the Arabic original of the Ibn Muʿadh's *Jaen Tables* has not yet been discovered, the Arabic text of the chapter dealing with the local direction of Mecca has been found and was very kindly translated by Prof. Julio Samsó.

Mathematically, one way of looking at the problem of determining the *qibla* from a given locality is shown in Fig. V-3-1, in which one is looking at the earth from somewhere out in space where one can see one's own city (Z) as well as Mecca (M). Point P is the celestial north pole, the great circle $WKFN$ is the equator and arcs PZK and PMF are meridians of one's locality and of Mecca, respectively. If we think of MZ as the great circle arc joining our locality and Mecca, then the angle PZM is the angle we must turn from true north to be facing along the great circle route to Mecca. Since one's latitude is measured by the arc KZ, and that of Mecca is measured by arc FM, one can calculate arcs ZP and MP as complements of the latitudes, so those two sides of the spherical triangle PZM are known. And the angle ZPM is just the difference of longitudes of one's locality and that of Mecca. Since two sides and the included angle of the triangle are known, one can "solve" the triangle, and, in particular, find angle PZM, which solves the problem.

[48]The basic train of reasoning that Ibn Muʿādh follows in this passage can be summarized thus: (1) The illuminated portion of the earth, represented by arc LFN is, as previously determined, $180°27'52''$. (2) Angle LHF = 1/2 angle $LFN = 90°13'56''$, and angle $BHF = 19°$ by construction. (3) Therefore, angle LHB = angle LHF – angle $BHF = 90°13'56'' - 19° = 71°13'56''$. (4) Therefore, angle EHL = angle EHB – angle $LHB = 90° - 71°13'56'' = 18°46'4''$. (5) Angle $QHE = 1/2$ angle $EHL = 9°23'2''$. (6) Therefore, angle HQE = angle KEH – angle $QHE = 90° - 9°23'2'' = 80°36'58''$. (7) Sine of angle $HQE = EH/QH = .986618 = 59;11,49$ (text has 59;11,48). (8) Since $ZH = EH$ (both being radii of circle EFN), then $ZH/QH = .986618 = 59;11,49$. (9) Therefore, $QH = ZH/.986618 = ZH/59;11,49$. (10) Normalizing ZH to 1 (decimal) or 60 (sexagesimal), we get $QH = 1/.986618$ or 60/59;11,49 = 1.013564.... (11) Therefore, $QZ = QH—ZH = 00;48,50$ or .813889 parts out of 60. (12) Given that the earth has a circumference of 24,000 miles and assuming a value of 22/7 for π, then we can calculate ZH, the earth's radius r, according to the formulation $r = 24{,}000/2\pi = 3818.181818$ or 3818;10,55 miles. (13) Divided into 60 radial parts, that value yields 63.636363 or 63:38,11 miles per part. 14) Therefore, $QZ = 63.636367 \times .813889$ or $63;38,11 \times 00;48,50 = 51.792737$ or $51;47,34 + 6/11$ miles, which is the maximum distance above the earth's surface that rising vapors can reach into the air.

[49]See [Goldstein, 1977, p. 110] for an English translation of the text omitted here by the Latin translator.

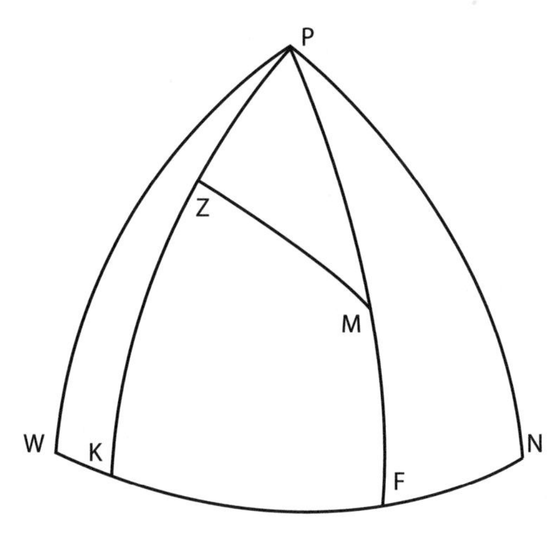

Fig. V-3-1.

The reader who follows Ibn Muʿādh's solution of the problem will see that he assumes as known the same quantities as does the method above, namely, two latitudes and the difference in the longitudes. However, he uses a different method than that described above, one that al-Bīrūnī called "the method of the *zijes*."[50] But, in the end, he finds the arc *ZM*, which measures the distance between one's locality and Mecca, and then the supplement of angle *PZM*, namely, *KZM*.

Fig. V-3-2 illustrates the case in which one (whose zenith is Z) is NE of Mecca (whose zenith is M) and one's longitude differs by less than 90° from that of Mecca. One is looking down at the horizon circle of one's locality, *ABGD*. The equator, *AKG*, cuts the horizon, and its north pole, *P*, is on one's N–S meridian, *BPZD*. (*B* is the north point and *D* the south point.) The meridian of Mecca is *PMF*. We assume known, as we said above, our latitude, arc *ZK*, the latitude of Mecca, arc *MF*, and the difference of our longitude and that of Mecca, namely arc *KF*. To aid in following Ibn Muʿ ādh's computation, the diagram includes the latitude circle of Mecca, *CLN*, which is parallel to the equator, as well as half of a great circle, *AOG*, through one's East and West points (*A* and *G*) and Mecca, *M*. The treatise now explains the necessary calculations, as follows.

As for the determination of the four cardinal points[51] and the *qibla* according to the method of the *qāḍī* Abū ʿAbd Allah b. Muʿādh: you should know the longitude of Mecca which is sixty-seven degrees from the Maghrib,[52] although, sometimes, half a degree is added [to this amount]. Its latitude is twenty-one and one-half, although some add half a degree also. You should also establish the longitude and the latitude of your locality.

[50] A *zīj* is an astronomical manual with tables and rules for using them in astronomy.

[51] We have not included the determination of the four cardinal points of the compass. Ibn Muʿādh uses a method originating in ancient India, known as the "Indian circle method."

[52] In the context here "Maghrib" (literally, "the West") probably means the traditional prime meridian set by Ptolemy, through the Canary Islands. The longitude given here for Mecca is exactly that given by al-Khwārizmī and a number of other authors who used Ptolemy's prime meridian.

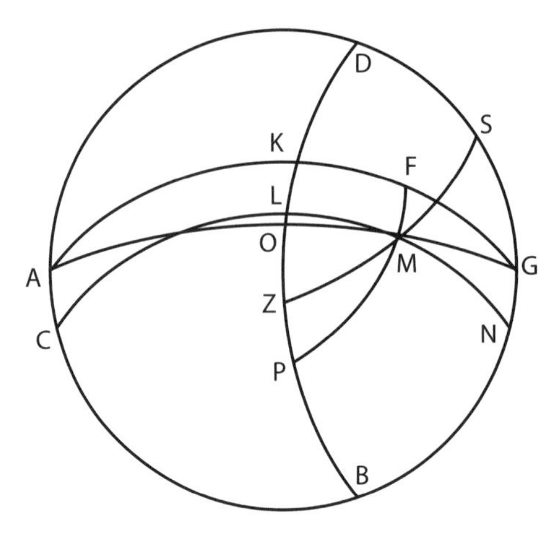

Fig. V-3-2.

Then, multiply the Cosine[53] of the latitude of Mecca [*FM*] by the Sine of the difference between the two longitudes [i.e., that of your locality and that of Mecca, *KF*] and divide the product by 60. Find the arc[Sine] corresponding to the result and the result [*MO*] will be the "longitudinal excess" (*al-fāḍla al-ṭūiyya*) which is [also called] the "perpendicular" (*al-ʿamūd*).[54]

Keep it and obtain its Cosine which will be used as a "divisor" (*imām*). Multiply the Sine of the latitude of Mecca by 60 and divide the result by the "divisor" (*imām*). Find the arc[Sine] of the result and the arc [*KO*] will be the "distance from the equator" (*al-buʿd min muʿaddil al-nahār*).[55] Keep it.

If the difference between the two longitudes is less than 90°,[56] consider the latitude of your locality and the "distance" [from the equator] which you have just noted and subtract the smaller one from the greater one [in this case $ZK - KO = ZO$]. If [the difference between the two longitudes] exceeds 90°, add it to the latitude of your locality. The result [of the subtraction or of the addition], whichever of the two operations you have performed, is the "distance of the locality" (*buʾd al-balad*).

Obtain the Cosine [of this "distance"] and multiply it by the Cosine of the "perpendicular." Divide the product by 60 and find the arc[Sine] of the result. Subtract this arc from 90° and the difference [*MZ*] will be the arc of the "distance between your locality and Mecca" (*masāfa mā bayna baladi-ka wa-Makka*).[57]

[53]Sine and Cosine are used in this translation because in the medieval world they were both side lengths in a certain right triangle. In Arabic texts the Cosine of an arc is called "the Sine of the complement" of the arc.

[54]In this step, the law of sines is applied to the right triangle *OMP*: $\text{Sin } PM : \text{Sin } O = \text{Sin } MO : \text{Sin } P$. (Note that arc *KF* is measured by angle *P*.)

[55]Here, apply the cosine law for right triangle *OMP*: $\text{Cos } PM \text{ Sin } O = \text{Cos } MO \text{ Cos } OP$. (Note that $\text{Cos } OP = \text{Sin } KO$.)

[56]Within 90° of longitude east and west of Mecca one would include most of China in the east and all of the world to the west known in medieval Islam.

[57]Here, apply the cosine law to right triangle *ZMO*: $\text{Cos } MZ \text{ Sin } O = \text{Cos } ZO \text{ Cos } MO$.

Obtain the Sine of this "distance" and use it as a divisor (*imām*). Then, multiply 60 by the Sine of the "perpendicular" (*ʿamūd*) and divide the result by the divisor *(imām).* You will obtain the Sine of the arc of the azimuth. Find the corresponding arc[Sine] and the result will be the arc of the azimuth [*DS*], the azimuth of Mecca.[58]

Then, consider [the following cases]: if the difference between the longitude of your locality and the longitude of Mecca is smaller than 90° and the latitude of your locality is greater than the "distance from the equator,"[59] the resulting azimuth should be counted from the south point towards the corresponding direction, east or west. If the latitude of your locality is smaller than the "distance of the equator" and the difference of longitudes exceeds 90°, the azimuth which you have obtained should be counted from the north point towards the direction of Mecca, may God increase her nobility and importance.

4. IBRĀHĪM IBN AL-ZARQĀLLUH, ON A UNIVERSAL ASTROLABE

Ibrāhīm ibn al-Zarqālluh was born in 1029 in Cordova and worked as an engraver of scientific instruments. He also taught astronomy and made astronomical observations in Toledo, where he died in 1099. Among other astronomical treatises, he wrote a work criticizing Ptolemy's model for the motion of Mercury, but our focus here is on a form of the astrolabe he invented, which, unlike the usual planispheric astrolabe, could be used for any latitude.

The following selections are from two treatises on the construction and use of the astrolabe[60] he invented. There are two variants of this instrument, the *shakkāziyya* and the *zarqāliyya*. The latter is of interest here, because the back of the instrument contains an orthographic projection of the celestial sphere instead of the stereographic projection used for the much more common planispheric astrolabe.[61] In an orthographic projection, points of a hemisphere bounded by some great circle are projected onto points in the plane containing that great circle by lines perpendicular to the plane.[62]

However, al-Zarqālluh discusses the plate entirely in terms of a geometrical construction with only vague references to spherical astronomy (see below). Nor does he introduce the concept of a pointwise projection. He simply lists the lines and curves one needs to draw and, in the case of ellipses, suggests a way to draw them.[63] All of these are contained within a circle whose circumference is graduated in steps of 5°, which he calls "the similar circle," since it is similar to a circle on the other face of the instrument.

This "similar circle" is divided into quadrants by two perpendicular diameters. The horizontal diameter is called the central meridian and that perpendicular to it is called the "greatest parallel." Lines parallel to this are drawn joining endpoints of the 5° intervals on the left hand rim of the similar circle to the corresponding endpoints on the other side. The greatest parallel is divided into 24 equal segments, and a semi-ellipse is drawn through each

[58]The sine law is applied to right triangle *OMZ*: Sin *OM* : Sin *Z* = Sin *MZ* : Sin *O*. Note that the arc *DS* is measured by angle *Z*. The supplement of this angle gives the *qibla*.

[59]This is the situation illustrated in Fig. V-3-2.

[60]The instrument is known as "the plate of the *zijes*" or "the plate of al-Zarqālluh."

[61]Further basic information on this instrument may be found in [Berggren, 1986, pp. 165–171].

[62][Berggren, 1982] drew attention to the fact that al-Bīrūnī described these projections in two treatises dealing with mappings. [Puig, 1996] discusses a third treatise by al-Bīrūnī, which describes the orthographic projection—this time as the basis for an astrolabe.

[63][Berggren, 1991] draws attention to this feature in Ptolemy's discussion of the grids in his map projections. One finds the same feature in al-Bīrūnī's treatise on a different projection of the celestial sphere [Berggren, 1982].

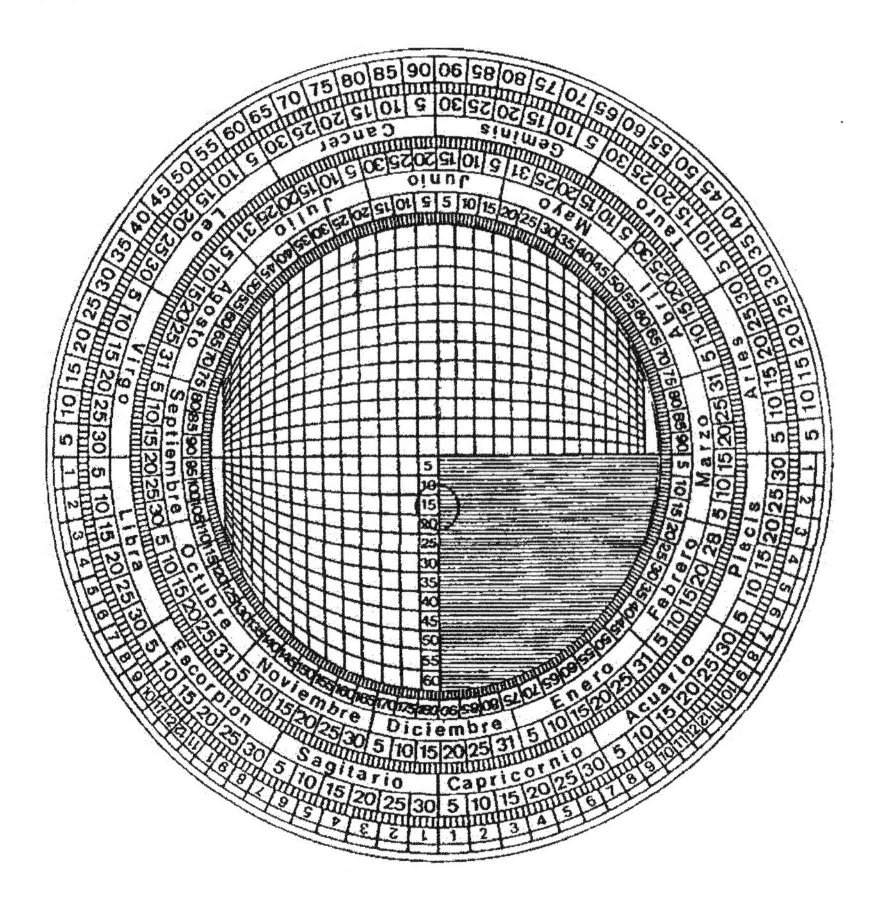

Fig. V-4-1. The back of the *ṣafīḥa zarqāliyya*, taken from [Puig, 1987a, p. 126], reprinted by permission of the University of Barcelona.

of these points and the endpoints of the central meridian, its major axis being the central meridian.

The feature that makes this projection so useful is that the same network of lines and elliptical arcs can be used to represent the orthographic projection of any one of the three principal coordinate systems of the celestial sphere (horizon, equator, and ecliptic). Hence it is ideal for any of the main problems of spherical astronomy, all of which involve transforming coordinates from one of these systems to another.

The first selection is taken from the *Treatise on the Use and Construction of al-Zarqālluh's Instrument* (*Risāla fī ʿamal ālat al-Zarqālluh wa-rasmi-hi*).[64] In it the author explains how to draw the image of the orthographic projection on the back of the *zarqālliyya* astrolabe, which is shown in Fig. V-4-1.

[64]The translation is by Julio Samsó, who has also translated the other selections. This text is also found in a treatise titled *Book on the Saphea*, a Castilian translation from Arabic, in Volume III of the astronomical compilation, *Libros del Saber*. This latter collection of works on astronomy was commissioned by King Alfonso X ("the Wise") during his reign as King of Castile, León, and Galicia from 1252 to 1284. It represented the collective efforts of Christian, Jewish, and Muslim scholars at his court.

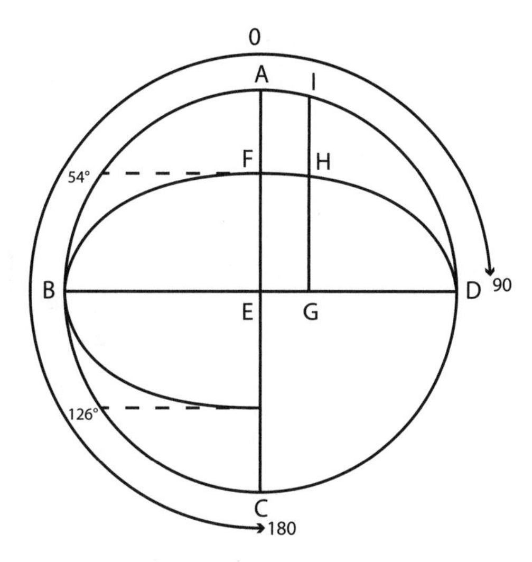

Fig. V-4-2. From [Puig, 1987b, p. 21].

Then we draw the circles of the quinary divisions inside [the circle of] the days as well as [the circle of] the degrees, as it has already been explained for the front of the *ṣafīḥa*. The smallest of these circles is circle *ABCD*[65] [Fig. V-4-2], where *A* is on the line joining the center with the point of suspension,[66] *B* is on the left of the person looking at the instrument, *C* is in the lower part of the diameter that passes through the point of suspension and *D* is at the end of the diameter *BD*, perpendicular to *AC*. The quinary divisions begin at *A* and reach 90 in *D*. The divisions of the other half [of the circle] begin at *A*, reach 90 in *B*, its maximum being 180 in *C*. Similarly, in the other quadrant, divisions begin in *D* and reach 90 in *C*.

Then, we divide diameter *AC* in 24 equal parts and we draw the lines of the ellipses.The major axis of each [ellipse] is segment *BD*, while half of its minor [axis] will be the distance between the center of the *ṣafīḥa* and the [corresponding] point of the divisions of [diameter] *AC*. The quadrant *CED* will not contain these lines. I have explained elsewhere the procedure to draw an ellipse.[67] [But] concerning this curve, if its major axis is the diameter of the circle and we have, in the circle, a chord that is parallel to its (the ellipse's) minor axis, the ratio of the minor semiaxis to the radius of that circle will be equal to the

[65] The letter C corresponds to *jīm* in the Arabic text and is often tranliterated "G." We use "C" here to keep agreement with Fig. V-4-2.

[66] This refers to a small protrusion on the rim of the instrument holding a ring from which the user can suspend the instrument. It is not shown in Fig. V-4-1.

[67] Another instance in which Ibn al-Zarqālluh describes an approximate procedure to draw an ellipse is in his treatise on an "analog computer" for finding planetary positions, known as an equatorium, where he says that the curve of Mercury's deferent, the result of the combination of two motions in Ptolemy's model, is an ellipse. To draw it in the instrument, he determines groups of three points of the deferent and traces arcs of a circle passing through these triples of points. The process continues with several sets of triples until the whole deferent is drawn. The resulting curve is, indeed, similar to an ellipse. See [Samsó and Mielgo, 1994].

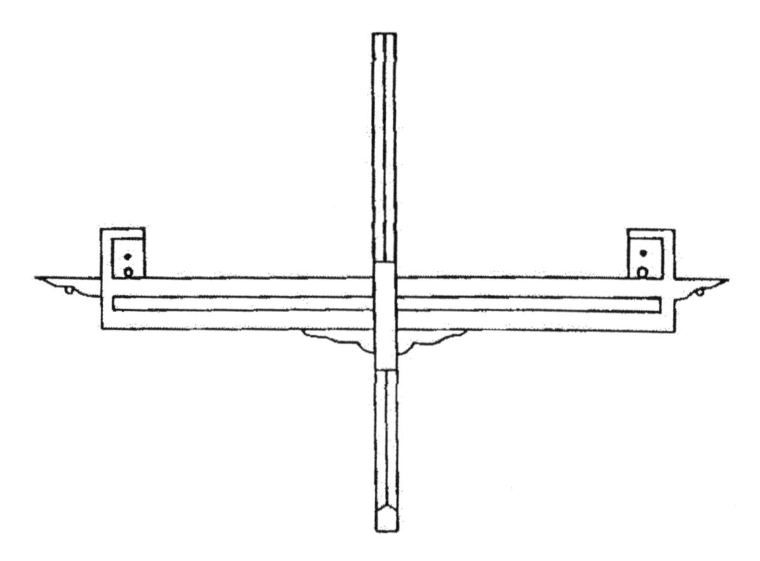

Fig. V-4-3. From [Puig, 1987b, p. 23].

ratio of half [of the segment] of the chord inside the ellipse to half of the [whole] chord.[68] If we divide each one of the lines parallel to the diameter *AC* according to the number of the divisions of *AC* and we join the points of [the first division] after diameter *BD*, the composite line obtained will be approximately an ellipse. Then we do the same with the points of the division that follow the previous one and the result will be another line [= ellipse]. We complete all points following this disposition. We have already seen that some people join three consecutive points with an arc of a circle, concealing how the line was traced except to those who know the procedure.

Then we will write the quinary divisions between the curved lines on one side of the diameter *AC*, ascending from the centre of the *ṣafīḥa* towards the point of suspension and descending towards the lower part of the *ṣafīḥa*.

I omit the description of how to draw the lines representing Sines in the lower right quadrant, which follows the text above. This is followed by an elaborate description of how to make an alidade with two sights. The piece (horizontal in Fig. V-4-3) rotates around the center of the instrument. A transversal ruler (*al-muʿtariḍa*) is perpendicular to the alidade and can slide along it. This piece is essential when one uses the back of the instrument.[69]

The above passage of the treatise on the construction of the instrument only uses the straight-line parallels to draw the half ellipses and gives the impression that they are only auxiliary lines that could be deleted after drawing the ellipses. Thus, we include here a

[68]In Fig. V-4-2, $FE:AE = GH:GI$. The result is found in Thābit ibn Qurra's treatise *On the Sections of a Cylinder and Its Surface*, Prop. 13, which Thābit proves on the basis of Apollonius's *Conics* I, 21. An analogous result is proposition 7 from the work of Ibn al-Samḥ (section IV-2).

[69]The Science Museum in London possesses an eighteenth-century Maghribī astrolabe (Inv. no. 1986-1190) containing two plates with the two kinds of *ṣafīḥa* (the *zarqāliyya* and *shakkāziyya*), which is the only specimen known of the alidade with transversal ruler. Prior to that our only evidence of how the alidade with transversal ruler looked was the illustrations in the Alfonsine translation.

passage from the first chapter of the treatise on the use of the instrument, where we find an explicit mention of the parallels, which will be used to solve the problems that employ the orthographic projection.

The next selection is from the description of the back of the *ṣafīḥa* in *Book on the Use of the* Saphea Zarqāliyya *Serving All Latitudes* (*Kitāb al-ʾamal bi l-ṣafīḥa al-zarqāliyya al-muʾadda li-jamīʿ al-ʾurūḍ*), the treatise on the use of the instrument.

Inside the circle of the months [see Fig. V-4-1] we have the circle of the quinary divisions and, within it, the circle of the degrees, which is similar to the meridian circle[70] on the face of the instrument. The number of the quinary divisions increases from the upper half of the "similar" circle towards right and left until it reaches 90 to the right of someone who is looking at the back when the instrument is hanging from the point of suspension. Then the number increases to the left of the viewer until it reaches 180 in the lower part of the "similar" circle. The diameter which goes from the point of suspension to the lower part of the *ṣafīḥa* is the greatest parallel, and the [straight] lines parallel to it [the diameter] which pass through the quinary divisions are the parallels. Those on the right of the greatest parallel are the southern parallels, while those on the left are the northern parallels. The diameter perpendicular to the greatest parallel is the middle meridian and the curved lines that pass through its two endpoints are the meridians.

The following selection, on using the instrument to determine the hour of the day, must be understood in the context the ancient and medieval worlds, where each period of daylight and each period of darkness was divided into 12 equal hours. (These hours are called "seasonal hours.") For example, noon in a given locality would have been the sixth hour of the day (when the sun would have been due south). Since the lengths of days and nights depend on both on the local latitude and the season of the year, the lengths of daytime hours and nighttime hours varied as well. Only for localities on the equator, or for all localities at the equinoxes, were the lengths of all 24 seasonal hours equal.

Chapter 25: Knowing the hour of the day on the basis of the [solar] altitude, and solar altitude from the hour of the day

Put the end of the alidade[71] that is nearer to the transversal ruler so that its edge [of the transversal ruler *aa′*] coincides with the number of degrees of the altitude established in the "similar" circle [Fig. V-4-4a].[72] Then move the end of the alidade in the upper northern quadrant until its edge coincides with the altitude of Aries [$90^\circ - \varphi$][73] in your locality, leaving the transversal ruler in its place without moving it from its position on the alidade [Fig. V-4-4b]. Enter in the parallels with the solar declination, in the [northern or southern] side according to the [sign of the] declination, and mark the place where the parallel [*pp′* in Fig. V-4-4b] meets the edge of the transversal ruler [*S* in Fig. V-4-4b]. Then

[70]It is concentric with the circle on the face.

[71]The alidade was a ruler, with sights at either end, which rotated around a pivot through its center and that of the astrolabe. In Fig. V-4-4a it is represented by the horizontal thick black line. One could use the sights and scale of degrees around the rim of the astrolabe to take the altitude of the sun, moon, or star above the horizon.

[72][Puig, 1987a, p. 130] remarks that the alidade should be placed on the central meridian.

[73]Since the beginning of the sign of Aries is on the equator, its altitude over the horizon (when it culminates on the local meridian) is the complement of the latitude.

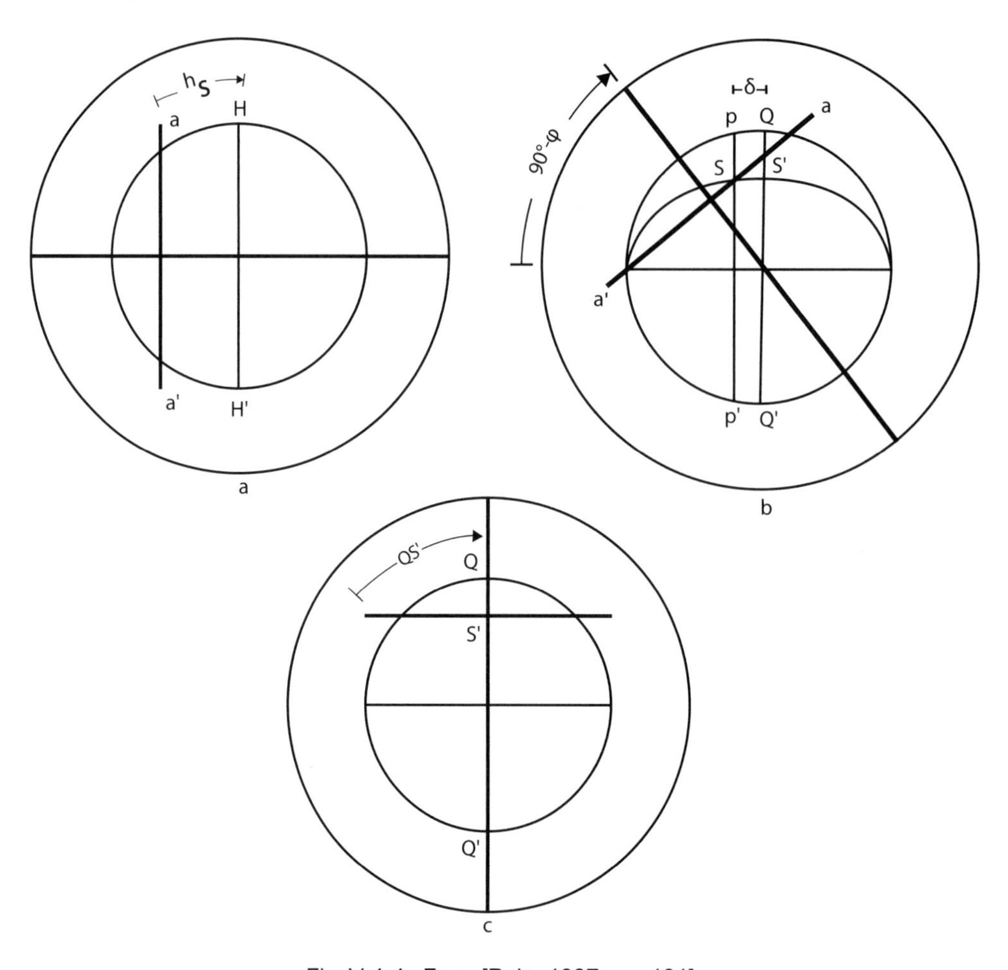

Fig. V-4-4. From [Puig, 1987a, p. 131].

note which meridian passes over the mark and make a second mark in the intersection of the aforementioned meridian with the greatest parallel [S' in Fig. V-4-4b].

Afterwards, place the edge of the alidade on the greatest parallel and move the transversal ruler until its edge coincides with the second mark [Fig. V-4-4c]. This edge will coincide with a particular number of degrees in the graduation of the "similar" circle in its northern side. Subtract this number from half the day arc[74] and divide the remainder by the number of degrees of one day-hour of that particular day. The result obtained will be the number of hours elapsed [since sunrise] of that day if [the computation] has been made [for a time] before midday. If it [the time] is after midday, divide the number of degrees marked by the edge of the transversal ruler in the graduation of the "similar"

[74]The day arc is based on the apparent fact that the sun follows, very nearly, a circular path across the sky during a day and night. The portion of that path (measured in degrees) above our horizon is called the "day circle" of the sun.

circle by the degrees of one day-hour and add six hours to the result. You will obtain the number of seasonal hours [since sunrise].

We follow Dr. Puig's explanation of this passage. In Fig. V-4-4a the alidade (thick straight line) is on the central meridian, the perpendicular diameter HH' represents the horizon, and the transversal ruler aa' represents the circle parallel to the horizon (*almucantar* in Arabic) corresponding to the solar altitude. Up to this point, we have been dealing with spherical coordinates based on the horizon. But in Fig. V-4-4b, where we rotate the alidade a number of degrees equivalent to the co-latitude of our locality, the system of coordinates becomes equatorial.[75] The intersection of the parallel of declination pp' with the *almucantar* (represented by the transversal ruler) determines the position of the Sun (S) at the time of the observation. S will be projected onto the equator (QQ'), the alidade will be superposed on the equator, and S' will be the endpoint of the arc of right ascension of the Sun. The "similar circle" will be the local meridian and the transversal ruler will mark, on the similar circle, the number of degrees of equatorial rotation before or after midday. (One must, of course, know whether it is morning or afternoon.)

5. ABŪ MUḤAMMAD JĀBIR IBN AFLAḤ, *CORRECTION OF THE ALMAGEST*

Abū Muḥammad Jābir ibn Aflaḥ lived during the twelfth century (and, perhaps, during the late eleventh century) in Seville. Little is known of his life, but his son was an acquaintance of the Jewish scholar Maimonides, who referred to Jābir's famous work on astronomy, his *Correction of the Almagest.*[76] This work was the first serious, technical correction of Ptolemy's work to be written in the Islamic West, and it reveals Jābir as an accomplished theoretical astronomer. The work, in its Hebrew and Latin translations, had an important effect on the development of astronomy in Western Europe,[77] and the trigonometric portions of the work clearly influenced Regiomontanus who, however, nowhere mentions Jābir. (See section II-5-6 in Chapter 1.)

The selections below are from Jābir's treatment of spherical trigonometry in his *Correction.* In his study of the tradition of Jābir's work, Professor Lorch states that the two manuscripts of that work in the Escorial may represent a revision of the work and states that there is some reason to believe that the revision may have been the work of Maimonides. The first selection contains Jābir's statement and proof of what has become known as the Rule of Four Quantities[78] and, in his work, replaced Menelaus's Theorem as the basic tool of spherical trigonometry. This is followed by translations of all of Jābir's theorems (and their proofs) that allow his reader to calculate unknown parts of spherical triangles given other, known parts.

As mentioned earlier in the chapter, the capitalized "Sine" is used to remind the reader that the medieval sine function was equal to the modern function multiplied by the hypotenuse of the reference right triangle. And, since the Sine function has the same value for θ and $180^\circ - \theta$, Jābir includes, at the end of the section, rules telling how, knowing the Sine of an angle

[75] See the initial remarks describing the grid of the projection.

[76] I thank Prof. José Bellver for sending me copies of the relevant pages of the manuscript, and a typed version of some of the marginal comments, as well as helpful comments. Professor Richard Lorch's summary of sections of the Latin version of Jābir's text was also of great help [Lorch, 1995]. Lorch says (p. 3) that in his summary he has, in the main, followed the printed (Latin) version of Jābir's work of 1533 but has consulted the Arabic texts as well.

[77] Jābir was known in the Latin West as "Geber."

[78] We do not know if Jābir referred to it as other than "the premise."

in a given spherical triangle, one can use knowledge of other parts of the triangle to decide whether the angle in question is acute or obtuse.

Jābir assumes throughout that his reader is familiar with such basic concepts as great circles on a sphere, their poles, and spherical triangles formed by arcs of great circles.[79]

In what follows, Jābir establishes the following results:

1. The Rule of Four Quantities: Let two great circles on a sphere intersect at A and let two points, B and C, be chosen, either both on one of the great circles, or one on each. From each of B and C, on whichever circle each happens to be, let great circle arcs, BD and CE, be drawn perpendicular to the other great circle. Then $\text{Sin}(AB)\text{:Sin}(BD) = \text{Sin}(AC)\text{:Sin}(CE)$.
2. The Sine Theorem: If ABC is a spherical triangle, then the ratio of the Sine of any angle to the Sine of the side opposite it is the same for all three angles.

The next two theorems both assume that ABC is a spherical triangle with a right angle at B, with a, b, c, designating the sides opposite angles A, B, C, respectively.

3. If B is the only right angle in the triangle then, $\text{Sin}(b)\text{:Sin}(a) = \text{Cos}(c)\text{:Cos}(C)$.
4. $\text{Cos}(b)\text{: Cos}(a) = \text{Cos}(c)\text{: Sin}(90°)$.

This last result is known as "Geber's Theorem," after the Latinized form of Jābir's name. Since the modern sine function has $\sin(90°) = 1$, it can be written as $\cos(b) = \cos(a)\cdot\cos(c)$. The selection begins with the statement of the Rule of Four Quantities.

If there are two great circles on a sphere [neither passing through the pole of the other], and if two points are marked on the circumference of one of them, or one point on the circumference of each of them, wherever it may be, and from each of them [the two points] is produced an arc of a great circle which contains with the arc of the second circle a right angle then the ratio of the Sine of the arc that is between one of the two points and one of the two points of the intersection to the Sine of the arc produced from that point to the second circle is as the ratio of the Sine of the arc that is between the second and between the other of the two points of the intersection to the Sine of the arc produced from that point to the second circle.

So, let the two circles *AGDB* and *AEZB* be great circles on the sphere and first of all let there be marked on the circumference of circle *ABGD* two points *G*, *D*, and let there be produced from them to the circumference of circle *AEZ* at two right angles [arcs *GE* and *DZ*]. Then I say that the ratio of the Sine of arc AG to the Sine of arc *GE* is as the ratio of the Sine of arc *AD* to the Sine of arc *DZ* [Fig. V-5-1].

Its proof is that if we produce from the two points *G* and *D* two perpendiculars onto the plane of circle *AEZ*, say, the two lines *GK* and *DT*, and likewise we also produce from them two perpendiculars to the diameter, *AB* (in the plane of circle *AGD*), namely, the two lines *GL* and *DM*, and we join its line[s] *KL* and *TM*, then because the two lines *KL* and *TM* are parallel, and similarly the two lines *KG* and *TD* are also parallel, the two angles

[79] The reader unfamiliar with these concepts may find a basic introduction in [Berggren, 1986, chapter 6] or a much fuller treatment in [Van Brummelen, 2013].

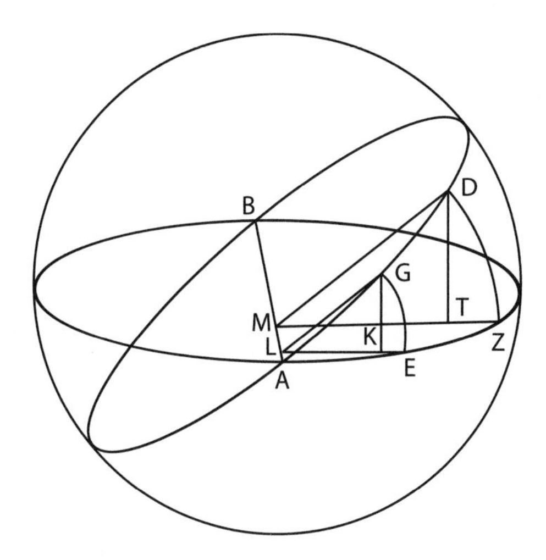

Fig. V-5-1.

KGL and *TDM* are equal. But each one of the two angles *GKL* and *DTM* are right, and so the two triangles *GKL* and *DTM* are similar to each other, and so the two sides *GL*, *GK* are proportional to the two sides *DM*, *DT*. But the line *GL* is the Sine of arc *AG* and, similarly, *GK* is the Sine of arc *GE* and line *DM* is the Sine of arc *AD* and the line *DT* is the Sine of arc *DZ*. And so, then, the ratio of the Sine of arc *AG* to the Sine of arc *GE* is as the Sine of arc *AD* to the Sine of arc *DZ*. End of its demonstration.

The proof of the case where the two points are on different circles is omitted here.

Now we want to prove that for every triangle [whose sides are] arcs of great circles, the ratio of the Sine of any of its sides to the Sine of the arc of the angle subtending it is a single ratio.

So let there be a triangle *ABG* of arcs of great circles. Then I say that the ratio of the Sine of each of its sides to the Sine of the arc of the angle subtending it is one and the same ratio. So first let triangle *ABG* be a right triangle and we will prove that about it first [Fig. V-5-2]. So let its angle *B* be right and let arc *GE* be a quadrant of a circle and let it end at point *E*. And [let] the pole of circle *EG* be on an arc of a great circle, and it is arc *ED*. And let arc *AG* be extended to point *D*. And because angle *E* is right and arc *EG* is a quadrant of a circle, the point *G* is the pole of arc *ED*, so arc *DG* is a quadrant of a circle. And each of the two angles *B* and *E* are right [and] there were two circles *AG*, *GB* on which two points were marked, *A* and *D*, from which two arcs, *AB* and *DE* were produced which contain with circle *BG* two right angles. And so, by what we premised, the ratio of the Sine of arc *AG* to the Sine of arc *AB* equals the ratio of the Sine of arc *GD* to the Sine of arc *DE* and so, *alternando,* the ratio of the Sine of arc *AG* to the Sine of arc *GD* equals the ratio of the Sine of arc *AB* to the Sine of arc *DE*. But arc *GD* is the arc of the right angle *B*, since it is a quadrant, and arc *DE* is the arc of angle *G*, because the point *G* is the pole of the arc *DE*. So, then, the ratio of the Sine of side *AG* to the Sine of the arc of the angle subtending it, *B*, [is equal to the ratio of the Sine of side *AB* to the Sine of the arc of the angle subtending it, *G*.][80]

[80] It seems this line was inadvertently dropped from the manuscript by the scribe copying it.

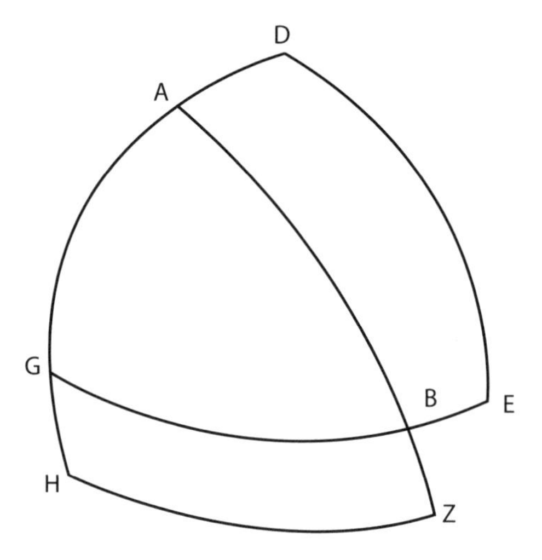

Fig. V-5-2.

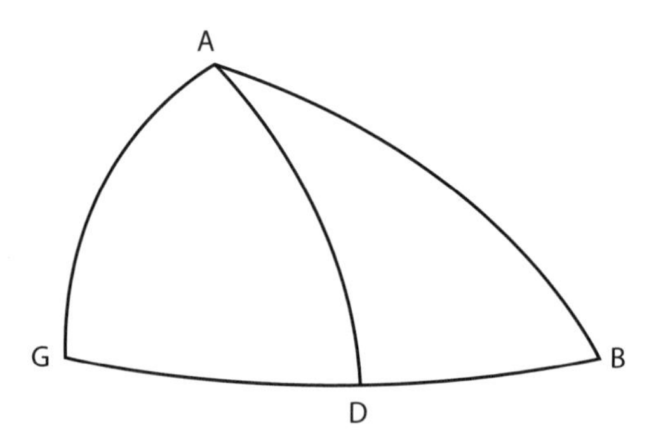

Fig. V-5-3.

And likewise it is shown if we make arc *AZ* a quadrant and draw[81] the arc of a great circle, namely arc *ZH*, which meets *AG* if it is produced at the point *H*, the ratio of the Sine of the side *AG* to the Sine of the arc of angle *B*, which it subtends, is as the ratio of the Sine of side *BG* to the Sine of the arc of angle *A* which subtends it. The proof [of the first case] ends.

And now let the triangle *ABG* have no right angles, and let arc *AD* contain, with arc *BG*, a right angle, angle *ADG* [Fig. V-5-3]. So because triangle *ADG* is right angled the ratio of the Sine of the side *AG* to the Sine of the arc of the right angle *D* subtending it is as the ratio of the Sine of the side *AD* to the Sine of the arc of the angle subtending it, *G*. And likewise, also, the ratio of the Sine of the right angle *D* of triangle *ABD* to the Sine of the side *AB* subtending it is as the ratio of the Sine of angle *B* to the Sine of the side *AD* subtending it. And *ex aequali* according to the perturbed ratio, the ratio of the Sine of the side *AG* to the Sine of angle *B* is as the ratio of the Sine of the side *AB* to the Sine of angle *G*. And, likewise, it is proved that the ratio of the Sine of the side *BG* to the Sine

[81]The text here has, "around point *Z* and around the pole of circle *ABZ*," which makes no sense.

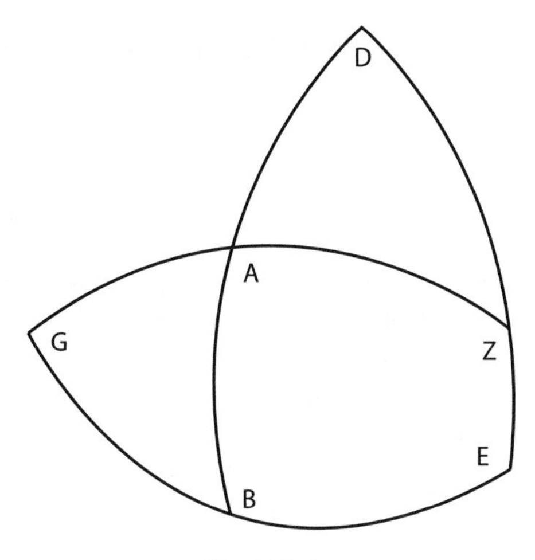

Fig. V-5-4.

of angle *A* equals the ratio of the Sine of the side *AG* to the Sine of angle *B*. And that is what we wanted to prove.

And likewise it is proved in the case where the perpendicular falls outside of triangle *ABG*, that is that one of the two angles *B*, *G* is obtuse since the Sine of an obtuse angle is itself the Sine of the acute angle adjoining it.

And likewise we prove also that in any triangle of arcs of great circles having only one right angle in it the ratio of the Sine of the side subtending the right angle to the Sine of one of the containing sides is as the ratio of the Sine of the complement of the other containing side to the Sine of the complement of the arc of the angle subtending it.

So let there be triangle *ABG* of arcs of great circles and a right angle only at point *B* of it. Then I say that the Sine of side *AG* to the Sine of side *BG* is as the ratio of the Sine of the complement of the side *AB* to the Sine of the complement of angle *G* subtending it [Fig. V-5-4].

Proof: We make arc *BD* a quadrant so that the point *D* is a pole of [the great] circle *BG*, and let an arc of a great circle, arc *DZE*, pass over the point *D* and over the pole of circle *AG*. And let it meet arc *AG* at point *Z* and arc *BG* at point *E*. Each of the angles *E* and *Z* are right, and so the two circles *DAB* and *ZAG* will be great circles and there are marked on them *D*, *G* from which are produced two arcs *DE* and *GB*. And each of the angles *E* and *B* are right. So, as we proved, Sine arc *AG* to Sine arc *BG* equals Sine arc *AD* to Sine arc *DZ*. But arc *AD* is the complement of side *AB* and arc *DZ* is the complement of the arc of angle *G*. And so, then, the ratio of the Sine of side *AG*, which subtends the right [angle], to the Sine of side *BG*, one of those containing it [the right angle], is as the ratio of the Sine of the complement of the remaining side, *AB*, to the Sine of the complement of angle *G* subtending it.

And so I demonstrated that the [ratio of the] Sine of the right angle *B* to the Sine of angle *A* is as the ratio of the Sine of the complement of the side *AB* on which are the two [non-right] angles to the Sine of the complement of the angle *G* subtending it, QED.

And likewise it is proved now that the Sine of the complement of the side subtending the right (angle) to the Sine of the complement of one of the two containing it is as the ratio of the Sine of the complement of the third side to the Sine of a quadrant. So, there are two [great] circles, *DAB* and *DZE*, and marked on the circumference of one of them are two points, *A* and *B*. And two arcs are produced from them, *AZ* and *BE*, containing with the circle *EZ* two right angles [Fig. V-5-4]. And so, by what we proved, Sine arc *AZ* to Sine arc *AD* equals Sine arc *BE* to Sine arc *DB*. But arc *AZ* equals the complement of arc *AG*, and arc *EB* equals the complement of arc *BG* and arc *AD* equals the complement of arc *AB*, since arc *DB* is a quadrant. And so, *alternando*, the ratio of the Sine of the complement of the side *AG*, which subtends the right [angle], to the Sine of the complement of the side *BG*, one of the two containing it [the right angle], is as the ratio of the Sine of the complement of the remaining side, *AB*, to the Sine of a quadrant. End of proof.

SOURCES

I. Arithmetic

1. Ibn al-Bannā, *Raising the Veil*, translated by J. L. Berggren from the edition and French translation in Mohamed Aballagh. 1988, *Rafʾu al-Hijāb dʾIbn al-Bannā: Edition critique, Traduction, Etude philosophique et Analyse mathématique*, dissertation, Université de Paris I; *Talkhis*, translated by J. L. Berggren from the edition and French translation in Ibn al-Bannāʾ. 1969. *Talkhīs aʾmāl al-ḥisāb*, Mohamed Souissi (ed. and trans.). Tunis: Université de Tunis.
2. Abū al-Ḥasan ibn ʿAlī al-Qalaṣādī, *Removing the Veil from the Science of Calculation*, translated by J. L. Berggren from the French in F. Woepcke. 1858–1859. "Traduction du traité d'arithmétique d'Aboūl Haçan Ali Ben Mohammad Alkalçādī," *Atti dell' Academia Pontificia de 'Nuovi Lincei* 12, 230–275.
3. Muḥammad ibn Muḥammad al-Fullānī Kishnāwī, On magic squares, translated by J. L. Berggren from the French translation in Jacques Sesiano. 1994. "Quelques methods arabes de construction des carrés magiques impairs," *Bulletin de la Société vaudoise des sciences naturelles* 83(1), 51–76.

II. Algebra

1. Ibn al-Bannāʾ, selections from the *Summary Account* and *Raising the Veil*, see I 1; *Book on Fundamentals and Premises for Algebra*, translated by J. L. Berggren from the French and Arabic in Ahmed Djebbar. 1990. "Le livre d'algèbre d'Ibn al-Bannāʾ," in Mathématiciens du Maghreb Médiéval (IXe–XVIe siècles): Contribution à l'étude des activités de l'Occident musulman, Ph.D. thesis, Université de Nantes/Université de Paris-Sud.
2. Ibn Badr, *An Abridgement of Algebra*, translated by J. L. Berggren from the edition in José Sánchez Pérez (ed.)1916. *Compendio de algebra de Abenbéder (Ibn Badr), Text árabe, traducción y studio por José A. Sánchez Pérez*. Madrid: Impr. Ibérica.

III. Combinatorics

1. Aḥmad Ibn Muʿnim al-ʿAbdari, *Fiqh al-Ḥisāb*, introduction to figured numbers, quoted (with minor changes) from the English translation in Ahmed Djebbar. Undated. Preprint, "Figured Numbers in the Mathematical Tradition of Andalusia and the Maghrib," footnotes 8 and 11; remainder translated by J. L. Berggren from the Arabic in Driss Lamrabet. 2006. *Fiqh al-Ḥisāb* (section 9 of chapter 1). Rabat: Dar al-Amane Publishers; *Fiqh al-Ḥisāb* (section 11 of chapter 1), translated by J. L. Berggren from the French and Arabic in Ahmed Djebbar. 1985. *L'analyse Combinatoire au Maghreb: L'Exemple d'ibn Munʿim (XIIe–XIIIe s.)*. Paris: Publications Mathématiques d'Orsay, Université de Paris-Sud.

2. Ibn al-Bannāʾ, on combinatorics, translated by Victor Katz from the French in Mohamed Aballagh. 1988. *Rafʿ al-Hijab d'Ibn al-Bannā: Edition critique, Traduction, Etude philosophique et Analyse mathématique*, dissertation, Université de Paris I, pp. 65–71.
3. Shihāb al-Dīn ibn-al-Majdī, "Enumerating Polynomial Equations," translated by J. L. Berggren (with the assistance of M. Bagheri) from the Arabic in *Enveloping the Core*, BM Add. MS 2469, folios 193b–194b.

IV. Geometry

1. Abū ʿAbd Allāh Muḥammad ibn ʿAbdūn, *On Measurement*, translated by J. L. Berggren from the Arabic in Ahmed Djebbar. 2005–2006. *Suhayl* 5, 7–68 (of the Arabic section) and *Suhayl* 6, 81–86.
2. Abū al-Qāsim ibn al-Samḥ, *The Plane Sections of a Cylinder and the Determination of Their Areas*, translated by Victor Katz from the French translation from the Hebrew by Tony Lévy in Roshdi Rashed. 1996. *Les mathématiques infinitésimals du IX*^{e} *au XI*^{e} *siècle*. London: Al-Furqān Islamic Heritage Foundation, vol. I, pp. 929–973.
3. Abū ʿAbd Allāh Muḥammad ibn al-Muʿādh al Jayyānī, *On Ratios*, translation from Edward B. Plooij. 1950. *Euclid's Conception of Ratio and His Definition of Proportional Magnitudes as Criticized by Arabian Commentators*. Rotterdam: W. J. Van Hengel.
4. Al-Muʾtaman ibn Hūd, selections from the *Kitāb al-Istikmāl*. 1. Heron's Theorem. Translated by J. L. Berggren from copies of Arabic text supplied by Jan Hogendijk in MS Copenhagen Royal Library, Or. 82, folio 41a. 2. "Projective Invariance of Cross Ratio." Translated by J. L. Berggren from copies of Arabic text (and a printed version) supplied by Jan Hogendijk in MS Copenhagen Royal Library, Or. 82, folio 42a–42b. 3. "Ceva's Theorem," translation from pp. 30–32 of Jan Hogendijk. 2004. "The Lost Geometrical Parts of the Istikmāl of Yūsuf al-Muʾtaman ibn Hūd (11th century) in the Redaction of ibn Sartāq (14th century): An Analytical Table of Contents," *Archives Internationales d'Histoire des Sciences* 53, 19–34. 4. Continued Proportion, translation from Jan Hogendijk. 1992. "Four Constructions of Two Mean Proportionals between Two Given Lines in the Book of Perfection (Istikmāl) of al-Muʾtaman Ibn Hūd," *Journal for History of Arabic Science* 10, 13–29. 5. Alhazen's Problems, translation from Jan Hogendijk. 1996. "Al-Muʾtaman's Simplified Lemmas for Solving 'Alhazen's Problem,'" in Josep Casulleras and Julio Samsó (eds.), *From Baghdad to Barcelona: Studies in the Islamic Exact Sciences in Honour of Prof. Juan Vernet*. Barcelona: Universitat de Barcelona (Anuari de Filologia XIX, B-2), vol. 1, 59–101, pp. 73–90.
5. Muḥyī al-Dīn ibn Abī al-Shukr al-Maghribī, *Recension of Euclid's Elements*. Material on parallel postulate slightly adapted from A. I. Sabra. 1969. "Simplicius on Euclid's Parallel Postulate," *Journal of the Warburg and Courtauld Institutes* 32, 1–24. Material on Book XV, translation from Jan Hogendijk. 1993. "An Arabic Text on the Comparison of the Five Regular Polyhedra: 'Book XV' of the *Revision of the Elements* by Muḥyi al-Dīn al-Maghribī," *Zeitschrift für Geschichte der arabisch-islamischen Wissenschaften* 8, 133–233, pp. 188–214.

V. Trigonometry and Astronomy

1. Abū ʿAbd Allāh Muḥammad ibn al-Muʿādh al-Jayyānī, *Book of Unknowns of Arcs of the Sphere*, translated by J. L. Berggren from the Arabic in the edition of M. V. Villuendas. 1979. *La Trigonometría Europea en el Siglo XI*. Barcelona: Instituto de Historia de la Ciencia de la Real Academia de Buenas Letras Barcelona.
2. Abū ʿAbd Allāh Muḥammad ibn al-Muʿādh al-Jayyānī, *On Twilight and the Rising of Clouds*, translation from A. Mark Smith. 1992, "The Latin Version of Ibn Muʿādh's Treatise 'On Twilight and the Rising of Clouds,'" *Arabic Sciences and Philosophy* 2, 83–132, pp. 117–132.
3. Abū ʿAbd Allāh Muḥammad ibn al-Muʿādh al-Jayyānī, on the *qibla*, translation by J. Samsó (private communication to author) from Arabic text published in J. Samsó and H. Mielgo. 1994. "Ibn Isḥāq al-Tūnisī and Ibn Muʿādh al-Jayyānī on the Qibla," in J. Samsó, *Islamic Astronomy and Medieval Spain*. Burlington, VT: Ashgate/Variorum, VI.
4. Ibrāhīm ibn al-Zarqālluh, on a universal astrolabe, translation by J. Samsó (private communication to author) from two Arabic manuscripts, *Treatise on the Use and Construction of al-Zarqālluh's Instrument* (*Risāla fī ʿamal ālat al-Zarqālluh wa-rasmi-hi*), BNF 4284, and *Book on the use of the* saphea zarqāliyya

serving all latitudes (*Kitāb al-ʿamal bi l-ṣafīḥa al-zarqāliyya al-muʾadda li-jamīʿ al-ʿurūḍ*), MS Escorial 962.

5. Abū Muḥammad Jābir ibn Aflaḥ, *Correction of the Almagest*, translated by J. L. Berggren from the Arabic manuscript 910 in the Escorial Library, in San Lorenzo de El Escorial, Spain.

REFERENCES

Bearman, P. J., et al. (eds.). (1986). *Encyclopædia of Islam*, second edition, 12 vols. Leiden: Brill.

Berggren, J. Lennart. 1982. "Al- Bīrūnīʾ on Plane Maps of the Sphere,"*Journal for the History of Arabic Science* 6, 47–96.

———. 1986. *Episodes in the Mathematics of Medieval Islam*. New York: Springer.

———. 1991. "Ptolemy's Maps of Earth and the Heavens: A New Interpretation," *Archive for History of Exact Sciences* 43, 133–144.

———. 2007. "Mathematics in Medieval Islam," in Victor J. Katz (ed.), *The Mathematics of Egypt, Mesopotamia, China, India, and Islam: A Sourcebook*. Princeton, NJ: Princeton University Press, pp. 515–675.

De Slane, W. M. 1925. *Kitāb al-ʿibar*. Paris.

Djebbar, Ahmed. 1992. "The Treatment of Fractions in the Arab Mathematical Tradition of the Maghreb," in Paul Benoit, Karine Chemla, and Jim Ritter (eds.), *Histoire de fractions, fractions d'histoire*. Basel: Birkhäuser, pp. 223–246.

———. 1995. "Les Mathématiques dans le Maghreb Médiéval." *Newsletter of the African Mathematical Union Commission on the History of Mathematics in Africa* (*AMUCHMA*) *Bulletin* 15, 3-45.

———. 2005. *L'algèbre arabe: Genèse d'un art*. Paris: Vuibert-Adapt.

Djebbar, A., and Y. Guergour. 2013, "La numeration *rūmī* dans des écrits mathématiques d'al-Andalus et du Maghreb avec l'édition d'une épître d'Ibn al-Bannāʾ, " *Suhayl* (in Arabic) 12, 7–74.

Doncel, M. G. 1982. "Quadratic Interpolations in Ibn Muʿadh," *Archives internationales d'histoire des sciences* 32, 68–77.

Goldstein, B. R. 1977. "Ibn Muʿādh's Treatise on Twilight and the Height of the Atmosphere," *Archive for History of Exact Sciences* 17, 97–118.

Heath, T. L. 1913. *Aristarchus of Samos*. Oxford: Oxford University Press.

Hogendijk, Jan. 1991. "The Geometrical Parts of the *Istikmāl* of Yūsuf al-Muʾtaman ibn Hūd (11th century): An Analytical Table of Contents," *Archives internationales d'histoire des sciences* 41, 209–281.

Ibn al-Bannāʾ. 1969. *Talkhīs aʿmāl al-ḥisāb*, edited and translated by Mohamed Souissi. Tunis: Université de Tunis.

Imhausen, Annette. 2007. "Egyptian Mathematics," in Victor J. Katz (ed.), *The Mathematics of Egypt, Mesopotamia, China, India, and Islam: A Sourcebook*. Princeton, NJ: Princeton University Press, pp. 7–56.

Lévi-Provençal, E. 1986. "ʿAbd al-Raḥmān," in P. J. Bearman et al. (eds.), *Encyclopædia of Islam*, second edition. Leiden: Brill, Vol. I, pp. 82^b–83^a.

Lorch, Richard. 1995. *Arabic Mathematical Sciences*. Burlington, VT: Ashgate/Variorum, articles VI (originally printed in *Centaurus*, 19, 1975), VII and VIII (first printings).

Montelle, Clemency. 2014. "Hypsicles of Alexandria and the Fourteenth Book of Euclid's *Elements*." (to be published soon)

Puig, Roser. 1987a. "La proyección ortográfica en el *Libro de la açafeha alfonsí*," in Mercè Comes, Roser Puig, and Julio Samsó (eds.), *De Astronomia Alphonsi Regis*. Barcelona: University of Barcelona.

———. 1987b. *Los tratados de construcción y uso de la azafea de Azarquiel*. Madrid: Instituto Hispano-Arabe de Cultura.

———. 1996. "On the Eastern Sources of Ibn al-Zarqālluh's Orthographic Projection," in J. Casulleras and J. Samsó (eds.), *From Baghdad to Barcelona: Studies in the Islamic Exact Sciences in Honour of Prof. Juan Vernet*. Barcelona: University of Barcelona, pp. 737–753.

Risner, F. 1572. *Opticae Thesaurus, Alhazeni Arabis libri septem.* Basel. Reprint, New York: Johnson Reprint, 1972.

Sabra, A. I. 1969. "Simplicius on Euclid's Parallel Postulate," *Journal of the Warburg and Courtauld Institutes* 32, 1–24.

———. 1982. "Ibn al-Haytham's Lemmas for Solving 'Al-Hazen's Problem,'" *Archive for History of Exact Sciences* 26, 299–324.

———. 1986. "Ilm al-ḥisāb (The Science of Reckoning)," in *Encyclopaedia of Islam,* new edition. Leiden: Brill, Vol. III, 1138a–1141b.

Ṣāʾid al-Andalusī. 1991. *Science in the Medieval World: Book of the Categories of Nations,* translated and edited by S. I. Salem and Alok Kumar. Austin: University of Texas Press.

Samsó, J., and H. Mielgo. 1994. "Ibn al-Zarqālluh on Mercury," *Journal for the History of Astronomy* 25, 289–296. Reprinted in J. Samsó. 2007. *Astronomy and Astrology in al-Andalus and the Maghrib.* Aldershot: Ashgate-Variorum, no. IV.

Sesiano, Jacques. 1994, "Quelques methods arabes de construction des carrés magiques impairs," *Bulletin de la Société vaudoise des sciences naturelles* 83(1), 51–76.

Smith, A. Mark. 1992. "The Latin Version of Ibn Muʿādh's Treatise 'On Twilight and the Rising of Clouds,'" *Arabic Sciences and Philosophy* 2, 83–132.

Souissi, Mohamed. 1968. *La langue des mathématiques en arabe.* Tunis: Impr. Officielle.

Van Brummelen, Glen. 2009. *The Mathematics of the Heavens and the Earth: The Early History of Trigonometry.* Princeton, NJ: Princeton University Press.

———. 2013. *Heavenly Mathematics: The Forgotten Art of Spherical Trigonometry.* Princeton, NJ: Princeton University Press.

Villuendas, M. V. 1979. *La Trigonometría Europea en el Siglo XI.* Barcelona: Instituto de Historia de la Ciencia de la Real Academia de Buenas Letras Barcelona.

Zemouli, T. 1993. "*Al-aʿmāl al-riyāḍiyya li-Ibn al-Yāsamīn* [The Mathematical Work of Ibn al-Yāsamīn]," master's thesis, Alger, École Normale Supérieure.

Appendices

APPENDIX 1: BYZANTINE MATHEMATICS

In order to better understand Byzantine arithmetic, one has to remember that Greek numerals are written with letters without using a symbol for zero: thus 201 is written $\sigma\alpha$, 1300 is written, $\alpha\tau$, 1308 is written, $\alpha\tau\eta$. Such a notation does not allow a positional arithmetic.[1] A symbol for zero existed, but was used only for astronomical tables and in the sexagesimal system. In the thirteenth century there was an attempt to introduce Indian numerals and a corresponding arithmetic, but without real success. We know that calculation was taught to children by means of finger reckoning, but we do not really know how ordinary people performed calculations of daily life. However, numerous calculations in the margins of the astronomical manuscripts show that ancient procedures were in use.

The most important books of arithmetic appear in the thirteenth and fourteenth centuries, with only a few treatises surviving from before this period. Of these, one of the most important is the mathematical papyrus of Akhmin (ancient Panopolis in Upper Egypt) (seventh century), which deals with fractions and problems in the Egyptian tradition. For the seventh and eighth centuries we know, thanks to some *Lives of the Saints*, that the education of young people included training in arithmetic, though no text on this survives from before the eleventh century when, around 1007–1008, one finds an anonymous *Quadrivium* in which the arithmetical part is based on Euclid and Nicomachus. It is mainly at the end of the thirteenth and during the first half of the fourteenth century that Byzantine scholars show a growing interest in arithmetic when the work of Diophantus is paraphrased in the *Quadrivium* of George Pachymeres (ca. 1242–1310) and studied by his contemporary Maximus Planudes (ca. 1255–1305). Two trends can be observed in the arithmetical manuals. On the one hand there are theoretical works often linked to astronomy, with many chapters devoted to the sexagesimal calculations; on the other, are practical manuals with problems useful in daily life. The learned element is based essentially on ancient authors (e.g., Euclid, Theon, Heron, and Nicomachus) and appears in an important treatise like the *Quadrivium* of George Pachymeres

[1] This introduction to Byzantine mathematics is taken from Anne Tihon. 2008. "Numeracy and Science," in Robin Cormack, John F. Haldon, and Elizabeth Jeffreys (eds.), *The Oxford Handbook of Byzantine Studies*, chapter III.17.4. Oxford: Oxford University Press.

(ca. 1300), book I of which was much used by his followers, or the *Stoicheiosis* (*Elements*) of Theodore Metochites (ca. 1300), an immense astronomical work that opens with a long arithmetical introduction, or the *Astronomical Tribiblos* of Theodore Meliteniotes (ca. 1352), of which an important part of book I is devoted to arithmetical procedures. Barlaam's *Logistic* is a purely abstract work based on Euclid, in which numbers are represented by lines and letters. The last representative of this learned strand is George of Trebizond. His *Introduction to the Almagest*, although written (ca. 1468?) after the fall of Constantinople, deserves to be mentioned in discussions of the later development of Byzantine arithmetic. George's treatise differs from the others in its criticism of Theon by the adoption of Arabo-Latin methods and also by an attempt to create a more abstract terminology for naming the terms of a ratio.

The practical strand appears in manuals on Indian calculation, which adopt the numerical Indian ciphers and associated calculations. Such methods had been introduced among the Arabs by al-Khwārizmī (ca. 825), among the Jews by ibn Ezra (ca. 1092–1167), and in the Western world by Leonardo of Pisa (ca. 1202–1238). They appear in Byzantium ca. 1252 in an anonymous treatise and in a manual based on it written by Maximus Planudes ca. 1300. But Indian numerals never knew a real success among Byzantine scholars, who remained deeply attached to the Greek tradition. Among the arithmetical works one must also refer to two *Arithmetical Letters* by Nicolas Aravasde Rhabdas (ca. 1340); the first explains finger reckoning, which had been widespread throughout the Mediterranean since antiquity. Nicolas Mesarites in his description of the school of the Holy Apostles in Constantinople ca. 1200 depicts terrorized children counting on their fingers while the blows of the rod of the whip of the master rain down over the least error. The *Treatise on Square Numbers* of Manuel Moschopoulos (ca. 1265–1315) deals with "magic squares," and the *Treatise on the Square Root* of Isaac Argyros (ca. 1368) tries to improve on the methods of Heron of Alexandria. There exist a number of anonymous manuals concerned with practical problems solved by algebraic procedures.

Geometry as such was not widely developed by Byzantine scholars, except in Justinian's time, when two great geometers—Anthemios of Tralles (d. ca. 534) and Isidores of Miletos (ca. 532)—were active. But we know from many accounts that Euclid was studied in each period of the Byzantine Empire, and the treatises on the quadrivium already mentioned included Euclidian geometry. Apart from the Euclidian tradition, there existed numerous small treatises on geodesy that explain procedures for surveying. This was illustrated by Michael Psellos (ca. 1018–1078), John Pediassimos (thirteenth century), and Isaac Argyros (ca. 1368), and also by anonymous treatises giving rather inexact nonscientific methods.

. . .

A survey of Byzantine scientific works related to the quadrivium demonstrates that access to them was available only to a restricted intellectual elite: scholars, astrologers, or physicians close to the imperial court or patronized by rich or powerful persons, civil servants, diplomats, the director of the Patriarchal School, or high members of the Orthodox Church. Scientific knowledge implied an excellent education, access to an important library, and a certain social status. It is difficult to form a clear understanding of the scientific education of an ordinary person. Compilations of basic elements of astronomy or geometry appear in many manuscripts, probably for the use of schoolteachers, but the ancient cosmology of Kosmas

Indikopleustes reappears, and absurd ideas explaining some astronomical phenomena can be read here and there in anonymous collections. Astrology and numerology had many supporters among Byzantine scholars but did not interfere with their scientific works.[2]

1. MAXIMUS PLANUDES, *THE GREAT CALCULATION ACCORDING TO THE INDIANS*

Maximus Planudes was born ca. 1255 in Nicomedia and died at Constantinople ca. 1305. He took the name Maximus, replacing his baptismal name of Manuel, when he became a monk, shortly before 1280. This work, *The Great Calculation According to the Indians*, introduces the eastern Arabic form of the Indian numerals to Constantinople along with a detailed study of algorithms for addition, subtraction, multiplication, division, and the extraction of square roots. The excerpts here are from the opening section in which Planudes explains the Hindu-Arabic numeral system in terms of Greek arithmetic, and then from a later part on taking square roots.

Since numbers continue without bound, but knowledge of the boundless is not possible, the more eminent of the astronomers invented certain signs and a method relating to them, so that the representation of those numbers they needed might be more easily and more clearly apprehended at a glance.

There are only nine signs required which are these: 1, 2, 3, 4, 5, 6, 7, 8, 9.[3] They also use a certain other sign which they call a *cipher*, which, according to the Indians, signifies "nothing." These nine signs are themselves of Indian origin and the cipher is written as 0.

When each of these 9 signs[4] stands alone by itself and in the first place beginning from the right-hand side, the symbol 1 indicates one, 2 indicates two, 3 three, 4 four, 5 five, 6 six, 7 seven, 8 eight and 9 nine. If, however, it is in the second place, then the symbol 1 indicates ten, 2 twenty, 3 thirty, and so on. In the third position 1 indicates a hundred,[5] 2 two hundred, 3 three hundred and so on. The pattern continues for the remaining places.

Thus, in the first position the signs are to be regarded as units, which, beginning from one, proceed to nine. (Two, three, four, up to nine will be reckoned as *monadic* numbers, since they are all bounded by ten, neither reaching nor exceeding it.) Hence any sign which occurs in the first position will be regarded as *monadic*, and in the second position as *decadic*, that is, between ten and ninety, and in the third position as *hecatontadic*, that is, between a hundred and nine hundred. So also a sign in the fourth position is regarded as a multiple of a thousand and in the fifth as myriads and in the sixth as tens of myriads, the seventh as hundreds of myriads, the eighth as thousands of myriads, and in the ninth as myriads of myriads. If one were to proceed even beyond this point, the tenth position counts as tens of myriads of myriads and the eleventh as hundreds of myriads of myriads, the twelfth as thousands of myriads of myriads and the thirteenth as myriads of myriads of myriads. Indeed one could proceed even further.

[2]Of course, astrology and numerology also appear in Latin, Hebrew, and Arabic works and were certainly part of the mathematical enterprise in the medieval period. For examples of works in numerology, see section I-2-3 in Chapter 1 and section II-1 in Chapter 2. For astrology, see section II-2 in Chapter 2.

[3]For ease of reading, I use modern forms for the symbols.

[4]I will use numerals when they are given in the Greek and names when Greek names for numbers are given.

[5]Note that this is one word in Greek, as with the other multiples of a hundred. This will be important in understanding Planudes's description of the four types later in the work.

Now to clearly illustrate what I have said by an example, suppose the given number is 8136274592, which occupies ten of these places. We begin from the right hand side, as has been said, and so the sign 2 in the first place indicates the number two, which is a monadic number. In the second place, the sign 9, is ninety, which is a decadic number, consisting in fact only of tens, just as the two before it, being monadic, consisted only of units. The sign 5 in the third place is five hundred, which is a hecatontadic number. The number 4, in the fourth place, is four thousand, which is a *chiliadic* number,[6] and 7 in the fifth represents seven myriads and is a *myriadic* number. The sign 2 in the sixth place is twenty myriads; it is a *decakismyriadic* number. 6 in the seventh place is six hundred myriads, a *hecatontakismyriadic* number. The sign 3 in the eighth is three thousand myriads, a *chiliontakismyriadic* number. The sign 1 in the ninth is a myriad of myriads, and is a *myriontakismyriadic* number. The sign 8 in the tenth place is eighty myriads of myriads, which is a *decakismyriontakismyriadic* number. The given number is thus read in its entirety as eighty myriads of myriads and a myriad of myriads and three thousand six hundred and twenty seven myriads, four thousand five hundred and ninety two.

. . .

The cipher is never placed at the left-hand end of the digits but can appear in the middle of the number or at the right-hand side, that is, at the extreme side before the smallest (non-zero) place digit. Not only one, but two, three, four or as many zeros as are required may be placed in the middle or in the other aforementioned place. Just as the (number of) places increases the size of the number, so too does the number of ciphers. For example, one cipher lying at the end makes the number decadic, so 50 is fifty. Two ciphers make it hecatontadic, thus 400 is four hundred, and so on in turn. If one cipher lies in the middle and there is only one symbol before it, it makes that number hecatontadic; thus 302 is three hundred and two, but if there are two such signs, the number is chiliadic, thus 6005 is six thousand and five. If there is a single cipher with two signs after it, this indicates a chiliadic number, thus 6043 is six thousand and forty three, but if there are two then the number is myriadic, thus 60043 is six myriads and forty three, and so on in turn. To put it simply, the number is to be understood by the order in which the symbols are placed.

On Finding the Square Root of Any Number

Since we have dealt with different problems arising from astronomical calculation, let us now deal with the quadrature of numbers which are not perfect squares, to show clearly that it is possible to find the square root of any given non-square number. This is achieved in the following way.

Take the square root of the nearest[7] perfect square and double it. Then take from the number whose square root you seek, the square you found nearest to it, and express the

[6] That is, a multiple of a thousand.

[7] Nearest, but less than.

remainder as a fraction of the number obtained by doubling the square root of the square.[8] For example, if eight were double the side of the square, take the fraction as eighths, if ten then as tenths, and so on.

Thus, suppose we wish to regard 18 as a square and find its root. Take a root of the perfect square nearest it, that is, of 16; its root is 4. Double this and obtain 8, and subtract 16 from 18 leaving 2; express this in terms of eighths and write that the square root of 18 is 4 and two eighths. Now two eighths equals a quarter and so the root is 4 and a quarter. To see that this is the correct answer, multiply 4 and a quarter by itself and so find the answer is 18. Here are the steps: We say then that four-times 4 is 16 and four-times two eighths, that is, one quarter four times, equals four quarters equals one. Furthermore [four-times] two eighths is one, and one and one is two. Combine this with the 16 and it makes 18. But this method is too simplistic and is incomplete, and it has a lack of accuracy, for this answer is not the root multiplied by itself. If we multiply also a quarter times itself, we get not just 18, but 18 and one sixteenth. In what follows a more accurate method will be outlined, which we claim to be our own discovery, with the help of God. Let us turn our attention to this method for a while, showing how it can be used even for large numbers, so that it might become easier for us to understand. We apply the aforementioned method to numbers from the unit to 99 and from a hundred to 9999, since the roots of all numbers between these two have two digits, and we should proceed by the following method.

Suppose we write the number 235 and seek to find its square root. Take then the number which when multiplied by itself exactly produces the first digit two or is nearest to it. 2 is not correct since when multiplied by itself it gives 4. The number produced by the multiplication must be either equal to 2 or less than it. Hence the product should be one times itself, and we say one times one is again one. Subtract this from 2 and one is left over. Write this in small print above in the space between the 2 and the 3 . . .

<table>
<tr><td></td><td>1</td><td></td><td>3</td><td></td><td></td></tr>
<tr><td>2</td><td></td><td>3</td><td></td><td>5</td><td></td></tr>
<tr><td>1</td><td>5</td><td></td><td></td><td></td><td>10</td></tr>
<tr><td></td><td></td><td>2</td><td></td><td>10</td><td></td></tr>
</table>

Now double the *one* which you found from the subtraction of one from 2. I refer here to the *one* which was the root not the square, for you should double that number which you found from the subtraction.[9] One and one is two. Write this *two* below the 3, but not so that it lies on the same line as the unit previously written below the 2, but below that again in the third row. Now find the number which when multiplied by this cancels with 13, but when multiplied by itself cancels with the residual.[10] I mean that the numbers must be less than or equal to 13 and the residual. This is what is always meant by "cancel." Six multiplied by two can be subtracted from 13, but when multiplied by itself produces

[8] In modern notation, $\sqrt{a^2 + b} \approx a + \frac{b}{2a}$.

[9] This is confusing. He means that you double the root, not the remainder.

[10] That is, 35.

a number which is greater than the residual. The remainder is 13 and the square is 36. Hence we pass over the 6 and take 5 and say five-times 2 is 10.[11] Write the 5 between the 3 and the 2 on the same row as the unit written below the 2. Taking the 10 from the 13 leaves 3. Write this in small print above in the space between the 3 and the 5. Again we multiply the 5 by itself giving 25 and subtract from 35 leaving 10. Write this outside of the row by itself. Now double the 5 and it becomes 10, just as at the beginning we doubled the unit below the 2. Write this in the third row in turn next to the two you wrote previously. Now combine the 20 and the 10 in the third row, since 2 lies in the tens column, this gives 30. Take half from it and we get 15, since 30 is double the root. The root then of 235 is 15 and ten thirtieths. The remainder is expressed in terms of that same thirtieth which is obtained by doubling the (integer part) of the square root. Now ten thirtieths is a third. Note that the (integer part) of the square root, that is 15, lies in the second row and its double is in the third. To check the result, proceed as follows: Multiply 15 by itself and by ten thirtieths and say, fifteen by 15 is 225, then fifteen times a third, that is ten thirtieths, gives 5, then 5 and 5 is 10. Combining this with 225 gives 235.

2. MANUEL MOSCHOPOULOS, *ON MAGIC SQUARES*

Manuel Moschopoulos (ca. 1265–1315) was a student of Maximus Planudes and took an interest in aspects of both science and letters. He was known both as a scholar and a teacher. The only mathematical work for which he is known is a treatise on magic squares, the first in the West to discuss this subject. Although the treatise deals with the construction of odd-order magic squares, this excerpt deals only with the construction of evenly-even squares, that is, squares whose side is a power of 2.

The Methods for Evenly-Even Squares

As before, two different methods for those squares formed from evenly-even numbers[12] have been found. The first of these is as follows: We draw up the cells for such a square, and then we place (certain) dots as shown. In the first such square we place the dots only in the cells along the diagonals, thus [Fig. A1-1].

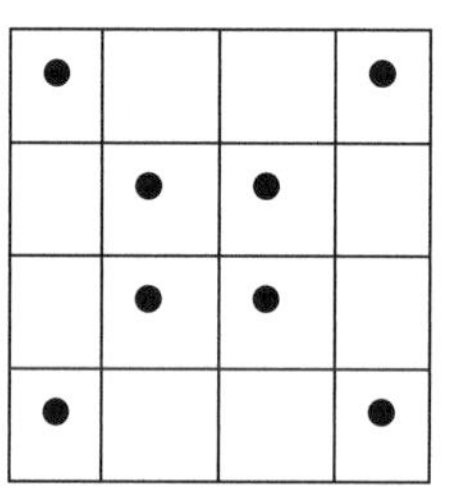

Fig. A1-1.

[11]Planudes has made the algorithm very complicated by not using place value. An easier way to express what he is doing is to say that we seek the largest integer x so that $(20 + x)x < 135$. Here $x = 6$ does not work, but $x = 5$ does.

[12]An evenly-even number is one of the form 2^n, for some positive integer n.

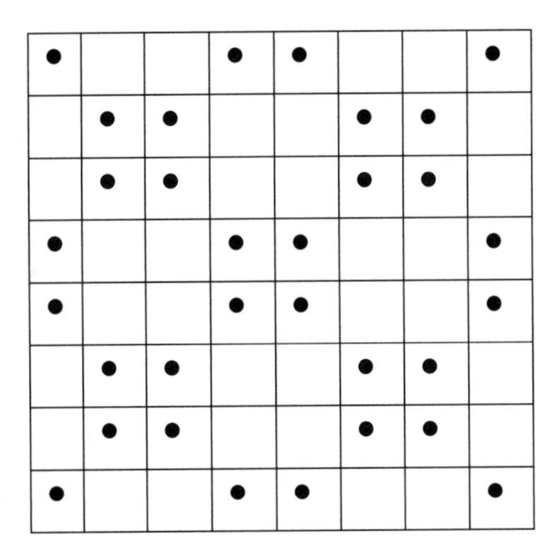

Fig. A1-2.

For each square in turn, [we place the dots] firstly along the diagonals and then proceed as follows. We count four places in turn to the right from the first of the cells in the topmost row, including the first one and three others, and we place a dot in the fourth cell, and another one in the cell next in turn directly on the right. We again count from this cell 4 places and we place a dot in the 4th cell and another one in the cell directly on the right immediately after it, and so on as far as possible. We continue this procedure also along the other sides of the square in a circle. We then place a series of dots from the 4th cell of the topmost (row of cells), counting from left to right, obliquely to the 4th cell of the left side (of the square), counting downwards, so that the dots meet to form an isosceles triangle with the corner of the square. (Likewise, we place a series of dots) from the 5th cell to the 5th cell on the right hand side, counting up (from the bottom). Then counting the 5th as the first, we join the 4th cell (to the 4th) cell at an angle, sloping to the left, then from the 5th cell (sloping) to the right. We continue this until we reach the end of the [remaining] cells on the top row, and then we turn the square around putting the top side to the bottom, and connect dots from that edge in a similar fashion, as one can see in Figs. A1-2 and A1-3.

After the placement of the dots as shown, we go through the numbers in turn, starting from the number one, and likewise through the (corresponding) cells in the given square, starting from the first of the cells in the topmost row from left to right. In those cells which have dots, we place the numbers corresponding to the cells, but where there are no dots, we pass over those cells and their corresponding numbers. We continue this process until the last of the cells of the entire square. Then again, beginning from the number one, we go through the numbers in turn and the cells of the square, beginning from the first cell in the bottom row from right to left. In those cells which are empty we place the corresponding numbers, but we pass over those cells which have numbers already and their corresponding numbers. We do this running through all the cells as far as the first cell in the topmost row, from which we began our descent.

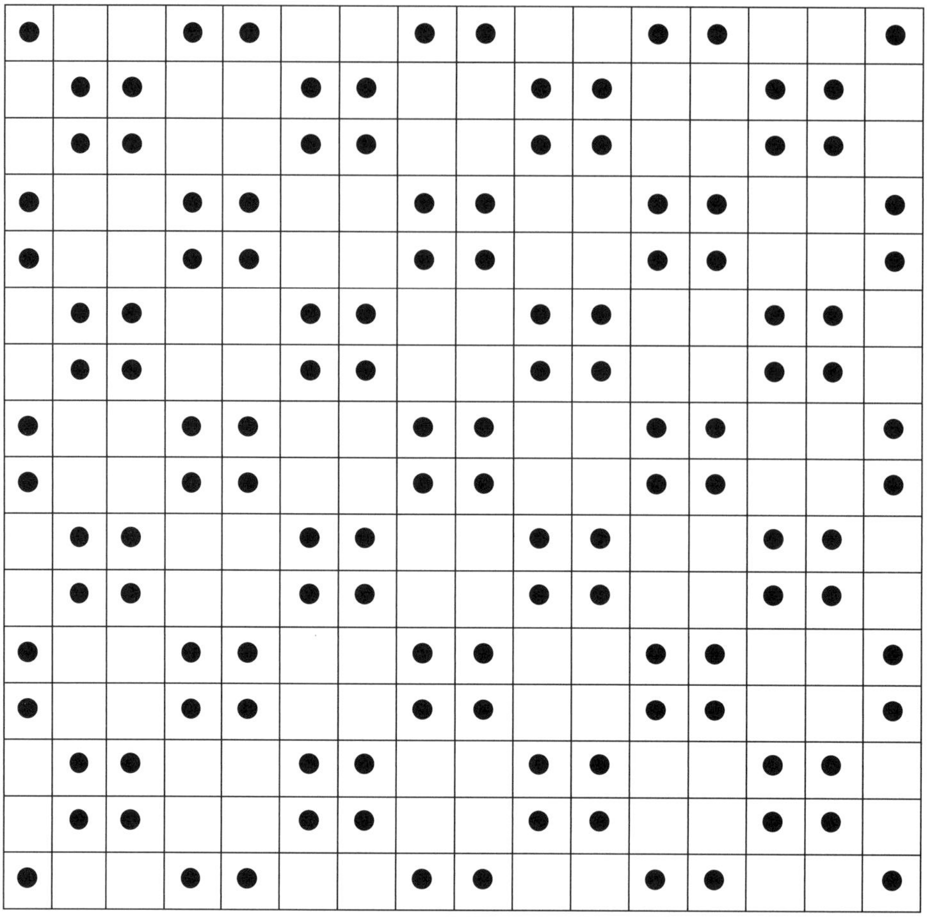

●			●	●			●	●			●	●			●
	●	●			●	●			●	●			●	●	
	●	●			●	●			●	●			●	●	
●			●	●			●	●			●	●			●
●			●	●			●	●			●	●			●
	●	●			●	●			●	●			●	●	
	●	●			●	●			●	●			●	●	
●			●	●			●	●			●	●			●
●			●	●			●	●			●	●			●
	●	●			●	●			●	●			●	●	
	●	●			●	●			●	●			●	●	
●			●	●			●	●			●	●			●
●			●	●			●	●			●	●			●
	●	●			●	●			●	●			●	●	
	●	●			●	●			●	●			●	●	
●			●	●			●	●			●	●			●

Fig. A1-3.

So that this might become clearer, let us practice (the method) on one such square. In particular then, suppose we are given the square with side 4, which we draw up and place the dots in the cells along the diameters, thus [Fig. A1-4]:

1	15	14	4
12	6	7	9
8	10	11	5
13	3	2	16

Fig. A1-4.

We then begin from the number one and from the 1st of the cells along the top, and we place straightaway the number one in that same 1st cell, since it contains a dot. Since there is no dot in the 2nd cell we pass over it and with it the number two which corresponds to it. Similarly we pass over the 3rd cell and with it the number three, but we place 4 in the 4th cell since it contains a dot. We pass over the 5th cell and with it the number five, and we place 6 in the 6th cell, and 7 in the 7th. We pass over 8 and the 8th cell and likewise 9 and the 9th, but we place 10 in the 10th and 11 in the 11th cell. We pass over the 12th and 12, but place 13 in the 13th cell. We pass over 14 and the 14th, and likewise 15 and the 15th cell, but place 16 in the 16th cell. Then we begin again from the number one, and take the first cell on the lowest row [from right to left] as the first cell in the square, counting to the left. We immediately pass over this square and with it the number one which corresponds to it, since it has a number in it. In the 2nd cell we place 2 since there is no number in it, and in the 3rd cell we place 3. We pass over 4 and the 4th cell and place 5 in the 5th. We pass over the 6th cell and 6, and likewise the 7th and 7, but we place 8 in the 8th and 9 in the 9th. The 10th and 10 we pass over and likewise 11 and the 11th, but we place 12 in the 12th and pass over 13 and the 13th. We place 14 in the 14th and 15 in the 15th, but pass over 16 and the 16th cell. All this is clear to see in the diagram. We use this procedure for squares of the same type. Thus the first method has been explained.

The Second Method:
The other method has the following format. I draw up the cells of the smallest possible square, that is to say, the one having side 4. I fill the cells with numbers as shown [Fig. A1-5].

1	14	11	8
12	7	2	13
6	9	16	3
15	4	5	10

Fig. A1-5.

I then use this square as model and archetype for each square of similar form in turn. For all such squares in turn can be subdivided into this one. The next square in turn after this one has double its side. Doubling the side always makes the [area of the] square four times as great as the area of the square whose side was doubled, therefore the square next in turn can be divided into four such squares. The next square again after this has double its side, and four times [the side of] the first square. Its area is four times as great as the second square, but, compared with the first, since its side is four times as great, it has area 16 times as great. It can therefore be divided into 16 such squares. We can easily find what multiple [the area of] one square is of another from the side, for we look to see what multiple one side is of the other, and we take the number by which one is a multiple of the other, and we multiply this number by itself and the number resulting from the multiplication is the ratio of [the area of] one square to the other. For example, suppose one side is four times the other. I take 4 and multiply it by itself and it makes 16. I have thus proven that the [larger] square is 16 times the [smaller] square. The other cases are similar.

We must now come to the positioning of the numbers which is as follows: We draw up the cells of another such square after the first one, which is already given, and we divide it up, using certain lines, into as many squares of the first kind as possible. Then we fill half the cells of the squares in turn starting from the top, looking to the first square and placing the numbers in the same order as in the first square. Then, beginning from the bottom, we reverse back to the top, filling the other half of the cells, those remaining in each square, looking again to the first and placing the numbers in the same order as in the first square.[13]

For greater clarity, let one such square be drawn, and let us show the placing (of the numbers) in it. Let it be, in fact, the square immediately after the first, which we draw up thus [Fig. A1-6].

1			8	9			16
	7	2			15	10	
6			3	14			11
	4	5			12	13	
17			24	25			32
	23	18			31	26	
22			19	30			27
	20	21			28	29	

Fig. A1-6.

We divide it by lines into as many squares of the first kind as possible. It can in fact be divided into 4. We fill half the cells as shown, beginning from the top and descending to the bottom.[14] We then reverse, in turn, beginning from the bottom, back to the top from whence we descended, filling those remaining cells in the same order as in the first square,[15] and when the whole thing is full, the *Sides* are equal in any direction [Fig. A1-7]. The procedure is the same in all other such squares.

Observe that in this arrangement, whenever you divide up the square into four sub-squares, the *Side* of each part equals that of the first—this was not the case with the

[13] Compared with his earlier descriptions, this section is remarkably terse and would be difficult to follow without an example.

[14] As shown, we move from left to right and add 8, 16, 24, respectively, to the entries in the basic square, as we do so.

[15] Adding, of course, 24, 32, 40, 48, respectively, to the basic square as we reverse back.

1	62	59	8	9	54	51	16
60	7	2	61	52	15	10	53
6	57	64	3	14	49	56	11
63	4	5	58	55	12	13	50
17	46	43	24	25	38	35	32
44	23	18	45	36	31	26	37
22	41	48	19	30	33	40	27
47	20	21	42	39	28	29	34

Fig. A1-7.

previous arrangement.[16] If the sides are divided in two, each part has the same sum (along the side)—this will happen in all cases except the first. The square has other elegant and graceful features, which the preceding account does not contain.

3. ISAAC ARGYROS, *ON SQUARE ROOTS*

Isaac Argyros (ca. 1300–1375) pursued his scientific activity in Constantinople from 1367 to 1373. He wrote two astronomical treatises based on Ptolemy's work, but for the meridian of Byzantium. He also wrote a *Treatise on the Date of Easter* and a *Treatise on the Astrolabe.* In the work excerpted here on square roots, he attempted to improve on the method of Heron. We present his method using ordinary fractions, but he also demonstrated how to calculate square roots in the sexagesimal system.

The study of square roots of numbers that are perfect squares is easy, since they are themselves integers. In fact, every number multiplied by itself forms a square. The root is the number which one multiplied by itself. The square numbers are formed beginning with unity by the addition of successive odd numbers. Thus, if one takes 1, 3, 5, 7, 9, 11, and so forth in sequence, the addition of unity and of 3 gives four, which is square. The addition of unity and 3 and 5 gives 9, which is also a square; the addition of the numbers up to 7 gives 16; the addition of the numbers through 9 gives 25, and adding 11 to the preceding numbers gives 36, and so forth. The roots of these numbers are the numbers in the sequence of consecutive numbers beginning with the unit. Thus the root of 4 is 2, since two times 2 will be 4; the root of 9 is 3, since three times 3 are 9; the root of 16 is 4, that of 25 is 5, that of 36 is 6, and so on.

Mathematicians study equally the roots of other numbers, those which are not perfect squares. It is absolutely impossible to determine these exactly; one can only obtain values very close to the exact value. For this reason, those who have preceded us have established different methods that present great difficulties to solve this particular problem. It therefore appeared to be most useful to choose the easiest of these methods

[16]This also holds for the diagonal sum of each subsquare.

to explain; it comes from the first principles of the ancient methods, but it has been corrected, in order to get the greatest exactitude possible, and with the help of God we used these methods themselves. This method is the following: For every given number which is not a perfect square, we find the square root this way. We take the nearest square number, either more than the given one or less. The root of this square number is clear, since it is an integer; we take this value and write it down. Next, we multiply by two the root obtained and make this result the denominator of a fraction. We add to the root written down a fraction of which the numerator is the number of units by which the number whose root we are finding exceeds the nearest square number, if that number is greater. If it smaller, we subtract the fraction.

Suppose, for example, that we are looking for the square root of 6. The number 6 is between 4 and 9, but it is closer to 4 than to 9. It exceeds that one, in fact, by two units, and it is short of the other by three. If we had wanted to find the root of 7, we would have been closer to 9 than to 4. To find the root of 6, we have taken the square 4 as the closest square, and we write down the root, that is, 2. Next, we multiply the 2 by two, giving 4, and this value gives to the fractions the value of quarters. Since 6 exceeds 4 by two units, we add to the root written down, that is to say, to two units, two quarters, that is, a half. You should know, in fact, that if the added or subtracted fractions have a numerator greater than a unit and if it is possible to reduce this to a unit fraction, as here we have reduced two quarters to a unit fraction, that is, to a half, we must always do that. We find thus as a first approximation greater than the root of six, 2 units and a half. This approximation is greater, since if one multiplies two units by itself and by a half, and then by a half again, one obtains the number 6. But it is also necessary to multiply a half by itself. In performing this multiplication, one gets a quarter, and one obtains a number that exceeds 6. . . .

We have established this coarse method as a foundation; it will be corrected, with the help of God, by another method we have derived and one of the greatest exactitude. We must now describe this method. Recording the number from which is derived the denominator of the fraction or fractions, that which exceeds the number for which one is finding the root, we begin with the method previously explained. Multiply by two the root obtained and write down the result of the multiplication by two, then multiply the numbers together. Take the number obtained as the denominator of a fraction, then subtract from the aforesaid root a unit fraction with this number as denominator. After this subtraction, we have a new root of the number for which we are finding the root, closer to the exact root. We know definitively that this root will be closer, since if one multiplies by itself this root that we have called exact, a part of the number obtained adjusted exactly with the proposed number, but the fraction subtracted according to the proposed method . . . multiplied by itself forms a fraction infinitely smaller than the preceding, which one finds newly in excess by the ratio to the number for which we are seeking the root. Since this fraction is an essentially negligible quantity because of its smallness, we do not consider it. This calculation will be clearer by an example.

To find the root of 6, we have found this to be 2 units and a half, and if one multiplies this by itself, it makes $6\frac{1}{4}$. This 1/4 is the amount that the square exceeds 6. We write the number 4, which is the denominator of the fraction one quarter. We multiply by two the [first] root $2\frac{1}{2}$ and we write down the result, 5. We multiply the 5 by the 4 to get 20, and we set this number as the denominator of a fraction 1/20. We subtract this fraction from $2\frac{1}{2}$. The root nearer to the exact root is two units and 9 twentieths. The half of a unit is in fact 10 twentieths. If one subtracts one twentieth, there remains 9 twentieths. The product of this number by itself is 6 and one four hundredth. We can neglect this fraction as very small and imperceptible.

But if we consider that it is not sufficient to stop here, and we want to continue, we may find a more precise value by beginning again in the same fashion by writing 400. Then, we multiply the second root by two, that is, the 2 units and 9 twentieths, and then multiply the result, 4 and 18 twentieths by 400, and we set the result 1960 as the denominator of the fraction one nineteen hundred sixtieth. We subtract this from the second root, that is, from two and 9 twentieths, and we find a value more precise than the second, since that one had an excess [when squared] of one four hundredth, which, we have said, may be neglected because of its smallness, but this latter one has an excess of one nineteen hundred sixtieth of one nineteen hundred sixtieth, that is, approximately one three thousand eight hundred thousandths. Since this value will continue to increase if we continue the process further, this is why in beginning our proposal we have said that it is impossible to find the exact root of a number which is not a perfect square.

4. ANONYMOUS FIFTEENTH-CENTURY MANUSCRIPT ON ARITHMETIC

This arithmetic manuscript was brought to Vienna from Constantinople in the mid-sixteenth century, but the evidence indicates that it was written in the Byzantine Empire before the fall of the capital in 1453. The book is a collection of problems, many with solutions, accomplished by various methods. Many of the problems are old and have appeared in other problem compilations from such diverse times and locales as China at the beginning of our era, fifth-century India, and medieval Islam.

18. Two people each want to buy silk fabric. They walked together to the market and found two single pieces at a merchant, for 40 florins for both. They negotiated to buy one piece without the other, but he would only sell as a whole. He told them that one was worth a certain amount and the other $1\frac{1}{2}$ times that and still 8 florins more. You now want to know how much each piece is worth. If one subtracts the 8 from the 40, there remains 32; then use the rule of three. The first amount and the second (the $1\frac{1}{2}$ times) together give $2\frac{1}{2}$. Then say, if the $2\frac{1}{2}$ gives 32, what will $1\frac{1}{2}$ give? It gives $19\frac{1}{5}$ florins. Now add the 8, and it gives you that one piece of silk fabric was worth $27\frac{1}{5}$, the other just $12\frac{4}{5}$ florins. Check that there are 40 florins in total.

45. A man became ill and he made a will, and his fortune was found to be in the amount of 1000 florins. His wife was pregnant, and he said, if my wife bears a boy, then the child shall have one share of my fortune and my wife two shares; if she, however, bears a girl, then my wife shall have one share of my fortune and the daughter two. She gave birth to twins, one boy and one girl. I want you to find the share of the inheritance coming to each, according to the available information. It is necessary that for the boy, you make a half of the share of the mother and for the daughter twice as much as the mother. Therefore take for the share of the son 1, and of the mother 2 and of the daughter 4. Add these, the 1 and 2 and 4, and you have 7 parts, and this is the divisor. Multiply the 2 parts with the sum, namely the 1000 florins, and it will be 2000 and divide by the 7, and thus the mother's share is $285\frac{5}{7}$. Multiply 1000 by the 4 shares of the daughter and you get 4000;

divide by 7, and there will be $571\frac{3}{7}$. Finally, multiply the one share of the son by 1000, and it remains as 1000. Then divide this by 7, and the son gets $142\frac{6}{7}$.

46. In a meadow there were girls dancing, and a man went by and greeted them and said: There are 100 girls dancing. And one of those answered and said: We are not 100, but if we were again as much as we are, and then a half and a quarter more, and with you together, there would be 100. I want to know how many there were. Do it thus: Take the smallest number, which contains a half and a quarter, and that is 4. Now make 4 the basis. Also, since one said again as many as we are and the half and the quarter, so do this with the 4 and take another 4 and half of 4 and a quarter of 4, and this gives 11. But she also said, with him there will be 100, so there remain 99, and then you will later add him. So divide the 99 by 11, and there results 9. Then multiply the 9 by the 4, and there results 36, so there were 36 girls.

53. A man told his servant that he should buy birds of three types, pigeons, turtledoves, and sparrows. Each pigeon cost 4 coins, each turtledove 2 coins, and one can get three sparrows for one coin. He gave him 100 coins that he should buy 100 birds, neither more nor fewer. I want to know, how many of each type of bird he will buy. Do it thus: begin with the birds of smallest value, namely the sparrows, and say: Suppose that he took 100 sparrows for $33\frac{1}{3}$ coins, and there remain to 100, $66\frac{2}{3}$ coins. Make everything in thirds, and there are 200/3. Look at how much money more the turtledoves cost than the sparrows, and you need 5/3 more for each, and these 5 are the pigeons. Find now also the turtledoves. Divide the 200 by 5, and this is 40. Then look at how much more the pigeons cost than the sparrows, and they are 11/3 more, and these 11 must be subtracted from 40, and there remain 29, and these 29 are the turtledoves; and there remain 66 sparrows.

59. A man wishes to build a house, and a builder said that he could do this in 24 days. A second builder said that he could do it in 12 days; another one said, I can build it in 6 days. In how many days can the three builders do this together? The first builds a house in 24 days, the second can build 2 houses in 24 days, and the third can in 24 days build 4 houses. It follows that they can build the house together in $3\frac{3}{7}$ days. Do it thus: If they can build 7 houses in 24 days, in how many days can one build 1 house? Thus it happens that the three builders can do this in $3\frac{3}{7}$ days.

71. We want to find a number, that if you multiply it by 12 and then divide it by 36, then the result is 64. Do it thus: Multiply the 36 by the 64, and you get 2304, and this number is such that if you divide it by 36, then the quotient is 64. Further you say, that you divide this by 12. Thus divide the 2304 by 12 and the result is 192, and this is the very number which you can multiply by 12 and then divide by 36, and get the quotient 64.

SOURCES

1. Maximus Planudes, *The Great Calculation According to the Indians*, translation in Peter G. Brown. 2006. "The Great Calculation According to the Indians, of Maximus Planudes," *Convergence*, 3, http://www.maa.org/publications/periodicals/convergence/the-great-calculation-according-to-the-indians-of-maximus-planudes-introduction.

2. Manuel Moschopoulos, *On Magic Squares*, translation in Peter G. Brown. 2005, "The Magic Squares of Manuel Moschopoulos," *Convergence* 2, http://www.maa.org/publications/periodicals/convergence/the-magic-squares-of-manuel-moschopoulos-introduction.
3. Isaac Argyros, *On Square Roots*, translated by Victor Katz from the French translation in André Allard. 1978, "Le petit traité d'Isaac Argyre sur la racine carrée," *Centaurus* 22, 1–43.
4. Anonymous fifteenth-century Byzantine manuscript, translated by Victor Katz from the German translation in Herbert Hunger and Kurt Vogel. 1963. *Ein Byzantinisches Rechenbuch des 15. Jahrhunderts: Text, Übersetzung und Kommentar*, Vienna: Herman Böhlaus Nachf.

APPENDIX 2

Diophantus *Arithmetica*, Book I, #24

To find three numbers such that, if each receives a given fraction of the sum of the other two, the results are equal.

Let the first receive 1/3 of (second + third), the second 1/4 of (third + first); and the third 1/5 of (first + second). Assume the first is x, and, for convenience sake, take the sum of the second and third as a number of units divisible by 3, say 3.

Then the sum of the three is $x + 3$, and the first $+1/3$ (second + third) is $x + 1$. Therefore, the second + 1/4(third + first) is $x + 1$; hence 3 times the second plus the sum of all is $4x + 4$, and therefore the second is $x + 1/3$. Lastly, the third $+1/5$(first + second) is $x + 1$, or 4 times the third plus the sum of all is $5x + 5$, and the third is $x + 1/2$. Therefore $x + (x + 1/3) + (x + 1/2) = x + 3$, and $x = 13/12$. The numbers, after multiplying all by the common denominator, are 13, 17, 19.

Source

From Thomas L. Heath. 1964. *Diophantus of Alexandria: A Study in the History of Greek Algebra*. New York: Dover (reprint of 1910 edition published by Cambridge University Press).

APPENDIX 3

From the *Ganitasārasangraha* of Mahavira

From Chapter VI
The rule for arriving at the value of the money contents of a purse which, when added to what is on hand with each of certain persons, becomes a specified multiple of the sum of what is on hand with the others:

233–235: The quantities obtained by adding one to each of the specified multiple numbers in the problem, and then multiplying these sums with each other, giving up in each case the sum relating to the particular specified multiple, are to be reduced to their lowest terms by the removal of common factors. These reduced quantities are then to be added. Thereafter, the square root[1] of this resulting sum is to be obtained, from which one is

[1]This makes no sense mathematically. In terms of the problem in verses 236–237, the value needed is 3.

to be subsequently subtracted. Then the reduced quantities referred to above are to be multiplied by this square root as diminished by one. Then these are to be separately subtracted from the sum of those same reduced quantities. Thus the moneys on hand with each of the several persons are arrived at. These quantities measuring the moneys on hand have to be added to one another, excluding from the addition in each case the value of the money on the hand of one of the persons; and the several sums so obtained are to be written down separately. These are then to be respectively multiplied by the specified multiple quantities mentioned above; from the several products so obtained the already found out values of the moneys on hand are to be separately subtracted. Then the same value of the money in the purse is obtained separately in relation to each of the several moneys on hand.

An example in illustration thereof:
236–237: Three merchants saw dropped on the way a purse containing money. One of them said to the others, "If I secure this purse, I shall become twice as rich as both of you with your moneys on hand." Then the second of them said, "I shall become three times as rich." Then the other, the third, said, "I shall become five times as rich." What is the value of the money in the purse, as also the money on hand with each of the three merchants?

Source

From Mahāvīra, 1912, *Ganitāsarangraha*, edited and translated by M. Rangācārya. Madras: Government Press.

APPENDIX 4

Time Line

What follows is a listing of the dates of every identifiable author whose work we have used in this sourcebook. Note that most dates are approximate.

Claudius Ptolemy	100–170
Martianus Capella	370–440
Boethius	480–524
Aurelius Cassiodorus	490–583
Isidore of Seville	560–636
Bede the Venerable	672–735
Alcuin of York	735–804
Muḥammad al-Khwārizmī	780–850
Banū Mūsā ibn Shākir	ninth century
Abū ʿAbdallah ibn ʿAbdun	923–976
Gerbert of Aurillac	945–1003
Ibn al-Samh	984–1035
Ibn Muʿādh al-Jayyānī	mid-eleventh century
Al-Muʾtaman ibn Hūd	d. 1085

Franco of Liége	d. 1083
Pandulf of Capua	late eleventh century
Ibrahim ibn al-Zarqālluh	1029–1099
Shlomo Ishaqi	1040–1105
Abraham bar Ḥiyya	1065–1145
Adelard of Bath	1080–1152
Plato of Tivoli	early twelfth century
Abraham ibn Ezra	1089–1167
Hugh of St. Victor	1096–1141
Hermann of Carinthia	1100–1160
Gerard of Cremona	1114–1187
Robert of Chester	mid-twelfth century
Domingo Gundisalvo	mid-twelfth century
Jabir ibn Aflaḥ	mid-twelfth century
Aḥmad ibn Munʿim al- ʿAbdari	d. 1228
Leonardo of Pisa	1170–1240
Robert Grosseteste	1175–1253
John of Sacrobosco	1195–1236
Jordanus de Nemore	early thirteenth century
William of Lunis	early thirteenth century
William of Moerbeke	1215–1286
Muḥyī al-Dīn ibn Abī al-Shukr al-Maghribī	d. 1290
Roger Bacon	1219–1292
Campanus of Novara	1220–1296
Witelo	1230–1300
Ibn Badr	thirteenth century(?)
Maximus Planudes	1255–1305
Aḥmad ibn al-Bannāʾ	1256–1321
Manuel Moschopoulos	1265–1315
John Duns Scotus	1266–1307
Abner of Burgos	1270–1348
Giovanni Villani	1278–1348
Qalonymos ben Qalonymos	1287–1329
Levi ben Gershon	1288–1344
Thomas Bradwardine	1290–1349
Richard of Wallingford	1291–1336
Isaac Argyros	1300–1375
Immanuel ben Jacob Bonfils	1300–1377
Johannes de Lineriis	early fourteenth century
John of Muris	early fourteenth century
William Heytesbury	1313–1373
Nicole Oresme	1320–1382
Dominicus de Clavasio	mid-fourteenth century
Giovanni di Casali	mid-fourteenth century
Paolo Girardi	mid-fourteenth century

Jacobo da Firenze	mid-fourteenth century
Master Dardi	mid-fourteenth century
Geoffrey Chaucer	1343–1400
Isaac ibn al-Aḥdab	1350–1430
Simon ben Ṣemaḥ	1361–1444
Don Benveniste ben Lavi	late fourteenth century
Jacob Canpanṭon	late fourteenth century
Ibn al-Majdī	d. 1447
Aaron ben Isaac	mid-fifteenth century
ʿAli al-Qalaṣādī	mid-fifteenth century
Simon Moṭoṭ	mid-fifteenth century
Regiomontanus	1436–1476
Elijah Mizraḥi	1450–1526
Solomon ben Isaac	early sixteenth century
Muḥammad al-Kishnāwī	d. 1741

Editors and Contributors

EDITORS

VICTOR J. KATZ is Professor of Mathematics Emeritus at the University of the District of Columbia. He has long been interested in the history of mathematics and its use in teaching. His most recent book, written with Karen Parshall, is *Taming the Unknown: A History of Algebra from Antiquity to the Early Twentieth Century*, which appeared in 2014. The third edition of his well-regarded textbook, *A History of Mathematics: An Introduction*, appeared in 2008. Katz is also the editor of *The Mathematics of Egypt, Mesopotamia, China, India and Islam: A Sourcebook*, published in 2007. Professor Katz has edited three books dealing with the use of the history of mathematics in the teaching of the subject as well as two collections of historical articles taken from journals of the Mathematical Association of America in the past 90 years. He has directed two NSF-sponsored projects to help college teachers learn the history of mathematics and its use in teaching and also involved secondary school teachers in writing materials demonstrating this use in the high school curriculum. These materials, *Historical Modules for the Teaching and Learning of Mathematics,* were published on a CD in 2005. Professor Katz was also the founding editor of *Convergence*, the Mathematical Association of America's online magazine on the history of mathematics and its use in teaching.

MENSO FOLKERTS studied mathematics and classical philology at the University of Göttingen and received his Ph.D. in 1967 with a dissertation on a medieval mathematical text. In 1973 he received his "Habilitation" in the history of exact sciences and technology at the Technical University of Berlin. He was Associate Professor of Mathematics and Its History at the University of Oldenburg (1976–1980) and Full Professor of History of Science at the University of Munich (1980–2008). His main research area is mathematics in the Middle Ages and in early modern times. Folkerts has edited numerous medieval mathematical texts in Latin and has published ten books and about 200 articles in the history of science. He is the editor of two series and co-editor of six journals in the history of science. He has been the president of the German Society of the History of Medicine, Science and Technology, member of the German National Committee in the International Union of the History and Philosophy of Science, and member of the Executive Committee of the International

Commission on the History of Mathematics. He is member of four German academies, including the German National Academy of Sciences Leopoldina, and of the Académie Internationale d'Histoire des Sciences. In 2013 he was awarded the Kenneth O. May Medal and Prize for outstanding contributions to the history of mathematics.

BARNABAS B. HUGHES, O.F.M., began his interest in the history of mathematics as an aid to the teaching of mathematics to high school students. Both they and he found the subjects more interesting because of historical anecdotes from the history of mathematics and its originators. During his 40 years at California State University, Northridge, he published numerous articles and half a dozen books on topics from its history, the last being this cooperative work on sources in medieval mathematics.

ROI WAGNER has a mathematics Ph.D. (1997) and a philosophy Ph.D. (2007) from Tel Aviv University. He taught mathematics and philosophy in Tel Aviv University, the Hebrew University of Jerusalem, and the Academic College of Tel Aviv Jaffa, and held postdoc and visiting positions in Paris 6, Cambridge University, the Max Planck Institute for the History of Science, and the Buber Society at the Hebrew University. He publishes on the history and philosophy of mathematics (usually applying semiotic perspectives) as well as political philosophy (focusing on resistance studies). His first book, *S(zp,zp): Post Structural Readings of Gödel's Proof*, was published in 2009, and his new book, *Making and Breaking Mathematical Sense: Histories and Philosophies of Mathematical Practice* will be published in 2016 by Princeton University Press. Roi Wagner is currently Professor for the History and Philosophy of Mathematical Sciences at the ETH Zurich.

J. LENNART BERGGREN is Emeritus Professor in the Department of Mathematics at Simon Fraser University, Canada, where he taught mathematics and its history for forty years. He has held visiting positions in the Mathematics Institute at the University of Warwick, and the History of Science Departments at Yale and Harvard Universities. He has published sixty refereed papers and authored or co-authored eight books on the history of mathematics. His special interest is the history of mathematical sciences in ancient Greece and medieval Islam, including cartography, spherical astronomy, and such mathematical instruments as sundials and astrolabes. Among his books are *Episodes in the Mathematics of Medieval Islam* (1986), which has been translated into Farsi and German, *Euclid's Phænomena* (with Robert Thomas, 1996; reprinted, 2006), and *Pi: A Sourcebook* (third edition, 2004, with J. Borwein and P. Borwein). He is a member of the Canadian Society for the History and Philosophy of Science (of which he has served three two-year terms as President).

CONTRIBUTORS

IMMO WARNTJES is lecturer in Irish Medieval History at Queen's University Belfast, United Kingdom. He is a graduate of the University of Göttingen (Germany) in both History and Mathematics and holds a Ph.D. in Medieval Studies from the National University of Ireland, Galway. His primary research interest is early medieval scientific thought, and he is a specialist in the science of computus (medieval time-reckoning) in the Latin West, ca. AD 400–1200.

NAOMI ARADI received her Ph.D. in 2016 from the Hebrew University of Jerusalem, Department of History and Philosophy of Science. Her dissertation is on the Arithmetic of the Jews in the Middle Ages.

AVINOAM BARANESS was born and lives in Israel. His closest family members serve as models and inspiration for his studies, especially his grandfather, who dedicated himself to the Torah and meditated on the Bible and Talmud day and night; his mother, who devoted herself to the Bible and Hebrew literature; and his father, who studies and teaches mathematics and Hebrew grammar. Avinoam attended the Hebrew University of Jerusalem and Lifschitz College and earned his bachelor's degree in Talmud and mathematics, with a minor in musicology. He holds an M.S. in mathematics teaching from the Hebrew University, in which he is currently a Ph.D. student in the department of the history and philosophy of science. His doctoral research is focused on Medieval and early Modern Hebrew treatises dealing with geometry. He has a number of publications on the topic of Talmud, which appear in *Ha'Me'ir La'aretz,* the yearbook of Ha'me'iry Yeshiva. Currently he is working with Professor Ruth Glasner on an English commented translation of *Rectifying the Curve* by Abner of Burgos, a section of which appears in this *Sourcebook.*

DAVID GARBER graduated from Bar-Ilan University in 1994, and finished his Ph.D. in Mathematics in Bar-Ilan University in 2001. He is now a Senior Lecturer at Holon Institute of Technology, Israel. His main interests are geometric topology (combinatorics and topology of plane curves), algebraic combinatorics, and algebraic cryptography. His additional area of interest is mathematical discussions in Rabbinical writings, and he has written numerous papers on this topic in both Hebrew and English.

STELA SEGEV studied Mathematics and Science Education at the Hebrew University of Jerusalem (B.S. and M.S.). She taught mathematics in high school, instructed mathematics teachers, and wrote several brochures and manuals on mathematics for secondary high schools. Later she became interested in the history of mathematics and in particular in Hebrew mathematicians in the Middle Ages. She received a Ph.D. from the Hebrew University of Jerusalem. The subject of her dissertation was *The Book of the Number*, written by Elija Mizraḥi at the beginning of the sixteenth century in Constantinople. Currently she is the head of the Mathematics Department at Herzog College (near Jerusalem). She is also continuing her research with the goal of publishing a critical edition of Mizraḥi's book.

SHAI SIMONSON is currently professor of computer science at Stonehill College. His research interests are in theoretical computer science (algorithms and complexity theory, grammars, and automata), computer science education, math education, and history of mathematics. Shai is the author of the text *Java, From the Ground Up* and the director of ArsDigita University, which offers free computer science lectures to students all over the world, especially in developing countries. Shai has a particular interest in mathematics education at the middle school and high school levels. He designed an innovative mathematics curriculum for the South Area Solomon Schechter Day School in Norwood, which led to his book *Rediscovering Mathematics.* Sponsored by the NSF, Shai spent 1999 at the Hebrew University working on Ralbag and *Ma'ase Hoshev.*

ILANA WARTENBERG studied mathematics (B.S., M.S.), linguistics (M.A.), as well as history and philosophy of science (Ph.D.) in Tel Aviv and Paris. She taught mathematics and logic at universities in Israel. She also worked as a linguist in the hi-tech industry in Israel. In the past fourteen years she has specialized in medieval Hebrew science, mainly in mathematics and the reckoning of the Jewish and other calendars. She also researches the connection between Hebrew and Arabic science and the creation and evolution of the Hebrew scientific terminology. Her past and future publications focus on the treatises of Isaac ibn al-Aḥdab (*Iggeret ha-Mispar/The Epistle of the Number*, composed in Sicily at the end of the fourteenth century), Jacob ben Samson (*Sod ha-Ibbur/The Calculation of the Jewish Calendar*, composed in northern France in 1123), Abraham bar Ḥiyya (*Sefer ha-Ibbur/The Book on the Jewish Calendar*, composed in southern France in 1123), Isaac Israeli (*Yesod Olam/The Foundation of the World*, composed in Toledo in 1310), and various shorter tracts and Genizah fragments in Hebrew and Judaeo-Arabic. She works as a research associate at the Department of Hebrew and Jewish Studies at University College London.

Index

Page numbers in *italics* refer to figures and tables.